MECHANICS

THIRD EDITION

KEITH R. SYMON

University of Wisconsin

ADDISON-WESLEY PUBLISHING COMPANY

Reading, Massachusetts

Menlo Park, California · London · Amsterdam · Don Mills, Ontario · Sydney

This book is in the
ADDISON-WESLEY SERIES IN PHYSICS

 Printed in the United States of America. Published simultaneously in Canada. Library of Congress Catalog Card No. 75–128910.

ISBN 0-201-07392-7
UVWXYZ-MA-89

PREFACE

This text is intended as the basis for an intermediate course in mechanics at the undergraduate level. Such a course, as essential preparation for advanced work in physics, has several major objectives. It must develop in the student a thorough understanding of the fundamental principles of mechanics. It should treat in detail certain specific problems of primary importance in physics, for example, the harmonic oscillator and the motion of a particle under a central force. The problems suggested and those worked out in the text have been chosen with regard to their interest and importance in physics, as well as to their instructive value.

The choice of topics and their treatment throughout the book are intended to emphasize the modern point of view. Applications to atomic physics are made wherever possible, with an indication as to the extent of the validity of the results of classical mechanics. The inadequacies in classical mechanics are carefully pointed out, and the points of departure for quantum mechanics and for the theory of relativity are indicated. The last two chapters then develop special relativistic mechanics. The development, except for the last six chapters, proceeds directly from Newton's laws of motion, which form a suitable basis from which to attack most mechanical problems. More advanced methods, using Lagrange's equations and tensor algebra, are introduced in Chapters 8 to 12.

An important objective of a first course in mechanics is to train the student to think about physical phenomena in mathematical terms. Most students have a fairly good intuitive feeling for mechanical phenomena in a qualitative way. The study of mechanics should aim at developing an almost equally intuitive feeling for the precise mathematical formulation of physical problems and for the physical interpretation of the mathematical solutions. The examples treated in the text have been worked out so as to integrate, as far as possible, the mathematical treatment with the physical interpretation. After working an assigned problem, the student should study it until he is sure he understands the physical interpretation of every feature of the mathematical treatment. He should decide whether the result agrees with his physical intuition about the problem. If not, then either his solution or his intuition should be appropriately corrected. If the answer is fairly complicated, he should try to see whether it can be simplified in certain special or limiting cases. He should try to formulate and solve similar problems on his own.

Only a knowledge of differential and integral calculus has been presupposed. Mathematical concepts beyond those treated in the first year of calculus are introduced and explained as needed. A previous course in elementary differential equations or vector analysis may be helpful, but it is the author's experience that students with an adequate preparation in algebra and calculus are able to handle the vector analysis and differential equations needed for this course with the explanations provided herein. A physics student is likely to get more out of his advanced courses in mathematics if he has previously encountered these concepts in physics.

Since the book contains more than enough material for a two-semester undergraduate course, an opportunity is provided for selection of topics to suit a particular class. The first seven chapters provide a basis for a one-semester course in intermediate mechanics. By omitting some topics in earlier chapters the author has found it possible in a one-semester course to include also parts of Chapters 8 or 9.

The text has been written so as to afford maximum flexibility in the selection and arrangement of topics to be covered. With certain obvious exceptions, many sections or groups of sections can be postponed or omitted without prejudice to the understanding of the remaining material. Where particular topics presented earlier are needed in later parts of the book, references to section and equation numbers make it easy to locate the earlier material needed.

Later chapters depend primarily upon the material covered in Chapters 2, 3, and 4, in Sections 1, 2, and 6 of Chapter 5, and Sections 1 and 2 of Chapter 7. When this material has been covered, most of the material in later chapters will be understandable. Section 5.11 should of course precede the material on fluid motion in Chapter 8 if that is to be covered. Chapters 9 and 10 should be covered before Chapter 11 or 12. The last two chapters on relativity may be taken up any time after Chapter 7. Chapter 14 contains a few references, examples, and extensions of material from Chapters 8, 9 and 10, but these may be omitted without affecting the continuity of the remaining material.

In the first chapter the basic concepts of mechanics are reviewed, and the laws of mechanics and of gravitation are formulated and applied to a few simple examples. The second chapter undertakes a fairly thorough study of the problem of one-dimensional motion. The chapter concludes with a study of the harmonic oscillator as probably the most important example of one-dimensional motion. Use is made of complex numbers to represent oscillating quantities. The last section, on the principle of superposition, makes some use of Fourier series and provides a basis for certain parts of Chapters 8 and 12. If these chapters are not to be covered, Section 2.11 may be omitted or, better, skimmed to get a brief indication of the significance of the principle of superposition and the way in which Fourier series are used to treat the problem of an arbitrary applied force function.

Chapter 3 begins with a development of vector algebra and its use in describing motions in a plane or in space. Boldface letters are used for vectors. Section 3.6 is a brief introduction to vector analysis, which is used very little in this book except in Chapter 8, and it may be omitted or skimmed if Chapter 8 and a few proofs in some other chapters are omitted. The author feels there is some advantage in introducing the student to the concepts and notation of vector analysis at this stage, where the level of treatment is fairly easy; in later courses where the physical concepts and mathematical treatment become more difficult, it will be well if the notations are already familiar. The theorems stating the time rates of change of momentum, energy, and angular momentum are derived for a moving particle, and several problems are discussed, of which motion under central forces receives major attention. Examples are taken from astronomical and from atomic problems.

In Chapter 4 the conservation laws of energy, momentum, and angular momentum are derived, with emphasis on their position as cornerstones of present-day physics. They are then applied to typical problems, particularly collision problems. The two-body problem is solved, and the motion of two coupled harmonic oscillators is worked out. The general theory of coupled oscillations is best treated by means of linear transformations in vector spaces, as in Chapter 12, but the behavior of coupled oscillating systems is too important to be omitted altogether from even a one-semester course. The section on two coupled oscillators can be omitted or postponed until Chapter 12. The rigid body is discussed in Chapter 5 as a special kind of system of particles. Only rotation about a fixed axis is treated; the more general study of the motion of a rigid body is left to a later chapter, where more advanced methods are used. The section on statics treats the problem of the reduction of a system of forces to an equivalent simpler system. Elementary treatments of the equilibrium of beams, flexible strings, and of fluids are given in Sections 5.9, 5.10, and 5.11. With the exception of Sections 1, 2, 6, and 8, each of which forms the basis for the material which follows, the various sections in Chapter 5 are independent of one another and can be taken up in any order or omitted as desired.

The theory of gravitation is studied in some detail in Chapter 6. The last section, on the gravitational field equations, may be omitted without disturbing the continuity of the remaining material. The laws of motion in moving coordinate systems are worked out in Chapter 7, and applied to motion on the rotating earth and to the motion of a system of charged particles in a magnetic field. Particular attention is paid to the status in Newtonian mechanics of the "fictitious forces" which appear when moving coordinate systems are introduced, and to the role to be played by such forces in the general theory of relativity.

In Chapter 8 an introductory treatment of vibrating strings and of the motion of fluids is presented, with emphasis on the fundamental concepts and mathematical methods used in treating the mechanics of continuous media. Chapter 9 on Lagrange's equations is intended as an introduction to the methods of advanced dynamics. Hamilton's equations and the concept of phase space are presented, since they are prerequisite to any later course in quantum mechanics or statistical mechanics, but the theory of canonical transformations and the use of variational principles are beyond the scope of this book. Chapter 10 develops the algebra of tensors, including orthogonal coordinate transformations, which are required in Chapters 11 and 12. The inertia tensor and the stress tensor are described in some detail as examples. Section 10.6 on the stress tensor will enable the reader to extend the discussion of ideal fluids in Chapter 8 to a solid or viscous medium. The methods developed in Chapters 9 and 10 are applied in Chapter 11 to the general rotation of a rigid body about a point, and in Chapter 12 to the study of small vibrations of a physical system about a state of equilibrium or of steady motion.

Chapter 13 is an introduction to the special theory of relativity. It presents the basic concepts and physical principles. The chapter concludes with a derivation and discussion of the Lorentz transformation. Chapter 14 contains a fairly complete development of relativistic particle mechanics. The mathematical tools needed for most applications of the theory are presented. The reader wishing to acquire a working mastery of the special theory of relativity will want to cover the entire chapter. If only a basic understanding of relativistic dynamics is required, then Sections 14.1, 14.2, 14.4, and 14.8 should suffice.

The problems at the end of each chapter are arranged in the order in which the material is covered in the chapter, for convenience in assignment. An attempt has been made to include a sufficient variety of problems to guarantee that anyone who can solve them has mastered the material in the text. The converse is not necessarily true, since most problems require more or less physical ingenuity in addition to an understanding of the text. Many of the problems are fairly easy and should be tractable for anyone who has understood the material presented. A few are probably too difficult for most college juniors or seniors to solve without some assistance. Those problems which are particularly difficult or time-consuming are marked with an asterisk.

This third edition differs from the second edition of this text primarily in the addition of the final two chapters on the theory of relativity, and in the addition to the first seven chapters of some additional problems, similar to those contained in the earlier editions but generally of a lower average level of difficulty.

Grateful acknowledgment is made to Professor Francis W. Sears of Dartmouth College and to Professor George H. Vineyard of Brookhaven National Laboratory for their many helpful suggestions, to Mr. Charles Vittitoe and Mr. Donald Roiseland for a critical reading of Chapters 8 through 12, and to Professor Robert March and Mrs. Vernita Aigner for a critical reading of Chapters 13 and 14. The author is particularly grateful to the many teachers and students who have offered corrections and suggestions for improvement which have been incorporated in this third edition. While space does not permit mentioning individuals here, I hope that each may find my thanks expressed in the changes that have been made in this edition.

Madison, Wisconsin
May, 1971

K.R.S.

CONTENTS

To my Father

CHAPTER 1

ELEMENTS OF NEWTONIAN MECHANICS

1.1 MECHANICS, AN EXACT SCIENCE

When we say that physics is an exact science, we mean that its laws are expressed in the form of mathematical equations which describe and predict the results of precise quantitative measurements. The advantage in a quantitative physical theory is not only the practical one that it gives us the power accurately to predict and to control natural phenomena. By a comparison of the results of accurate measurements with the numerical predictions of the theory, we can gain considerable confidence that the theory is correct, and we can determine in what respects it needs to be modified. It is often possible to explain a given phenomenon in several rough qualitative ways, and if we are content with that, it may be impossible to decide which theory is correct. But if a theory can be given which predicts correctly the results of measurements to four or five (or even two or three) significant figures, the theory can hardly be very far wrong. Rough agreement might be a coincidence, but close agreement is unlikely to be. Furthermore, there have been many cases in the history of science when small but significant discrepancies between theory and accurate measurements have led to the development of new and more far-reaching theories. Such slight discrepancies would not even have been detected if we had been content with a merely qualitative explanation of the phenomena.

The symbols which are to appear in the equations that express the laws of a science must represent quantities which can be expressed in numerical terms. Hence the concepts in terms of which an exact science is to be developed must be given precise numerical meanings. If a definition of a quantity (mass, for example) is to be given, the definition must be such as to specify precisely how the value of the quantity is to be determined in any given case. A qualitative remark about its meaning may be helpful, but is not sufficient as a definition. As a matter of fact, it is probably not possible to give an ideally precise definition of every concept appearing in a physical theory. Nevertheless, when we write down a mathematical equation, the presumption is that the symbols appearing in it have precise meanings, and we should strive to make our ideas as clear and precise as possible and to recognize at what points there is a lack of precision or clarity. Sometimes a new concept can be defined in terms of others whose meanings are known, in which case there is no problem. For example,

$$\text{momentum} = \text{mass} \times \text{velocity}$$

gives a perfectly precise definition of "momentum" provided "mass" and "velocity" are assumed to be precisely defined already. But this kind of definition will not do for all terms in a theory, since we must start somewhere with a set of basic concepts or "primitive" terms whose meanings are assumed known. The first concepts to be introduced in a theory cannot be defined in the above way, since at first we have nothing to put on the right side of the equation. The meanings of these primitive terms must be made clear by some means that lies outside of the physical theories being set up. We might, for example, simply use the terms over and over until their meanings become clear. This is the way babies learn a language, and probably, to some extent, freshman physics students learn the same way. We might define all primitive terms by stating their meaning in terms of observation and experiment. In particular, nouns designating measurable quantities, like force, mass, etc., may be defined by specifying the operational process for measuring them. One school of thought holds that all physical terms should be defined in this way. Or we might simply state what the primitive terms are, with a rough indication of their physical meaning, and then let the meaning be determined more precisely by the laws and postulates we lay down and the rules that we give for interpreting theoretical results in terms of experimental situations. This is the most convenient and flexible way, and is the way physical theories are usually set up. It has the disadvantage that we are never sure that our concepts have been given a precise meaning. It is left to experience to decide not only whether our laws are correct, but even whether the concepts we use have a precise meaning. The modern theories of relativity and quanta arise as much from fuzziness in classical concepts as from inaccuracies in classical laws.

Historically, mechanics was the earliest branch of physics to be developed as an exact science. The laws of levers and of fluids in static equilibrium were known to Greek scientists in the third century B.C. The tremendous development of physics in the last three centuries began with the discovery of the laws of mechanics by Galileo and Newton. The laws of mechanics, as formulated by Isaac Newton in the middle of the seventeenth century, and the laws of electricity and magnetism, as formulated by James Clerk Maxwell about two hundred years later, are the two basic theories of classical physics. Relativistic physics, which began with the work of Einstein in 1905, and quantum physics, as based upon the work of Heisenberg and Schroedinger in 1925–1926, require a modification and reformulation of mechanics and electrodynamics in terms of new physical concepts. Nevertheless, modern physics builds on the foundations laid by classical physics, and a clear understanding of the principles of classical mechanics and electrodynamics is still essential in the study of relativistic and quantum physics. Furthermore, in the vast majority of practical applications of mechanics to the various branches of engineering and to astronomy, the laws of classical mechanics can still be applied. Except when bodies travel at speeds approaching the speed of light, or when enormous masses or enormous distances are involved, relativistic mechanics gives the same results as classical mechanics; indeed, it must, since we know from experience that classical mechanics gives correct results in ordinary applications.

Similarly, quantum mechanics should and does agree with classical mechanics except when applied to physical systems of molecular size or smaller. Indeed, one of the chief guiding principles in formulating new physical theories is the requirement that they must agree with the older theories when applied to those phenomena where the older theories are known to be correct.

Mechanics is the study of the motions of material bodies. Mechanics may be divided into three subdisciplines, *kinematics*, *dynamics*, and *statics*. Kinematics is the study and description of the possible motions of material bodies. Dynamics is the study of the laws which determine, among all possible motions, which motion will actually take place in any given case. In dynamics we introduce the concept of force. The central problem of dynamics is to determine for any physical system the motions which will take place under the action of given forces. Statics is the study of forces and systems of forces, with particular reference to systems of forces which act on bodies at rest.

We may also subdivide the study of mechanics according to the kind of physical system to be studied. This is, in general, the basis for the outline of the present book. The simplest physical system, and the one we shall study first, is a single particle. Next we shall study the motion of a system of particles. A rigid body may be treated as a special kind of system of particles. Finally, we shall study the motions of continuous media, elastic and plastic substances, solids, liquids, and gases.

A great many of the applications of classical mechanics may be based directly on Newton's laws of motion. All of the problems studied in this book, except in Chapters 9 to 14, are treated in this way. There are, however, a number of other ways of formulating the principles of classical mechanics. The equations of Lagrange and of Hamilton are examples. They are not new physical theories, for they may be derived from Newton's laws, but they are different ways of expressing the same physical theory. They use more advanced mathematical concepts, they are in some respects more elegant than Newton's formulation, and they are in some cases more powerful in that they allow the solutions of some problems whose solution based directly on Newton's laws would be very difficult. The more different ways we know to formulate a physical theory, the better chance we have of learning how to modify it to fit new kinds of phenomena as they are discovered. This is one of the main reasons for the importance of the more advanced formulations of mechanics. They are a starting point for the newer theories of relativity and quanta.

1.2 KINEMATICS, THE DESCRIPTION OF MOTION

Mechanics is the science which studies the motions of physical bodies. We must first describe motions. Easiest to describe are the motions of a *particle*, that is, an object whose size and internal structure are negligible for the problem with which we are concerned. The earth, for example, could be regarded as a particle for most problems in planetary motion, but certainly not for terrestrial problems. We can

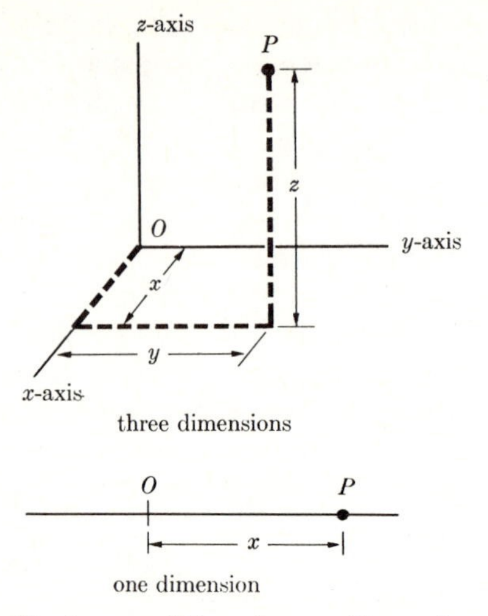

Fig. 1.1 Rectangular coordinates specifying the position of a particle P relative to an origin O.

describe the position of a particle by specifying a point in space. This may be done by giving three coordinates. Usually, rectangular coordinates are used. For a particle moving along a straight line (Chapter 2) only one coordinate need be given. To describe the motion of a particle, we specify the coordinates as functions of time:

$$\begin{aligned} &\text{one dimension: } x(t), \\ &\text{three dimensions: } x(t),\ y(t),\ z(t). \end{aligned} \tag{1.1}$$

The basic problem of classical mechanics is to find ways to determine functions like these which specify the positions of objects as functions of time, for any mechanical situation. The physical meaning of the function $x(t)$ is contained in the rules which tell us how to measure the coordinate x of a particle at a time t. Assuming we know the meaning of $x(t)$, or at least that it has a meaning (this assumption, which we make in classical mechanics, is not quite correct according to quantum mechanics), we can define the x-component of velocity v_x at time t as*

$$v_x = \dot{x} = \frac{dx}{dt}, \tag{1.2}$$

and, similarly,

$$v_y = \dot{y} = \frac{dy}{dt}, \qquad v_z = \dot{z} = \frac{dz}{dt}.$$

*We shall denote a time derivative either by d/dt or by a dot. Both notations are given in Eq. (1.2).

We now define the components of acceleration a_x, a_y, a_z as the derivatives of the velocity components with respect to time (we list several equivalent notations which may be used):

$$
\begin{aligned}
a_x &= \dot{v}_x = \frac{dv_x}{dt} = \ddot{x} = \frac{d^2x}{dt^2}, \\
a_y &= \dot{v}_y = \frac{dv_y}{dt} = \ddot{y} = \frac{d^2y}{dt^2}, \qquad (1.3) \\
a_z &= \dot{v}_z = \frac{dv_z}{dt} = z = \frac{d^2z}{dt^2}.
\end{aligned}
$$

For many purposes some other system of coordinates may be more convenient for specifying the position of a particle. When other coordinate systems are used, appropriate formulas for components of velocity and acceleration must be worked out. Spherical, cylindrical, and plane polar coordinates will be discussed in Chapter 3. For problems in two and three dimensions, the concept of a vector is very useful as a means of representing positions, velocities, and accelerations. A systematic development of vector algebra will be given in Section 3.1.

To describe a system of particles, we may specify the coordinates of each particle in any convenient coordinate system. Or we may introduce other kinds of coordinates, for example, the coordinates of the center of mass, or the distance between two particles. If the particles form a rigid body, the three coordinates of its center of mass and three angular coordinates specifying its orientation in space are sufficient to specify its position. To describe the motion of continuous matter, for example a fluid, we would need to specify the density $\rho(x, y, z, t)$ at any point (x, y, z) in space at each instant t in time, and the velocity vector $\mathbf{v}\,(x, y, z, t)$ with which the matter at the point (x, y, z) is moving at time t. Appropriate devices for describing the motion of physical systems will be introduced as needed.

1.3 DYNAMICS. MASS AND FORCE

Experience leads us to believe that the motions of physical bodies are controlled by interactions between them and their surroundings. Observations of the behavior of projectiles and of objects sliding across smooth, well-lubricated surfaces suggest the idea that changes in the velocity of a body are produced by interaction with its surroundings. A body isolated from all interactions would have a constant velocity. Hence, in formulating the laws of dynamics, we focus our attention on accelerations.

Let us imagine two bodies interacting with each other and otherwise isolated from interaction with their surroundings. As a rough approximation to this situation, imagine two boys, not necessarily of equal size, engaged in a tug of war over a rigid pole on smooth ice. Although no two actual bodies can ever be isolated completely from interactions with all other bodies, this is the simplest kind of situation to think about and one for which we expect the simplest mathematical laws. Careful experiments with actual bodies lead us to conclusions as to what we should observe if we could achieve ideal isolation of two bodies. We should observe that the two bodies are always accelerated in opposite directions, and that the ratio of their accelerations is constant for any particular pair of bodies no matter how strongly they may be pushing or pulling each other. If we measure the coordinates x_1 and x_2 of the two bodies along the line of their accelerations, then

$$\ddot{x}_1/\ddot{x}_2 = -k_{12}, \tag{1.4}$$

where k_{12} is a positive constant characteristic of the two bodies concerned. The negative sign expresses the fact that the accelerations are in opposite directions.

Furthermore, we find that in general the larger or heavier or more massive body is accelerated the least. We find, in fact, that the ratio k_{12} is proportional to the ratio of the weight of body 2 to that of body 1. The accelerations of two interacting bodies are inversely proportional to their weights. This suggests the possibility of a dynamical definition of what we shall call the *masses* of bodies in terms of their mutual accelerations. We choose a standard body as a unit mass. The mass of any other body is defined as the ratio of the acceleration of the unit mass to the acceleration of the other body when the two are in interaction:

$$m_i = k_{1i} = -\ddot{x}_1/\ddot{x}_i, \tag{1.5}$$

where m_i is the mass of body i, and body 1 is the standard unit mass.

In order that Eq. (1.5) may be a useful definition, the ratio k_{12} of the mutual accelerations of two bodies must satisfy certain requirements. If the mass defined by Eq. (1.5) is to be a measure of what we vaguely call the amount of matter in a body, then the mass of a body should be the sum of the masses of its parts, and this turns out to be the case to a very high degree of precision. It is not essential, in order to be useful in scientific theories, that physical concepts for which we give precise definitions should correspond closely to any previously held common-sense ideas. However, most precise physical concepts have originated from more or less vague common-sense ideas, and mass is a good example. Later, in the theory of relativity, the concept of mass is somewhat modified, and it is no longer exactly true that the mass of a body is the sum of the masses of its parts.

One requirement which is certainly essential is that the concept of mass be independent of the particular body which happens to be chosen as having unit mass, in the sense that the ratio of two masses will be the same no matter what unit of mass may be chosen. This will be true because of the following relation, which is found experimentally, between the mutual acceleration ratios defined by Eq. (1.4) of any three bodies:

$$k_{12}k_{23}k_{31} = 1. \tag{1.6}$$

Suppose that body 1 is the unit mass. Then if bodies 2 and 3 interact with each other, we find, using Eqs. (1.4), (1.6), and (1.5),

$$\begin{aligned}\ddot{x}_2/\ddot{x}_3 &= -k_{23}\\ &= -1/(k_{12}k_{31})\\ &= -k_{13}/k_{12}\\ &= -m_3/m_2. \end{aligned} \tag{1.7}$$

The final result contains no explicit reference to body 1, which was taken to be the standard unit mass. Thus the ratio of the masses of any two bodies is the negative inverse of the ratio of their mutual accelerations, independently of the unit of mass chosen.

By Eq. (1.7), we have, for two interacting bodies,

$$m_2\ddot{x}_2 = -m_1\ddot{x}_1. \tag{1.8}$$

This suggests that the quantity (*mass* $\times$ *acceleration*) will be important, and we call this quantity the *force* acting on a body. The acceleration of a body in space has three components, and the three components of force acting on the body are

$$F_x = m\ddot{x}, \qquad F_y = m\ddot{y}, \qquad F_z = m\ddot{z}. \tag{1.9}$$

The forces which act on a body are of various kinds, electric, magnetic, gravitational, etc., and depend on the behavior of other bodies. In general, forces due to several sources may act on a given body, and it is found that the total force given by Eqs. (1.9) is the vector sum of the forces which would be present if each source were present alone.

The theory of electromagnetism is concerned with the problem of determining the electric and magnetic forces exerted by electrical charges and currents upon one another. The theory of gravitation is concerned with the problem of determining the gravitational forces exerted by masses upon one another. The fundamental problem of mechanics is to determine the motions of any mechanical system, given the forces acting on the bodies which make up the system.

1.4 NEWTON'S LAWS OF MOTION

Isaac Newton was the first to give a complete formulation of the laws of mechanics. Newton stated his famous three laws as follows:*

1. Every body continues in its state of rest or of uniform motion in a straight line unless it is compelled to change that state by forces impressed upon it.
2. Rate of change of momentum is proportional to the impressed force, and is in the direction in which the force acts.

---

*Isaac Newton, *Mathematical Principles of Natural Philosophy* and his *System of the World*, tr. by F. Cajori (p. 13). Berkeley: University of California Press, 1934.

3. To every action there is always opposed an equal reaction.

In the second law, momentum is to be defined as the product of the mass and the velocity of the particle. Momentum, for which we use the symbol p, has three components, defined along x-, y-, and z-axes by the equations

$$p_x = mv_x, \qquad p_y = mv_y, \qquad p_z = mv_z. \tag{1.10}$$

The first two laws, together with the definition of momentum, Eqs. (1.10), and the fact that the mass is constant by Eq. (1.4),* are equivalent to Eqs. (1.9), which express them in mathematical form. The third law states that when two bodies interact, the force exerted on body 1 by body 2 is equal and opposite in direction to that exerted on body 2 by body 1. This law expresses the experimental fact given by Eq. (1.4), and can easily be derived from Eq. (1.4) and from Eqs. (1.5) and (1.9).

The status of Newton's first two laws, or of Eqs. (1.9), is often the subject of dispute. We may regard Eqs. (1.9) as defining force in terms of mass and acceleration. In this case, Newton's first two laws are not laws at all but merely definitions of a new concept to be introduced in the theory. The physical laws are then the laws of gravitation, electromagnetism, etc., which tell us what the forces are in any particular situation. Newton's discovery was not that force equals mass times acceleration, for this is merely a definition of "force." What Newton discovered was that the laws of physics are most easily expressed in terms of the concept of force defined in this way. Newton's third law is still a legitimate physical law expressing the experimental result given by Eq. (1.4) in terms of the concept of force. This point of view toward Newton's first two laws is convenient for many purposes and is often adopted. Its chief disadvantage is that Eqs. (1.9) define only the total force acting on a body, whereas we often wish to speak of the total force as a (vector) sum of component forces of various kinds due to various sources. The whole science of statics, which deals with the forces acting in structures at rest, would be unintelligible if we took Eqs. (1.9) as our definition of force, for all accelerations are zero in a structure at rest.

We may also take the laws of electromagnetism, gravitation, etc., together with the parallelogram law of addition, as defining "force." Equations (1.9) then become a law connecting previously defined quantities. This has the disadvantage that the definition of force changes whenever a new kind of force (e.g., nuclear force) is discovered, or whenever modifications are made in electromagnetism or in gravitation. Probably the best plan, the most flexible at least, is to take force as a primitive concept of our theory, perhaps defined operationally in terms of measurements with a spring balance. Then Newton's laws are laws, and so

*In the theory of relativity, the mass of a body is not constant, but depends on its velocity. In this case, law (2) and Eqs. (1.9) are not equivalent, and it turns out that law (2) is the correct formulation. Force should then be equated to time rate of change of momentum. The simple definition (1.5) of mass is not correct according to the theory of relativity unless the particles being accelerated move at low velocities.

are the laws of theories of special forces like gravitation and electromagnetism.

Aside from the question of procedure in regard to the definition of force, there are other difficulties in Newton's mechanics. The third law is not always true. It fails to hold for electromagnetic forces, for example, when the interacting bodies are far apart or rapidly accelerated and, in fact, it fails for any forces which propagate from one body to another with finite velocities. Fortunately, most of our development is based on the first two laws. Whenever the third law is used, its use will be explicitly noted and the results obtained will be valid only to the extent that the third law holds.

Another difficulty is that the concepts of Newtonian mechanics are not perfectly clear and precise, as indeed no concepts can probably ever be for any theory, although we must develop the theory as if they were. An outstanding example is the fact that no specification is made of the coordinate system with respect to which the accelerations mentioned in the first two laws are to be measured. Newton himself recognized this difficulty but found no very satisfactory way of specifying the correct coordinate system to use. Perhaps the best way to formulate these laws is to say that there is a coordinate system with respect to which they hold, leaving it to experiment to determine the correct coordinate system. It can be shown that if these laws hold in any coordinate system, they hold also in any coordinate system moving uniformly with respect to the first. This is called the principle of Newtonian relativity, and will be proved in Section 7.1, although the reader should find little difficulty in proving it for himself.

Two assumptions which are made throughout classical physics are that the behavior of measuring instruments is unaffected by their state of motion so long as they are not rapidly accelerated, and that it is possible, in principle at least, to devise instruments to measure any quantity with as small an error as we please. These two assumptions fail in extreme cases, the first at very high velocities, the second when very small magnitudes are to be measured. The failure of these assumptions forms the basis of the theory of relativity and the theory of quantum mechanics, respectively. However, for a very wide range of phenomena, Newton's mechanics is correct to a very high degree of accuracy, and forms the starting point at which the modern theories begin. Not only the laws but also the concepts of classical physics must be modified according to the modern theories. However, an understanding of the concepts of modern physics is made easier by a clear understanding of the concepts of classical physics. These difficulties are pointed out here in order that the reader may be prepared to accept later modifications in the theory. This is not to say that Newton himself (or the reader either at this stage) ought to have worried about these matters before setting up his laws of motion. Had he done so, he probably never would have developed his theory at all. It was necessary to make whatever assumptions seemed reasonable in order to get started. Which assumptions needed to be altered, and when, and in what way, could only be determined later by the successes and failures of the theory in predicting experimental results.

1.5 GRAVITATION

Although there had been previous suggestions that the motions of the planets and of falling bodies on earth might be due to a property of physical bodies by which they attract one another, the first to formulate a mathematical theory of this phenomenon was Isaac Newton. Newton showed, by methods to be considered later, that the motions of the planets could be quantitatively accounted for if he assumed that with every pair of bodies is associated a force of attraction proportional to their masses and inversely proportional to the square of the distance between them. In symbols,

$$F = \frac{Gm_1m_2}{r^2}, \tag{1.11}$$

where m_1, m_2 are the masses of the attracting bodies, r is the distance between them, and G is a universal constant whose value according to experiment is*

$$G = (6.670 \pm 0.005) \times 10^{-8} \text{ cm}^3\text{-sec}^{-2}\text{-g}^{-1}. \tag{1.12}$$

For a spherically symmetrical body, we shall show later (Section 6.2) that the force can be computed as if all the mass were at the center. For a small body of mass m at the surface of the earth, the force of gravitation is therefore

$$F = mg, \tag{1.13}$$

where

$$g = \frac{GM}{R^2} = 980.2 \text{ cm-sec}^{-2}, \tag{1.14}$$

and M is the mass of the earth and R its radius. The quantity g has the dimensions of an acceleration, and we can readily show by Eqs. (1.9) and (1.13) that any freely falling body at the surface of the earth is accelerated downward with an acceleration g.

The fact that the gravitational force on a body is proportional to its mass, rather than to some other constant characterizing the body (e.g., its electric charge), is more or less accidental from the point of view of Newton's theory. This fact is fundamental in the general theory of relativity. The proportionality between gravitational force and mass is probably the reason why the theory of gravitation is ordinarily considered a branch of mechanics, while theories of other kinds of force are not.

Equation (1.13) gives us a more convenient practical way of measuring mass than that contemplated in the original definition (1.5). We may measure a mass by measuring the gravitational force on it, as in a spring balance, or by comparing the gravitational force on it with that on a standard mass, as in the beam or platform balance; in other words, by weighing it.

---

*_Smithsonian Physical Tables_, 9th ed., 1954.

1.6 UNITS AND DIMENSIONS

In setting up a system of units in terms of which to express physical measurements, we first choose arbitrary standard units for a certain set of "fundamental" physical quantities (e.g., mass, length, and time) and then define further derived units in terms of the fundamental units (e.g., the unit of velocity is one unit length per unit of time). It is customary to choose mass, length, and time as the fundamental quantities in mechanics, although there is nothing sacred in this choice. We could equally well choose some other three quantities, or even more or fewer than three quantities, as fundamental.

There are three systems of units in common use, the centimeter-gram-second or cgs system, the meter-kilogram-second or mks system, and the foot-pound-second or English system, the names corresponding to the names of the three fundamental units in each system.* Units for other kinds of physical quantities are obtained from their defining equations by substituting the units for the fundamental quantities which occur. For example, velocity, by Eq. (1.2),

$$v_x = \frac{dx}{dt},$$

is defined as a distance divided by a time. Hence the units of velocity are cm/sec, m/sec, and ft/sec in the three above-mentioned systems, respectively.

Similarly, the reader can show that the units of force in the three systems as given by Eqs. (1.9) are g-cm-sec^{-2}, kg-m-sec^{-2}, lb-ft-sec^{-2}. These units happen to have the special names *dyne*, *newton*, and *poundal*, respectively. *Gravitational units* of force are sometimes defined by replacing Eqs. (1.9) by the equations

$$F_x = m\ddot{x}/g, \qquad F_y = m\ddot{y}/g, \qquad F_z = m\ddot{z}/g, \tag{1.15}$$

where $g = 980.2$ cm-sec^{-2} $= 9.802$ m-sec^{-2} $= 32.16$ ft-sec^{-2} is the standard acceleration of gravity at the earth's surface. Unit force is then that force exerted by the standard gravitational field on unit mass. The names gram-weight, kilogram-weight, pound-weight are given to the gravitational units of force in the three systems. In the present text, we shall write the fundamental law of mechanics in the form (1.9) rather than (1.15); hence we shall be using the *absolute* units for force and not the gravitational units.

Henceforth the question of units will rarely arise, since nearly all our examples will be worked out in algebraic form. It is assumed that the reader is sufficiently familiar with the units of measurement and their manipulation to be able to work out numerical examples in any system of units should the need arise.

In any physical equation, the dimensions or units of all additive terms on both sides of the equation must agree when reduced to fundamental units. As an

*In the mks system, there is a fourth fundamental unit, the coulomb of electrical charge, which enters into the definitions of electrical units. Electrical units in the cgs system are all defined in terms of centimeters, grams, and seconds. Electrical units in the English system are practically never used.

example, we may check that the dimensions of the gravitational constant in Eq. (1.11) are correctly given in the value quoted in Eq. (1.12):

$$F = \frac{Gm_1 m_2}{r^2}. \tag{1.11}$$

We substitute for each quantity the units in which it is expressed:

$$(\text{g-cm-sec}^{-2}) = \frac{(\text{cm}^3\text{-sec}^{-2}\text{-g}^{-1})\,(\text{g})\,(\text{g})}{(\text{cm}^2)} = (\text{g-cm-sec}^{-2}). \tag{1.16}$$

The check does not depend on which system of units we use so long as we use absolute units of force, and we may check dimensions without any reference to units, using symbols l, m, t for length, mass, time:

$$(mlt^{-2}) = \frac{(l^3 t^{-2} m^{-1})\,(m)\,(m)}{(l^2)} = (mlt^{-2}). \tag{1.17}$$

When constant factors like G are introduced, we can, of course, always make the dimensions agree in any particular equation by choosing appropriate dimensions for the constant. If the units in the terms of an equation do not agree, the equation is certainly wrong. If they do agree, this does not guarantee that the equation is right. However, a check on dimensions in a result will reveal most of the mistakes that result from algebraic errors. The reader should form the habit of mentally checking the dimensions of his formulas at every step in a derivation. When constants are introduced in a problem, their dimensions should be worked out from the first equation in which they appear, and used in checking subsequent steps.

1.7 SOME ELEMENTARY PROBLEMS IN MECHANICS

Before beginning a systematic development of mechanics based on the laws introduced in this chapter, we shall review a few problems from elementary mechanics in order to fix these laws clearly in mind.

One of the simplest mechanical problems is that of finding the motion of a body moving in a straight line, and acted upon by a constant force. If the mass of the body is m and the force is F, we have, by Newton's second law,

$$F = ma. \tag{1.18}$$

The acceleration is then constant:

$$a = \frac{dv}{dt} = \frac{F}{m}. \tag{1.19}$$

If we multiply Eq. (1.19) by dt, we obtain an expression for the change in velocity dv occurring during the short time dt:

$$dv = \frac{F}{m}\,dt. \tag{1.20}$$

Integrating, we find the total change in velocity during the time t:

$$\int_{v_0}^{v} dv = \int_0^t \frac{F}{m}\, dt, \tag{1.21}$$

$$v - v_0 = \frac{F}{m}\, t, \tag{1.22}$$

where v_0 is the velocity at $t = 0$. If x is the distance of the body from a fixed origin, measured along its line of travel, then

$$v = \frac{dx}{dt} = v_0 + \frac{F}{m}\, t. \tag{1.23}$$

We again multiply by dt and integrate to find x:

$$\int_{x_0}^{x} dx = \int_0^t \left(v_0 + \frac{F}{m}\, t\right) dt, \tag{1.24}$$

$$x = x_0 + v_0 t + \tfrac{1}{2} \frac{F}{m}\, t^2, \tag{1.25}$$

where x_0 represents the position of the body at $t = 0$. We now have a complete description of the motion. We can calculate from Eqs. (1.25) and (1.22) the velocity of the body at any time t, and the distance it has traveled. A body falling freely near the surface of the earth is acted upon by a constant force given by Eq. (1.13), and by no other force if air resistance is negligible. In this case, if x is the height of the body above some reference point, we have

$$F = -mg. \tag{1.26}$$

The negative sign appears because the force is downward and the positive direction of x is upward. Substituting in Eqs. (1.19), (1.22), and (1.25), we have the familiar equations

$$a = -g, \tag{1.27}$$

$$v = v_0 - gt, \tag{1.28}$$

$$x = x_0 + v_0 t - \tfrac{1}{2} g t^2. \tag{1.29}$$

In applying Newton's law of motion, Eq. (1.18), it is essential to decide first to what body the law is to be applied, then to insert the mass m of that body and the total force F acting on it. Failure to keep in mind this rather obvious point is the source of many difficulties, one of which is illustrated by the horse-and-wagon dilemma. A horse pulls upon a wagon, but according to Newton's third law the wagon pulls back with an equal and opposite force upon the horse. How then can either the wagon or the horse move? The reader who can solve Problem 6 at the end of this chapter will have no difficulty answering this question.

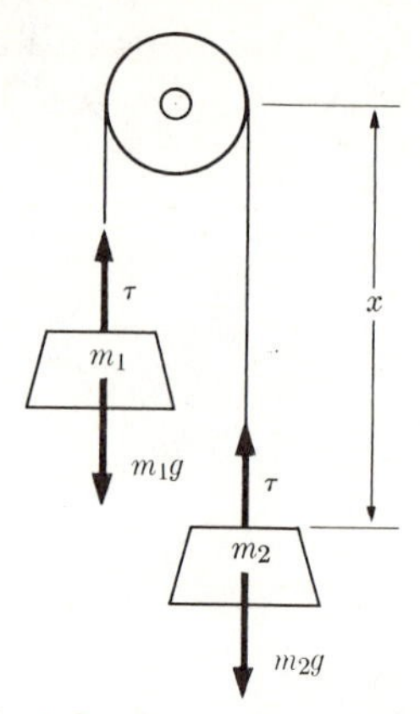

Fig. 1.2 Atwood's machine.

Consider the motion of the system illustrated in Fig. 1.2. Two masses m_1 and m_2 hang from the ends of a rope over a pulley, and we will suppose that m_2 is greater than m_1. We take x as the distance from the pulley to m_2. Since the length of the rope is constant, the coordinate x fixes the positions of both m_1 and m_2. Both move with the same velocity

$$v = \frac{dx}{dt}, \tag{1.30}$$

the velocity being positive when m_1 is moving upward and m_2 is moving downward. If we neglect friction and air resistance, the forces on m_1 and m_2 are

$$F_1 = -m_1 g + \tau, \tag{1.31}$$

$$F_2 = m_2 g - \tau, \tag{1.32}$$

where τ is the tension in the rope. The forces are taken as positive when they tend to produce a positive velocity dx/dt. Note that the terms involving τ in these equations satisfy Newton's third law. The equations of motion of the two masses are

$$-m_1 g + \tau = m_1 a, \tag{1.33}$$

$$m_2 g - \tau = m_2 a, \tag{1.34}$$

where a is the acceleration dv/dt, and is the same for both masses. By adding Eqs. (1.33) and (1.34), we can eliminate τ and solve for the acceleration:

$$a = \frac{d^2x}{dt^2} = \frac{(m_2 - m_1)}{(m_1 + m_2)} g. \tag{1.35}$$

The acceleration is constant and the velocity v and position x can be found at any time t as in the preceding example. We can substitute for a from Eq. (1.35) in either Eq. (1.33) or (1.34) and solve for the tension:

$$\tau = \frac{2m_1 m_2}{m_1 + m_2} g. \tag{1.36}$$

As a check, we note that if $m_1 = m_2$, then $a = 0$ and

$$\tau = m_1 g = m_2 g, \tag{1.37}$$

as it should if the masses are in static equilibrium. As a matter of interest, note that if $m_2 \gg m_1$, then

$$a \doteq g, \tag{1.38}$$

$$\tau \doteq 2m_1 g. \tag{1.39}$$

The reader should convince himself that these two results are to be expected in this case.

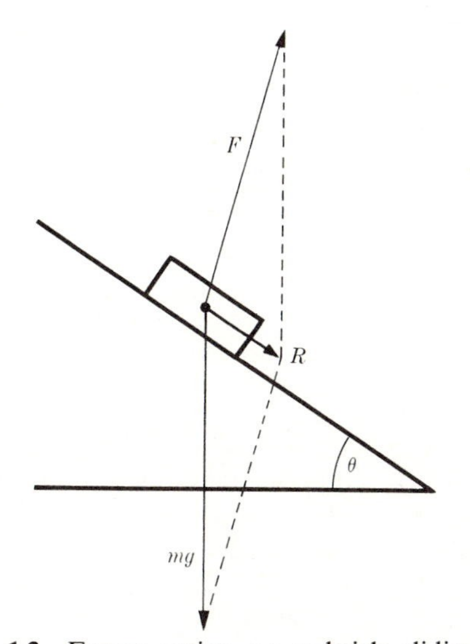

Fig. 1.3 Forces acting on a brick sliding down an incline.

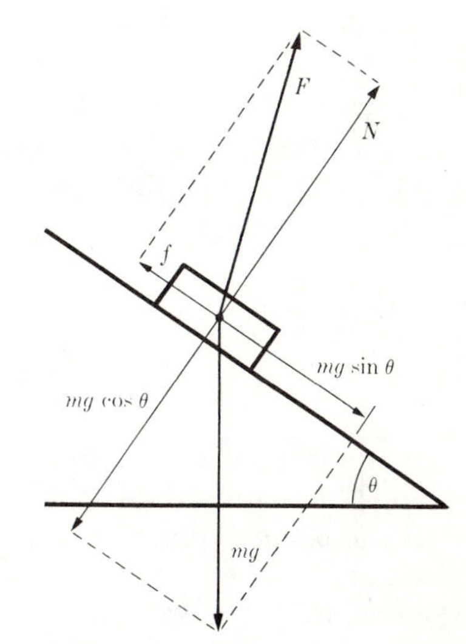

Fig. 1.4 Resolution of forces into components parallel and perpendicular to the incline.

When several forces act on a body, its acceleration is determined by the vector sum of the forces which act. Conversely, any force can be resolved in any convenient manner into vector components whose vector sum is the given force, and these components can be treated as separate forces acting on the body.* As an example, we consider a brick of mass m sliding down an incline, as shown in

*A systematic development of vector algebra will be given in Chapter 3. Only an understanding of the parallelogram law for vector addition is needed for the present discussion.

Fig. 1.3. The two forces which act on the brick are the weight mg and the force F with which the plane acts on the brick. These two forces are added according to the parallelogram law to give a resultant R which acts on the brick:

$$R = ma. \tag{1.40}$$

Since the brick is accelerated in the direction of the resultant force, it is evident that if the brick slides down the incline without jumping off or penetrating into the inclined plane, the resultant force R must be directed along the incline. In order to find R, we resolve each force into components parallel and perpendicular to the incline, as in Fig. 1.4. The force F exerted on the brick by the plane is resolved in Fig. 1.4 into two components, a force N normal to the plane preventing the brick from penetrating the plane, and a force f parallel to the plane, and opposed to the motion of the brick, arising from the friction between the brick and the plane. Adding parallel components, we obtain

$$R = mg \sin \theta - f, \tag{1.41}$$

and

$$0 = N - mg \cos \theta. \tag{1.42}$$

If the frictional force f is proportional to the normal force N, as is often approximately true for dry sliding surfaces, then

$$f = \mu N = \mu mg \cos \theta, \tag{1.43}$$

where μ is the coefficient of friction. Using Eqs. (1.43), (1.41), and (1.40), we can calculate the acceleration:

$$a = g(\sin \theta - \mu \cos \theta). \tag{1.44}$$

The velocity and position can now be found as functions of the time t, as in the first example. Equation (1.44) holds only when the brick is sliding down the incline. If it is sliding up the incline, the force f will oppose the motion, and the second term in Eq. (1.44) will be positive. This could only happen if the brick were given an initial velocity up the incline. If the brick is at rest, the frictional force f may have any value up to a maximum $\mu_s N$:

$$f \leq \mu_s N, \tag{1.45}$$

where μ_s, the coefficient of static friction, is usually greater than μ. In this case R is zero, and

$$f = mg \sin \theta \leq \mu_s mg \cos \theta. \tag{1.46}$$

According to Eq. (1.46), the angle θ of the incline must not be greater than a limiting value θ_r, the *angle of repose:*

$$\tan \theta \leq \tan \theta_r = \mu_s. \tag{1.47}$$

If θ is greater than θ_r, the brick cannot remain at rest.

If a body moves with constant speed v around a circle of radius r, its acceleration is toward the center of the circle, as we shall prove in Chapter 3, and is of magnitude

$$a = \frac{v^2}{r}. \tag{1.48}$$

Such a body must be acted on by a constant force toward the center. This centripetal force is given by

$$F = ma = \frac{mv^2}{r}. \tag{1.49}$$

Note that mv^2/r is not a "centrifugal force" directed away from the center, but is mass times acceleration and is directed toward the center, as is the centripetal force F. As an example, the moon's orbit around the earth is nearly circular, and if we assume that the earth is at rest at the center, then, by Eq. (1.11), the force on the moon is

$$F = \frac{GMm}{r^2}, \tag{1.50}$$

where M is the mass of the earth and m that of the moon. We can express this force in terms of the radius R of the earth and the acceleration g of gravity at the earth's surface by substituting for GM from Eq. (1.14):

$$F = \frac{mg\,R^2}{r^2}. \tag{1.51}$$

The speed v of the moon is

$$v = \frac{2\pi r}{T}, \tag{1.52}$$

where T is the period of revolution. Substituting Eqs. (1.51) and (1.52) in Eq. (1.49), we can find r:

$$r^3 = \frac{g\,R^2T^2}{4\pi^2}. \tag{1.53}$$

This equation was first worked out by Isaac Newton in order to check his inverse square law of gravitation.* It will not be quite accurate because the moon's orbit is not quite circular, and also because the earth does not remain at rest at the center of the moon's orbit, but instead wobbles slightly due to the attraction of the moon. By Newton's third law, this attractive force is also given by Eq. (1.51). Since the earth is much heavier than the moon, its acceleration is much smaller, and Eq. (1.53) will not be far wrong. The exact treatment of this problem is given in Section 4.7. Another small error is introduced by the fact that g, as determined

*Isaac Newton, *op. cit.*, p. 407.

experimentally, includes a small effect due to the earth's rotation. (See Section 7.3.) If we insert the measured values,

$$g = 980.2 \text{ cm-sec}^{-2},$$

$$R = 6{,}368 \text{ km},$$

$$T = 27\tfrac{1}{3} \text{ days},$$

we obtain, from Eq. (1.53),

$$r = 383{,}000 \text{ km}.$$

The mean distance to the moon according to modern measurements is

$$r = 385{,}000 \text{ km}.$$

The values of r and R available to Newton would not have given such close agreement.

PROBLEMS

1. Compute the gravitational force of attraction between an electron and a proton at a separation of 0.5 Å (1 Å = 10^{-8} cm). Compare with the electrostatic force of attraction at the same distance.

2. The coefficient of viscosity η is defined by the equation

$$\frac{F}{A} = \eta \frac{dv}{ds},$$

where F is the frictional force acting across an area A in a moving fluid, and dv is the difference in velocity parallel to A between two layers of fluid a distance ds apart, ds being measured perpendicular to A. Find the units in which the viscosity η would be expressed in the foot-pound-second, cgs, and mks systems. Find the three conversion factors for converting coefficients of viscosity from one of these systems to another.

3. A fluid flows through a cylindrical pipe of length l and radius a. A pressure difference ΔP (force per unit area) causes a flux Φ (volume per second) to flow through the pipe. Assume that ΔP is proportional to l and depends otherwise only on Φ, on the radius a of the pipe, and on the viscosity η and where η is defined in Problem 2. Show from dimensional considerations that ΔP must also be proportional to η and to Φ and inversely proportional to a^4.

4. A system of units often used by mechanical engineers chooses, in addition to the foot and the second, a third fundamental unit of force, the *pound-weight* (usually just called *pound*). The unit of mass is then a derived unit, based on Eqs. (1.9), and is called the *slug*. Express the slug in terms of the fundamental units (ft, lb-wt, sec). Express the slug in terms of pounds in the foot-pound-second system. Find the gravitational constant G in the foot-pound-weight-second system.

5. A motorist is approaching a green traffic light with speed v_0, when the light turns to amber.

a) If his reaction time is τ, during which he makes his decision to stop and applies his foot to the brake, and if his maximum braking deceleration is a, what is the minimum distance s_{min} from the intersection at the moment the light turns to amber in which he can bring his car to a stop?

b) If the amber light remains on for a time t before turning red, what is the maximum distance s_{max} from the intersection at the moment the light turns to amber such that he can continue into the intersection at speed v_0 without running the red light?

c) Show that if his initial speed v_0 is greater than

$$v_{0\,max} = 2a(t-\tau),$$

there will be a range of distances from the intersection such that he can neither stop in time nor continue through without running the red light.

d) Make some reasonable estimates of τ, t, and a, and calculate $v_{0\,max}$ in miles per hour. If $v_0 = \frac{2}{3}v_{0\,max}$, calculate s_{min} and s_{max}.

6. A boy of mass m pulls (horizontally) a sled of mass M. The coefficient of friction between sled and snow is μ.

a) Draw a diagram showing all forces acting on the boy and on the sled.

b) Find the horizontal and vertical components of each force at a moment when boy and sled each have an acceleration a.

c) If the coefficient of static friction between the boy's feet and the ground is μ_s, what is the maximum acceleration he can give to himself and the sled, assuming traction to be the limiting factor?

7. A floor mop of mass m is pushed with a force F directed along the handle, which makes an angle θ with the vertical. The coefficient of friction with the floor is μ.

a) Draw a diagram showing all forces acting on the mop.

b) For given θ, μ, find the force F required to slide the mop with uniform velocity across the floor.

c) Show that if θ is less than the angle of repose [as defined by Eq. (1.47)], the mop cannot be started across the floor by pushing along the handle. Neglect the mass of the mop handle.

8. A box of mass m slides across a horizontal table with coefficient of friction μ. The box is connected by a rope which passes over a pulley to a body of mass M hanging alongside the table. Find the acceleration of the system and the tension in the rope.

9. The brick shown in Figs. 1.3 and 1.4 is given an initial velocity v_0 up the incline. The angle θ is greater than the angle of repose. Find the distance the brick moves up the incline, and the time required for it to slide up and back to its original position.

10. A curve in a highway of radius of curvature r is banked at an angle θ with the horizontal. If the coefficient of friction is μ_s, what is the maximum speed with which a car can round the curve without skidding?

11. Assuming the earth moves in a circle of radius 93,000,000 miles, with a period of revolution of one year, find the mass of the sun in tons.

12. a) Compute the mass of the earth from its radius and the values of g and G.

b) Look up the masses and distances of the sun and moon and compute the force of attraction between earth and sun and between earth and moon. Check your results by making a rough estimate of the ratio of these two forces from a consideration of the fact that the former causes the earth to revolve about the sun once a year, whereas the latter causes the earth to wobble in a small circle, approximately once a month, about the common center of gravity of the earth-moon system.

13. The sun is about 25,000 light years from the center of the galaxy, and travels approximately in a circle at a speed of 175 mi/sec. Find the approximate mass of the galaxy by assuming that the gravitational force on the sun can be calculated as if all the mass of the galaxy were at its center. Express the result as a ratio of the galactic mass to the sun's mass. You do not need to look up either G or the sun's mass to do this problem if you compare the revolution of the sun around the galactic center with the revolution of the earth about the sun.

14. A neutron star is a collection of neutrons bound together by their mutual gravitation with a density comparable to that of an atomic nucleus (approximately 10^{12} g/cm^3). Assume that the neutron star is a sphere and show that the maximum frequency with which it may rotate, if mass is not to fly off at the equator, is $f = (\rho G/3\pi)^{1/2}$, where ρ is the density. Calculate f for a density of 10^{12} g/cm^3. It has been suggested that pulsars, which emit regular bursts of radiation at repetition rates up to about 30/sec, are rotating neutron stars.

CHAPTER 2

MOTION OF A PARTICLE IN ONE DIMENSION

2.1 MOMENTUM AND ENERGY THEOREMS

In this chapter, we study the motion of a particle of mass m along a straight line, which we will take to be the x-axis, under the action of a force F directed along the x-axis. The discussion will be applicable, as we shall see, to other cases where the motion of a mechanical system depends on only one coordinate, or where all but one coordiante can be eliminated from the problem.

The motion of the particle is governed, according to Eqs. (1.9), by the equation

$$m\frac{d^2x}{dt^2} = F. \tag{2.1}$$

Before considering the solution of Eq. (2.1), we shall define some concepts which are useful in discussing mechanical problems and prove some simple general theorems about one-dimensional motion. The linear momentum p, according to Eq. (1.10), is defined as

$$p = mv = m\frac{dx}{dt}. \tag{2.2}$$

From Eq. (2.1), using Eq. (2.2) and the fact that m is constant, we obtain

$$\frac{dp}{dt} = F. \tag{2.3}$$

This equation states that the time rate of change of momentum is equal to the applied force, and is, of course, just Newton's second law. We may call it the (differential) momentum theorem. If we multiply Eq. (2.3) by dt and integrate from t_1 to t_2, we obtain an integrated form of the momentum theorem:

$$p_2 - p_1 = \int_{t_1}^{t_2} F\,dt. \tag{2.4}$$

Equation (2.4) gives the change in momentum due to the action of the force F between the times t_1 and t_2. The integral on the right is called the *impulse* delivered by the force F during this time; F must be known as a function of t alone in order to evaluate the integral. If F is given as $F(x, v, t)$, then the impulse can be computed for any particular given motion $x(t)$, $v(t)$.

A quantity which will turn out to be of considerable importance is the *kinetic energy*, defined (in classical mechanics) by the equation

$$T = \tfrac{1}{2}mv^2. \tag{2.5}$$

If we multiply Eq. (2.1) by v, we obtain

$$mv\frac{dv}{dt} = Fv,$$

or

$$\frac{d}{dt}\left(\tfrac{1}{2}mv^2\right) = \frac{dT}{dt} = Fv. \tag{2.6}$$

Equation (2.6) gives the rate of change of kinetic energy, and may be called the (differential) energy theorem. If we multiply by dt and integrate from t_1 to t_2, we obtain the integrated form of the energy theorem:

$$T_2 - T_1 = \int_{t_1}^{t_2} Fv\,dt. \tag{2.7}$$

Equation (2.7) gives the change in energy due to the action of the force F between the times t_1 and t_2. The integral on the right is called the *work* done by the force during this time. The integrand Fv on the right is the time rate of doing work, and is called the *power* supplied by the force F. In general, when F is given as $F(x, v, t)$, the work can only be computed for a particular specified motion $x(t)$, $v(t)$. Since $v = dx/dt$, we can rewrite the work integral in a form which is convenient when F is known as a function of x:

$$T_2 - T_1 = \int_{x_1}^{x_2} F\,dx. \tag{2.8}$$

2.2 DISCUSSION OF THE GENERAL PROBLEM OF ONE-DIMENSIONAL MOTION

If the force F is known, the equation of motion (2.1) becomes a second-order ordinary differential equation for the unknown function $x(t)$. The force F may be known as a function of any or all of the variables t, x, and v. For any given motion of a dynamical system, all dynamical variables (x, v, F, p, T, etc.) associated with the system are, of course, functions of the time t, that is, each has a definite value at any particular time t. However, in many cases a dynamical variable such as the force may be known to bear a certain functional relationship to x, or to v, or to any combination of x, v, and t. As an example, the gravitational force acting on a body falling from a great height above the earth is known as a function of the height above the earth. The frictional drag on such a body would depend on its speed and on the density of the air and hence on the height above the earth; if atmospheric conditions are changing, it would also depend on t. If F is given as $F(x, v, t)$, then when $x(t)$ and $v(t)$ are known, these functions can be substituted to give F as a function of the time t alone; however, in general, this cannot be done until after Eq. (2.1) has been solved, and even then the function $F(t)$ may be different for different possible motions of the particle. In any case, if F is given as $F(x, v, t)$

(where F may depend on any or all of these variables), then Eq. (2.1) becomes a definite differential equation to be solved:

$$\frac{d^2x}{dt^2} = \frac{1}{m} F(x, \dot{x}, t). \tag{2.9}$$

This is the most general type of second-order ordinary differential equation, and we shall be concerned in this chapter with studying its solutions and their applications to mechanical problems.

Equation (2.9) applies to all possible motions of the particle under the action of the specified force. In general, there will be many such motions, for Eq. (2.9) prescribes only the acceleration of the particle at every instant in terms of its position and velocity at that instant. If we know the position and velocity of a particle at a certain time, we can determine its position a short time later (or earlier). Knowing also its acceleration, we can find its velocity a short time later. Equation (2.9) then gives the acceleration a short time later. In this manner, we can trace out the past or subsequent positions and velocities of a particle if its position x_0 and velocity v_0 are known at any one time t_0. Any pair of values of x_0 and v_0 will lead to a possible motion of the particle. We call t_0 the *initial instant*, although it may be any moment in the history of the particle, and the values of x_0 and v_0 at t_0 we call the *initial conditions*. Instead of specifying initial values for x and v, we could specify initial values of any two quantities from which x and v can be determined; for example, we may specify x_0 and the initial momentum $p_0 = mv_0$. These initial conditions, together with Eq. (2.9), then represent a perfectly definite problem whose solution should be a unique function $x(t)$ representing the motion of the particle under the specified conditions.

The mathematical theory of second-order ordinary differential equations leads to results in agreement with what we expect from the nature of the physical problem in which the equation arises. The theory asserts that, ordinarily, an equation of the form (2.9) has a unique continuous solution $x(t)$ which takes on given values x_0 and v_0 of x and $\dot{x}$ at any chosen initial value t_0 of t. "Ordinarily" here means, as far as the beginning mechanics student is concerned, "in all cases of physical interest."* The properties of differential equations like Eq. (2.9) are derived in most treatises on differential equations. We know that any physical problem must always have a unique solution, and therefore any force function $F(x, \dot{x}, t)$ which can occur in a physical problem will necessarily satisfy the required conditions for those values of x, $\dot{x}$, t of physical interest. Thus ordinarily we do not need to worry about whether a solution exists. However, most mechanical problems involve some simplification of the actual physical situation, and it is possible to oversimplify or otherwise distort a physical problem in such a way

---

*For a rigorous mathematical statement of the conditions for the existence of a solution of Eq. (2.9), see a text on differential equations, e.g. W. Leighton, *An Introduction to the Theory of Differential Equations*. New York: McGraw-Hill, 1952. (Appendix 1.)

that the resulting mathematical problem no longer possesses a unique solution. The general practice of physicists in mechanics and elsewhere is to proceed, ignoring questions of mathematical rigor. On those fortunately rare occasions when we run into difficulty, we then consult our physical intuition, or check our lapses of rigor, until the source of the difficulty is discovered. Such a procedure may bring shudders to the mathematician, but it is the most convenient and rapid way to apply mathematics to the solution of physical problems. The physicist, while he may proceed in a nonrigorous fashion, should nevertheless be acquainted with the rigorous treatment of the mathematical methods which he uses.

The existence theorem for Eq. (2.9) guarantees that there is a unique mathematical solution to this equation for all cases which will arise in practice. In some cases the exact solution can be found by elementary methods. Most of the problems considered in this text will be of this nature. Fortunately, many of the most important mechanical problems in physics can be solved without too much difficulty. In fact, one of the reasons why certain problems are considered important is that they can be easily solved. The physicist is concerned with discovering and verifying the laws of physics. In checking these laws experimentally, he is free, to a large extent, to choose those cases where the mathematical analysis is not too difficult to carry out. The engineer is not so fortunate, since his problems are selected not because they are easy to solve, but because they are of practical importance. In engineering, and often also in physics, many cases arise where the exact solution of Eq. (2.9) is difficult or impossible to obtain. In such cases various methods are available for obtaining at least approximate answers. The reader is referred to courses and texts on differential equations for a discussion of such methods.* From the point of view of theoretical mechanics, the important point is that a solution always does exist and can be found as accurately as desired. We shall restrict our attention to examples which can be treated by simple methods.

2.3 APPLIED FORCE DEPENDING ON THE TIME

If the force F is given as a function of the time, then the equation of motion (2.9) can be solved in the following manner. Multiplying Eq. (2.9) by dt and integrating from an initial instant t_0 to any later (or earlier) instant t, we obtain Eq. (2.4), which in this case we write in the form

$$mv - mv_0 = \int_{t_0}^{t} F(t)\, dt. \tag{2.10}$$

Since $F(t)$ is a known function of t, the integral on the right can, at least in principle, be evaluated and the right member is then a function of t (and t_0). We solve for v:

$$v = \frac{dx}{dt} = v_0 + \frac{1}{m}\int_{t_0}^{t} F(t)\, dt. \tag{2.11}$$

---

*W. E. Milne, *Numerical Calculus*. Princeton: Princeton University Press, 1949. (Chapter 5.) H. Levy and E. A. Baggott, *Numerical Solutions of Differential Equations*. New York: Dover Publications, 1950.

Now multiply by dt and integrate again from t_0 to t:

$$x - x_0 = v_0(t - t_0) + \frac{1}{m}\int_{t_0}^{t}\left[\int_{t_0}^{t} F(t)\,dt\right]dt. \tag{2.12}$$

To avoid confusion, we may rewrite the variable of integration as t' in the first integral and t'' in the second:

$$x = x_0 + v_0(t - t_0) + \frac{1}{m}\int_{t_0}^{t} dt'' \int_{t_0}^{t''} F(t')\,dt'. \tag{2.13}$$

This gives the required solution $x(t)$ in terms of two integrals which can be evaluated when $F(t)$ is given. A definite integral can always be evaluated. If an explicit formula for the integral cannot be found, then at least it can always be computed as accurately as we please by numerical methods. For this reason, in the discussion of a general type of problem such as the one above, we ordinarily consider the problem solved when the solution has been expressed in terms of one or more definite integrals. In a practical problem, the integrals would have to be evaluated to obtain the final solution in usable form.*

Problems in which F is given as a function of t usually arise when we seek to find the behavior of a mechanical system under the action of some external influence. As an example, we consider the motion of a free electron of charge $-e$ when subject to an oscillating electric field along the x-axis:

$$E_x = E_0 \cos(\omega t + \theta). \tag{2.14}$$

The force on the electron is

$$F = -eE_x = -eE_0 \cos(\omega t + \theta). \tag{2.15}$$

The equation of motion is

$$m\frac{dv}{dt} = -eE_0 \cos(\omega t + \theta). \tag{2.16}$$

We multiply by dt and integrate, taking $t_0 = 0$:

$$v = \frac{dx}{dt} = v_0 + \frac{eE_0 \sin\theta}{m\omega} - \frac{eE_0}{m\omega}\sin(\omega t + \theta). \tag{2.17}$$

---

*The reader who has studied differential equations may be disturbed by the appearance of three constants, t_0, v_0, and x_0, in the solution (2.13), whereas the general solution of a second-order differential equation should contain only two arbitrary constants. Mathematically, there are only two independent constants in Eq. (2.13), an additive constant containing the terms $x_0 - v_0 t_0$ plus a term from the lower limit of the last integral, and a constant multiplying t containing the term v_0 plus a term from the lower limit of the first integral. Physically, we can take any initial instant t_0, and then just two parameters x_0 and v_0 are required to specify one out of all possible motions subject to the given force.

Integrating again, we obtain

$$x = x_0 - \frac{eE_0 \cos \theta}{m\omega^2} + \left(v_0 + \frac{eE_0 \sin \theta}{m\omega} \right) t + \frac{eE_0}{m\omega^2} \cos (\omega t + \theta). \tag{2.18}$$

If the electron is initially at rest at $x_0 = 0$, this becomes

$$x = -\frac{eE_0 \cos \theta}{m\omega^2} + \frac{eE_0 \sin \theta}{m\omega} t + \frac{eE_0}{m\omega^2} \cos (\omega t + \theta). \tag{2.19}$$

It is left to the reader to explain physically the origin of the constant term and the term linear in t in Eq. (2.19) in terms of the phase of the electric field at the initial instant. How do the terms in Eq. (2.19) depend on e, m, E_0, and ω? Explain physically. Why does the oscillatory term turn out to be out of phase with the applied force?

The problem considered here is of interest in connection with the propagation of radio waves through the ionosphere, which contains a high density of free electrons. Associated with a radio wave of angular frequency ω is an electric field which may be given by Eq. (2.14). The oscillating term in Eq. (2.18) has the same frequency ω and is independent of the initial conditions. This coherent oscillation of the free electrons modifies the propagation of the wave. The nonoscillating terms in Eq. (2.18) depend on the initial conditions, and hence on the detailed motion of each electron as the wave arrives. These terms cannot contribute to the propagation characteristics of the wave, since they do not oscillate with the frequency of the wave, although they may affect the leading edge of the wave which arrives first. We see that the oscillatory part of the displacement x is 180° out of phase with the applied force due to the electric field. Since the electron has a negative charge, the resulting electric polarization is 180° out of phase with the electric field. The result is that the dielectric coefficient of the ionosphere is less than one. (In an ordinary dielectric at low frequencies, the charges are displaced in the direction of the electric force on them, and the dielectric coefficient is greater than one.) Since the velocity of light is

$$v = c(\varepsilon\mu/\varepsilon_0\mu_0)^{-1/2}, \tag{2.20}$$

where $c = 3 \times 10^8$ m/sec and $\varepsilon/\varepsilon_0$ and μ/μ_0 are the relative dielectric and magnetic coefficients respectively, and since $\mu = \mu_0$ here, the (phase) velocity v of radio waves in the ionosphere is greater than the velocity c of electromagnetic waves in empty space. Thus waves entering the ionosphere at an angle are bent back toward the earth. The effect is seen to be inversely proportional to ω^2, so that for high enough frequencies, the waves do not return to the earth but pass out through the ionosphere.

Only a slight knowledge of electromagnetic theory is required to carry this discussion through mathematically.* The dipole moment of the electron displaced from its equilibrium position is

$$-ex = -\frac{e^2}{m\omega^2} E_0 \cos(\omega t + \theta) = -\frac{e^2}{m\omega^2} E_x \tag{2.21}$$

if we consider only the oscillating term. If there are N electrons per cm^3, the total dipole moment per unit volume is

$$P_x = -\frac{Ne^2}{m\omega^2} E_x. \tag{2.22}$$

The electric displacement is

$$D_x = \varepsilon_0 E_x + P_x = \varepsilon_0 \left(1 - \frac{Ne^2}{m\omega^2}\right) E_x \text{ (mks units).} \tag{2.23}$$

Since the electric permittivity is defined by

$$D_x = \varepsilon E_x, \tag{2.24}$$

we conclude that

$$\varepsilon/\varepsilon_0 = 1 - \frac{Ne^2}{m\omega^2}, \tag{2.25}$$

and since $\mu = \mu_0$,

$$v = c\left(1 - \frac{Ne^2}{m\omega^2}\right)^{-1/2}. \tag{2.26}$$

2.4 DAMPING FORCE DEPENDING ON THE VELOCITY

Another type of force which allows an easy solution of Eq. (2.9) is the case when F is a function of v alone:

$$m\frac{dv}{dt} = F(v). \tag{2.27}$$

To solve, we multiply by $[mF(v)]^{-1}\,dt$ and integrate from t_0 to t:

$$\int_{v_0}^{v} \frac{dv}{F(v)} = \frac{t - t_0}{m}. \tag{2.28}$$

The integral on the left can be evaluated, in principle at least, when $F(v)$ is given, and an equation containing the unknown v results. If this equation is solved for v (we assume in general discussions that this can always be done), we will have an equation of the form

$$v = \frac{dx}{dt} = \varphi\left(v_0 \; \frac{t - t_0}{m}\right). \tag{2.29}$$

---

*See, e.g., F. W. Sears and M. W. Zemansky, *University Physics*, 3rd ed., Reading, Mass.: Addison-Wesley, 1964. (Sections 26.7, 27.7, 27.9.)

The solution for x is then

$$x = x_0 + \int_{t_0}^{t} \varphi\left(v_0, \frac{t-t_0}{m}\right) dt. \tag{2.30}$$

In the case of one-dimensional motion, the only important kinds of forces which depend on the velocity are frictional forces. The force of sliding or rolling friction between dry solid surfaces is nearly constant for a given pair of surfaces with a given normal force between them, and depends on the velocity only in that its direction is always opposed to the velocity. The force of friction between lubricated surfaces or between a solid body and a liquid or gaseous medium depends on the velocity in a complicated way, and the function $F(v)$ can usually be given only in the form of a tabulated summary of experimental data. In certain cases and over certain ranges of velocity, the frictional force is proportional to some fixed power of the velocity:

$$F = (\mp) b v^n. \tag{2.31}$$

If n is an odd integer, the negative sign should be chosen in the above equation. Otherwise the sign must be chosen so that the force has the opposite sign to the velocity v. The frictional force is always opposed to the velocity, and therefore does negative work, i.e., absorbs energy from the moving body. A velocity-dependent force in the same direction as the velocity would represent a source of energy; such cases do not often occur.

As an example, we consider the problem of a boat traveling with initial velocity v_0, which shuts off its engines at $t_0 = 0$ when it is at the position $x_0 = 0$. We assume the force of friction given by Eq. (2.31) with $n = 1$:

$$m \frac{dv}{dt} = -bv. \tag{2.32}$$

We solve Eq. (2.32), following the steps outlined above [Eqs. (2.27) through (2.30)]:

$$\int_{v_0}^{v} \frac{dv}{v} = -\frac{b}{m} t,$$

$$\ln \frac{v}{v_0} = -\frac{b}{m} t,$$

$$v = v_0 e^{-bt/m}. \tag{2.33}$$

We see that as $t \to \infty$, $v \to 0$, as it should, but that the boat never comes completely to rest in any finite time. The solution for x is

$$x = \int_0^t v_0 e^{-bt/m}\, dt$$

$$= \frac{mv_0}{b}(1 - e^{-bt/m}). \tag{2.34}$$

As $t \to \infty$, x approaches the limiting value

$$x_s = \frac{mv_0}{b}. \tag{2.35}$$

Thus we can specify a definite distance that the boat travels in stopping. Although according to the above result, Eq. (2.33), the velocity never becomes exactly zero, when t is sufficiently large the velocity becomes so small that the boat is practically stopped. Let us choose some small velocity v_s such that when $v < v_s$ we are willing to regard the boat as stopped (say, for example, the average random speed given to an anchored boat by the waves passing by it). Then we can define the time t_s required for the boat to stop by

$$v_s = v_0 e^{-bt_s/m}, \qquad t_s = \frac{m}{b} \ln \frac{v_0}{v_s}. \tag{2.36}$$

Since the logarithm is a slowly changing function, the stopping time t_s will not depend to any great extent on precisely what value of v_s we choose so long as it is much smaller than v_0. It is often instructive to expand solutions in a Taylor series in t. If we expand the right side of Eqs. (2.33) and (2.34) in power series in t, we obtain*

$$v = v_0 - \frac{bv_0}{m} t + \cdots, \tag{2.37}$$

$$x = v_0 t - \frac{1}{2} \frac{bv_0}{m} t^2 + \cdots. \tag{2.38}$$

Note that the first two terms in the series for v and x are just the formulas for a particle acted on by a constant force $-bv_0$, which is the initial value of the frictional force in Eq. (2.32). This is to be expected, and affords a fairly good check on the algebra which led to the solution (2.34). Series expansions are a very useful means of obtaining simple approximate formulas valid for a short range of time.

The characteristics of the motion of a body under the action of a frictional force as given by Eq. (2.31) depend on the exponent n. In general, a large exponent n

---

*The reader who has not already done so should memorize the Taylor series for a few simple functions like

$$e^x = 1 + x + \frac{x^2}{2} + \frac{x^3}{2 \cdot 3} + \frac{x^4}{2 \cdot 3 \cdot 4} + \cdots,$$

$$\ln(1+x) = x - \frac{x^2}{2} + \frac{x^3}{3} - \frac{x^4}{4} + \cdots,$$

$$(1+x)^n = 1 + nx + \frac{n(n-1)}{2} x^2 + \frac{n(n-1)(n-2)}{2 \cdot 3} x^3 + \cdots.$$

These three series are extremely useful in obtaining approximations to complicated formulas, valid when x is small.

will result in rapid initial slowing but slow final stopping, and vice versa, as one can see by sketching graphs of F versus v for various values of n. For small enough values of n, the velocity comes to zero in a finite time. For large values of n, the body not only requires an infinite time, but travels an infinite distance before stopping. This disagrees with ordinary experience, an indication that while the exponent n may be large at high velocities, it must become smaller at low velocities. The exponent $n = 1$ is often assumed in problems involving friction, particularly when friction is only a small effect to be taken into account approximately. The reason for taking $n = 1$ is that this gives easy equations to solve, and is often a fairly good approximation when the frictional force is small, provided b is properly chosen.

2.5 CONSERVATIVE FORCE DEPENDING ON POSITION. POTENTIAL ENERGY

One of the most important types of motion occurs when the force F is a function of the coordinate x alone:

$$m\frac{dv}{dt} = F(x). \tag{2.39}$$

We have then, by the energy theorem (2.8),

$$\tfrac{1}{2}mv^2 - \tfrac{1}{2}mv_0^2 = \int_{x_0}^{x} F(x)\,dx. \tag{2.40}$$

The integral on the right is the work done by the force when the particle goes from x_0 to x. We now define the *potential energy* $V(x)$ as the work done by the force when the particle goes from x to some chosen standard point x_s:

$$V(x) = \int_{x}^{x_s} F(x)\,dx = -\int_{x_s}^{x} F(x)\,dx. \tag{2.41}$$

The reason for calling this quantity potential energy will appear shortly. In terms of $V(x)$, we can write the integral in Eq. (2.40) as follows:

$$\int_{x_0}^{x} F(x)\,dx = -V(x) + V(x_0). \tag{2.42}$$

With the help of Eq. (2.42), Eq. (2.40) can be written

$$\tfrac{1}{2}mv^2 + V(x) = \tfrac{1}{2}mv_0^2 + V(x_0). \tag{2.43}$$

The quantity on the right depends only on the initial conditions and is therefore constant during the motion. It is called the *total energy* E, and we have the law of conservation of kinetic plus potential energy, which holds, as we can see, only when the force is a function of position alone:

$$\tfrac{1}{2}mv^2 + V(x) = T + V = E. \tag{2.44}$$

Solving for v, we obtain

$$v = \frac{dx}{dt} = \sqrt{\frac{2}{m}}[E - V(x)]^{1/2}. \tag{2.45}$$

The function $x(t)$ is to be found by solving for x the equation

$$\sqrt{\frac{m}{2}} \int_{x_0}^{x} [E - V(x)]^{-1/2}\, dx = t - t_0. \tag{2.46}$$

In this case, the initial conditions are expressed in terms of the constants E and x_0.

In applying Eq. (2.46), and in taking the indicated square root in the integrand, care must be taken to use the proper sign, depending on whether the velocity v given by Eq. (2.45) is positive or negative. In cases where v is positive during some parts of the motion and negative during other parts, it may be necessary to carry out the integration in Eq. (2.46) separately for each part of the motion.

From the definition (2.41) we can express the force in terms of the potential energy:

$$F = -\frac{dV}{dx}. \tag{2.47}$$

This equation can be taken as expressing the physical meaning of the potential energy. The potential energy is a function whose negative derivative gives the force. The effect of changing the coordinate of the standard point x_s is to add a constant to $V(x)$. Since it is the derivative of V which enters into the dynamical equations as the force, the choice of standard point x_s is immaterial. A constant can always be added to the potential $V(x)$ without affecting the physical results. (The same constant must, of course, be added to E.)

As an example, we consider the problem of a particle subject to a linear restoring force, for example, a mass fastened to a spring:

$$F = -kx. \tag{2.48}$$

The potential energy, if we take $x_s = 0$, is

$$\begin{aligned} V(x) &= -\int_0^x (-kx)\, dx \\ &= \tfrac{1}{2}kx^2. \end{aligned} \tag{2.49}$$

Equation (2.46) becomes, for this case, with $t_0 = 0$,

$$\sqrt{\frac{m}{2}} \int_{x_0}^{x} (E - \tfrac{1}{2}kx^2)^{-1/2}\, dx = t. \tag{2.50}$$

Now make the substitutions

$$\sin\theta = x\sqrt{\frac{k}{2E}}, \tag{2.51}$$

$$\omega = \sqrt{\frac{k}{m}}, \tag{2.52}$$

so that

$$\sqrt{\frac{m}{2}}\int_{x_0}^{x}(E-\tfrac{1}{2}kx^2)^{-1/2}\,dx = \frac{1}{\omega}\int_{\theta_0}^{\theta} d\theta = \frac{1}{\omega}(\theta-\theta_0),$$

and, by Eq. (2.50),

$$\theta = \omega t + \theta_0 .$$

We can now solve for x in Eq. (2.51):

$$x = \sqrt{\frac{2E}{k}}\sin\theta = A\sin(\omega t+\theta_0), \tag{2.53}$$

where

$$A = \sqrt{\frac{2E}{k}}. \tag{2.54}$$

Thus the coordinate x oscillates harmonically in time, with amplitude A and frequency $\omega/2\pi$. The initial conditions are here determined by the constants A and θ_0, which are related to E and x_0 by

$$E = \tfrac{1}{2}kA^2, \tag{2.55}$$

$$x_0 = A\sin\theta_0 . \tag{2.56}$$

Notice that in this example we meet the sign difficulty in taking the square root in Eq. (2.50) by replacing $(1-\sin^2\theta)^{-1/2}$ by $(\cos\theta)^{-1}$, a quantity which can be made either positive or negative as required by choosing θ in the proper quadrant.

A function of the dependent variable and its first derivative which is constant for all solutions of a second-order differential equation, is called a *first integral* of the equation. The function $\frac{1}{2}m\dot{x}^2+V(x)$ is called the *energy integral* of Eq. (2.39). An integral of the equations of motion of a mechanical system is also called a *constant of the motion.* In general, any mechanical problem can be solved if we can find enough first integrals, or constants of the motion.

Even in cases where the integral in Eq. (2.46) cannot easily be evaluated or the resulting equation solved to give an explicit solution for $x(t)$, the energy integral,

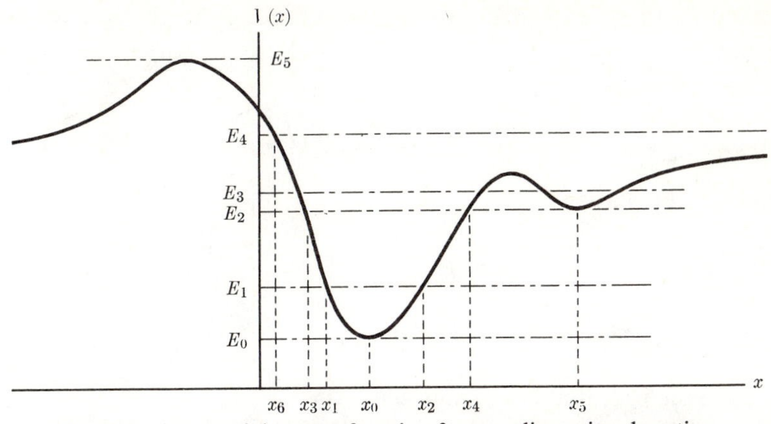

Fig. 2.1 A potential-energy function for one-dimensional motion.

Eq. (2.44), gives us useful information about the solution. For a given energy E, we see from Eq. (2.45) that the particle is confined to those regions on the x-axis where $V(x) \leq E$. Furthermore, the velocity is proportional to the square root of the difference between E and $V(x)$. Hence, if we plot $V(x)$ versus x, we can give a good qualitative description of the kinds of motion that are possible. For the potential-energy function shown in Fig. 2.1 we note that the least energy possible is E_0. At this energy, the particle can only be at rest at x_0. With a slightly higher energy E_1, the particle can move between x_1 and x_2; its velocity decreases as it approaches x_1 or x_2, and it stops and reverses its direction when it reaches either x_1 or x_2, which are called *turning points* of the motion. With energy E_2, the particle may oscillate between turning points x_3 and x_4, or remain at rest at x_5. With energy E_3, there are four turning points and the particle may oscillate in either of the two potential valleys. With energy E_4, there is only one turning point; if the particle is initially traveling to the left, it will turn at x_6 and return to the right, speeding up over the valleys at x_0 and x_5, and slowing down over the hill between. At energies above E_5, there are no turning points and the particle will move in one direction only, varying its speed according to the depth of the potential at each point.

When a particle is oscillating near a point of stable equilibrium, we can find an approximate solution for its motion. Let $V(x)$ have a minimum at $x = x_0$, and expand the function $V(x)$ in a Taylor series about this point:

$$V(x) = V(x_0) + \left(\frac{dV}{dx}\right)_{x_0}(x-x_0) + \frac{1}{2}\left(\frac{d^2V}{dx^2}\right)_{x_0}(x-x_0)^2 + \frac{1}{6}\left(\frac{d^3V}{dx^3}\right)_{x_0}(x-x_0)^3 + \cdots. \tag{2.57}$$

The constant $V(x_0)$ can be dropped without affecting the physical results. Since x_0 is a minimum point,

$$\left(\frac{dV}{dx}\right)_{x_0} = 0, \qquad \left(\frac{d^2V}{dx^2}\right)_{x_0} \geq 0. \tag{2.58}$$

Making the abbreviations

$$k = \left(\frac{d^2V}{dx^2}\right)_{x_0}, \tag{2.59}$$

$$x' = x - x_0, \tag{2.60}$$

we can write the potential function in the form

$$V(x') = \tfrac{1}{2}kx'^2 \cdots. \tag{2.61}$$

For sufficiently small values of x', provided $k \neq 0$, we can neglect the terms represented by dots, and Eq. (2.61) becomes identical with Eq. (2.49). Hence, for small oscillations about any potential minimum, except in the exceptional case $k = 0$, the motion is that of a harmonic oscillator, with frequency given by Eqs. (2.52) and (2.59).

A point where $V(x)$ has a minimum is called a point of *stable equilibrium.* A particle at rest at such a point will remain at rest. If displaced a slight distance, it will experience a restoring force tending to return it, and it will oscillate about the equilibrium point. A point where $V(x)$ has a maximum is called a point of *unstable* equilibrium. In theory, a particle at rest there can remain at rest, since the force is zero, but if it is displaced the slightest distance, the force acting on it will push it farther away from the unstable equilibrium position. A region where $V(x)$ is constant is called a region of *neutral* equilibrium, since a particle can be displaced slightly without suffering either a restoring or a repelling force.

This kind of qualitative discussion, based on the energy integral, is simple and very useful. Study this example until you understand it well enough to be able to see at a glance, for any potential energy curve, the types of motion that are possible.

It may be that only part of the force on a particle is derivable from a potential function $V(x)$. Let F' be the remainder of the force:

$$F = -\frac{dV}{dx} + F'. \tag{2.62}$$

In this case the energy $(T+V)$ is no longer constant. If we substitute F from Eq. (2.62) in Eq. (2.1), and multiply by dx/dt, we have, after rearranging terms,

$$\frac{d}{dt}(T+V) = F'v. \tag{2.63}$$

The time rate of change of kinetic plus potential energy is equal to the power delivered by the additional force F'.

2.6 FALLING BODIES

One of the simplest and most commonly occurring types of one-dimensional motion is that of falling bodies. We take up this type of motion here as an illustration of the principles discussed in the preceding sections.

A body falling near the surface of the earth, if we neglect air resistance, is subject to a constant force

$$F = -mg, \tag{2.64}$$

where we have taken the positive direction as upward. The equation of motion is

$$m\frac{d^2x}{dt^2} = -mg. \tag{2.65}$$

The solution may be obtained by any of the three methods discussed in Sections 2.3, 2.4, and 2.5, since a constant force may be considered as a function of either t, v, or x. The reader will find it instructive to solve the problem by all three methods. We have already obtained the result in Chapter 1 [Eqs. (1.28) and (1.29)].

In order to include the effect of air resistance, we may assume a frictional force proportional to v, so that the total force is

$$F = -mg-bv. \tag{2.66}$$

The constant b will depend on the size and shape of the falling body, as well as on the viscosity of the air. The problem must now be treated as a case of $F(v)$:

$$m\frac{dv}{dt} = -mg-bv. \tag{2.67}$$

Taking $v_0 = 0$ at $t = 0$, we proceed as in Section 2.4 [Eq. (2.28)]:

$$\int_0^v \frac{dv}{v+(mg/b)} = -\frac{bt}{m}. \tag{2.68}$$

We integrate and solve for v:

$$v = -\frac{mg}{b}(1-e^{-bt/m}). \tag{2.69}$$

We may obtain a formula useful for short times of fall by expanding the exponential function in a power series:

$$v = -gt+\frac{1}{2}\frac{bg}{m}t^2+\cdots. \tag{2.70}$$

Thus for a short time ($t \ll m/b$), $v = -gt$, approximately, and the effect of air resistance can be neglected. After a long time, we see from Eq. (2.69) that

$$v \doteq -\frac{mg}{b}, \qquad \text{if} \qquad t \gg \frac{m}{b}.$$

The velocity mg/b is called the *terminal velocity* of the falling body in question. The body reaches within $1/e$ of its terminal velocity in a time $t = m/b$. We could use the experimentally determined terminal velocity to find the constant b. We now integrate Eq. (2.69), taking $x_0 = 0$:

$$x = \frac{m^2 g}{b^2}\left(1 - \frac{bt}{m} - e^{-bt/m}\right). \tag{2.71}$$

By expanding the exponential function in a power series, we obtain

$$x = -\tfrac{1}{2}gt^2 + \tfrac{1}{6}\frac{bg}{m}t^3 + \cdots. \tag{2.72}$$

If $t \ll m/b$, $x \doteq -\frac{1}{2}gt^2$, as in Eq. (1.29). When $t \gg m/b$,

$$x \doteq \left(\frac{m^2 g}{b^2} - \frac{mg}{b}t\right).$$

This result is easily interpreted in terms of terminal velocity. Why is the positive constant present?

It is worth noting that we may obtain the series solution (2.70) directly from the differential equation (2.67) without solving it exactly. Let us first neglect altogether the term involving b, so that the solution is

$$v^{(0)} = -gt.$$

Substitute this result in the last term in Eq. (2.67) and integrate again:

$$v^{(1)} = -gt + \tfrac{1}{2}\frac{bg}{m}t^2.$$

This result agrees with the first two terms in Eq. (2.70). If we put $v = v^{(1)}$ into the last term in Eq. (2.67) and integrate, we get a better approximation $v^{(2)}$, good to order b^2, and so on. This method of successive approximations is often useful in solving an equation containing a small term which in a zero-order approximation may be neglected. A similar method can be used to solve by successive approximations an algebraic equation containing one or more small terms.

For small heavy bodies with large terminal velocities, a better approximation may be

$$F = bv^2. \tag{2.73}$$

The reader should be able to show that with the frictional force given by Eq. (2.73), the result (taking $x_0 = v_0 = 0$ at $t_0 = 0$) is

$$v = -\sqrt{\frac{mg}{b}} \tanh\left(\sqrt{\frac{bg}{m}}\, t\right)$$

$$\doteq \begin{cases} -gt, & \text{if} \quad t \ll \sqrt{\dfrac{m}{bg}}, \\[2ex] -\sqrt{\dfrac{mg}{b}}, & \text{if} \quad t \gg \sqrt{\dfrac{m}{bg}}, \end{cases} \tag{2.74}$$

$$x = -\frac{m}{b} \ln \cosh\left(\sqrt{\frac{bg}{m}}\, t\right)$$

$$\doteq \begin{cases} -\frac{1}{2}gt^2, & \text{if} \quad t \ll \sqrt{\dfrac{m}{bg}}, \\[2ex] \dfrac{m}{b} \ln 2 - \sqrt{\dfrac{mg}{b}}\, t, & \text{if} \quad t \gg \sqrt{\dfrac{m}{bg}}. \end{cases} \tag{2.75}$$

Again there is a terminal velocity, given this time by $(mg/b)^{1/2}$. The terminal velocity can always be found as the velocity at which the frictional force equals the gravitational force, and will exist whenever the frictional force becomes sufficiently large at high velocities.

In the case of bodies falling from a great height, the variation of the gravitational force with height should be taken into account. In this case, we neglect air resistance (in order to be able to use the energy method), and measure x from the center of the earth. Then if M is the mass of the earth and m the mass of the falling body,

$$F = -\frac{mMG}{x^2}, \tag{2.76}$$

and

$$V(x) = -\int_{\infty}^{x} F\, dx = -\frac{mMG}{x}, \tag{2.77}$$

where we have taken $x_s = \infty$ in order to avoid a constant term in $V(x)$. Equation (2.45) becomes

$$v = \frac{dx}{dt} = \pm\sqrt{\frac{2}{m}}\left(E + \frac{mMG}{x}\right)^{1/2}. \tag{2.78}$$

The plus sign refers to ascending motion, the minus sign to descending motion.

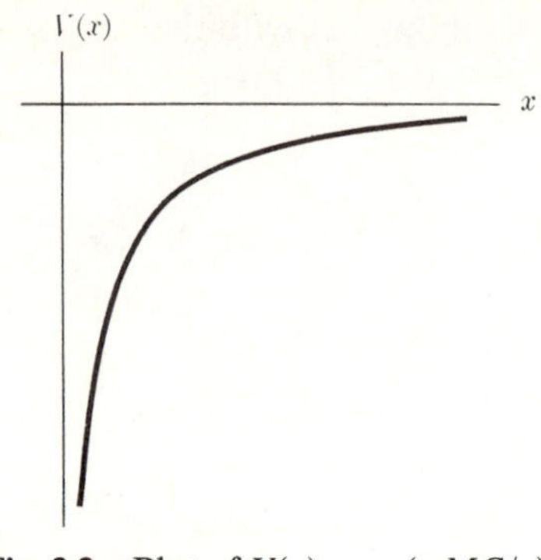

Fig. 2.2 Plot of $V(x) = -(mMG/x)$.

The function $V(x)$ is plotted in Fig. 2.2. We see that there are two types of motion, depending on whether E is positive or negative. When E is positive, there is no turning point, and if the body is initially moving upward, it will continue to move upward forever, with decreasing velocity, approaching the limiting velocity

$$v_l = \sqrt{\frac{2E}{m}}. \tag{2.79}$$

When E is negative, there is a turning point at a height

$$x_T = \frac{mMG}{-E}. \tag{2.80}$$

If the body is initially moving upward, it will come to a stop at x_T, and fall back to the earth. The dividing case between these two types of motion occurs when the initial position and velocity are such that $E = 0$. The turning point is then at infinity, and the body moves upward forever, approaching the limiting velocity $v_l = 0$. If $E = 0$, then at any height x, the velocity will be

$$v_e = \sqrt{\frac{2MG}{x}}. \tag{2.81}$$

This is called the *escape velocity* for a body at distance x from the center of the earth, because a body moving upward at height x with velocity v_e will just have sufficient energy to travel upward indefinitely (if there is no air resistance).

To find $x(t)$, we must evaluate the integral

$$\int_{x_0}^{x} \frac{dx}{\pm\left(E + \frac{mMG}{x}\right)^{1/2}} = \sqrt{\frac{2}{m}}\, t, \tag{2.82}$$

where x_0 is the height at $t = 0$. To solve for the case when E is negative, we substitute

$$\cos\theta = \sqrt{\frac{-Ex}{mMG}}. \tag{2.83}$$

Equation (2.82) then becomes

$$\frac{mMG}{(-E)^{3/2}}\int_{\theta_0}^{\theta} 2\cos^2\theta\, d\theta = \sqrt{\frac{2}{m}}\,t. \tag{2.84}$$

(We choose a positive sign for the integrand so that θ will increase when t increases.) We can, without loss of generality, take x_0 to be at the turning point x_T, since the body will at some time in its past or future career pass through x_T if no force except gravity acts upon it, provided $E < 0$. Then $\theta_0 = 0$, and

$$\frac{mMG}{(-E)^{3/2}}(\theta+\sin\theta\cos\theta) = \sqrt{\frac{2}{m}}\,t,$$

or

$$\theta+\tfrac{1}{2}\sin 2\theta = \sqrt{\frac{2MG}{x_T^3}}\,t, \tag{2.85}$$

and

$$x = x_T\cos^2\theta. \tag{2.86}$$

This pair of equations cannot be solved explicitly for $x(t)$. A numerical solution can be obtained by choosing a sequence of values of θ and finding the corresponding values of x and t from Eqs. (2.85) and (2.86). That part of the motion for which x is less than the radius of the earth will, of course, not be correctly given, since Eq. (2.76) assumes all the mass of the earth concentrated at $x = 0$ (not to mention the fact that we have omitted from our equation of motion not only air resistance, but also the forces which would act on the body when it collides with the earth).

The solution can be obtained in a similar way for the cases when E is positive or zero.

2.7 THE SIMPLE HARMONIC OSCILLATOR

The most important problem in one-dimensional motion, and fortunately one of the easiest to solve, is the harmonic or linear oscillator. The simplest example is that of a mass m fastened to a spring whose constant is k (Fig. 2.3). If we measure x from the relaxed position of the spring, then the spring exerts a restoring force

$$F = -kx. \tag{2.87}$$

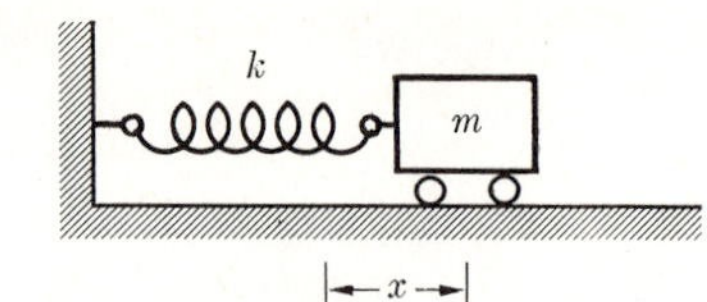

Fig. 2.3 Model of a simple harmonic oscillator.

The potential energy associated with this force is

$$V(x) = \tfrac{1}{2}kx^2. \tag{2.88}$$

The equation of motion, if we assume no other force acts, is

$$m\frac{d^2x}{dt^2} + kx = 0. \tag{2.89}$$

Equation (2.89) describes the free harmonic oscillator. Its solution was obtained in Section 2.5. The motion is a simple sinusoidal oscillation about the point of equilibrium. In all physical cases there will be some frictional force acting, though it may often be very small. As a good approximation in most cases, particularly when the friction is small, we can assume that the frictional force is proportional to the velocity. Since this is almost the only kind of frictional force for which the problem can easily be solved, we shall restrict our attention to this case. If we use Eq. (2.31) for the frictional force with $n = 1$, the equation of motion then becomes

$$m\frac{d^2x}{dt^2} + b\frac{dx}{dt} + kx = 0. \tag{2.90}$$

This equation describes the damped harmonic oscillator. Its motion, at least for small damping, consists of a sinusoidal oscillation of gradually decreasing amplitude, as we shall show later. If the oscillator is subject to an additional impressed force $F(t)$, its motion will be given by

$$m\frac{d^2x}{dt^2} + b\frac{dx}{dt} + kx = F(t). \tag{2.91}$$

If $F(t)$ is a sinusoidally varying force, Eq. (2.91) leads to the phenomenon of resonance, where the amplitude of oscillation becomes very large when the frequency of the impressed force equals the natural frequency of the undamped free oscillator.

The importance of the harmonic oscillator problem lies in the fact that equations of the same form as Eqs. (2.89) through (2.91) turn up in a wide variety of physical problems. In almost every case of one-dimensional motion where the potential energy function $V(x)$ has one or more minima, the motion of the particle for small oscillations about the minimum point follows Eq. (2.89), as we have shown in Section 2.5.

When a solid is deformed, it resists the deformation with a force proportional to the amount of deformation, provided the deformation is not too great. This statement is called *Hooke's law*. It follows from the fact that the undeformed solid is at a potential-energy minimum and that the potential energy may be expanded in a Taylor series in the coordinate describing the deformation. If a solid is deformed beyond a certain point, called its *elastic limit*, it will remain permanently deformed; that is, its structure is altered so that its undeformed shape for minimum potential energy is changed. It turns out in most cases that the higher-order terms in the series (2.57) are negligible almost up to the elastic limit, so that Hooke's law holds almost up to the elastic limit. When the elastic limit is exceeded and plastic flow takes place, the forces depend in a complicated way not only on the shape of the material, but also on the velocity of deformation and even on its previous history, so that the forces can no longer be specified in terms of a potential-energy function.

Thus practically any problem involving mechanical vibrations reduces to that of the harmonic oscillator at small amplitudes of vibration, that is, so long as the elastic limits of the materials involved are not exceeded. The motions of stretched strings and membranes, and of sound vibrations in an enclosed gas or in a solid, result in a number of so-called normal modes of vibration, each mode behaving in many ways like an independent harmonic oscillator. An electric circuit containing inductance L, resistance R, and capacitance C in series, and subject to an applied electromotive force $E(t)$, satisfies the equation

$$L\frac{d^2q}{dt^2}+R\frac{dq}{dt}+\frac{q}{C} = E(t), \tag{2.92}$$

where q is the charge on the condenser and dq/dt is the current. This equation is identical in form with Eq. (2.91). Early work on electrical circuits was often carried out by analogy with the corresponding mechanical problem. Today the situation is often reversed, and the mechanical and acoustical engineers are able to make use of the simple and effective methods developed by electrical engineers for handling vibration problems. The theory of electrical oscillations in a transmission line or in a cavity is similar mathematically to the problem of the vibrating string or resonating air cavity. The quantum-mechanical theory of an atom can be put in a form which is identical mathematically with the theory of a system of harmonic oscillators.

2.8 LINEAR DIFFERENTIAL EQUATIONS WITH CONSTANT COEFFICIENTS

Equations (2.89) to (2.91) are examples of second-order linear differential equations. The *order* of a differential equation is the order of the highest derivative that occurs in it. Most equations of mechanics are of second order. (Why?) A *linear* differential equation is one in which there are no terms of higher than first degree in the dependent variable (in this case x) and its derivatives. Thus the most general type of linear differential equation of order n would be

$$a_n(t)\frac{d^n x}{dt^n}+a_{n-1}(t)\frac{d^{n-1}x}{dt^{n-1}}+\cdots+a_1(t)\frac{dx}{dt}+a_0(t)x = b(t). \tag{2.93}$$

If $b(t) = 0$, the equation is said to be *homogeneous;* otherwise it is *inhomogeneous.* Linear equations are important because there are simple general methods for solving them, particularly when the coefficients $a_0, a_1, \ldots, a_n$ are constants, as in Eqs. (2.89) to (2.91). In the present section, we shall solve the problem of the free harmonic oscillator [Eq. (2.89)], and at the same time develop a general method of solving any linear homogeneous differential equation with constant coefficients. This method is applied in Section 2.9 to the damped harmonic oscillator equation (2.90). In Section 2.10 we shall study the behavior of a harmonic oscillator under a sinusoidally oscillating impressed force. In Section 2.11 a theorem is developed which forms the basis for attacking Eq. (2.91) with any impressed force $F(t)$, and the methods of attack are discussed briefly.

The solution of Eq. (2.89), which we obtained in Section 2.5, we now write in the form

$$x = A\sin(\omega_0 t+\theta), \qquad \omega_0 = \sqrt{k/m}. \tag{2.94}$$

This solution depends on two "arbitrary" constants A and θ. They are called arbitrary because no matter what values are given to them, the solution (2.94) will satisfy Eq. (2.89). They are not arbitrary in a physical problem, but depend on the initial conditions. It can be shown that the general solution of any second-order differential equation depends on two arbitrary constants. By this we mean that we can write the solution in the form

$$x = x(t;\, C_1, C_2), \tag{2.95}$$

such that for every value of C_1 and C_2, or every value within a certain range, $x(t;\, C_1, C_2)$ satisfies the equation and, furthermore, practically every solution of the equation is included in the function $x(t;\, C_1, C_2)$ for some value of C_1 and C_2.* If we can find a solution containing two arbitrary constants which satisfies a second-order differential equation, then we can be sure that practically every solution will be included in it. The methods of solution of the differential equations studied in previous sections have all been such as to lead directly to a solution corresponding to the initial conditions of the physical problem. In the present and subsequent sections of this chapter, we shall consider methods which lead to a general solution containing two arbitrary constants. These constants must then

*The only exceptions are certain "singular" solutions which may occur in regions where the mathematical conditions for a unique solution (Section 2.2) are not satisfied.

be given the proper values to fit the initial conditions of the physical problem; the fact that a solution with two arbitrary constants is the general solution guarantees that we can always satisfy the initial conditions by proper choice of the constants.

We now state two theorems regarding linear homogeneous differential equations:

Theorem 1. *If $x = x_1(t)$ is any solution of a linear homogeneous differential equation, and C is any constant, then $x = Cx_1(t)$ is also a solution.*

Theorem 2. *If $x = x_1(t)$ and $x = x_2(t)$ are solutions of a linear homogeneous differential equation, then $x = x_1(t)+x_2(t)$ is also a solution.*

We prove these theorems only for the case of a second-order equation:

$$a_2(t)\frac{d^2x}{dt^2}+a_1(t)\frac{dx}{dt}+a_0(t)x = 0. \tag{2.96}$$

The proof can easily be generalized to higher-order equations. Assume that $x = x_1(t)$ satisfies Eq. (2.96). Then

$$a_2(t)\frac{d^2(Cx_1)}{dt^2}+a_1(t)\frac{d(Cx_1)}{dt}+a_0(t)(Cx_1) = C\left[a_2(t)\frac{d^2x_1}{dt^2}+a_1(t)\frac{dx_1}{dt}+a_0(t)x_1\right] = 0.$$

Hence $x = Cx_1(t)$ also satisfies Eq. (2.96). If $x_1(t)$ and $x_2(t)$ both satisfy Eq. (2.96), then

$$\begin{aligned}a_2(t)\frac{d^2(x_1+x_2)}{dt^2}&+a_1(t)\frac{d(x_1+x_2)}{dt}+a_0(t)(x_1+x_2)\\&=\left[a_2(t)\frac{d^2x_1}{dt^2}+a_1(t)\frac{dx_1}{dt}+a_0(t)x_1\right]\\&\quad+\left[a_2(t)\frac{d^2x_2}{dt^2}+a_1(t)\frac{dx_2}{dt}+a_0(t)x_2\right] = 0.\end{aligned}$$

Hence $x = x_1(t)+x_2(t)$ also satisfies Eq. (2.96). The problem of finding the general solution of Eq. (2.96) thus reduces to that of finding any two independent "particular" solutions $x_1(t)$ and $x_2(t)$, for then Theorems 1 and 2 guarantee that

$$x = C_1x_1(t)+C_2x_2(t) \tag{2.97}$$

is also a solution. Since this solution contains two arbitrary constants, it must be the general solution. The requirement that $x_1(t)$ and $x_2(t)$ be independent means in this case that one is not a multiple of the other. If $x_1(t)$ were a constant multiple of $x_2(t)$, then Eq. (2.97) would really contain only one arbitrary constant. The right member of Eq. (2.97) is called a *linear combination* of x_1 and x_2.

In the case of equations like (2.89) and (2.90), where the coefficients are constant, a solution of the form $x = e^{pt}$ always exists. To show this, assume that a_0, a_1, and a_2 are all constant in Eq. (2.96) and substitute

$$x = e^{pt}, \qquad \frac{dx}{dt} = pe^{pt}, \qquad \frac{d^2x}{dt^2} = p^2 e^{pt}. \tag{2.98}$$

We then have

$$(a_2 p^2 + a_1 p + a_0) e^{pt} = 0. \tag{2.99}$$

Canceling out e^{pt}, we have an algebraic equation of second degree in p. Such an equation has, in general, two roots. If they are different, this gives two independent functions e^{pt} satisfying Eq. (2.96) and our problem is solved. If the two roots for p should be equal, we have found only one solution, but then, as we shall show in the next section, the function

$$x = te^{pt} \tag{2.100}$$

also satisfies the differential equation. The linear homogeneous equation of nth order with constant coefficients can also be solved by this method.

Let us apply the method to Eq. (2.89). Making the substitution (2.98), we have

$$mp^2 + k = 0, \tag{2.101}$$

whose solution is

$$p = \pm\sqrt{-\frac{k}{m}} = \pm i\omega_0, \qquad \omega_0 = \sqrt{\frac{k}{m}}. \tag{2.102}$$

This gives, as the general solution,

$$x = C_1 e^{i\omega_0 t} + C_2 e^{-i\omega_0 t}. \tag{2.103}$$

In order to interpret this result, we remember that

$$e^{i\theta} = \cos\theta + i\sin\theta. \tag{2.104}$$

If we allow complex numbers x as solutions of the differential equation, then the arbitrary constants C_1 and C_2 must also be complex in order for Eq. (2.103) to be the general solution. The solution of the physical problem must be real, hence we must choose C_1 and C_2 so that x turns out to be real. The sum of two complex numbers is real if one is the complex conjugate of the other. If

$$C = a + ib, \tag{2.105}$$

and

$$C^* = a - ib, \tag{2.106}$$

then

$$\begin{aligned} C + C^* &= 2a, \\ C - C^* &= 2ib. \end{aligned} \tag{2.107}$$

Now $e^{i\omega_0 t}$ is the complex conjugate of $e^{-i\omega_0 t}$, so that if we set $C_1 = C$, $C_2 = C^*$, then x will be real:

$$x = Ce^{i\omega_0 t} + C^*e^{-i\omega_0 t}. \tag{2.108}$$

We could evaluate x by using Eqs. (2.104), (2.105), and (2.106), but the algebra is simpler if we make use of the polar representation of a complex number:

$$C = a + ib = re^{i\theta}, \tag{2.109}$$

$$C^* = a - ib = re^{-i\theta}, \tag{2.110}$$

where

$$r = (a^2 + b^2)^{1/2}, \qquad \tan\theta = b/a, \tag{2.111}$$

$$a = r\cos\theta, \qquad b = r\sin\theta. \tag{2.112}$$

The reader should verify that these equations follow algebraically from Eq. (2.104). If we represent C as a point in the complex plane, then a and b are its rectangular coordinates, and r and θ are its polar coordinates. Using the polar representation of C, Eq. (2.108) becomes (we set $r = \frac{1}{2}A$)

$$\begin{aligned} x &= \tfrac{1}{2}Ae^{i(\omega_0 t + \theta)} + \tfrac{1}{2}Ae^{-i(\omega_0 t + \theta)} \\ &= A\cos(\omega_0 t + \theta). \end{aligned} \tag{2.113}$$

This is the general real solution of Eq. (2.89). It differs from the solution (2.94) only by a shift of $\pi/2$ in the phase constant θ.

Setting $B_1 = A\cos\theta$, $B_2 = -A\sin\theta$, we can write our solution in another form:

$$x = B_1\cos\omega_0 t + B_2\sin\omega_0 t. \tag{2.114}$$

The constants A, θ, or B_1, B_2, are to be obtained in terms of the initial values x_0, v_0 at $t = 0$ by setting

$$x_0 = A\cos\theta = B_1, \tag{2.115}$$

$$v_0 = -\omega_0 A\sin\theta = \omega_0 B_2. \tag{2.116}$$

The solutions are easily obtained:

$$A = \left(x_0^2 + \frac{v_0^2}{\omega_0^2}\right)^{1/2}, \tag{2.117}$$

$$\tan\theta = -\frac{v_0}{x_0\omega_0}, \tag{2.118}$$

or

$$B_1 = x_0, \tag{2.119}$$

$$B_2 = \frac{v_0}{\omega_0}. \tag{2.120}$$

Another way of handling Eq. (2.103) would be to notice that, since Eq. (2.89) contains only real coefficients, a complex function can satisfy it only if both real and imaginary parts satisfy it separately. (The proof of this statement is a matter of substituting $x = u + iw$ and carrying out a little algebra.) Hence if a solution is (we set $r = A$)

$$x = Ce^{i\omega_0 t} = Ae^{i(\omega_0 t + \theta)}$$

$$= A \cos(\omega_0 t + \theta) + iA \sin(\omega_0 t + \theta), \tag{2.121}$$

then both the real and imaginary parts of this solution must separately be solutions, and we have either solution (2.113) or (2.94). We can carry through the solutions of linear equations like this, and perform any algebraic operations we please on them in their complex form (so long as we do not multiply two complex numbers together), with the understanding that at each step what we are really concerned with is only the real part or only the imaginary part. This procedure is often useful in the treatment of problems involving harmonic oscillations, and we shall use it in Section 2.10.

It is often very convenient to represent a sinusoidal function as a complex exponential:

$$\cos\theta = \text{real part of } e^{i\theta} = \frac{e^{i\theta} + e^{-i\theta}}{2}, \tag{2.122}$$

$$\sin\theta = \text{imaginary part of } e^{i\theta} = \frac{e^{i\theta} - e^{-i\theta}}{2i}. \tag{2.123}$$

Exponential functions are easier to handle algebraically than sines and cosines. The reader will find the relations (2.122), (2.123), and (2.104) useful in deriving trigonometric formulas. The power series for the sine and cosine functions are readily obtained by expanding $e^{i\theta}$ in a power series and separating the real and imaginary parts. The trigonometric rule for $\sin(A+B)$ and $\cos(A+B)$ can be easily obtained from the algebraic rule for adding exponents. Many other examples could be cited.

2.9 THE DAMPED HARMONIC OSCILLATOR

The equation of motion for a particle subject to a linear restoring force and a frictional force proportional to its velocity is [Eq. (2.90)]

$$m\ddot{x} + b\dot{x} + kx = 0, \tag{2.124}$$

where the dots stand for time derivatives. Applying the method of Section 2.8, we make the substitution (2.98) and obtain

$$mp^2 + bp + k = 0. \tag{2.125}$$

The solution is

$$p = -\frac{b}{2m} \pm \left[\left(\frac{b}{2m}\right)^2 - \frac{k}{m}\right]^{1/2}. \tag{2.126}$$

We distinguish three cases: (a) $k/m > (b/2m)^2$, (b) $k/m < (b/2m)^2$, and (c) $k/m = (b/2m)^2$.

In case (a), we make the substitutions

$$\omega_0 = \sqrt{\frac{k}{m}}, \tag{2.127}$$

$$\gamma = \frac{b}{2m}, \tag{2.128}$$

$$\omega_1 = (\omega_0^2 - \gamma^2)^{1/2}, \tag{2.129}$$

where γ is called the damping coefficient and $(\omega_0/2\pi)$ is the natural frequency of the undamped oscillator. There are now two solutions for p:

$$p = -\gamma \pm i\omega_1. \tag{2.130}$$

The general solution of the differential equation is therefore

$$x = C_1 e^{-\gamma t + i\omega_1 t} + C_2 e^{-\gamma t - i\omega_1 t}. \tag{2.131}$$

Setting

$$C_1 = \tfrac{1}{2}Ae^{i\theta}, \qquad C_2 = \tfrac{1}{2}Ae^{-i\theta}, \tag{2.132}$$

we have

$$x = Ae^{-\gamma t}\cos(\omega_1 t + \theta). \tag{2.133}$$

This corresponds to an oscillation of frequency $(\omega_1/2\pi)$ with an amplitude $Ae^{-\gamma t}$ which decreases exponentially with time (Fig. 2.4). The constants A and θ depend

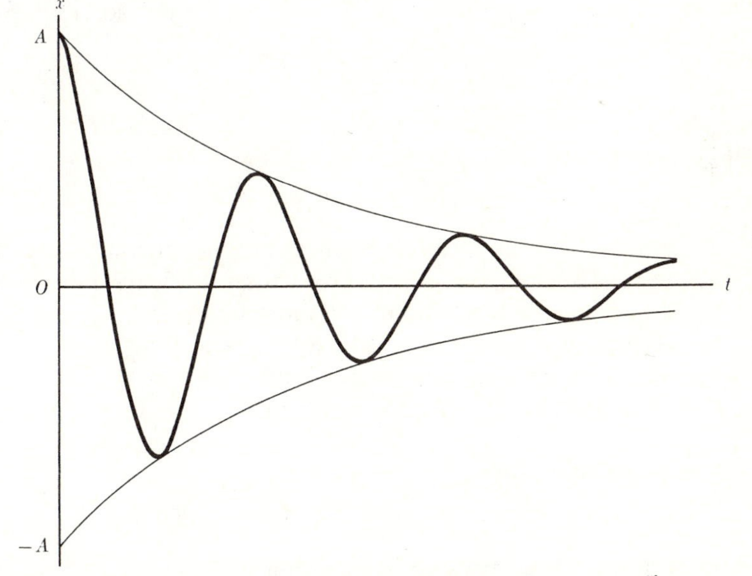

Fig. 2.4 Motion of damped harmonic oscillator. Heavy curve: $x = Ae^{-\gamma t}\cos\omega t$, $\gamma = \omega/8$. Light curve: $x = \pm Ae^{-\gamma t}$.

upon the initial conditions. The frequency of oscillation is less than without damping. The solution (2.133) can also be written

$$x = e^{-\gamma t}(B_1 \cos \omega_1 t + B_2 \sin \omega_1 t). \tag{2.134}$$

In terms of the constants ω_0 and γ, Eq. (2.124) can be written

$$\ddot{x} + 2\gamma\dot{x} + \omega_0^2 x = 0. \tag{2.135}$$

This form of the equation is often used in discussing mechanical oscillations.

The total energy (kinetic plus potential) of the oscillator is

$$E = \tfrac{1}{2}m\dot{x}^2 + \tfrac{1}{2}kx^2. \tag{2.136}$$

It is no longer constant; the friction $-b\dot{x}$ plays the role of F' in Eq. (2.63). In the important case of small damping, $\gamma \ll \omega_0$, we can set $\omega_1 \doteq \omega_0$ and neglect γ compared with ω_0, and we have for the energy corresponding to the solution (2.133), approximately,

$$E \doteq \tfrac{1}{2}kA^2 e^{-2\gamma t} = E_0 e^{-2\gamma t}. \tag{2.137}$$

Thus the energy falls off exponentially at twice the rate at which the amplitude decays. The fractional rate of decline or *logarithmic derivative* of E is

$$\frac{1}{E}\frac{dE}{dt} = \frac{d \ln E}{dt} = -2\gamma. \tag{2.138}$$

We now consider case (b), ($\omega_0 < \gamma$). In this case, the two solutions for p are

$$\begin{aligned} p &= -\gamma_1 = -\gamma - (\gamma^2 - \omega_0^2)^{1/2}, \\ p &= -\gamma_2 = -\gamma + (\gamma^2 - \omega_0^2)^{1/2}. \end{aligned} \tag{2.139}$$

The general solution is

$$x = C_1 e^{-\gamma_1 t} + C_2 e^{-\gamma_2 t}. \tag{2.140}$$

These two terms both decline exponentially with time, one at a faster rate than the other. The constants C_1 and C_2 may be chosen to fit the initial conditions. The reader should determine them for two important cases: $x_0 \neq 0$, $v_0 = 0$ and $x_0 = 0$, $v_0 \neq 0$, and draw curves $x(t)$ for the two cases.

In case (c), ($\omega_0 = \gamma$), we have only one solution for p:

$$p = -\gamma. \tag{2.141}$$

The corresponding solution for x is

$$x = e^{-\gamma t}. \tag{2.142}$$

We now show that, in this case, another solution is

$$x = te^{-\gamma t}. \tag{2.143}$$

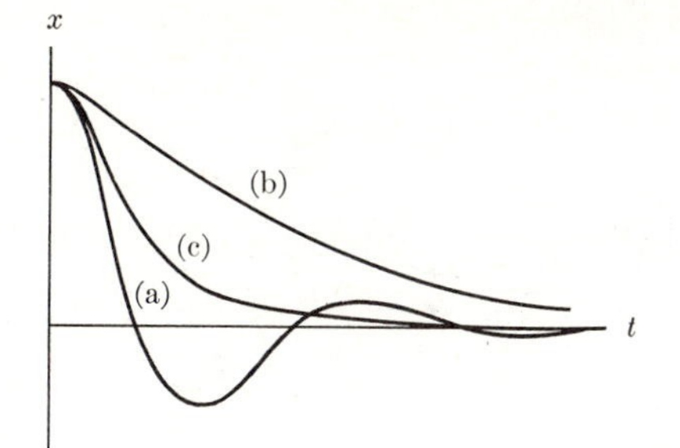

Fig. 2.5 Return of harmonic oscillator to equilibrium. (a) Underdamped. (b) Overdamped. (c) Critically damped.

To prove this, we compute

$$\begin{aligned} \dot{x} &= e^{-\gamma t} - \gamma t e^{-\gamma t}, \\ \ddot{x} &= -2\gamma e^{-\gamma t} + \gamma^2 t e^{-\gamma t}. \end{aligned} \tag{2.144}$$

The left side of Eq. (2.135) is, for this x,

$$\ddot{x} + 2\gamma\dot{x} + \omega_0^2 x = (\omega_0^2 - \gamma^2) t e^{-\gamma t}. \tag{2.145}$$

This is zero if $\omega_0 = \gamma$. Hence the general solution in case $\omega_0 = \gamma$ is

$$x = (C_1 + C_2 t)e^{-\gamma t}. \tag{2.146}$$

If we keep either ω_0 or γ fixed and let the other vary, we see from Eq. (2.139) that

$$\gamma_1 > \gamma_c > \gamma_2, \tag{2.147}$$

where γ_c is the value when $\gamma = \omega_0$. The solution (2.146) therefore declines exponentially at a rate intermediate between that of the two exponential terms in Eq. (2.140). Hence, for fixed γ or ω_0, the solution (2.146) falls to zero faster after a sufficiently long time than the solution (2.140), except in the case $C_2 = 0$ in Eq. (2.140). Cases (a), (b), and (c) are important in problems involving mechanisms which approach an equilibrium position under the action of a frictional damping force, e.g., pointer reading meters, hydraulic and pneumatic spring returns for doors, etc. In most cases, it is desired that, if displaced, the mechanism move quickly and smoothly back to its equilibrium position. For a given damping coefficient γ, or for a given ω_0, this is accomplished in the shortest time without overshoot if $\omega_0 = \gamma$ [case (c)].* This case is called *critical damping*. If $\omega_0 < \gamma$, the system is said to be *overdamped;* it behaves sluggishly and does not return as quickly to $x = 0$ as for critical damping. If $\omega_0 > \gamma$, the system is said to be *underdamped;* the coordinate x then overshoots the value $x = 0$ and oscillates. Note that at critical damping, $\omega_1 = 0$, so that the period of oscillation becomes infinite.

*Note, however, Problem 41 at the end of this chapter.

The behavior is shown in Fig. 2.5 for the case of a system displaced from equilibrium and released ($x_0 \neq 0$, $v_0 = 0$). The reader should draw similar curves for the case where the system is given a sharp blow at $t = 0$ (i.e., $x_0 = 0$, $v_0 \neq 0$).

2.10 THE FORCED HARMONIC OSCILLATOR

The harmonic oscillator subject to an external applied force is governed by Eq. (2.91). In order to simplify the problem of solving this equation, we state the following theorem:

Theorem 3. *If $x_i(t)$ is a solution of an inhomogeneous linear equation [e.g., Eq. (2.91)], and $x_h(t)$ is a solution of the corresponding homogeneous equation [e.g., Eq. (2.90)], then $x(t) = x_i(t)+x_h(t)$ is also a solution of the inhomogeneous equation.*

This theorem applies whether the coefficients in the equation are constants or functions of t. The proof is a matter of straightforward substitution, and is left to the reader. In consequence of Theorem 3, if we know the general solution x_h of the homogeneous equation (2.90) (we found this in Section 2.9), then we need find only one particular solution x_i of the inhomogeneous equation (2.91). For we can add x_i to x_h and obtain a solution of Eq. (2.91) which contains two arbitrary constants and is therefore the general solution.

The most important case is that of a sinusoidally oscillating applied force. If the applied force oscillates with angular frequency ω and amplitude F_0, the equation of motion is

$$m\frac{d^2x}{dt^2}+b\frac{dx}{dt}+kx = F_0 \cos(\omega t+\theta_0), \tag{2.148}$$

where θ_0 is a constant specifying the phase of the applied force. There are, of course, many solutions of Eq. (2.148), of which we need find only one. From physical considerations, we expect that one solution will be a steady oscillation of the coordinate x at the same frequency as the applied force:

$$x = A_s \cos(\omega t+\theta_s). \tag{2.149}$$

The amplitude A_s and phase θ_s of the oscillations in x will have to be determined by substituting Eq. (2.149) in Eq. (2.148). This procedure is straightforward and leads to the correct answer. The algebra is simpler, however, if we write the force as the real part of a complex function:*

$$F(t) = \mathrm{Re}(\mathbf{F}_0 e^{i\omega t}), \tag{2.150}$$

$$\mathbf{F}_0 = F_0 e^{i\theta_0}. \tag{2.151}$$

*Note the use of bold face type ($\mathbf{F}$, $\mathbf{x}$) to distinguish complex quantities from the corresponding real quantities (F, x).

Thus if we can find a solution $\mathbf{x}(t)$ of

$$m\frac{d^2\mathbf{x}}{dt^2}+b\frac{d\mathbf{x}}{dt}+k\mathbf{x} = \mathbf{F}_0 e^{i\omega t}, \tag{2.152}$$

then, by splitting the equation into real and imaginary parts, we can show* that the real part of $\mathbf{x}(t)$ will satisfy Eq. (2.148). We assume a solution of the form

$$\mathbf{x} = \mathbf{x}_0 e^{i\omega t},$$

so that

$$\dot{\mathbf{x}} = i\omega \mathbf{x}_0 e^{i\omega t}, \qquad \ddot{\mathbf{x}} = -\omega^2 \mathbf{x}_0 e^{i\omega t}. \tag{2.153}$$

Substituting in Eq. (2.152), we solve for $\mathbf{x}_0$:

$$\mathbf{x}_0 = \frac{\mathbf{F}_0/m}{\omega_0^2-\omega^2+2i\gamma\omega}. \tag{2.154}$$

The solution of Eq. (2.152) is therefore

$$\mathbf{x} = \mathbf{x}_0 e^{i\omega t} = \frac{(\mathbf{F}_0/m)e^{i\omega t}}{\omega_0^2-\omega^2+2i\gamma\omega}. \tag{2.155}$$

The simplest way to write Eq. (2.155) is to express the denominator in polar form [Eq. (2.109)]:

$$\omega_0^2-\omega^2+2i\gamma\omega = [(\omega_0^2-\omega^2)^2+4\gamma^2\omega^2]^{1/2} \exp\left(i \tan^{-1}\frac{2\gamma\omega}{\omega_0^2-\omega^2}\right). \tag{2.156}$$

It is convenient to define the angle

$$\beta = \frac{\pi}{2} - \tan^{-1}\frac{2\gamma\omega}{\omega_0^2-\omega^2} = \tan^{-1}\frac{\omega_0^2-\omega^2}{2\gamma\omega}, \tag{2.157}$$

$$\sin\beta = \frac{\omega_0^2-\omega^2}{[(\omega_0^2-\omega^2)^2+4\gamma^2\omega^2]^{1/2}}, \tag{2.158}$$

$$\cos\beta = \frac{2\gamma\omega}{[(\omega_0^2-\omega^2)^2+4\gamma^2\omega^2]^{1/2}}. \tag{2.159}$$

This definition is purely a matter of taste, and is arranged so that $\beta = 0$ when $\omega = \omega_0$ and $\beta \to \pm\pi/2$ as $\omega \to \pm\infty$. (See Fig. 2.6.) This definition also makes our treatment parallel to the customary treatment of Eq. (2.92) in electrical engineering. If we use Eqs. (2.156) and (2.157) and the fact that

$$i = e^{i\pi/2}, \tag{2.160}$$

---

*The assertion "we can show that . . . " throughout this book will mean that the reader who has followed the discussion to this point should be able to supply the proof himself. (In this case, put $\mathbf{x} = x+iy$ and the result falls out.) Long or tricky proofs will either be given in the text, or a reference cited, or the reader will be warned that it is not easy.

we may rewrite Eq. (2.155) in the form

$$\mathbf{x} = \frac{F_0}{im[(\omega_0^2 - \omega^2)^2 + 4\gamma^2\omega^2]^{1/2}} e^{i(\omega t + \theta_0 + \beta)} . \tag{2.161}$$

The complex velocity is

$$\dot{\mathbf{x}} = i\omega\mathbf{x} = \frac{\omega F_0}{m[(\omega_0^2 - \omega^2)^2 + 4\gamma^2\omega^2]^{1/2}} e^{i(\omega t + \theta_0 + \beta)}. \tag{2.162}$$

The real position and velocity are then

$$\begin{aligned} x &= \mathrm{Re}(\mathbf{x}) \\ &= \frac{F_0}{m} \frac{1}{[(\omega_0^2 - \omega^2)^2 + 4\gamma^2\omega^2]^{1/2}} \sin(\omega t + \theta_0 + \beta), \end{aligned} \tag{2.163}$$

and

$$\begin{aligned} \dot{x} &= \mathrm{Re}(\dot{\mathbf{x}}) \\ &= \frac{F_0}{m} \frac{\omega}{[(\omega_0^2 - \omega^2)^2 + 4\gamma^2\omega^2]^{1/2}} \cos(\omega t + \theta_0 + \beta). \end{aligned} \tag{2.164}$$

This is a particular solution of Eq. (2.148) containing no arbitrary constants. By Theorem 3 and Eq. (2.133), the general solution (for the underdamped oscillator) is

$$x = Ae^{-\gamma t} \cos(\omega_1 t + \theta) + \frac{F_0/m}{[(\omega_0^2 - \omega^2)^2 + 4\gamma^2\omega^2]^{1/2}} \sin(\omega t + \theta_0 + \beta). \tag{2.165}$$

This solution contains two arbitrary constants A, θ, whose values are determined by the initial values x_0, v_0 at $t = 0$. The first term dies out exponentially in time and is called the *transient*. The second term is called the *steady state*, and oscillates with constant amplitude. The transient depends on the initial conditions. The steady state which remains after the transient dies away is independent of the initial conditions. (When there is no damping, $\gamma = 0$, the "transient" does not die away, but we may still define it as that part of the solution which has the natural frequency $\omega_1 = \omega_0$; the term "transient" is not very descriptive in this case.)

In the steady state, the rate at which work is done on the oscillator by the applied force is

$$\begin{aligned} \dot{x}F(t) &= \frac{F_0^2}{m} \frac{\omega}{[(\omega^2 - \omega_0^2)^2 + 4\gamma^2\omega^2]^{1/2}} \cos(\omega t + \theta_0) \cos(\omega t + \theta_0 + \beta) \\ &= \frac{F_0^2}{m} \frac{\omega \cos\beta \cos^2(\omega t + \theta_0)}{[(\omega^2 - \omega_0^2)^2 + 4\gamma^2\omega^2]^{1/2}} - \frac{F_0^2}{2m} \frac{\omega \sin\beta \sin 2(\omega t + \theta_0)}{[(\omega^2 - \omega_0^2)^2 + 4\gamma^2\omega^2]^{1/2}}. \end{aligned} \tag{2.166}$$

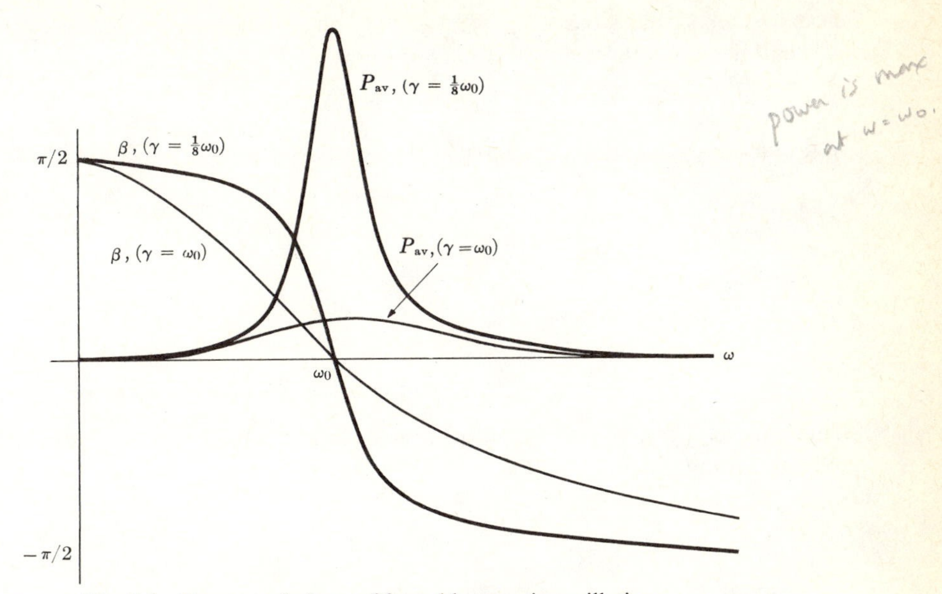

Fig. 2.6 Power and phase of forced harmonic oscillations.

The last term on the right is zero on the average, while the average value of $\cos^2 (\omega t + \theta_0)$ over a complete cycle is $\frac{1}{2}$. Hence the average power delivered by the applied force is

$$P_{av} = \langle \dot{x}F(t) \rangle_{av} = \frac{F_0^2 \cos \beta}{2m} \frac{\omega}{[(\omega^2 - \omega_0^2)^2 + 4\gamma^2\omega^2]^{1/2}}, \tag{2.167}$$

or

$$P_{av} = \tfrac{1}{2} F_0 \dot{x}_m \cos \beta, \tag{2.168}$$

where $\dot{x}_m$ is the maximum value of $\dot{x}$. A similar relation holds for power delivered to an electrical circuit. The factor $\cos \beta$ is called the *power factor*. In the electrical case, β is the phase angle between the current and the applied emf. Using formula (2.162) for $\cos \beta$, we can rewrite Eq. (2.167):

$$P_{av} = \frac{F_0^2}{m} \frac{\gamma\omega^2}{(\omega^2 - \omega_0^2)^2 + 4\gamma^2\omega^2}. \tag{2.169}$$

It is easy to show that in the steady state power is supplied to the oscillator at the same average rate that power is being dissipated by friction, as of course it must be. The power P_{av} has a maximum for $\omega = \omega_0$. In Fig. 2.6, the power P_{av} (in arbitrary units) and the phase β of steady-state forced oscillations are plotted against ω for two values of γ. The heavy curves are for small damping; the light curves are for greater damping. Formula (2.169) can be simplified somewhat in case $\gamma \ll \omega_0$. In this case, P_{av} is large only near the resonant frequency ω_0, and we shall deduce a formula valid near $\omega = \omega_0$. Defining

$$\Delta\omega = \omega - \omega_0, \tag{2.170}$$

and assuming $\Delta\omega \ll \omega_0$, we have

$$(\omega^2 - \omega_0^2) = (\omega + \omega_0)\,\Delta\omega \doteq 2\omega_0\,\Delta\omega\,, \tag{2.171}$$

$$\omega^2 \doteq \omega_0^2. \tag{2.172}$$

Hence

$$P_{av} \doteq \frac{F_0^2}{4m} \frac{\gamma}{(\Delta\omega)^2 + \gamma^2}\,. \tag{2.173}$$

This simple formula gives a good approximation to P_{av} near resonance. The corresponding formula for β is

$$\cos\beta \doteq \frac{\gamma}{[(\Delta\omega)^2 + \gamma^2]^{1/2}}, \qquad \sin\beta \doteq \frac{-\Delta\omega}{[(\Delta\omega)^2 + \gamma^2]^{1/2}}\,. \tag{2.174}$$

When $\omega \ll \omega_0$, $\beta \doteq \pi/2$, and Eq. (2.164) becomes

$$x \doteq \frac{F_0}{\omega_0^2 m} \cos(\omega t + \theta_0) = \frac{F(t)}{k}. \tag{2.175}$$

This result is easily interpreted physically; when the force varies slowly, the particle moves in such a way that the applied force is just balanced by the restoring force. When $\omega \gg \omega_0$, $\beta \doteq -\pi/2$, and Eq. (2.164) becomes

$$x \doteq -\frac{F_0}{\omega^2 m} \cos(\omega t + \theta_0) = -\frac{F(t)}{\omega^2 m}\,. \tag{2.176}$$

The motion now depends only on the mass of the particle and on the frequency of the applied force, and is independent of the friction and the restoring force. This result is, in fact, identical with that obtained in Section 2.3 [see Eqs. (2.15) and (2.19)] for a free particle subject to an oscillating force.

We can apply the result (2.165) to the case of an electron bound to an equilibrium position $x = 0$ by an elastic restoring force, and subject to an oscillating electric field:

$$E_x = E_0 \cos\omega t, \tag{2.177}$$

$$F = -eE_0 \cos\omega t. \tag{2.178}$$

The motion will be given by

$$x = Ae^{-\gamma t}\cos(\omega_1 t + \theta) - \frac{eE_0}{m} \frac{\sin(\omega t + \beta)}{[(\omega^2 - \omega_0^2)^2 + 4\gamma^2\omega^2]^{1/2}}\,. \tag{2.179}$$

The term of interest here is the second one, which is independent of the initial conditions and oscillates with the frequency of the electric field. Expanding the second term, we get

$$
\begin{aligned}
x &= -\frac{eE_0}{m}\frac{\sin\beta\cos\omega t}{[(\omega^2-\omega_0^2)^2+4\gamma^2\omega^2]^{1/2}}-\frac{eE_0}{m}\frac{\cos\beta\sin\omega t}{[(\omega^2-\omega_0^2)^2+4\gamma^2\omega^2]^{1/2}} \\
&= \frac{-eE_0\cos\omega t}{m}\frac{\omega_0^2-\omega^2}{[(\omega^2-\omega_0^2)^2+4\gamma^2\omega^2]} \\
&\quad -\frac{eE_0\sin\omega t}{m}\frac{2\gamma\omega}{[(\omega^2-\omega_0^2)^2+4\gamma^2\omega^2]}.
\end{aligned}
\tag{2.180}
$$

The first term represents an oscillation of x in phase with the applied force at low frequencies, 180° out of phase at high frequencies. The second term represents an oscillation of x that is 90° out of phase with the applied force, the velocity $\dot{x}$ for this term being in phase with the applied force. Hence the second term corresponds to an absorption of energy from the applied force. The second term contains a factor γ and is therefore small, if $\gamma \ll \omega_0$, except near resonance. If we imagine a dielectric medium consisting of electrons bound by elastic forces to positions of equilibrium, then the first term in Eq. (2.180) will represent an electric polarization proportional to the applied oscillating electric field, while the second term will represent an absorption of energy from the electric field. Near the resonant frequency, the dielectric medium will absorb energy, and will be opaque to electromagnetic radiation. Above the resonant frequency, the displacement of the electrons is out of phase with the applied force, and the resulting electric polarization will be out of phase with the applied electric field. The dielectric constant and index of refraction will be less than one. For very high frequencies, the first term of Eq. (2.180) approaches the last term of Eq. (2.18), and the electrons behave as if they were free. Below the resonant frequency, the electric polarization will be in phase with the applied electric field, and the dielectric constant and index of refraction will be greater than one.

Computing the dielectric constant from the first term in Eq. (2.180), in the same manner as for a free electron [see Eqs. (2.20) to (2.26)], we find, for N electrons per

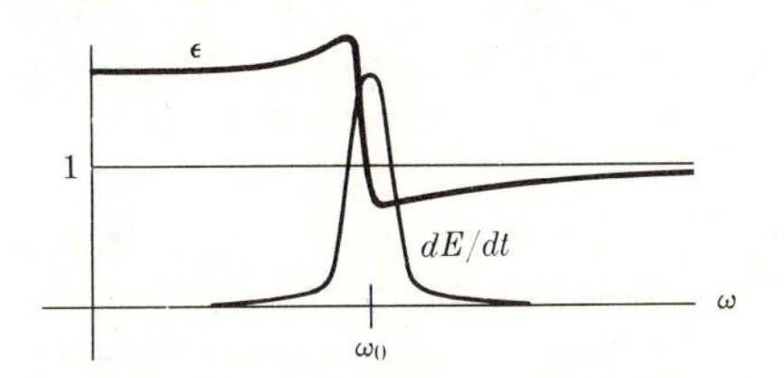

Fig. 2.7 Dielectric constant and energy absorption for medium containing harmonic oscillators.

unit volume:

$$\frac{\varepsilon}{\varepsilon_0} = 1 + \frac{Ne^2}{m} \frac{\omega_0^2 - \omega^2}{(\omega_0^2 - \omega^2)^2 + 4\gamma^2\omega^2}. \tag{2.181}$$

The index of refraction for electromagnetic waves ($\mu = \mu_0$) is

$$\begin{aligned} n = \frac{c}{v} &= \left(\frac{\mu\varepsilon}{\mu_0\varepsilon_0}\right)^{1/2} \\ &= \left(\frac{\varepsilon}{\varepsilon_0}\right)^{1/2}. \end{aligned} \tag{2.182}$$

For very high or low frequencies, Eq. (2.181) becomes

$$\frac{\varepsilon}{\varepsilon_0} \doteq 1 + \frac{Ne^2}{m\omega_0^2}, \quad \omega \ll \omega_0, \tag{2.183}$$

$$\frac{\varepsilon}{\varepsilon_0} \doteq 1 - \frac{Ne^2}{m\omega^2}, \quad \omega \gg \omega_0. \tag{2.184}$$

The mean rate of energy absorption per unit volume is given by Eq. (2.169):

$$\frac{dE}{dt} = \frac{Ne^2E_0^2}{m} \frac{\gamma\omega^2}{(\omega^2 - \omega_0^2)^2 + 4\gamma^2\omega^2}. \tag{2.185}$$

The resulting dielectric constant and energy absorption versus frequency are plotted in Fig. 2.7. Thus the dielectric constant is constant and greater than one at low frequencies, increases as we approach the resonant frequency, falls to less than one in the region of "anomalous dispersion" where there is strong absorption of electromagnetic radiation, and then rises, approaching one at high frequencies. The index of refraction will follow a similar curve. This is precisely the sort of behavior which is exhibited by matter in all forms. Glass, for example, has a constant dielectric constant at low frequencies; in the region of visible light its index of refraction increases with frequency; and it becomes opaque in a certain band in the ultraviolet. X-rays are transmitted with an index of refraction very slightly less than one. A more realistic model of a transmitting medium would result from assuming several different resonant frequencies corresponding to electrons bound with various values of the spring constant k. This picture is then capable of explaining most of the features in the experimental curves for ε or n versus frequency. Not only is there qualitative agreement, but the formulas (2.181) to (2.185) agree quantitatively with experimental results, provided the constants N, ω_0 and γ are properly chosen for each material. Likewise the shapes of the absorption lines in the spectra of gases fit Formula (2.173). The success of this theory was one of the reasons for the adoption, until the year 1913, of the "jelly model" of the atom, in which electrons were imagined embedded in a positively charged jelly in which they oscillated as harmonic oscillators. The experiments of

Rutherford in 1913 forced physicists to adopt the "planetary" model of the atom, but this model was unable to explain even qualitatively the optical and electromagnetic properties of matter until the advent of quantum mechanics. The result of the quantum-mechanical treatment is that, for the interaction of matter and radiation, the simple oscillator picture gives essentially correct results when the constants are properly chosen.*

We now consider an applied force $F(t)$ which is large only during a short time interval δt and is zero or negligible at all other times. Such a force is called an impulse, and corresponds to a sudden blow. We assume the oscillator initially at rest at $x = 0$, and we assume the time δt so short that the mass moves only a negligibly small distance while the force is acting. According to Eq. (2.4), the momentum just after the force is applied will equal the impulse delivered by the force:

$$mv_0 = p_0 = \int F\,dt, \tag{2.186}$$

where v_0 is the velocity just after the impulse, and the integral is taken over the time interval δt during which the force acts. After the impulse, the applied force is zero, and the oscillator must move according to Eq. (2.133) if the damping is less than critical. We are assuming δt so small that the oscillator does not move appreciably during this time, hence we choose $\theta = -(\pi/2) - \omega_1 t_0$, in order that $x = 0$ at $t = t_0$, where t_0 is the instant at which the impulse occurs:

$$x = Ae^{-\gamma t} \sin [\omega_1(t - t_0)]. \tag{2.187}$$

The velocity at $t = t_0$ is

$$v_0 = \omega_1 A e^{-\gamma t_0}. \tag{2.188}$$

Thus

$$A = \frac{v_0}{\omega_1} e^{\gamma t_0}. \tag{2.189}$$

The solution when an impulse p_0 is delivered at $t = t_0$ to an oscillator at rest is therefore

$$x = \begin{cases} 0, & t \leq t_0, \\ \dfrac{p_0}{m\omega_1} e^{-\gamma(t - t_0)} \sin [\omega_1(t - t_0)], & t > t_0. \end{cases} \tag{2.190}$$

Here we have neglected the short time δt during which the force acts.

We see that the result of an impulse-type force depends only on the total impulse p_0 delivered, and is independent of the particular form of the function $F(t)$,

*See John C. Slater, *Quantum Theory of Matter*. New York: McGraw-Hill Book Co., 1951. (Page 378.)

provided only that $F(t)$ is negligible except during a very short time interval δt. Several possible forms of $F(t)$ which have this property are listed below:

$$F(t) = \begin{cases} 0, & t < t_0, \\ p_0/\delta t, & t_0 \leq t \leq t_0 + \delta t, \\ 0, & t > t_0 + \delta t, \end{cases} \tag{2.191}$$

$$F(t) = \frac{p_0\,\delta t}{\pi} \frac{1}{(t-t_0)^2 + (\delta t)^2}, \qquad -\infty < t < \infty, \tag{2.192}$$

$$F(t) = \frac{p_0}{\delta t \sqrt{\pi}} \exp\left[-\frac{(t-t_0)^2}{(\delta t)^2}\right], \qquad -\infty < t < \infty. \tag{2.193}$$

The reader may verify that each of these functions is negligible except within an interval of the order of δt around t_0, and that the total impulse delivered by each is p_0. The exact solution of Eq. (2.91) with $F(t)$ given by any of the above expressions must reduce to Eq. (2.190) when $\delta t \to 0$ (see Problem 55).

2.11 THE PRINCIPLE OF SUPERPOSITION. HARMONIC OSCILLATOR WITH ARBITRARY APPLIED FORCE

An important property of the harmonic oscillator is that its motion $x(t)$, when subject to an applied force $F(t)$ which can be regarded as the sum of two or more other forces $F_1(t)$, $F_2(t)$, . . . , is the sum of the motions $x_1(t)$, $x_2(t)$, . . . , which it would have if each of the forces $F_n(t)$ were acting separately. This principle applies to small mechanical vibrations, electrical vibrations, sound waves, electromagnetic waves, and all physical phenomena governed by linear differential equations. The principle is expressed in the following theorem.

Theorem 4. *Let the (finite or infinite*) set of functions $x_n(t)$, $n = 1, 2, 3, \ldots$, be solutions of the equations*

$$m\ddot{x}_n + b\dot{x}_n + kx_n = F_n(t), \tag{2.194}$$

and let

$$F(t) = \sum_n F_n(t). \tag{2.195}$$

Then the function

$$x(t) = \sum_n x_n(t) \tag{2.196}$$

satisfies the equation

$$m\ddot{x} + b\dot{x} + kx = F(t). \tag{2.197}$$

---

*When the set of functions is infinite, there are certain mathematical restrictions which need not concern us here.

To prove this theorem, we substitute Eq. (2.196) in the left side of Eq. (2.197):

$$\begin{aligned} m\ddot{x}+b\dot{x}+kx &= m\sum_n \ddot{x}_n+b\sum_n \dot{x}_n+k\sum_n x_n \\ &= \sum_n (m\ddot{x}_n+b\dot{x}_n+kx_n) \\ &= \sum_n F_n(t) \\ &= F(t). \end{aligned}$$

This theorem enables us to find a solution of Eq. (2.197) whenever the force $F(t)$ can be expressed as a sum of forces $F_n(t)$ for which the solutions of the corresponding equations (2.194) can be found. In particular, whenever $F(t)$ can be written as a sum of sinusoidally oscillating terms:

$$F(t) = \sum_n C_n \cos(\omega_n t+\theta_n), \tag{2.198}$$

a particular solution of Eq. (2.197) will be, by Theorem 4 and Eq. (2.163),

$$x = \sum_n \frac{C_n}{m}\frac{1}{[(\omega_n^2-\omega_0^2)^2+4\gamma^2\omega_n^2]^{1/2}}\sin(\omega_n t+\theta_n+\beta_n), \tag{2.199}$$

$$\beta_n = \tan^{-1}\frac{\omega_0^2-\omega_n^2}{2\gamma\omega_n}.$$

The general solution is then

$$x = Ae^{-\gamma t}\cos(\omega_1 t+\theta)+\sum_n \frac{C_n}{m}\frac{\sin(\omega_n t+\theta_n+\beta_n)}{[(\omega_n^2-\omega_0^2)^2+4\gamma^2\omega_n^2]^{1/2}}, \tag{2.200}$$

where A and θ are, as usual, to be chosen to make the solution (2.200) fit the initial conditions.

We can write Eqs. (2.198) and (2.199) in a different form by setting

$$A_n = C_n\cos\theta_n, \qquad B_n = -C_n\sin\theta_n. \tag{2.201}$$

Then

$$F(t) = \sum_n (A_n\cos\omega_n t+B_n\sin\omega_n t), \tag{2.202}$$

and

$$x = \sum_n \frac{A_n\sin(\omega_n t+\beta_n)-B_n\cos(\omega_n t+\beta_n)}{m[(\omega_n^2-\omega_0^2)^2+4\gamma^2\omega_n^2]^{1/2}}. \tag{2.203}$$

An important case of this kind is that of a periodic force $F(t)$, that is, a force such that

$$F(t+T) = F(t), \tag{2.204}$$

where T is the period of the force. For any continuous function $F(t)$ satisfying Eq. (2.204) (and, in fact, even for only piecewise continuous functions), it can be shown that $F(t)$ can always be written as a sum of sinusoidal functions:

$$F(t) = \tfrac{1}{2}A_0 + \sum_{n=1}^{\infty}\left(A_n \cos\frac{2\pi nt}{T} + B_n \sin\frac{2\pi nt}{T}\right), \tag{2.205}$$

where

$$\begin{aligned} A_n &= \frac{2}{T}\int_0^T F(t)\cos\frac{2\pi nt}{T}\,dt, \qquad n = 0, 1, 2, \ldots, \\ B_n &= \frac{2}{T}\int_0^T F(t)\sin\frac{2\pi nt}{T}\,dt, \qquad n = 1, 2, 3, \ldots. \end{aligned} \tag{2.206}$$

This result enables us, at least in principle, to solve the problem of the forced oscillator for any periodically varying force. The sum in Eq. (2.205) is called a *Fourier series.** The actual computation of the solution by this method is in most cases rather laborious, particularly the fitting of the constants A, θ in Eq. (2.200) to the initial conditions. However, the knowledge that such a solution exists is often useful in itself. Note also that the transient part of the solution, which depends on the initial conditions, dies out eventually if the oscillator is damped, and we are left after a long time with the steady-state solution (2.199). If any of the frequencies $2\pi n/T$ coincides with the natural frequency ω_0 of the oscillator, then the corresponding terms in the series in Eqs. (2.199) or (2.203) will be relatively much larger than the rest. Thus a force which oscillates nonsinusoidally at half the frequency ω_0 may cause the oscillator to perform a nearly sinusoidal oscillation at its natural frequency ω_0.

A generalization of the Fourier series theorem [Eqs. (2.205) and (2.206)] applicable to nonperiodic forces is the Fourier integral theorem, which allows us to represent any continuous (or piecewise continuous) function $F(t)$, subject to certain limitations, as a superposition of harmonically oscillating forces. By means of Fourier series and integrals, we may solve Eq. (2.197) for almost any physically reasonable force $F(t)$. We shall not pursue the subject further here. Suffice it to say that while the methods of Fourier series and Fourier integrals are of considerable practical value in solving vibration problems, their greatest

*For a proof of the above statements and a more complete discussion of Fourier series, see Dunham Jackson, *Fourier Series and Orthogonal Polynomials.* Menasha, Wisconsin: George Banta Pub. Co., 1941. (Chapter 1.)

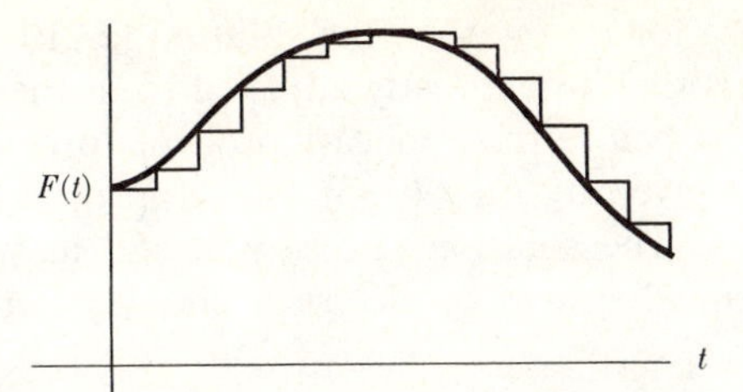

Fig. 2.8 Representation of a force as a sum of impulses. Heavy curve: $F(t)$. Light curve: $\sum_n F_n(t)$.

importance in physics probably lies in the fact that in principle such a solution exists. Many important results can be deduced without ever actually evaluating the series or integrals at all.

A method of solution known as Green's method is based on the solution (2.190) for an impulse-type force. We can think of any force $F(t)$ as the sum of a series of impulses, each acting during a short time δt and delivering an impulse $F(t)\,\delta t$:

$$F(t) \doteq \sum_{n=-\infty}^{\infty} F_n(t), \tag{2.207}$$

$$F_n(t) = \begin{cases} 0, & \text{if} \quad t < t_n, \quad \text{where} \quad t_n = n\,\delta t, \\ F(t_n), & \text{if} \quad t_n \le t \le t_{n+1}, \\ 0, & \text{if} \quad t > t_{n+1}. \end{cases} \tag{2.208}$$

As $\delta t \to 0$, the sum of all the impulse forces $F_n(t)$ will approach $F(t)$. (See Fig. 2.8.) According to Theorem 4 and Eq. (2.190), a solution of Eq. (2.197) for a force given by Eq. (2.207) is

$$x(t) \doteq \sum_{n=-\infty}^{n_0} \frac{F(t_n)\,\delta t}{m\omega_1} e^{-\gamma(t-t_n)} \sin\left[\omega_1(t-t_n)\right], \tag{2.209}$$

where $t_{n_0} \le t < t_{n_0+1}$. If we let $\delta t \to 0$ and write $t_n = t'$, Eq. (2.209) becomes

$$x(t) = \int_{-\infty}^{t} \frac{F(t')}{m\omega_1} e^{-\gamma(t-t')} \sin\left[\omega_1(t-t')\right] dt'. \tag{2.210}$$

The function

$$G(t, t') = \begin{cases} 0, & \text{if} \quad t' > t, \\ \dfrac{e^{-\gamma(t-t')}}{m\omega_1} \sin\left[\omega_1(t-t')\right], & \text{if} \quad t' \le t, \end{cases} \tag{2.211}$$

is called the *Green's function* for Eq. (2.197). In terms of Green's function,

$$x(t) = \int_{-\infty}^{\infty} G(t, t')F(t')\,dt'. \tag{2.212}$$

If the force $F(t)$ is zero for $t < t_0$, then the solution (2.210) will give $x(t) = 0$ for $t < t_0$. This solution is therefore already adjusted to fit the initial condition that the oscillator be at rest before the application of the force. For any other initial condition, a transient given by Eq. (2.133), with appropriate values of A and θ, will have to be added. The solution (2.210) is useful in studying the transient behavior of a mechanical system or electrical circuit when subject to forces of various kinds.

PROBLEMS

1. a) A certain jet engine at its maximum rate of fuel intake develops a constant thrust (force) of 3000 lb-wt. Given that it is operated at maximum thrust during take-off, calculate the power (in horsepower) delivered to the airplane by the engine when the airplane's velocity is 20 mph, 100 mph, and 300 mph (1 horsepower = 746 watts).

b) A piston engine at its maximum rate of fuel intake develops a constant power of 500 horsepower. Calculate the force it applies to the airplane during take-off at 20 mph, 100 mph, and 300 mph.

2. A particle of mass m is subject to a constant force F. At $t = 0$ it has zero velocity. Use the momentum theorem to find its velocity at any later time t. Calculate the energy of the particle at any later time from both Eqs. (2.7) and (2.8) and check that the results agree.

3. A particle of mass m is subject to a force given by Eq. (2.192). (In Eq. (2.192), δt is a fixed small time interval.) Find the total impulse delivered by the force during the time $-\infty < t < \infty$. If its initial velocity (at $t \to -\infty$) is v_0, what is its final velocity (as $t \to \infty$)? Use the momentum theorem.

4. A high-speed proton of electric charge e moves with constant speed v_0 in a straight line past an electron of mass m and charge $-e$, initially at rest. The electron is at a distance a from the path of the proton.

a) Assume that the proton passes so quickly that the electron does not have time to move appreciably from its initial position until the proton is far away. Show that the component of force in a direction perpendicular to the line along which the proton moves is

$$F = \frac{e^2 a}{4\pi\varepsilon_0 (a^2 + v_0^2 t^2)^{3/2}}, \text{ (mks units)}$$

where $t = 0$ when the proton passes closest to the electron.

b) Calculate the impulse delivered by this force.

c) Write the component of the force in a direction parallel to the proton velocity and show that the net impulse in that direction is zero.

d) Using these results, calculate the (approximate) final momentum and final kinetic energy of the electron.

e) Show that the condition for the original assumption in part (a) to be valid is $(e^2/4\pi\varepsilon_0) \ll \frac{1}{2}mv_0^2$.

5. A particle of mass m at rest at $t = 0$ is subject to a force $F(t) = F_0 \sin^2 \omega t$.

a) Sketch the form you expect for $v(t)$ and $x(t)$, for several periods of oscillation of the force.

b) Find $v(t)$ and $x(t)$ and compare with your sketch.

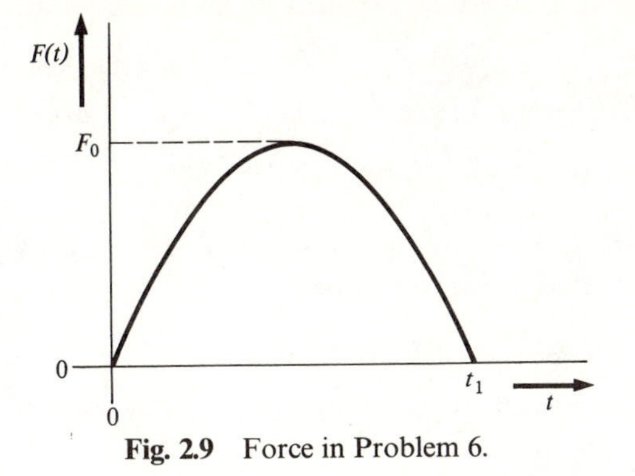

Fig. 2.9 Force in Problem 6.

6. A particle of mass m, initial velocity v_0 is subject beginning at $t = 0$ to a force $F(t)$ as sketched in Fig. 2.9.

a) Make a sketch showing $F(t)$ and the expected form of $v(t)$ and $x(t)$.

b) Devise a simple function $F(t)$ having this form, and find $x(t)$ and $v(t)$.

7. A particle which had originally a velocity v_0 is subject to a force given by Eq. (2.191).

a) Find $v(t)$ and $x(t)$.

b) Show that as $\delta t \to 0$, the motion approaches motion at constant velocity with an abrupt change in velocity at $t = t_0$ of amount p_0/m. (δt is a fixed time interval.)

8. A microphone contains a diaphragm of mass m and area A, suspended so that it can move freely in a direction perpendicular to the diaphragm. A sound wave impinges on the diaphragm so that the pressure on its front face is

$$p = p_0 + p' \sin \omega t.$$

Assume that the pressure on its back face remains constant at the atmospheric pressure p_0. Neglecting all other forces except that due to the pressure difference across the diaphragm, find its motion. In an actual microphone there is a restoring force on the diaphragm which keeps it from moving too far. Since this force is neglected here, nothing prevents the diaphragm from drifting away with a constant velocity. Avoid this difficulty by choosing the initial velocity so that the motion is purely oscillatory. If the output voltage of the microphone is to be proportional to the sound pressure p' and independent of ω, how must it depend upon the amplitude and frequency of the motion of the diaphragm?

9. A tug of war is held between two teams of five men each. Each man weighs 160 lb and can initially pull on the rope with a force of 200 lb-wt. At first the teams are evenly matched, but as the men tire, the force with which each man pulls decreases according to the formula

$$F = (200 \text{ lb-wt})\, e^{-t/\tau},$$

where the mean tiring time τ is 10 sec for one team and 20 sec for the other. Find the motion. Assume the men do not change their grip on the rope. ($g = 32 \text{ ft-sec}^{-2}$.) What is the final velocity of the two teams? Which of our assumptions is responsible for this unreasonable result?

10. A particle initially at rest is subject, beginning at $t = 0$, to a force

$$F = F_0 e^{-\gamma t} \cos(\omega t + \theta).$$

a) Find its motion.
b) How does the final velocity depend on θ, and on ω? [*Hint:* The algebra is simplified by writing $\cos(\omega t + \theta)$ in terms of complex exponential functions.]

11. A boat with initial velocity v_0 is slowed by a frictional force

$$F = -be^{\alpha v}.$$

a) Find its motion.
b) Find the time and the distance required to stop.

12. A boat is slowed by a frictional force $F(v)$. Its velocity decreases according to the formula

$$v = C(t - t_1)^2,$$

where C is a constant and t_1 is the time at which it stops. Find the force $F(v)$.

13. A jet engine which develops a constant maximum thrust F_0 is used to power a plane with a frictional drag proportional to the square of the velocity. If the plane starts at $t = 0$ with a negligible velocity and accelerates with maximum thrust, find its velocity $v(t)$.

14. Assume that the engines of a propeller-driven airplane of mass m deliver a constant power P at full throttle. Find the force $F(v)$. Neglecting friction use the method of Section 2.4 to find the velocity and position of the plane as it accelerates down the runway, starting from rest at $t = 0$. Check your result for the velocity using the energy theorem. In what ways are the assumptions in this problem physically unrealistic? In what ways would the answer be changed by more realistic assumptions?

15. The engine of a racing car of mass m delivers a constant power P at full throttle. Assuming that the friction is proportional to the velocity, find an expression for $v(t)$ if the car accelerates from a standing start at full throttle. Does your solution behave correctly as $t \to \infty$?

16. a) A body of mass m slides on a rough horizontal surface. The coefficient of static friction is μ_s, and the coefficient of sliding friction is μ. Devise an analytic function $F(v)$ to represent the frictional force which has the proper constant value at appreciable velocities and reduces to the static value at very low velocities.
b) Find the motion under the force you have devised if the body starts with an initial velocity v_0.

17. Find $v(t)$ and $x(t)$ for a particle of mass m which starts at $x_0 = 0$ with velocity v_0, subject to a force given by Eq. (2.31) with $n \neq 1$. Find the time to stop, and the distance required to stop, and verify the remarks in the last paragraph of Section 2.4.

18. A particle of mass m is subject to a force

$$F = -kx + kx^3/a^2$$

where k, a are constants.

a) Find $V(x)$ and discuss the kinds of motion which can occur.

b) Show that if $E = \frac{1}{4}ka^2$ the integral in Eq. (2.46) can be evaluated by elementary methods. Find $x(t)$ for this case, choosing x_0, t_0 in any convenient way. Show that your result agrees with the qualitative discussion in part (a) for this particular energy.

19. A particle of mass m is repelled from the origin by a force inversely proportional to the cube of its distance from the origin. Set up and solve the equation of motion if the particle is initially at rest at a distance x_0 from the origin.

20. A mass m is connected to the origin with a spring of constant k, whose length when relaxed is l. The restoring force is very nearly proportional to the amount the spring has been stretched or compressed so long as it is not stretched or compressed very far. However, when the spring is compressed too far, the force increases very rapidly, so that it is impossible to compress the spring to less than half its relaxed length. When the spring is stretched more than about twice its relaxed length, it begins to weaken, and the restoring force becomes zero when it is stretched to very great lengths.

a) Devise a force function $F(x)$ which represents this behavior. (Of course a real spring is deformed if stretched too far, so that F becomes a function of its previous history, but you are to assume here that F depends only on x.)

b) Find $V(x)$ and describe the types of motion which may occur.

21. A particle of mass m is acted on by a force whose potential energy is

$$V = ax^2 - bx^3.$$

a) Find the force.

b) The particle starts at the origin $x = 0$ with velocity v_0. Show that, if $|v_0| < v_c$, where v_c is a certain critical velocity, the particle will remain confined to a region near the origin. Find v_c.

22. An alpha particle in a nucleus is held by a potential having the shape shown in Fig. 2.10.

a) Describe the kinds of motion that are possible.

b) Devise a function $V(x)$ having this general form and having the values $-V_0$ and V_1 at $x = 0$ and $x = \pm x_1$, and find the corresponding force.

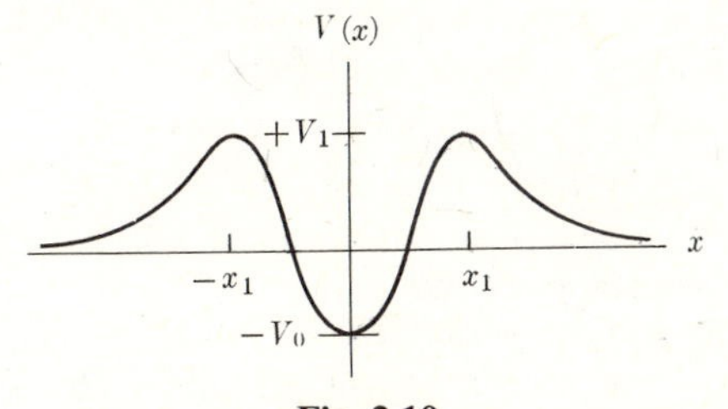

Fig. 2.10

23. A particle is subject to a force

$$F = -kx + \frac{a}{x^3}.$$

a) Find the potential $V(x)$, describe the nature of the solutions, and find the solution $x(t)$.
b) Can you give a simple interpretation of the motion when $E^2 \gg ka$?

24. A particle of mass m is subject to a force given by

$$F = B\left(\frac{a^2}{x^2} - \frac{28a^5}{x^5} + \frac{27a^8}{x^8}\right).$$

The particle moves only along the positive x-axis.
a) Find and sketch the potential energy. (B and a are positive.)
b) Describe the types of motion which may occur. Locate all equilibrium points and determine the frequency of small oscillations about any which are stable.
c) A particle starts at $x = 3a/2$ with a velocity $v = -v_0$, where v_0 is positive. What is the smallest value of v_0 for which the particle may eventually escape to a very large distance? Describe the motion in that case. What is the maximum velocity the particle will have? What velocity will it have when it is very far from its starting point?

25. The potential energy for the force between two atoms in a diatomic molecule has the approximate form:

$$V(x) = -\frac{a}{x^6} + \frac{b}{x^{12}},$$

where x is the distance between the atoms and a, b are positive constants.
a) Find the force.
b) Assuming one of the atoms is very heavy and remains at rest while the other moves along a straight line, describe the possible motions.
c) Find the equilibrium distance and the period of small oscillations about the equilibrium position if the mass of the lighter atom is m.

26. Find the solution for the motion of a body subject to a linear repelling force $F = kx$. Show that this is the type of motion to be expected in the neighborhood of a point of unstable equilibrium.

27. A particle of mass m moves in a potential well given by

$$V(x) = \frac{-V_0 a^2(a^2+x^2)}{8a^4+x^4}.$$

a) Sketch $V(x)$ and $F(x)$.
b) Discuss the motions which may occur. Locate all equilibrium points and determine the frequency of small oscillations about any that are stable.
c) A particle starts at a great distance from the potential well with velocity v_0 toward the well. As it passes the point $x = a$, it suffers a collision with another particle, during which it loses a fraction α of its kinetic energy. How large must α be in order that the particle thereafter

remains trapped in the well? How large must α be in order that the particle be trapped in one side of the well? Find the turning points of the new motion if $\alpha = 1$.

28. Solve Eq. (2.65) by each of the three methods discussed in Sections 2.3, 2.4, and 2.5.

29. Derive the solutions (2.74) and (2.75) for a falling body subject to a frictional force proportional to the square of the velocity.

30. A body of mass m falls from rest through a medium which exerts a frictional drag (force) $be^{\alpha|v|}$.

a) Find its velocity $v(t)$.
b) What is the terminal velocity?
c) Expand your solution in a power series in t, keeping terms up to t^2.
d) Why does the solution fail to agree with Eq. (1.28) even for short times t?

31. A projectile is fired vertically upward with an initial velocity v_0. Find its motion, assuming a frictional drag proportional to the square of the velocity. (Constant g.)

32. Derive equations analogous to Eqs. (2.85) and (2.86) for the motion of a body whose velocity is greater than the escape velocity. [*Hint:* Set $\sinh \beta = (Ex/mMG)^{1/2}$.]

33. Find the motion of a body projected upward from the earth with a velocity equal to the escape velocity. Neglect air resistance.

34. Starting with $e^{2i\theta} = (e^{i\theta})^2$, obtain formulas for $\sin 2\theta$, $\cos 2\theta$ in terms of $\sin \theta$, $\cos \theta$.

35. By writing $\cos \theta$ in the form (2.122) derive the formula

$$\cos^3\theta = \tfrac{1}{4} \cos 3\theta + \tfrac{3}{4} \cos \theta.$$

36. Find the general solutions of the equations:

$$\text{a)} \qquad m\ddot{x} + b\dot{x} - kx = 0,$$

$$\text{b)} \qquad m\ddot{x} - b\dot{x} + kx = 0.$$

Discuss the physical interpretation of these equations and their solutions, assuming that they are the equations of motion of a particle.

37. Show that when $\omega_0^2 - \gamma^2$ is very small, the underdamped solution (2.133) is approximately equal to the critically damped solution (2.146), for a short time interval. What is the relation between the constants C_1, C_2 and A, θ? This result suggests how one might discover the additional solution (2.143) in the critical case.

38. A freely rolling freight car weighing 10^4 kg arrives at the end of its track with a speed of 2 m/sec. At the end of the track is a snubber consisting of a firmly anchored spring with $k = 1.6 \times 10^4$ kg/sec^2. The car compresses the spring. If the friction is proportional to the velocity, find the damping constant b_c for critical damping. Sketch the motion $x(t)$ and find the maximum distance by which the spring is compressed (for $b = b_c$). Show that if $b \geq b_c$, the car will come to a stop, but if $b \leq b_c$, the car will rebound and roll back down the track. (Note that the car is not fastened to the spring. As long as it pushes on the spring, it moves according to the harmonic oscillator equation, but instead of pulling on the spring, it will simply roll back down the track.)

39. A mass m subject to a linear restoring force $-kx$ and damping $-b\dot{x}$ is displaced a distance x_0 from equilibrium and released with zero initial velocity. Find the motion in the underdamped, critically damped, and overdamped cases.

40. Solve Problem 39 for the case when the mass starts from its equilibrium position with an initial velocity v_0. Sketch the motion for the three cases.

41. Solve Problem 39 for the case when the mass has an initial displacement x_0 and an initial velocity v_0 directed back toward the equilibrium point. Show that if $|v_0| > |\gamma_1 x_0|$, the mass will overshoot the equilibrium in the critically damped and overdamped cases so that the remarks at the end of Section 2.9 do not apply. Sketch the motion in these cases.

42. It is desired to design a bathroom scale with a platform deflection of one inch under a 200-lb man. If the motion is to be critically damped, find the required spring constant k and the damping constant b. Show that the motion will then be overdamped for a lighter person. If a 200-lb man steps on the scale, what is the maximum upward force exerted by the scale platform against his feet while the platform is coming to rest?

43. A mass of 1000 kg drops from a height of 10 m on a platform of negligible mass. It is desired to design a spring and dashpot on which to mount the platform so that the platform will settle to a new equilibrium position 0.2 m below its original position as quickly as possible after the impact *without overshooting*.

a) Find the spring constant k and the damping constant b of the dashpot. Be sure to examine your proposed solution $x(t)$ to make sure that it satisfies the correct initial conditions and does not overshoot.

b) Find, to two significant figures, the time required for the platform to settle within 1 mm of its final position.

44. A force $F_0 e^{-at}$ acts on a harmonic oscillator of mass m, spring constant k, and damping constant b. Find a particular solution of the equation of motion by starting from the guess that there should be a solution with the same time dependence as the applied force.

45. a) Find the motion of a damped harmonic oscillator subject to a constant applied force F_0, by guessing a "steady-state" solution of the inhomogeneous equation (2.91) and adding a solution of the homogeneous equation.

b) Solve the same problem by making the substitution $x' = x - a$, and choosing the constant a so as to reduce the equation in x' to the homogeneous equation (2.90). Hence show that the effect of the application of a constant force is merely to shift the equilibrium position without affecting the nature of the oscillations.

46. An underdamped harmonic oscillator is subject to an applied force

$$F = F_0 e^{-at} \cos(\omega t + \theta).$$

Find a particular solution by expressing F as the real part of a complex exponential function and looking for a solution for x having the same exponential time dependence.

47. An undamped harmonic oscillator ($b = 0$), initially at rest, is subject beginning at $t = 0$ to an applied force $F_0 \sin \omega t$. Find the motion $x(t)$.

48. An undamped harmonic oscillator ($b = 0$) is subject to an applied force $F_0 \cos \omega t$. Show that if $\omega = \omega_0$, there is no steady-state solution. Find a particular solution by starting with a solution for $\omega = \omega_0 + \varepsilon$, and passing to the limit $\varepsilon \to 0$. [*Hint:* If you start with the steady-state solution and let $\varepsilon \to 0$, it will blow up. Try starting with a solution which fits the initial condition $x_0 = 0$, so that it cannot blow up at $t = 0$.]

49. A critically damped harmonic oscillator with mass m and spring constant k, is subject to an applied force $F_0 \cos \omega t$. If, at $t = 0$, $x = x_0$ and $v = v_0$, what is $x(t)$?

50. A force $F_0 \cos (\omega t + \theta_0)$ acts on a damped harmonic oscillator beginning at $t = 0$.
a) What must be the initial values of x and v in order that there be no transient?
b) If instead $x_0 = v_0 = 0$, find the amplitude A and phase θ of the transient in terms of F_0, θ_0.

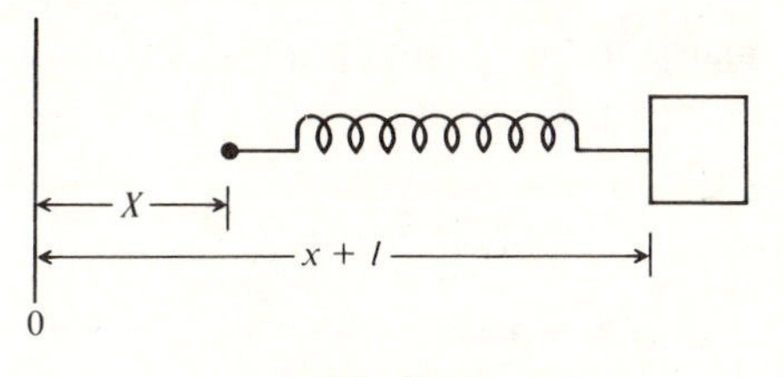

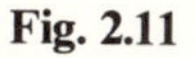
Fig. 2.11

51. A mass m is attached to a spring with force constant k, relaxed length l, as shown in Fig. 2.11. The left end of the spring is not fixed, but is instead made to oscillate with amplitude a, frequency ω, so that $X = a \sin \omega t$, where X is measured from a fixed reference point 0. Write the equation of motion, and show that it is equivalent to Eq. (2.148) with an applied force $ka \sin \omega t$, if the friction is given by Eq. (2.31). Show that, if the friction comes instead from a dashpot connected between the ends of the spring, so that the frictional force is $-b(\dot{x} - \dot{X})$, then the equation of motion has an additional applied force $\omega ba \cos \omega t$.

52. An automobile weighing one ton (2000 lb, including passengers but excluding wheels and everything else below the springs) settles one inch closer to the road for every 200 lb of passengers. It is driven at 20 mph over a washboard road with sinusoidal undulations having a distance between bumps of 1 ft and an amplitude of 2 in (height of bumps and depth of holes from mean road level). Find the amplitude of oscillation of the automobile, assuming it moves vertically as a simple harmonic oscillator without damping (no shock absorbers). (Neglect the mass of wheels and springs.) If shock absorbers are added to provide damping, is the ride better or worse? (Use the result of Problem 51.)

53. An undamped harmonic oscillator of mass m, natural frequency ω_0, is initially at rest and is subject at $t = 0$ to a blow so that it starts from $x_0 = 0$ with initial velocity v_0 and oscillates freely until $t = 3\pi/2\omega_0$. From this time on, a force $F = B \cos(\omega t + \theta)$ is applied. Find the motion.

54. Find the motion of a mass m subject to a restoring force $-kx$, and to a damping force $(\pm)\mu mg$ due to dry sliding friction. Show that the oscillations are isochronous (period independent of amplitude) with the amplitude of oscillation decreasing by $2\mu g/\omega_0^2$ during each half-cycle until the mass comes to a stop. [*Hint:* Use the result of Problem 45. When the force has a different algebraic form at different times during the motion, as here, where the sign of the damping force must be chosen so that the force is always opposed to the velocity, it is necessary to solve the equation of motion separately for each interval of time during which a particular expression for the force is to be used, and to choose as initial conditions for each time interval the final position and velocity of the preceding time interval.]

55. An undamped harmonic oscillator ($\gamma = 0$), initially at rest, is subject to a force given by Eq. (2.191).

a) Find $x(t)$.

b) For a fixed p_0, for what value of δt is the final amplitude of oscillation greatest?

c) Show that as $\delta t \to 0$, your solution approaches that given by Eq. (2.190).

56. Find the solution analogous to Eq. (2.190) for a critically damped harmonic oscillator subject to an impulse p_0 delivered at $t = t_0$.

57. a) Find, using the principle of superposition, the motion of an underdamped oscillator $[\gamma = (1/3)\omega_0]$ initially at rest and subject, after $t = 0$, to a force

$$F = A \sin \omega_0 t + B \sin 3\omega_0 t,$$

where ω_0 is the natural frequency of the oscillator.

b) What ratio of B to A is required in order for the forced oscillation at frequency $3\omega_0$ to have the same amplitude as that at frequency ω_0?

58. A force $F_0(1 - e^{-at})$ acts on a harmonic oscillator which is at rest at $t = 0$. The mass is m, the spring constant $k = 4ma^2$, and $b = ma$. Find the motion. Sketch $x(t)$.

***59.** Solve Problem 58 for the case $k = ma^2$, $b = 2ma$.

60. Find, by the Fourier-series method, the steady-state solution for the damped harmonic oscillator subject to a force

$$F(t) = \begin{cases} 0, & \text{if } nT < t \le (n+\frac{1}{2})T, \\ F_0, & \text{if } (n+\frac{1}{2})T < t \le (n+1)T, \end{cases}$$

where n is any integer, and $T = 6\pi/\omega_0$, where ω_0 is the resonance frequency of the oscillator. Show that if $\gamma \ll \omega_0$, the motion is nearly sinusoidal with period $T/3$.

*An asterisk is used, as explained in the Preface, to indicate problems which may be particularly difficult.

61. Find, by the Fourier-series method, the steady-state solution for an undamped harmonic oscillator subject to a force having the form of a rectified sine-wave:

$$F(t) = F_0 \left|\sin \omega_0 t\right|,$$

where ω_0 is the natural frequency of the oscillator.

62. Solve Problem 58 by using Green's solution (2.210).

63. An underdamped oscillator initially at rest is acted upon, beginning at $t = 0$, by a force given by Eq. (2.191). Find its motion by using Green's solution (2.210).

64. Using the result of Problem 56, find by Green's method the motion of a critically damped oscillator initially at rest and subject to a force $F(t)$.

CHAPTER 3

MOTION OF A PARTICLE IN TWO OR THREE DIMENSIONS

3.1 VECTOR ALGEBRA

The discussion of motion in two or three dimensions is vastly simplified by the introduction of the concept of a vector. A *vector* is defined geometrically as a physical quantity characterized by a magnitude and a direction in space. Examples are velocity, force, and position with respect to a fixed origin. Schematically, we represent a vector by an arrow whose length and direction represent the magnitude and direction of the vector. We shall represent a vector by a letter in boldface italic type. The same letter in ordinary italics will represent the magnitude of the vector. (See Fig. 3.1.) The magnitude of a vector may also be represented by vertical bars enclosing the vector symbol:

$$A = |\boldsymbol{A}|. \tag{3.1}$$

Two vectors are equal if they have the same magnitude and direction; the concept of a vector itself makes no reference to any particular location.*

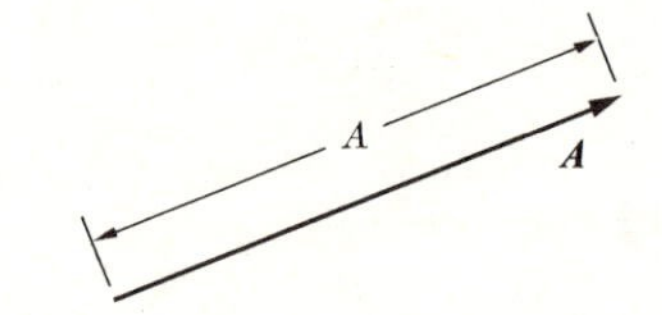

Fig. 3.1 A vector $\boldsymbol{A}$ and its magnitude A.

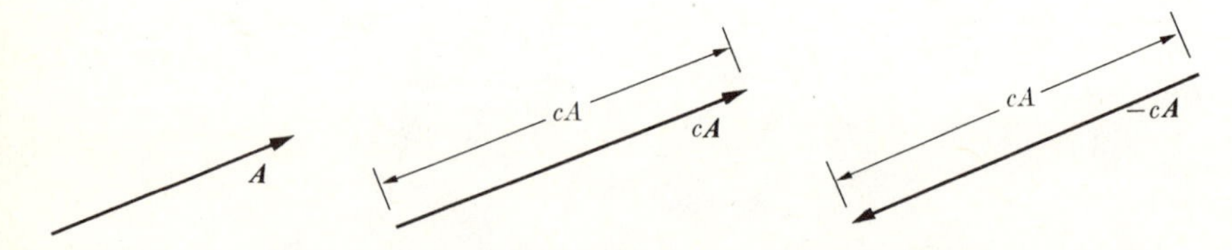

Fig. 3.2 Definition of multiplication of a vector by a scalar ($c > 0$).

*A distinction is sometimes made between "free" vectors, which have no particular location in space; "sliding" vectors, which may be located anywhere along a line; and "fixed" vectors, which must be located at a definite point in space. We prefer here to regard the vector as distinguished by its magnitude and direction alone, so that two vectors may be regarded as equal if they have the same magnitude and direction, regardless of position in space. In the case of a vector quantity like force, it may then be necessary to specify not only the *vector* which describes its magnitude and direction, but also the *location* at which it is applied.

A quantity represented by an ordinary (positive or negative) number is often called a *scalar*, to distinguish it from a vector. We define a product of a vector $\boldsymbol{A}$ and a positive scalar c as a vector $c\boldsymbol{A}$ in the same direction as $\boldsymbol{A}$ of magnitude cA. If c is negative, we define $c\boldsymbol{A}$ as having the magnitude $|c|A$ and a direction opposite to $\boldsymbol{A}$. (See Fig. 3.2.) It follows from this definition that

$$|c\boldsymbol{A}| = |c|\,|\boldsymbol{A}|. \tag{3.2}$$

It is also readily shown, on the basis of this definition, that multiplication by a scalar is associative in the following sense:

$$(cd)\boldsymbol{A} = c(d\boldsymbol{A}). \tag{3.3}$$

It is sometimes convenient to be able to write the scalar to the right of the vector, and we define $\boldsymbol{A}c$ as meaning the same vector as $c\boldsymbol{A}$:

$$\boldsymbol{A}c = c\boldsymbol{A}. \tag{3.4}$$

We define the sum $(\boldsymbol{A}+\boldsymbol{B})$ of two vectors $\boldsymbol{A}$ and $\boldsymbol{B}$ as the vector which extends from the tail of $\boldsymbol{A}$ to the tip of $\boldsymbol{B}$ when $\boldsymbol{A}$ is drawn with its tip at the tail of $\boldsymbol{B}$, as in Fig. 3.3. This definition is equivalent to the usual parallelogram rule, and is more convenient to use. It is readily extended to the sum of any number of vectors, as in Fig. 3.4.

On the basis of the definition given in Fig. 3.3, we can readily prove that vector addition is commutative and associative:

$$\boldsymbol{A}+\boldsymbol{B} = \boldsymbol{B}+\boldsymbol{A}, \tag{3.5}$$

$$(\boldsymbol{A}+\boldsymbol{B})+\boldsymbol{C} = \boldsymbol{A}+(\boldsymbol{B}+\boldsymbol{C}). \tag{3.6}$$

According to Eq. (3.6), we may omit parentheses in writing a vector sum, since the order of adding does not matter. From the definitions given by Figs. 3.2 and 3.3, we can also prove the following distributive laws:

$$c(\boldsymbol{A}+\boldsymbol{B}) = c\boldsymbol{A}+c\boldsymbol{B}, \tag{3.7}$$

$$(c+d)\boldsymbol{A} = c\boldsymbol{A}+d\boldsymbol{A}. \tag{3.8}$$

These statements can be proved by drawing diagrams representing the right and left members of each equation according to the definitions given. For example,

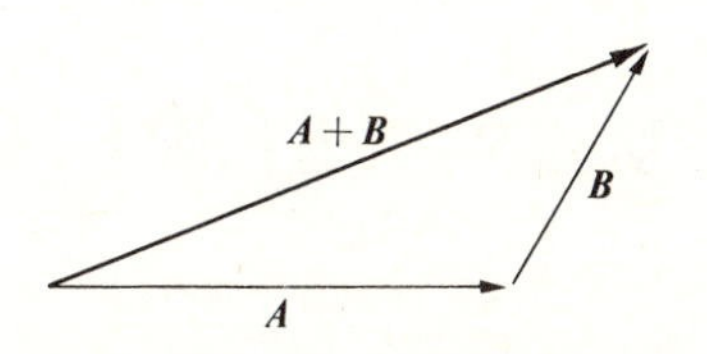

Fig. 3.3 Definition of addition of two vectors.

Fig. 3.4 Addition of several vectors.

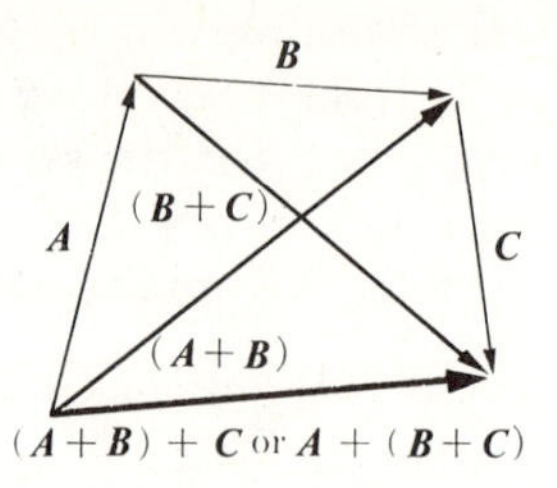

Fig. 3.5 Proof of Eq. (3.6).

the diagram in Fig. 3.5 makes it evident that the result of adding $\boldsymbol{C}$ to $(\boldsymbol{A}+\boldsymbol{B})$ is the same as the result of adding $(\boldsymbol{B}+\boldsymbol{C})$ to $\boldsymbol{A}$.

According to Eqs. (3.3) through (3.8), the sum and product we have defined have most of the algebraic properties of sums and products of ordinary numbers. This is the justification for calling them sums and products. Thus it is unnecessary to commit these results to memory. We need only remember that we can manipulate these sums and products just as we manipulate numbers in ordinary algebra, provided we remember that the product defined by Fig. 3.2 can be formed only between a scalar and a vector, and the result is a vector, and that the sum defined by Fig. 3.3 can be formed only between two vectors, and the result is a vector.

A vector may be represented algebraically in terms of its *components* or *projections* along a set of coordinate axes. Drop perpendiculars from the tail and tip of the vector onto the coordinate axes as in Fig. 3.6. Then the component of the vector along any axis is defined as the length of the segment cut off on the axis by these perpendiculars. The component is taken as positive or negative according to whether the projection of the tip of the vector lies in the positive or negative direction along the axis from the projection of the tail. The components

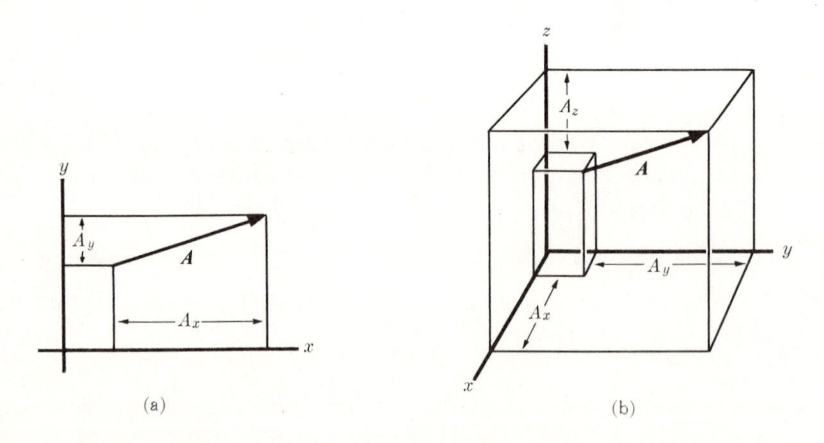

Fig. 3.6 (a) Components of a vector in a plane. (b) Components of a vector in space.

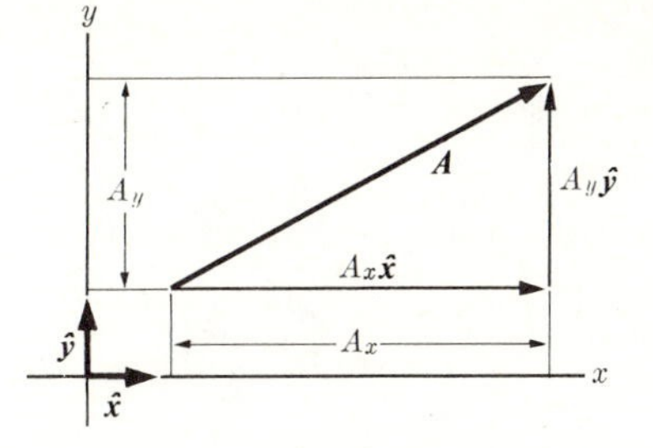

Fig. 3.7 Diagrammatic proof of the formula $\boldsymbol{A} = A_x\hat{\boldsymbol{x}}+A_y\hat{\boldsymbol{y}}$.

of a vector $\boldsymbol{A}$ along x-, y-, and z-axes will be written A_x, A_y, and A_z. The notation (A_x, A_y, A_z) will sometimes be used to represent the vector $\boldsymbol{A}$:

$$\boldsymbol{A} = (A_x, A_y, A_z). \tag{3.9}$$

If we define vectors $\hat{\boldsymbol{x}}$, $\hat{\boldsymbol{y}}$, $\hat{\boldsymbol{z}}$ of unit length along the x-, y-, z-axes respectively, then we can write any vector as a sum of products of its components with $\hat{\boldsymbol{x}}$, $\hat{\boldsymbol{y}}$, $\hat{\boldsymbol{z}}$:*

$$\boldsymbol{A} = A_x\hat{\boldsymbol{x}}+A_y\hat{\boldsymbol{y}}+A_z\hat{\boldsymbol{z}}. \tag{3.10}$$

The correctness of this formula can be made evident by drawing a diagram in which the three vectors on the right, which are parallel to the three axes, are added to give $\boldsymbol{A}$. Figure 3.7 shows this construction for the two-dimensional case.

We now have two equivalent ways of defining a vector: geometrically as a quantity with a magnitude and direction in space, or algebraically as a set of three numbers (A_x, A_y, A_z), which we call its components.† The operations of addition and multiplication by a scalar, which are defined geometrically in Figs. 3.2 and 3.3 in terms of the lengths and directions of the vectors involved, can also be defined algebraically as operations on the components of the vectors. Thus $c\boldsymbol{A}$ is the vector whose components are the components of $\boldsymbol{A}$, each multiplied by c:

$$c\boldsymbol{A} = (cA_x, cA_y, cA_z), \tag{3.11}$$

and $\boldsymbol{A}+\boldsymbol{B}$ is the vector whose components are obtained by adding the components of $\boldsymbol{A}$ and $\boldsymbol{B}$:

$$\boldsymbol{A}+\boldsymbol{B} = (A_x+B_x, A_y+B_y, A_z+B_z). \tag{3.12}$$

---

*We will use the caret over a boldface italic letter to denote a vector of unit length.

†These two ways of defining a vector are not quite equivalent as given here, for the algebraic definition requires that a coordinate system be set up, whereas the geometric definition does not refer to any particular set of axes. This flaw can be remedied by making the algebraic definition also independent of any particular set of axes. This is done by studying how the components change when the axes are changed, and defining a vector algebraically as a set of three quantities which transform in a certain way when the axes are changed. This refinement will not concern us in this chapter.

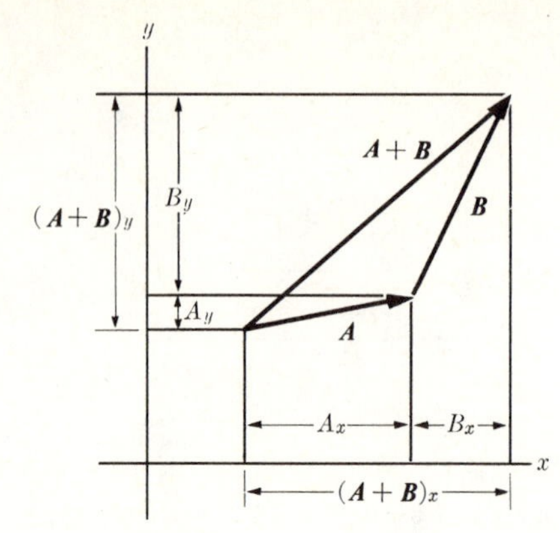

Fig. 3.8 Proof of equivalence of algebraic and geometric definitions of vector addition.

The equivalence of the definitions (3.11) and (3.12) to the corresponding geometrical definitions can be demonstrated by drawing suitable diagrams. Figure 3.8 constitutes a proof of Eq. (3.12) for the two-dimensional case. All vectors are drawn in Fig. 3.8 so that their components are positive; for a complete proof, similar diagrams should be drawn for the cases where one or both components of either vector are negative. The length of a vector can be defined algebraically as follows:

$$|\boldsymbol{A}| = (A_x^2 + A_y^2 + A_z^2)^{1/2}, \tag{3.13}$$

where the positive square root is to be taken.

We can now give algebraic proofs of Eqs. (3.2), (3.3), (3.5), (3.6), (3.7), and (3.8), based on the definitions (3.11), (3.12), and (3.13). For example, to prove Eq. (3.7), we show that each component of the left side agrees with each component on the right. For the x-component, the proof runs:

$$\begin{aligned}
[c(\boldsymbol{A}+\boldsymbol{B})]_x &= c(\boldsymbol{A}+\boldsymbol{B})_x && \text{[by Eq. (3.11)]} \\
&= c(A_x + B_x) && \text{[by Eq. (3.12)]} \\
&= cA_x + cB_x \\
&= (c\boldsymbol{A})_x + (c\boldsymbol{B})_x && \text{[by Eq. (3.11)]} \\
&= (c\boldsymbol{A} + c\boldsymbol{B})_x. && \text{[by Eq. (3.12)]}
\end{aligned}$$

Since all components are treated alike in the definitions (3.11), (3.12), (3.13), the same proof holds for the y- and z-components, and hence the vectors on the left and right sides of Eq. (3.7) are equal.

In view of the equivalence of the geometrical and algebraic definitions of the vector operations, it is unnecessary, for geometrical applications, to give both an algebraic and a geometric proof of each formula of vector algebra. Either a geo-

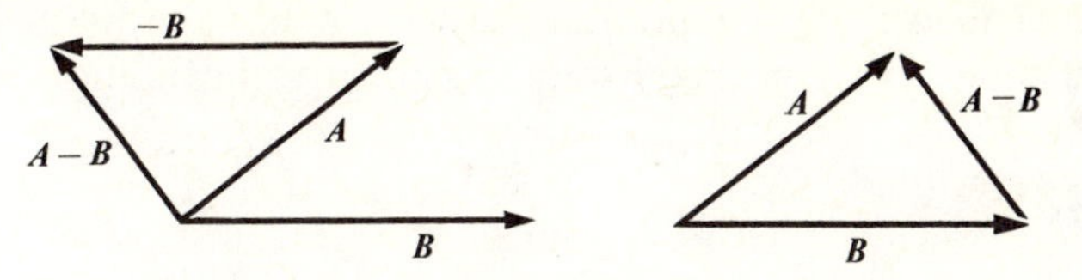

Fig. 3.9 Two methods of subtraction of vectors.

metric or an algebraic proof, whichever is easiest, will suffice. However, there are important cases in physics where we have to consider sets of quantities which behave algebraically like the components of vectors although they cannot be interpreted geometrically as quantities with a magnitude and direction in ordinary space. In order that we may apply the rules of vector algebra in such applications, it is important to know that all of these rules can be proved purely algebraically from the algebraic definitions of the vector operations. The geometric approach has the advantage of enabling us to visualize the meanings of the various vector notations and formulas. The algebraic approach simplifies certain proofs, and has the further advantage that it makes possible wide applications of the mathematical concept of a vector, including many cases where the ordinary geometric meaning is no longer retained.

We may define subtraction of vectors in terms of addition and multiplication by -1:

$$\boldsymbol{A}-\boldsymbol{B} = \boldsymbol{A}+(-\boldsymbol{B}) = (A_x-B_x,\ A_y-B_y,\ A_z-B_z). \tag{3.14}$$

The difference $\boldsymbol{A}-\boldsymbol{B}$ may be found geometrically according to either of the two schemes shown in Fig. 3.9. Subtraction of vectors may be shown to have all the algebraic properties to be expected by analogy with subtraction of numbers.

It is useful to define a *scalar product* $(\boldsymbol{A}\cdot\boldsymbol{B})$ of two vectors $\boldsymbol{A}$ and $\boldsymbol{B}$ as the product of their magnitudes times the cosine of the angle between them (Fig. 3.10):

$$\boldsymbol{A}\cdot\boldsymbol{B} = AB\cos\theta. \tag{3.15}$$

The scalar product is a scalar or number. It is also called the *dot product* or *inner product*, and can also be defined as the product of the magnitude of either vector times the projection of the other along it. An example of its use is the expression for the work done when a force $\boldsymbol{F}$ acts through a distance $\boldsymbol{s}$ not necessarily parallel to it:

$$W = Fs\cos\theta = \boldsymbol{F}\cdot\boldsymbol{s}.$$

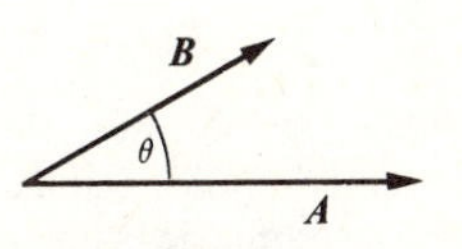

Fig. 3.10 Angle between two vectors.

We are entitled to call $\boldsymbol{A}\cdot\boldsymbol{B}$ a product because it has the following algebraic properties which are easily proved from the geometrical definition (3.15):*

$$(c\boldsymbol{A})\cdot\boldsymbol{B} = \boldsymbol{A}\cdot(c\boldsymbol{B}) = c(\boldsymbol{A}\cdot\boldsymbol{B}), \tag{3.16}$$

$$\boldsymbol{A}\cdot(\boldsymbol{B}+\boldsymbol{C}) = \boldsymbol{A}\cdot\boldsymbol{B}+\boldsymbol{A}\cdot\boldsymbol{C}, \tag{3.17}$$

$$\boldsymbol{A}\cdot\boldsymbol{B} = \boldsymbol{B}\cdot\boldsymbol{A}, \tag{3.18}$$

$$\boldsymbol{A}\cdot\boldsymbol{A} = A^2. \tag{3.19}$$

These equations mean that we can treat the dot product algebraically like a product in the algebra of ordinary numbers, provided we keep in mind that the two factors must be vectors and the resulting product is a scalar. The following statements are also consequences of the definition (3.15), where $\hat{\boldsymbol{x}}$, $\hat{\boldsymbol{y}}$, and $\hat{\boldsymbol{z}}$ are the unit vectors along the three coordinate axes:

$$\begin{aligned} \hat{\boldsymbol{x}}\cdot\hat{\boldsymbol{x}} &= \hat{\boldsymbol{y}}\cdot\hat{\boldsymbol{y}} = \hat{\boldsymbol{z}}\cdot\hat{\boldsymbol{z}} = 1, \\ \hat{\boldsymbol{x}}\cdot\hat{\boldsymbol{y}} &= \hat{\boldsymbol{y}}\cdot\hat{\boldsymbol{z}} = \hat{\boldsymbol{z}}\cdot\hat{\boldsymbol{x}} = 0. \end{aligned} \tag{3.20}$$

$$\boldsymbol{A}\cdot\boldsymbol{B} = AB, (-AB), \quad \text{when } \boldsymbol{A} \text{ is parallel (anti-parallel) to } \boldsymbol{B}, \tag{3.21}$$

$$\boldsymbol{A}\cdot\boldsymbol{B} = 0, \quad \text{when } \boldsymbol{A} \text{ is perpendicular to } \boldsymbol{B}. \tag{3.22}$$

Notice that, according to Eq. (3.22), the dot product of two vectors is zero if they are perpendicular, even though neither vector is of zero length.

The dot product can also be defined algebraically in terms of components:

$$\boldsymbol{A}\cdot\boldsymbol{B} = A_xB_x+A_yB_y+A_zB_z. \tag{3.23}$$

To prove that Eq. (3.23) is equivalent to the geometric definition (3.15), we write $\boldsymbol{A}$ and $\boldsymbol{B}$ in the form given by Eq. (3.10), and make use of Eqs. (3.16), (3.17), (3.18), and (3.20), which follow from Eq. (3.15):

$$\begin{aligned} \boldsymbol{A}\cdot\boldsymbol{B} &= (\hat{\boldsymbol{x}}A_x+\hat{\boldsymbol{y}}A_y+\hat{\boldsymbol{z}}A_z)\cdot(\hat{\boldsymbol{x}}B_x+\hat{\boldsymbol{y}}B_y+\hat{\boldsymbol{z}}B_z) \\ &= (\hat{\boldsymbol{x}}\cdot\hat{\boldsymbol{x}})A_xB_x+(\hat{\boldsymbol{x}}\cdot\hat{\boldsymbol{y}})A_xB_y+(\hat{\boldsymbol{x}}\cdot\hat{\boldsymbol{z}})A_xB_z+(\hat{\boldsymbol{y}}\cdot\hat{\boldsymbol{x}})A_yB_x+(\hat{\boldsymbol{y}}\cdot\hat{\boldsymbol{y}})A_yB_y+(\hat{\boldsymbol{y}}\cdot\hat{\boldsymbol{z}})A_yB_z \\ &\qquad +(\hat{\boldsymbol{z}}\cdot\hat{\boldsymbol{x}})A_zB_x+(\hat{\boldsymbol{z}}\cdot\hat{\boldsymbol{y}})A_zB_y+(\hat{\boldsymbol{z}}\cdot\hat{\boldsymbol{z}})A_zB_z \\ &= A_xB_x+A_yB_y+A_zB_z. \end{aligned}$$

This proves Eq. (3.23). The properties (3.16) to (3.20) can all be proved readily from the algebraic definition (3.23) as well as from the geometric definition (3.15). We can regard Eqs. (3.21) and (3.22) as algebraic definitions of *parallel* and *perpendicular*.

---

*Note that we call Eq. (3.15) a geometrical definition, even though it is expressed as an equation, because it refers only to the magnitudes and directions of $\boldsymbol{A}$ and $\boldsymbol{B}$ and not to their components along any particular set of axes. Likewise a geometric proof may utilize algebraic manipulations as long as it does not refer to components along coordinate axes.

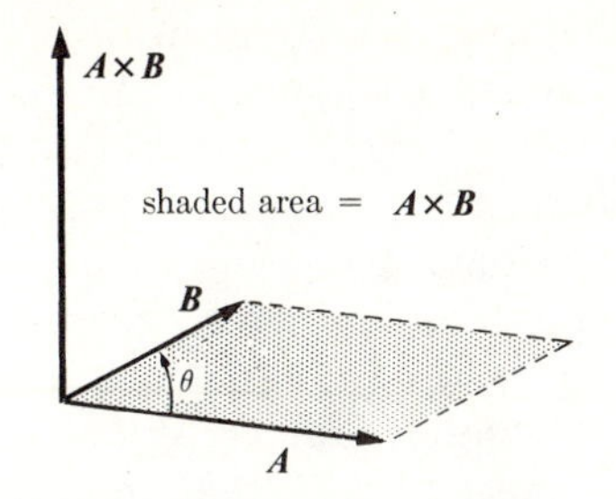

Fig. 3.11 Definition of vector product.

Another product convenient to define is the *vector product*, also called the *cross product* or *outer product*. The cross product $(\boldsymbol{A} \times \boldsymbol{B})$ of two vectors $\boldsymbol{A}$ and $\boldsymbol{B}$ is defined as a vector perpendicular to the plane of $\boldsymbol{A}$ and $\boldsymbol{B}$ whose magnitude is the area of the parallelogram having $\boldsymbol{A}$ and $\boldsymbol{B}$ as sides. The sense or direction of $(\boldsymbol{A} \times \boldsymbol{B})$ is defined as the direction of advance of a right-hand screw rotated from $\boldsymbol{A}$ toward $\boldsymbol{B}$. (See Fig. 3.11.) The length of $(\boldsymbol{A} \times \boldsymbol{B})$, in terms of the angle θ between the two vectors, is given by

$$|\boldsymbol{A} \times \boldsymbol{B}| = AB \sin \theta. \tag{3.24}$$

Note that the scalar product of two vectors is a scalar or number, while the vector product is a new vector. The vector product has the following algebraic properties which can be proved from the definition given in Fig. 3.11:*

$$\boldsymbol{A} \times \boldsymbol{B} = -\boldsymbol{B} \times \boldsymbol{A}, \tag{3.25}$$

$$(c\boldsymbol{A}) \times \boldsymbol{B} = \boldsymbol{A} \times (c\boldsymbol{B}) = c(\boldsymbol{A} \times \boldsymbol{B}), \tag{3.26}$$

$$\boldsymbol{A} \times (\boldsymbol{B}+\boldsymbol{C}) = (\boldsymbol{A} \times \boldsymbol{B})+(\boldsymbol{A} \times \boldsymbol{C}), \tag{3.27}$$

$$\boldsymbol{A} \times \boldsymbol{A} = \boldsymbol{0}, \tag{3.28}$$

$$\boldsymbol{A} \times \boldsymbol{B} = \boldsymbol{0}, \quad \text{when } \boldsymbol{A} \text{ is parallel to } \boldsymbol{B}, \tag{3.29}$$

$$|\boldsymbol{A} \times \boldsymbol{B}| = AB, \quad \text{when } \boldsymbol{A} \text{ is perpendicular to } \boldsymbol{B}, \tag{3.30}$$

$$\begin{gathered} \hat{\boldsymbol{x}} \times \hat{\boldsymbol{x}} = \hat{\boldsymbol{y}} \times \hat{\boldsymbol{y}} = \hat{\boldsymbol{z}} \times \hat{\boldsymbol{z}} = \boldsymbol{0}, \\ \hat{\boldsymbol{x}} \times \hat{\boldsymbol{y}} = \hat{\boldsymbol{z}}, \quad \hat{\boldsymbol{y}} \times \hat{\boldsymbol{z}} = \hat{\boldsymbol{x}}, \quad \hat{\boldsymbol{z}} \times \hat{\boldsymbol{x}} = \hat{\boldsymbol{y}}. \end{gathered} \tag{3.31}$$

Hence the cross product can be treated algebraically like an ordinary product with the exception that the order of multiplication must not be changed, and provided we keep in mind that the two factors must be vectors and the result is a vector. Switching the order of factors in a cross product changes the sign. This is the first unexpected deviation of the rules of vector algebra from those of ordinary

*Here $\boldsymbol{0}$ stands for the vector of zero length, sometimes called the *null vector*. It has no particular direction in space. It has the properties:

$$\boldsymbol{A}+\boldsymbol{0} = \boldsymbol{A}, \quad \boldsymbol{A}\cdot\boldsymbol{0} = 0, \quad \boldsymbol{A}\times\boldsymbol{0} = \boldsymbol{0}, \quad \boldsymbol{A}-\boldsymbol{A} = \boldsymbol{0}, \quad \boldsymbol{0} = (0, 0, 0).$$

algebra. The reader should therefore memorize Eq. (3.25). Equations (3.29) and (3.30), as well as the analogous Eqs. (3.21) and (3.22), are also worth remembering. (It goes without saying that all geometrical and algebraic definitions should be memorized.) In a repeated vector product like $(\boldsymbol{A} \times \boldsymbol{B}) \times (\boldsymbol{C} \times \boldsymbol{D})$, the parentheses cannot be omitted or rearranged, for the result of carrying out the multiplications in a different order is not, in general, the same. [See, for example, Eqs. (3.35) and (3.36).] Notice that according to Eq. (3.29) the cross product of two vectors may be null without either vector being the null vector.*

From Eqs. (3.25) to (3.31), using Eq. (3.10) to represent $\boldsymbol{A}$ and $\boldsymbol{B}$, we can prove that the geometric definition (Fig. 3.11) is equivalent to the following algebraic definition of the cross product:

$$\boldsymbol{A} \times \boldsymbol{B} = (A_y B_z - A_z B_y,\ A_z B_x - A_x B_z,\ A_x B_y - A_y B_x). \tag{3.32}$$

We can also write $\boldsymbol{A} \times \boldsymbol{B}$ as a determinant:

$$\boldsymbol{A} \times \boldsymbol{B} = \begin{vmatrix} \hat{\boldsymbol{x}} & \hat{\boldsymbol{y}} & \hat{\boldsymbol{z}} \\ A_x & A_y & A_z \\ B_x & B_y & B_z \end{vmatrix}. \tag{3.33}$$

Expansion of the right side of Eq. (3.33) according to the ordinary rules for determinants yields Eq. (3.32). Again the properties (3.25) to (3.31) follow also from the algebraic definition (3.32).

The following useful identities can be proved:

$$\boldsymbol{A} \cdot (\boldsymbol{B} \times \boldsymbol{C}) = (\boldsymbol{A} \times \boldsymbol{B}) \cdot \boldsymbol{C}, \tag{3.34}$$

$$\boldsymbol{A} \times (\boldsymbol{B} \times \boldsymbol{C}) = \boldsymbol{B}(\boldsymbol{A} \cdot \boldsymbol{C}) - \boldsymbol{C}(\boldsymbol{A} \cdot \boldsymbol{B}), \tag{3.35}$$

$$(\boldsymbol{A} \times \boldsymbol{B}) \times \boldsymbol{C} = \boldsymbol{B}(\boldsymbol{A} \cdot \boldsymbol{C}) - \boldsymbol{A}(\boldsymbol{B} \cdot \boldsymbol{C}), \tag{3.36}$$

$$\hat{\boldsymbol{x}} \cdot (\hat{\boldsymbol{y}} \times \hat{\boldsymbol{z}}) = 1. \tag{3.37}$$

The first three of these should be committed to memory. Equation (3.34) allows us to interchange dot and cross in the scalar triple product. The quantity $\boldsymbol{A} \cdot (\boldsymbol{B} \times \boldsymbol{C})$ can be shown to be the volume of the parallelepiped whose edges are $\boldsymbol{A}$, $\boldsymbol{B}$, $\boldsymbol{C}$, with positive or negative sign depending on whether $\boldsymbol{A}$, $\boldsymbol{B}$, $\boldsymbol{C}$ are in the same relative orientation as $\hat{\boldsymbol{x}}$, $\hat{\boldsymbol{y}}$, $\hat{\boldsymbol{z}}$, that is, depending on whether a right-hand screw rotated from $\boldsymbol{A}$ toward $\boldsymbol{B}$ would advance along $\boldsymbol{C}$ in the positive or negative direction. The triple vector product formulas (3.35) and (3.36) are easy to remember if we note that the positive term on the right in each case is the middle vector ($\boldsymbol{B}$) times the scalar product ($\boldsymbol{A} \cdot \boldsymbol{C}$) of the other two, while the negative term is the other vector within the parentheses times the scalar product of the other two.

*The mathematically inclined reader may be interested to know that there is no way of defining a vector product in three dimensions without giving up some of the rules of ordinary algebra.

As an example of the use of the vector product, the rule for the force exerted by a magnetic field of induction $\boldsymbol{B}$ on a moving electric charge q (esu) can be expressed as

$$\boldsymbol{F} = \frac{q}{c}\boldsymbol{v}\times\boldsymbol{B},$$

where c is the speed of light and $\boldsymbol{v}$ is the velocity of the charge. This equation gives correctly both the magnitude and direction of the force. The reader will remember that the subject of electricity and magnetism is full of right- and left-hand rules. Vector quantities whose directions are determined by right- or left-hand rules generally turn out to be expressible as cross products.

3.2 APPLICATIONS TO A SET OF FORCES ACTING ON A PARTICLE

According to the principles set down in Section 1.3, if a set of forces $\boldsymbol{F}_1, \boldsymbol{F}_2, \ldots, \boldsymbol{F}_n$ act on a particle, the total force $\boldsymbol{F}$, which determines its acceleration, is to be obtained by taking the vector sum of the forces $\boldsymbol{F}_1, \boldsymbol{F}_2, \ldots, \boldsymbol{F}_n$:

$$\boldsymbol{F} = \boldsymbol{F}_1+\boldsymbol{F}_2+\cdots+\boldsymbol{F}_n. \tag{3.38}$$

The forces $\boldsymbol{F}_1, \boldsymbol{F}_2, \ldots, \boldsymbol{F}_n$ are often referred to as *component* forces, and $\boldsymbol{F}$ is called their *resultant*. The term *component* is here used in a more general sense than in the preceding section, where the components of a vector were defined as the projections of the vector on a set of coordinate axes. When *component* is meant in this sense as one of a set of vectors whose sum is $\boldsymbol{F}$, we shall use the term (*vector*) *component*. In general, unless otherwise indicated, the term *component* of a vector $\boldsymbol{F}$ in a certain direction will mean the perpendicular projection of the vector $\boldsymbol{F}$ on a line in that direction. In symbols, the component of $\boldsymbol{F}$ in the direction of the unit vector $\hat{\boldsymbol{n}}$ is

$$F_n = \hat{\boldsymbol{n}}\cdot\boldsymbol{F}. \tag{3.39}$$

In this sense, the component of $\boldsymbol{F}$ is not a vector, but a number. The components of $\boldsymbol{F}$ along the x-, y-, and z-axes are the components in the sense of Eq. (3.39) in the directions $\hat{\boldsymbol{x}}$, $\hat{\boldsymbol{y}}$, and $\hat{\boldsymbol{z}}$.

If the forces $\boldsymbol{F}_1, \boldsymbol{F}_2, \ldots, \boldsymbol{F}_n$ are given, the sum may be determined graphically by drawing a careful scale diagram according to the definition of Fig. 3.3 or 3.4. The sum may also be determined analytically by drawing a rough sketch of the sum diagram and using trigonometry to calculate the magnitude and direction of the vector $\boldsymbol{F}$. If, for example, two vectors are to be added, the sum can be found by using the cosine and sine laws. In Fig. 3.12, F_1, F_2, and θ are given, and the magnitude and direction of the sum $\boldsymbol{F}$ are calculated from

$$F^2 = F_1^2+F_2^2-2F_1F_2\cos\theta, \tag{3.40}$$

$$\frac{F_1}{\sin\beta} = \frac{F_2}{\sin\alpha} = \frac{F}{\sin\theta}. \tag{3.41}$$

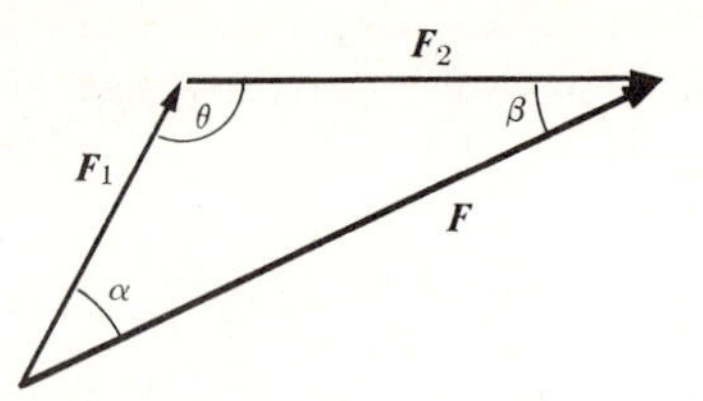

Fig. 3.12 Sum of two forces.

Note that the first of these equations can be obtained by squaring, in the sense of the dot product, the equation

$$\boldsymbol{F} = \boldsymbol{F}_1 + \boldsymbol{F}_2. \tag{3.42}$$

Taking the dot product of each member of this equation with itself, we obtain

$$\begin{aligned} \boldsymbol{F}\cdot\boldsymbol{F} = F^2 &= \boldsymbol{F}_1\cdot\boldsymbol{F}_1 + 2\boldsymbol{F}_1\cdot\boldsymbol{F}_2 + \boldsymbol{F}_2\cdot\boldsymbol{F}_2 \\ &= F_1^2 + F_2^2 - 2F_1F_2 \cos\theta. \end{aligned}$$

(Note that θ in Fig. 3.12 is the supplement of the angle between $\boldsymbol{F}_1$ and $\boldsymbol{F}_2$ as defined by Fig. 3.10.) This technique can be applied to obtain directly the magnitude of the sum of any number of vectors in terms of their lengths and the angles between them. Simply square Eq. (3.38), and split up the right side according to the laws of vector algebra into a sum of squares and dot products of the component forces. The angle between $\boldsymbol{F}$ and any of the component forces can be found by crossing or dotting the component vector into Eq. (3.38). For example, in the case of a sum of two forces, we cross $\boldsymbol{F}_1$ into Eq. (3.42):

$$\boldsymbol{F}_1 \times \boldsymbol{F} = \boldsymbol{F}_1 \times \boldsymbol{F}_1 + \boldsymbol{F}_1 \times \boldsymbol{F}_2.$$

We take the magnitude of each side, using Eqs. (3.28) and (3.24):

$$F_1F \sin\alpha = F_1F_2 \sin\theta, \qquad \text{or} \qquad \frac{F}{\sin\theta} = \frac{F_2}{\sin\alpha}.$$

When a sum of more than two vectors is involved, it is usually simpler to take the dot product of the component vector with each side of Eq. (3.38).

The vector sum in Eq. (3.38) can also be obtained by adding separately the components of $\boldsymbol{F}_1, \ldots, \boldsymbol{F}_n$ along any convenient set of axes:

$$\begin{aligned} F_x &= F_{1x} + F_{2x} + \cdots + F_{nx}, \\ F_y &= F_{1y} + F_{2y} + \cdots + F_{ny}, \\ F_z &= F_{1z} + F_{2z} + \cdots + F_{nz}, \end{aligned} \tag{3.43}$$

When a sum of a large number of vectors is to be found, this is likely to be the quickest method. The reader should use his ingenuity in combining and modifying these methods to suit the problem at hand. Obviously, if a set of vectors is to be added which contains a group of parallel vectors, it will be simpler to add these parallel vectors first before trying to apply the methods of the preceding paragraph.

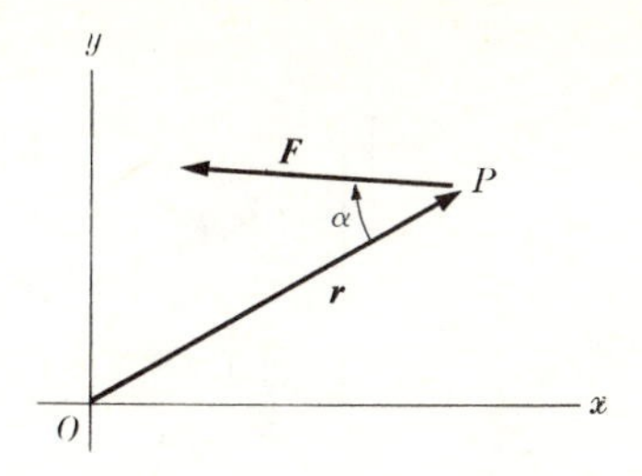

Fig. 3.13 Force F acting at point P.

Just as the various forces acting on a particle are to be added vectorially to give the total force, so, conversely, the total force, or any individual force, acting on a particle may be resolved in any convenient manner into a sum of (vector) component forces which may be considered as acting individually on the particle. Thus in the problem discussed in Section 1.7 (Fig. 1.4), the reaction force F exerted by the plane on the brick is resolved into a normal component N and a frictional component f. The effect of the force F on the motion of the brick is the same as that of the forces N and f acting together. If it is desired to resolve a force $\boldsymbol{F}$ into a sum of (vector) component forces in two or three perpendicular directions, this can be done by taking the perpendicular projections of $\boldsymbol{F}$ in these directions, as in Fig. 3.6. The magnitudes of the vector components of $\boldsymbol{F}$, along a set of perpendicular directions, are just the ordinary components of $\boldsymbol{F}$ in these directions in the sense of Eq. (3.39).

If a force $\boldsymbol{F}$ in the xy-plane acts on a particle at the point P, we define the *torque*, or *moment* of the force $\boldsymbol{F}$ about the origin O (Fig. 3.13) as the product of the distance $\overline{OP}$ and the component of $\boldsymbol{F}$ perpendicular to $\boldsymbol{r}$:

Torque —
$$N_O = rF \sin \alpha. \tag{3.44}$$

The moment N_O of the force $\boldsymbol{F}$ about the point O is defined as positive when $\boldsymbol{F}$ acts in a counterclockwise direction about O as in Fig. 3.13, and negative when $\boldsymbol{F}$ acts in a clockwise direction. We can define in a similar way the moment about O of any vector quantity located at the point P. The concept of moment will be found useful in our study of the mechanics of particles and rigid bodies. The geometrical and algebraic properties of torques will be studied in detail in Chapter 5. Notice that torque can be defined in terms of the vector product:

$$N_O = \pm|\boldsymbol{r} \times \boldsymbol{F}|, \tag{3.45}$$

where the + or − sign is used according to whether the vector $\boldsymbol{r} \times \boldsymbol{F}$ points in the positive or negative direction along the z-axis.

We can generalize the above definition of torque to the three-dimensional case by defining the torque or moment of a force $\boldsymbol{F}$, acting at a point P, about an

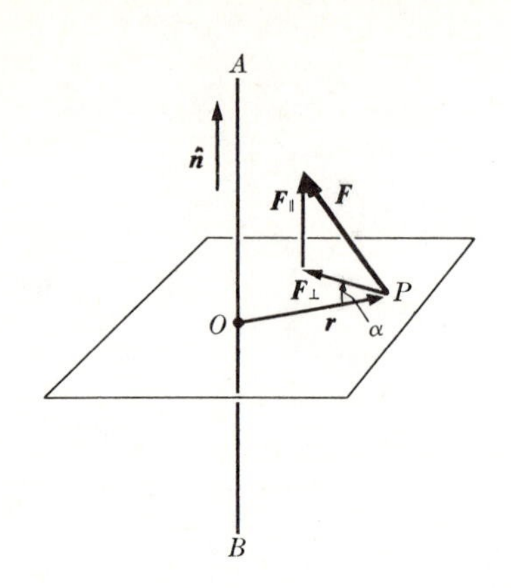

Fig. 3.14 Moment of a force about an axis in space.

axis $\overline{AB}$ (Fig. 3.14). Let $\hat{\boldsymbol{n}}$ be a unit vector in the direction of $\overline{AB}$, and let $\boldsymbol{F}$ be resolved into vector components parallel and perpendicular to $\overline{AB}$:

$$\boldsymbol{F} = \boldsymbol{F}_{\parallel} + \boldsymbol{F}_{\perp}, \tag{3.46}$$

where

$$\begin{aligned} \boldsymbol{F}_{\parallel} &= \hat{\boldsymbol{n}}(\hat{\boldsymbol{n}} \cdot \boldsymbol{F}), \\ \boldsymbol{F}_{\perp} &= \boldsymbol{F} - \boldsymbol{F}_{\parallel}. \end{aligned} \tag{3.47}$$

We now define the moment of $\boldsymbol{F}$ about the axis $\overline{AB}$ as the moment, defined by Eq. (3.44) or (3.45), of the force $\boldsymbol{F}_{\perp}$, in a plane through the point P perpendicular to $\overline{AB}$, about the point O at which the axis $\overline{AB}$ passes through this plane:

$$N_{AB} = \pm r F_{\perp} \sin \alpha = \pm |\boldsymbol{r} \times \boldsymbol{F}_{\perp}|, \tag{3.48}$$

where the + or − sign is used, depending on whether $\boldsymbol{r} \times \boldsymbol{F}_{\perp}$ is in the same or opposite direction to $\hat{\boldsymbol{n}}$. According to this definition, a force like $\boldsymbol{F}_{\parallel}$ parallel to $\overline{AB}$ has no torque or moment about $\overline{AB}$. Since $\boldsymbol{r} \times \boldsymbol{F}_{\parallel}$ is perpendicular to $\boldsymbol{n}$,

$$\begin{aligned} \boldsymbol{n} \cdot (\boldsymbol{r} \times \boldsymbol{F}) &= \hat{\boldsymbol{n}} \cdot [\boldsymbol{r} \times (\boldsymbol{F}_{\parallel} + \boldsymbol{F}_{\perp})] \\ &= \hat{\boldsymbol{n}} \cdot (\boldsymbol{r} \times \boldsymbol{F}_{\parallel}) + \boldsymbol{n} \cdot (\boldsymbol{r} \times \boldsymbol{F}_{\perp}) \\ &= \hat{\boldsymbol{n}} \cdot (\boldsymbol{r} \times \boldsymbol{F}_{\perp}) \\ &= \pm |\boldsymbol{r} \times \boldsymbol{F}_{\perp}|. \end{aligned}$$

Hence we can define N_{AB} in a neater way as follows:

$$N_{AB} = \hat{\boldsymbol{n}} \cdot (\boldsymbol{r} \times \boldsymbol{F}). \tag{3.49}$$

This definition automatically includes the proper sign, and does not require a resolution of $\boldsymbol{F}$ into $\boldsymbol{F}_{\parallel}$ and $\boldsymbol{F}_{\perp}$. Furthermore, $\boldsymbol{r}$ can now be drawn to P from any point on the axis $\overline{AB}$, since a component of $\boldsymbol{r}$ parallel to $\overline{AB}$, like a component of $\boldsymbol{F}$ parallel to $\overline{AB}$, gives a component in the cross product perpendicular to $\hat{\boldsymbol{n}}$ which disappears from the dot product.

Equation (3.49) suggests the definition of a *vector torque* or *vector moment*, about a point O, of a force $\boldsymbol{F}$ acting at a point P, as follows:

$$\boldsymbol{N}_O = \boldsymbol{r} \times \boldsymbol{F}, \tag{3.50}$$

where $\boldsymbol{r}$ is the vector from O to P. The vector torque $\boldsymbol{N}_O$ has, according to Eq. (3.49), the property that its component in any direction is the torque, in the previous sense, of the force $\boldsymbol{F}$ about an axis through O in that direction. Hereafter the term torque will usually mean the vector torque defined by Eq. (3.50). Torque about an axis $\overline{AB}$ in the previous sense will be called the component of torque along $\overline{AB}$. We can define the vector moment of any vector located at a point P, about a point O, by an equation analogous to Eq. (3.50).

3.3 DIFFERENTIATION AND INTEGRATION OF VECTORS

A vector $\boldsymbol{A}$ may be a function of a scalar quantity, say t, in the sense that with each value of t a certain vector $\boldsymbol{A}(t)$ is associated, or algebraically in the sense that its components may be functions of t:

$$\boldsymbol{A} = \boldsymbol{A}(t) = [A_x(t), A_y(t), A_z(t)]. \tag{3.51}$$

The most common example is that of a vector function of the time; for example, the velocity of a moving particle is a function of the time: $\boldsymbol{v}(t)$. Other cases also occur, however; for example, in Eq. (3.76), the vector $\hat{\boldsymbol{r}}$ is a function of the angle θ. We may define the derivative of the vector $\boldsymbol{A}$ with respect to t in analogy with the usual definition of the derivative of a scalar function (see Fig. 3.15):

$$\frac{d\boldsymbol{A}}{dt} = \lim_{\Delta t \to 0} \frac{\boldsymbol{A}(t+\Delta t) - \boldsymbol{A}(t)}{\Delta t}. \tag{3.52}$$

(Division by Δt here means multiplication by $1/\Delta t$.) We may also define the vector derivative algebraically in terms of its components:

$$\frac{d\boldsymbol{A}}{dt} = \left(\frac{dA_x}{dt}, \frac{dA_y}{dt}, \frac{dA_z}{dt}\right) = \hat{\boldsymbol{x}}\frac{dA_x}{dt} + \hat{\boldsymbol{y}}\frac{dA_y}{dt} + \hat{\boldsymbol{z}}\frac{dA_z}{dt}. \tag{3.53}$$

As an example, if $\boldsymbol{v}(t)$ is the vector velocity of a particle, its vector acceleration $\boldsymbol{a}$ is

$$\boldsymbol{a} = d\boldsymbol{v}/dt.$$

Examples of the calculation of vector derivatives based on either definition (3.52) or (3.53) will be given in Sections 3.4 and 3.5.

The following properties of vector differentiation can be proved by straightforward calculation from the algebraic definition (3.53), or they may be proved from the definition (3.52) in the same way the analogous properties are proved for differentiation of a scalar function:

$$\frac{d}{dt}(\boldsymbol{A}+\boldsymbol{B}) = \frac{d\boldsymbol{A}}{dt}+\frac{d\boldsymbol{B}}{dt}, \tag{3.54}$$

$$\frac{d}{dt}(f\boldsymbol{A}) = \frac{df}{dt}\boldsymbol{A}+f\frac{d\boldsymbol{A}}{dt}, \tag{3.55}$$

$$\frac{d}{dt}(\boldsymbol{A}\cdot\boldsymbol{B}) = \frac{d\boldsymbol{A}}{dt}\cdot\boldsymbol{B}+\boldsymbol{A}\cdot\frac{d\boldsymbol{B}}{dt}, \tag{3.56}$$

$$\frac{d}{dt}(\boldsymbol{A}\times\boldsymbol{B}) = \frac{d\boldsymbol{A}}{dt}\times\boldsymbol{B}+\boldsymbol{A}\times\frac{d\boldsymbol{B}}{dt}. \tag{3.57}$$

These results imply that differentiation of vector sums and products obeys the same algebraic rules as differentiation of sums and products in ordinary calculus, except, however, that the order of factors in the cross product must not be changed [Eq. (3.57)]. To prove Eq. (3.55), for example, from the definition (3.53), we simply show by direct calculation that the corresponding components on both sides of the equation are equal, making use of the definitions and properties of the vector operations introduced in the preceding section. For the x-component, the proof runs:

$$\left[\frac{d}{dt}(f\boldsymbol{A})\right]_x = \frac{d}{dt}(f\boldsymbol{A})_x \qquad \text{[by Eq. (3.53)]}$$

$$= \frac{d}{dt}(fA_x) \qquad \text{[by Eq. (3.11)]}$$

$$= \frac{df}{dt}A_x+f\frac{dA_x}{dt} \qquad \text{[standard rule of ordinary calculus]}$$

$$= \frac{df}{dt}A_x+f\left(\frac{d\boldsymbol{A}}{dt}\right)_x \qquad \text{[by Eq. (3.53)]}$$

$$= \left(\frac{df}{dt}\boldsymbol{A}\right)_x+\left(f\frac{d\boldsymbol{A}}{dt}\right)_x \qquad \text{[by Eq. (3.11)]}$$

$$= \left(\frac{df}{dt}\boldsymbol{A}+f\frac{d\boldsymbol{A}}{dt}\right)_x. \qquad \text{[by Eq. (3.12)]}$$

As another example, to prove Eq. (3.56) from the definition (3.52), we proceed as in the proof of the corresponding theorem for products of ordinary scalar functions. We shall use the symbol Δ to stand for the increment in the values of any function between t and $t+\Delta t$; the increment $\Delta\boldsymbol{A}$ of a vector $\boldsymbol{A}$ is defined in Fig. 3.15. Using

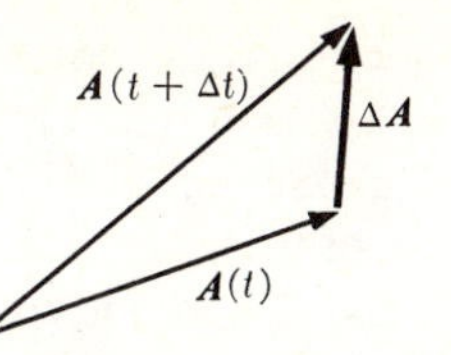

Fig. 3.15 Vector increment $\Delta A = A(t+\Delta t)-A(t)$.

this definition of Δ, and the rules of vector algebra given in the preceding section, we have

$$\begin{aligned}\frac{\Delta(\boldsymbol{A}\cdot\boldsymbol{B})}{\Delta t} &= \frac{(\boldsymbol{A}+\Delta\boldsymbol{A})\cdot(\boldsymbol{B}+\Delta\boldsymbol{B})-\boldsymbol{A}\cdot\boldsymbol{B}}{\Delta t}\\ &= \frac{(\Delta\boldsymbol{A})\cdot\boldsymbol{B}+\boldsymbol{A}\cdot(\Delta\boldsymbol{B})+(\Delta\boldsymbol{A})\cdot(\Delta\boldsymbol{B})}{\Delta t}\\ &= \frac{(\Delta\boldsymbol{A})\cdot\boldsymbol{B}}{\Delta t}+\frac{\boldsymbol{A}\cdot(\Delta\boldsymbol{B})}{\Delta t}+\frac{(\Delta\boldsymbol{A})\cdot(\Delta\boldsymbol{B})}{\Delta t}\\ &= \frac{\Delta\boldsymbol{A}}{\Delta t}\cdot\boldsymbol{B}+\boldsymbol{A}\cdot\frac{\Delta\boldsymbol{B}}{\Delta t}+\frac{(\Delta\boldsymbol{A})\cdot(\Delta\boldsymbol{B})}{\Delta t}.\end{aligned} \tag{3.58}$$

When $\Delta t \to 0$, the left side of Eq. (3.58) approaches the left side of Eq. (3.56), and the first two terms on the right side of Eq. (3.58) approach the two terms on the right of Eq. (3.56), while the last term on the right of Eq. (3.58) vanishes. The rigorous justification of this limit process is exactly similar to the justification required for the corresponding process in ordinary calculus.

In treating motions in three-dimensional space, we often meet scalar and vector quantities which have a definite value at every point in space. Such quantities are functions of the space coordinates, commonly x, y, and z. They may also be thought of as functions of the position vector $\boldsymbol{r}$ from the origin to the point x, y, z (Fig. 3.16). We thus distinguish scalar point functions

$$u(\boldsymbol{r}) = u(x, y, z),$$

and vector point functions

$$\boldsymbol{A}(\boldsymbol{r}) = \boldsymbol{A}(x, y, z) = [A_x(x, y, z), A_y(x, y, z), A_z(x, y, z)].$$

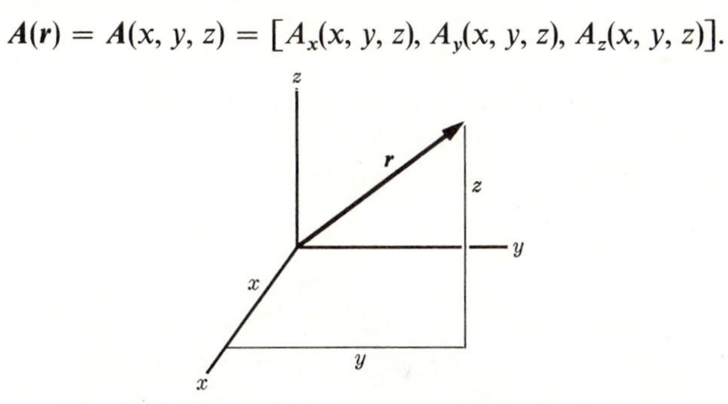

Fig. 3.16 The position vector $\boldsymbol{r}$ of the point (x, y, z).

An example of a scalar point function is the potential energy $V(x, y, z)$ of a particle moving in three dimensions. An example of a vector point function is the electric field intensity $\boldsymbol{E}(x, y, z)$. Scalar and vector point functions are often functions of the time t as well as of the point x, y, z in space.

If we are given a curve C in space, and a vector function $\boldsymbol{A}$ defined at points along this curve, we may consider the *line integral* of $\boldsymbol{A}$ along C:

$$\int_C \boldsymbol{A} \cdot d\boldsymbol{r}.$$

To define the line integral, imagine the curve C divided into small segments, and let any segment be represented by a vector $d\boldsymbol{r}$ in the direction of the segment and of length equal to the length of the segment. Then the curve consists of the successive vectors $d\boldsymbol{r}$ laid end to end. Now for each segment, form the product $\boldsymbol{A} \cdot d\boldsymbol{r}$, where $\boldsymbol{A}$ is the value of the vector function at the position of that segment. The line integral above is defined as the limit of the sums of the products $\boldsymbol{A} \cdot d\boldsymbol{r}$ as the number of segments increases without limit, while the length $|d\boldsymbol{r}|$ of every segment approaches zero. As an example, the work done by a force $\boldsymbol{F}$, which may vary from point to point, on a particle which moves along a curve C is

$$W = \int_C \boldsymbol{F} \cdot d\boldsymbol{r},$$

which is a generalization, to the case of a varying force and an arbitrary curve C, of the formula

$$W = \boldsymbol{F} \cdot \boldsymbol{s},$$

for a constant force acting on a body moving along a straight line segment $\boldsymbol{s}$. The reason for using the symbol $d\boldsymbol{r}$ to represent a segment of the curve is that if $\boldsymbol{r}$ is the position vector from the origin to a point on the curve, then $d\boldsymbol{r}$ is the increment in $\boldsymbol{r}$ (see Fig. 3.15) from one end to the other of the corresponding segment. If we write $\boldsymbol{r}$ in the form

$$\boldsymbol{r} = \hat{\boldsymbol{x}}x + \hat{\boldsymbol{y}}y + \hat{\boldsymbol{z}}z, \tag{3.59}$$

then

$$d\boldsymbol{r} = \hat{\boldsymbol{x}}\, dx + \hat{\boldsymbol{y}}\, dy + \hat{\boldsymbol{z}}\, dz, \tag{3.60}$$

where dx, dy, dz are the differences in the coordinates of the two ends of the segment. If s is the distance measured along the curve from some fixed point, we may express the line integral as an ordinary integral over the coordinate s:

$$\int_C \boldsymbol{A} \cdot d\boldsymbol{r} = \int A \cos\theta\, ds, \tag{3.61}$$

where θ is the angle between $\boldsymbol{A}$ and the tangent to the curve at each point. (See Fig. 3.17.) This formula may be used to evaluate the integral if we know A and $\cos\theta$ as functions of s. We may also write the integral, using Eq. (3.60), as

$$\int_C \boldsymbol{A} \cdot d\boldsymbol{r} = \int_C (A_x\, dx + A_y\, dy + A_z\, dz). \tag{3.62}$$

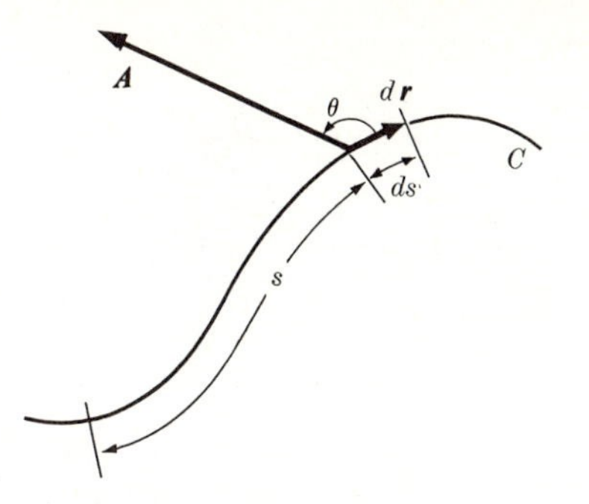

Fig. 3.17 Elements involved in the line integral.

One of the most convenient ways to represent a curve in space is to give the three coordinates (x, y, z) or, equivalently, the position vector $\boldsymbol{r}$, as functions of a parameter s which has a definite value assigned to each point of the curve. The parameter s is often, though not necessarily, the distance measured along the curve from some reference point, as in Fig. 3.17 and in Eq. (3.61). The parameter s may also be the time at which a moving particle arrives at any given point on the curve. If we know $\boldsymbol{A}(\boldsymbol{r})$ and $\boldsymbol{r}(s)$, then the line integral can be evaluated from the formula

$$\begin{aligned}\int_C \boldsymbol{A}\cdot d\boldsymbol{r} &= \int\left(\boldsymbol{A}\cdot\frac{d\boldsymbol{r}}{ds}\right)ds\\ &= \int\left(A_x\frac{dx}{ds}+A_y\frac{dy}{ds}+A_z\frac{dz}{ds}\right)ds.\end{aligned} \tag{3.63}$$

The right member of this equation is an ordinary integral over the variable s.

As an example of the calculation of a line integral, let us compute the work done on a particle moving in a semicircle of radius a about the origin in the xy-plane, by a force attracting the particle toward the point $(x = a,\ y = 0)$ and proportional to the distance of the particle from the point $(a, 0)$. Using the notation indicated in Fig. 3.18, we can write down the following relations:

$$\beta = \tfrac{1}{2}(\pi-\alpha), \qquad \theta = \frac{\pi}{2}-\beta = \tfrac{1}{2}\alpha,$$

$$D^2 = 2a^2(1-\cos\alpha), \qquad D = 2a\sin\frac{\alpha}{2},$$

$$\boldsymbol{F} = -k\boldsymbol{D}, \qquad F = kD = 2ka\sin\frac{\alpha}{2},$$

$$s = a(\pi-\alpha).$$

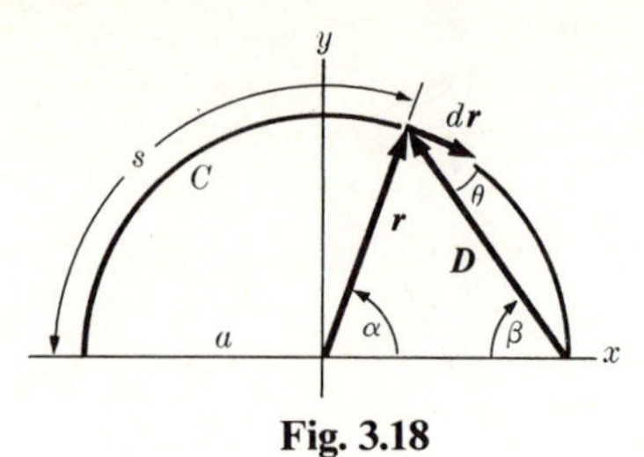

Fig. 3.18

Using these relations, we can evaluate the work done, using Eq. (3.61):

$$
\begin{aligned}
W &= \int_C \boldsymbol{F}\cdot d\boldsymbol{r} \\
&= \int_{s=0}^{\pi a} F \cos\theta \, ds \\
&= -\int_{\alpha=\pi}^{0} 2ka^2 \sin\frac{\alpha}{2}\cos\frac{\alpha}{2}\, d\alpha \\
&= 2ka^2.
\end{aligned}
$$

In order to calculate the same integral from Eq. (3.63), we express $\boldsymbol{r}$ and $\boldsymbol{F}$ along the curve as functions of the parameter α:

$$x = a\cos\alpha, \qquad y = a\sin\alpha,$$

$$F_x = kD\cos\beta = 2ka\sin^2\frac{\alpha}{2} = ka(1-\cos\alpha),$$

$$F_y = -kD\sin\beta = -2ka\sin\frac{\alpha}{2}\cos\frac{\alpha}{2} = -ka\sin\alpha.$$

The work is now, according to Eq. (3.63),

$$
\begin{aligned}
W &= \int_C \boldsymbol{F}\cdot d\boldsymbol{r} \\
&= \int_{\alpha=\pi}^{0} \left(F_x\frac{dx}{d\alpha} + F_y\frac{dy}{d\alpha}\right) d\alpha \\
&= \int_{\pi}^{0} [-ka^2(1-\cos\alpha)\sin\alpha - ka^2\sin\alpha\cos\alpha]\, d\alpha \\
&= ka^2\int_0^{\pi}\sin\alpha\, d\alpha \\
&= 2ka^2.
\end{aligned}
$$

3.4 KINEMATICS IN A PLANE

Kinematics is the science which describes the possible motions of mechanical systems without regard to the dynamical laws that determine which motions actually occur. In studying the kinematics of a particle in a plane, we shall be concerned with methods for describing the position of a particle, and the path followed by the particle, and with methods for finding the various components of its velocity and acceleration.

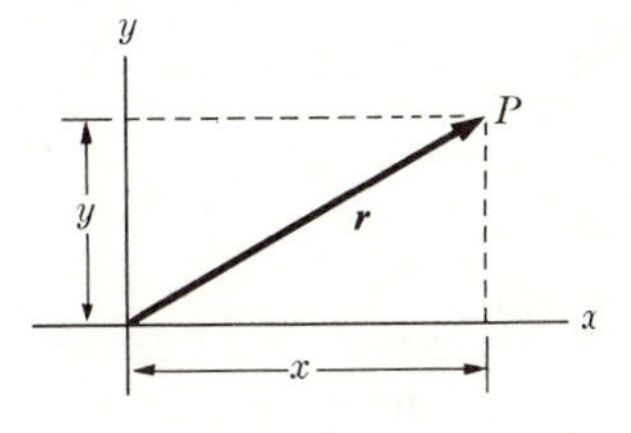

Fig. 3.19 Position vector and rectangular coordinates of a point P in a plane.

The simplest method of locating a particle in a plane is to set up two perpendicular axes and to specify any position by its rectangular coordinates x, y with respect to these axes (Fig. 3.19). Equivalently, we may specify the position vector $\boldsymbol{r} = (x, y)$ from the origin to the position of the particle. If we locate a position by specifying (in any convenient manner) the vector $\boldsymbol{r}$, then we need to specify in addition only the origin O from which the vector is drawn. If we specify the coordinates x, y, then we must also specify the coordinate axes from which x, y are measured.

Having set up a coordinate system, we next wish to describe the path of a particle in the plane. A curve in the xy-plane may be specified by giving y as a function of x along the curve, or vice versa:

$$y = y(x), \tag{3.64}$$

or

$$x = x(y). \tag{3.65}$$

Forms (3.64) and (3.65), however, are not convenient in many cases, for example when the curve doubles back on itself. We may also specify the curve by giving a relation between x and y,

$$f(x, y) = 0, \tag{3.66}$$

such that the curve consists of those points whose coordinates satisfy this relation. An example is the equation of a circle:

$$x^2 + y^2 - a^2 = 0.$$

One of the most convenient ways to represent a curve is in terms of a parameter s:

$$x = x(s), \qquad y = y(s), \tag{3.67}$$

or

$$\boldsymbol{r} = \boldsymbol{r}(s).$$

The parameter s has a unique value at each point of the curve. As s varies, the point $[x(s), y(s)]$ traces out the curve. The parameter s may, for example, be the distance measured along the curve from some fixed point. The equations of a circle can be expressed in terms of a parameter θ in the form

$$x = a \cos \theta,$$

$$y = a \sin \theta,$$

where θ is the angle between the x-axis and the radius a to the point (x, y) on the circle. In terms of the distance s measured around the circle,

$$x = a \cos \frac{s}{a},$$

$$y = a \sin \frac{s}{a}.$$

In mechanical problems, the parameter is usually the time, in which case Eqs. (3.67) specify not only the path of the particle, but also the rate at which the particle traverses the path. If a particle travels with constant speed v around a circle, its position at any time t may be given by

$$x = a \cos \frac{vt}{a},$$

$$y = a \sin \frac{vt}{a}.$$

If a particle moves along the path given by Eq. (3.67), we may specify its motion by giving $s(t)$, or by specifying directly

$$x = x(t), \qquad y = y(t), \tag{3.68}$$

or

$$\boldsymbol{r} = \boldsymbol{r}(t). \tag{3.69}$$

The velocity and acceleration, and their components, are given by

$$\begin{aligned} \boldsymbol{v} &= \frac{d\boldsymbol{r}}{dt} = \hat{\boldsymbol{x}}\frac{dx}{dt} + \hat{\boldsymbol{y}}\frac{dy}{dt}, \\ v_x &= \frac{dx}{dt}, \qquad v_y = \frac{dy}{dt}, \end{aligned} \tag{3.70}$$

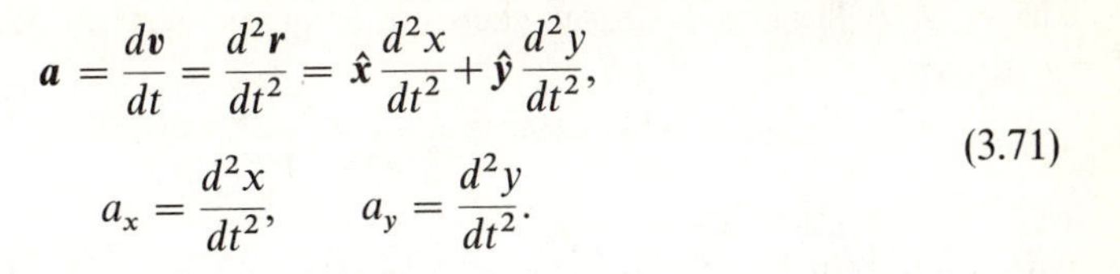

$$\boldsymbol{a} = \frac{d\boldsymbol{v}}{dt} = \frac{d^2\boldsymbol{r}}{dt^2} = \hat{\boldsymbol{x}}\frac{d^2x}{dt^2} + \hat{\boldsymbol{y}}\frac{d^2y}{dt^2},$$

$$a_x = \frac{d^2x}{dt^2}, \qquad a_y = \frac{d^2y}{dt^2}. \tag{3.71}$$

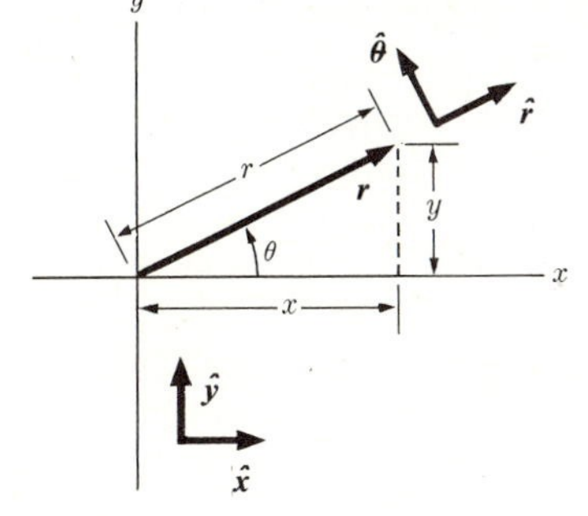

Fig. 3.20 Plane polar coordinates.

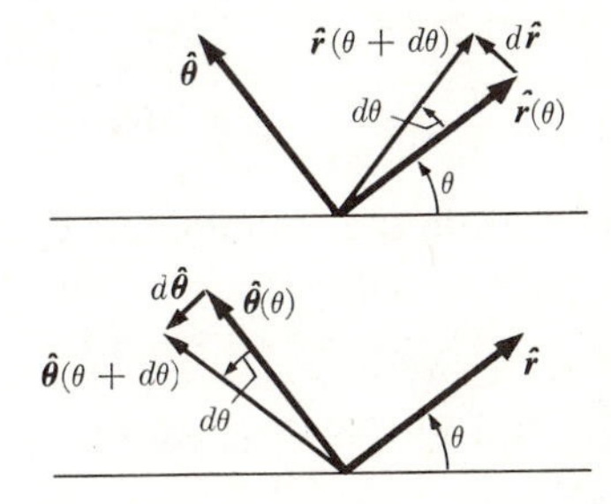

Fig. 3.21 Increments in the vectors $\hat{\boldsymbol{r}}$ and $\hat{\boldsymbol{\theta}}$.

Polar coordinates, shown in Fig. 3.20, are convenient in many problems. The coordinates r, θ are related to x, y by the following equations:

$$x = r\cos\theta, \qquad y = r\sin\theta, \tag{3.72}$$

and

$$r = (x^2+y^2)^{1/2},$$

$$\theta = \tan^{-1}\frac{y}{x} = \sin^{-1}\frac{y}{(x^2+y^2)^{1/2}} = \cos^{-1}\frac{x}{(x^2+y^2)^{1/2}}. \tag{3.73}$$

We define unit vectors $\hat{\boldsymbol{r}}$, $\hat{\boldsymbol{\theta}}$ in the directions of increasing r and θ, respectively, as shown. The vectors $\hat{\boldsymbol{r}}$, $\hat{\boldsymbol{\theta}}$ are functions of the angle θ, and are related to $\hat{\boldsymbol{x}}$, $\hat{\boldsymbol{y}}$ by the equations

$$\hat{\boldsymbol{r}} = \hat{\boldsymbol{x}}\cos\theta + \hat{\boldsymbol{y}}\sin\theta,$$

$$\hat{\boldsymbol{\theta}} = -\hat{\boldsymbol{x}}\sin\theta + \hat{\boldsymbol{y}}\cos\theta. \tag{3.74}$$

Equations (3.74) follow by inspection of Fig. 3.20. Differentiating, we obtain the important formulas

$$\frac{d\hat{\boldsymbol{r}}}{d\theta} = \hat{\boldsymbol{\theta}}, \qquad \frac{d\hat{\boldsymbol{\theta}}}{d\theta} = -\hat{\boldsymbol{r}}. \tag{3.75}$$

Formulas (3.75) can also be obtained by studying Fig. 3.21 (remembering that

$|\hat{\boldsymbol{r}}| = |\hat{\boldsymbol{\theta}}| = 1$). The position vector $\boldsymbol{r}$ is given very simply in terms of polar coordinates:

$$\boldsymbol{r} = r\hat{\boldsymbol{r}}(\theta). \tag{3.76}$$

We may describe the motion of a particle in polar coordinates by specifying $r(t)$, $\theta(t)$, thus determining the position vector $\boldsymbol{r}(t)$. The velocity vector is

$$\boldsymbol{v} = \frac{d\boldsymbol{r}}{dt} = \frac{dr}{dt}\hat{\boldsymbol{r}} + r\frac{d\hat{\boldsymbol{r}}}{d\theta}\frac{d\theta}{dt} = \dot{r}\hat{\boldsymbol{r}} + r\dot{\theta}\hat{\boldsymbol{\theta}}. \tag{3.77}$$

Thus we obtain the components of velocity in the $\hat{\boldsymbol{r}}$, $\hat{\boldsymbol{\theta}}$ directions:

$$v_r = \dot{r}, \qquad v_\theta = r\dot{\theta}. \tag{3.78}$$

The acceleration vector is

$$\begin{aligned} \boldsymbol{a} &= \frac{d\boldsymbol{v}}{dt} = \ddot{r}\hat{\boldsymbol{r}} + \dot{r}\frac{d\hat{\boldsymbol{r}}}{d\theta}\frac{d\theta}{dt} + \dot{r}\dot{\theta}\hat{\boldsymbol{\theta}} + r\ddot{\theta}\hat{\boldsymbol{\theta}} + r\dot{\theta}\frac{d\hat{\boldsymbol{\theta}}}{d\theta}\frac{d\theta}{dt} \\ &= (\ddot{r} - r\dot{\theta}^2)\hat{\boldsymbol{r}} + (r\ddot{\theta} + 2\dot{r}\dot{\theta})\hat{\boldsymbol{\theta}}. \end{aligned} \tag{3.79}$$

The components of acceleration are

$$a_r = \ddot{r} - r\dot{\theta}^2, \qquad a_\theta = r\ddot{\theta} + 2\dot{r}\dot{\theta}. \tag{3.80}$$

The term $r\dot{\theta}^2 = v_\theta^2/r$ is called the *centripetal acceleration* arising from motion in the θ direction. If $\ddot{r} = \dot{r} = 0$, the path is a circle, and $a_r = -v_\theta^2/r$. This result is familiar from elementary physics. The term $2\dot{r}\dot{\theta}$ is sometimes called the coriolis acceleration.

3.5 KINEMATICS IN THREE DIMENSIONS

The development in the preceding section for kinematics in two dimensions utilizing rectangular coordinates can be extended immediately to the three-dimensional case. A point is specified by its coordinates x, y, z, with respect to chosen rectangular axes in space, or by its position vector $\boldsymbol{r} = (x, y, z)$ with respect to a chosen origin. A path in space may be represented in the form of two equations in x, y, and z:

$$f(x, y, z) = 0, \qquad g(x, y, z) = 0. \tag{3.81}$$

Each equation represents a surface. The path is the intersection of the two surfaces. A path may also be represented parametrically:

$$x = x(s), \qquad y = y(s), \qquad z = z(s). \tag{3.82}$$

Velocity and acceleration are again given by

$$\boldsymbol{v} = \frac{d\boldsymbol{r}}{dt} = \hat{\boldsymbol{x}}v_x + \hat{\boldsymbol{y}}v_y + \hat{\boldsymbol{z}}v_z, \tag{3.83}$$

$$v_x = \frac{dx}{dt}, \qquad v_y = \frac{dy}{dt}, \qquad v_z = \frac{dz}{dt}, \tag{3.84}$$

and

$$\boldsymbol{a} = \frac{d\boldsymbol{v}}{dt} = \hat{\boldsymbol{x}}a_x + \hat{\boldsymbol{y}}a_y + \hat{\boldsymbol{z}}a_z, \tag{3.85}$$

$$a_x = \frac{d^2x}{dt^2}, \qquad a_y = \frac{d^2y}{dt^2}, \qquad a_z = \frac{d^2z}{dt^2}. \tag{3.86}$$

Many coordinate systems other than cartesian are useful for special problems. Perhaps the most widely used are spherical polar coordinates and cylindrical polar coordinates. Cylindrical polar coordinates (ρ, φ, z) are defined as in Fig. 3.22, or by the equations

$$x = \rho \cos\varphi, \qquad y = \rho \sin\varphi, \qquad z = z, \tag{3.87}$$

and, conversely,

$$\begin{aligned} \rho &= (x^2+y^2)^{1/2}, \\ \varphi &= \tan^{-1}\frac{y}{x} = \sin^{-1}\frac{y}{(x^2+y^2)^{1/2}} = \cos^{-1}\frac{x}{(x^2+y^2)^{1/2}}, \\ z &= z. \end{aligned} \tag{3.88}$$

A system of unit vectors $\hat{\boldsymbol{\rho}}$, $\hat{\boldsymbol{\varphi}}$, $\hat{\boldsymbol{z}}$, in the directions of increasing ρ, φ, z, respectively,

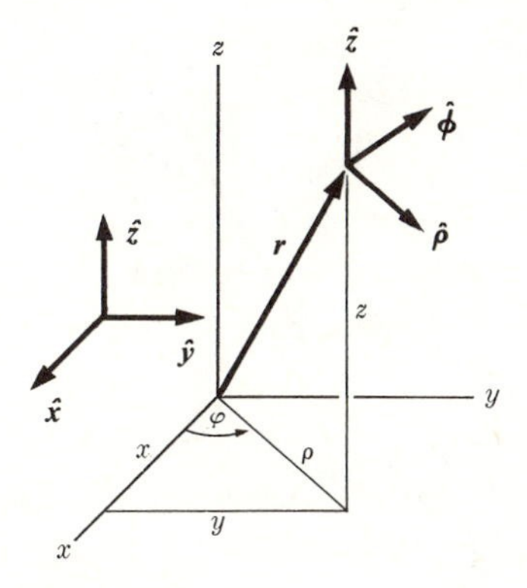

Fig. 3.22 Cylindrical polar coordinates.

is shown in Fig. 3.22. $\hat{\boldsymbol{z}}$ is constant, but $\hat{\boldsymbol{\varphi}}$ and $\hat{\boldsymbol{\rho}}$ are functions of φ, just as in plane polar coordinates:

$$\hat{\boldsymbol{\rho}} = \hat{\boldsymbol{x}} \cos \varphi + \hat{\boldsymbol{y}} \sin \varphi, \qquad \hat{\boldsymbol{\varphi}} = -\hat{\boldsymbol{x}} \sin \varphi + \hat{\boldsymbol{y}} \cos \varphi, \tag{3.89}$$

and, likewise,

$$\frac{d\hat{\boldsymbol{\rho}}}{d\varphi} = \hat{\boldsymbol{\varphi}}, \qquad \frac{d\hat{\boldsymbol{\varphi}}}{d\varphi} = -\hat{\boldsymbol{\rho}}. \tag{3.90}$$

The position vector $\boldsymbol{r}$ can be expressed in cylindrical coordinates in the form

$$\boldsymbol{r} = \rho\hat{\boldsymbol{\rho}} + z\hat{\boldsymbol{z}}. \tag{3.91}$$

Differentiating, we obtain for velocity and acceleration, using Eq. (3.90):

$$\boldsymbol{v} = \frac{d\boldsymbol{r}}{dt} = \dot{\rho}\hat{\boldsymbol{\rho}} + \rho\dot{\varphi}\hat{\boldsymbol{\varphi}} + \dot{z}\hat{\boldsymbol{z}}, \tag{3.92}$$

$$\boldsymbol{a} = \frac{d\boldsymbol{v}}{dt} = (\ddot{\rho} - \rho\dot{\varphi}^2)\hat{\boldsymbol{\rho}} + (\rho\ddot{\varphi} + 2\dot{\rho}\dot{\varphi})\hat{\boldsymbol{\varphi}} + \ddot{z}\hat{\boldsymbol{z}}. \tag{3.93}$$

Since $\hat{\boldsymbol{z}}$, $\hat{\boldsymbol{\varphi}}$, $\hat{\boldsymbol{\rho}}$ form a set of mutually perpendicular unit vectors, any vector $\boldsymbol{A}$ can be expressed in terms of its components along $\hat{\boldsymbol{z}}$, $\hat{\boldsymbol{\varphi}}$, $\hat{\boldsymbol{\rho}}$:

$$\boldsymbol{A} = A_\rho\hat{\boldsymbol{\rho}} + A_\varphi\hat{\boldsymbol{\varphi}} + A_z\hat{\boldsymbol{z}}. \tag{3.94}$$

It must be noted that since $\hat{\boldsymbol{\rho}}$ and $\hat{\boldsymbol{\varphi}}$ are functions of φ, the set of components (A_ρ, A_φ, A_z) refers in general to a specific point in space at which the vector $\boldsymbol{A}$ is to be located, or at least to a specific value of the coordinate φ. Thus the components of a vector in cylindrical coordinates, and in fact in all systems of curvilinear coordinates, depend not only on the vector itself, but also on its location in space. If $\boldsymbol{A}$ is a function of a parameter, say t, then we may compute its derivative by differentiating Eq. (3.94), but we must be careful to take account of the variation

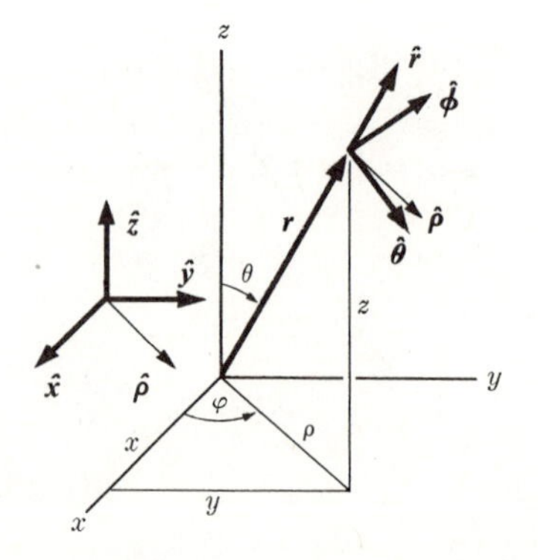

Fig. 3.23 Spherical polar coordinates.

of $\hat{\boldsymbol{\rho}}$ and $\hat{\boldsymbol{\varphi}}$ if the location of the vector is also changing with t (e.g., if $\boldsymbol{A}$ is the force acting on a moving particle):

$$\frac{d\boldsymbol{A}}{dt} = \left(\frac{dA_\rho}{dt} - A_\varphi \frac{d\varphi}{dt}\right)\hat{\boldsymbol{\rho}} + \left(\frac{dA_\varphi}{dt} + A_\rho \frac{d\varphi}{dt}\right)\hat{\boldsymbol{\varphi}} + \frac{dA_z}{dt}\hat{\boldsymbol{z}}. \tag{3.95}$$

Formulas (3.92) and (3.93) are special cases of Eq. (3.95). A formula for $d\boldsymbol{A}/dt$ could have been worked out also for the case of polar coordinates in two dimensions considered in the preceding section, and would, in fact, have been exactly analogous to Eq. (3.95) except that the last term would be missing.

Spherical polar coordinates (r, θ, φ) are defined as in Fig. 3.23 or by the equations

$$x = r \sin\theta \cos\varphi, \qquad y = r \sin\theta \sin\varphi, \qquad z = r\cos\theta. \tag{3.96}$$

The expressions for x and y follow if we note that $\rho = r\sin\theta$, and use Eq. (3.87); the formula for z is evident from the diagram. Conversely,

$$\begin{aligned} r &= (x^2+y^2+z^2)^{1/2}, \\ \theta &= \tan^{-1}\frac{(x^2+y^2)^{1/2}}{z}, \\ \varphi &= \tan^{-1}\frac{y}{x}. \end{aligned} \tag{3.97}$$

Unit vectors $\hat{\boldsymbol{r}}$, $\hat{\boldsymbol{\theta}}$, $\hat{\boldsymbol{\varphi}}$ appropriate to spherical coordinates are indicated in Fig. 3.23, where $\hat{\boldsymbol{\varphi}}$ is the same vector as in cylindrical coordinates. The unit vector $\hat{\boldsymbol{\rho}}$ is also useful in obtaining relations involving $\hat{\boldsymbol{r}}$ and $\hat{\boldsymbol{\theta}}$. We note that $\hat{\boldsymbol{z}}$, $\hat{\boldsymbol{\rho}}$, $\hat{\boldsymbol{r}}$, $\hat{\boldsymbol{\theta}}$, all lie in one vertical plane. From the figure, and Eq. (3.89), we have

$$\begin{aligned} \hat{\boldsymbol{r}} &= \hat{\boldsymbol{z}}\cos\theta + \hat{\boldsymbol{\rho}}\sin\theta = \hat{\boldsymbol{z}}\cos\theta + \hat{\boldsymbol{x}}\sin\theta\cos\varphi + \hat{\boldsymbol{y}}\sin\theta\sin\varphi, \\ \hat{\boldsymbol{\theta}} &= -\hat{\boldsymbol{z}}\sin\theta + \hat{\boldsymbol{\rho}}\cos\theta = -\hat{\boldsymbol{z}}\sin\theta + \hat{\boldsymbol{x}}\cos\theta\cos\varphi + \hat{\boldsymbol{y}}\cos\theta\sin\varphi, \\ \hat{\boldsymbol{\varphi}} &= -\hat{\boldsymbol{x}}\sin\varphi + \hat{\boldsymbol{y}}\cos\varphi. \end{aligned} \tag{3.98}$$

By differentiating these formulas, or more easily by inspection of the diagram (as in Fig. 3.21), noting that variation of θ, with φ and r fixed, corresponds to rotation in the $\hat{\boldsymbol{z}}, \hat{\boldsymbol{r}}, \hat{\boldsymbol{\rho}}, \hat{\boldsymbol{\theta}}$ plane, while variation of φ, with θ and r fixed, corresponds to rotation around the z-axis, we find

$$\begin{aligned} &\frac{\partial\hat{\boldsymbol{r}}}{\partial\theta} = \hat{\boldsymbol{\theta}}, &&\frac{\partial\hat{\boldsymbol{r}}}{\partial\varphi} = \hat{\boldsymbol{\varphi}}\sin\theta, \\ &\frac{\partial\hat{\boldsymbol{\theta}}}{\partial\theta} = -\hat{\boldsymbol{r}}, &&\frac{\partial\hat{\boldsymbol{\theta}}}{\partial\varphi} = \hat{\boldsymbol{\varphi}}\cos\theta, \\ &\frac{\partial\hat{\boldsymbol{\varphi}}}{\partial\theta} = 0, &&\frac{\partial\hat{\boldsymbol{\varphi}}}{\partial\varphi} = -\hat{\boldsymbol{\rho}} = -\hat{\boldsymbol{r}}\sin\theta - \hat{\boldsymbol{\theta}}\cos\theta. \end{aligned} \tag{3.99}$$

In spherical coordinates the position vector is simply

$$\boldsymbol{r} = r\hat{\boldsymbol{r}}(\theta, \varphi). \tag{3.100}$$

Differentiating and using Eqs. (3.99), we obtain the velocity and acceleration:

$$\boldsymbol{v} = \frac{d\boldsymbol{r}}{dt} = \dot{r}\hat{\boldsymbol{r}} + r\dot{\theta}\hat{\boldsymbol{\theta}} + (r\dot{\varphi}\sin\theta)\hat{\boldsymbol{\varphi}}, \tag{3.101}$$

$$\boldsymbol{a} = \frac{d\boldsymbol{v}}{dt} = (\ddot{r} - r\dot{\theta}^2 - r\dot{\varphi}^2\sin^2\theta)\hat{\boldsymbol{r}} + (r\ddot{\theta} + 2\dot{r}\dot{\theta} - r\dot{\varphi}^2\sin\theta\cos\theta)\hat{\boldsymbol{\theta}}$$

$$+ (r\ddot{\varphi}\sin\theta + 2\dot{r}\dot{\varphi}\sin\theta + 2r\dot{\theta}\dot{\varphi}\cos\theta)\hat{\boldsymbol{\varphi}}. \tag{3.102}$$

Again, $\hat{\boldsymbol{r}}$, $\hat{\boldsymbol{\theta}}$, $\hat{\boldsymbol{\varphi}}$ form a set of mutually perpendicular unit vectors, and any vector $\boldsymbol{A}$ may be represented in terms of its spherical components:

$$\boldsymbol{A} = A_r\hat{\boldsymbol{r}} + A_\theta\hat{\boldsymbol{\theta}} + A_\varphi\hat{\boldsymbol{\varphi}}. \tag{3.103}$$

Here again the components depend not only on $\boldsymbol{A}$ but also on its location. If $\boldsymbol{A}$ and its location are functions of t, then

$$\frac{d\boldsymbol{A}}{dt} = \left(\frac{dA_r}{dt} - A_\theta\frac{d\theta}{dt} - A_\varphi\sin\theta\frac{d\varphi}{dt}\right)\hat{\boldsymbol{r}}$$

$$+ \left(\frac{dA_\theta}{dt} + A_r\frac{d\theta}{dt} - A_\varphi\cos\theta\frac{d\varphi}{dt}\right)\hat{\boldsymbol{\theta}}$$

$$+ \left(\frac{dA_\varphi}{dt} + A_r\sin\theta\frac{d\varphi}{dt} + A_\theta\cos\theta\frac{d\varphi}{dt}\right)\hat{\boldsymbol{\varphi}}. \tag{3.104}$$

3.6 ELEMENTS OF VECTOR ANALYSIS

A scalar function $u(x, y, z)$ has three derivatives, which may be thought of as the components of a vector point function called the *gradient* of u:

$$\text{grad } u = \left(\frac{\partial u}{\partial x}, \frac{\partial u}{\partial y}, \frac{\partial u}{\partial z}\right) = \hat{\boldsymbol{x}}\frac{\partial u}{\partial x} + \hat{\boldsymbol{y}}\frac{\partial u}{\partial y} + \hat{\boldsymbol{z}}\frac{\partial u}{\partial z}. \tag{3.105}$$

We may also define grad u geometrically as a vector whose direction is the direction in which u increases most rapidly and whose magnitude is the *directional derivative* of u, i.e., the rate of increase of u per unit distance in that direction. That this geometrical definition is equivalent to the algebraic definition (3.105) can be seen by taking the differential of u:

$$du = \frac{\partial u}{\partial x}dx + \frac{\partial u}{\partial y}dy + \frac{\partial u}{\partial z}dz. \tag{3.106}$$

Equation (3.106) has the form of a scalar product of grad u with the vector $\boldsymbol{dr}$ whose components are dx, dy, dz:

$$du = \boldsymbol{dr}\cdot\text{grad } u. \tag{3.107}$$

Geometrically, du is the change in u when we move from the point $\boldsymbol{r} = (x, y, z)$ to a nearby point $\boldsymbol{r}+\boldsymbol{dr} = (x+dx, y+dy, z+dz)$. By Eq. (3.15):

$$du = |\boldsymbol{dr}|\,|\text{grad } u| \cos\theta, \tag{3.108}$$

where θ is the angle between $\boldsymbol{dr}$ and grad u. Thus at a fixed small distance $|\boldsymbol{dr}|$ from the point $\boldsymbol{r}$, the change in u is a maximum when $\boldsymbol{dr}$ is in the same direction as grad u, and then:

$$|\text{grad } u| = \frac{du}{|\boldsymbol{dr}|}.$$

This confirms the geometrical description of grad u given above. An alternative geometrical definition of grad u is that it is a vector such that the change in u, for an arbitrary small change of position $\boldsymbol{dr}$, is given by Eq. (3.107).

In a purely symbolic way, the right member of Eq. (3.105) can be thought of as the "product" of a "vector":

$$\boldsymbol{\nabla} = \left(\frac{\partial}{\partial x}, \frac{\partial}{\partial y}, \frac{\partial}{\partial z}\right) = \hat{\boldsymbol{x}}\frac{\partial}{\partial x} + \hat{\boldsymbol{y}}\frac{\partial}{\partial y} + \hat{\boldsymbol{z}}\frac{\partial}{\partial z}, \tag{3.109}$$

with the scalar function u:

$$\text{grad } u = \boldsymbol{\nabla} u. \tag{3.110}$$

The symbol $\boldsymbol{\nabla}$ is pronounced "del." $\boldsymbol{\nabla}$ itself is not a vector in the geometrical sense, but an operation on a function u which gives a vector $\boldsymbol{\nabla} u$. However, algebraically, $\boldsymbol{\nabla}$ has properties nearly identical with those of a vector. The reason is that the differentiation symbols ($\partial/\partial x$, $\partial/\partial y$, $\partial/\partial z$) have algebraic properties like those of ordinary numbers except when they act on a product of functions:

$$\frac{\partial}{\partial x}(u+v) = \frac{\partial u}{\partial x} + \frac{\partial v}{\partial x}, \qquad \frac{\partial}{\partial x}\frac{\partial}{\partial y}u = \frac{\partial}{\partial y}\frac{\partial}{\partial x}u, \tag{3.111}$$

and

$$\frac{\partial}{\partial x}(au) = a\frac{\partial u}{\partial x}, \tag{3.112}$$

provided a is constant. However,

$$\frac{\partial}{\partial x}(uv) = \frac{\partial u}{\partial x}v + u\frac{\partial v}{\partial x}. \tag{3.113}$$

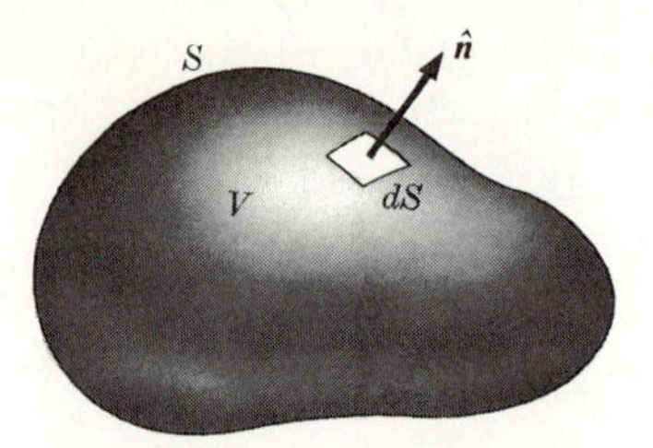

Fig. 3.24 A volume V bounded by a surface S.

In this one respect differentiation operators differ algebraically from ordinary numbers. If $\partial/\partial x$ were a number, $\partial/\partial x(uv)$ would equal either $u(\partial/\partial x)v$ or $v(\partial/\partial x)u$. Thus we may say that $\partial/\partial x$ behaves algebraically as a number except that when it operates on a product, the result is a sum of terms in which each factor is differentiated separately, as in Eq. (3.113). A similar remark applies to the symbol $\boldsymbol{\nabla}$. It behaves algebraically as a vector, except that when it operates on a product it must be treated also as a differentiation operation. This rule enables us to write down a large number of identities involving the $\boldsymbol{\nabla}$ symbol, based on vector identities. We shall require very few of these in this text, and shall not list them here.*

We can form the scalar product of $\boldsymbol{\nabla}$ with a vector point function $\boldsymbol{A}(x, y, z)$. This is called the *divergence* of $\boldsymbol{A}$:

$$\operatorname{div} \boldsymbol{A} = \boldsymbol{\nabla}\cdot\boldsymbol{A} = \frac{\partial A_x}{\partial x}+\frac{\partial A_y}{\partial y}+\frac{\partial A_z}{\partial z}. \tag{3.114}$$

The geometrical meaning of div $\boldsymbol{A}$ is given by the following theorem, called the divergence theorem, or Gauss' theorem:

$$\iiint_V \boldsymbol{\nabla}\cdot\boldsymbol{A}\, dV = \iint_S \hat{\boldsymbol{n}}\cdot\boldsymbol{A}\, dS, \tag{3.115}$$

where V is a given volume, S is the surface bounding the volume V, and $\hat{\boldsymbol{n}}$ is a unit vector perpendicular to the surface S pointing out from the volume at each point of S (Fig. 3.24). Thus $\hat{\boldsymbol{n}}\cdot\boldsymbol{A}$ is the component of $\boldsymbol{A}$ normal to S, and Eq. (3.115) says that the "total amount of $\boldsymbol{\nabla}\cdot\boldsymbol{A}$ inside V" is equal to the "total flux of $\boldsymbol{A}$ outward through the surface S." If $\boldsymbol{v}$ represents the velocity of a moving fluid at any point in space, then

$$\iint_S \hat{\boldsymbol{n}}\cdot\boldsymbol{v}\, dS$$

represents the volume of fluid flowing across S per second. If the fluid is incompressible, then according to Eq. (3.115),

$$\iiint_V \boldsymbol{\nabla}\cdot\boldsymbol{v}\, dV$$

*For a more complete treatment of vector analysis, see H. B. Phillips, *Vector Analysis*. New York: John Wiley & Sons, 1933.

would represent the total volume of fluid being produced within the volume V per second. Hence $\nabla \cdot \boldsymbol{v}$ would be positive at sources from which the fluid is flowing, and negative at "sinks" into which it is flowing. We omit the proof of Gauss' theorem [Eq. (3.115)]; it may be found in any book on vector analysis.*

We can also form a cross product of ∇ with a vector point function $\boldsymbol{A}(x, y, z)$. This is called the *curl* of $\boldsymbol{A}$:

$$\operatorname{curl} \boldsymbol{A} = \nabla \times \boldsymbol{A} = \hat{\boldsymbol{x}}\left(\frac{\partial A_z}{\partial y} - \frac{\partial A_y}{\partial z}\right) + \hat{\boldsymbol{y}}\left(\frac{\partial A_x}{\partial z} - \frac{\partial A_z}{\partial x}\right) + \hat{\boldsymbol{z}}\left(\frac{\partial A_y}{\partial x} - \frac{\partial A_x}{\partial y}\right). \tag{3.116}$$

The geometrical meaning of the curl is given by Stokes' theorem:

$$\iint_S \hat{\boldsymbol{n}} \cdot (\nabla \times \boldsymbol{A})\, dS = \int_C \boldsymbol{A} \cdot d\boldsymbol{r}, \tag{3.117}$$

where S is any surface in space, $\hat{\boldsymbol{n}}$ is the unit vector normal to S, and C is the curve bounding S, $d\boldsymbol{r}$ being taken in that direction in which a man would walk around C if his left hand were on the inside and his head in the direction of $\hat{\boldsymbol{n}}$. (See Fig. 3.25.) According to Eq. (3.117), curl $\boldsymbol{A}$ at any point is a measure of the extent to which the vector function $\boldsymbol{A}$ circles around that point. A good example is the magnetic field around a wire carrying an electric current, where the curl of the magnetic field intensity is proportional to the current density. We omit the proof of Stokes' theorem [Eq. (3.117)].†

The reader should not be bothered by the difficulty of fixing these ideas in his mind. Understanding of new mathematical concepts like these comes to most people only slowly, as they are put to use. The definitions are recorded here for future use. One cannot be expected to be familiar with them until he has seen how they are used in physical problems.

‡The symbolic vector ∇ can also be expressed in cylindrical coordinates in

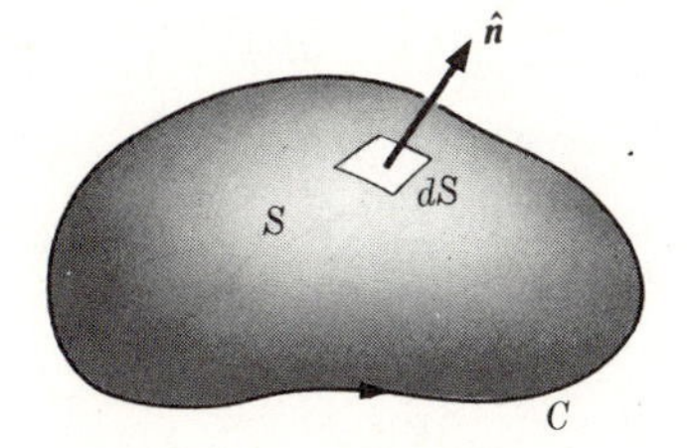

Fig. 3.25 A surface S bounded by a curve C.

*See, e.g., Phillips, *op. cit.* Chapter 3, Section 32.

†For the proof see Phillips, *op. cit.* Chapter 3, Section 29.

‡The remainder of this section can be omitted on first reading.

terms of its components along $\hat{\boldsymbol{\rho}}$, $\hat{\boldsymbol{\varphi}}$, $\hat{\boldsymbol{z}}$. (See Fig. 3.22.) We note that if $u = u(\rho, \varphi, z)$,

$$du = \frac{\partial u}{\partial \rho} d\rho + \frac{\partial u}{d\varphi} d\varphi + \frac{\partial u}{\partial z} dz, \tag{3.118}$$

and, from Eqs. (3.91) and (3.90),

$$d\boldsymbol{r} = \hat{\boldsymbol{\rho}}\, d\rho + \hat{\boldsymbol{\varphi}}\rho\, d\varphi + \hat{\boldsymbol{z}}\, dz, \tag{3.119}$$

a result whose geometric significance will be evident from Fig. 3.22. Hence, if we write

$$\boldsymbol{\nabla} = \hat{\boldsymbol{\rho}} \frac{\partial}{\partial \rho} + \frac{\hat{\boldsymbol{\varphi}}}{\rho} \frac{\partial}{\partial \varphi} + \hat{\boldsymbol{z}} \frac{\partial}{\partial z}, \tag{3.120}$$

we will have, since $\hat{\boldsymbol{\rho}}$, $\hat{\boldsymbol{\varphi}}$, $\hat{\boldsymbol{z}}$ are a set of mutually perpendicular unit vectors,

$$du = d\boldsymbol{r} \cdot \boldsymbol{\nabla} u, \tag{3.121}$$

as required by the geometrical definition of $\boldsymbol{\nabla} u = \text{grad } u$. [See the remarks following Eq. (3.107).] A formula for $\boldsymbol{\nabla}$ could have been worked out also for the case of polar coordinates in two dimensions and would have been exactly analogous to Eq. (3.120) except that the term in z would be missing. In applying the symbol $\boldsymbol{\nabla}$ to expressions involving vectors expressed in cylindrical coordinates [Eq. (3.94)], it must be remembered that the unit vectors $\hat{\boldsymbol{\rho}}$ and $\hat{\boldsymbol{\varphi}}$ are functions of φ and subject to differentiation when they occur after $\partial/\partial\varphi$.

We may also find the vector $\boldsymbol{\nabla}$ in spherical coordinates (Fig. 3.23) by noting that

$$du = \frac{\partial u}{dr} dr + \frac{\partial u}{\partial \theta} d\theta + \frac{\partial u}{\partial \varphi} d\varphi, \tag{3.122}$$

and

$$d\boldsymbol{r} = \hat{\boldsymbol{r}}\, dr + \hat{\boldsymbol{\theta}} r\, d\theta + \hat{\boldsymbol{\varphi}} r \sin\theta\, d\varphi. \tag{3.123}$$

Hence

$$\boldsymbol{\nabla} = \hat{\boldsymbol{r}} \frac{\partial}{\partial r} + \frac{\hat{\boldsymbol{\theta}}}{r} \frac{\partial}{\partial \theta} + \frac{\hat{\boldsymbol{\varphi}}}{r \sin\theta} \frac{\partial}{\partial \varphi}, \tag{3.124}$$

in order that Eq. (3.121) may hold. Again we caution that in working with Eq. (3.124), the dependence of $\hat{\boldsymbol{r}}$, $\hat{\boldsymbol{\theta}}$, $\hat{\boldsymbol{\varphi}}$ on θ, φ must be kept in mind. For example, the divergence of a vector function $\boldsymbol{A}$ expressed in spherical coordinates [Eq. (3.103)] is

$$\begin{aligned}\nabla\cdot\boldsymbol{A} &= \hat{\boldsymbol{r}}\cdot\frac{\partial \boldsymbol{A}}{\partial r}+\frac{\hat{\boldsymbol{\theta}}}{r}\cdot\frac{\partial \boldsymbol{A}}{\partial \theta}+\frac{\hat{\boldsymbol{\varphi}}}{r\sin\theta}\cdot\frac{\partial \boldsymbol{A}}{\partial \varphi}\\ &= \frac{\partial A_r}{\partial r}+\frac{1}{r}\left(\frac{\partial A_\theta}{\partial \theta}+A_r\right)+\frac{1}{r\sin\theta}\left(\frac{\partial A_\varphi}{\partial\varphi}+A_r\sin\theta+A_\theta\cos\theta\right)\\ &= \frac{\partial A_r}{\partial r}+\frac{2A_r}{r}+\frac{1}{r}\frac{\partial A_\theta}{\partial\theta}+\frac{A_\theta}{r\tan\theta}+\frac{1}{r\sin\theta}\frac{\partial A_\varphi}{\partial\varphi}\\ &= \frac{1}{r^2}\frac{\partial}{\partial r}(r^2A_r)+\frac{1}{r\sin\theta}\frac{\partial}{\partial\theta}(\sin\theta\, A_\theta)+\frac{1}{r\sin\theta}\frac{\partial A_\varphi}{\partial\varphi}.\end{aligned}$$

(In the above calculation, we use the fact that $\hat{\boldsymbol{r}}$, $\hat{\boldsymbol{\theta}}$, $\hat{\boldsymbol{\varphi}}$ are a set of mutually perpendicular unit vectors.)

3.7 MOMENTUM AND ENERGY THEOREMS

Newton's second law, as formulated in Chapter 1, leads, in two or three dimensions, to the vector equation

$$m\frac{d^2\boldsymbol{r}}{dt^2} = \boldsymbol{F}. \tag{3.125}$$

In two dimensions this is equivalent to two component equations, in three dimensions to three, which are in cartesian coordinates

$$m\frac{d^2x}{dt^2} = F_x, \qquad m\frac{d^2y}{dt^2} = F_y, \qquad m\frac{d^2z}{dt^2} = F_z. \tag{3.126}$$

In this section, we prove, using Eq. (3.125), some theorems for motion in two or three dimensions which are the vector analogs to those proved in Section 2.1 for one-dimensional motion.

The linear momentum vector $\boldsymbol{p}$ of a particle is to be defined, according to Eq. (1.10), as follows:

$$\boldsymbol{p} = m\boldsymbol{v}. \tag{3.127}$$

Equations (3.125) and (3.126) can then be written

$$\frac{d}{dt}(m\boldsymbol{v}) = \frac{d\boldsymbol{p}}{dt} = \boldsymbol{F}, \tag{3.128}$$

or, in component form,

$$\frac{dp_x}{dt} = F_x, \qquad \frac{dp_y}{dt} = F_y, \qquad \frac{dp_z}{dt} = F_z. \tag{3.129}$$

If we multiply by dt, and integrate from t_1 to t_2, we obtain the change in momentum between t_1 and t_2:

$$\boldsymbol{p}_2 - \boldsymbol{p}_1 = m\boldsymbol{v}_2 - m\boldsymbol{v}_1 = \int_{t_1}^{t_2} \boldsymbol{F}\, dt. \tag{3.130}$$

The integral on the right is the impulse delivered by the force, and is a vector whose components are the corresponding integrals of the components of $\boldsymbol{F}$. In component form:

$$\begin{aligned} p_{x_2} - p_{x_1} &= \int_{t_1}^{t_2} F_x\, dt, \\ p_{y_2} - p_{y_1} &= \int_{t_1}^{t_2} F_y\, dt, \\ p_{z_2} - p_{z_1} &= \int_{t_1}^{t_2} F_z\, dt. \end{aligned} \tag{3.131}$$

In order to obtain an equation for the rate of change of kinetic energy, we proceed as in Section 2.1, multiplying Eqs. (3.126) by v_x, v_y, v_z, respectively, to obtain

$$\frac{d}{dt}\left(\tfrac{1}{2}mv_x^2\right) = F_x v_x, \qquad \frac{d}{dt}\left(\tfrac{1}{2}mv_y^2\right) = F_y v_y, \qquad \frac{d}{dt}\left(\tfrac{1}{2}mv_z^2\right) = F_z v_z. \tag{3.132}$$

Adding these equations, we have

$$\frac{d}{dt}\left[\tfrac{1}{2}m(v_x^2 + v_y^2 + v_z^2)\right] = F_x v_x + F_y v_y + F_z v_z,$$

or

$$\frac{d}{dt}\left(\tfrac{1}{2}mv^2\right) = \frac{dT}{dt} = \boldsymbol{F}\cdot\boldsymbol{v}. \tag{3.133}$$

This equation can also be deduced from the vector equation (3.125) by taking the dot product with $\boldsymbol{v}$ on each side, and noting that

$$\frac{d}{dt}(v^2) = \frac{d}{dt}(\boldsymbol{v}\cdot\boldsymbol{v}) = \frac{d\boldsymbol{v}}{dt}\cdot\boldsymbol{v} + \boldsymbol{v}\cdot\frac{d\boldsymbol{v}}{dt} = 2\boldsymbol{v}\cdot\frac{d\boldsymbol{v}}{dt}.$$

Thus, by Eq. (3.132),

$$\boldsymbol{F}\cdot\boldsymbol{v} = m\boldsymbol{v}\cdot\frac{d\boldsymbol{v}}{dt} = \tfrac{1}{2}m\frac{d(v^2)}{dt} = \frac{d}{dt}\left(\tfrac{1}{2}mv^2\right).$$

Multiplying Eq. (3.133) by dt, and integrating, we obtain the integrated form of the energy theorem:

$$T_2 - T_1 = \tfrac{1}{2}mv_2^2 - \tfrac{1}{2}mv_1^2 = \int_{t_1}^{t_2} \boldsymbol{F}\cdot\boldsymbol{v}\, dt. \tag{3.134}$$

Since $\boldsymbol{v}\, dt = d\boldsymbol{r}$, if $\boldsymbol{F}$ is given as a function of $\boldsymbol{r}$, we can write the right member of Eq. (3.134) as a line integral:

$$T_2 - T_1 = \int_{\boldsymbol{r}_1}^{\boldsymbol{r}_2} \boldsymbol{F}\cdot d\boldsymbol{r}, \tag{3.135}$$

where the integral is to be taken along the path followed by the particle between

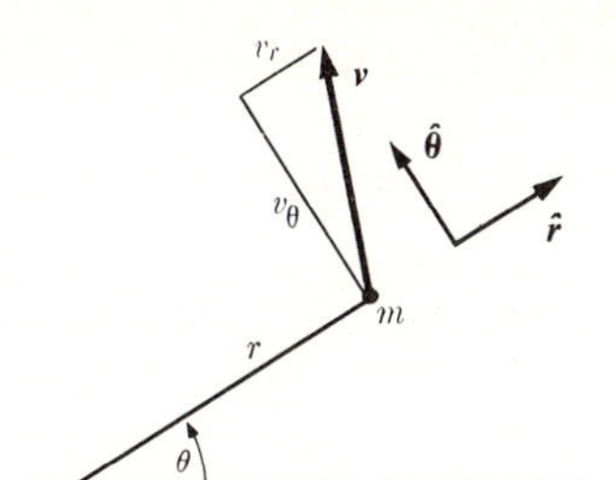

Fig. 3.26 Components of velocity in a plane.

the points $\boldsymbol{r}_1$ and $\boldsymbol{r}_2$. The integral on the right in Eqs. (3.134) and (3.135) is the work done on the particle by the force between the times t_1 and t_2. Note how the vector notation brings out the analogy between the one- and the two- or three-dimensional cases of the momentum and energy theorems.

3.8 PLANE AND VECTOR ANGULAR MOMENTUM THEOREMS

If a particle moves in a plane, we define its angular momentum L_O about a point O as the moment of its momentum vector about the point O, that is, as the product of its distance from O times the component of momentum perpendicular to the line joining the particle to O. The subscript $_O$ will usually be omitted, except when moments about more than one origin enter into the discussion, but it must be remembered that angular momentum, like torque, refers to a particular origin about which moments are taken. The angular momentum L is taken as positive when the particle is moving in a counter-clockwise sense with respect to O. It is expressed most simply in terms of polar coordinates with O as origin. Let the particle have mass m. Then its momentum is $m\boldsymbol{v}$, and the component of momentum perpendicular to the radius vector from O is mv_θ (Fig. 3.26), so that, if we use Eq. (3.78),

$$L = rmv_\theta = mr^2\dot{\theta}. \tag{3.136}$$

If we write the force in terms of its polar components,

$$\boldsymbol{F} = \hat{\boldsymbol{r}}F_r + \hat{\boldsymbol{\theta}}F_\theta, \tag{3.137}$$

then in plane polar coordinates the equation of motion, Eq. (3.125), becomes, by Eq. (3.80),

$$ma_r = m\ddot{r} - mr\dot{\theta}^2 = F_r, \tag{3.138}$$

$$ma_\theta = mr\ddot{\theta} + 2m\dot{r}\dot{\theta} = F_\theta. \tag{3.139}$$

We now note that

$$\frac{dL}{dt} = 2mr\dot{r}\dot{\theta} + mr^2\ddot{\theta}.$$

Thus, multiplying Eq. (3.139) by r, we have

$$\frac{dL}{dt} = \frac{d}{dt}(mr^2\dot{\theta}) = rF_\theta = N. \tag{3.140}$$

The quantity rF_θ is the torque exerted by the force $\boldsymbol{F}$ about the point O. Integrating Eq. (3.140), we obtain the integrated form of the angular momentum theorem for motion in a plane:

$$L_2 - L_1 = mr_2^2\dot{\theta}_2 - mr_1^2\dot{\theta}_1 = \int_{t_1}^{t_2} rF_\theta \, dt. \tag{3.141}$$

We can generalize the definition of angular momentum to apply to three-dimensional motion by defining the angular momentum of a particle about an axis in space as the moment of its momentum vector about this axis, just as in Section 3.2 we defined the moment of a force about an axis. The development is most easily carried out in cylindrical coordinates with the z-axis as the axis about which moments are to be taken. The generalization of theorems (3.140) and (3.141) to this case is then easily proved in analogy with the proof given above. This development is left as an exercise.

As a final generalization of the concept of angular momentum, we define the vector angular momentum $\boldsymbol{L}_O$ about a point O as the vector moment of the momentum vector about O:

$$\boldsymbol{L}_O = \boldsymbol{r} \times \boldsymbol{p} = m(\boldsymbol{r} \times \boldsymbol{v}), \tag{3.142}$$

where the vector $\boldsymbol{r}$ is taken from the point O as origin to the position of the particle of mass m. Again we shall omit the subscript $_O$ when no confusion can arise. The component of the vector $\boldsymbol{L}$ in any direction is the moment of the momentum vector $\boldsymbol{p}$ about an axis in that direction through O.

By taking the cross product of $\boldsymbol{r}$ with both members of the vector equation of motion [Eq. (3.125)], we obtain

$$\boldsymbol{r} \times \left(m\frac{d\boldsymbol{v}}{dt}\right) = \boldsymbol{r} \times \boldsymbol{F}. \tag{3.143}$$

By the rules of vector algebra and vector calculus,

$$\begin{aligned}
\frac{d\boldsymbol{L}}{dt} &= \frac{d}{dt}[\boldsymbol{r} \times (m\boldsymbol{v})] \\
&= \boldsymbol{r} \times \frac{d}{dt}(m\boldsymbol{v}) + \frac{d\boldsymbol{r}}{dt} \times (m\boldsymbol{v}) \\
&= \boldsymbol{r} \times \frac{d}{dt}(m\boldsymbol{v}) + \boldsymbol{v} \times (m\boldsymbol{v}) \\
&= \boldsymbol{r} \times \left(m\frac{d\boldsymbol{v}}{dt}\right).
\end{aligned}$$

We substitute this result in Eq. (3.143):

$$dL/dt = r \times F = N. \tag{3.144}$$

The time rate of change of the vector angular momentum of a particle is equal to the vector torque acting on it. The integral form of the angular momentum theorem is

$$L_2 - L_1 = \int_{t_1}^{t_2} N \, dt. \tag{3.145}$$

The theorems for plane angular momentum and for angular momentum about an axis follow from the vector angular momentum theorems by taking components in the appropriate direction.

3.9 DISCUSSION OF THE GENERAL PROBLEM OF TWO- AND THREE-DIMENSIONAL MOTION

If the force F is given, in general as a function $F(v, r, t)$ of position, velocity, and time, the equations of motion (3.126) become a set of three (or, in two dimensions, two) simultaneous second-order differential equations:

$$\begin{aligned} m\frac{d^2x}{dt^2} &= F_x(\dot{x}, \dot{y}, \dot{z}, x, y, z, t), \\ m\frac{d^2y}{dt^2} &= F_y(\dot{x}, \dot{y}, \dot{z}, x, y, z, t), \\ m\frac{d^2z}{dt^2} &= F_z(\dot{x}, \dot{y}, \dot{z}, x, y, z, t). \end{aligned} \tag{3.146}$$

If we are given the position $r_0 = (x_0, y_0, z_0)$, and the velocity $v_0 = (v_{x_0}, v_{y_0}, v_{z_0})$ at any instant t_0, Eqs. (3.146) give us d^2r/dt^2, and from r, $\dot{r}$, $\ddot{r}$, at time t, we can determine r, $\dot{r}$ a short time later or earlier at $t+dt$, thus extending the functions r, $\dot{r}$, $\ddot{r}$, into the past and future with the help of Eqs. (3.146). This argument can be made mathematically rigorous, and leads to an existence theorem guaranteeing the existence of a unique solution of these equations for any given position and velocity at an initial instant t_0. We note that the general solution of Eqs. (3.146) involves the six "arbitrary" constants x_0, y_0, z_0, v_{x_0}, v_{y_0}, v_{z_0}. Instead of these six constants, we might specify any other six quantities from which they can be determined. (In two dimensions, we will have two second-order differential equations and four initial constants.)

In general, the solution of the three simultaneous equations (3.146) will be much more difficult than the solution of the single equation (2.9) for one-dimensional motion. The reason for the greater difficulty is that, in general, all the variables x, y, z and their derivatives are involved in all three equations, which makes the problem of the same order of difficulty as a single sixth-order differential equation. [In fact, the set of Eqs. (3.146) can be shown to be equivalent to a single sixth-order equation.] If each force component involved only the corresponding coordinate and its derivatives,

$$F_x = F_x(\dot{x}, x, t),$$
$$F_y = F_y(\dot{y}, y, t), \tag{3.147}$$
$$F_z = F_z(\dot{z}, z, t),$$

then the three equations (3.146) would be independent of one another. We could solve for $x(t)$, $y(t)$, $z(t)$ separately as three independent problems in one-dimensional motion. The most important example of this case is probably when the force is given as a function of time only:

$$\boldsymbol{F} = \boldsymbol{F}(t) = [F_x(t), F_y(t), F_z(t)]. \tag{3.148}$$

The x, y, and z equations of motion can then each be solved separately by the method given in Section 2.3. The case of a frictional force proportional to the velocity will also be an example of the type (3.147). Other cases will sometimes occur, for example, the three-dimensional harmonic oscillator (e.g., a baseball in a tubful of gelatine, or an atom in a crystal lattice), for which the force is

$$F_x = -k_x x,$$
$$F_y = -k_y y, \tag{3.149}$$
$$F_z = -k_z z,$$

when the axes are suitably chosen. The problem now splits into three separate linear harmonic oscillator problems in x, y, and z. In most cases, however, we are not so fortunate, and Eq. (3.147) does not hold. Special methods are available for solving certain classes of two- and three-dimensional problems. Some of these will be developed in this chapter. Problems not solvable by such methods are always, in principle, solvable by various numerical methods of integrating sets of equations like Eqs. (3.146) to get approximate solutions to any required degree of accuracy. Such methods are even more tedious in the three-dimensional case than in the one-dimensional case, and are usually impractical unless one has the services of one of the large automatic computing machines.

When we try to extend the idea of potential energy to two or three dimensions, we will find that having the force given as $\boldsymbol{F}(\boldsymbol{r})$, a function of $\boldsymbol{r}$ alone, is not sufficient to guarantee the existence of a potential-energy function $V(\boldsymbol{r})$. In the one-dimensional case, we found that if the force is given as a function of position alone, a potential-energy function can always be defined by Eq. (2.41). Essentially, the reason is that in one dimension, a particle which travels from x_1 to x_2 and returns to x_1 must return by the same route, so that if the force is a function of position alone, the work done by the force on the particle during its return trip must necessarily be the same as that expended against the force in going from x_1 to x_2. In three dimensions, a particle can travel from $\boldsymbol{r}_1$ to $\boldsymbol{r}_2$ and return by a different route, so that even if $\boldsymbol{F}$ is a function of $\boldsymbol{r}$, the particle may be acted on by a different force on the return trip and the work done may not be the same. In Section 3.12 we shall formulate a criterion to determine when a potential energy $V(\boldsymbol{r})$ exists.

When $V(\boldsymbol{r})$ exists, a conservation of energy theorem still holds, and the total energy $(T+V)$ is a constant of the motion. However, whereas in one dimension the energy integral is always sufficient to enable us to solve the problem at least in principle (Section 2.5), in two and three dimensions this is no longer the case. If x is the only coordinate, then if we know a relation $(T+V = E)$ between x and $\dot{x}$, we can solve for $\dot{x} = f(x)$ and reduce the problem to one of carrying out a single integration. But with coordinates x, y, z, one relation between $x, y, z, \dot{x}, \dot{y}, \dot{z}$ is not enough. We would need to know five such relations, in general, in order to eliminate, for example, x, y, $\dot{x}$, and $\dot{y}$, and find $\dot{z} = f(z)$. In the two-dimensional case, we would need three relations between $x, y, \dot{x}, \dot{y}$ to solve the problem by this method. To find four more relations like the energy integral from Eqs. (3.146) (or two more in two dimensions) is hopeless in most cases. In fact, such relations do not usually exist. Often, however, we can find other quantities (e.g., the angular momentum) which are constants of the motion, and thus obtain one or two more relations between $x, y, z, \dot{x}, \dot{y}, \dot{z}$, which in many cases will be enough to allow a solution of the problem. Examples will be given later.

3.10 THE HARMONIC OSCILLATOR IN TWO AND THREE DIMENSIONS

In this section and the next, we consider a few simple problems in which the force has the form of Eqs. (3.147), so that the equations of motion separate into independent equations in x, y, and z. Mathematically, we then simply have three separate problems, each of the type considered in Chapter 2. The only new feature will be the interpretation of the three solutions $x(t)$, $y(t)$, $z(t)$ as representing a motion in three-dimensional space.

We first consider briefly the solution of the problem of the three-dimensional harmonic oscillator without damping, whose equations of motion are

$$\begin{aligned} m\ddot{x} &= -k_x x, \\ m\ddot{y} &= -k_y y, \\ m\ddot{z} &= -k_z z. \end{aligned} \tag{3.150}$$

An approximate model could be constructed by suspending a mass between three perpendicular sets of springs (Fig. 3.27). The solutions of these equations, we know from Section 2.8:

$$\begin{aligned} x &= A_x \cos(\omega_x t + \theta_x), \qquad & \omega_x^2 &= k_x/m, \\ y &= A_y \cos(\omega_y t + \theta_y), \qquad & \omega_y^2 &= k_y/m, \\ z &= A_z \cos(\omega_z t + \theta_z), \qquad & \omega_z^2 &= k_z/m. \end{aligned} \tag{3.151}$$

The six constants $(A_x, A_y, A_z, \theta_x, \theta_y, \theta_z)$ depend on the initial values x_0, y_0, z_0,

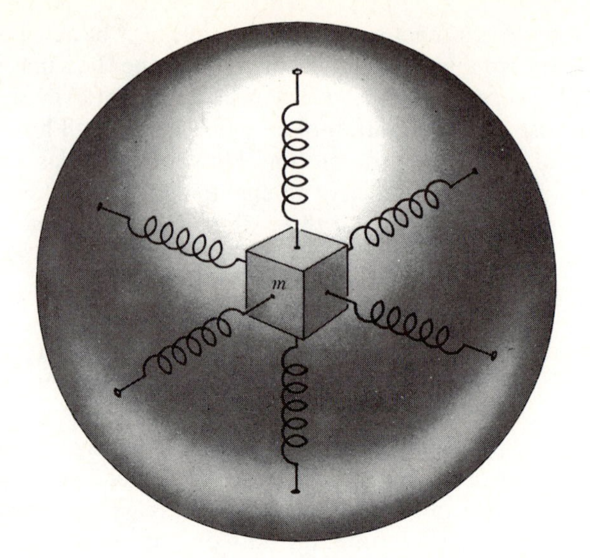

Fig. 3.27 Model of a three-dimensional harmonic oscillator.

$\dot{x}_0, \dot{y}_0, \dot{z}_0$. Each coordinate oscillates independently with simple harmonic motion at a frequency depending on the corresponding restoring force coefficient, and on the mass. The resulting motion of the particle takes place within a rectangular box of dimensions $2A_x \times 2A_y \times 2A_z$ about the origin. If the angular frequencies ω_x, ω_y, ω_z are commensurable, that is, if for some set of integers (n_x, n_y, n_z),

$$\frac{\omega_x}{n_x} = \frac{\omega_y}{n_y} = \frac{\omega_z}{n_z}, \tag{3.152}$$

then the path of the mass m in space is closed, and the motion is periodic. If (n_x, n_y, n_z) are chosen so that they have no common integral factor, then the period of the motion is

$$\tau = \frac{2\pi n_x}{\omega_x} = \frac{2\pi n_y}{\omega_y} = \frac{2\pi n_z}{\omega_z}. \tag{3.153}$$

During one period, the coordinate x makes n_x oscillations, the coordinate y makes n_y oscillations, and the coordinate z makes n_z oscillations, so that the particle returns at the end of the period to its initial position and velocity. In the two-dimensional case, if the path of the oscillating particle is plotted for various combinations of frequencies ω_x and ω_y, and various phases θ_x and θ_y, many interesting and beautiful patterns are obtained. Such patterns are called *Lissajous figures* (Fig. 3.28), and may be produced mechanically by a mechanism designed to move a pencil or other writing device according to Eqs. (3.151). Similar patterns may be obtained electrically on a cathode-ray oscilloscope by sweeping horizon-

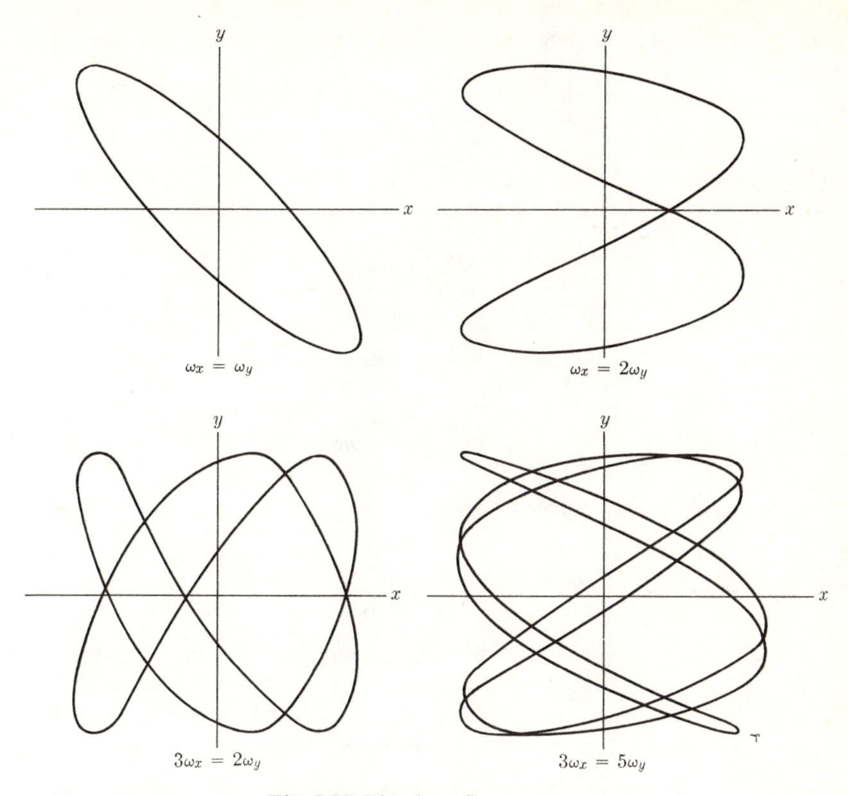

Fig. 3.28 Lissajous figures.

tally and vertically with suitable oscillating voltages. If the frequencies ω_x, ω_y, ω_z are incommensurable, so that Eq. (3.152) does not hold for any set of integers, the motion is not periodic, and the path fills the entire box $2A_x \times 2A_y \times 2A_z$, in the sense that the particle eventually comes arbitrarily close to every point in the box. The discussion can readily be extended to the cases of damped and forced oscillations in two and three dimensions.

If the three constants k_x, k_y, k_z are all equal, the oscillator is said to be *isotropic*, that is, the same in all directions. In this case, the three frequencies ω_x, ω_y, ω_z are all equal and the motion is periodic, with each coordinate executing one cycle of oscillation in a period. The path can be shown to be an ellipse, a straight line, or a circle, depending on the amplitudes and phases (A_x, A_y, A_z, θ_x, θ_y, θ_z).

3.11 PROJECTILES

An important problem in the history of the science of mechanics is that of determining the motion of a projectile. A projectile moving under the action of gravity near the surface of the earth moves, if air resistance is neglected, according to the equation

$$m\frac{d^2\boldsymbol{r}}{dt^2} = -mg\hat{\boldsymbol{z}}, \tag{3.154}$$

where the z-axis is taken in the vertical direction. In component form:

$$m\frac{d^2x}{dt^2} = 0, \tag{3.155}$$

$$m\frac{d^2y}{dt^2} = 0, \tag{3.156}$$

$$m\frac{d^2z}{dt^2} = -mg. \tag{3.157}$$

The solutions of these equations are

$$x = x_0 + v_{x_0}t, \tag{3.158}$$

$$y = y_0 + v_{y_0}t, \tag{3.159}$$

$$z = z_0 + v_{z_0}t - \tfrac{1}{2}gt^2, \tag{3.160}$$

or, in vector form,

$$\boldsymbol{r} = \boldsymbol{r}_0 + \boldsymbol{v}_0 t - \tfrac{1}{2}gt^2\hat{\boldsymbol{z}}. \tag{3.161}$$

We assume the projectile starts from the origin (0, 0, 0), with its initial velocity in the xz-plane, so that $v_{y_0} = 0$. This is no limitation on the motion of the projectile, but merely corresponds to a convenient choice of coordinate system. Equations (3.158), (3.159), (3.160) then become

$$x = v_{x_0}t, \tag{3.162}$$

$$y = 0, \tag{3.163}$$

$$z = v_{z_0}t - \tfrac{1}{2}gt^2. \tag{3.164}$$

These equations give a complete description of the motion of the projectile. Solving the first equation for t and substituting in the third, we have an equation for the path in the xz-plane:

$$z = \frac{v_{z_0}}{v_{x_0}}x - \frac{1}{2}\frac{g}{v_{x_0}^2}x^2. \tag{3.165}$$

This can be rewritten in the form

$$\left(x - \frac{v_{z_0}v_{x_0}}{g}\right)^2 = -2\frac{v_{x_0}^2}{g}\left(z - \frac{v_{z_0}^2}{2g}\right). \tag{3.166}$$

This is a parabola, concave downward, whose maximum altitude occurs at

$$z_m = \frac{v_{z0}^2}{2g}, \tag{3.167}$$

and which crosses the horizontal plane $z = 0$ at the origin and at the point

$$x_m = 2\frac{v_{z0}v_{x0}}{g}. \tag{3.168}$$

If the surface of the earth is horizontal, x_m is the range of the projectile.

Let us now take account of air resistance by assuming a frictional force proportional to the velocity:

$$m\frac{d^2\mathbf{r}}{dt^2} = -mg\hat{\mathbf{z}} - b\frac{d\mathbf{r}}{dt}. \tag{3.169}$$

In component notation, if we assume that the motion takes place in the xz-plane,

$$\begin{aligned} m\frac{d^2x}{dt^2} &= -b\frac{dx}{dt}, \\ m\frac{d^2z}{dt^2} &= -mg - b\frac{dz}{dt}. \end{aligned} \tag{3.170}$$

It should be pointed out that the actual resistance of the air against a moving projectile is a complicated function of velocity, so that the solutions we obtain will be only approximate, although they indicate the general nature of the motion. If the projectile starts from the origin at $t = 0$, the solutions of Eqs. (3.170) are (see Sections 2.4 and 2.6)

$$v_x = v_{x0}e^{-bt/m}, \tag{3.171}$$

$$x = \frac{mv_{x0}}{b}(1 - e^{-bt/m}), \tag{3.172}$$

$$v_z = \left(\frac{mg}{b} + v_{z0}\right)e^{-bt/m} - \frac{mg}{b}, \tag{3.173}$$

$$z = \left(\frac{m^2g}{b^2} + \frac{mv_{z0}}{b}\right)(1 - e^{-bt/m}) - \frac{mg}{b}t. \tag{3.174}$$

Solving Eq. (3.172) for t and substituting in Eq. (3.174), we obtain an equation for the trajectory:

$$z = \left(\frac{mg}{bv_{x0}} + \frac{v_{z0}}{v_{x0}}\right)x - \frac{m^2g}{b^2}\ln\left(\frac{mv_{x0}}{mv_{x0} - bx}\right). \tag{3.175}$$

For low air resistance, or short distances, when $(bx)/(mv_{x0}) \ll 1$, we may expand in powers of $(bx)/(mv_{x0})$ to obtain

$$z = \frac{v_{z0}}{v_{x0}} x - \frac{1}{2}\frac{g}{v_{x0}^2} x^2 - \frac{1}{3}\frac{bg}{mv_{x0}^3} x^3 - \cdots. \tag{3.176}$$

Thus the trajectory starts out as a parabola, but for larger values of x (taking v_{x0} as positive), z falls more rapidly than for a parabola. The fact that the first two terms agree with Eq. (3.165) and that the third term clearly has the right sign is a pretty good check on the algebra leading to Eq. (3.176); expansion in a power series in a small parameter is a simple and useful way of checking a result, and in addition often gives a simple approximate formula which is easy to interpret. According to Eq. (3.175), as x approaches the value $(mv_{x0})/b$, z approaches minus infinity, i.e., the trajectory ends as a vertical drop at $x = (mv_{x0})/b$. From Eq. (3.173), we see that the vertical fall at the end of the trajectory takes place at the terminal velocity $-mg/b$. (The projectile may, of course, return to earth before reaching this part of its trajectory.)

If we set $z = 0$ in Eq. (3.176), it has, in addition to the obvious solution $x = 0$, a solution for the range x_m which we may find by successive approximations. We first rewrite the equation in the following way:

$$x_m = \frac{2v_{x0}v_{z0}}{g} - \frac{2}{3}\frac{b}{v_{x0}} x_m^2 - \cdots. \tag{3.177}$$

If we neglect the second term, we get as a first approximation,

$$x_m \doteq \frac{2v_{x0}v_{z0}}{g},$$

which agrees with Eq. (3.168). Let us substitute this solution in the second term of Eq. (3.177), to obtain a second approximation

$$x_m \doteq \frac{2v_{x0}v_{z0}}{g} - \frac{8}{3}\frac{bv_{z0}^2 v_{x0}}{mg^2}. \tag{3.178}$$

The second term gives the first-order correction to the range due to air resistance, and the first two terms will give a good approximation when the effect of air resistance is small. Higher-order terms could be calculated by repeating the process of substituting approximate solutions back in Eq. (3.177). We thus obtain successive terms for x_m as a power series in b. The extreme opposite case, when air resistance is predominant in determining range (Fig. 3.29), occurs when the vertical drop at $x = (mv_{x0})/b$ begins above the horizontal plane $z = 0$. The range is then, approximately,

$$x_m \doteq \frac{mv_{x0}}{b}, \qquad \left(\frac{bv_{z0}}{mg} \gg 1\right). \tag{3.179}$$

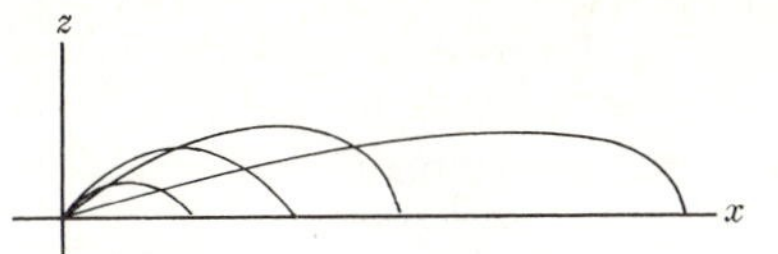

Fig. 3.29 Trajectories for maximum range for projectiles with various muzzle velocities.

We can treat (approximately) the problem of the effect of wind on the projectile by assuming the force of air resistance to be proportional to the relative velocity of the projectile with respect to the air:

$$m\frac{d^2\boldsymbol{r}}{dt^2} = -mg\hat{\boldsymbol{z}} - b\left(\frac{d\boldsymbol{r}}{dt} - \boldsymbol{v}_w\right), \tag{3.180}$$

where $\boldsymbol{v}_w$ is the wind velocity. If $\boldsymbol{v}_w$ is constant, the term $b\boldsymbol{v}_w$ in Eq. (3.180) behaves as a constant force added to $-mg\hat{\boldsymbol{z}}$, and the problem is easily solved by the method above, the only difference being that there may be constant forces in addition to frictional forces in all three directions x, y, z.

The air resistance to a projectile decreases with altitude, so that a better form for the equation of motion of a projectile which rises to appreciable altitudes would be

$$m\frac{d^2\boldsymbol{r}}{dt^2} = -mg\hat{\boldsymbol{z}} - be^{-z/h}\frac{d\boldsymbol{r}}{dt}, \tag{3.181}$$

where h is the height (say about five miles) at which the air resistance falls to $1/e$ of its value at the surface of the earth. In component form,

$$\begin{gathered} m\ddot{x} = -b\dot{x}e^{-z/h}, \qquad m\ddot{y} = -b\dot{y}e^{-z/h}, \\ m\ddot{z} = -mg - b\dot{z}e^{-z/h}. \end{gathered} \tag{3.182}$$

These equations are much harder to solve. Since z appears in the x and y equations, we must first solve the z equation for $z(t)$ and substitute in the other two equations. The z equation is not of any of the simple types discussed in Chapter 2. The importance of this problem was brought out during the First World War, when it was discovered accidentally that aiming a cannon at a much higher elevation than that which had previously been believed to give maximum range resulted in a great increase in the range of the shell. The reason is that the reduction in air resistance, at altitudes of several miles, more than makes up for the loss in horizontal component of muzzle velocity resulting from aiming the gun higher.

3.12 POTENTIAL ENERGY

If the force $\boldsymbol{F}$ acting on a particle is a function of its position $\boldsymbol{r} = (x, y, z)$, then the work done by the force when the particle moves from $\boldsymbol{r}_1$ to $\boldsymbol{r}_2$ is given by the line integral

$$\int_{\boldsymbol{r}_1}^{\boldsymbol{r}_2} \boldsymbol{F}\cdot d\boldsymbol{r}.$$

It is suggested that we try to define a potential energy $V(\boldsymbol{r}) = V(x, y, z)$ in analogy with Eq. (2.41) for one-dimensional motion, as the work done by the force on the particle when it moves from $\boldsymbol{r}$ to some chosen standard point $\boldsymbol{r}_s$:

$$V(\boldsymbol{r}) = -\int_{\boldsymbol{r}_s}^{\boldsymbol{r}} \boldsymbol{F}(\boldsymbol{r})\cdot d\boldsymbol{r}. \tag{3.183}$$

Such a definition implies, however, that the function $V(\boldsymbol{r})$ shall be a function only of the coordinates (x, y, z) of the point $\boldsymbol{r}$ (and of the standard point $\boldsymbol{r}_s$, which we regard as fixed), whereas in general the integral on the right depends upon the path of integration from $\boldsymbol{r}_s$ to $\boldsymbol{r}$. Only if the integral on the right is independent of the path of integration will the definition be legitimate.

Let us assume that we have a force function $\boldsymbol{F}(x, y, z)$ such that the line integral in Eq. (3.183) is independent of the path of integration from $\boldsymbol{r}_s$ to any point $\boldsymbol{r}$. The value of the integral then depends only on $\boldsymbol{r}$ (and on $\boldsymbol{r}_s$), and Eq. (3.183) defines a potential energy function $V(\boldsymbol{r})$. The change in V when the particle moves from $\boldsymbol{r}$ to $\boldsymbol{r}+d\boldsymbol{r}$ is the negative of the work done by the force $\boldsymbol{F}$:

$$dV = -\boldsymbol{F}\cdot d\boldsymbol{r}. \tag{3.184}$$

Comparing Eq. (3.184) with the geometrical definition [Eq. (3.107)] of the gradient, we see that

$$\begin{aligned} -\boldsymbol{F} &= \text{grad } V, \\ \boldsymbol{F} &= -\nabla V. \end{aligned} \tag{3.185}$$

Equation (3.185) may be regarded as the solution of Eq. (3.183) for $\boldsymbol{F}$ in terms of V. In component form,

$$F_x = -\frac{\partial V}{\partial x}, \qquad F_y = -\frac{\partial V}{\partial y}, \qquad F_z = -\frac{\partial V}{\partial z}, \tag{3.186}$$

In seeking a condition to be satisfied by the function $\boldsymbol{F}(\boldsymbol{r})$ in order that the integral in Eq. (3.183) be independent of the path, we note that, since Eq. (3.28) can be proved from the algebraic definition of the cross product, it must hold also for the vector symbol ∇:

$$\nabla \times \nabla = \boldsymbol{0}. \tag{3.187}$$

Applying $(\nabla \times \nabla)$ to the function V, we have

$$\nabla \times \nabla V = \text{curl (grad } V) = \boldsymbol{0}. \tag{3.188}$$

Equation (3.188) can readily be verified by direct computation. From Eqs. (3.188) and (3.185), we have

$$\nabla \times \boldsymbol{F} = \text{curl } \boldsymbol{F} = \boldsymbol{0}. \tag{3.189}$$

Since Eq. (3.189) has been deduced on the assumption that a potential function exists, it represents a necessary condition which must be satisfied by the force

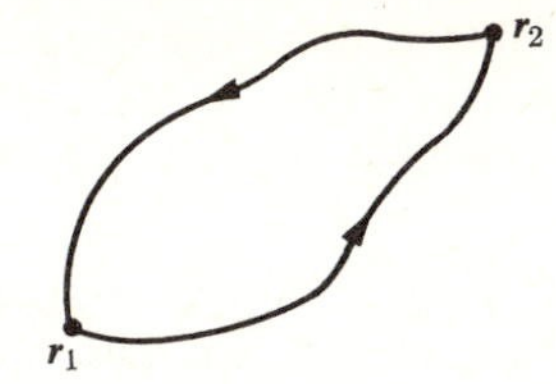

Fig. 3.30 Two paths between $\boldsymbol{r}_1$ and $\boldsymbol{r}_2$, forming a closed path.

function $\boldsymbol{F}(x, y, z)$ before a potential function can be defined. We can show that Eq. (3.189) is also a sufficient condition for the existence of a potential by making use of Stokes' theorem [Eq. (3.117)]. By Stokes' theorem, if we consider any closed path C in space, the work done by the force $\boldsymbol{F}(\boldsymbol{r})$ when the particle travels around this path is

$$\int_C \boldsymbol{F}\cdot d\boldsymbol{r} = \iint_S \hat{\boldsymbol{n}}\cdot(\nabla\times\boldsymbol{F})\, dS, \tag{3.190}$$

where S is a surface in space bounded by the closed curve C. If now Eq. (3.189) is assumed to hold, the integral on the right is zero, and we have, for any closed path C,

$$\int_C \boldsymbol{F}\cdot d\boldsymbol{r} = \boldsymbol{0}. \tag{3.191}$$

But if the work done by the force $\boldsymbol{F}$ around any closed path is zero, then the work done in going from $\boldsymbol{r}_1$ to $\boldsymbol{r}_2$ will be independent of the path followed. Consider any two paths between $\boldsymbol{r}_1$ and $\boldsymbol{r}_2$, and let a closed path C be formed going from $\boldsymbol{r}_1$ to $\boldsymbol{r}_2$ by one path and returning to $\boldsymbol{r}_1$ by the other (Fig. 3.30). Since the work done around C is zero, the work in going from $\boldsymbol{r}_1$ to $\boldsymbol{r}_2$ must be equal and opposite to that on the return trip, hence the work in going from $\boldsymbol{r}_1$ to $\boldsymbol{r}_2$ by either path is the same. Applying this argument to the integral on the right in Eq. (3.183), we see that the result is independent of the path of integration from $\boldsymbol{r}_s$ to $\boldsymbol{r}$, and therefore the integral is a function $V(\boldsymbol{r})$ of the upper limit alone, when the lower limit $\boldsymbol{r}_s$ is fixed. Thus Eq. (3.189) is both necessary and sufficient for the existence of a potential function $V(\boldsymbol{r})$ when the force is given as a function of position $\boldsymbol{F}(\boldsymbol{r})$.

When curl $\boldsymbol{F}$ is zero, we can express the work done by the force when the particle moves from $\boldsymbol{r}_1$ to $\boldsymbol{r}_2$ as the difference between the values of the potential energy at these points:

$$\begin{aligned} \int_{\boldsymbol{r}_1}^{\boldsymbol{r}_2} \boldsymbol{F}\cdot d\boldsymbol{r} &= \int_{\boldsymbol{r}_1}^{\boldsymbol{r}_s} \boldsymbol{F}\cdot d\boldsymbol{r} + \int_{\boldsymbol{r}_s}^{\boldsymbol{r}_2} \boldsymbol{F}\cdot d\boldsymbol{r} \\ &= V(\boldsymbol{r}_1) - V(\boldsymbol{r}_2). \end{aligned} \tag{3.192}$$

Combining Eq. (3.192) with the energy theorem (3.135), we have for any two times t_1 and t_2:

$$T_1 + V(\boldsymbol{r}_1) = T_2 + V(\boldsymbol{r}_2). \tag{3.193}$$

Hence the total energy $(T+V)$ is again constant, and we have an energy integral for motion in three dimensions:

$$T+V = \tfrac{1}{2}m(\dot{x}^2+\dot{y}^2+\dot{z}^2)+V(x, y, z) = E. \tag{3.194}$$

A force which is a function of position alone, and whose curl vanishes, is said to be *conservative*, because it leads to the theorem of conservation of kinetic plus potential energy [Eq. (3.194)].

In some cases, a force may be a function of both position and time $\boldsymbol{F}(\boldsymbol{r}, t)$. If at any time t the curl of $\boldsymbol{F}(\boldsymbol{r}, t)$ vanishes, then a potential-energy function $V(\boldsymbol{r}, t)$ can be defined as

$$V(\boldsymbol{r}, t) = -\int_{\boldsymbol{r}_s}^{\boldsymbol{r}} \boldsymbol{F}(\boldsymbol{r}, t)\cdot d\boldsymbol{r}, \tag{3.195}$$

and we will have, for any time t such that $\nabla \times \boldsymbol{F}(\boldsymbol{r}, t) = \boldsymbol{0}$,

$$\boldsymbol{F}(\boldsymbol{r}, t) = -\nabla V(\boldsymbol{r}, t). \tag{3.196}$$

However, the conservation law of energy can no longer be proved, for Eq. (3.192) no longer holds. It is no longer true that the change in potential energy equals the negative of the work done on the particle, for the integral which defines the potential energy at time t is computed from the force function at that time, whereas the integral that defines the work is computed using at each point the force function at the time the particle passed through that point. Consequently, the energy $T+V$ is not a constant when $\boldsymbol{F}$ and V are functions of time, and such a force is not to be called a conservative force.

When the forces acting on a particle are conservative, Eq. (3.194) enables us to compute its speed as a function of its position. The energy E is fixed by the initial conditions of the motion. Equation (3.194), like Eq. (2.44), gives no information as to the direction of motion. This lack of knowledge of direction is much more serious in two and three dimensions, where there is an infinity of possible directions, than in one dimension, where there are only two opposite directions in which the particle may move. In one dimension, there is only one path along which the particle may move. In two or three dimensions, there are many paths, and unless we know the path of the particle, Eq. (3.194) alone allows us to say very little about the motion except that it can occur only in the region where $V(x, y, z) \leq E$. As an example, the potential energy of an electron in the attractive electric field of two protons (ionized hydrogen molecule $H_2{}^+$) is

$$V = -\frac{e^2}{r_1}-\frac{e^2}{r_2}, \text{ (esu)} \tag{3.197}$$

where r_1, r_2 are the distances of the electron from the two protons. The function $V(x, y)$ (for motion in the xy-plane only) is plotted in Fig. 3.31 as a contour map, where the two protons are 2 A apart at the points $y = 0$, $x = \pm 1$ A, and the figures on the contours of constant potential energy are the corresponding potential energies in units of 10^{-12} erg. So long as $E < -46\times 10^{-12}$ erg, the

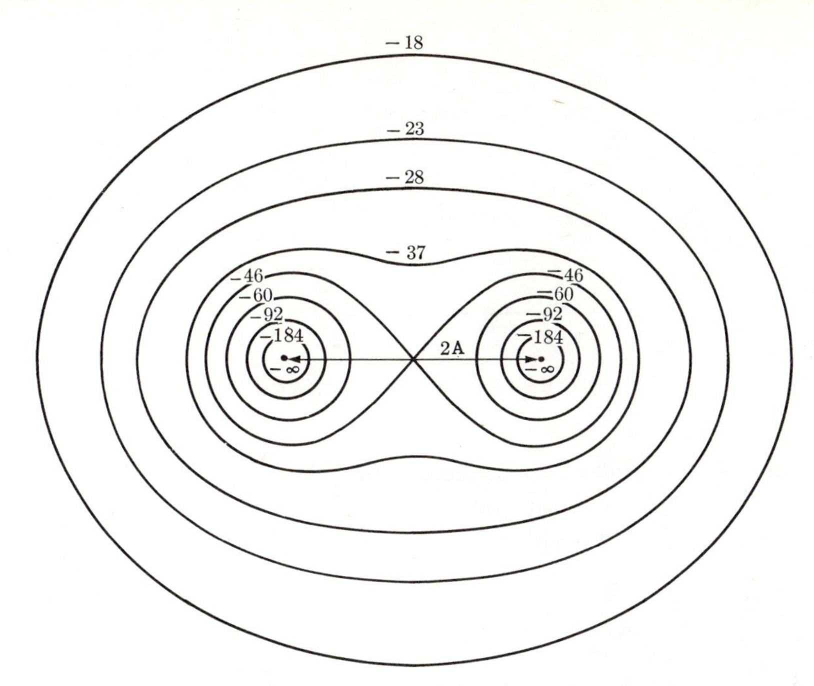

Fig. 3.31 Potential energy of electron in electric field of two protons 2 A apart. (Potential energies in units of 10^{-12} erg.)

electron is confined to a region around one proton or the other, and we expect its motion will be either an oscillation through the attracting center or an orbit around it, depending on initial conditions. (These comments on the expected motion require some physical insight or experience in addition to what we can say from the energy integral alone.) For $0 > E > -46 \times 10^{-12}$ erg, the electron is confined to a region which includes both protons, and a variety of motions are possible. For $E > 0$, the electron is not confined to any finite region in the plane. For $E \ll -46 \times 10^{-12}$ erg, the electron is confined to a region where the equipotentials are nearly circles about one proton, and its motion will be practically the same as if the other proton were not there. For $E < 0$, but $|E| \ll 46 \times 10^{-12}$ erg, the electron *may* circle in an orbit far from the attracting centers, and its motion then will be approximately that of an electron bound to a single attracting center of charge $2e$, as the equipotential lines far from the attracting centers are again very nearly circles.

Given a potential energy function $V(x, y, z)$, Eq. (3.186) enables us to compute the components of the corresponding force at any point. Conversely, given a force $\boldsymbol{F}(x, y, z)$, we may compute its curl to determine whether a potential energy function exists for it. If all components of curl $\boldsymbol{F}$ are zero within any region of space, then within that region $\boldsymbol{F}$ may be represented in terms of a potential-energy function as $-\nabla V$. The potential energy is to be computed from Eq. (3.183). Furthermore,

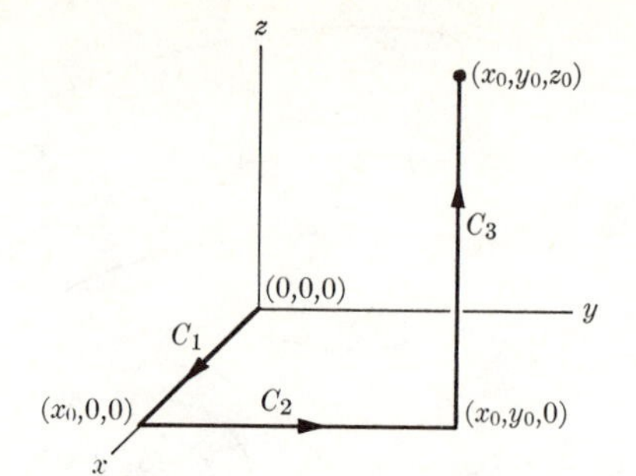

Fig. 3.32 A path of integration from (0, 0, 0) to (x_0, y_0, z_0).

since curl $\boldsymbol{F} = \mathbf{0}$, the result is independent of the path of integration, and we may compute the integral along any convenient path. As an example, consider the following two force functions:

$$\text{a)}\ F_x = axy, \qquad F_y = -az^2, \qquad F_z = -ax^2,$$

$$\text{b)}\ F_x = ay(y^2-3z^2), \quad F_y = 3ax(y^2-z^2), \quad F_z = -6axyz,$$

where a is a constant. We compute the curl in each case:

$$\text{a)}\ \nabla\times\boldsymbol{F} = \hat{\boldsymbol{x}}\left(\frac{\partial F_z}{\partial y}-\frac{\partial F_y}{\partial z}\right)+\hat{\boldsymbol{y}}\left(\frac{\partial F_x}{\partial z}-\frac{\partial F_z}{\partial x}\right)+\hat{\boldsymbol{z}}\left(\frac{\partial F_y}{\partial x}-\frac{\partial F_x}{\partial y}\right)$$

$$= (2az)\hat{\boldsymbol{x}}+(2ax)\hat{\boldsymbol{y}}-(ax)\hat{\boldsymbol{z}},$$

$$\text{b)}\ \nabla\times\boldsymbol{F} = \mathbf{0}.$$

In case (a) no potential energy exists. In case (b) there is a potential energy function, and we proceed to find it. Let us take $\boldsymbol{r}_s = \mathbf{0}$, that is, take the potential as zero at the origin. Since the components of force are given as functions of x, y, z, the simplest path of integration from (0, 0, 0) to (x_0, y_0, z_0) along which to compute the integral in Eq. (3.183) is one which follows lines parallel to the coordinate axes, for example as shown in Fig. 3.32:

$$V(x_0, y_0, z_0) = -\int_{(0,0,0)}^{(x_0,y_0,z_0)} \boldsymbol{F}\cdot d\boldsymbol{r} = -\int_{C_1} \boldsymbol{F}\cdot d\boldsymbol{r}-\int_{C_2} \boldsymbol{F}\cdot d\boldsymbol{r}-\int_{C_3} \boldsymbol{F}\cdot d\boldsymbol{r}.$$

Now along C_1, we have

$$y = z = 0, \qquad F_x = F_y = F_z = 0, \qquad d\boldsymbol{r} = \hat{\boldsymbol{x}}\,dx.$$

Thus

$$\int_{C_1} \boldsymbol{F}\cdot d\boldsymbol{r} = \int_0^{x_0} F_x\,dx = 0.$$

Along C_2,

$$x = x_0, \qquad z = 0,$$

$$F_x = ay^3, \qquad F_y = 3ax_0y^2, \qquad F_z = 0,$$

$$d\boldsymbol{r} = \hat{\boldsymbol{y}}\, dy.$$

Thus

$$\int_{C_2} \boldsymbol{F}\cdot d\boldsymbol{r} = \int_0^{y_0} F_y\, dy = ax_0y_0^3.$$

Along C_3,

$$x = x_0, \qquad y = y_0,$$

$$F_x = ay_0(y_0^2 - 3z^2), \qquad F_y = 3ax_0(y_0^2 - z^2), \qquad F_z = -6ax_0y_0z,$$

$$d\boldsymbol{r} = \hat{\boldsymbol{z}}\, dz.$$

Thus

$$\int_{C_3} \boldsymbol{F}\cdot d\boldsymbol{r} = \int_0^{z_0} F_z\, dz = -3ax_0y_0z_0^2.$$

Thus the potential energy, if the subscript zero is dropped, is

$$V(x, y, z) = -axy^3 + 3axyz^2.$$

It is readily verified that the gradient of this function is the force given by (b) above. In fact, one way to find the potential energy, which is often faster than the above procedure, is simply to try to guess a function whose gradient will give the required force.

An important case of a conservative force is the central force, a force directed always toward or away from a fixed center O, and whose magnitude is a function only of the distance from O. In spherical coordinates, with O as origin,

$$\boldsymbol{F} = \hat{\boldsymbol{r}}F(r). \tag{3.198}$$

The cartesian components of a central force are (since $\hat{\boldsymbol{r}} = \boldsymbol{r}/r$)

$$F_x = \frac{x}{r}F(r), \qquad F_y = \frac{y}{r}F(r), \qquad [r = (x^2 + y^2 + z^2)^{1/2}], \qquad F_z = \frac{z}{r}F(r). \tag{3.199}$$

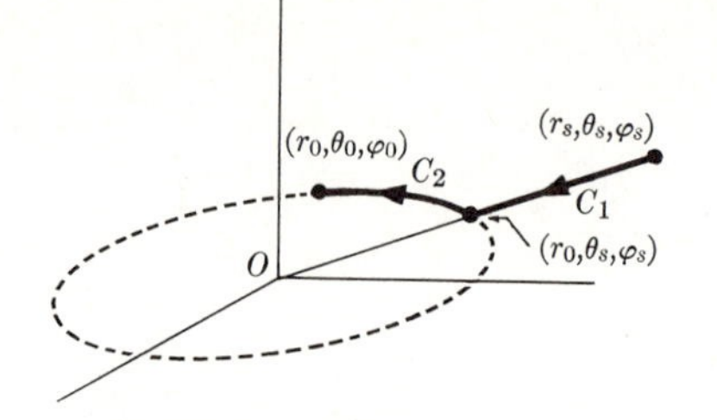

Fig. 3.33 Path of integration for a central force.

The curl of this force can be shown by direct computation to be zero, no matter what the function $F(r)$ may be. For example, we find

$$\frac{\partial F_x}{\partial y} = x\frac{d}{dr}\left(\frac{F(r)}{r}\right)\frac{\partial r}{\partial y} = \frac{xy}{r}\frac{d}{dr}\left(\frac{F(r)}{r}\right),$$

$$\frac{\partial F_y}{\partial x} = y\frac{d}{dr}\left(\frac{F(r)}{r}\right)\frac{\partial r}{\partial x} = \frac{xy}{r}\frac{d}{dr}\left(\frac{F(r)}{r}\right).$$

Therefore the z-component of curl $\boldsymbol{F}$ vanishes, and so, likewise, do the other two components. To compute the potential energy, we choose any standard point $\boldsymbol{r}_s$, and integrate from $\boldsymbol{r}_s$ to $\boldsymbol{r}_0$ along a path (Fig. 3.33) following a radius (C_1) from $\boldsymbol{r}_s$, whose coordinates are $(r_s, \theta_s, \varphi_s)$, to the point $(r_0, \theta_s, \varphi_s)$, then along a circle ($C_2$) of radius r_0 about the origin to the point $(r_0, \theta_0, \varphi_0)$. Along C_1,

$$d\boldsymbol{r} = \hat{\boldsymbol{r}}\, dr,$$

$$\int_{C_1} \boldsymbol{F}\cdot d\boldsymbol{r} = \int_{r_s}^{r_0} F(r)\, dr.$$

Along C_2,

$$d\boldsymbol{r} = \hat{\boldsymbol{\theta}} r\, d\theta + \hat{\boldsymbol{\varphi}} r \sin\theta\, d\varphi,$$

$$\int_{C_2} \boldsymbol{F}\cdot d\boldsymbol{r} = 0.$$

Thus

$$V(\boldsymbol{r}_0) = -\int_{\boldsymbol{r}_s}^{\boldsymbol{r}_0} \boldsymbol{F}\cdot d\boldsymbol{r} = -\int_{C_1} \boldsymbol{F}\cdot d\boldsymbol{r} - \int_{C_2} \boldsymbol{F}\cdot d\boldsymbol{r}$$

$$= -\int_{r_s}^{r_0} F(r)\, dr.$$

The potential energy is a function of r alone:

$$V(\boldsymbol{r}) = V(r) = -\int_{r_s}^{r} F(r)\, dr. \tag{3.200}$$

3.13 MOTION UNDER A CENTRAL FORCE

A central force is a force of the form given by Eq. (3.198). Physically, such a force represents an attraction [if $F(r) < 0$] or repulsion [if $F(r) > 0$] from a fixed point located at the origin $r = 0$. In most cases where two particles interact with each other, the force between them is (at least primarily) a central force; that is, if either particle be located at the origin, the force on the other is given by Eq. (3.198). Examples of attractive central forces are the gravitational force acting on a planet due to the sun, or the electrical attraction acting on an electron due to the nucleus of an atom. The force between a proton or an alpha particle and another nucleus is a repulsive central force. In the most important cases, the force $F(r)$ is inversely proportional to r^2. This case will be treated in the next section. Other forms of the function $F(r)$ occur occasionally; for example, in some problems involving the structure and interactions of nuclei, complex atoms, and molecules. In this section, we present the general method of attack on the problem of a particle moving under the action of a central force.

Since in all these examples neither of the two interacting particles is actually fastened to a fixed position, the problem we are solving, like most problems in physics, represents an idealization of the actual problem, valid when one of the particles can be regarded as practically at rest at the origin. This will be the case if one of the particles is much heavier than the other. Since the forces acting on the two particles have the same magnitude by Newton's third law, the acceleration of the heavy one will be much smaller than that of the lighter one, and the motion of the heavy particle can be neglected in comparison with the motion of the lighter one. We shall discover later, in Section 4.7, that, with a slight modification, our solution can be made to yield an exact solution to the problem of the motion of two interacting particles, even when their masses are equal.

We may note that the vector angular momentum of a particle under the action of a central force is constant, since the torque is

$$\boldsymbol{N} = \boldsymbol{r} \times \boldsymbol{F} = (\boldsymbol{r} \times \hat{\boldsymbol{r}})F(r) = \boldsymbol{0}. \tag{3.201}$$

Therefore, by Eq. (3.144),

$$\frac{d\boldsymbol{L}}{dt} = \boldsymbol{0}. \tag{3.202}$$

As a consequence, the angular momentum about any axis through the center of force is constant. It is because many physical forces are central forces that the concept of angular momentum is of importance.

In solving for the motion of a particle acted on by a central force, we first show that the path of the particle lies in a single plane containing the center of force. To show this, let the position $\boldsymbol{r}_0$ and velocity $\boldsymbol{v}_0$ be given at any initial time t_0, and choose the x-axis through the initial position $\boldsymbol{r}_0$ of the particle, and the z-axis perpendicular to the initial velocity $\boldsymbol{v}_0$. Then we have initially:

$$x_0 = |\boldsymbol{r}_0|, \qquad y_0 = z_0 = 0, \tag{3.203}$$

$$v_{x_0} = \boldsymbol{v}_0 \cdot \hat{\boldsymbol{x}}, \qquad v_{y_0} = \boldsymbol{v}_0 \cdot \hat{\boldsymbol{y}}, \qquad v_{z_0} = 0. \tag{3.204}$$

The equations of motion in rectangular coordinates are, by Eqs. (3.199),

$$m\ddot{x} = \frac{x}{r}F(r), \qquad m\ddot{y} = \frac{y}{r}F(r), \qquad m\ddot{z} = \frac{z}{r}F(r). \tag{3.205}$$

A solution of the z-equation which satisfies the initial conditions on z_0 and v_{z_0} is

$$z(t) = 0. \tag{3.206}$$

Hence the motion takes place entirely in the xy-plane. We can see physically that if the force on a particle is always toward the origin, the particle can never acquire any component of velocity out of the plane in which it is initially moving. We can also regard this result as a consequence of the conservation of angular momentum. By Eq. (3.202), the vector $\boldsymbol{L} = m(\boldsymbol{r} \times \boldsymbol{v})$ is constant; therefore both $\boldsymbol{r}$ and $\boldsymbol{v}$ must always lie in a fixed plane perpendicular to $\boldsymbol{L}$.

We have now reduced the problem to one of motion in a plane with two differential equations and four initial conditions remaining to be satisfied. If we choose polar coordinates r, θ in the plane of the motion, the equations of motion in the r and θ directions are, by Eqs. (3.80) and (3.198),

$$m\ddot{r} - mr\dot{\theta}^2 = F(r), \tag{3.207}$$

$$mr\ddot{\theta} + 2m\dot{r}\dot{\theta} = 0. \tag{3.208}$$

Multiplying Eq. (3.208) by r, as in the derivation of the (plane) angular momentum theorem, we have

$$\frac{d}{dt}(mr^2\dot{\theta}) = \frac{dL}{dt} = 0. \tag{3.209}$$

This equation expresses the conservation of angular momentum about the origin and is a consequence also of Eq. (3.202) above. It may be integrated to give the angular momentum integral of the equations of motion:

$$mr^2\dot{\theta} = L = \text{a constant}. \tag{3.210}$$

The constant L is to be evaluated from the initial conditions. Another integral of Eqs. (3.207) and (3.208), since the force is conservative, is

$$T + V = \tfrac{1}{2}m\dot{r}^2 + \tfrac{1}{2}mr^2\dot{\theta}^2 + V(r) = E, \tag{3.211}$$

where $V(r)$ is given by Eq. (3.200) and E is the energy constant, to be evaluated from the initial conditions. If we substitute for $\dot{\theta}$ from Eq. (3.210), the energy becomes

$$\tfrac{1}{2}m\dot{r}^2 + \frac{L^2}{2mr^2} + V(r) = E. \tag{3.212}$$

We can solve for $\dot{r}$:

$$\dot{r} = \sqrt{\frac{2}{m}}\left(E - V(r) - \frac{L^2}{2mr^2}\right)^{1/2}. \tag{3.213}$$

Therefore

$$\int_{r_0}^{r} \frac{dr}{\left(E - V(r) - \frac{L^2}{2mr^2}\right)^{1/2}} = \sqrt{\frac{2}{m}}\, t. \tag{3.214}$$

The integral is to be evaluated and the resulting equation solved for $r(t)$. We then obtain $\theta(t)$ from Eq. (3.210):

$$\theta = \theta_0 + \int_0^t \frac{L}{mr^2}\, dt. \tag{3.215}$$

We thus obtain the solution of Eqs. (3.207) and (3.208) in terms of the four constants L, E, r_0, θ_0, which can be evaluated when the initial position and velocity in the plane are given.

It will be noted that our treatment based on Eq. (3.212) is analogous to our treatment of the one-dimensional problem based on the energy integral [Eq. (2.44)]. The coordinate r here plays the role of x, and the $\dot{\theta}$ term in the kinetic energy, when $\dot{\theta}$ is eliminated by Eq. (3.210), plays the role of an addition to the potential energy. We may bring out this analogy further by substituting from Eq. (3.210) into Eq. (3.207):

$$m\ddot{r} - \frac{L^2}{mr^3} = F(r). \tag{3.216}$$

If we transpose the term $-L^2/mr^3$ to the right side, we obtain

$$m\ddot{r} = F(r) + \frac{L^2}{mr^3}. \tag{3.217}$$

This equation has exactly the form of an equation of motion in one dimension for a particle subject to the actual force $F(r)$ plus a "centrifugal force" L^2/mr^3. The centrifugal force is not really a force at all but a part of the mass times acceleration, transposed to the right side of the equation in order to reduce the equation for r to an equation of the same form as for one-dimensional motion. We may call it a "fictitious force." If we treat Eq. (3.217) as a problem in one-dimensional motion, the effective "potential energy" corresponding to the "force" on the right is

$$\begin{aligned} 'V'(r) &= -\int F(r)\, dr - \int \frac{L^2}{mr^3}\, dr \\ &= V(r) + \frac{L^2}{2mr^2}. \end{aligned} \tag{3.218}$$

The second term in 'V' is the "potential energy" associated with the "centrifugal force." The resulting energy integral is just Eq. (3.212). The reason why we have been able to obtain a complete solution to our problem based on only two integrals, or constants of the motion (L and E), is that the equations of motion do not contain the coordinate θ, so that the constancy of L is sufficient to enable us to eliminate θ entirely from Eq. (3.207) and to reduce the problem to an equivalent problem in one-dimensional motion.

The integral in Eq. (3.214) sometimes turns out rather difficult to evaluate in practice, and the resulting equation difficult to solve for $r(t)$. It is sometimes easier to find the path of the particle in space than to find its motion as a function of time. We can describe the path of the particle by giving $r(\theta)$. The resulting equation is somewhat simpler if we make the substitution

$$u = \frac{1}{r}, \qquad r = \frac{1}{u}. \tag{3.219}$$

Then we have, using Eq. (3.210),

$$\begin{aligned} \dot{r} &= -\frac{1}{u^2}\frac{du}{d\theta}\dot{\theta} = -r^2\dot{\theta}\frac{du}{d\theta} \\ &= -\frac{L}{m}\frac{du}{d\theta}, \end{aligned} \tag{3.220}$$

$$\ddot{r} = -\frac{L}{m}\frac{d^2u}{d\theta^2}\dot{\theta} = -\frac{L^2u^2}{m^2}\frac{d^2u}{d\theta^2}. \tag{3.221}$$

Substituting for r and $\ddot{r}$ in Eq. (3.217), and multiplying by $-m/(L^2u^2)$, we have a differential equation for the path or orbit in terms of $u(\theta)$:

$$\frac{d^2u}{d\theta^2} = -u - \frac{m}{L^2u^2}F\left(\frac{1}{u}\right). \tag{3.222}$$

In case $L = 0$, Eq. (3.222) blows up, but we see from Eq. (3.210) that in this case θ is constant, and the path is a straight line through the origin.

Even in cases where the explicit solutions of Eqs. (3.214) and (3.215), or Eq. (3.222), are difficult to carry through, we can obtain qualitative information about the r motion from the effective potential 'V' given by Eq. (3.218), just as in the one-dimensional case discussed in Section 2.5. By plotting 'V'(r), we can decide for any total energy E whether the motion in r is periodic or aperiodic, we can locate the turning points, and we can describe roughly how the velocity $\dot{r}$ varies during the motion. If 'V'(r) has a minimum at a point r_0, then for energy E slightly greater than 'V'(r_0), r may execute small, approximately harmonic oscillations about r_0 with angular frequency given by

$$\omega^2 = \frac{1}{m}\left(\frac{d^2\text{'}V\text{'}}{dr^2}\right)_{r_0}. \tag{3.223}$$

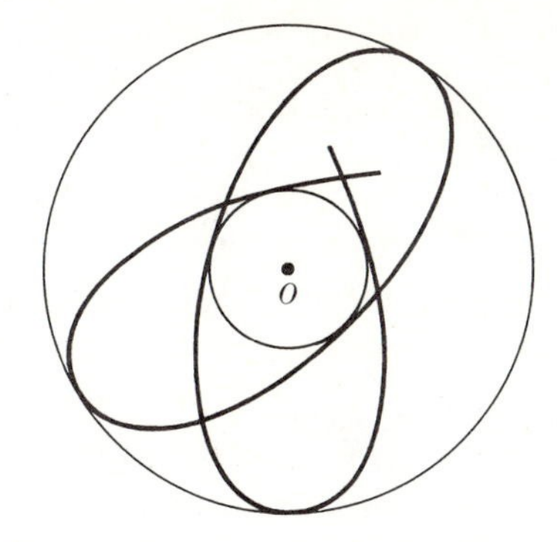

Fig. 3.34 An aperiodic bounded orbit.

[See the discussion in Section 2.7 concerning Eq. (2.57).] We must remember, of course, that at the same time the particle is revolving around the center of force with an angular velocity

$$\dot{\theta} = \frac{L}{mr^2}. \tag{3.224}$$

The rate of revolution decreases as r increases. In cases where the r motion is not periodic, then $\dot{\theta} \to 0$ as $r \to \infty$, and the particle may or may not perform one or more complete revolutions as it moves toward $r = \infty$, depending on how rapidly r increases. When the r motion is periodic, the period of the r motion is not, in general, the same as the period of revolution, so that the orbit may not be closed, although it is confined to a finite region of space. (See Fig. 3.34.) If the ratio of the period of the r motion to the period of revolution is a rational number, the orbit will be closed. If this ratio is an integer, the orbit is a simple closed curve whose period is related to the area of the orbit. This can be seen as follows. The area swept out by the radius from the origin to the particle when the particle moves through a small angle $d\theta$ is approximately (Fig. 3.35)

$$dS = \tfrac{1}{2}r^2\, d\theta. \tag{3.225}$$

Hence the rate at which area is swept out by the radius is, by Eq. (3.210),

$$\frac{dS}{dt} = \tfrac{1}{2}r^2\dot{\theta} = \frac{L}{2m}. \tag{3.226}$$

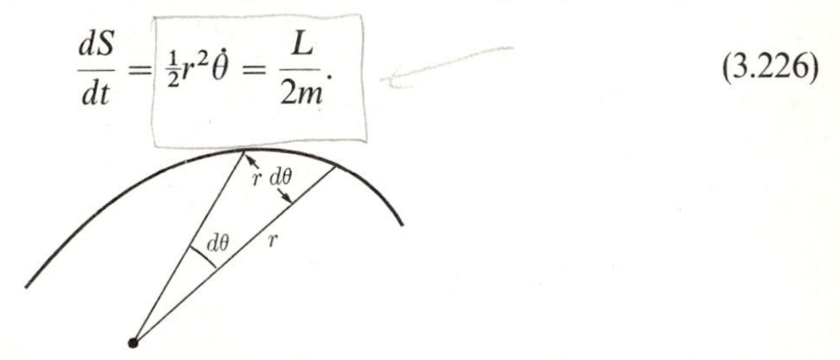

Fig. 3.35 Area swept out by radius vector.

This result is true for any particle moving under the action of a central force. If the motion is periodic, then, integrating over a complete period τ of the motion, we have for the area of the orbit

$$S = \frac{L\tau}{2m}. \tag{3.227}$$

If the orbit is known, the period of revolution can be calculated from this formula.

3.14 THE CENTRAL FORCE INVERSELY PROPORTIONAL TO THE SQUARE OF THE DISTANCE

The most important problem in three-dimensional motion is that of a mass moving under the action of a central force inversely proportional to the square of the distance from the center:

$$\boldsymbol{F} = \frac{K}{r^2}\hat{\boldsymbol{r}}, \tag{3.228}$$

for which the potential energy is

$$V(r) = \frac{K}{r}, \tag{3.229}$$

where the standard radius r_s is taken to be infinite in order to avoid an additional constant term in $V(r)$. As an example, the gravitational force (Section 1.5) between two masses m_1 and m_2 a distance r apart is given by Eq. (3.228) with

$$K = -Gm_1m_2, \qquad G = 6.67 \times 10^{-8} \text{ dyne-g}^{-2}\text{-cm}^2, \tag{3.230}$$

where K is negative, since the gravitational force is attractive. Another example is

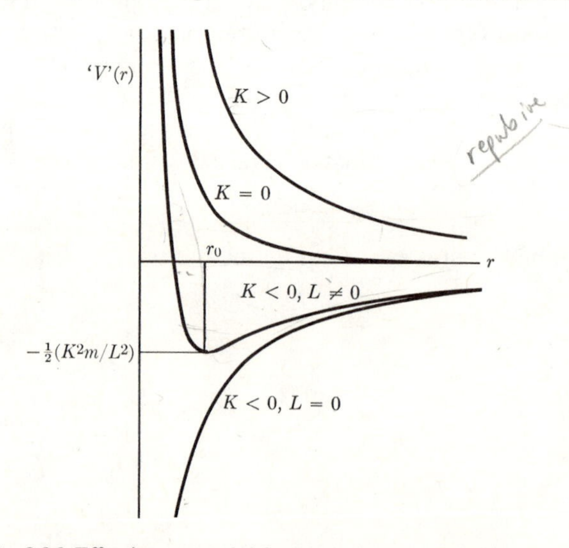

Fig. 3.36 Effective potential for central inverse square law of force.

the electrostatic force between two electric charges q_1 and q_2 a distance r apart, given by Eq. (3.228) with

$$K = q_1 q_2, \tag{3.231}$$

where the charges are in electrostatic units, and the force is in dynes. The electrostatic force is repulsive when q_1 and q_2 have the same sign, otherwise attractive. Historically, the first problems to which Newton's mechanics was applied were problems involving the motion of the planets under the gravitational attraction of the sun, and the motion of satellites around the planets. The success of the theory in accounting for such motions was responsible for its initial acceptance.

We first determine the nature of the orbits given by the inverse square law of force. In Fig. 3.36 is plotted the effective potential

$$\text{`}V\text{'}(r) = \frac{K}{r} + \frac{L^2}{2mr^2}. \tag{3.232}$$

For a repulsive force ($K > 0$), there are no periodic motions in r; only positive total energies E are possible, and the particle comes in from $r = \infty$ to a turning point and travels out to infinity again. For a given energy and angular momentum, the turning point occurs at a larger value of r than for $K = 0$ (no force), for which the orbit would be a straight line. For an attractive force ($K < 0$) with $L \neq 0$, the motion is also unbounded if $E > 0$, but in this case the turning point occurs at a smaller value of r than for $K = 0$. Hence the orbits are as indicated in Fig. 3.37. The light lines in Fig. 3.37 represent the turning point radius or perihelion distance measured from the point of closest approach of the particle to the attracting or repelling center. For $K < 0$, and $-\frac{1}{2}K^2m/L^2 < E < 0$, the coordinate r oscillates between two turning points. For $E = -\frac{1}{2}K^2m/L^2$, the particle moves in a circle of radius $r_0 = L^2/(-Km)$. Computation shows (see Problem 44 at the end of this

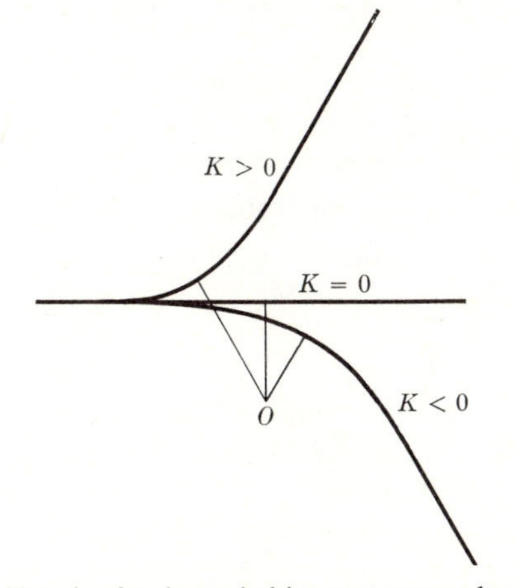

Fig. 3.37 Sketch of unbounded inverse square law orbits.

chapter) that the period of small oscillations in r is the same as the period of revolution, so that for E near $-\frac{1}{2}K^2m/L^2$, the orbit is a closed curve with the origin slightly off center (at least in the approximation of small oscillations). We shall show later that the orbit is, in fact, an ellipse for all negative values of E if $L \neq 0$. If $L = 0$, the problem reduces to the one-dimensional motion of a falling body, discussed in Section 2.6.

To evaluate the integrals in Eqs. (3.214) and (3.215) for the inverse square law of force is rather laborious. We shall find that we can obtain all the essential information about the motion more simply by starting from Eq. (3.222) for the orbit. Equation (3.222) for the orbit becomes, in this case,

$$\frac{d^2u}{d\theta^2}+u = -\frac{mK}{L^2}. \tag{3.233}$$

This equation has the same form as that of a harmonic oscillator (of unit frequency) subject to a constant force, where θ here plays the role of t. The homogeneous equation and its general solution are

$$\frac{d^2u}{d\theta^2}+u = 0, \tag{3.234}$$

$$u = A\cos(\theta-\theta_0), \tag{3.235}$$

where A, θ_0 are arbitrary constants. An obvious particular solution of the inhomogeneous equation (3.233) is the constant solution

$$u = -\frac{mK}{L^2}. \tag{3.236}$$

Hence the general solution of Eq. (3.233) is

$$u = \frac{1}{r} = -\frac{mK}{L^2}+A\cos(\theta-\theta_0). \tag{3.237}$$

This is the equation of a conic section (ellipse, parabola, or hyperbola) with focus at $r = 0$, as we shall presently show. The constant θ_0 determines the orientation of the orbit in the plane. The constant A, which may be taken as positive (since θ_0 is arbitrary), determines the turning points of the r motion, which are given by

$$\frac{1}{r_1} = -\frac{mK}{L^2}+A, \qquad \frac{1}{r_2} = -\frac{mK}{L^2}-A. \tag{3.238}$$

If $A > -mK/L^2$ (as it necessarily is for $K > 0$), then there is only one turning point, r_1, since r cannot be negative. We cannot have $A < mK/L^2$, since r could then not be positive for any value of θ. For a given E, the turning points are solutions of the equation

$$\text{‘}V\text{’}(r) = \frac{K}{r}+\frac{L^2}{2mr^2} = E. \tag{3.239}$$

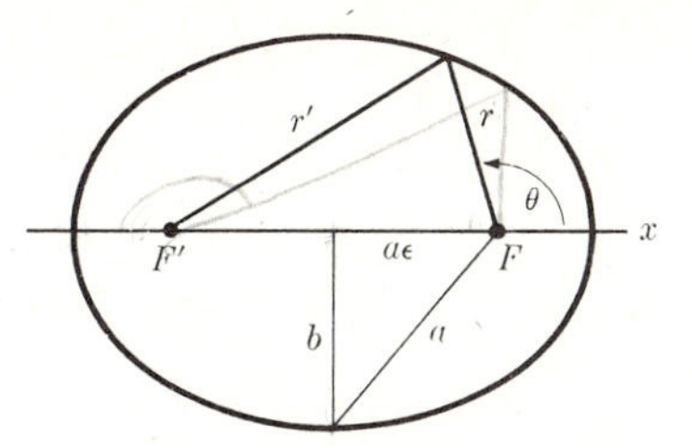

Fig. 3.38 Geometry of the ellipse.

The solutions are

$$\frac{1}{r_1} = -\frac{mK}{L^2}+\left[\left(\frac{mK}{L^2}\right)^2+\frac{2mE}{L^2}\right]^{1/2},$$
$$\frac{1}{r_2} = -\frac{mK}{L^2}-\left[\left(\frac{mK}{L^2}\right)^2+\frac{2mE}{L^2}\right]^{1/2}. \tag{3.240}$$

Comparing Eq. (3.238) with Eq. (3.240), we see that the value of A in terms of the energy and angular momentum is given by

$$A^2 = \frac{m^2K^2}{L^4}+\frac{2mE}{L^2}. \tag{3.241}$$

The orbit is now determined in terms of the initial conditions.

An ellipse is defined as the curve traced by a particle moving so that the sum of its distances from two fixed points F, F' is constant.* The points F, F' are called the *foci* of the ellipse. Using the notation indicated in Fig. 3.38, we have

$$r'+r = 2a, \tag{3.242}$$

where a is half the largest diameter (major axis) of the ellipse. In terms of polar coordinates with center at the focus F and with the negative x-axis through the focus F', the cosine law gives

$$r'^2 = r^2+4a^2\varepsilon^2+4ra\varepsilon \cos\theta, \tag{3.243}$$

where $a\varepsilon$ is the distance from the center of the ellipse to the focus; ε is called the *eccentricity* of the ellipse. If $\varepsilon = 0$, the foci coincide and the ellipse is a circle. As $\varepsilon \to 1$, the ellipse degenerates into a parabola or straight line segment, depending on whether the focus F' recedes to infinity or remains a finite distance from F. Substituting r' from Eq. (3.242) in Eq. (3.243), we find

$$r = \frac{a(1-\varepsilon^2)}{1+\varepsilon\cos\theta}. \tag{3.244}$$

---

*For a more detailed treatment of conic sections, see W. F. Osgood and W. C. Graustein, *Plane and Solid Analytic Geometry*. New York: Macmillan, 1938. (Chapters 6, 7, 8, 10.)

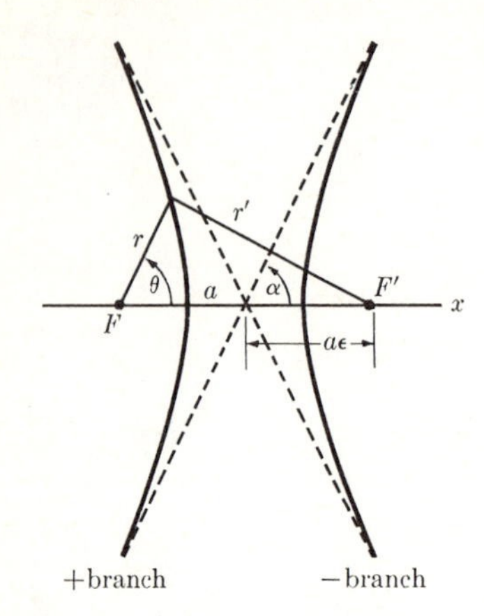

Fig. 3.39 Geometry of the hyperbola.

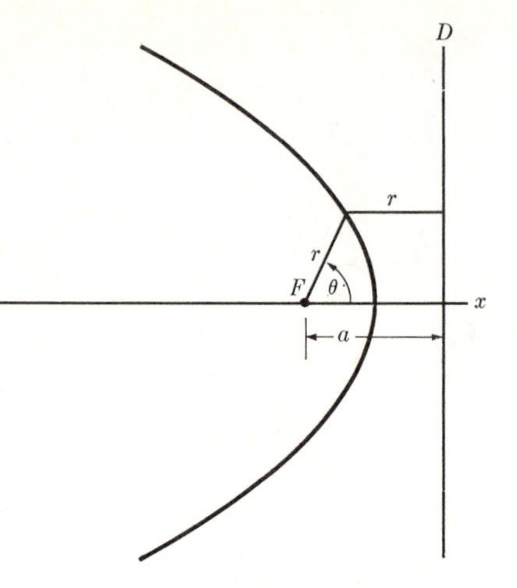

Fig. 3.40 Geometry of the parabola.

This is the equation of an ellipse in polar coordinates with the origin at one focus. If b is half the smallest diameter (minor axis), we have, from Fig. 3.38,

$$b = a(1-\varepsilon^2)^{1/2}. \tag{3.245}$$

The area of the ellipse can be obtained in a straightforward way by integration:

$$S = \pi ab. \tag{3.246}$$

A hyperbola is defined as the curve traced by a particle moving so that the difference of its distances from two fixed foci F, F' is constant (Fig. 3.39). A hyperbola has two branches defined by

$$\begin{aligned} r'-r &= 2a \quad (+\text{branch}), \\ r'-r &= -2a \quad (-\text{branch}). \end{aligned} \tag{3.247}$$

We shall call the branch which encircles F the $+$branch (left branch in the figure), and the branch which avoids F the $-$branch (right branch in the figure). Equation (3.243) holds also for the hyperbola, but the eccentricity ε is now greater than one. The equation of the hyperbola becomes in polar coordinates:

$$r = \frac{a(\varepsilon^2-1)}{\pm 1+\varepsilon \cos \theta}. \tag{3.248}$$

(The $+$ sign refers to the $+$branch, the $-$ sign to the $-$branch.) The asymptotes of the hyperbola (dotted lines in Fig. 3.39) make an angle α with the axis through the foci, where α is the value of θ for which r is infinite:

$$\cos \alpha = \pm\frac{1}{\varepsilon}. \tag{3.249}$$

A parabola is the curve traced by a particle moving so that its distance from a fixed line D (the *directrix*) equals its distance from a fixed focus F. From Fig. 3.40, we have

$$r = \frac{a}{1+\cos\theta}, \tag{3.250}$$

where a is the distance from the focus F to the directrix D.

We can write the equations for all three conic sections in the standard form

$$\frac{1}{r} = B + A\cos\theta, \tag{3.251}$$

where A is positive, and B and A are given as follows:

$B > A$, ellipse,

$$B = \frac{1}{a(1-\varepsilon^2)}, \qquad A = \frac{\varepsilon}{a(1-\varepsilon^2)}; \tag{3.252}$$

$B = A$, parabola,

$$B = \frac{1}{a}, \qquad A = \frac{1}{a}; \tag{3.253}$$

$0 < B < A$, hyperbola, $+$branch,

$$B = \frac{1}{a(\varepsilon^2-1)}, \qquad A = \frac{\varepsilon}{a(\varepsilon^2-1)}; \tag{3.254}$$

$-A < B < 0$, hyperbola, $-$branch,

$$B = -\frac{1}{a(\varepsilon^2-1)}, \qquad A = \frac{\varepsilon}{a(\varepsilon^2-1)}. \tag{3.255}$$

The case $B < -A$ cannot occur, since r would then not be positive for any value of θ. If we allow an arbitrary orientation of the curve with respect to the x-axis, then Eq. (3.251) becomes

$$\frac{1}{r} = B + A\cos(\theta-\theta_0), \tag{3.256}$$

where θ_0 is the angle between the x-axis and the line from the origin to the perihelion (point of closest approach of the curve to the origin). It will be noted that in all cases

$$\varepsilon = \frac{A}{|B|}. \tag{3.257}$$

For an ellipse or hyperbola,

$$a = \left|\frac{B}{A^2 - B^2}\right|. \tag{3.258}$$

Equation (3.237) for the orbit of a particle under an inverse square law force has the form of Eq. (3.256) for a conic section, with [if we use Eq. (3.241)]

$$B = -\frac{mK}{L^2},$$
$$A = \left(B^2 + \frac{2mE}{L^2}\right)^{1/2}. \tag{3.259}$$

The eccentricity of the orbit, by Eq. (3.257), is

$$\varepsilon = \left(1 + \frac{2EL^2}{mK^2}\right)^{1/2}. \tag{3.260}$$

For an attractive force ($K < 0$), the orbit is an ellipse, parabola, or hyperbola, depending on whether $E < 0$, $E = 0$, or $E > 0$; if a hyperbola, it is the $+$branch. For a repulsive force ($K > 0$), we must have $E > 0$, and the orbit can only be the $-$branch of a hyperbola. These results agree with our preliminary qualitative discussion. For elliptic and hyperbolic orbits, the semimajor axis a is given by

$$a = \left|\frac{K}{2E}\right|. \tag{3.261}$$

It is curious that this relation does not involve the eccentricity or the angular momentum; the energy E depends only on the semimajor axis a, and vice versa. Equations (3.260) and (3.261) may be obtained directly from Eq. (3.239) for the turning points of the r motion. If we solve this equation for r, we obtain the turning points

$$r_{1,2} = \frac{K}{2E} \pm \left[\left(\frac{K}{2E}\right)^2 + \frac{L^2}{2mE}\right]^{1/2}. \tag{3.262}$$

The maximum and minimum radii for an ellipse are

$$r_{1,2} = a(1 \pm \varepsilon), \tag{3.263}$$

and the minimum radius for a hyperbola is

$$r_1 = a(\varepsilon \mp 1), \tag{3.264}$$

where the upper sign is for the $+$branch and the lower sign for the $-$branch. Comparing Eqs. (3.263) and (3.264) with Eq. (3.262), we can read off the values of a and ε. Thus if we know that the path is an ellipse or hyperbola, we can find the size and shape from Eq. (3.239), which follows from the simple energy method of treatment, without going through the exact solution of the equation for the orbit. This is a useful point to remember.

3.15 ELLIPTIC ORBITS. THE KEPLER PROBLEM

Early in the seventeenth century, before Newton's discovery of the laws of motion, Kepler announced the following three laws describing the motion of the planets, deduced from the extensive and accurate observations of planetary motions by Tycho Brahe:

1. The planets move in ellipses with the sun at one focus.
2. Areas swept out by the radius vector from the sun to a planet in equal times are equal.
3. The square of the period of revolution is proportional to the cube of the semimajor axis.

The second law is expressed by our Eq. (3.226) and is a consequence of the conservation of angular momentum; it shows that the force acting on the planet is directed toward the sun. The first law follows, as we have shown, from the fact that the force is inversely proportional to the square of the distance. The third law follows from the fact that the gravitational force is proportional to the mass of the planet, as we now show.

In the case of an elliptical orbit, we can find the period of the motion from Eqs. (3.227) and (3.246):

$$\tau = \frac{2m}{L}\pi ab = \frac{2m}{L}\pi a^2(1-\varepsilon^2)^{1/2} = \left(\frac{\pi^2 K^2 m}{2|E|^3}\right)^{1/2}, \tag{3.265}$$

or, using Eq. (3.261),

$$\tau^2 = 4\pi^2 a^3 \left|\frac{m}{K}\right|. \tag{3.266}$$

In the case of a small body of mass m moving under the gravitational attraction [Eq. (3.230)] of a large body of mass M, this becomes

$$\tau^2 = \frac{4\pi^2}{MG}a^3. \tag{3.267}$$

The coefficient of a^3 is now a constant for all planets, in agreement with Kepler's third law. Equation (3.267) allows us to "weigh" the sun, if we know the value of G, by measuring the period and major axis of any planetary orbit. This has already been worked out in Chapter 1, Problem 11, for a circular orbit. Equation (3.267) now shows that the result applies also to elliptical orbits if the semimajor axis is substituted for the radius.

We have shown that Kepler's laws follow from Newton's laws of motion and the law of gravitation. The converse problem, to deduce the law of force from Kepler's laws and the law of motion, is an easier problem, and a very important one historically, for it was in this way that Newton deduced the law of gravitation. We expect that the motions of the planets should show slight deviations from

Kepler's laws, in view of the fact that the central force problem which was solved in the last section represents an idealization of the actual physical problem. In the first place, as pointed out in Section 3.13, we have assumed that the sun is stationary, whereas actually it must wobble slightly due to the attraction of the planets going around it. This effect is very small, even in the case of the largest planets, and can be corrected for by the methods explained later in Section 4.7. In the second place, a given planet, say the earth, is acted on by the gravitational pull of the other planets, as well as by the sun. Since the masses of even the heaviest planets are only a few percent of the mass of the sun, this will produce small but measurable deviations from Kepler's laws. The expected deviations can be calculated, and they agree with the very precise astronomical observations. In fact, the planets Neptune and Pluto were discovered as a result of their effects on the orbits of the other planets. Observations of the planet Uranus for about sixty years after its discovery in 1781 showed unexplained deviations from the predicted orbit, even after corrections were made for the gravitational effects of the other known planets. By a careful and elaborate mathematical analysis of the data, Adams and Leverrier were able to show that the deviations could be accounted for by assuming an unknown planet beyond Uranus, and they calculated the position of the unknown planet. The planet Neptune was promptly discovered in the predicted place.

The orbits of the comets, which are occasionally observed to move in around the sun and out again, are, at least in some cases, very elongated ellipses. It is not at present known whether any of the comets come from beyond the solar system, in which case they would, at least initially, have parabolic or hyperbolic orbits. Even those comets whose orbits are known to be elliptical have rather irregular periods due to the perturbing gravitational pull of the larger planets near which they occasionally pass. Between close encounters with the larger planets, a comet will follow fairly closely a path given by Eq. (3.256), but during each such encounter, its motion will be disturbed, so that afterwards the constants A, B, and θ_0 will have values different from those before the encounter.

As noted in Section 3.13, we expect in general that the bounded orbits arising from an attractive central force $F(r)$ will not be closed (Fig. 3.34). Closed orbits (except for circular orbits) arise only where the period of radial oscillations is equal to, or is an exact rational multiple of, the period of revolution. Only for certain special forms of the function $F(r)$, of which the inverse square law is one, will the orbits be closed. Any change in the inverse square law, either a change in the exponent of r or an addition to $F(r)$ of a term not inversely proportional to r^2, will be expected to lead to orbits that are not closed. However, if the change is very small, then the orbits ought to be approximately elliptical. The period of revolution will then be only slightly greater or slightly less than the period of radial oscillations, and the orbit will be approximately an ellipse whose major axis rotates slowly about the center of force. As a matter of fact, a slow precession of the major axis of the orbit of the planet Mercury has been observed, with an angular velocity of 41 seconds of arc per century, over and above the perturbations accounted for by the gravitational effects of the other planets. It was once thought that this could be accounted for by the gravitational effect of dust in the solar system, but it can be shown that the amount of dust is far too small to account for the effect. It is now fairly certain that the effect is due to slight corrections to Newton's theory of planetary motion required by the theory of relativity.*

The problem of the motion of electrons around the nucleus of an atom would be the same as that of the motion of planets around the sun, if Newtonian mechanics were applicable. Actually, the motion of electrons must be calculated from the laws of quantum mechanics. Before the discovery of quantum mechanics, Bohr was able to give a fair account of the behavior of atoms by assuming that the electrons revolve in certain orbits specially selected from among those given by Newtonian mechanics. Bohr's theory is still useful as a rough picture of atomic structure.†

3.16 HYPERBOLIC ORBITS. THE RUTHERFORD PROBLEM. SCATTERING CROSS SECTION

The hyperbolic orbits are of interest in connection with the motion of particles around the sun which may come from or escape to outer space, and also in connection with the collisions of two charged particles. If a light particle of charge q_1 encounters a heavy particle of charge q_2 at rest, the light particle will follow a hyperbolic trajectory past the heavy particle, according to the results obtained in Section 3.14. In the case of collisions of atomic particles, the region in which the trajectory bends from one asymptote to the other is very small (a few angstrom units or less), and what is observed is the deflection angle $\Theta = \pi - 2\alpha$ (Fig. 3.41) between the paths of the incident particle before and after the collision. Figure 3.41 is drawn for the case of a repelling center of force at F. By Eqs. (3.249) and (3.260),

$$\tan\frac{\Theta}{2} = \cot\alpha = (\varepsilon^2 - 1)^{-1/2} = \left(\frac{mK^2}{2EL^2}\right)^{1/2}. \tag{3.268}$$

Let the particle have an initial speed v_0, and let it be traveling in such a direction that, if undeflected, it would pass a distance s from the center of force (F). The distance s is called the *impact parameter* for the collision. We can readily compute the energy and angular momentum in terms of the speed and impact parameter:

$$E = \tfrac{1}{2}mv_0^2, \tag{3.269}$$

$$L = mv_0 s. \tag{3.270}$$

---

*A. Einstein and L. Infeld, *The Evolution of Physics*. New York: Simon and Schuster, 1938. (Page 253.) For a mathematical discussion, see R. C. Tolman, *Relativity, Thermodynamics, and Cosmology*. Oxford: Oxford University Press, 1934. (Section 83.)

†M. Born, *Atomic Physics*, tr. by John Dougall. New York: Stechert, 1936. (Chapter 5.)

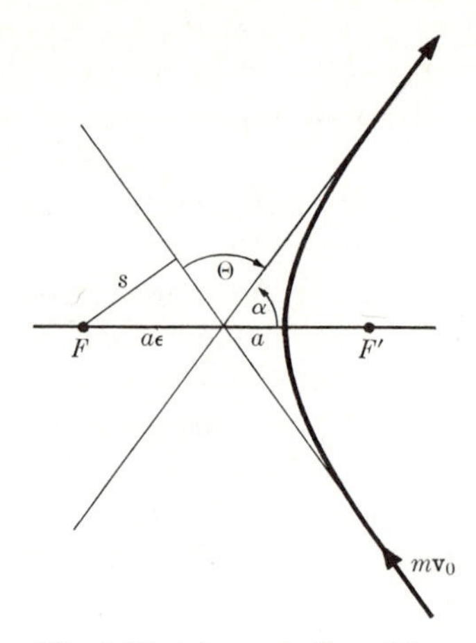

Fig. 3.41 A hyperbolic orbit.

Substituting in Eq. (3.268), we have for the scattering angle Θ:

$$\tan\frac{\Theta}{2} = \frac{K}{msv_0^2}. \tag{3.271}$$

If a light particle of positive charge q_1 collides with a heavy particle of positive charge q_2, this is, by Eq. (3.231),

$$\tan\frac{\Theta}{2} = \frac{q_1q_2}{msv_0^2}. \tag{3.272}$$

In a typical scattering experiment, a stream of charged particles may be shot in a definite direction through a thin foil. Many of the particles emerge from the foil in a different direction, after being deflected or scattered through an angle Θ by a collision with a particle within the foil. To put Eq. (3.272) in a form in which it can be compared with experiment, we must eliminate the impact parameter s, which cannot be determined experimentally. In the experiment, the fraction of incident particles scattered through various angles Θ is observed. It is customary to express the results in terms of a *cross section* defined as follows. If N incident particles strike a thin foil containing n scattering centers per unit area, the average number dN of particles scattered through an angle between Θ and $\Theta+d\Theta$ is given in terms of the cross section $d\sigma$ by the formula

$$\frac{dN}{N} = n\,d\sigma. \tag{3.273}$$

$d\sigma$ is called the cross section for scattering through an angle between Θ and $\Theta+d\Theta$, and can be thought of as the effective area surrounding the scattering center which the incident particle must hit in order to be scattered through an

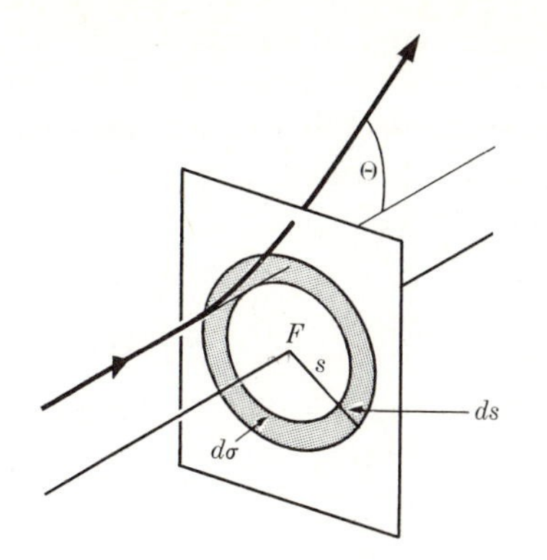

Fig. 3.42 Cross section for scattering.

angle between Θ and $\Theta + d\Theta$. For if there is a "target area" $d\sigma$ around each scattering center, then the total target area in a unit area is $n\, d\sigma$. If N particles strike one unit area, the average number striking the target area is $Nn\, d\sigma$, and this, according to Eq. (3.273), is just dN, the number of particles scattered through an angle between Θ and $\Theta + d\Theta$.

Now consider an incident particle approaching a scattering center F as in Figs. 3.41 and 3.42. If the impact parameter is between s and $s + ds$, the particle will be scattered through an angle between Θ and $\Theta + d\Theta$, where Θ is given by Eq. (3.272), and $d\Theta$ is given by the differential of Eq. (3.272):

$$\frac{1}{2\cos^2(\Theta/2)}\, d\Theta = -\frac{q_1 q_2}{m s^2 v_0^2}\, ds. \tag{3.274}$$

The area of the ring around F of inner radius s, outer radius $s + ds$, at which the incident particle must be aimed in order to be scattered through an angle between Θ and $\Theta + d\Theta$, is

$$d\sigma = 2\pi s\, ds. \tag{3.275}$$

Substituting for s from Eq. (3.272), and for ds from Eq. (3.274) (omitting the negative sign), we obtain

$$d\sigma = \left(\frac{q_1 q_2}{2 m v_0^2}\right)^2 \frac{2\pi \sin\Theta}{\sin^4(\Theta/2)}\, d\Theta. \tag{3.276}$$

This formula can be compared with $d\sigma$ determined experimentally as given by Eq. (3.273). Formula (3.276) was deduced by Rutherford and used in interpreting his experiments on the scattering of alpha particles by thin metal foils. He was able to show that the formula agrees with his experiments with $q_1 = 2e$ (charge on alpha particle),* and $q_2 = Ze$ (charge on atomic nucleus), so long as the perihelion

*Here e stands for the magnitude of the electronic charge.

distance ($a+a\varepsilon$ in Fig. 3.41) is larger than about 10^{-12} cm, which shows that the positive charge on the atom must be concentrated within a region of radius less than 10^{-12} cm. This was the origin of the nuclear theory of the atom. The perihelion distance can be computed from formula (3.262) or by using the conservation laws for energy and angular momentum, and is given by

$$r_1 = \frac{q_1 q_2}{2E}\left[1+\left(1+\frac{2EL^2}{mq_1^2q_2^2}\right)^{1/2}\right]. \tag{3.277}$$

The smallest perihelion distance for incident particles of a given energy occurs when $L = 0$ ($s = 0$), and has the value

$$r_{1\,\mathrm{min}} = \frac{q_1 q_2}{E}. \tag{3.278}$$

Hence if there is a deviation from Coulomb's law of force when the alpha particle grazes or penetrates the nucleus, it should show up first as a deviation from Rutherford's law [Eq. (3.276)] at large angles of deflection Θ, and should show up when the energy E is large enough so that

$$E > \frac{q_1 q_2}{r_0}, \tag{3.279}$$

where r_0 is the radius of the nucleus. The earliest measurements of nuclear radii were made in this way by Rutherford, and turn out to be of the order of 10^{-12} cm.

The above calculation of the cross section is strictly correct only when the alpha particle impinges on a nucleus much heavier than itself, since the scattering center is assumed to remain fixed. This restriction can be removed by methods to be discussed in Section 4.8. Alpha particles also collide with electrons, but the electron is so light that it cannot appreciably deflect the alpha particle. The collision of an alpha particle with a nucleus should really be treated by the methods of quantum mechanics. The concept of a definite trajectory with a definite impact parameter s is no longer valid in quantum mechanics. The concept of cross section is still valid in quantum mechanics, however, as it should be, since it is defined in terms of experimentally determined quantities. The final result for the scattering cross section turns out the same as our formula (3.276).* It is a fortunate coincidence in the history of physics that classical mechanics gives the right answer to this problem.

3.17 MOTION OF A PARTICLE IN AN ELECTROMAGNETIC FIELD

The laws determining the electric and magnetic fields due to various arrangements of electric charges and currents are the subject matter of electromagnetic theory. The determination of the motions of charged particles under given electric and

*D. Bohm, *Quantum Theory*. New York: Prentice-Hall, 1951. (Page 537.)

magnetic forces is a problem in mechanics. The electric force on a particle of charge q located at a point $\boldsymbol{r}$ is

$$\boldsymbol{F} = q\boldsymbol{E}(\boldsymbol{r}), \tag{3.280}$$

where $\boldsymbol{E}(\boldsymbol{r})$ is the electric field intensity at the point $\boldsymbol{r}$. The electric field intensity may be a function of time as well as of position in space. The force exerted by a magnetic field on a charged particle at a point $\boldsymbol{r}$ depends on the velocity $\boldsymbol{v}$ of the particle, and is given in terms of the magnetic induction $\boldsymbol{B}(\boldsymbol{r})$ by the equation*

$$\boldsymbol{F} = \frac{q}{c}\boldsymbol{v} \times \boldsymbol{B}(\boldsymbol{r}), \tag{3.281}$$

where $c = 3 \times 10^{10}$ cm/sec is the velocity of light, and all quantities are in Gaussian units, i.e., q is in electrostatic units, $\boldsymbol{B}$ in electromagnetic units (Gauss), and $\boldsymbol{v}$ and $\boldsymbol{F}$ are in cgs units. In mks units, the equation reads

$$\boldsymbol{F} = q\boldsymbol{v} \times \boldsymbol{B}(\boldsymbol{r}). \tag{3.282}$$

Equation (3.280) holds for either Gaussian or mks units. We shall base our discussion on Eq. (3.281) (Gaussian units), but the results are readily transcribed into mks units by omitting c wherever it occurs. The total electromagnetic force acting on a particle due to an electric field intensity $\boldsymbol{E}$ and a magnetic induction $\boldsymbol{B}$ is

$$\boldsymbol{F} = q\boldsymbol{E} + \frac{q}{c}\boldsymbol{v} \times \boldsymbol{B}. \tag{3.283}$$

If an electric charge moves near the north pole of a magnet, the magnet will exert a force on the charge given by Eq. (3.281); and by Newton's third law the charge should exert an equal and opposite force on the magnet. This is indeed found to be the case, at least when the velocity of the particle is small compared with the speed of light, if the magnetic field due to the moving charge is calculated and the force on the magnet computed. However, since the magnetic induction $\boldsymbol{B}$ is directed radially away from the pole, and the force $\boldsymbol{F}$ is perpendicular to $\boldsymbol{B}$, the forces on the charge and on the pole are not directed along the line joining them, as in the case of a central force. Newton's third law is sometimes stated in the "strong" form in which action and reaction are not only equal and opposite, but are directed along the line joining the interacting particles. For magnetic forces, the law holds only in the "weak" form in which nothing is said about the directions of the two forces except that they are opposite. This is true not only of the forces between magnets and moving charges, but also of the magnetic forces exerted by moving charges on one another.

---

*G. P. Harnwell, *Principles of Electricity and Electromagnetism*, 2nd ed. New York: McGraw-Hill, 1949. (Page 302.)

If the magnetic field is constant in time, then the electric field intensity can be shown to satisfy the equation

$$\nabla \times \boldsymbol{E} = \mathbf{0}. \tag{3.284}$$

The proof of this statement belongs to electromagnetic theory and need not concern us here.* We note, however, that this implies that for static electric and magnetic fields, the electric force on a charged particle is conservative. We can therefore define an electric potential

$$\phi(\boldsymbol{r}) = -\int_{\boldsymbol{r}_s}^{\boldsymbol{r}} \boldsymbol{E} \cdot d\boldsymbol{r}, \tag{3.285}$$

such that

$$\boldsymbol{E} = -\nabla\phi. \tag{3.286}$$

Since $\boldsymbol{E}$ is the force per unit charge, ϕ will be the potential energy per unit charge associated with the electric force:

$$V(\boldsymbol{r}) = q\phi(\boldsymbol{r}). \tag{3.287}$$

Furthermore, since the magnetic force is perpendicular to the velocity, it can do no work on a charged particle. Consequently, the law of conservation of energy holds for a particle in a static electromagnetic field:

$$T + q\phi = E, \tag{3.288}$$

where E is a constant.

A great variety of problems of practical and theoretical interest arise involving the motion of charged particles in electric and magnetic fields. In general, special methods of attack must be devised for each type of problem. We shall discuss two special problems which are of interest both for the results obtained and for the methods of obtaining those results.

We first consider the motion of a particle of mass m, charge q, in a uniform constant magnetic field. Let the z-axis be chosen in the direction of the field, so that

$$\boldsymbol{B}(\boldsymbol{r}, t) = B\hat{\boldsymbol{z}}, \tag{3.289}$$

where B is a constant. The equations of motion are then, by Eq. (3.281),

$$m\ddot{x} = \frac{qB}{c}\dot{y}, \qquad m\ddot{y} = -\frac{qB}{c}\dot{x}, \qquad m\ddot{z} = 0. \tag{3.290}$$

According to the last equation, the z-component of velocity is constant, and we shall consider the case when $v_z = 0$ and the motion is entirely in the xy-plane. The first two equations are not hard to solve, but we can avoid solving them directly by making use of the energy integral, which in this case reads

$$\tfrac{1}{2}mv^2 = E. \tag{3.291}$$

*Harnwell, *op. cit.* (Page 340.)

The force is given by

$$\boldsymbol{F} = \frac{qB}{c}\boldsymbol{v}\times\hat{\boldsymbol{z}}, \tag{3.292}$$

$$F = \frac{qBv}{c}. \tag{3.293}$$

The force, and consequently the acceleration, is therefore of constant magnitude and perpendicular to the velocity. A particle moving with constant speed v and constant acceleration a perpendicular to its direction of motion moves in a circle of radius r given by Eq. (3.80):

$$a = r\dot{\theta}^2 = \frac{v^2}{r} = \frac{F}{m}. \tag{3.294}$$

We substitute for F from Eq. (3.293) and solve for r:

$$r = \frac{cmv}{qB}. \tag{3.295}$$

The product Br is therefore proportional to the momentum and inversely proportional to the charge.

This result has many practical applications. If a bubble chamber is placed in a uniform magnetic field, one can measure the momentum of a charged particle by measuring the radius of curvature of its track. The same principle is used in a beta-ray spectrometer to measure the momentum of a fast electron by the curvature of its path in a magnetic field. In a mass spectrometer, a particle is accelerated through a known difference of electric potential, so that, by Eq. (3.288), its kinetic energy is

$$\tfrac{1}{2}mv^2 = q(\phi_0 - \phi_1). \tag{3.296}$$

It is then passed through a uniform magnetic field B. If q is known, and r, B, $(\phi_0 - \phi_1)$ are measured, we can eliminate v between Eqs. (3.295) and (3.296), and solve for the mass:

$$m = \frac{qB^2r^2}{2c^2(\phi_0 - \phi_1)}. \tag{3.297}$$

There are many variations of this basic idea. The historic experiments of J. J. Thomson which demonstrated the existence of the electron were essentially of this type, and by them Thomson succeeded in showing that the path traveled by a cathode ray is that which would be followed by a stream of charged particles, all with the same ratio q/m.

In a cyclotron, charged particles travel in circles in a uniform magnetic field, and receive increments in energy twice per revolution by passing through an alternating electric field. The radius r of the circles therefore increases, according to Eq. (3.295), until a maximum radius is reached, at which radius the particles emerge in a beam of definite energy determined by Eq. (3.295). The frequency ν of the alternating electric field must be the same as the frequency ν of revolution of the particles, which is given by

$$v = 2\pi r\nu. \tag{3.298}$$

Combining this equation with Eq. (3.295), we have

$$\nu = \frac{qB}{2\pi mc}. \tag{3.299}$$

Thus if B is constant, ν is independent of r, and this is the fundamental principle on which the operation of the cyclotron is based.*

In the betatron, electrons travel in circles, and the magnetic field within the circle is made to increase. Since B is changing with time, $\nabla \times \boldsymbol{E}$ is no longer zero; the changing magnetic flux induces a voltage around the circle such that a net amount of work is done on the electrons by the electric field as they travel around the circle. The betatron is so designed that the increase of B at the electron orbit is proportional to the increase of mv, so that r remains constant.

Finally, we consider a particle of mass m, charge q, moving in a uniform constant electric field intensity $\boldsymbol{E}$ and a uniform constant magnetic induction $\boldsymbol{B}$. Again let the z-axis be chosen in the direction of $\boldsymbol{B}$, and let the y-axis be chosen so that $\boldsymbol{E}$ is parallel to the yz-plane:

$$\boldsymbol{B} = B\hat{\boldsymbol{z}}, \qquad \boldsymbol{E} = E_y\hat{\boldsymbol{y}} + E_z\hat{\boldsymbol{z}}, \tag{3.300}$$

where B, E_y, E_z are constants. The equations of motion, by Eq. (3.283), are

$$m\ddot{x} = \frac{qB}{c}\dot{y}, \tag{3.301}$$

$$m\ddot{y} = -\frac{qB}{c}\dot{x} + qE_y, \tag{3.302}$$

$$m\ddot{z} = qE_z. \tag{3.303}$$

The z-component of the motion is uniformly accelerated:

$$z = z_0 + \dot{z}_0 t + \frac{1}{2}\frac{qE_z}{m}t^2. \tag{3.304}$$

---

*According to the theory of relativity, the mass of a particle increases with velocity at velocities near the speed of light, and consequently the cyclotron cannot accelerate particles to such speeds unless ν is reduced or B is increased as the particle velocity increases. [It turns out that Eq. (3.295) still holds in relativity theory.]

To solve the x and y equations, we differentiate Eq. (3.301) and substitute in Eq. (3.302) in order to eliminate $\ddot{y}$.

$$\frac{m^2 c}{qB}\dddot{x} = -\frac{qB}{c}\dot{x} + qE_y. \tag{3.305}$$

By making the substitutions

$$\omega = \frac{qB}{mc}, \tag{3.306}$$

$$a = \frac{qE_y}{m}, \tag{3.307}$$

we can write Eq. (3.305) in the form

$$\frac{d^2\dot{x}}{dt^2} + \omega^2 \dot{x} = a\omega. \tag{3.308}$$

This equation has the same form as the equation for a harmonic oscillator with angular frequency ω subject to a constant applied "force" $a\omega$, except that $\dot{x}$ appears in place of the coordinate. The corresponding oscillator problem was considered in Chapter 2, Problem 45. The solution in this case will be

$$\dot{x} = \frac{a}{\omega} + A_x \cos(\omega t + \theta_x), \tag{3.309}$$

where A_x and θ_x are arbitrary constants to be determined. By eliminating $\ddot{x}$ from Eqs. (3.301) and (3.302), in a similar way, we obtain a solution for $\dot{y}$:

$$\dot{y} = A_y \cos(\omega t + \theta_y). \tag{3.310}$$

We get x and y by integrating Eqs. (3.309) and (3.310):

$$x = C_x + \frac{at}{\omega} + \frac{A_x}{\omega}\sin(\omega t + \theta_x), \tag{3.311}$$

$$y = C_y + \frac{A_y}{\omega}\sin(\omega t + \theta_y). \tag{3.312}$$

Now a difficulty arises, for we have six constants A_x, A_y, θ_x, θ_y, C_x, and C_y to be determined, and only four initial values x_0, y_0, $\dot{x}_0$, $\dot{y}_0$ to determine them. The trouble is that we obtained the solutions (3.311) and (3.312) by differentiating the original equations, and differentiating an equation may introduce new solutions that do not satisfy the original equation. Consider, for example, the very simple equation

$$x = 3.$$

Differentiating, we get

$$\dot{x} = 0,$$

whose solution is

$$x = C.$$

Now only for one particular value of the constant C will this satisfy the original equation. Let us substitute Eqs. (3.311) and (3.312) or, equivalently, Eqs. (3.309) and (3.310) into the original Eqs. (3.301) and (3.302), using Eqs. (3.306) and (3.307):

$$-\frac{qB}{c} A_x \sin(\omega t+\theta_x) = \frac{qB}{c} A_y \cos(\omega t+\theta_y), \tag{3.313}$$

$$-\frac{qB}{c} A_y \sin(\omega t+\theta_y) = -\frac{qB}{c} A_x \cos(\omega t+\theta_x). \tag{3.314}$$

These two equations will hold only if A_x, A_y, θ_x, and θ_y are chosen so that

$$A_x = A_y, \tag{3.315}$$

$$\sin(\omega t+\theta_x) = -\cos(\omega t+\theta_y), \tag{3.316}$$

$$\cos(\omega t+\theta_x) = \sin(\omega t+\theta_y). \tag{3.317}$$

The latter two equations are satisfied if

$$\theta_y = \theta_x + \frac{\pi}{2}. \tag{3.318}$$

Let us set

$$A_x = A_y = \omega A, \tag{3.319}$$

$$\theta_x = \theta, \tag{3.320}$$

$$\theta_y = \theta + \frac{\pi}{2}. \tag{3.321}$$

Then Eqs. (3.311) and (3.312) become

$$x = C_x + A \sin(\omega t+\theta) + \frac{at}{\omega}, \tag{3.322}$$

$$y = C_y + A \cos(\omega t+\theta). \tag{3.323}$$

There are now only four constants, A, θ, C_x, C_y, to be determined by the initial values x_0, y_0, $\dot{x}_0$, $\dot{y}_0$. The z-motion is, of course, given by Eq. (3.304). If $E_y = 0$, the xy-motion is in a circle of radius A with angular velocity ω about the point (C_x, C_y); this is the motion considered in the previous example. The effect of E_y is to add to this uniform circular motion a uniform translation in the *x-direction*! The resulting path in the xy-plane will be a cycloid having loops, cusps, or ripples, depending on the initial conditions and on the magnitude of E_y (Fig. 3.43). This problem is of interest in connection with the design of magnetrons. The translation

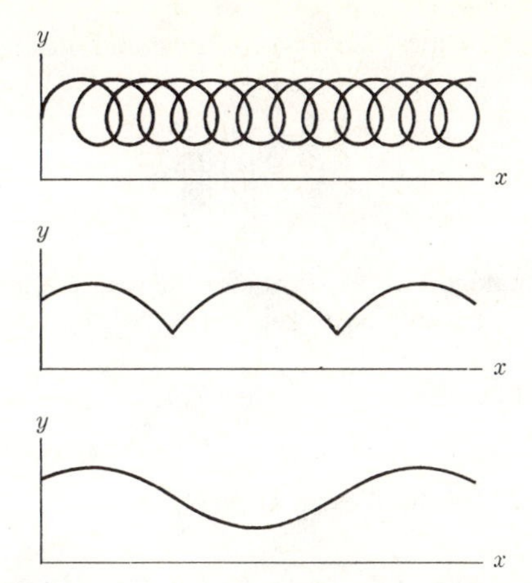

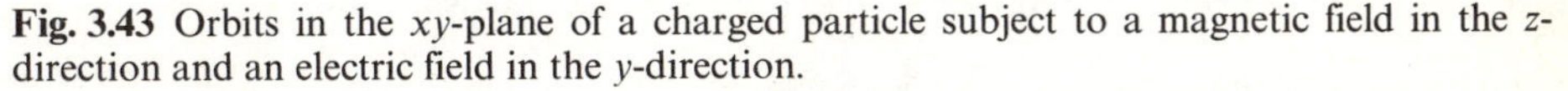

Fig. 3.43 Orbits in the xy-plane of a charged particle subject to a magnetic field in the z-direction and an electric field in the y-direction.

has the velocity

$$v_D = a/\omega = E_y\, c/B.$$

$$\boldsymbol{v}_D = c\boldsymbol{E} \times \boldsymbol{B}/B^2. \tag{3.324}$$

This drift velocity of a charged particle in crossed electric and magnetic fields is of fundamental importance in the theory of plasmas.

PROBLEMS

1. Prove, on the basis of the geometric definitions of the operations of vector algebra, the following equations. In many cases a diagram will suffice. (a) Eq. (3.7), (b) Eq. (3.17), (c) Eq. (3.26), (d) Eq. (3.27), *(e) Eq. (3.35).

2. Prove, on the basis of the algebraic definitions of the operations of vector algebra in terms of components, the following equations: (a) Eq. (3.8), (b) Eq. (3.17), (c) Eq. (3.27), (d) Eq. (3.34), (e) Eq. (3.35).

3. Derive Eq. (3.32) by direct calculation, using Eq. (3.10) to represent $\boldsymbol{A}$ and $\boldsymbol{B}$, and making use of Eqs. (3.25) to (3.31).

4. a) Prove that $\boldsymbol{A}\cdot(\boldsymbol{B}\times\boldsymbol{C})$ is the volume of the parallelepiped whose edges are $\boldsymbol{A}$, $\boldsymbol{B}$, $\boldsymbol{C}$ with positive or negative sign according to whether a right-hand screw rotated from $\boldsymbol{A}$ toward $\boldsymbol{B}$ would advance along $\boldsymbol{C}$ in the positive or negative direction. $\boldsymbol{A}$, $\boldsymbol{B}$, $\boldsymbol{C}$ are any three vectors not lying in a single plane.

b) Use this result to prove Eq. (3.34) geometrically. Verify that the right and left members of Eq. (3.34) are equal in sign as well as in magnitude.

5. Prove the following inequalities. Give a geometric and an algebraic proof (in terms of components) for each:

a) $$|\boldsymbol{A}+\boldsymbol{B}| \leq |\boldsymbol{A}|+|\boldsymbol{B}|.$$
b) $$|\boldsymbol{A}\cdot\boldsymbol{B}| \leq |\boldsymbol{A}|\,|\boldsymbol{B}|.$$
c) $$|\boldsymbol{A}\times\boldsymbol{B}| \leq |\boldsymbol{A}|\,|\boldsymbol{B}|.$$

6. a) Obtain a formula analogous to Eq. (3.40) for the magnitude of the sum of three forces $\boldsymbol{F}_1, \boldsymbol{F}_2, \boldsymbol{F}_3$, in terms of F_1, F_2, F_3, and the angles $\theta_{12}, \theta_{23}, \theta_{31}$ between pairs of forces. [Use the suggestions following Eq. (3.40).]

b) Obtain a formula in the same terms for the angle α_1, between the total force and the component force $\boldsymbol{F}_1$.

7. Prove Eqs. (3.54) and (3.55) from the definition (3.52) of vector differentiation.

8. Prove Eqs. (3.56) and (3.57) from the algebraic definition (3.53) of vector differentiation.

9. Give suitable definitions, analogous to Eqs. (3.52) and (3.53), for the integral of a vector function $\boldsymbol{A}(t)$ with respect to a scalar t:

$$\int_{t_1}^{t_2} \boldsymbol{A}(t)\, dt.$$

Write a set of equations like Eqs. (3.54)–(3.57) expressing the algebraic properties you would expect such an integral to have. Prove that on the basis of either definition

$$\frac{d}{dt}\int_0^t \boldsymbol{A}(t)\, dt = \boldsymbol{A}(t).$$

10. A 45° isosceles right triangle ABC has a hypotenuse AB of length $4a$. A particle is acted on by a force attracting it toward a point O on the hypotenuse a distance a from the point A. The force is equal in magnitude to k/r^2, where r is the distance of the particle from the point O. Calculate the work done by this force when the particle moves from A to C to B along the two legs of the triangle. Make the calculation by both methods, that based on Eq. (3.61) and that based on Eq. (3.63).

11. A particle moves around a semicircle of radius R, from one end A of a diameter to the other B. It is attracted toward its starting point A by a force proportional to its distance from A. When the particle is at B, the force toward A is F_0. Calculate the work done against this force when the particle moves around the semicircle from A to B.

12. A particle is acted on by a force whose components are

$$F_x = ax^3 + bxy^2 + cz,$$

$$F_y = ay^3 + bx^2y,$$

$$F_z = cx.$$

Calculate the work done by this force when the particle moves along a straight line from the origin to the point (x_0, y_0, z_0).

13. a) A particle in the xy-plane is attracted toward the origin by a force $F = k/y$, inversely proportional to its distance from the x-axis. Calculate the work done by the force when the particle moves from the point $x = 0$, $y = a$ to the point $x = 2a$, $y = 0$ along a path which follows the sides of a rectangle consisting of a segment parallel to the x-axis from $x = 0$, $y = a$ to $x = 2a$, $y = a$, and a vertical segment from the latter point to the x-axis.

b) Calculate the work done by the same force when the particle moves along an ellipse of semiaxes a, $2a$. [*Hint:* Set $x = 2a \sin \theta$, $y = a \cos \theta$.]

14. Find the r- and θ-components of $d\boldsymbol{a}/dt$ in plane polar coordinates, where $\boldsymbol{a}$ is the acceleration of a particle.

15. Find the components of $d^2\boldsymbol{A}/dt^2$ in cylindrical polar coordinates, where the vector $\boldsymbol{A}$ is a function of t and is located at a moving point.

16. Find the components of $d^3\boldsymbol{r}/dt^3$ in spherical coordinates.

***17.** a) Plane parabolic coordinates f, h are defined in terms of cartesian coordinates x, y by the equations

$$x = f - h, \qquad y = 2(fh)^{1/2},$$

where f and h are never negative. Find f and h in terms of x and y. Let unit vectors $\hat{\boldsymbol{f}}$, $\hat{\boldsymbol{h}}$ be defined in the directions of increasing f and h respectively. That is, $\hat{\boldsymbol{f}}$ is a unit vector in the direction in which a point would move if its f-coordinate increases slightly while its h-coordinate remains constant. Show that $\hat{\boldsymbol{f}}$ and $\hat{\boldsymbol{h}}$ are perpendicular at every point. [*Hint:* $\hat{\boldsymbol{f}} = (\hat{\boldsymbol{x}}\, dx + \hat{\boldsymbol{y}}\, dy)[(dx)^2 + (dy)^2]^{-1/2}$, when $df > 0$, $dh = 0$. Why?]

b) Show that $\hat{\boldsymbol{f}}$ and $\hat{\boldsymbol{h}}$ are functions of f, h, and find their derivatives with respect to f and h. Show that $\boldsymbol{r} = f^{1/2}(f+h)^{1/2}\hat{\boldsymbol{f}} + h^{1/2}(f+h)^{1/2}\hat{\boldsymbol{h}}$. Find the components of velocity and acceleration in parabolic coordinates.

18. A particle moves along the parabola

$$y^2 = 4f_0^2 - 4f_0 x,$$

where f_0 is a constant. Its speed v is constant. Find its velocity and acceleration components in rectangular and in polar coordinates. Show that the equation of the parabola in polar coordinates is

$$r \cos^2 \frac{\theta}{2} = f_0 .$$

What is the equation of this parabola in parabolic coordinates (Problem 17)?

19. A particle moves with varying speed along an arbitrary curve lying in the xy-plane. The position of the particle is to be specified by the distance s the particle has traveled along the curve from some fixed point on the curve. Let $\hat{\boldsymbol{\tau}}(s)$ be a unit vector tangent to the curve at the point s in the direction of increasing s. Show that

$$\frac{d\hat{\boldsymbol{\tau}}}{ds} = \frac{\hat{\boldsymbol{\nu}}}{r}\text{,"}$$

where $\hat{\boldsymbol{\nu}}(s)$ is a unit vector normal to the curve at the point s, and $r(s)$ is the radius of curvature at the point s, defined as the distance from the curve to the point of intersection of two nearby normals. Hence derive the following formulas for the velocity and acceleration of the particle:

$$\boldsymbol{v} = \dot{s}\hat{\boldsymbol{\tau}}, \qquad \boldsymbol{a} = \ddot{s}\hat{\boldsymbol{\tau}} + \frac{\dot{s}^2}{r}\,\hat{\boldsymbol{\nu}}.$$

20. Using the properties of the vector symbol $\boldsymbol{\nabla}$, derive the vector identities:

$$\text{curl (curl } \boldsymbol{A}) = \text{grad (div } \boldsymbol{A}) - \boldsymbol{\nabla}^2 \boldsymbol{A},$$

$$u \text{ grad } v = \text{grad } (uv) - v \text{ grad } u.$$

Then write out the x-components of each side of these equations and prove by direct calculation that they are equal in each case. (One must be very careful, in using the first identity in curvilinear coordinates, to take proper account of the dependence of the unit vectors on the coordinates.)

21. Calculate curl $\boldsymbol{A}$ in cylindrical coordinates.

22. If the particle in Problem 12 moves with a constant velocity $\boldsymbol{v}$, what is the impulse delivered to it by the given force?

23. a) Given that the particle in Problem 11 moves with a constant speed v around the semicircle, find the rectangular components $F_x(t)$, $F_y(t)$ of the additional force which must act on it besides the force given in Problem 11. Take the x-axis along the diameter AB.

b) Calculate the impulse delivered by this additional force.

24. A particle of mass m moves with constant speed v around a circle of radius r, starting at $t = 0$ from a point P on the circle. Find the angular momentum about the point P at any time t, the force, and the torque about P, and verify that the angular momentum theorem (3.140) is satisfied.

25. A particle of mass m moves according to the equations

$$x = x_0 + at^2,$$

$$y = bt^3,$$

$$z = ct.$$

Find the angular momentum $\boldsymbol{L}$ at any time t. Find the force $\boldsymbol{F}$ and from it the torque $\boldsymbol{N}$ acting on the particle. Verify that the angular momentum theorem (3.144) is satisfied.

26. Give a suitable definition of the angular momentum of a particle about an axis in space. Taking the specified axis as the z-axis, express the angular momentum in terms of cylindrical coordinates. If the force acting on the particle has cylindrical components F_z, F_ρ, F_φ, prove that the time rate of change of angular momentum about the z-axis is equal to the torque about that axis.

27. A moving particle of mass m is located by spherical coordinates $r(t)$, $\theta(t)$, $\varphi(t)$. The force acting on it has spherical components F_r, F_θ, F_φ. Calculate the spherical components of the angular momentum vector and of the torque vector about the origin, and verify by direct calculation that the equation

$$\frac{d\boldsymbol{L}}{dt} = \boldsymbol{N}$$

follows from Newton's equation of motion.

28. The solutions plotted in Fig. 3.28 correspond to the first two of Eqs. (3.151). If $\theta_x = 0$, estimate θ_y for the case $\omega_x = 2\omega_y$ as drawn. Sketch the corresponding figure for the case $\theta_x = \theta_y$. Sketch a typical figure for the case $4\omega_x = 3\omega_y$.

29. Find a lowest order correction to Eq. (3.179) by putting $x_m = (mv_{x0}/b)(1-\delta)$ and solving Eq. (3.175) for δ, assuming $\delta \ll 1$ and $bv_{z0}/mg \gg 1$. [*Hint:* The algebra is not difficult, but you must think carefully about which are the most important terms in this limiting case.]

30. Find the maximum height z_{max} reached by a projectile whose equation of motion is Eq. (3.169). Expand your result in a power series in b, keeping terms in z_{max} up to first order in b, and check the lowest order term against Eq. (3.167).

31. A projectile is fired from the origin with initial velocity $\boldsymbol{v}_0 = (v_{x0}, v_{y0}, v_{z0})$. The wind velocity is $\boldsymbol{v}_w = w\hat{\boldsymbol{y}}$. Solve the equations of motion (3.180) for x, y, z as functions of t. Find the point x_1, y_1 at which the projectile will return to the horizontal plane, keeping only first-order terms in b. Show that if air resistance and wind velocity are neglected in aiming the gun, air resistance alone will cause the projectile to fall short of its target a fraction $4bv_{z0}/3mg$ of the target distance, and that the wind causes an additional miss in the y-coordinate of amount $2bwv_{z0}^2/(mg^2)$.

32. Solve for the next term beyond those given in Eqs. (3.176) and (3.178).

33. A projectile is to be fired from the origin in the xz-plane (z-axis vertical) with muzzle velocity v_0 to hit a target at the point $x = x_0$, $z = 0$. (a) Neglecting air resistance, find the correct angle of elevation of the gun. Show that, in general, there are two such angles unless the target is at or beyond the maximum range.

b) Find the first-order correction to the angle of elevation due to air resistance.

34. Show that the forces in Problems 11 and 12 are conservative, find the potential energy, and use it to find the work done in each case.

35. Determine which of the following forces are conservative, and find the potential energy for those which are:

a) $F_x = 6abz^3y - 20bx^3y^2$, $\quad F_y = 6abxz^3 - 10bx^4y$, $\quad F_z = 18abxz^2y$.

b) $F_x = 18abyz^3 - 20bx^3y^2$, $\quad F_y = 18abxz^3 - 10bx^4y$, $\quad F_z = 6abxyz^2$.

c) $\boldsymbol{F} = \hat{\boldsymbol{x}}F_x(x) + \hat{\boldsymbol{y}}F_y(y) + \hat{\boldsymbol{z}}F_z(z)$.

36. Determine the potential energy for each of the following forces which is conservative:

a) $F_x = 2ax(z^3+y^3), \quad F_y = 2ay(z^3+y^3)+3ay^2(x^2+y^2), \quad F_z = 3az^2(x^2+y^2).$
b) $F_\rho = a\rho^2 \cos\varphi, \quad F_\varphi = a\rho^2 \sin\varphi, \quad F_z = 2az^2.$
c) $F_r = -2ar \sin\theta \cos\varphi, \quad F_\theta = -ar \cos\theta \cos\varphi, \quad F_\varphi = ar \sin\theta \sin\varphi.$

37. Determine the potential energy for each of the following forces which is conservative:

a) $F_x = axe^{-R}, \quad F_y = bye^{-R}, \quad F_z = cze^{-R}$, where $R = ax^2+by^2+cz^2$.
b) $\boldsymbol{F} = \boldsymbol{A}f(\boldsymbol{A}\cdot\boldsymbol{r})$, where $\boldsymbol{A}$ is a constant vector and $f(s)$ is any suitable function of $s = \boldsymbol{A}\cdot\boldsymbol{r}$.
c) $\boldsymbol{F} = (\boldsymbol{r}\times\boldsymbol{A})f(\boldsymbol{A}\cdot\boldsymbol{r})$.

38. A particle is attracted toward the z-axis by a force $\boldsymbol{F}$ proportional to the square of its distance from the xy-plane and inversely proportional to its distance from the z-axis. Add an additional force perpendicular to $\boldsymbol{F}$ in such a way as to make the total force conservative, and find the potential energy. Be sure to write expressions for the forces and potential energy which are dimensionally consistent.

39. Show that $\boldsymbol{F} = \hat{\boldsymbol{r}}F(r)$ is a conservative force by showing by direct calculation that the integral

$$\int_{\boldsymbol{r}_1}^{\boldsymbol{r}_2} \boldsymbol{F}\cdot d\boldsymbol{r}$$

along any path between $\boldsymbol{r}_1$ and $\boldsymbol{r}_2$ depends only on r_1 and r_2. [*Hint:* Express $\boldsymbol{F}$ and $d\boldsymbol{r}$ in spherical coordinates.]

40. Find the components of force for the following potential-energy functions:

a) $V = axy^2z^3.$
b) $V = \frac{1}{2}kr^2.$
c) $V = \frac{1}{2}k_x x^2 + \frac{1}{2}k_y y^2 + \frac{1}{2}k_z z^2.$

41. Find the force on the electron in the hydrogen molecule ion for which the potential is

$$V = -\frac{e^2}{r_1} - \frac{e^2}{r_2},$$

see p. 118

where r_1 is the distance from the electron to the point $y = z = 0, x = -a$, and r_2 is the distance from the electron to the point $y = z = 0, x = a$.

42. Devise a potential-energy function which vanishes as $r \to \infty$, and which yields a force $\boldsymbol{F} \doteq -k\boldsymbol{r}$ when $r \to 0$. Find the force. Verify by doing the appropriate line integrals that the work done by this force on a particle going from $\boldsymbol{r} = 0$ to $\boldsymbol{r} = \boldsymbol{r}_0$ is the same if the particle travels in a straight line as it is if it follows the path shown in Fig. 3.32.

43. The potential energy for an isotropic harmonic oscillator is

$$V = \tfrac{1}{2}kr^2.$$

Plot the effective potential energy for the r-motion when a particle of mass m moves with this potential energy and with angular momentum L about the origin. Discuss the types of motion that are possible, giving as complete a description as is possible without carrying out the solution. Find the frequency of revolution for circular motion and the frequency of small radial oscillations about this circular motion. Hence describe the nature of the orbits which differ slightly from circular orbits.

44. Find the frequency of small radial oscillations about steady circular motion for the effective potential given by Eq. (3.232) for an attractive inverse square law force, and show that it is equal to the frequency of revolution.

45. Find $r(t)$, $\theta(t)$ for the orbit of the particle in Problem 43. Compare with the orbits found in Section 3.10 for the three-dimensional harmonic oscillator.

46. A particle of mass m moves under the action of a central force whose potential is

$$V(r) = Kr^4, \qquad K > 0.$$

For what energy and angular momentum will the orbit be a circle of radius a about the origin? What is the period of this circular motion? If the particle is slightly disturbed from this circular motion, what will be the period of small radial oscillations about $r = a$?

47. According to Yukawa's theory of nuclear forces, the attractive force between a neutron and a proton has the potential

$$V(r) = \frac{Ke^{-\alpha r}}{r}, \qquad K < 0.$$

a) Find the force, and compare it with an inverse square law of force.

b) Discuss the types of motion which can occur if a particle of mass m moves under such a force.

c) Discuss how the motions will be expected to differ from the corresponding types of motion for an inverse square law of force.

d) Find L and E for motion in a circle of radius a.

e) Find the period of circular motion and the period of small radial oscillations.

f) Show that the nearly circular orbits are almost closed when a is very small.

48. Solve the orbital equation (3.222) for the case $F = 0$. Show that your solution agrees with Newton's first law.

49. It will be shown in Chapter 6 (Problem 7) that the effect of a uniform distribution of dust of density ρ about the sun is to add to the gravitational attraction of the sun on a planet of mass m an additional attractive central force

$$F' = -mkr,$$

where

$$k = \frac{4\pi}{3}\rho G.$$

a) If the mass of the sun is M, find the angular velocity of revolution of the planet in a circular orbit of radius r_0, and find the angular frequency of small radial oscillations. Hence show that if F' is much less than the attraction due to the sun, a nearly circular orbit will be approximately an ellipse whose major axis precesses slowly with angular velocity

$$\omega_p = 2\pi\rho\left(\frac{r_0^3 G}{M}\right)^{1/2}.$$

b) Does the axis precess in the same or in the opposite direction to the orbital angular velocity? Look up M and the radius of the orbit of Mercury, and calculate the density of dust required to cause a precession of 41 seconds of arc per century.

50. a) Discuss by the method of the effective potential the types of motion to be expected for an attractive central force inversely proportional to the cube of the radius:

$$F(r) = -\frac{K}{r^3}, \qquad K > 0.$$

b) Find the ranges of energy and angular momentum for each type of motion.
c) Solve the orbital equation (3.222), and show that the solution is one of the forms:

$$\frac{1}{r} = A\cos[\beta(\theta-\theta_0)], \tag{1}$$

$$\frac{1}{r} = A\cosh[\beta(\theta-\theta_0)], \tag{2}$$

$$\frac{1}{r} = A\sinh[\beta(\theta-\theta_0)], \tag{3}$$

$$\frac{1}{r} = A(\theta-\theta_0), \tag{4}$$

$$\frac{1}{r} = \frac{1}{r_0}e^{\pm\beta\theta}. \tag{5}$$

d) For what values of L and E does each of the above types of motion occur? Express the constants A and β in terms of E and L for each case.
e) Sketch a typical orbit of each type.

51. (a) Discuss the types of motion that can occur for a central force

$$F(r) = -\frac{K}{r^2}+\frac{K'}{r^3}.$$

Assume that $K > 0$, and consider both signs for K'.

b) Solve the orbital equation, and show that the bounded orbits have the form (if $L^2 > -mK'$)

$$r = \frac{a(1-\varepsilon^2)}{1+\varepsilon\cos\alpha\theta}.$$

c) Show that this is a precessing ellipse, determine the angular velocity of precession, and state whether the precession is in the same or in the opposite direction to the orbital angular velocity.

52. Sputnik I had a perigee (point of closest approach to the earth) 227 km above the earth's surface, at which point its speed was 28,710 km/hr. Find its apogee (maximum) distance from the earth's surface and its period of revolution. (Assume the earth is a sphere, and neglect air resistance. You need only look up g and the earth's radius to do this problem.)

53. Explorer I had a perigee 360 km and an apogee 2,549 km above the earth's surface. Find its distance above the earth's surface when it passed over a point 90° around the earth from its perigee.

54. A comet is observed a distance of 1.00×10^8 km from the sun, traveling toward the sun with a velocity of 51.6 km per second at an angle of 45° with the radius from the sun. Work out an equation for the orbit of the comet in polar coordinates with origin at the sun and x-axis through the observed position of the comet. (The mass of the sun is 2.00×10^{30} kg.)

55. It can be shown (Chapter 6, Problems 17 and 21) that the correction to the potential energy of a mass m in the earth's gravitational field, due to the oblate shape of the earth, is approximately, in spherical coordinates, relative to the polar axis of the earth,

$$V' = -\frac{\eta m M G R^2}{5r^3}(1-3\cos^2\theta),$$

where M is the mass of the earth and $2R$, $2R(1-\eta)$ are the equatorial and polar diameters of the earth. Calculate the rate of precession of the perigee (point of closest approach) of an earth satellite moving in a nearly circular orbit in the equatorial plane. Look up the equatorial and polar diameters of the earth, and estimate the rate of precession in degrees per revolution for a satellite 400 miles above the earth.

***56.** Calculate the torque on an earth satellite due to the oblateness potential energy correction given in Problem 55. A satellite moves in a circular orbit of radius r whose plane is inclined so that its normal makes an angle α with the polar axis. Assume that the orbit is very little affected in one revolution, and calculate the average torque during a revolution. Show that the effect of such a torque is to make the normal to the orbit precess in a cone of half angle α about the polar axis, and find a formula for the rate of precession in degrees per revolution. Calculate the rate for a satellite 400 miles above the earth, using suitable values for M, η, and R.

57. It can be shown that the orbit given by the special theory of relativity for a particle of mass m moving under a potential energy $V(r)$ is the same as the orbit which the particle would follow according to Newtonian mechanics if the potential energy were

$$V(r) - \frac{[E-V(r)]^2}{2mc^2},$$

where E is the energy (kinetic plus potential), and c is the speed of light. Discuss the nature of the orbits for an inverse square law of force according to the theory of relativity. Show by comparing the orbital angular velocity with the frequency of radial oscillations for nearly circular motion that the nearly circular orbits, when the relativistic correction is small, are precessing ellipses, and calculate the angular velocity of precession. [See Eq. (14.101).]

58. Mars has a perihelion (closest) distance from the sun of 2.06×10^8 km, and an aphelion (maximum) distance of 2.485×10^8 km. Assume that the earth moves in the same plane in a circle of radius 1.49×10^8 km with a period of one year. From this data alone, find the speed of Mars at perihelion. Assume that a Mariner space probe is launched so that its perihelion is at the earth's orbit and its aphelion at the perihelion of Mars. Find the velocity of the Mariner relative to Mars at the point where they meet. Which has the higher velocity? Which has the higher average angular velocity during the period of the flight?

59. Mariner 4 left the earth on an orbit whose perihelion distance from the sun was approximately the distance of the earth (1.49×10^8 km), and whose aphelion distance was approximately the distance of Mars from the sun (2.2×10^8 km). With what velocity did it leave relative to the earth? With what velocity must it leave the earth (relative to the earth) in order to escape altogether from the sun's gravitational pull? (You need no further data to answer this problem except the length of the year, if you assume the earth moves in a circle.)

60. a) A satellite is to be launched from the surface of the earth. Assume the earth is a sphere of radius R, and neglect friction with the atmosphere. The satellite is to be launched at an angle α with the vertical, with a velocity v_0, so as to coast without power until its velocity is horizontal at an altitude h_1 above the earth's surface. A horizontal thrust is then applied by the last stage rocket so as to add an additional velocity Δv_1 to the velocity of the satellite. The final orbit is to be an ellipse with perigee h_1 (point of closest approach) and apogee h_2 (point farthest away) measured from the earth's surface. Find the required initial velocity v_0 and additional velocity Δv_1, in terms of R, α, h_1, h_2, and g, the acceleration of gravity at the earth's surface.

b) Write a formula for the change δh_1 in perigee height due to a small error $\delta\beta$ in the final thrust direction, to order $(\delta\beta)^2$.

61. Two planets move in the same plane in circles of radii r_1, r_2 about the sun. A space probe is to be launched from planet 1 with velocity v_1 relative to the planet, so as to reach the orbit of planet 2. (The velocity v_1 is the relative velocity after the probe has escaped from the gravitational field of the planet.) Show that v_1 is a minimum for an elliptical orbit whose perihelion and aphelion are r_1 and r_2. In that case, find v_1, and the relative velocity v_2 between the space probe and planet 2 if the probe arrives at radius r_2 at the proper time to intercept planet 2. Express your results in terms of r_1, r_2, and the length of the year Y_1 of planet 1. Look up the appropriate values of r_1 and r_2, and estimate v_1 for trips to Venus and Mars from the earth.

62. A rocket is in an elliptical orbit around the earth, perigee r_1, apogee r_2, measured from the center of the earth. At a certain point in its orbit, its engine is fired for a short time so as to give a velocity increment Δv in order to put the rocket on an orbit which escapes from the earth with a final velocity v_0 relative to the earth. (Neglect any effects due to the sun and moon.) Show that Δv is a minimum if the thrust is applied at perigee, parallel to the orbital velocity. Find Δv in that case in terms of the elliptical orbit parameters ε, a, the acceleration g at a distance R from the earth's center, and the final velocity v_0. Can you explain physically why Δv is smaller for larger ε?

63. A satellite moves around the earth in an orbit which passes across the poles. The time at which it crosses each parallel of latitude is measured so that the function $\theta(t)$ is known. Show how to find the perigee, the semimajor axis, and the eccentricity of its orbit in terms of $\theta(t)$, and the value of g at the surface of the earth. Assume the earth is a sphere of radius R.

64. A particle of mass m moves in an elliptical orbit of major axis $2a$, eccentricity ε, in such a way that the radius to the particle from the center of the ellipse sweeps out area at a constant rate

$$\frac{dS}{dt} = C,$$

and with period τ independent of a and ε. (a) Write out the equation of the ellipse in polar coordinates with origin at the center of the ellipse.

b) Show that the force on the particle is a central force, and find $F(r)$ in terms of m, τ.

65. Show that the Rutherford cross-section formula (3.276) holds also when one of the charges is negative.

66. A particle is reflected from the surface of a hard sphere of radius R in such a way that the incident and reflected lines of travel lie in a common plane with the radius to the point of impact and make equal angles with the radius. Find the cross-section $d\sigma$ for scattering through an angle between Θ and $\Theta + d\Theta$. Integrate $d\sigma$ over all angles and show that the total cross-section has the expected value πR^2.

67. Exploit the analogy u, $\theta \leftrightarrow x$, t between Eqs. (3.222) and (2.39) in order to develop a solution of Eq. (3.222) analogous to the solution (2.46) of Eq. (2.39). Use your solution to show that the scattering angle Θ (Fig. 3.42) for a particle subject to a central force $F(r)$ is given by

$$\Theta = \left|\pi - 2s \int_0^{u_0} \left[1 - s^2u^2 - V(u^{-1})/(\tfrac{1}{2}mv_0^2)\right]^{-1/2} du\right|,$$

where $V(r = u^{-1})$ is the potential energy,

$$V(r) = \int_r^\infty F(r)\, dr,$$

s is the impact parameter, and u_0 is the value of u at which the quantity in square brackets vanishes. [This problem is not difficult if you keep clearly in mind the physical and geometrical significance of the various quantities involved at each step in the solution.]

68. Show that a hard sphere as defined in Problem 66 can be represented as a limiting case of a central force where

$$V(r) = \begin{matrix} 0, \text{ if } r > R, \\ \infty, \text{ if } r < R, \end{matrix}$$

that is, show that such a potential gives the same law of reflection as specified in Problem 66. Hence use the result of Problem 67 to solve Problem 66.

69. Use the result of Problem 67 to derive the Rutherford cross-section formula (3.276).

70. A rocket moves with initial velocity v_0 toward the moon of mass M, radius r_0. Find the cross-section σ for striking the moon. Take the moon to be at rest, and ignore all other bodies.

71. Show that for a repulsive central force inversely proportional to the cube of the radius,

$$F(r) = \frac{K}{r^3}, \qquad K > 0,$$

the orbits are of the form (1) given in Problem 50, and express β in terms of K, E, L, and the mass m of the incident particle. Show that the cross-section for scattering through an angle between Θ and $\Theta + d\Theta$ for a particle subject to this force is

$$d\sigma = \frac{2\pi^3 K}{mv_0^2} \frac{\pi - \Theta}{\Theta^2(2\pi - \Theta)^2} d\Theta.$$

72. A particle of charge q, mass m at rest in a constant, uniform magnetic field $\boldsymbol{B} = B_0\hat{\boldsymbol{z}}$ is subject, beginning at $t = 0$, to an oscillating electric field

$$\boldsymbol{E} = E_0\hat{\boldsymbol{x}} \sin \omega t.$$

Find its motion.

73. Solve Problem 72 for the case $\omega = qB_0/mc$.

74. A charged particle moves in a constant, uniform electric and magnetic field. Show that if we introduce a new variable

$$\boldsymbol{r}' = \boldsymbol{r} - \frac{\boldsymbol{E} \times \boldsymbol{B}}{B^2} ct,$$

the equation of motion for $\boldsymbol{r}'$ is the same as that for $\boldsymbol{r}$ except that the component of $\boldsymbol{E}$ perpendicular to $\boldsymbol{B}$ has been eliminated.

75. A particle of charge q in a cylindrical magnetron moves in a uniform magnetic field

$$\boldsymbol{B} = B\hat{\boldsymbol{z}},$$

and an electric field, directed radially outward or inward from a central wire along the z-axis,

$$\boldsymbol{E} = \frac{a}{\rho}\hat{\boldsymbol{\rho}},$$

where ρ is the distance from the z-axis, and $\hat{\boldsymbol{\rho}}$ is a unit vector directed radially outward from the z-axis. The constants a and B may be either positive or negative.

a) Set up the equations of motion in cylindrical coordinates.

b) Show that the quantity

$$m\rho^2\dot{\varphi} + \frac{qB}{2c}\rho^2 = K$$

is a constant of the motion.

c) Using this result, give a qualitative discussion, based on the energy integral, of the types of motion that can occur. Consider all cases, including all values of a, B, K, and E.

d) Under what conditions can circular motion about the axis occur?

e) What is the frequency of small radial oscillations about this circular motion?

76. A velocity selector for a beam of charged particles of mass m, charge e, is to be designed to select particles of a particular velocity v_0. The velocity selector utilizes a uniform electric field E in the x-direction and a uniform magnetic field B in the y-direction. The beam emerges from a narrow slit along the y-axis and travels in the z-direction. After passing through the crossed fields for a distance l, the beam passes through a second slit parallel to the first and also in the yz-plane. The fields E and B are chosen so that particles with the proper velocity moving parallel to the z-axis experience no net force.

a) If a particle leaves the origin with a velocity v_0 at a small angle with the z-axis, find the point at which it arrives at the plane $z = l$. Assume that the initial angle is small enough so that second-order terms in the angle may be neglected.

b) What is the best choice of E, B in order that as large a fraction as possible of the particles with velocity v_0 arrive at the second slit, while particles of other velocities miss the slit as far as possible?

c) If the slit width is h, what is the maximum velocity deviation δv from v_0 for which a particle moving initially along the z-axis can pass through the second slit? Assume that E, B have the values chosen in part (b).

CHAPTER 4

THE MOTION OF A SYSTEM OF PARTICLES

4.1 CONSERVATION OF LINEAR MOMENTUM. CENTER OF MASS

We consider in this chapter the behavior of mechanical systems containing two or more particles acted upon by *internal* forces exerted by the particles upon one another, and by *external* forces exerted upon particles of the system by agents not belonging to the system. We assume the particles to be point masses each specified by its position (x, y, z) in space, like the single particle whose motion was studied in the preceding chapter.

Let the system we are studying contain N particles, and let them be numbered $1, 2, \ldots, N$. The masses of the particles we designate by $m_1, m_2, \ldots, m_N$. The total force acting on the kth particle will be the sum of the internal forces exerted on particle k by all the other $(N-1)$ particles in the system, plus any external force which may be applied to particle k. Let the sum of the internal forces on particle k be $\boldsymbol{F}_k^i$, and let the total external force on particle k be $\boldsymbol{F}_k^e$. Then the equation of motion of the kth particle will be

$$m_k\ddot{\boldsymbol{r}}_k = \boldsymbol{F}_k^e + \boldsymbol{F}_k^i, \qquad k = 1, 2, \ldots, N. \tag{4.1}$$

The N equations obtained by letting k in Eqs. (4.1) run over the numbers $1, \ldots, N$ are the equations of motion of our system. Since each of these N equations is itself a vector equation, we have in general a set of $3N$ simultaneous second-order differential equations to be solved. The solution will be a set of functions $\boldsymbol{r}_k(t)$ specifying the motion of each particle in the system. The solution will depend on $6N$ "arbitrary" constants specifying the initial position and velocity of each particle. The problem of solving the set of equations (4.1) is very difficult, except in certain special cases, and no general methods are available for attacking the N-body problem, even in the case where the forces between the bodies are central forces. The two-body problem can often be solved, as we shall see, and some general theorems are available when the internal forces satisfy certain conditions.

If $\boldsymbol{p}_k = m_k\boldsymbol{v}_k$ is the linear momentum of the kth particle, we can write Eqs. (4.1) in the form

$$\frac{d\boldsymbol{p}_k}{dt} = \boldsymbol{F}_k^e + \boldsymbol{F}_k^i, \qquad k = 1, \ldots, N. \tag{4.2}$$

Summing the right and left sides of these equations over all the particles, we have

$$\sum_{k=1}^{N} \frac{d\boldsymbol{p}_k}{dt} = \frac{d}{dt}\sum_{k=1}^{N} \boldsymbol{p}_k = \sum_{k=1}^{N} \boldsymbol{F}_k^e + \sum_{k=1}^{N} \boldsymbol{F}_k^i. \tag{4.3}$$

We designate by $\boldsymbol{P}$ the total linear momentum of the particles, and by $\boldsymbol{F}$ the total external force:

$$\boldsymbol{P} = \sum_{k=1}^{N} \boldsymbol{p}_k = \sum_{k=1}^{N} m_k \boldsymbol{v}_k, \tag{4.4}$$

$$\boldsymbol{F} = \sum_{k=1}^{N} \boldsymbol{F}_k^e. \tag{4.5}$$

We now make the assumption, to be justified below, that the sum of the internal forces acting on all the particles is zero:

$$\sum_{k=1}^{N} \boldsymbol{F}_k^i = \boldsymbol{0}. \tag{4.6}$$

When Eqs. (4.4), (4.5), and (4.6) are substituted in Eq. (4.3), it becomes

$$\frac{d\boldsymbol{P}}{dt} = \boldsymbol{F}. \tag{4.7}$$

This is the momentum theorem for a system of particles. It states that the time rate of change of the total linear momentum is equal to the total external force. An immediate corollary is the conservation theorem for linear momentum, which states that the total momentum $\boldsymbol{P}$ is constant when no external forces act.

We now try to justify the assumption (4.6). Our first proof is based on Newton's third law. We assume that the force $\boldsymbol{F}_k^i$ acting on particle k due to all the other particles can be represented as a sum of separate forces due to each of the other particles:

$$\boldsymbol{F}_k^i = \sum_{l \neq k} \boldsymbol{F}_{l \to k}^i, \tag{4.8}$$

where $\boldsymbol{F}_{l \to k}^i$ is the force on particle k due to particle l. According to Newton's third law, the force exerted by particle l on particle k is equal and opposite to that exerted by k on l:

$$\boldsymbol{F}_{k \to l}^i = -\boldsymbol{F}_{l \to k}^i. \tag{4.9}$$

Equation (4.9) expresses Newton's third law in what we may call the weak form; that is, it says that the forces are equal and opposite, but does not imply that the forces act along the line joining the two particles. If we now consider the sum in Eq. (4.6), we have

$$\sum_{k=1}^{N} \boldsymbol{F}_k^i = \sum_{k=1}^{N} \sum_{l \neq k} \boldsymbol{F}_{l \to k}^i. \tag{4.10}$$

The sum on the right is over all forces acting between all pairs of particles in the system. Since for each pair of particles k, l, two forces $\boldsymbol{F}_{k \to l}^i$ and $\boldsymbol{F}_{l \to k}^i$ appear in the total sum, and by Eq. (4.9) the sum of each such pair is zero, the total sum on the right in Eq. (4.10) vanishes, and Eq. (4.6) is proved.

Thus Newton's third law, in the form (4.9), is sufficient to guarantee the conservation of linear momentum for a system of particles, and it was for this purpose that the law was introduced. The law of conservation of momentum has, however, a more general validity than Newton's third law, as we shall see later. We can derive assumption (4.6) on the basis of a somewhat weaker assumption than Newton's third law. We do not need to assume that the particles interact in pairs. We assume only that the internal forces are such that they would do no net work if every particle in the system should be displaced the same small distance $\delta \boldsymbol{r}$ from its position at any particular instant. An imagined motion of all the particles in the system is called a *virtual displacement*. The motion described, in which every particle moves the same small distance $\delta \boldsymbol{r}$, is called a *small virtual translation* of the system. We assume, then, that in any small virtual translation $\delta \boldsymbol{r}$ of the entire system, the internal forces would do no net work. From the point of view of the general idea of conservation of energy, this assumption amounts to little more than assuming that space is homogeneous. If we move the system to a slightly different position in space without otherwise disturbing it, the internal state of the system should be unaffected, hence in particular the distribution of various kinds of energy within it should remain the same and no net work can have been done by the internal forces. Let us use this idea to prove Eq. (4.6). The work done by the force $\boldsymbol{F}_k^i$ in a small virtual translation $\delta \boldsymbol{r}$ is

$$\delta W_k = \boldsymbol{F}_k^i \cdot \delta \boldsymbol{r}. \tag{4.11}$$

The total work done by all the internal forces is

$$\delta W = \sum_{k=1}^{N} \delta W_k = \delta \boldsymbol{r} \cdot \left(\sum_{k=1}^{N} \boldsymbol{F}_k^i \right), \tag{4.12}$$

where we have factored out $\delta \boldsymbol{r}$ from the sum, since it is the same for all particles. Assuming that $\delta W = 0$, we have

$$\delta \boldsymbol{r} \cdot \left(\sum_{k=1}^{N} \boldsymbol{F}_k^i \right) = 0. \tag{4.13}$$

Since Eq. (4.13) must hold for any $\delta \boldsymbol{r}$, Eq. (4.6) follows.

We can put Eq. (4.7) in an illuminating form by introducing the concept of center of mass of the system of particles. The vector $\boldsymbol{R}$ which locates the center of mass is defined by the equation

$$M\boldsymbol{R} = \sum_{k=1}^{N} m_k \boldsymbol{r}_k, \tag{4.14}$$

where M is the total mass:

$$M = \sum_{k=1}^{N} m_k. \tag{4.15}$$

The coordinates of the center of mass are given by the components of Eq. (4.14):

$$X = \frac{1}{M} \sum_{k=1}^{N} m_k x_k, \qquad Y = \frac{1}{M} \sum_{k=1}^{N} m_k y_k, \qquad Z = \frac{1}{M} \sum_{k=1}^{N} m_k z_k. \tag{4.16}$$

The total momentum defined by Eq. (4.4) is, in terms of the center of mass,

$$\boldsymbol{P} = \sum_{k=1}^{N} m_k \dot{\boldsymbol{r}}_k = M\dot{\boldsymbol{R}}, \tag{4.17}$$

so that Eq. (4.7) can be written

$$M\ddot{\boldsymbol{R}} = \boldsymbol{F}. \tag{4.18}$$

This equation has the same form as the equation of motion of a particle of mass M acted on by a force $\boldsymbol{F}$. We thus have the important theorem that [when Eq. (4.6) holds] *the center of mass of a system of particles moves like a single particle, whose mass is the total mass of the system, acted on by a force equal to the total external force acting on the system.*

4.2 CONSERVATION OF ANGULAR MOMENTUM

Let us calculate the time rate of change of the total angular momentum of a system of N particles relative to a point Q not necessarily fixed in space. The vector angular momentum of particle k about a point Q, not necessarily the origin, is to be defined according to Eq. (3.142):

$$\boldsymbol{L}_{kQ} = m_k(\boldsymbol{r}_k - \boldsymbol{r}_Q) \times (\dot{\boldsymbol{r}}_k - \dot{\boldsymbol{r}}_Q), \tag{4.19}$$

where $\boldsymbol{r}_Q$ is the position vector of the point Q, and $(\boldsymbol{r}_k - \boldsymbol{r}_Q)$ is the vector from Q to particle k. Note that in place of the velocity $\dot{\boldsymbol{r}}_k$ we have written the velocity $(\dot{\boldsymbol{r}}_k - \dot{\boldsymbol{r}}_Q)$ relative to the point Q as origin, so that $\boldsymbol{L}_{kQ}$ is the angular momentum of m_k calculated as if Q were a fixed origin. This is the most useful way to define the angular momentum about a moving point Q. Taking the cross product of $(\boldsymbol{r}_k - \boldsymbol{r}_Q)$ with the equation of motion (4.2) for particle k, as in the derivation of Eq. (3.144), we obtain

$$(\boldsymbol{r}_k - \boldsymbol{r}_Q) \times \frac{d\boldsymbol{p}_k}{dt} = (\boldsymbol{r}_k - \boldsymbol{r}_Q) \times \boldsymbol{F}_k^e + (\boldsymbol{r}_k - \boldsymbol{r}_Q) \times \boldsymbol{F}_k^i. \tag{4.20}$$

We now differentiate Eq. (4.19):

$$\frac{d\boldsymbol{L}_{kQ}}{dt} = (\boldsymbol{r}_k - \boldsymbol{r}_Q) \times \frac{d\boldsymbol{p}_k}{dt} + m_k(\dot{\boldsymbol{r}}_k - \dot{\boldsymbol{r}}_Q) \times (\dot{\boldsymbol{r}}_k - \dot{\boldsymbol{r}}_Q) - m_k(\boldsymbol{r}_k - \boldsymbol{r}_Q) \times \ddot{\boldsymbol{r}}_Q. \tag{4.21}$$

The second term on the right vanishes. Therefore, by Eq. (4.20),

$$\frac{d\boldsymbol{L}_{kQ}}{dt} = (\boldsymbol{r}_k - \boldsymbol{r}_Q) \times \boldsymbol{F}_k^e + (\boldsymbol{r}_k - \boldsymbol{r}_Q) \times \boldsymbol{F}_k^i - m_k(\boldsymbol{r}_k - \boldsymbol{r}_Q) \times \ddot{\boldsymbol{r}}_Q. \tag{4.22}$$

The total angular momentum and total external torque about the point Q are defined as follows:

$$\boldsymbol{L}_Q = \sum_{k=1}^{N} \boldsymbol{L}_{kQ}, \tag{4.23}$$

$$\boldsymbol{N}_Q = \sum_{k=1}^{N} (\boldsymbol{r}_k - \boldsymbol{r}_Q) \times \boldsymbol{F}_k^e. \tag{4.24}$$

Summed over all particles, Eq. (4.22) becomes, if we use Eq. (4.14),

$$\frac{d\boldsymbol{L}_Q}{dt} = \boldsymbol{N}_Q + \sum_{k=1}^{N} (\boldsymbol{r}_k - \boldsymbol{r}_Q) \times \boldsymbol{F}_k^i - M(\boldsymbol{R} - \boldsymbol{r}_Q) \times \ddot{\boldsymbol{r}}_Q. \tag{4.25}$$

The last term will vanish if the acceleration of the point Q is zero or is along the line joining Q with the center of mass. We shall restrict the discussion to moments about a point Q satisfying this condition:

$$(\boldsymbol{R} - \boldsymbol{r}_Q) \times \ddot{\boldsymbol{r}}_Q = \boldsymbol{0}. \tag{4.26}$$

The most important applications will be to cases where Q is at rest, or where Q is the center of mass. If we also assume that the total internal torque vanishes:

$$\sum_{k=1}^{N} (\boldsymbol{r}_k - \boldsymbol{r}_Q) \times \boldsymbol{F}_k^i = \boldsymbol{0}, \tag{4.27}$$

then Eq. (4.25) becomes

$$\frac{d\boldsymbol{L}_Q}{dt} = \boldsymbol{N}_Q. \tag{4.28}$$

This is the angular momentum theorem for a system of particles. An immediate corollary is the conservation theorem for angular momentum, which states that the total angular momentum of a system of particles is constant if there is no external torque on the system.

In order to prove Eq. (4.27) from Newton's third law, we need to assume a stronger version of the law than that needed in the preceding section, namely, that the force $\boldsymbol{F}_{k \to l}^i$ is not only equal and opposite to $\boldsymbol{F}_{l \to k}^i$, but that these forces act along the line joining the two particles; that is, the two particles can only attract or repel each other. We shall assume, as in the previous section, that $\boldsymbol{F}_k^i$ is the sum of forces due to each of the other particles:

$$\begin{aligned}\sum_{k=1}^{N} (\boldsymbol{r}_k - \boldsymbol{r}_Q) \times \boldsymbol{F}_k^i &= \sum_{k=1}^{N} \sum_{l \neq k} (\boldsymbol{r}_k - \boldsymbol{r}_Q) \times \boldsymbol{F}_{l \to k}^i \\ &= \sum_{k=1}^{N} \sum_{l=1}^{k-1} [(\boldsymbol{r}_k - \boldsymbol{r}_Q) \times \boldsymbol{F}_{l \to k}^i + (\boldsymbol{r}_l - \boldsymbol{r}_Q) \times \boldsymbol{F}_{k \to l}^i].\end{aligned} \tag{4.29}$$

In the second step, the sum of torques has been rearranged as a sum of pairs of torques due to pairs of forces which, according to Newton's third law, are equal and opposite [Eq. (4.9)], so that

$$\sum_{k=1}^{N} (\boldsymbol{r}_k - \boldsymbol{r}_Q) \times \boldsymbol{F}_k^i = \sum_{k=1}^{N} \sum_{l=1}^{k-1} [(\boldsymbol{r}_k - \boldsymbol{r}_Q) - (\boldsymbol{r}_l - \boldsymbol{r}_Q)] \times \boldsymbol{F}_{l \to k}^i$$
$$= \sum_{k=1}^{N} \sum_{l=1}^{k-1} (\boldsymbol{r}_k - \boldsymbol{r}_l) \times \boldsymbol{F}_{l \to k}^i. \tag{4.30}$$

The vector $(\boldsymbol{r}_k - \boldsymbol{r}_l)$ has the direction of the line joining particle l with particle k. If $\boldsymbol{F}_{l \to k}^i$ acts along this line, the cross product in Eq. (4.30) vanishes. Hence if we assume Newton's third law in the strong form, then assumption (4.27) can be proved.

Alternatively, by assuming that no net work is done by the internal forces in a small virtual rotation about any axis through the point Q, we can show that the component of total internal torque in any direction is zero, and hence justify Eq. (4.27).

As an application of Eq. (4.28), we consider the action of a gyroscope or top. A gyroscope is a rigid system of particles symmetrical about an axis and rotating about that axis. The reader can convince himself that when the gyroscope is rotating about a fixed axis, the angular momentum vector of the gyroscope about a point Q on the axis of rotation is directed along the axis of rotation, as in Fig. 4.1. The symmetry about the axis guarantees that any component of the angular momentum $\boldsymbol{L}_k$ of particle k that is perpendicular to the axis will be compensated by an equal and opposite component due to the diametrically opposite particle. Let us choose the point Q where the gyroscope axis rests on its support. If now a

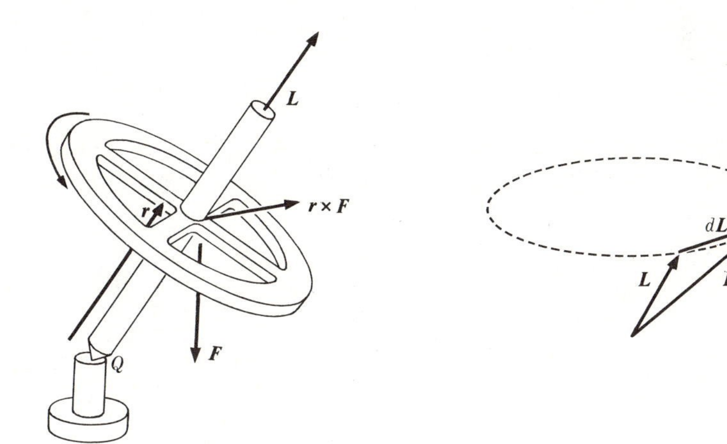

Fig. 4.1 Motion of a simple gyroscope.

force $\boldsymbol{F}$ is applied downward on the gyroscope axis (e.g., the force of gravity), the torque $(\boldsymbol{r} \times \boldsymbol{F})$ due to $\boldsymbol{F}$ will be directed perpendicular to $\boldsymbol{r}$ and to $\boldsymbol{L}$, as shown in Fig. 4.1. By Eq. (4.28) the vector $d\boldsymbol{L}/dt$ is in the same direction, as shown in the figure, and the vector $\boldsymbol{L}$ tends to precess around the figure in a cone under the action of the force $\boldsymbol{F}$. Now the statement that $\boldsymbol{L}$ is directed along the gyroscope axis is strictly true only if the gyroscope is simply rotating about its axis. If the gyroscope axis itself is changing its direction, then this latter motion will contribute an additional component of angular momentum. If, however, the gyroscope is spinning very rapidly, then the component of angular momentum along its axis will be much greater than the component due to the motion of the axis, and $\boldsymbol{L}$ will be very nearly parallel to the gyroscope axis. Therefore the gyroscope axis must also precess around the vertical, remaining essentially parallel to $\boldsymbol{L}$. A careful analysis of the off-axis components of $\boldsymbol{L}$ shows that, if the gyroscope axis is initially stationary in a certain direction and is released, it will wobble slightly down and up as it precesses around the vertical. This will be shown in Chapter 11. The gyroscope does not "resist any change in its direction," as is sometimes asserted, for the rate of change in its angular momentum is always equal to the applied torque, just as the rate of change of linear momentum is always equal to the applied force. We can make the gyroscope turn in any direction we please by applying the appropriate torque. The importance of the gyroscope as a directional stabilizer arises from the fact that the angular momentum vector $\boldsymbol{L}$ remains constant when no torque is applied. The changes in direction of a well-made gyroscope are small because the applied torques are small and $\boldsymbol{L}$ is very large, so that a small $d\boldsymbol{L}$ gives no appreciable change in direction. Furthermore, a gyroscope only changes direction while a torque is applied; if it shifts slightly due to occasional small frictional torques in its mountings, it stops shifting when the torque stops. A large nonrotating mass, if mounted like a gyroscope, would acquire only small angular velocities due to frictional torques, but once set in motion by a small torque, it would continue to rotate, and the change in position might eventually become large.

4.3 CONSERVATION OF ENERGY

In many cases, the total force acting on any particle in a system of particles depends only on the positions of the particles in the system:

$$\boldsymbol{F}_k = \boldsymbol{F}_k(\boldsymbol{r}_1, \boldsymbol{r}_2, \ldots, \boldsymbol{r}_N), \qquad k = 1, 2, \ldots, N. \tag{4.31}$$

The external force $\boldsymbol{F}_k^e$, for example, might depend on the position $\boldsymbol{r}_k$ of particle k, and the internal force $\boldsymbol{F}_k^i$ might depend on the positions of the other particles relative to particle k. It may be that a potential function $V(\boldsymbol{r}_1, \boldsymbol{r}_2, \ldots, \boldsymbol{r}_N)$ exists such that

$$F_{kx} = -\frac{\partial V}{\partial x_k}, \qquad F_{ky} = -\frac{\partial V}{\partial y_k}, \qquad F_{kz} = -\frac{\partial V}{\partial z_k}, \quad k = 1, \ldots, N. \tag{4.32}$$

Conditions to be satisfied by the force functions $\boldsymbol{F}_k(\boldsymbol{r}_1, \ldots, \boldsymbol{r}_N)$ in order for a potential V to exist can be worked out, analogous to the condition (3.189) for a single particle. The general result is rather unwieldy and of little practical importance, and we omit this development here. The special case when $\boldsymbol{F}_k^i$ is a sum of forces between pairs of particles is of great practical importance and will be treated later. If a potential energy exists, we can derive a conservation of energy theorem as follows. By Eq. (4.32), the equations of motion of the kth particle are

$$m_k \frac{dv_{kx}}{dt} = -\frac{\partial V}{\partial x_k}, \qquad m_k \frac{dv_{ky}}{dt} = -\frac{\partial V}{\partial y_k}, \qquad m_k \frac{dv_{kz}}{dt} = -\frac{\partial V}{\partial z_k}. \tag{4.33}$$

Multiplying Eqs. (4.33) by v_{kx}, v_{ky}, v_{kz}, respectively, and adding, we have for each k:

$$\frac{d}{dt}(\tfrac{1}{2}m_k v_k^2) + \frac{\partial V}{\partial x_k}\frac{dx_k}{dt} + \frac{\partial V}{\partial y_k}\frac{dy_k}{dt} + \frac{\partial V}{\partial z_k}\frac{dz_k}{dt} = 0, \quad k = 1, \ldots, N. \tag{4.34}$$

This is to be summed over all values of k:

$$\frac{d}{dt}\sum_{k=1}^{N}(\tfrac{1}{2}m_k v_k^2) + \sum_{k=1}^{N}\left(\frac{\partial V}{\partial x_k}\frac{dx_k}{dt} + \frac{\partial V}{\partial y_k}\frac{dy_k}{dt} + \frac{\partial V}{\partial z_k}\frac{dz_k}{dt}\right) = 0. \tag{4.35}$$

The second term in Eq. (4.35) is dV/dt:

$$\frac{dV}{dt} = \sum_{k=1}^{N}\left(\frac{\partial V}{\partial x_k}\frac{dx_k}{dt} + \frac{\partial V}{\partial y_k}\frac{dy_k}{dt} + \frac{\partial V}{\partial z_k}\frac{dz_k}{dt}\right), \tag{4.36}$$

and the first term is the time derivative of the total kinetic energy

$$T = \sum_{k=1}^{N} \tfrac{1}{2}m_k v_k^2. \tag{4.37}$$

Consequently, Eq. (4.35) can be written

$$\frac{d}{dt}(T + V) = 0. \tag{4.38}$$

Hence we again have a conservation of energy theorem,

$$T + V = E, \tag{4.39}$$

where E is constant. If the internal forces are derivable from a potential-energy function V, as in Eq. (4.32), but the external forces are not, the energy theorem will be

$$\frac{d}{dt}(T + V) = \sum_{k=1}^{N} \boldsymbol{F}_k^e \cdot \boldsymbol{v}_k. \tag{4.40}$$

Suppose the internal force acting on any particle k can be regarded as the sum of forces due to each of the other particles, where the force $\boldsymbol{F}_{l\to k}^i$ on k due to

l depends only on the relative position $(\boldsymbol{r}_k - \boldsymbol{r}_l)$ of particle k with respect to particle l:

$$\boldsymbol{F}_k^i = \sum_{l \neq k} \boldsymbol{F}_{l \to k}^i(\boldsymbol{r}_k - \boldsymbol{r}_l). \tag{4.41}$$

It may be that the vector function $\boldsymbol{F}_{l \to k}^i(\boldsymbol{r}_k - \boldsymbol{r}_l)$ is such that we can define a potential-energy function

$$V_{kl}(\boldsymbol{r}_{kl}) = -\int_{\boldsymbol{r}_s}^{\boldsymbol{r}_{kl}} \boldsymbol{F}_{l \to k}^i(\boldsymbol{r}_{kl}) \cdot d\boldsymbol{r}_{kl}, \tag{4.42}$$

$$\boldsymbol{r}_{kl} = \boldsymbol{r}_k - \boldsymbol{r}_l. \tag{4.43}$$

This will be true if $\boldsymbol{F}_{l \to k}^i$ is a conservative force in the sense of Chapter 3, that is, if

$$\text{curl } \boldsymbol{F}_{l \to k}^i = \boldsymbol{0}, \tag{4.44}$$

where the derivatives are with respect to x_{kl}, y_{kl}, z_{kl}. The gravitational and electrostatic forces between pairs of particles are examples of conservative forces. If $\boldsymbol{F}_{l \to k}^i$ is conservative, so that V_{kl} can be defined, then*

$$\begin{aligned} \boldsymbol{F}_{l \to k}^i &= -\hat{\boldsymbol{x}} \frac{\partial V_{kl}}{\partial x_{kl}} - \hat{\boldsymbol{y}} \frac{\partial V_{kl}}{\partial y_{kl}} - \hat{\boldsymbol{z}} \frac{\partial V_{kl}}{\partial z_{kl}} \\ &= -\hat{\boldsymbol{x}} \frac{\partial V_{kl}}{\partial x_k} - \hat{\boldsymbol{y}} \frac{\partial V_{kl}}{\partial y_k} - \hat{\boldsymbol{z}} \frac{\partial V_{kl}}{\partial z_k}. \end{aligned} \tag{4.45}$$

If Newton's third law (weak form) holds, then

$$\begin{aligned} \boldsymbol{F}_{k \to l}^i = -\boldsymbol{F}_{l \to k}^i &= \hat{\boldsymbol{x}} \frac{\partial V_{kl}}{\partial x_{kl}} + \hat{\boldsymbol{y}} \frac{\partial V_{kl}}{\partial y_{kl}} + \hat{\boldsymbol{z}} \frac{\partial V_{kl}}{\partial z_{kl}} \\ &= -\hat{\boldsymbol{x}} \frac{\partial V_{kl}}{\partial x_l} - \hat{\boldsymbol{y}} \frac{\partial V_{kl}}{\partial y_l} - \hat{\boldsymbol{z}} \frac{\partial V_{kl}}{\partial z_l}. \end{aligned} \tag{4.46}$$

Thus V_{kl} will also serve as the potential-energy function for the force $\boldsymbol{F}_{k \to l}^i$. We can now define the total internal potential energy V^i for the system of particles as the sum of V_{kl} over all the pairs of particles:

$$V^i(\boldsymbol{r}_1, \ldots, \boldsymbol{r}_N) = \sum_{k=1}^{N} \sum_{l=1}^{k-1} V_{kl}(\boldsymbol{r}_k - \boldsymbol{r}_l). \tag{4.47}$$

It follows from Eqs. (4.41), (4.45), and (4.46), that the internal forces are given by

$$\boldsymbol{F}_k^i = -\hat{\boldsymbol{x}} \frac{\partial V^i}{\partial x_k} - \hat{\boldsymbol{y}} \frac{\partial V^i}{\partial y_k} - \hat{\boldsymbol{z}} \frac{\partial V^i}{\partial z_k}, \qquad k = 1, \ldots, N. \tag{4.48}$$

---

*Note that $V(\boldsymbol{r}_{kl}) = V(x_{kl}, y_{kl}, z_{kl}) = V(x_k - x_l, y_k - y_l, z_k - z_l)$, so that $\partial V / \partial x_k = \partial V / \partial x_{kl} = -\partial V / \partial x_l$, etc.

In particular, if the forces between pairs of particles are central forces, the potential energy $V_{kl}(r_{kl})$ for each pair of particles depends only on the distance r_{kl} between them, and is given by Eq. (3.200); the internal forces of the system are then conservative, and Eq. (4.48) holds. The energy theorem (4.40) will be valid for such a system of particles. If the external forces are also conservative, their potential energy can be added to V^i, and the total energy is constant.

If there is internal friction, as is often the case, the internal frictional forces depend on the relative velocities of the particles, and the conservation law of potential plus kinetic energy no longer holds.

4.4 CRITIQUE OF THE CONSERVATION LAWS

We may divide the phenomena to which the laws of mechanics have been applied into three major classes. The motions of celestial bodies—stars, satellites, planets—are described with extremely great precision by the laws of classical mechanics. It was in this field that the theory had many of its important early successes. The motions of the bodies in the solar system can be predicted with great accuracy for periods of thousands of years. The theory of relativity predicts a few slight deviations from the classically predicted motion, but these are too small to be observed except in the case of the orbit of Mercury, where relativity and observation agree in showing an otherwise unaccounted for slow precession of the axis of the elliptical orbit around the sun at an angular velocity of about 0.01 degree per century.

The motion of terrestrial bodies of macroscopic and microscopic size constitutes the second major division of phenomena. Motions in this class are properly described by Newtonian mechanics, without any significant corrections, but the laws of force are usually very complicated, and often not precisely known, so that the beautifully precise calculations of celestial mechanics cannot be duplicated here.

The third class of phenomena is the motion of "atomic" particles: molecules, atoms, electrons, nuclei, protons, neutrons, etc. Early attempts to describe the motions of such particles were based on classical mechanics, and many phenomena in this class can be understood and predicted on this basis. However, the finer details of the behavior of atomic particles can only be properly described in terms of quantum mechanics and, for high velocities, relativistic quantum mechanics must be introduced. We might add a fourth class of phenomena, having to do with the intrinsic structure of the elementary particles themselves (protons, neutrons, electrons, etc.). Even quantum mechanics fails to describe such phenomena correctly, and physics is now struggling to produce a new theory which will describe this class of phenomena.

The conservation law for linear momentum holds for systems of celestial bodies as well as for bodies of macroscopic and microscopic size. The gravitational and mechanical forces acting between such bodies satisfy Newton's third law, at least to a high degree of precision. Linear momentum is also conserved in most interactions of particles of atomic size, except when high velocities or rapid accelerations are involved. The electrostatic forces between electric charges at rest satisfy Newton's third law, but when the charges are in motion, their electric fields propogate with the velocity of light, so that if two charges are in rapid relative motion, the forces between them may not at any instant be exactly equal and opposite. When electric charges accelerate, they may emit electromagnetic radiation and lose momentum in so doing. It turns out that the law of conservation of momentum can be preserved also in such cases, but only by associating momentum with the electromagnetic field as well as with moving particles. Such a redefinition of momentum goes beyond the original limits of Newtonian mechanics.

Celestial bodies and bodies of macroscopic or microscopic size are obviously not really particles, since they have a structure which for many purposes is not adequately represented by merely giving to the body three position coordinates x, y, z. Nevertheless, the motion of such bodies, in problems where their structure can be neglected, is correctly represented by the law of motion of a single particle,

$$m\ddot{\boldsymbol{r}} = \boldsymbol{F}. \tag{4.49}$$

This is often justified by regarding the macroscopic body as a system of smaller particles satisfying Newton's third law. For such a system, the linear momentum theorem holds, and can be written in the form of Eq. (4.18), which has the same form as Eq. (4.49). This is a very convenient way of justifying the application of Eq. (4.49) to bodies of macroscopic or astronomical size, provided our conscience is not troubled by the fact that according to modern ideas it does not make sense. If the particles of which the larger body is composed are taken as atoms and molecules, then in the first place Newton's third law does not invariably hold for such particles, and in the second place we should apply quantum mechanics, not classical mechanics, to their motion. The momentum theorem (4.18) can be derived for bodies made up of atoms by using the laws of electrodynamics and quantum mechanics, but this lies outside the scope of Newtonian mechanics. Hence, for the present, we must take the law of motion (4.49), as applied to macroscopic and astronomical bodies, as a fundamental postulate in itself, whose justification is based on experimental grounds or on the results of deeper theories. The theorems proved in Section 4.1 show that this postulate gives a consistent theory of mechanics in the sense that if, from bodies satisfying this postulate, we construct a composite body, the latter body will also satisfy the postulate.

The law of conservation of angular momentum, as formulated in Section 4.2 for a system of particles, holds for systems of celestial bodies (regarded as particles) and for systems of bodies of macroscopic size whenever effects due to rotation of the individual bodies can be neglected. When rotations of the individual bodies enter into the motion, then a conservation law for angular momentum still holds, provided we include the angular momentum associated with such rotations; the bodies are then no longer regarded as particles of the simple type considered in the preceding sections whose motions are completely described simply by specifying the function $\boldsymbol{r}(t)$ for each particle. The total angular momentum of the solar system is very nearly constant, even if the sun, planets, and satellites are regarded as simple particles whose rotations can be neglected. Tidal forces, however, convert some rotational angular momentum into orbital angular momentum of the planets and satellites, and so rotational angular momentum must be included if the law of conservation of angular momentum is to hold precisely. Some change in angular momentum occurs due to friction with interplanetary dust and rocks, but the effect is too small to be observed, and could in any case be included by adding the angular momentum of the interplanetary matter to the total.

The law of conservation of total angular momentum, including rotation, of astronomical and terrestrial bodies can be justified by regarding each body as a system of smaller particles whose mutual forces satisfy Newton's third law (strong form). The argument of Section 4.2 then gives the law of conservation of total angular momentum, the rotational angular momentum of a body appearing as ordinary orbital angular momentum ($\boldsymbol{r} \times \boldsymbol{p}$) of the particles of which it is composed. This argument is subject to the same criticism as applied above to the case of linear momentum. If the "particles" of which a body is composed are atoms and molecules, then Newton's third law does not always hold, particularly in its strong form; moreover, the laws of quantum mechanics apply to such particles; and in addition atoms and molecules also possess rotational angular momentum which must be taken into account. Even the elementary particles—electrons, protons, neutrons, etc.—possess an intrinsic angular momentum which is not associated with their orbital motion. This angular momentum is called *spin* angular momentum from its analogy with the intrinsic angular momentum of rotation of a macroscopic body, and must be included if the total is to satisfy a conservation law. Thus we never arrive at the ideal simple particle of Newtonian mechanics, described by its position $\boldsymbol{r}(t)$ alone. We are left with the choice of accepting the conservation law of angular momentum as a basic postulate, or appealing for its justification to theories which go beyond classical mechanics.

The gravitational forces acting between astronomical bodies are conservative, so that the principle of conservation of mechanical energy holds very accurately in astronomy. In principle, there is a small loss of mechanical energy in the solar system due to friction with interplanetary dust and rocks, but the effect is too small to produce any observable effects on planetary motion, even with the high precision with which astronomical events are predicted and observed. There is also a very gradual but measurable loss of rotational energy of planets and satellites due to tidal friction. For terrestrial bodies of macroscopic or microscopic size, friction usually plays an important part, and only in certain special cases where friction may be neglected can the principle of conservation of energy in the form (4.39) or even (4.40) be applied. However, it was discovered by Joule that we can associate energy with heat in such a way that the law of conservation of energy of a system of bodies still applies to the total kinetic plus potential plus heat energy. If we regard a body as composed of atoms and molecules, its heat energy turns out to be kinetic and potential energy of random motion of its atoms and molecules. The electromagnetic forces on moving charged particles are not conservative, and an electromagnetic energy must be associated with the electromagnetic field in order to preserve the conservation law of energy. Such extensions of the concept of energy to include heat and electromagnetic energy are, of course, outside the domain of mechanics. When the definition of energy is suitably extended to include not only kinetic energy, but energy associated with the electromagnetic fields and any other force fields which may act, then a law of conservation of energy holds quite generally, in classical, relativistic, and quantum physics.

The conservation laws of energy, momentum, and angular momentum are the cornerstones of present-day physics, being generally valid in all physical theories. It seems at present an idle exercise to attempt to prove them for material bodies within the framework of classical mechanics by appealing to an outmoded picture of matter as made up of simple Newtonian particles exerting central forces upon one another. The conservation laws are in a sense not laws at all, but postulates which we insist must hold in any physical theory. If, for example, for moving charged particles, we find that the total energy, defined as $(T + V)$, is not constant, we do not abandon the law, but change its meaning by redefining energy to include electromagnetic energy in such a way as to preserve the law. We prefer always to look for quantities which are conserved, and agree to apply the names "total energy," "total momentum," "total angular momentum" only to such quantities. The conservation of these quantities is then not a physical fact, but a consequence of our determination to define them in this way. It is, of course, a statement of physical fact, which may or may not be true, to assert that such definitions of energy, momentum, and angular momentum can always be found. This assertion, has so far been true; a deeper justification will be suggested at the end of Section 9.6.

4.5 ROCKETS, CONVEYOR BELTS, AND PLANETS

There are many problems that can be solved by appropriate applications of the conservation laws of linear momentum, angular momentum, and energy. In solving such problems, it is necessary to decide which conservation laws are appropriate. The conservation laws of linear and angular momentum or, rather, the theorems (4.7) and (4.28) of which they are corollaries, are always applicable to any physical system provided all external forces and torques are taken into account, and application of one or the other is appropriate whenever the external forces or torques are known. The law of conservation of kinetic plus potential energy is applicable only when there is no conversion of mechanical energy into other forms of energy. We cannot use the law of conservation of energy when there is friction, for example, unless there is a way to determine the amount of heat energy produced.

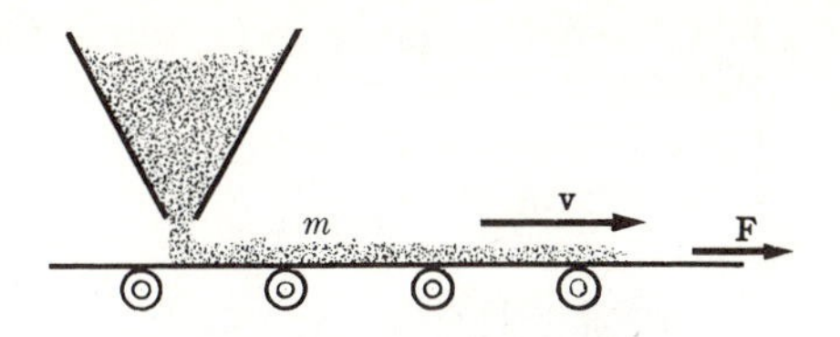

Fig. 4.2 A conveyor belt.

The conservation laws of energy, momentum, and angular momentum refer always to a definite fixed system of particles. In applying the conservation laws, care must be taken to decide just how much is included in the system to which they are to be applied, and to include all the energy and momentum of this system in writing down the equations. One may choose the system arbitrarily, including and excluding whatever particles may be convenient, but if any forces act from outside the system on particles in the system, these must be taken into account.

A typical problem in which the linear momentum theorem is applicable is the conveyor belt problem. Material is dropped continuously from a hopper onto a moving belt, and it is required to find the force F required to keep the belt moving at constant velocity v (Fig. 4.2). Let the rate at which mass is dropped on the belt be dm/dt. If m is the mass of material on the belt, and M is the mass of the belt (which really does not figure in the problem), the total momentum of the system, belt plus material on the belt and in the hopper, is

$$P = (m+M)v. \tag{4.50}$$

We assume that the hopper is at rest; otherwise the momentum of the hopper and its contents must be included in Eq. (4.50). The linear momentum theorem requires that

$$F = \frac{dP}{dt} = v\frac{dm}{dt}. \tag{4.51}$$

This gives the force applied to the belt. The power supplied by the force is

$$Fv = v^2\frac{dm}{dt} = \frac{d}{dt}(mv^2) = \frac{d}{dt}[(m+M)v^2]. \tag{4.52}$$

This is twice the rate at which the kinetic energy is increasing, so that the conservation theorem of mechanical energy (4.40) does not apply here. Where is the excess half of the power going?

The equation of motion of a rocket can be obtained from the law of conservation of momentum. Let the mass of the rocket at any given instant be M, and let its speed be $\boldsymbol{v}$ relative to some fixed coordinate system. If material is shot out of the rocket motor with an exhaust velocity $\boldsymbol{u}$ relative to the rocket, the velocity of the exhaust relative to the fixed coordinate system is $\boldsymbol{v}+\boldsymbol{u}$. If an external force

F also acts on the rocket, then the linear momentum theorem (4.7) reads in this case:

$$\frac{d}{dt}(M\boldsymbol{v})-\frac{dM}{dt}(\boldsymbol{v}+\boldsymbol{u}) = \boldsymbol{F}. \tag{4.53}$$

The first term is the time rate of change of momentum of the rocket. The second term represents the rate at which momentum is appearing in the rocket exhaust, where $-(dM/dt)$ is the rate at which matter is being exhausted. The conservation law applies to a definite fixed system of particles. Since fuel in the rocket appears later in the exhaust, we must include the exhaust gases in the system when we apply the conservation law of momentum. The equation can be rewritten:

$$M\frac{d\boldsymbol{v}}{dt} = \boldsymbol{u}\frac{dM}{dt}+\boldsymbol{F}. \tag{4.54}$$

The first term on the right is called the *thrust* of the rocket motor. Since dM/dt is negative, the thrust is opposite in direction to the exhaust velocity. The force $\boldsymbol{F}$ may represent air resistance, or a gravitational force, or any other force acting on the rocket.

Let us first consider the case of a rocket motor fastened to a fixed support, as in a "bench test." The left side of Eq. (4.54) is then zero, and the force F exerted by the restraining support is

$$F = -\boldsymbol{u}\frac{dM}{dt}.$$

To measure the thrust of the rocket engine, we measure the force F required to hold it stationary.

Let us now solve Eq. (4.54) for the special case where there is no external force:

$$M\frac{d\boldsymbol{v}}{dt} = \boldsymbol{u}\frac{dM}{dt}. \tag{4.55}$$

We multiply by dt/M and integrate, assuming that $\boldsymbol{u}$ is constant:

$$\boldsymbol{v}-\boldsymbol{v}_0 = -\boldsymbol{u}\ln\frac{M_0}{M}. \tag{4.56}$$

The change of speed in any interval of time depends only on the exhaust velocity and on the fraction of mass exhausted during that time interval. This result is independent of any assumption as to the rate at which mass is exhausted.

Problems in which the law of conservation of angular momentum is useful turn up frequently in astronomy. The angular momentum of the galaxy of stars, or of the solar system, remains constant during the course of its development provided no material is ejected from the system. The effect of lunar tides is gradually to slow down the rotation of the earth. As the angular momentum of the rotating earth decreases, the angular momentum of the moon must increase. The magnitude of the (orbital) angular momentum of the moon is

$$L = mr^2\omega, \tag{4.57}$$

where m is the mass, ω is the angular velocity, and r is the radius of the orbit of the moon. We can equate the mass times the centripetal acceleration to the gravitational force, to obtain the relation

$$mr\omega^2 = \frac{GMm}{r^2}, \tag{4.58}$$

where M is the mass of the earth. Solving this equation for ω and substituting in Eq. (4.57), we obtain

$$L = (GMm^2r)^{1/2}. \tag{4.59}$$

Therefore, as the moon's angular momentum increases, it moves farther away from the earth. (In attempting to determine the rate of recession of the moon by equating the change of L to the change of the earth's rotational angular momentum, it would be necessary to determine how much of the slowing down of the earth's rotation by tidal friction is due to the moon and how much to the sun. The angular momentum of the moon plus the rotational angular momentum of the earth is not constant because of the tidal friction due to the sun. The total angular momentum of the earth-moon system about the sun is very nearly constant except for the very small effect of tides raised on the sun by the earth.)

4.6 COLLISION PROBLEMS

Many questions concerning collisions of particles can be answered by applying the conservation laws. Since the conservation laws are valid also in quantum mechanics,* results obtained with their use are valid for particles of atomic and subatomic size, as well as for macroscopic particles. In most collision problems, the colliding particles are moving at constant velocity, free of any force, for some time before and after the collision, while during the collision they are under the action of the forces which they exert on one another. If the mutual forces during the collision satisfy Newton's third law, then the total linear momentum of the particles is the same before and after the collision. If Newton's third law holds in the strong form, the total angular momentum is conserved also. If the forces are conservative, kinetic energy is conserved (since the potential energy before and after the collision is the same). In any case, the conservation laws are always valid if we take into account all the energy, momentum, and angular momentum, including that associated with any radiation which may be emitted and including any energy which is converted from kinetic energy into other forms, or vice versa.

*P. A. M. Dirac, *The Principles of Quantum Mechanics*, 4th ed. Oxford: Oxford University Press, 1958. (Page 115.)

We consider first a collision between two particles, 1 and 2, in which the total kinetic energy and linear momentum are known to be conserved. Such a collision is said to be *elastic*. If we designate by subscripts 1 and 2 the two particles, and by subscripts I and F the values of kinetic energy and momentum before and after the collision respectively, the conservation laws require

$$\boldsymbol{p}_{1I}+\boldsymbol{p}_{2I} = \boldsymbol{p}_{1F}+\boldsymbol{p}_{2F}, \tag{4.60}$$

$$T_{1I}+T_{2I} = T_{1F}+T_{2F}. \tag{4.61}$$

Equation (4.61) can be rewritten in terms of the momenta and masses of the particles:

$$\frac{p_{1I}^2}{2m_1}+\frac{p_{2I}^2}{2m_2} = \frac{p_{1F}^2}{2m_1}+\frac{p_{2F}^2}{2m_2}. \tag{4.62}$$

To specify any momentum vector $\boldsymbol{p}$, we must specify three quantities, which may be either its three components along any set of axes, or its magnitude and direction (the latter specified perhaps by spherical angles θ, φ). Thus Eqs. (4.60) and (4.62) represent four equations involving the ratio of the two masses and twelve quantities required to specify the momenta involved. If nine of these quantities are given, the equations can be solved for the remaining four. In a typical case, we might be given the masses and initial momenta of the two particles, and the final direction of motion of one of the particles, say particle 1. We could then find the final momentum $\boldsymbol{p}_{2F}$ of particle 2, and the magnitude of the final momentum $\boldsymbol{p}_{1F}$ (or equivalently, the energy) of particle 1. In many important cases, the mass of one of the particles is unknown, and can be computed from Eqs. (4.60) and (4.62) if enough is known about the momenta and energies before and after the collision. Note that the initial conditions alone are not enough to determine the outcome of the collision from Eqs. (4.60) and (4.62); we must know something about the motion after the collision. The initial conditions alone would determine the outcome if we could solve the equations of motion of the system.

Consider a collision of a particle of mass m_1, momentum $\boldsymbol{p}_{1I}$, with a particle of mass m_2 at rest. This is a common case. (There is actually no loss of generality in this problem, since, as we pointed out in Section 1.4 and will show in Section 7.1, if m_2 is initially moving with a uniform velocity $\boldsymbol{v}_{2I}$, Newton's laws are equally applicable in a coordinate system moving with uniform velocity $\boldsymbol{v}_{2I}$, in which m_2 is initially at rest.) Let m_1 be "scattered" through an angle ϑ_1; that is, let ϑ_1 be the angle between its final and its initial direction of motion (Fig. 4.3). The momentum $\boldsymbol{p}_{2F}$ must lie in the same plane as $\boldsymbol{p}_{1I}$ and $\boldsymbol{p}_{1F}$ since there is no component of momentum perpendicular to this plane before the collision, and there must be none after. Let $\boldsymbol{p}_{2F}$ make an angle ϑ_2 with the direction of $\boldsymbol{p}_{1I}$. We write out Eq. (4.60) in components along and perpendicular to $\boldsymbol{p}_{1I}$:

$$p_{1I} = p_{1F}\cos\vartheta_1+p_{2F}\cos\vartheta_2, \tag{4.63}$$

$$0 = p_{1F}\sin\vartheta_1-p_{2F}\sin\vartheta_2. \tag{4.64}$$

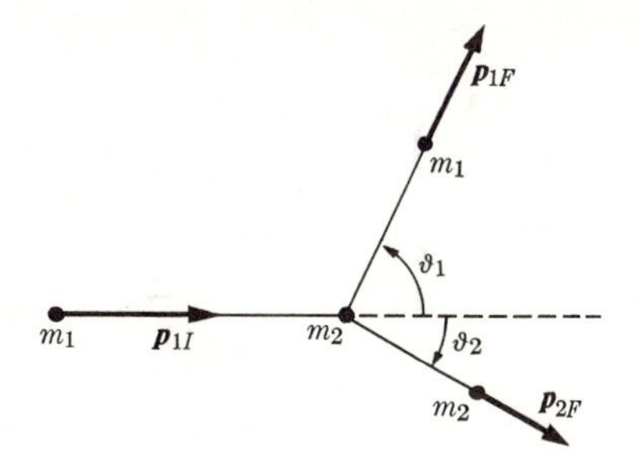

Fig. 4.3 Collision of particle m_1 with particle m_2 at rest.

Equation (4.62) becomes, in the present case,

$$\frac{p_{1I}^2 - p_{1F}^2}{m_1} = \frac{p_{2F}^2}{m_2}. \tag{4.65}$$

If two of the quantities

$$(p_{1F}/p_{1I},\ p_{2F}/p_{1I},\ \vartheta_1,\ \vartheta_2,\ m_1/m_2)$$

are known, the remaining three can be found. If the masses, the initial momentum p_{1I}, and the angle ϑ_1 are known, for example, we can solve for p_{1F}, p_{2F}, ϑ_2 as follows. Transposing the first term on the right to the left side in Eqs. (4.63) and (4.64), squaring, and adding, we eliminate ϑ_2:

$$p_{1I}^2 + p_{1F}^2 - 2p_{1I}p_{1F}\cos\vartheta_1 = p_{2F}^2. \tag{4.66}$$

After substituting this in Eq. (4.65), we can solve for p_{1F}:

$$\frac{p_{1F}}{p_{1I}} = \frac{m_1}{m_1+m_2}\cos\vartheta_1 \pm \left[\left(\frac{m_1}{m_1+m_2}\right)^2 \cos^2\vartheta_1 + \frac{m_2-m_1}{m_1+m_2}\right]^{1/2}, \tag{4.67}$$

and p_{2F} can now be found from Eq. (4.66), and ϑ_2 from Eq. (4.63). If $m_1 > m_2$, the quantity under the radical is zero for $\vartheta_1 = \vartheta_m$, where ϑ_m is given by

$$\cos^2\vartheta_m = 1 - \frac{m_2^2}{m_1^2}, \qquad 0 \le \vartheta_m \le \frac{\pi}{2}. \tag{4.68}$$

If $\vartheta_1 > \vartheta_m$ (and $\vartheta_1 \le \pi$), then p_{1F}/p_{1I} is either imaginary or negative, neither of which is allowable physically, so that ϑ_m represents the maximum angle through which m_1 can be scattered. If $m_1 \gg m_2$, this angle is very small, as we know from experience. For $\vartheta_1 < \vartheta_m$, there are two values of p_{1F}/p_{1I}, the larger corresponding to a glancing collision, the smaller to a more nearly head-on collision; ϑ_2 will be different for these two cases. The case $\vartheta_1 = 0$ may represent either no collision at all ($p_{1F} = p_{1I}$) or a head-on collision. In the latter case,

$$\frac{p_{1F}}{p_{1I}} = \frac{m_1 - m_2}{m_1 + m_2}, \qquad \vartheta_2 = 0, \qquad \frac{p_{2F}}{p_{1I}} = \frac{2m_2}{m_1 + m_2}. \tag{4.69}$$

If $m_1 = m_2$, Eqs. (4.67), (4.66), and (4.64) reduce to

$$\frac{p_{1F}}{p_{1I}} = \cos \vartheta_1, \qquad \frac{p_{2F}}{p_{1I}} = \sin \vartheta_1, \qquad \vartheta_2 = \left(\frac{\pi}{2} - \vartheta_1\right). \tag{4.70}$$

ϑ_1 now varies from $\vartheta_1 = 0$ for no collision to $\vartheta_1 = \pi/2$ for a head-on collision in which the entire momentum is transferred to particle 2. (Actually, ϑ_1 is undefined if $p_{1F} = 0$, but $\vartheta_1 \to \pi/2$ and $p_{1F} \to 0$ as the collision approaches a head-on collision.) If $m_1 < m_2$, all values of ϑ_1 from 0 to π are possible, and give a positive value for p_{1F}/p_{1I} if the plus sign is chosen in Eq. (4.67). The minus sign cannot be chosen, since it leads to a negative value for p_{1F}/p_{1I}. If $\vartheta_1 = 0$, then $p_{1F} = p_{1I}$; this is the case when there is no collision. The case $\vartheta_1 = \pi$ corresponds to a head-on collision, for which

$$\begin{aligned} \frac{p_{1F}}{p_{1I}} &= \frac{m_2 - m_1}{m_1 + m_2}, \qquad \vartheta_1 = \pi, \\ \frac{p_{2F}}{p_{1I}} &= \frac{2m_2}{m_1 + m_2}, \qquad \vartheta_2 = 0. \end{aligned} \tag{4.71}$$

If m_1 is unknown, but either p_{1I} or T_{1I} can be measured or calculated, observation of the final momentum of particle 2 (whose mass is assumed known) is sufficient to determine m_1. As an example, if $T_{1I} = p_{1I}^2/2m_1$ is known, and T_{2F} is measured for a head-on collision, m_1 is given by Eq. (4.69) or (4.71):

$$\frac{m_1}{m_2} = \frac{2T_{1I}}{T_{2F}} - 1 \pm \left[\left(\frac{2T_{1I}}{T_{2F}} - 1\right)^2 - 1\right]^{1/2}. \tag{4.72}$$

We thus determine m_1 to within one of two possible values. If results for a collision with another particle of different mass m_2, or for a different scattering angle, are known, m_1 is determined uniquely. Essentially this method was used by Chadwick to establish the existence of the neutron.* Unknown neutral particles created in a nuclear reaction were allowed to impinge on matter containing various nuclei of known masses. The energies of two kinds of nuclei of different masses m_2, m_2' projected forward by head-on collisions were measured. By writing Eq. (4.72) for both cases, the unknown energy T_{1I} could be eliminated, and the mass m_1 was found to be practically equal to that of the proton.

We have seen that if we know the initial momenta of two colliding particles of known masses, and the angle of scattering ϑ_1 (or ϑ_2), all other quantities involved in the collision can be calculated from the conservation laws. To predict the angles

*J. Chadwick, *Nature*, **129**, 312 (1932).

of scattering, we must know not only the initial momenta and the initial trajectories, but also the law of force between the particles. An example is the collision of two particles acted on by a central inverse square law of force, to be treated in Section 4.8. Such predictions can be made for collisions of macroscopic or astronomical bodies under suitable assumptions as to the law of force. For atomic particles, which obey quantum mechanics, this cannot be done, although we can predict the probabilities of observing various angles ϑ_1 (or ϑ_2) for given initial conditions; that is, we can predict cross sections. In all cases where energy is conserved, the relationships between energies, momenta, and angles of scattering developed above are valid except at particle velocities comparable with the velocity of light. In the latter case, Eqs. (4.60), (4.61), (4.63), and (4.64) are still valid, but the relativistic relationships between mass, momentum, and energy must be used, instead of Eq. (4.62). We will derive in Chapter 14 the relativistic expressions for kinetic energy and momentum:

$$T = mc^2\left(\frac{1}{\sqrt{1-(v^2/c^2)}}-1\right), \tag{4.73}$$

$$p = \frac{mv}{\sqrt{1-(v^2/c^2)}}, \tag{4.74}$$

where c is the speed of light, v is the velocity, and m is the rest mass of the particle, that is, the mass when the particle is at rest. These reduce to the classical expressions (2.5) and (3.127) when $v \ll c$. We easily deduce the following useful relation between momentum and energy:

$$\frac{p^2}{2m} = T + \frac{T^2}{2mc^2}. \tag{4.75}$$

Unless v is nearly equal to c, the second term on the right in Eq. (4.75) is much smaller than the first, and this equation reduces to the classical one. With the help of Eq. (4.75), the conservation laws can be applied to collisions involving velocities near the speed of light.

Atoms, molecules, and nuclei possess internal potential and kinetic energy associated with the motion of their parts, and may absorb or release energy on collision. Such inelastic collisions between atomic particles are said to be of the *first kind*, or *endoergic*, if kinetic energy of translational motion is absorbed, and of the *second kind*, or *exoergic*, if kinetic energy is released in the process. It may also happen that in an atomic or nuclear collision, the final particles after the collision are not the same as the initial particles before collision. For example, a proton may collide with a nucleus and be absorbed while a neutron is released and

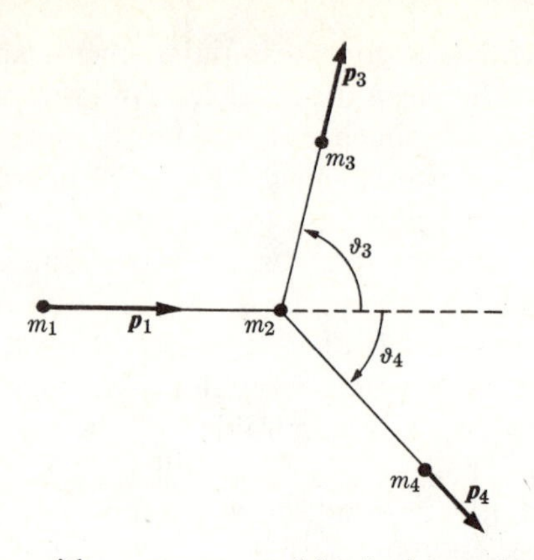

Fig. 4.4 Collision of m_1 with m_2 at rest, resulting in the production of m_3 and m_4.

flies away. There are a great many possible types of such processes. Two particles may collide and stick together to form a single particle or, conversely, a single particle may suddenly break up into two particles which fly apart. Two particles may collide and form two other particles which fly apart. Or three or more particles may be formed in the process and fly apart after the collision. In all these cases, the law of conservation of momentum holds, and the law of conservation of energy also if we take into account the internal energy of the atoms and molecules. We consider here a case in which a particle of mass m_1 collides with a particle of mass m_2 at rest (Fig. 4.4). Particles of masses m_3 and m_4 leave the scene of the collision at angles ϑ_3 and ϑ_4 with respect to the original direction of motion of m_1. Let kinetic energy Q be produced in the process ($Q > 0$ for an exoergic collision; $Q = 0$ for an elastic collision; $Q < 0$ for an endoergic collision). Then, applying the conservation laws of energy and momentum, we write

$$p_1 = p_3 \cos \vartheta_3 + p_4 \cos \vartheta_4, \tag{4.76}$$

$$0 = p_3 \sin \vartheta_3 - p_4 \sin \vartheta_4, \tag{4.77}$$

$$T_1 + Q = T_3 + T_4. \tag{4.78}$$

Since kinetic energy can be expressed in terms of momentum, if the masses are known, we may find any three of the quantities p_1, p_3, p_4, ϑ_3, ϑ_4, Q in terms of the other three. In many cases p_1 is known, p_3 and ϑ_3 are measured, and it is desired to calculate Q. By eliminating ϑ_4 from Eqs. (4.76) and (4.77), as in the previous example, we obtain

$$p_4^2 = p_1^2 + p_3^2 - 2p_1 p_3 \cos \vartheta_3. \tag{4.79}$$

This may now be substituted in Eq. (4.78) to give Q in terms of known quantities:

$$Q = T_3+T_4-T_1 = \frac{p_3^2}{2m_3}+\frac{p_1^2+p_3^2-2p_1p_3\cos\vartheta_3}{2m_4}-\frac{p_1^2}{2m_1}, \tag{4.80}$$

or

$$Q = T_3\left(1+\frac{m_3}{m_4}\right)-T_1\left(1-\frac{m_1}{m_4}\right)-2\left(\frac{m_1m_3T_1T_3}{m_4^2}\right)^{1/2}\cos\vartheta_3. \tag{4.81}$$

Every step up to the substitution for T_1, T_3, and T_4 is valid also for particles moving at velocities of the order of the velocity of light. At high velocities, the relativistic relation (4.75) between T and p should be used in the last step. Equation (4.81) is useful in obtaining Q for a nuclear reaction in which an incident particle m_1 of known energy collides with a nucleus m_2, with the result that a particle m_3 is emitted whose energy and direction of motion can be observed. Equation (4.81) allows us to determine Q from these known quantities, taking into account the effect of the slight recoil of the residual nucleus m_4, which is usually difficult to observe directly.

Collisions of inert macroscopic bodies are always inelastic and endoergic, kinetic energy being converted to heat by frictional forces during the impact. Kinetic energy of translation may also be converted into kinetic energy of rotation, and conversely. (Exchanges of rotational energy are included in Q in the previous analysis.) Such collisions range from the nearly elastic collisions of hard steel balls, to which the above analysis of elastic collisions applies when rotation is not involved, to completely inelastic collisions in which the two bodies stick together after the collision. Let us consider a completely inelastic collision in which a bullet of mass m_1, velocity $\boldsymbol{v}_1$ strikes and sticks in an object of mass m_2 at rest. Let the velocity of the two after the collision be $\boldsymbol{v}_2$. Evidently the conservation of momentum implies that $\boldsymbol{v}_2$ be in the same direction as $\boldsymbol{v}_1$, and we have:

$$m_1\boldsymbol{v}_1 = (m_1+m_2)\boldsymbol{v}_2. \tag{4.82}$$

The velocity after the collision is

$$\boldsymbol{v}_2 = \frac{m_1}{m_1+m_2}\boldsymbol{v}_1. \tag{4.83}$$

Energy is not conserved in such a collision. The amount of energy converted into heat is

$$Q = \tfrac{1}{2}m_1v_1^2-\tfrac{1}{2}(m_1+m_2)v_2^2 = \tfrac{1}{2}m_1v_1^2\left(\frac{m_2}{m_1+m_2}\right). \tag{4.84}$$

In a head-on collision of two bodies in which rotation is not involved, it was found experimentally by Isaac Newton that the ratio of relative velocity after impact to relative velocity before impact is roughly constant for any two given bodies. Let bodies m_1, m_2, traveling with initial velocities v_{1I}, v_{2I} along the x-axis, collide and rebound along the same axis with velocities v_{1F}, v_{2F}. Then the experimental result is expressed by the equation*

$$v_{2F}-v_{1F} = e(v_{1I}-v_{2I}), \tag{4.85}$$

where the constant e is called the *coefficient of restitution*, and has a value between 0 and 1. If $e = 1$, the collision is perfectly elastic; if $e = 0$, it is completely inelastic. Conservation of momentum yields, in any case,

$$m_1v_{1I}+m_2v_{2I} = m_1v_{1F}+m_2v_{2F}. \tag{4.86}$$

Equations (4.85) and (4.86) enable us to find the final velocities v_{1F} and v_{2F} for a head-on collision when the initial velocities are known.

4.7 THE TWO-BODY PROBLEM

We consider in this section the motion of a system of two particles acted on by internal forces satisfying Newton's third law (weak form), and by no external forces, or by external forces satisfying a rather specialized condition to be introduced later. We shall find that this problem can be separated into two single-particle problems. The motion of the center of mass is governed by an equation (4.18) of the same form as that for a single particle. In addition, we shall find that the motion of either particle, with respect to the other as origin, is the same as the motion with respect to a fixed origin, of a single particle of suitably chosen mass acted on by the same internal force. This result will allow application of the results of Section 3.14 to cases where the motion of the attracting center cannot be neglected.

Let the two particles have masses m_1 and m_2, and let them be acted on by external forces $\boldsymbol{F}_1^e$, $\boldsymbol{F}_2^e$, and internal forces $\boldsymbol{F}_1^i$, $\boldsymbol{F}_2^i$ exerted by each particle on the other, and satisfying Newton's third law:

$$\boldsymbol{F}_1^i = -\boldsymbol{F}_2^i. \tag{4.87}$$

The equations of motion for the system are then

$$m_1\ddot{\boldsymbol{r}}_1 = \boldsymbol{F}_1^i+\boldsymbol{F}_1^e, \tag{4.88}$$

$$m_2\ddot{\boldsymbol{r}}_2 = \boldsymbol{F}_2^i+\boldsymbol{F}_2^e. \tag{4.89}$$

We now introduce a change of coordinates:

$$\boldsymbol{R} = \frac{m_1\boldsymbol{r}_1+m_2\boldsymbol{r}_2}{m_1+m_2}, \tag{4.90}$$

$$\boldsymbol{r} = \boldsymbol{r}_1-\boldsymbol{r}_2. \tag{4.91}$$

*More recent experiments show that e is not really constant, but depends on the initial velocities, on the medium in which the collision takes place, and on the past history of the bodies. For a more complete discussion with references, see G. Barnes, "Study of Collisions," *Am. J. Phys.* **26**, 5 (January, 1958).

The inverse transformation is

$$\boldsymbol{r}_1 = \boldsymbol{R} + \frac{m_2}{m_1+m_2}\boldsymbol{r}, \tag{4.92}$$

$$\boldsymbol{r}_2 = \boldsymbol{R} - \frac{m_1}{m_1+m_2}\boldsymbol{r}, \tag{4.93}$$

where $\boldsymbol{R}$ is the coordinate of the center of mass, and $\boldsymbol{r}$ is the relative coordinate of m_1 with respect to m_2. (See Fig. 4.5.) Adding Eqs. (4.88) and (4.89) and using Eq. (4.87), we obtain the equation of motion for $\boldsymbol{R}$:

$$(m_1+m_2)\ddot{\boldsymbol{R}} = \boldsymbol{F}_1^e + \boldsymbol{F}_2^e. \tag{4.94}$$

Multiplying Eq. (4.89) by m_1, and subtracting from Eq. (4.88) multiplied by m_2, using Eq. (4.87), we obtain the equation of motion for $\boldsymbol{r}$:

$$m_1 m_2 \ddot{\boldsymbol{r}} = (m_1+m_2)\boldsymbol{F}_1^i + m_1 m_2\left(\frac{\boldsymbol{F}_1^e}{m_1} - \frac{\boldsymbol{F}_2^e}{m_2}\right). \tag{4.95}$$

We now assume that

$$\frac{\boldsymbol{F}_1^e}{m_1} = \frac{\boldsymbol{F}_2^e}{m_2}, \tag{4.96}$$

and introduce the abbreviations

$$M = m_1 + m_2, \tag{4.97}$$

$$\mu = \frac{m_1 m_2}{m_1+m_2}, \tag{4.98}$$

$$\boldsymbol{F} = \boldsymbol{F}_1^e + \boldsymbol{F}_2^e. \tag{4.99}$$

Equations (4.94) and (4.95) then take the form of single-particle equations of motion:

$$M\ddot{\boldsymbol{R}} = \boldsymbol{F}, \tag{4.100}$$

$$\mu\ddot{\boldsymbol{r}} = \boldsymbol{F}_1^i. \tag{4.101}$$

Equation (4.100) is the familiar equation for the motion of the center of mass. Equation (4.101) is the equation of motion for a particle of mass μ acted on by the

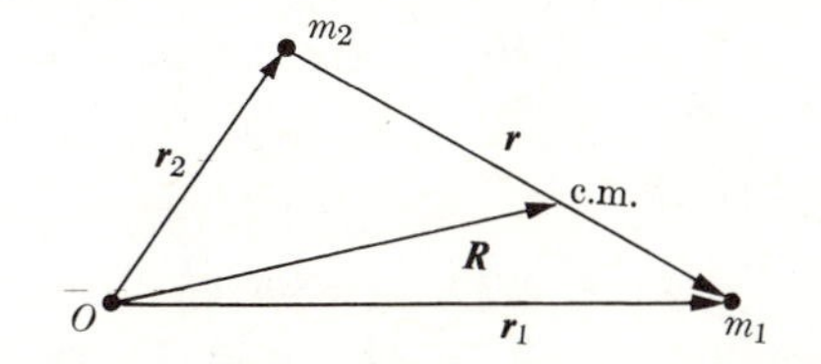

Fig. 4.5 Coordinates for the two-body problem.

internal force $\boldsymbol{F}_1^i$ that particle 2 exerts on particle 1. Thus the motion of particle 1 as viewed from particle 2 is the same as if particle 2 were fixed and particle 1 had a mass μ (μ is called the *reduced mass*). If one particle is much heavier than the other, μ is slightly less than the mass of the lighter particle. If the particles are of equal mass, μ is half the mass of either. We may now apply the results of Section 3.14 to any two-body problem in which the two particles exert an inverse square law attraction or repulsion on each other, provided the external forces are either zero or are proportional to the masses, as required by Eq. (4.96).

Equation (4.96) is satisfied if the external forces are gravitational forces exerted by masses whose distances from the two bodies m_1 and m_2 are much greater than the distance r form m_1 to m_2. As an example, the motion of the earth-moon system can be treated, to a good approximation, by the method of this section, since the moon is much closer to the earth than either is to the sun (or to the other planets). Atomic particles are acted on by electrical forces proportional to their charges, and hence Eq. (4.96) holds ordinarily only if the external forces are zero. There is also the less important case where the two particles have the same ratio of charge to mass, and are acted on by external forces due to distant charges. We may remark here that although Eqs. (4.88) and (4.89) are not the correct equations for describing the motions of atomic particles, the introduction of the coordinates $\boldsymbol{R}$, $\boldsymbol{r}$, and the reduction of the two-body problem to two one-body problems can be carried out in the quantum-mechanical treatment in a way exactly analogous to the above classical treatment, under the same assumptions about the forces.

It is worth noticing that the kinetic energy of the two-body system can be separated into two parts, one associated with each of the two one-body problems into which we have separated the two-body problem. The center-of-mass velocity and the relative velocity are, according to Eqs. (4.90)–(4.93), related to the particle velocities by

$$\boldsymbol{V} = \dot{\boldsymbol{R}} = \frac{m_1\boldsymbol{v}_1+m_2\boldsymbol{v}_2}{m_1+m_2}, \tag{4.102}$$

$$\boldsymbol{v} = \dot{\boldsymbol{r}} = \boldsymbol{v}_1-\boldsymbol{v}_2, \tag{4.103}$$

or

$$\boldsymbol{v}_1 = \boldsymbol{V}+\frac{\mu}{m_1}\boldsymbol{v}, \tag{4.104}$$

$$\boldsymbol{v}_2 = \boldsymbol{V}-\frac{\mu}{m_2}\boldsymbol{v}. \tag{4.105}$$

The total kinetic energy is

$$\begin{aligned} T &= \tfrac{1}{2}m_1v_1^2+\tfrac{1}{2}m_2v_2^2 \\ &= \tfrac{1}{2}MV^2+\tfrac{1}{2}\mu v^2. \end{aligned} \tag{4.106}$$

The angular momentum can similarly be separated into two parts:

$$\begin{aligned} \boldsymbol{L} &= m_1(\boldsymbol{r}_1 \times \boldsymbol{v}_1) + m_2(\boldsymbol{r}_2 \times \boldsymbol{v}_2) \\ &= M(\boldsymbol{R} \times \boldsymbol{V}) + \mu(\boldsymbol{r} \times \boldsymbol{v}). \end{aligned} \tag{4.107}$$

The total linear momentum is, however, just

$$\boldsymbol{P} = m_1\boldsymbol{v}_1 + m_2\boldsymbol{v}_2 = M\boldsymbol{V}. \tag{4.108}$$

There is no term $\mu\boldsymbol{v}$ in the total linear momentum.

4.8 CENTER-OF-MASS COORDINATES. RUTHERFORD SCATTERING BY A CHARGED PARTICLE OF FINITE MASS

By making use of the results of the preceding section, we can solve a two-body scattering problem completely, if we know the interaction force between the two particles, by solving the one-body equation of motion for the coordinate $\boldsymbol{r}$. The result, however, is not in a very convenient form for application. The solution $\boldsymbol{r}(t)$ describes the motion of particle 1 with respect to particle 2 as origin. Since particle 2 itself will be moving along some orbit, this is not usually a very convenient way of interpreting the motion. It would be better to describe the motion of both particles by means of coordinates $\boldsymbol{r}_1(t)$, $\boldsymbol{r}_2(t)$ referred to some fixed origin. Usually one of the particles is initially at rest; we shall take it to be particle 2, and call it the *target* particle. Particle 1, approaching the target with an initial velocity $\boldsymbol{v}_{1I}$, we shall call the *incident* particle. The two particles are to be located by vectors $\boldsymbol{r}_1$ and $\boldsymbol{r}_2$ relative to an origin with respect to which the target particle is initially at rest. We shall call the coordinates $\boldsymbol{r}_1$, $\boldsymbol{r}_2$ the *laboratory coordinate system.*

The translation from the coordinates $\boldsymbol{R}$, $\boldsymbol{r}$ to laboratory coordinates is most conveniently carried out in two steps. We first introduce a *center-of-mass coordinate system* in which the particles are located by vectors $\boldsymbol{r}_1^i$, $\boldsymbol{r}_2^i$ with respect to the center of mass as origin:

$$\boldsymbol{r}_1^i = \boldsymbol{r}_1 - \boldsymbol{R}, \qquad \boldsymbol{r}_2^i = \boldsymbol{r}_2 - \boldsymbol{R}, \tag{4.109}$$

and, conversely,

$$\boldsymbol{r}_1 = \boldsymbol{r}_1^i + \boldsymbol{R}, \qquad \boldsymbol{r}_2 = \boldsymbol{r}_2^i + \boldsymbol{R}. \tag{4.110}$$

The relation between the center-of-mass coordinates and the relative coordinate $\boldsymbol{r}$ is obtained from Eqs. (4.92) and (4.93):

$$\begin{aligned} \boldsymbol{r}_1^i &= \frac{m_2}{m_1+m_2}\boldsymbol{r} = \frac{\mu}{m_1}\boldsymbol{r}, \\ \boldsymbol{r}_2^i &= -\frac{m_1}{m_1+m_2}\boldsymbol{r} = -\frac{\mu}{m_2}\boldsymbol{r}. \end{aligned} \tag{4.111}$$

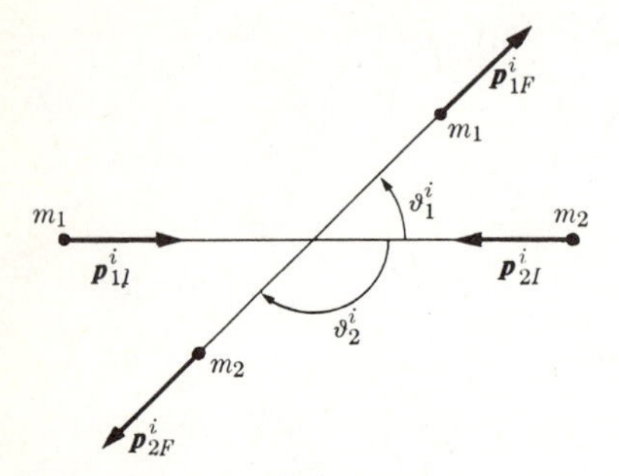

Fig. 4.6 Two-particle collision in center-of-mass coordinates.

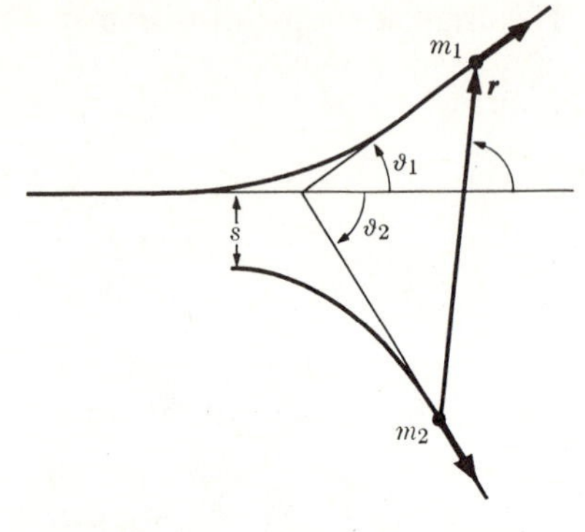

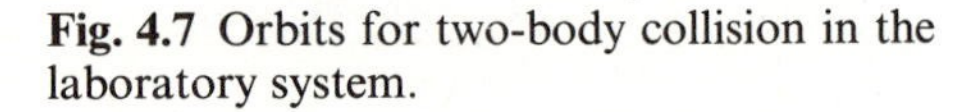

Fig. 4.7 Orbits for two-body collision in the laboratory system.

The position vectors of the particles relative to the center of mass are constant multiples of the relative coordinate $\boldsymbol{r}$. The center of mass has the advantage over particle 2, as an origin of coordinates, in that it moves with uniform velocity in collision problems where no external forces are assumed to act.

In the center-of-mass coordinate system the total linear momentum is zero, and the momenta $\boldsymbol{p}_1^i$ and $\boldsymbol{p}_2^i$ of the two particles are always equal and opposite. The scattering angles ϑ_1^i and ϑ_2^i between the two final directions of motion and the initial direction of motion of particle 1 are the supplements of each other, as shown in Fig. 4.6.

We now determine the relation between the scattering angle Θ in the equivalent one-body problem and the scattering angle ϑ_1 in the laboratory coordinate system (Fig. 4.7). The velocity of the incident particle in the center-of-mass system is related to the relative velocity in the one-body problem, according to Eq. (4.111), by

$$\boldsymbol{v}_1^i = \frac{\mu}{m_1}\boldsymbol{v}. \tag{4.112}$$

Since these two velocities are always parallel, the angle of scattering ϑ_1^i of the incident particle in the center-of-mass system is equal to the angle of scattering Θ in the one-body problem. The incident particle velocities in the center-of-mass and laboratory systems are related by [Eq. (4.110)]

$$\boldsymbol{v}_1 = \boldsymbol{v}_1^i + \boldsymbol{V}, \tag{4.113}$$

where the constant velocity of the center of mass can be expressed in terms of the initial velocity in the laboratory system by Eq. (4.102):

$$\boldsymbol{V} = \frac{m_1}{m_1+m_2}\boldsymbol{v}_{1I} = \frac{\mu}{m_2}\boldsymbol{v}_{1I}. \tag{4.114}$$

The relation expressed by Eq. (4.113) is shown in Fig. 4.8, from which the relation between $\vartheta_1^i = \Theta$ and ϑ_1 can be determined:

$$\tan \vartheta_1 = \frac{v_{1F}^i \sin \Theta}{v_{1F}^i \cos \Theta + V}, \tag{4.115}$$

or, with the help of Eqs. (4.112) and (4.114),

$$\tan \vartheta_1 = \frac{\sin \Theta}{\cos \Theta + (m_1 v_I / m_2 v_F)}, \tag{4.116}$$

where v_I and v_F are the initial and final relative speeds, and we have substituted v_I for v_{1I}, since initially the relative velocity is just the velocity of the incident particle. If the collision is elastic, the initial and final relative speeds are the same and Eq. (4.116) reduces to:

$$\tan \vartheta_1 = \frac{\sin \Theta}{\cos \Theta + (m_1/m_2)}. \tag{4.117}$$

A similar relation for ϑ_2 can be worked out.

If the incident particle is much heavier than the target particle, then ϑ_1 will be very small, no matter what value Θ may have. This corresponds to the result obtained in Section 4.6, that ϑ_1 can never be larger than ϑ_m given by Eq. (4.68), if $m_1 > m_2$. If $m_1 = m_2$, then Eq. (4.117) is easily solved for ϑ_1:

$$\tan \vartheta_1 = \frac{\sin \Theta}{\cos \Theta + 1} = \frac{2 \sin (\Theta/2) \cos (\Theta/2)}{2 \cos^2 (\Theta/2)} = \tan \frac{\Theta}{2},$$

$$\vartheta_1 = \tfrac{1}{2}\Theta. \tag{4.118}$$

Since Θ may always have any value between 0 and π without violating the conservation laws in the center-of-mass system, the maximum value of ϑ_1 in this case is $\pi/2$, in agreement with the corresponding result of Section 4.6. If the target mass m_2 is much larger than the incident mass m_1, then $\tan \vartheta_1 \doteq \tan \Theta$; this justifies rigorously our application to this case of Eq. (3.276) for the Rutherford cross section, deduced in Chapter 3 for the one-body scattering problem with an inverse square law force.

According to the above developments, Eq. (3.276) applies also to the two-body

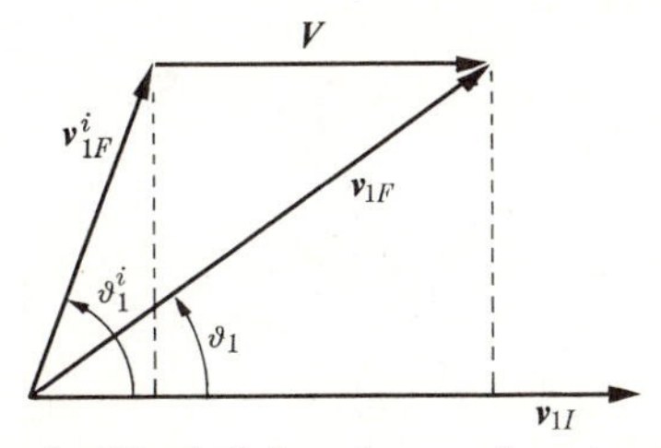

Fig. 4.8 Relation between velocities in laboratory and center-of-mass coordinate systems.

problem for any ratio m_1/m_2 of incident mass to target mass, but Θ must be interpreted as the angle of scattering in terms of relative coordinates, or else in terms of center-of-mass coordinates. That is, $d\sigma$ in Eq. (3.276) is the cross section for a scattering process in which the relative velocity $\boldsymbol{v}$ after the collision makes an angle between Θ and $\Theta+d\Theta$ with the initial velocity. Since it is the laboratory scattering angle ϑ_1 that is ordinarily measured, we must substitute for Θ and $d\Theta$ in Eq. (3.276) their values in terms of ϑ_1 and $d\vartheta_1$ as determined from Eq. (4.117). This is most easily done in case $m_1 = m_2$, when, by Eq. (4.118), the Rutherford scattering cross section [Eq. (3.276)] becomes

$$d\sigma = \left(\frac{q_1 q_2}{2\mu v_0^2}\right)^2 \frac{4\cos\vartheta_1}{\sin^4\vartheta_1} 2\pi \sin\vartheta_1 \, d\vartheta_1. \tag{4.119}$$

4.9 THE *N*-BODY PROBLEM

It would be very satisfactory if we could arrive at a general method of solving the problem of any number of particles moving under the forces which they exert on one another, analogous to the method given in Section 4.7 by which the two-body problem was reduced to two separate one-body problems. Unfortunately no such general method is available for systems of more than two particles. This does not mean that such problems cannot be solved. The extremely accurate calculations of the motions of the planets represent a solution of a problem involving the gravitational interactions of a considerable number of bodies. However, these solutions are not general solutions of the equations of motion, like the system of orbits we have obtained for the two-body case, but are numerical solutions obtained by elaborate calculations for specified initial conditions and holding over certain periods of time. Even the three-body problem admits of no general reduction, say, to three one-body problems, or to any other manageable set of equations.

However, we can partially separate the problem of the motion of a system of particles into two problems: first, to find the motion of the center of mass, and

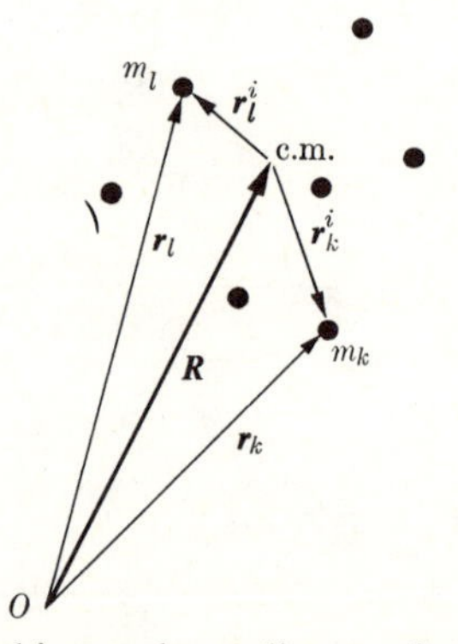

Fig. 4.9 Center-of-mass and internal coordinates of a system of particles.

second, to find the internal motion of the system, that is, the motion of its particles relative to the center of mass. Let us define the internal coordinate vector $\boldsymbol{r}_k^i$ of the kth particle as the vector from the center of mass to the kth particle (Fig. 4.9):

$$\boldsymbol{r}_k^i = \boldsymbol{r}_k - \boldsymbol{R}, \qquad k = 1, \ldots, N, \tag{4.120}$$

$$\boldsymbol{r}_k = \boldsymbol{R} + \boldsymbol{r}_k^i, \qquad k = 1, \ldots, N. \tag{4.121}$$

In view of the definition (4.14) of the center of mass, the internal coordinates $\boldsymbol{r}_k^i$ satisfy the equation

$$\sum_{k=1}^{N} m_k \boldsymbol{r}_k^i = \boldsymbol{0}. \tag{4.122}$$

We define the center-of-mass velocity and the internal velocities:

$$\boldsymbol{V} = \dot{\boldsymbol{R}}, \tag{4.123}$$

$$\boldsymbol{v}_k^i = \dot{\boldsymbol{r}}_k^i = \boldsymbol{v}_k - \boldsymbol{V}. \tag{4.124}$$

The total internal momentum of a system of particles (i.e., the momentum relative to the center of mass) vanishes by Eq. (4.122):

$$\sum_{k=1}^{N} m_k \boldsymbol{v}_k^i = \boldsymbol{0}. \tag{4.125}$$

We first show that the total kinetic energy, momentum, and angular momentum can each be split up into a part depending on the total mass M and the motion of the center of mass, and an internal part depending only on the internal coordinates and velocities. The total kinetic energy of the system of particles is

$$T = \sum_{k=1}^{N} \tfrac{1}{2} m_k v_k^2. \tag{4.126}$$

By substituting for $\boldsymbol{v}_k$ from Eq. (4.124), and making use of Eq. (4.125), we can split T into two parts:

$$\begin{aligned} T &= \sum_{k=1}^{N} \tfrac{1}{2} m_k (V^2 + 2\boldsymbol{V} \cdot \boldsymbol{v}_k^i + v_k^{i2}) \\ &= \sum_{k=1}^{N} \tfrac{1}{2} m_k V^2 + \sum_{k=1}^{N} \tfrac{1}{2} m_k v_k^{i2} + \sum_{k=1}^{N} m_k \boldsymbol{V} \cdot \boldsymbol{v}_k^i \\ &= \tfrac{1}{2} M V^2 + \sum_{k=1}^{N} \tfrac{1}{2} m_k v_k^{i2} + \boldsymbol{V} \cdot \sum_{k=1}^{N} m_k \boldsymbol{v}_k^i \\ &= \tfrac{1}{2} M V^2 + \sum_{k=1}^{N} \tfrac{1}{2} m_k v_k^{i2}. \end{aligned} \tag{4.127}$$

The total linear momentum is, if we make use of Eqs. (4.124) and (4.125),

$$\begin{aligned} \boldsymbol{P} &= \sum_{k=1}^{N} m_k \boldsymbol{v}_k \\ &= \sum_{k=1}^{N} m_k \boldsymbol{V} + \sum_{k=1}^{N} m_k \boldsymbol{v}_k^i \\ &= M\boldsymbol{V}. \end{aligned} \tag{4.128}$$

The internal linear momentum is zero.

The total angular momentum about the origin is, if we use Eqs. (4.121), (4.122), (4.124), and (4.125),

$$\begin{aligned} \boldsymbol{L} &= \sum_{k=1}^{N} m_k(\boldsymbol{r}_k \times \boldsymbol{v}_k) \\ &= \sum_{k=1}^{N} m_k(\boldsymbol{R} \times \boldsymbol{V} + \boldsymbol{r}_k^i \times \boldsymbol{V} + \boldsymbol{R} \times \boldsymbol{v}_k^i + \boldsymbol{r}_k^i \times \boldsymbol{v}_k^i) \\ &= \sum_{k=1}^{N} m_k(\boldsymbol{R} \times \boldsymbol{V}) + \left(\sum_{k=1}^{N} m_k \boldsymbol{r}_k^i\right) \times \boldsymbol{V} + \boldsymbol{R} \times \left(\sum_{k=1}^{N} m_k \boldsymbol{v}_k^i\right) + \sum_{k=1}^{N} m_k(\boldsymbol{r}_k^i \times \boldsymbol{v}_k^i) \\ &= M(\boldsymbol{R} \times \boldsymbol{V}) + \sum_{k=1}^{N} m_k(\boldsymbol{r}_k^i \times \boldsymbol{v}_k^i). \end{aligned} \tag{4.129}$$

Notice that the internal angular momentum depends only on the internal coordinates and velocities and is independent of the origin about which $\boldsymbol{L}$ is being computed (and from which the vector $\boldsymbol{R}$ is drawn).

The position of particle k with respect to particle l is specified by the vector

$$\boldsymbol{r}_k - \boldsymbol{r}_l = \boldsymbol{r}_k^i - \boldsymbol{r}_l^i. \tag{4.130}$$

The relative positions of the particles with respect to each other depend only on the internal coordinates $\boldsymbol{r}_k^i$, and likewise the relative velocities, so that the internal forces $\boldsymbol{F}_k^i$ will be expected to depend only on the internal coordinates $\boldsymbol{r}_k^i$, and possibly on the internal velocities. If there is a potential energy associated with the internal forces, it likewise will depend only on the internal coordinates.

Although the forces, energy, momentum, and angular momentum can each be split into two parts, a part associated with the motion of the center of mass and an internal part depending only on the internal coordinates and velocities, it must not be supposed that the internal motion and the center-of-mass motion are two completely separate problems. The motion of the center of mass, as governed by Eq. (4.18), is a separate one-body problem *when the external force* $\boldsymbol{F}$ *is given.* However, in most cases $\boldsymbol{F}$ will depend to some extent on the internal motion of the system. The internal equations of motion contain the external forces except in special cases and, furthermore, they also depend on the motion of the center of mass. If we substitute Eqs. (4.121) in Eqs. (4.1), and rearrange, we have

$$m_k\ddot{\boldsymbol{r}}_k^i = \boldsymbol{F}_k^i + \boldsymbol{F}_k^e - m_k\ddot{\boldsymbol{R}}. \tag{4.131}$$

There are many cases, however, in which a group of particles forms a system which seems to have some identity of its own independent of other particles and systems of particles. An atomic nucleus, made up of neutrons and protons, is an example, as is an atom, made up of nucleus and electrons, or a molecule, composed of nuclei and electrons, or the collection of particles which make up a baseball. In all such cases, it turns out that the internal forces are much stronger than the external ones, and the acceleration $\ddot{\boldsymbol{R}}$ is small, so that the internal equations of motion (4.131) depend essentially only on the internal forces, and their solutions represent internal motions which are nearly independent of the external forces and of the motion of the system as a whole. The system viewed externally then behaves like a single particle with coordinate vector $\boldsymbol{R}$, mass M, acted on by the (external) force $\boldsymbol{F}$, but a particle which has, in addition to its "orbital" energy, momentum, and angular momentum associated with the motion of its center of mass, an intrinsic or internal energy and angular momentum associated with its internal motion. The external force $\boldsymbol{F}$ may depend on the internal state of motion of the system. The orbital and intrinsic parts of the energy, momentum, and angular momentum can be identified in Eqs. (4.127), (4.128), and (4.129). The internal angular momentum is usually called *spin* and is independent of the position or velocity of the center of mass relative to the origin about which the total angular momentum is to be computed. So long as the external forces are small, this approximate representation of the system as a single particle is valid. Whenever the external forces are strong enough to affect appreciably the internal motion, the separation into problems of internal and of orbital motions breaks down and the system begins to lose its individuality. Some of the central problems at the frontiers of present-day physical theories are concerned with bridging the gap between a loose collection of particles and a system with sufficient individuality to be treated as a single particle.

4.10 TWO COUPLED HARMONIC OSCILLATORS

A very commonly occurring type of mechanical system is one in which several harmonic oscillators interact with one another. As a typical example of such a system, consider the mechanical system shown in Fig. 4.10, consisting of two

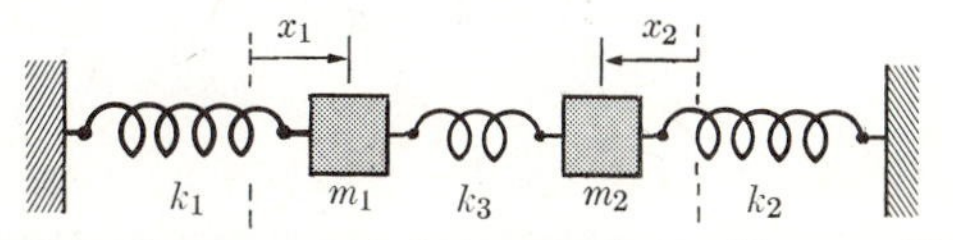

Fig. 4.10 A simple model of two coupled harmonic oscillators. Dashed lines mark the equilibrium positions.

masses m_1, m_2 fastened to fixed supports by springs whose elastic constants are k_1, k_2, and connected by a third spring of elastic constant k_3. We suppose the masses are free to move only along the x-axis; they may, for example, slide along a rail. If spring k_3 were not present, the two masses would vibrate independently in simple harmonic motion with angular frequencies (neglecting damping)

$$\omega_{10}^0 = \sqrt{\frac{k_1}{m_1}}, \qquad \omega_{20}^0 = \sqrt{\frac{k_2}{m_2}}. \tag{4.132}$$

We wish to investigate the effect of coupling these two oscillators together by means of the spring k_3. We describe the positions of the two masses by specifying the distances x_1 and x_2 that the springs k_1 and k_2 have been stretched from their equilibrium positions, as shown in Fig. 4.10. For reasons of mathematical taste, we have chosen to define the coordinates x_1 and x_2 so that they measure the amounts by which the respective springs have been stretched. This choice provides a symmetry in the physical meanings of the coordinates, which is reflected in an algebraic symmetry in the resulting equations in that they remain the same if we interchange subscripts '$_1$' and '$_2$'. This symmetry makes the algebra a little easier to work out and to check. Our choice has the disadvantage that the distances x_1 and x_2 are not measured from a common origin, and moreover, they are measured in opposite directions! Thus in interpreting our results geometrically, we must remember that the positive direction for m_1 is to the right, while the positive direction for m_2 is to the left. Other ways of defining x_1 and x_2 are equally permissible. We assume for simplicity that when springs k_1 and k_2 are relaxed ($x_1 = x_2 = 0$), spring k_3 is also relaxed. The amount by which spring k_3 is compressed is then (x_1+x_2). The equations of motion for the masses m_1, m_2 (neglecting friction) are

$$m_1\ddot{x}_1 = -k_1x_1 - k_3(x_1+x_2), \tag{4.133}$$

$$m_2\ddot{x}_2 = -k_2x_2 - k_3(x_1+x_2). \tag{4.134}$$

We rewrite these in the form

$$m_1\ddot{x}_1 + k_1'x_1 + k_3x_2 = 0, \tag{4.135}$$

$$m_2\ddot{x}_2 + k_2'x_2 + k_3x_1 = 0, \tag{4.136}$$

where

$$k_1' = k_1 + k_3, \tag{4.137}$$

$$k_2' = k_2 + k_3. \tag{4.138}$$

We have two second-order linear differential equations to solve simultaneously. If the third terms were not present, the equations would be independent of one another, and we would have independent harmonic vibrations of x_1 and x_2 at frequencies

$$\omega_{10} = \sqrt{\frac{k_1'}{m_1}}, \tag{4.139}$$

$$\omega_{20} = \sqrt{\frac{k_2'}{m_2}}. \tag{4.140}$$

These are the frequencies with which each mass would vibrate if the other were held fixed. Thus the first effect of the coupling spring is simply to change the frequency of independent vibration of each mass, due to the fact that each mass is now held in position by two springs instead of one. The third terms in Eqs. (4.135) and (4.136) give rise to a coupling between the motions of the two masses, so that they no longer move independently.

We may solve Eqs. (4.135), (4.136) by an extension of the method of Section 2.8 applicable to any set of simultaneous linear differential equations with constant coefficients. We assume that

$$x_1 = C_1 e^{pt}, \tag{4.141}$$

$$x_2 = C_2 e^{pt}, \tag{4.142}$$

where C_1, C_2 are constants. Note that the same time dependence is assumed for both x_1 and x_2, in order that the factor e^{pt} will cancel out when we substitute in Eqs. (4.135) and (4.136):

$$(m_1 p^2 + k_1')C_1 + k_3 C_2 = 0, \tag{4.143}$$

$$(m_2 p^2 + k_2')C_2 + k_3 C_1 = 0. \tag{4.144}$$

We now have two algebraic equations in the three unknown quantities C_1, C_2, p. We note that either Eq. (4.143) or (4.144) can be solved for the ratio C_2/C_1:

$$\frac{C_2}{C_1} = -\frac{m_1 p^2 + k_1'}{k_3} = -\frac{k_3}{m_2 p^2 + k_2'}. \tag{4.145}$$

The two values of C_2/C_1 must be equal, and we have an equation for p:

$$\frac{m_1 p^2 + k_1'}{k_3} = \frac{k_3}{m_2 p^2 + k_2'}, \tag{4.146}$$

which may be rearranged as a quadratic equation in p^2, called the *secular equation:*

$$m_1 m_2 p^4 + (m_2 k_1' + m_1 k_2')p^2 + (k_1' k_2' - k_3^2) = 0, \tag{4.147}$$

whose solutions are

$$p^2 = -\frac{1}{2}\left(\frac{k_1'}{m_1}+\frac{k_2'}{m_2}\right) \pm \left[\frac{1}{4}\left(\frac{k_1'}{m_1}+\frac{k_2'}{m_2}\right)^2 - \frac{k_1'k_2'}{m_1m_2}+\frac{k_3^2}{m_1m_2}\right]^{1/2}$$

$$= -\frac{1}{2}(\omega_{10}^2+\omega_{20}^2) \pm \left[\frac{1}{4}(\omega_{10}^2-\omega_{20}^2)^2+\frac{k_3^2}{m_1m_2}\right]^{1/2}. \tag{4.148}$$

It is not hard to show that the quantity in brackets is less than the square of the first term, so that we have two negative solutions for p^2. If we assume that $\omega_{10} \geq \omega_{20}$, the solutions for p^2 are

$$\begin{aligned} p^2 &= -\omega_1^2 = -(\omega_{10}^2+\tfrac{1}{2}\Delta\omega^2), \\ p^2 &= -\omega_2^2 = -(\omega_{20}^2-\tfrac{1}{2}\Delta\omega^2), \end{aligned} \tag{4.149}$$

where

$$\Delta\omega^2 = (\omega_{10}^2-\omega_{20}^2)\left[\left(1+\frac{4\kappa^4}{(\omega_{10}^2-\omega_{20}^2)^2}\right)^{1/2}-1\right], \tag{4.150}$$

with the abbreviation

$$\kappa^2 = \frac{k_3}{\sqrt{m_1m_2}}, \tag{4.151}$$

where κ is the coupling constant. If $\omega_{10} = \omega_{20}$, Eq. (4.150) reduces to

$$\Delta\omega^2 = 2\kappa^2. \tag{4.152}$$

The four solutions for p are

$$p = \pm i\omega_1,\ \pm i\omega_2. \tag{4.153}$$

If $p^2 = -\omega_1^2$, Eq. (4.145) can be written

$$\frac{C_2}{C_1} = \frac{m_1}{k_3}(\omega_1^2-\omega_{10}^2) = \frac{\Delta\omega^2}{2\kappa^2}\sqrt{\frac{m_1}{m_2}}, \tag{4.154}$$

and if $p^2 = -\omega_2^2$, it can be written

$$\frac{C_1}{C_2} = \frac{m_2}{k_3}(\omega_2^2-\omega_{20}^2) = -\frac{\Delta\omega^2}{2\kappa^2}\sqrt{\frac{m_2}{m_1}}. \tag{4.155}$$

By substituting from Eq. (4.153) in Eqs. (4.141), (4.142), we get four solutions of Eqs. (4.135) and (4.136) provided the ratio C_2/C_1 is chosen according to Eq. (4.154) or (4.155). Each of these solutions involves one arbitrary constant (C_1 or C_2). Since the equations (4.135), (4.136) are linear, the sum of these four solutions will also be a solution, and is in fact the general solution, for it will contain four arbitrary constants (say C_1, C_1', C_2, C_2'):

$$x_1 = C_1 e^{i\omega_1 t} + C_1' e^{-i\omega_1 t} - \frac{\Delta\omega^2}{2\kappa^2}\sqrt{\frac{m_2}{m_1}}\, C_2 e^{i\omega_2 t} - \frac{\Delta\omega^2}{2\kappa^2}\sqrt{\frac{m_2}{m_1}}\, C_2' e^{-i\omega_2 t}, \tag{4.156}$$

$$x_2 = \frac{\Delta\omega^2}{2\kappa^2}\sqrt{\frac{m_1}{m_2}}\, C_1 e^{i\omega_1 t} + \frac{\Delta\omega^2}{2\kappa^2}\sqrt{\frac{m_1}{m_2}}\, C_1' e^{-i\omega_1 t} + C_2 e^{i\omega_2 t} + C_2' e^{-i\omega_2 t}. \tag{4.157}$$

In order to make x_1 and x_2 real, we choose

$$C_1 = \tfrac{1}{2}A_1 e^{i\theta_1}, \qquad C_1' = \tfrac{1}{2}A_1 e^{-i\theta_1}, \tag{4.158}$$

$$C_2 = \tfrac{1}{2}A_2 e^{i\theta_2}, \qquad C_2' = \tfrac{1}{2}A_2 e^{-i\theta_2}, \tag{4.159}$$

so that

$$x_1 = A_1 \cos(\omega_1 t + \theta_1) - \frac{\Delta\omega^2}{2\kappa^2}\sqrt{\frac{m_2}{m_1}}\, A_2 \cos(\omega_2 t + \theta_2), \tag{4.160}$$

$$x_2 = \frac{\Delta\omega^2}{2\kappa^2}\sqrt{\frac{m_1}{m_2}}\, A_1 \cos(\omega_1 t + \theta_1) + A_2 \cos(\omega_2 t + \theta_2). \tag{4.161}$$

This is the general solution, involving the four arbitrary constants A_1, A_2, θ_1, θ_2. We see that the motion of each coordinate is a superposition of two harmonic vibrations at frequencies ω_1 and ω_2. The oscillation frequencies are the same for both coordinates, but the relative amplitudes are different, and are given by Eqs. (4.154) and (4.155).

If A_1 or A_2 is zero, only one frequency of oscillation appears. The resulting motion is called a *normal mode of vibration.* The normal mode of highest frequency is given by

$$x_1 = A_1 \cos(\omega_1 t + \theta_1), \tag{4.162}$$

$$x_2 = \frac{\Delta\omega^2}{2\kappa^2}\sqrt{\frac{m_1}{m_2}}\, A_1 \cos(\omega_1 t + \theta_1), \tag{4.163}$$

$$\omega_1^2 = \omega_{10}^2 + \tfrac{1}{2}\Delta\omega^2. \tag{4.164}$$

The frequency of oscillation is higher than ω_{10}. By referring to Fig. 4.10, we see that in this mode of oscillation the two masses m_1 and m_2 are oscillating out of phase; that is, their displacements are in opposite directions. The mode of oscillation of lower frequency is given by

$$x_1 = -\frac{\Delta\omega^2}{2\kappa^2}\sqrt{\frac{m_2}{m_1}}\, A_2 \cos(\omega_2 t + \theta_2), \tag{4.165}$$

$$x_2 = A_2 \cos(\omega_2 t + \theta_2), \tag{4.166}$$

$$\omega_2^2 = \omega_{20}^2 - \tfrac{1}{2}\Delta\omega^2. \tag{4.167}$$

In this mode, the two masses oscillate in phase at a frequency lower than ω_{20}. The most general motion of the system is given by Eqs. (4.160), (4.161), and is a superposition of the two normal modes of vibration.

The effect of coupling is thus to cause both masses to participate in the oscillation at each frequency, and to raise the highest frequency and lower the lowest frequency of oscillation. Even when both frequencies are initially equal, the coupling results in two frequencies of vibration, one higher and one lower than the frequency without coupling. When the coupling is very weak, i.e., when

$$\kappa^2 \ll \tfrac{1}{2}(\omega_{10}^2 - \omega_{20}^2), \tag{4.168}$$

then Eq. (4.150) becomes

$$\Delta\omega^2 \doteq \frac{2\kappa^4}{\omega_{10}^2 - \omega_{20}^2}. \tag{4.169}$$

For the highest frequency mode of vibration, the ratio of the amplitude of vibration of mass m_2 to that of mass m_1 is then

$$\frac{x_2}{x_1} = \frac{\Delta\omega^2}{2\kappa^2}\sqrt{\frac{m_1}{m_2}} \doteq \frac{\kappa^2}{\omega_{10}^2 - \omega_{20}^2}\sqrt{\frac{m_1}{m_2}}. \tag{4.170}$$

Thus, unless $m_2 \ll m_1$, the mass m_2 oscillates at much smaller amplitude than m_1. Similarly, it can be shown that for the low-frequency mode of vibration, m_1 oscillates at much smaller amplitude than m_2. If two oscillators of different frequency are weakly coupled together, there are two normal modes of vibration of the system. In one mode, the oscillator of higher frequency oscillates at a frequency slightly higher than without coupling, and the other oscillates weakly out of phase at the same frequency. In the other mode, the oscillator of lowest frequency oscillates at a frequency slightly lower than without coupling, and the other oscillates weakly and in phase at the same frequency. At (or very near) resonance, when the two natural frequencies ω_{10} and ω_{20} are equal, the condition for weak coupling [Eq. (4.168)] is not satisfied even when the coupling constant is very small. $\Delta\omega^2$ is then given by Eq. (4.152), and we find for the two normal modes of vibration:

$$\frac{x_2}{x_1} = \pm\sqrt{\frac{m_1}{m_2}}, \tag{4.171}$$

$$\omega^2 = \omega_{10}^2 \pm \kappa^2. \tag{4.172}$$

The two oscillators oscillate in or out of phase with an amplitude ratio depending only on their mass ratio, and with a frequency lower or higher than the uncoupled frequency by an amount depending on the coupling constant.

An interesting special case is the case of two identical oscillators ($m_1 = m_2$,

$k_1 = k_2$) coupled together. The general solution (4.160), (4.161) is, in this case,

$$x_1 = A_1 \cos(\omega_1 t + \theta_1) - A_2 \cos(\omega_2 t + \theta_2), \tag{4.173}$$

$$x_2 = A_1 \cos(\omega_1 t + \theta_1) + A_2 \cos(\omega_2 t + \theta_2), \tag{4.174}$$

where ω_1 and ω_2 are given by Eq. (4.172). If $A_2 = 0$, we have the high-frequency normal mode of vibration, and if $A_1 = 0$, we have the low-frequency normal mode. Let us suppose that initially m_2 is at rest in its equilibrium position, while m_1 is displaced a distance A from equilibrium and released at $t = 0$. The choice of constants which fits these initial conditions is

$$\begin{aligned} \theta_1 &= \theta_2 = 0, \\ A_1 &= -A_2 = \tfrac{1}{2}A, \end{aligned} \tag{4.175}$$

so that Eqs. (4.173), (4.174) become

$$x_1 = \tfrac{1}{2}A(\cos \omega_1 t + \cos \omega_2 t), \tag{4.176}$$

$$x_2 = \tfrac{1}{2}A(\cos \omega_1 t - \cos \omega_2 t), \tag{4.177}$$

which can be rewritten in the form

$$x_1 = A \cos\left(\frac{\omega_1 - \omega_2}{2} t\right) \cos\left(\frac{\omega_1 + \omega_2}{2} t\right), \tag{4.178}$$

$$x_2 = -A \sin\left(\frac{\omega_1 - \omega_2}{2} t\right) \sin\left(\frac{\omega_1 + \omega_2}{2} t\right). \tag{4.179}$$

If the coupling is small, ω_1 and ω_2 are nearly equal, and x_1 and x_2 oscillate rapidly at the angular frequency $(\omega_1 + \omega_2)/2 \doteq \omega_1 \doteq \omega_2$, with an amplitude which varies sinusoidally at angular frequency $(\omega_1 - \omega_2)/2$. The motion of each oscillator is a superposition of its two normal-mode motions, which leads to beats, the beat frequency being the difference between the two normal-mode frequencies. This is illustrated in Fig. 4.11, where oscillograms of the motion of x_2 are shown: (a) when the high-frequency normal mode alone is excited, (b) when the low-frequency normal mode is excited, and (c) when oscillator m_1 alone is initially displaced. In Fig. 4.12, oscillograms of x_1 and x_2 as given by Eqs. (4.178), (4.179) are shown. It can be seen that the oscillators periodically exchange their energy, due to the coupling between them. Figure 4.13 shows the same motion when the springs k_1 and k_2 are not exactly equal. In this case, oscillator m_1 does not give up all its energy to m_2 during the beats. Figure 4.14 shows that the effect of increasing the coupling is to increase the beat frequency $\omega_1 - \omega_2$ [Eq. (4.172)].

If a frictional force acts on each oscillator, the equations of motion (4.135) and (4.136) become

$$m_1 \ddot{x}_1 + b_1 \dot{x}_1 + k_1' x_1 + k_3 x_2 = 0, \tag{4.180}$$

$$m_2 \ddot{x}_2 + b_2 \dot{x}_2 + k_2' x_2 + k_3 x_1 = 0, \tag{4.181}$$

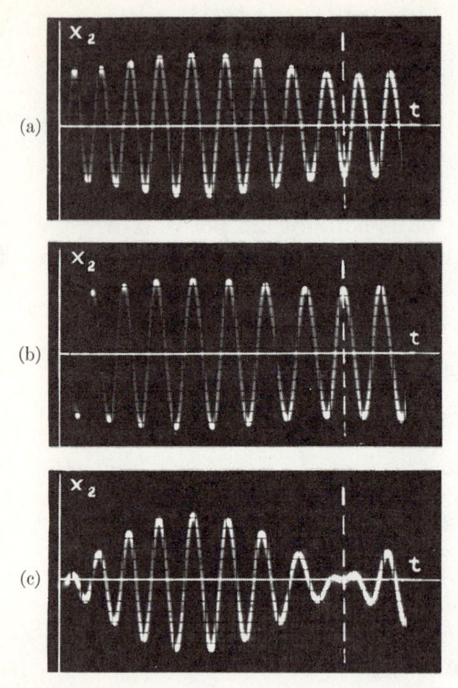

Fig. 4.11 Motion of coupled harmonic oscillators. (a) High-frequency normal mode. (b) Low-frequency normal mode. (c) m_1 initially displaced.

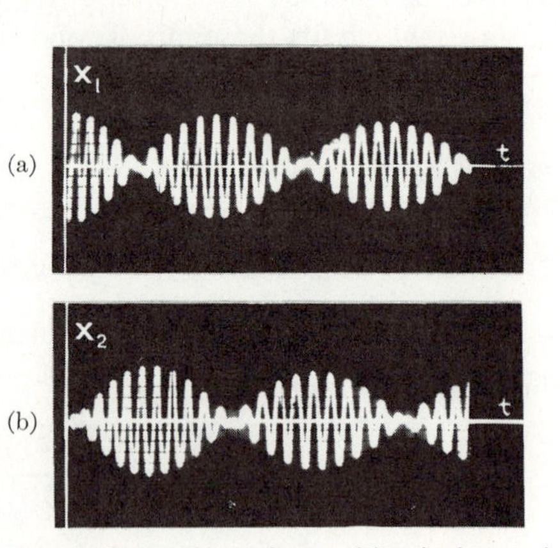

Fig. 4.12 Motion of two identical coupled oscillators.

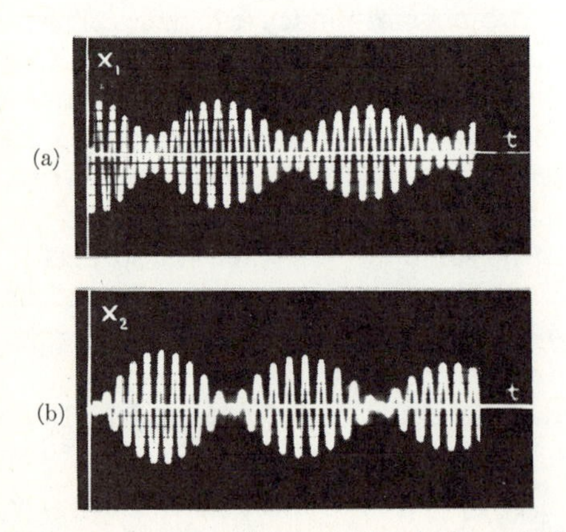

Fig. 4.13 Motion of two nonidentical coupled oscillators.

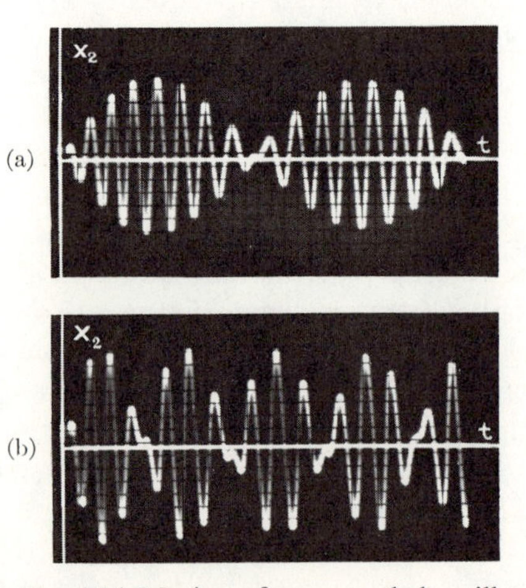

Fig. 4.14 Motion of two coupled oscillators. (a) Weak coupling. (b) Strong coupling.

where b_1 and b_2 are the respective friction coefficients. The substitution (4.141), (4.142) leads to a fourth-degree secular equation for p:

$$m_1m_2p^4+(m_2b_1+m_1b_2)p^3+(m_2k_1'+m_1k_2'+b_1b_2)p^2$$
$$+(b_1k_2'+b_2k_1')p+(k_1'k_2'-k_3^2) = 0. \quad (4.182)$$

This equation cannot be solved so easily as Eq. (4.147). The four roots for p are, in general, complex, and have the form (if b_1 and b_2 are not too large)

$$p = -\gamma_1 \pm i\omega_1,$$
$$p = -\gamma_2 \pm i\omega_2. \quad (4.183)$$

That the roots have this form with γ_1 and γ_2 positive can be shown (though not easily) algebraically from a study of the coefficients in Eq. (4.182). Physically, it is evident that the roots have the form (4.183), since this will lead to damped vibrations, the expected result of friction. If b_1 and b_2 are large enough, one or both of the pairs of complex roots may become a pair of real negative roots, the corresponding normal mode or modes being overdamped. A practical solution of Eq. (4.182) can, in general, be obtained only by numerical methods when numerical values for the constants are given, although an approximate algebraic solution can be found when the damping is very small.

The problem of the motion of a system of two coupled harmonic oscillators subject to a harmonically oscillating force applied to either mass can be solved by methods similar to those which apply to a single harmonic oscillator. A steady-state solution can be found in which both oscillators oscillate at the frequency of the applied force with definite amplitudes and phases, depending on their masses, the spring constants, the damping, and the amplitude and phase of the applied force. The system is in resonance with the applied force when its frequency corresponds to either of the two normal modes of vibration, and the masses then vibrate at large amplitudes limited only by the damping. The general solution consists of the steady-state solution plus the general solution of the unforced problem. A superposition principle can be proved according to which, if a number of forces act on either or both masses, the solution is the sum of the solutions with each force acting separately. This theorem can be used to treat the problem of arbitrary forces acting on the two masses.

Other types of coupling between the oscillators are possible in addition to coupling by means of a spring as in the example above. The oscillators may be coupled by frictional forces. A simple example would be the case where one mass

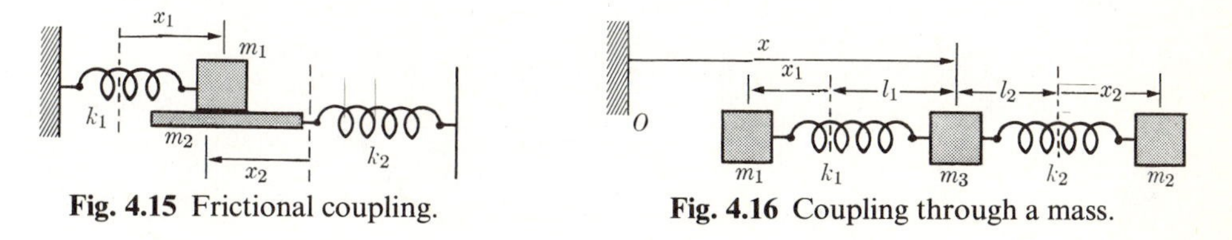

Fig. 4.15 Frictional coupling.

Fig. 4.16 Coupling through a mass.

slides over the other, as in Fig. 4.15. We assume that the force of friction is proportional to the relative velocity of the two masses. The equations of motion of m_1 and m_2 are then

$$m_1\ddot{x}_1 = -k_1x_1 - b(\dot{x}_1 + \dot{x}_2), \tag{4.184}$$

$$m_2\ddot{x}_2 = -k_2x_2 - b(\dot{x}_2 + \dot{x}_1), \tag{4.185}$$

or

$$m_1\ddot{x}_1 + b\dot{x}_1 + k_1x_1 + b\dot{x}_2 = 0, \tag{4.186}$$

$$m_2\ddot{x}_2 + b\dot{x}_2 + k_2x_2 + b\dot{x}_1 = 0. \tag{4.187}$$

The coupling is expressed in Eqs. (4.186), (4.187) by a term in the equation of motion of each oscillator depending on the velocity of the other. The oscillators may also be coupled by a mass, as in Fig. 4.16. It is left to the reader to set up the equations of motion. (See Problem 40 at the end of this chapter.)

Two oscillators may be coupled in such a way that the force acting on one depends on the position, velocity, or acceleration of the other, or on any combination of these. In general, all three types of coupling occur to some extent; a spring, for example, has always some mass, and is subject to some internal friction. Thus the most general pair of equations for two coupled harmonic oscillators is of the form

$$m_1\ddot{x}_1 + b_1\dot{x}_1 + k_1x_1 + m_c\ddot{x}_2 + b_c\dot{x}_2 + k_cx_2 = 0, \tag{4.188}$$

$$m_2\ddot{x}_2 + b_2\dot{x}_2 + k_2x_2 + m_c\ddot{x}_1 + b_c\dot{x}_1 + k_cx_1 = 0. \tag{4.189}$$

These equations can be solved by the method described above, with similar results. Two normal modes of vibration appear, if the frictional forces are not too great.

Equations of the form (4.188), (4.189), or the simpler special cases considered in the preceding discussions, arise not only in the theory of coupled mechanical oscillators, but also in the theory of coupled electrical circuits. Applying Kirchhoff's second law to the two meshes of the circuit shown in Fig. 4.17, with mesh currents

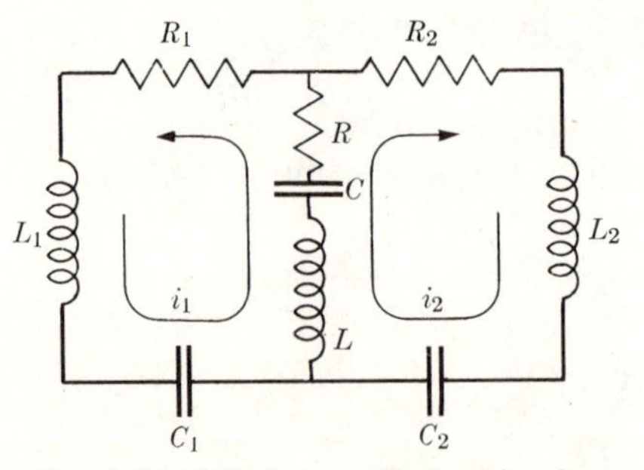

Fig. 4.17 Coupled oscillating circuits.

i_1, i_2 around the two meshes as shown, we obtain

$$(L+L_1)\ddot{q}_1+(R+R_1)\dot{q}_1+\left(\frac{1}{C}+\frac{1}{C_1}\right)q_1+L\ddot{q}_2+R\dot{q}_2+\frac{1}{C}q_2=0, \tag{4.190}$$

and

$$(L+L_2)\ddot{q}_2+(R+R_2)\dot{q}_2+\left(\frac{1}{C}+\frac{1}{C_2}\right)q_2+L\ddot{q}_1+R\dot{q}_1+\frac{1}{C}q_1=0, \tag{4.191}$$

where q_1 and q_2 are the charges built up on C_1 and C_2 by the mesh currents i_1 and i_2. These equations have the same form as Eqs. (4.188), (4.189), and can be solved by similar methods. In electrical circuits, the damping is often fairly large, and finding the solution becomes a formidable task.

The discussion of this section can be extended to the case of any number of coupled mechanical or electrical harmonic oscillators, with analogous results. The algebraic details become almost prohibitive, however, unless we make use of more advanced mathematical techniques. We therefore postpone further discussion of this problem to Chapter 12.

All mechanical and electrical vibration problems reduce in the limiting case of small amplitudes of vibration to problems involving one or several coupled harmonic oscillators. Problems involving vibrations of strings, membranes, elastic solids, and electrical and acoustical vibrations in transmission lines, pipes, or cavities, can be reduced to problems of coupled oscillators, and exhibit similar normal modes of vibration. The treatment of the behavior of an atom or molecule according to quantum mechanics results in a mathematical problem identical with the problem of coupled harmonic oscillators, in which the energy levels play the role of oscillators, and external perturbing influences play the role of the coupling mechanism.

PROBLEMS

1. Formulate and prove a conservation law for the angular momentum about the origin of a system of particles confined to a plane.

2. Water is poured into a barrel at the rate of 120 lb per minute from a height of 16 ft. The barrel weighs 25 lb, and rests on a scale. Find the scale reading after the water has been pouring into the barrel for one minute.

3. A ballistic pendulum to be used to measure the speed of a bullet is constructed by suspending a block of wood of mass M by a cord of length l. The pendulum initially hangs vertically at rest. A bullet of mass m is fired into the block and becomes imbedded in it. The pendulum then begins to swing and rises until the cord makes a maximum angle θ with the vertical. Find the initial speed of the bullet in terms of M, m, l, and θ by applying appropriate conservation laws.

4. A box of mass m falls on a conveyor belt moving with constant speed v_0. The coefficient of sliding friction between the box and the belt is μ. How far does the box slide along the belt before it is moving with the same speed as the belt? What force F must be applied to the belt to keep it moving at constant speed after the box falls on it, and for how long? Calculate the impulse delivered by this force and check that momentum is conserved between the time before the box falls on the belt and the time when the box is moving with the belt. Calculate the work done by the force F in pulling the belt. Calculate the work dissipated in friction between the box and the belt. Check that the energy delivered to the belt by the force F is just equal to the kinetic energy increase of the box plus the energy dissipated in friction.

5. A scoop of mass m_1 is attached to an arm of length l and negligible weight. The arm is pivoted so that the scoop is free to swing in a vertical arc of radius l. At a distance l directly below the pivot is a pile of sand. The scoop is lifted until the arm is at a 45° angle with the vertical, and released. It swings down and scoops up a mass m_2 of sand. To what angle with the vertical does the arm of the scoop rise after picking up the sand? This problem is to be solved by considering carefully which conservation laws are applicable to each part of the swing of the scoop. Friction is to be neglected, except that required to keep the sand in the scoop.

6. a) A spherical satellite of mass m, radius a, moves with speed $\mathbf{v}$ through a tenuous atmosphere of density ρ. Find the frictional force on it, assuming that the speed of the air molecules can be neglected in comparison with $\mathbf{v}$, and that each molecule which is struck becomes embedded in the skin of the satellite. Do you think these assumptions are valid?
b) If the orbit is a circle 400 km above the earth (radius 6360 km), where $\rho = 10^{-11}$kg/m^{-3}, and if $a = 1$ m, $m = 100$ kg, find the change in altitude and the change in period of revolution in one week.

7. A lunar landing craft approaches the moon's surface. Assume that one-third of its weight is fuel, that the exhaust velocity from its rocket engine is 1500 m/sec, and that the acceleration of gravity at the lunar surface is one-sixth of that at the earth's surface. How long can the craft hover over the moon's surface before it runs out of fuel?

8. A toy rocket consists of a plastic bottle partly filled with water containing also air at a high pressure p. The water is ejected through a small nozzle of area A. Calculate the exhaust velocity v by assuming that frictional losses of energy are negligible, so that the kinetic energy of the escaping water is equal to the work done by the gas pressure in pushing it out. Show that the thrust of this rocket engine is then $2\,pA$. (Assume that the water leaves the nozzle of area A with velocity v.) If the empty rocket weighs 500 g, if it contains initially 500 g of water, and if $A = 5$ mm^2, what pressure is required in order that the rocket can just support itself against gravity? If it is then released so that it accelerates upward, what maximum velocity will it reach? Approximately how high will it go? What effects are neglected in the calculation, and how would each of them affect the final result?

***9.** A two-stage rocket is to be built capable of accelerating a 100-kg payload to a velocity of 6000 m/sec in free flight in empty space (no gravitational field). (In a two-stage rocket, the first stage is detached after exhausting its fuel, before the second stage is fired.) Assume that the fuel used can reach an exhaust velocity of 1500 m/sec, and that structural requirements imply that an empty rocket (without fuel or payload) will weigh 10% as much as the fuel it can carry. Find the optimum choice of masses for the two stages so that the total take-off weight is a minimum. Show that it is impossible to build a single-stage rocket which will do the job.

10. A rocket is to be fired vertically upward. The initial mass is M_0, the exhaust velocity $-u$ is constant, and the rate of exhaust $-(dM/dt) = A$ is constant. After a total mass ΔM is exhausted, the rocket engine runs out of fuel.

a) Neglecting air resistance and assuming that the acceleration g of gravity is constant, set up and solve the equation of motion.

*b) Show that if M_0, u, and ΔM are fixed, then the larger the rate of exhaust A, that is, the faster it uses up its fuel, the greater the maximum altitude reached by the rocket.

11. Assume that essentially all of the mass M of the gyroscope in Fig. 4.1 is concentrated in the rim of the wheel of radius R, and that the center of mass lies on the axis at a distance l from the pivot point Q. If the gyroscope rotates rapidly with angular velocity ω, show that the angular velocity of precession of its axis in a cone making an angle α with the vertical is approximately

$$\omega_p = gl/(R^2\omega^2).$$

12. A diver executing a $2\frac{1}{2}$ flip doubles up with his knees in his arms in order to increase his angular velocity. Estimate the ratio by which he thus increases his angular velocity relative to his angular velocity when stretched out straight with his arms over his head. Explain your reasoning.

13. A uniform spherical planet of radius a revolves about the sun in a circular orbit of radius r_0, and rotates about its axis with angular velocity ω_0, normal to the plane of the orbit. Due to tides raised on the planet by the sun, its angular velocity of rotation is decreasing. Find a formula expressing the orbit radius r as a function of angular velocity ω of rotation at any later or earlier time. [You will need formulas (5.9) and (5.91) from Chapter 5.] Apply your formula to the earth, neglecting the effect of the moon, and estimate how much farther the earth will be from the sun when the day has become equal to the present year. If the effect of the moon were taken into account, would the distance be greater or less?

***14.** A mass m of gas and debris surrounds a star of mass M. The radius of the star is negligible in comparison with the distances to the particles of gas and debris. The material surrounding the star has initially a total angular momentum $\mathbf{L}$, and a total kinetic and potential energy E. Assume that $m \ll M$, so that the gravitational fields due to the mass m are negligible in comparison with that of the star. Due to internal friction, the surrounding material continually loses mechanical energy. Show that there is a maximum energy ΔE which can be lost in this way, and that when this energy has been lost, the material must all lie on a circular ring around the star (but not necessarily uniformly distributed). Find ΔE and the radius of the ring. (You will need to use the method of Lagrange multipliers.)

15. A particle of mass m_1, energy T_{1I} collides elastically with a particle of mass m_2, at rest. If the mass m_2 leaves the collision at an angle ϑ_2 with the original direction of motion of m_1, what is the energy T_{2F} delivered to particle m_2? Show that T_{2F} is a maximum for a head-on collision, and that in this case the energy lost by the incident particle in the collision is

$$T_{1I} - T_{1F} = \frac{4m_1m_2}{(m_1+m_2)^2}\, T_{1I}.$$

16. A cloud-chamber picture shows the track of an incident particle which makes a collision and is scattered through an angle ϑ_1. The track of the target particle makes an angle ϑ_2 with the direction of the incident particle. Assuming that the collision was elastic and that the target particle was initially at rest, find the ratio m_1/m_2 of the two masses. (Assume small velocities so that the classical expressions for energy and momentum may be used.)

17. A proton of mass m_1 collides elastically with an unknown nucleus in a bubble chamber and is scattered through an angle ϑ_1. The ratio p_{1F}/p_{1I} is determined from the curvature of its initial and final tracks. Find the mass m_2 of the target nucleus. How might it be possible to determine whether the collision was indeed elastic?

18. In an experiment in which particles of mass m_1 collide elastically with stationary particles of mass m_2, it is desired to place a counter in a position where it will count particles which have lost half their initial momentum. At what angle ϑ_1 with the incident beam should the counter be placed? For what range of mass ratios m_1/m_2 does this problem have an answer?

19. Show that an elastic collision corresponds to a coefficient of restitution $e = 1$, that is, show that for a head-on elastic collision between two particles, Eq. (4.85) holds with $e = 1$.

20. Calculate the energy loss $-Q$ for a head-on collision between a particle of mass m_1, velocity v_1 with a particle of mass m_2 at rest, if the coefficient of restitution is e.

21. A particle of mass m_1, momentum p_{1I} collides elastically with a particle of mass m_2, momentum p_{2I} going in the opposite direction. If m_1 leaves the collision at an angle ϑ_1 with its original course, find its final momentum.

22. Find the relativistic corrections to Eq. (4.81) when the incident particle m_1 and the emitted particle m_3 move with speeds near the speed of light. Assume that the recoil particle m_4 is moving slowly enough so that the classical relation between energy and momentum can be used for it.

23. A particle of mass m_1, momentum p_1 collides with a particle of mass m_2 at rest. A reaction occurs from which two particles of masses m_3 and m_4 result, which leave the collision at angles ϑ_3 and ϑ_4 with the original path of m_1. Find the energy Q produced in the reaction in terms of the masses, the angles, and p_1.

24. A nuclear reaction whose Q is known occurs in a photographic plate in which the tracks of the incident particle m_1 and the two product particles m_3 and m_4 can be seen. Find the energy of the incident particle in terms of m_1, m_3, m_4, Q, and the measured angles ϑ_3 and ϑ_4 between the incident track and the two final tracks. What happens if $Q = 0$?

25. A billiard ball sliding on a frictionless table strikes an identical stationary ball. The balls leave the collision at angles $\pm\vartheta$ with the original direction of motion. Show that after the collision the balls must have a rotational energy equal to $1-\frac{1}{2}\cos^{-2}\vartheta$ of the initial kinetic energy, assuming that no energy is dissipated in friction.

26. A neutral particle of unknown momentum and direction produces a reaction in a bubble chamber in which two charged particles of masses m_3, m_4 emerge with momenta p_3, p_4. The angle between their tracks is α. Find the direction and momentum of the incident particle. If the mass m_1 of the incident particle is known or suspected, find the energy Q released in the reaction. (Nonrelativistic velocities.)

27. The Compton scattering of x-rays can be interpreted as the result of elastic collisions between x-ray photons and free electrons. According to quantum theory, a photon of wavelength λ has a kinetic energy hc/λ, and a linear momentum of magnitude h/λ, where h is Planck's constant and c is the speed of light. In the Compton effect, an incident beam of x-rays of known wavelength λ_I in a known direction is scattered in passing through matter, and the scattered radiation at an angle ϑ_1 to the incident beam is found to have a longer wavelength λ_F, which is a function of the angle ϑ_1. Assuming an elastic collision between an incident photon and an electron of mass m at rest, set up the equations expressing conservation of energy and momentum. Use the relativistic expressions for the energy and momentum of the electron. Show that the change in x-ray wavelength is

$$\lambda_F - \lambda_I = \frac{h}{mc}(1 - \cos\vartheta_1),$$

and that the ejected electron appears at an angle given by

$$\tan\vartheta_2 = \frac{\sin\vartheta_1}{[1 + (h/\lambda_I mc)]\,(1 - \cos\vartheta_1)}.$$

28. Work out a correction to Eq. (3.267) which takes into account the motion of the central mass M under the influence of the revolving mass m. A pair of stars revolve about each other, so close together that they appear in the telescope as a single star. It is determined from spectroscopic observations that the two stars are of equal mass and that each revolves in a circle with speed v and period τ under the gravitational attraction of the other. Find the mass m of each star by using your formula.

29. A space ship of mass m, initial velocity $\mathbf{v}_0$ approaches the moon and passes by it. The distance of closest approach is R (measured from the center of the moon). The velocity $\mathbf{v}_0$ is perpendicular to the orbital velocity $\mathbf{V}$ of the moon. Show that if the space ship passes behind the moon, it will gain energy and calculate the increase in its kinetic energy as it leaves the vicinity of the moon. Assume that $M \gg m$, where M is the mass of the moon.

30. A star of mass m, initial speed v_0, approaches a second star of mass $2m$ at rest. The first star travels initially along a line which if continued would pass the second star at a distance s. Find the final speed and direction of motion of each star.

31. Show that if the incident particle is much heavier than the target particle ($m_1 \gg m_2$), the Rutherford scattering cross section [Eq. (3.276)] in laboratory coordinates is approximately

$$d\sigma \doteq \left(\frac{q_1 q_2}{2m_2 v_0^2}\right)^2 \frac{4\gamma^2}{[1-(1-\gamma^2\vartheta_1^2)^{1/2}]^2\,(1-\gamma^2\vartheta_1^2)^{1/2}}\, 2\pi \sin\vartheta_1\, d\vartheta_1$$

if $\gamma\vartheta_1 < 1$, where $\gamma = m_1/m_2$. Otherwise, $d\sigma = 0$.

32. Find an expression analogous to Eq. (4.116) for the angle of recoil of the target particle (ϑ_2 in Fig. 4.7) in terms of the scattering angle Θ in the equivalent one-body problem. Show that, for an elastic collision,

$$\vartheta_2 = \tfrac{1}{2}(\pi - \Theta).$$

33. Assume that $m_2 \gg m_1$, and that $\Theta = \vartheta_1 + \delta$, in Eq. (4.117). Find a formula for δ in terms of ϑ_1. Show that the first-order correction to the Rutherford scattering cross section [Eq. (3.276)], due to the finite mass of m_2, vanishes.

34. An elastic sphere of radius a collides with an identical elastic sphere at rest. Assume that in the center of mass coordinate system, each sphere rebounds from the other so that the relative velocities before and after impact make equal angles with the normal to the spheres at the point of contact. Find the cross section for scattering the incident sphere through an angle ϑ_1.

35. A pair of masses m_1, m_2, connected by a spring of force constant k, slide without friction along the x-axis. Show that the center of mass moves with uniform velocity and that the masses oscillate with frequency $(k(m_1+m_2)/m_1m_2)^{1/2}$.

36. Set up the equations of motion for Fig. 4.10, assuming that the relaxed length of each spring is l, and that the distance between the walls is $3l+a$, so that the springs are stretched, even in the equilibrium position. Show that the equations can be put in the same form as Eqs. (4.135) and (4.136).

37. For the normal mode of vibration given by Eqs. (4.162) and (4.163), find the force exerted on m_1 through the coupling spring, and show that the motion of x_1 satisfies the equation for a simple harmonic oscillator subject to this driving force.

38. In Fig. 4.10, $m_1 = m_2 = m$, $k_1 = k$, $k_2 = 0.9k$, $k_3 = 0.1k$. Initially mass m_2 is held fixed at its equilibrium position and mass m_1 is pulled a distance A from its equilibrium position; then both masses are released. Find $x_1(t)$ and $x_2(t)$ and show that your result agrees qualitatively with Fig. 4.13.

39. Find the two normal modes of vibration for a pair of identical damped coupled harmonic oscillators [Eqs. (4.180), (4.181)]. That is, $m_1 = m_2$, $b_1 = b_2$, $k_1 = k_2$. [*Hint:* If $k_3 = 0$, you can certainly find the solution. You will find this point helpful in factoring the secular equation.]

40. Set up the equations of motion for the system shown in Fig. 4.16. The relaxed lengths of the two springs are l_1, l_2. Separate the problem into two problems, one involving the motion of the center of mass, and the other involving the "internal motion" described by the two coordinates x_1, x_2. Find the normal modes of vibration.

41. The system of coupled oscillators shown in Fig. 4.10 is subject to an applied force

$$F = F_0 \cos \omega t,$$

applied to mass m_1. Set up the equations of motion and find the steady-state solution. Sketch the amplitude and phase of the oscillations of each oscillator as functions of ω.

CHAPTER 5

RIGID BODIES. ROTATION ABOUT AN AXIS. STATICS

5.1 THE DYNAMICAL PROBLEM OF THE MOTION OF A RIGID BODY

In order to apply the theorems of the preceding chapter to the motion of a rigid body, we regard a rigid body as a system of many particles whose positions relative to one another remain fixed. We may define a rigid body as a system of particles whose mutual distances are all constant. The forces which hold the particles at fixed distances from one another are internal forces, and may be imagined as exerted by rigid weightless rods connected between all pairs of particles. Forces like this which maintain certain fixed relations between the particles of a system are called *forces of constraint*. Such forces of constraint can always be regarded as satisfying Newton's third law (strong form), since the constraints *could* be maintained by rigid rods fastened to the particles by frictionless universal joints. We may therefore apply the theorems of conservation of linear and angular momentum to the motion of a rigid body. For a perfectly rigid body, the theorem of conservation of mechanical energy holds also, since we can show by Newton's third law that the forces of constraint do no work in a rigid motion of the system of particles. The work done by the force exerted by a moving rod on a particle at one end is equal and opposite to the work done by the force exerted by the rod on a particle at the other end, since both particles have the same component of velocity in the direction of the rod (Fig. 5.1):

$$\begin{aligned} \boldsymbol{F}_{2\to1}\cdot\boldsymbol{v}_1+\boldsymbol{F}_{1\to2}\cdot\boldsymbol{v}_2 &= \boldsymbol{F}_{2\to1}\cdot\boldsymbol{v}_1-\boldsymbol{F}_{2\to1}\cdot\boldsymbol{v}_2 \\ &= \boldsymbol{F}_{2\to1}\cdot(\boldsymbol{v}_1-\boldsymbol{v}_2) \\ &= 0. \end{aligned} \tag{5.1}$$

We shall base our derivation of the equations of motion of a rigid body on these conservation laws. No actual solid body is ever perfectly rigid, so that our theory of the motion of rigid bodies will be an idealized approximation to the motion of actual bodies. However, in most applications the deviation of actual solid bodies from true rigidity is not significant. In a like spirit is our assumption that the ideal rigid body can be imagined as made up of ideal point particles held at fixed distances from one another.

A solid body of ordinary size is composed of such a large number of atoms and molecules that for most purposes it is more convenient to represent its

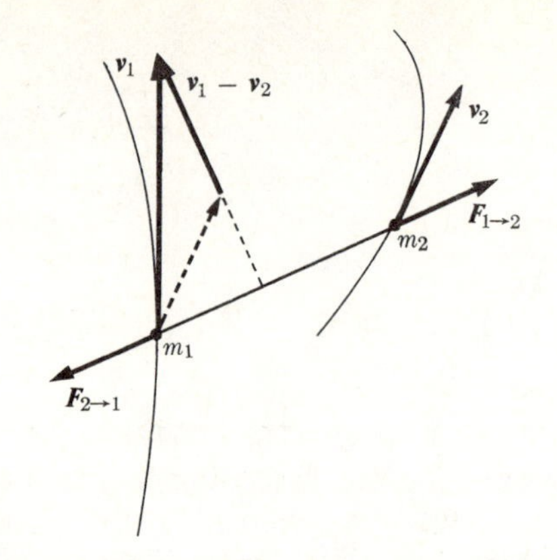

Fig. 5.1 Forces exerted by two particles connected by a rigid rod.

structure by specifying the average density ρ of mass per unit volume at each point in the body. The density is defined by

$$\rho = \frac{dM}{dV}, \tag{5.2}$$

where dM is the total mass in a volume dV which is to be chosen large enough to contain a large number of atoms, yet small enough so that the properties of the material are practically uniform within the volume dV. Only when a dV satisfying these two requirements can be chosen in the neighborhood of a point in the body can the density ρ be properly defined at that point. Sums over all the particles, such as occur in the expressions for total mass, total momentum, etc., can be replaced by integrals over the volume of the body. For example, the total mass is

$$M = \sum_i m_i = \iiint_{\text{(body)}} \rho \, dV. \tag{5.3}$$

Further examples will appear in the following sections.

In order to describe the position of a rigid body in space, six coordinates are needed. We may, for example, specify the coordinates (x_1, y_1, z_1) of some point P_1 in the body. Any other point P_2 of the body a distance r from P_1 will then lie somewhere on a sphere of radius r with center at (x_1, y_1, z_1). We can locate P_2 on this sphere with two coordinates, for example, the spherical coordinate angles θ_2, φ_2 with respect to a set of axes through the point (x_1, y_1, z_1). Any third point P_3 a distance $a \neq 0$ from the line through P_1 and P_2 must now lie on a circle of radius a about this line. We can locate P_3 on this circle with one coordinate. We thus require a total of six coordinates to locate the three points P_1, P_2, P_3 of the body, and when three noncollinear points are fixed, the locations of all points of a rigid body are fixed. There are many possible ways of choosing six coordinates by which the position of a body in space can be specified. Usually three of the six coordinates are used as above to locate some point in the body. The remaining three coordinates determine the orientation of the body about this point.

If a body is not connected to any supports, so that it is free to move in any manner, it is convenient to choose the center of mass as the point to be located by three coordinates (X, Y, Z), or by the vector $\boldsymbol{R}$. The motion of the center of mass $\boldsymbol{R}$ is then determined by the linear momentum theorem, which can be expressed in the form (4.18):

$$M\ddot{\boldsymbol{R}} = \boldsymbol{F}, \tag{5.4}$$

where M is the total mass and $\boldsymbol{F}$ is the total external force. The equation for the rotational motion about the center of mass is given by the angular momentum theorem (4.28):

$$\frac{d\boldsymbol{L}}{dt} = \boldsymbol{N}, \tag{5.5}$$

where $\boldsymbol{L}$ is the angular momentum and $\boldsymbol{N}$ is the torque about the point $\boldsymbol{R}$. If the force $\boldsymbol{F}$ is independent of the orientation of the body in space, as in the case of a body moving in a uniform gravitational field, the motion of the center of mass is independent of the rotational motion, and Eq. (5.4) is a separate equation which can be solved by the methods of Chapter 3. If the torque $\boldsymbol{N}$ is independent of the position $\boldsymbol{R}$ of the center of mass, or if $\boldsymbol{R}(t)$ is already known, so that $\boldsymbol{N}$ can be calculated as a function of time and of the orientation of the body, then the rotational motion about the center of mass may be determined from Eq. (5.5). In the more general case, when $\boldsymbol{F}$ and $\boldsymbol{N}$ each depend on both position and orientation, Eqs. (5.4) and (5.5) must be solved simultaneously as six coupled equations in some suitable set of coordinates; this case we shall not attempt to treat, although after the reader has studied Chapter 11, he will be able to set up for himself the six equations which must be solved.

If the body is constrained by external supports to rotate about a fixed point O, then moments and torques are to be computed about that point. We have to solve Eq. (5.5) for the rotation about the point O. In this case Eq. (5.4) serves only to determine the constraining force required to maintain the point O at rest.

The difficulty in applying Eq. (5.5) lies in the choice of three coordinates to describe the orientation of the body in space. The first thought that comes to mind is to choose a zero position for the body, and to specify any other orientation by specifying the angles of rotation φ_x, φ_y, φ_z, about three perpendicular axes, required to bring the body to this orientation. However, a little experimenting with a solid body will convince anyone that no suitable coordinates of this sort exist. Consider, for example, the position specified by $\varphi_x = 90°$, $\varphi_y = 90°$, $\varphi_z = 0$. If a body is first rotated 90° about the x-axis, and then 90° about the y-axis, the final position will be found to be different from that resulting from a 90° rotation about the y-axis followed by a 90° rotation about the x-axis. It turns out that no simple symmetric set of coordinates can be found to describe the orientation of a body, analogous to the coordinates x, y, z which locate the position of a point in space. We therefore postpone to Chapter 11 the treatment of the rather difficult problem of the rotation of a body around a point. We shall discuss here only the simple problem of rotation about a fixed axis.

5.2 ROTATION ABOUT AN AXIS

It requires only one coordinate to specify the orientation of a body which is free to rotate only about a fixed axis. Let the fixed axis be taken as the z-axis, and let a line $\overline{OA}$ in the body, through the axis and lying in (or parallel to) the xy-plane, be chosen. We fix the position of the body by specifying the angle θ between the line $\overline{OA}$ fixed in the body and the x-axis. Choosing cylindrical coordinates to locate each particle in the body, we now compute the total angular momentum about the z-axis. (See Fig. 5.2.) We shall write r_i instead of ρ_i to represent the distance of particle m_i from the z-axis, in order to avoid confusion with the density ρ:

$$L = \sum_i m_i r_i^2 \dot{\varphi}_i. \tag{5.6}$$

Let β_i be the angle between the direction of the line $\overline{OA}$ in the body and the direction of the radius from the z-axis to the particle m_i. Then, for a rigid body, β_i is constant, and

$$\varphi_i = \theta + \beta_i, \tag{5.7}$$

$$\dot{\varphi}_i = \dot{\theta}. \tag{5.8}$$

Substituting in Eq. (5.6), we have

$$\begin{aligned} L &= \sum_i m_i r_i^2 \dot{\theta} \\ &= \left(\sum_i m_i r_i^2\right) \dot{\theta} \\ &= I_z \dot{\theta}, \end{aligned} \tag{5.9}$$

where

$$I_z = \sum_i m_i r_i^2. \tag{5.10}$$

The quantity I_z is a constant for a given body rotating about a given axis, and is called the *moment of inertia* about that axis. We may also express I_z as an integral over the body:

$$I_z = \underset{\text{(body)}}{\iiint} \rho r^2 \, dV. \tag{5.11}$$

It is sometimes convenient to introduce the *radius of gyration* k_z defined by the equation

$$Mk_z^2 = I_z; \tag{5.12}$$

that is, k_z is a radius such that if all the mass of the body were situated a distance k_z from the axis, its moment of inertia would be I_z.

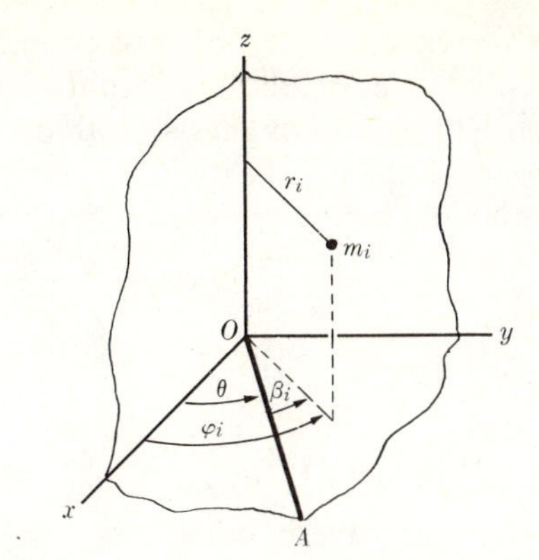

Fig. 5.2 Coordinates of a particle in a rigid body.

Using Eq. (5.9), we may write the component of Eq. (5.5) along the axis of rotation in the form

$$\frac{dL}{dt} = I_z\ddot{\theta} = N_z, \tag{5.13}$$

where N_z is the total external torque about the axis. Equation (5.13) is the equation of motion for rotation of a rigid body about a fixed axis. It has the same form as Eq. (2.1) for the motion of a particle along a straight line. The problem of rotation of a body about a fixed axis is therefore equivalent to the problem treated in Chapter 2. All methods and results of Chapter 2 can be extended directly to the present problem according to the following scheme of analogy:

Rectilinear motion		*Rotation about a fixed axis*	
position:	x	angular position:	θ
velocity:	$v = \dot{x}$	angular velocity:	$\omega = \dot{\theta}$
acceleration:	$a = \ddot{x}$	angular acceleration:	$\alpha = \ddot{\theta}$
force:	F	torque:	N_z
mass:	m	moment of inertia:	I_z
potential energy: $V(x) = -\int_{x_s}^{x} F(x)\,dx$, $F(x) = -\dfrac{dV}{dx}$		potential energy: $V(\theta) = -\int_{\theta_s}^{\theta} N_z(\theta)\,d\theta$, $N_z(\theta) = -\dfrac{dV}{d\theta}$	
kinetic energy:	$T = \frac{1}{2}m\dot{x}^2$	kinetic energy:	$T = \frac{1}{2}I_z\dot{\theta}^2$
linear momentum:	$p = m\dot{x}$	angular momentum:	$L = I_z\dot{\theta}$

The only mathematical difference between the two problems is that the moment of inertia I_z depends upon the location of the axis in the body, while the mass of a body does not depend on its position or on its motion. This does not affect the treatment of rotation about a single fixed axis.

The rotational potential and kinetic energies defined by the equations,

$$V(\theta) = -\int_{\theta_s}^{\theta} N_z(\theta)\, d\theta, \tag{5.14}$$

$$N_z = -\frac{dV}{d\theta}, \tag{5.15}$$

$$T = \tfrac{1}{2}I_z\dot{\theta}^2, \tag{5.16}$$

are not merely analogous to the corresponding quantities defined by Eqs. (2.41), (2.47), and (2.5) for linear motion. They are, in fact, equal to the potential and kinetic energies, defined in Chapters 2 and 4, of the system of particles making up the rigid body. The potential energy defined by Eq. (5.14), for example, is the work done against the forces whose torque is N_z, when the body is rotated through the angle $\theta - \theta_s$. The kinetic energy defined by Eq. (5.16) is just the sum of the ordinary kinetic energies of motion of the particles making up the body. The proof of these statements is left as an exercise.

5.3 THE SIMPLE PENDULUM

As an example of the treatment of rotational motion, we consider the motion of a simple pendulum, consisting of a mass m suspended from a fixed point O by a string or weightless rigid rod of length l. If a string supports the mass m, we must suppose that it remains taut, so that the distance l from m to O remains constant; otherwise we cannot treat the system as a rigid one. We consider only motions of the pendulum in one vertical plane, in order to be able to apply the simple theory of motion about a single fixed axis through O. We then have (Fig. 5.3)

$$I_z = ml^2, \tag{5.17}$$

$$N_z = -mgl \sin \theta, \tag{5.18}$$

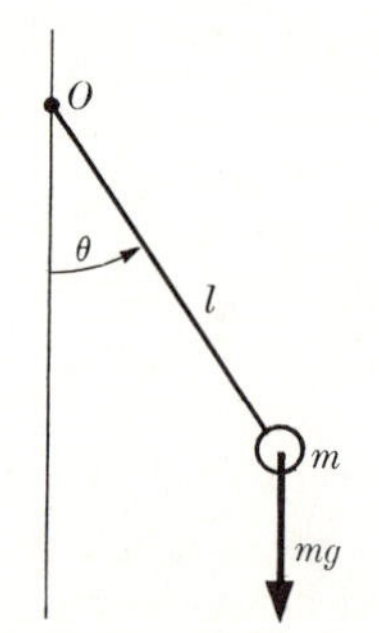

Fig. 5.3 The simple pendulum.

where the z-axis is an axis through O perpendicular to the plane in which the pendulum is swinging. The torque is taken as negative, since it acts in such a direction as to decrease the angle θ. Substituting in the equation of motion (5.13), we find

$$\ddot{\theta} = -\frac{g}{l}\sin\theta. \tag{5.19}$$

This equation is not easy to solve. If, however, we consider only small oscillations of the pendulum (say $\theta \ll \pi/2$), then $\sin\theta \doteq \theta$, and we can write

$$\ddot{\theta} + \frac{g}{l}\theta \doteq 0. \tag{5.20}$$

This is of the same form as Eq. (2.89) for the harmonic oscillator. Its solution is

$$\theta = \kappa \cos(\omega t + \beta), \tag{5.21}$$

where

$$\omega = \left(\frac{g}{l}\right)^{1/2}, \tag{5.22}$$

and κ and β are arbitrary constants which determine the amplitude and phase of the oscillation. Notice that the frequency of oscillation is independent of the amplitude, provided the amplitude is small enough so that Eq. (5.20) is a good approximation. This is the basis for the use of a pendulum to regulate the speed of a clock.

We can treat the problem of motion at large amplitudes by means of the energy integral. The potential energy associated with the torque given by Eq. (5.18) is

$$\begin{aligned} V(\theta) &= -\int_{\theta_s}^{\theta} -mgl\sin\theta\, d\theta \\ &= -mgl\cos\theta, \end{aligned} \tag{5.23}$$

where we have taken $\theta_s = \pi/2$ for convenience. We could have written down $V(\theta)$ right away as the gravitational potential energy of a mass m, referred to the horizontal plane through O as the level of zero potential energy. The energy integral is

$$\tfrac{1}{2}ml^2\dot{\theta}^2 - mgl\cos\theta = E. \tag{5.24}$$

We could prove that E is constant from the equation of motion (5.13), but we need not, since the analogy described in the preceding section guarantees that all theorems for one-dimensional linear motion will hold in their analogous forms for rotational motion about an axis. The potential energy $V(\theta)$ is plotted in Fig. 5.4. We see that for $-mgl < E < mgl$, the motion is an oscillating one, becoming simple harmonic motion for E slightly greater than $-mgl$. For $E > mgl$, the motion is nonoscillatory; θ steadily increases or steadily decreases, with $\dot{\theta}$ oscillating

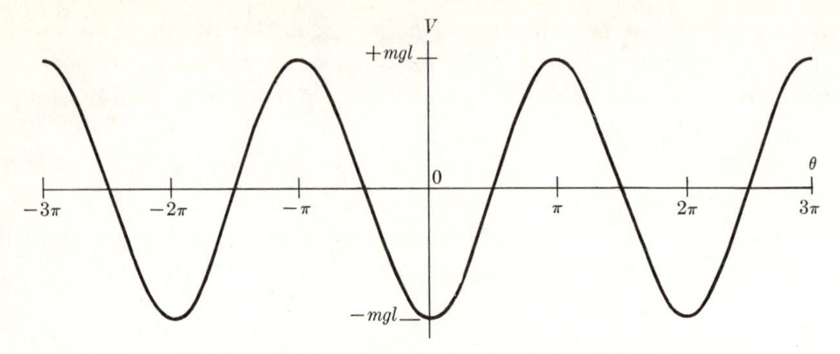

Fig. 5.4 Potential energy for simple pendulum.

between a maximum and minimum value. Physically, when $E > mgl$, the pendulum has enough energy to swing around in a complete circle. (In this case, of course, the mass must be held by a rigid rod instead of a string, unless $\dot{\theta}$ is very large.) This motion is still a periodic one, the pendulum making one complete revolution each time θ increases or decreases by 2π. In either case, the attempt to solve Eq. (5.24) for θ leads to the equation

$$\int_{\theta_0}^{\theta} \frac{d\theta}{(E/mgl+\cos\theta)^{1/2}} = \left(\frac{2g}{l}\right)^{1/2} t. \tag{5.25}$$

The integral on the left must be evaluated in terms of elliptic functions. The period of the motion can be obtained by integrating between appropriate limits. When the motion is oscillatory ($E < mgl$), the maximum value κ of θ is given, according to Eq. (5.24), by

$$E = -mgl\cos\kappa. \tag{5.26}$$

Equation (5.25) becomes, in this case,

$$\int_{\theta_0}^{\theta} \frac{d\theta}{(\cos\theta-\cos\kappa)^{1/2}} = \left(\frac{2g}{l}\right)^{1/2} t, \tag{5.27}$$

which can also be written

$$\int_{\theta_0}^{\theta} \frac{d\theta}{[\sin^2(\kappa/2)-\sin^2(\theta/2)]^{1/2}} = 2\left(\frac{g}{l}\right)^{1/2} t. \tag{5.28}$$

The angle θ oscillates between the limits $\pm\kappa$. We now introduce a new variable φ which runs from 0 to 2π for one cycle of oscillation of θ:

$$\sin\varphi = \frac{\sin\theta/2}{\sin\kappa/2} = \frac{1}{a}\sin\frac{\theta}{2}, \tag{5.29}$$

where

$$a = \sin\frac{\kappa}{2}. \tag{5.30}$$

With these substitutions, Eq. (5.28) can be written

$$\int_0^{\varphi} \frac{d\varphi}{(1-a^2\sin^2\varphi)^{1/2}} = \left(\frac{g}{l}\right)^{1/2} t, \tag{5.31}$$

where we have taken $\theta_0 = 0$, for convenience. The integral is now in a standard form for elliptic integrals. When a is small, the integrand can be expanded in a power series in a^2:

$$\int_0^{\varphi} [1+\tfrac{1}{2}a^2\sin^2\varphi+\cdots]\,d\varphi = \left(\frac{g}{l}\right)^{1/2} t. \tag{5.32}$$

This can be integrated term by term:

$$\varphi+\tfrac{1}{8}a^2(2\varphi-\sin 2\varphi)+\cdots = \left(\frac{g}{l}\right)^{1/2} t. \tag{5.33}$$

The period of the motion is obtained by setting $\varphi = 2\pi$:

$$\tau = 2\pi\left(\frac{l}{g}\right)^{1/2}\left(1+\frac{a^2}{4}+\cdots\right). \tag{5.34}$$

Thus as the amplitude of oscillation becomes large, the period becomes slightly longer than for small oscillations, a prediction which is readily verified experimentally by setting up two pendulums of equal length and setting them to swinging at unequal amplitudes. Equation (5.33) can be solved approximately for φ by successive approximations, and the result substituted in Eq. (5.29), which can be solved for θ by successive approximations. The result, to a second approximation, is

$$\theta \doteq \left(\kappa+\frac{\kappa^3}{192}\right)\sin\omega' t+\frac{\kappa^3}{192}\sin 3\omega' t, \tag{5.35}$$

where

$$\omega' = \frac{2\pi}{\tau} = \left(\frac{g}{l}\right)^{1/2}\left(1-\frac{\kappa^2}{16}+\cdots\right). \tag{5.36}$$

If we neglect terms in κ^2 and κ^3, this solution agrees with Eq. (5.21). At larger amplitudes in second approximation, the frequency is slightly lower than at small amplitudes, and the motion of θ contains a small third harmonic term.

5.4 THE COMPOUND PENDULUM

A rigid body suspended and free to swing about an axis is called a *compound pendulum*. We assume that the axis does not pass through the center of mass, and we specify the position of the body by the angle θ between a vertical line and a perpendicular line drawn from a point O on the axis, through the center of mass

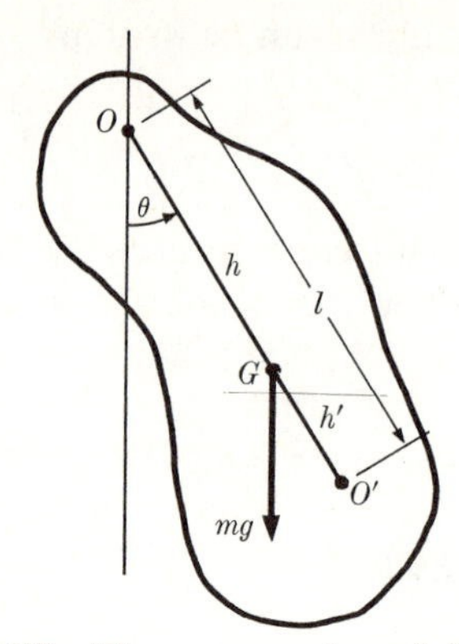

Fig. 5.5 The compound pendulum.

G (Fig. 5.5). In order to compute the total torque exerted by gravity, we anticipate a theorem, to be proved later [Eq. (5.103)], that the total torque is the same as if the total gravitational force were applied at the center of mass G. We then have, using Eqs. (5.12) and (5.13),

$$Mk_O^2\ddot{\theta} = -Mgh\sin\theta, \tag{5.37}$$

where h is the distance $\overline{OG}$. This equation is the same as Eq. (5.19) for a simple pendulum of length l, if we take

$$l = \frac{k_O^2}{h}. \tag{5.38}$$

The point O' a distance l from O along the line through the center of mass G is called the *center of oscillation*. If all the mass M were at O', the motion of the pendulum would be the same as its actual motion, for any given initial conditions. If the distance $\overline{O'G}$ is h', we have

$$l = h+h', \tag{5.39}$$

$$hh' = k_O^2-h^2. \tag{5.40}$$

It will be shown in the next section [theorem (5.81)] that the moment of inertia about any axis equals the moment of inertia about a parallel axis through the center of mass G plus Mh^2, where h is the distance from the axis to G. Let k_G be the radius of gyration about G. We then have

$$k_O^2 = k_G^2+h^2, \tag{5.41}$$

so that Eq. (5.40) becomes

$$hh' = k_G^2\,. \tag{5.42}$$

Since this equation is symmetrical in h and h', we conclude that if the body were

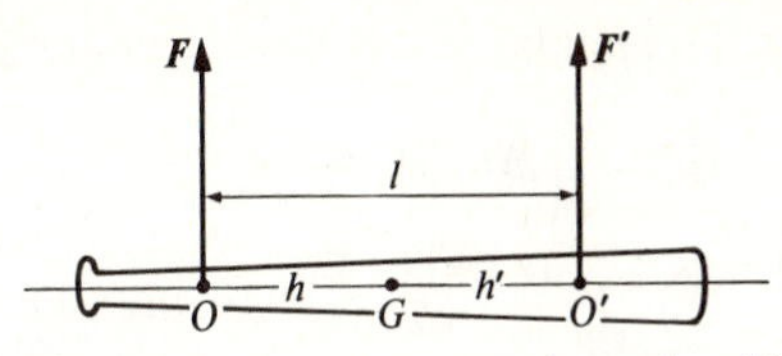

Fig. 5.6 Rigid body pivoted at O and struck a blow at O'.

suspended about a parallel axis through O', the center of oscillation would be at O. The acceleration g of gravity can be measured very accurately by measuring the period of small oscillations of a pendulum and using Eq. (5.22). If a compound pendulum is used, the radius of gyration must be known, or the period measured about two axes (preferably O, O'), so that the radius of gyration can be eliminated from the equations.

A problem closely related to the compound pendulum is the following. Consider a body (Fig. 5.6) free to rotate about a fixed axis through a point O. Let it be struck a blow at a point O' a distance l from the axis, the direction of the blow being perpendicular to the line $\overline{OO'}$ from the axis to O'. Place O' so that the line $\overline{OO'}$ passes through the center of mass G, and let h, h' be the distances $\overline{OG}$, $\overline{O'G}$. The impulse delivered at the point O' by the force F' during the blow is

$$J' = \int F'dt. \tag{5.43}$$

At the instant the blow is struck, a force F will, in general, have to be exerted on the body at the point O on the axis in order to keep O fixed. The impulse delivered to the body at O is

$$J = \int F\,dt. \tag{5.44}$$

Since we are concerned only with its motion shortly after the blow, we will neglect all other forces on the body except the two impulses F and F'. We wish to find the impulse J, and in particular, to find under what conditions $J = 0$. To do this, we use the linear and angular momentum theorems.

The momentum theorem for the component P' of linear momentum of the body in the direction of $\boldsymbol{F}$ is:

$$\frac{dP}{dt} = \frac{d}{dt}(Mh\dot{\theta}) = F+F', \tag{5.45}$$

where $\dot{\theta}$ is the angular velocity of the body about O. From this we have, for the momentum just after the blow,

$$Mh\dot{\theta} = J+J', \tag{5.46}$$

assuming that the body is initially at rest. The angular momentum theorem about O is:

$$\frac{dL}{dt} = \frac{d}{dt}(Mk_O^2\dot{\theta}) = F'l. \tag{5.47}$$

Integrating, we have, for the angular momentum just after the blow,

$$Mk_O^2\dot{\theta} = J'l. \tag{5.48}$$

We eliminate $\dot{\theta}$ between Eqs. (5.46) and (5.48):

$$hl = k_O^2\left\{1+\frac{J}{J'}\right\}. \tag{5.49}$$

We could now solve for J. By Newton's third law, an equal and opposite impulse $-J$ is delivered by the body to the axis at O. We now ask for the condition that no impulsive force be exerted on the axis at O at the instant of the blow, i.e., $J = 0$:

$$hl = k_O^2. \tag{5.50}$$

This equation is identical with Eq. (5.38) and may also be expressed in the symmetrical form [Eq. (5.42)]

$$hh' = k_G^2. \tag{5.51}$$

The point O' at which a blow must be struck in order that no impulse be felt at the point O is called the *center of percussion* relative to O. We see that the center of percussion is the same as the center of oscillation relative to O, and that O is the center of percussion relative to O'. This problem is of interest to a batter attempting to hit a baseball with his bat. He should try to hit the ball at the center of percussion (O') relative to his hands (at O). If the ball hits very far from the center of percussion, the blow is transmitted to his hands by the bat. If an unsupported body at rest, but free to move in space, is struck a blow J' at a point O', we can see that its initial motion must be a rotation about the center of percussion O relative to O'. For if it were to begin rotating about any other point O, an impulse J would be required at O. Its initial angular velocity $\dot{\theta}$ about O can be found from Eq. (5.46) (with $J = 0$). Of course, if the body is unsupported, its center of mass will move with the constant velocity $h\dot{\theta}$ after the blow (as long as gravity is neglected), and it will rotate with constant angular velocity $\dot{\theta}$. The point O will not remain fixed in this case although its initial velocity, just after the blow, is zero.

5.5 COMPUTATION OF CENTERS OF MASS AND MOMENTS OF INERTIA

We have given in Section 4.1 the following definition of center of mass for a system of particles:

$$\boldsymbol{R} = \frac{1}{M}\sum_i m_i\boldsymbol{r}_i. \tag{5.52}$$

For a solid body, the sum may be expressed as an integral:

$$\boldsymbol{R} = \frac{1}{M}\iiint \rho \boldsymbol{r}\, dV, \tag{5.53}$$

or, in component form,

$$X = \frac{1}{M}\iiint \rho x\, dV, \tag{5.54}$$

$$Y = \frac{1}{M}\iiint \rho y\, dV, \tag{5.55}$$

$$Z = \frac{1}{M}\iiint \rho z\, dV. \tag{5.56}$$

The integrals can be extended either over the volume of the body, or over all space, since $\rho = 0$ outside the body. These equations define a point G of the body whose coordinates are (X, Y, Z). We should first prove that the point G thus defined is independent of the choice of coordinate system. Since Eq. (5.52) or (5.53) is in vector form, and makes no reference to any particular set of axes, the definition of G certainly does not depend on any particular choice of directions for the axes. We should prove, however, that G is independent also of the choice of origin. Consider a system of particles, and let any particle m_i be located by vectors $\boldsymbol{r}_i$ and $\boldsymbol{r}'_i$ with respect to any two origins O and O'. If $\boldsymbol{a}$ is the vector from O to O', the relation between $\boldsymbol{r}_i$ and $\boldsymbol{r}'_i$ is (Fig. 5.7)

$$\boldsymbol{r}_i = \boldsymbol{r}'_i + \boldsymbol{a}. \tag{5.57}$$

The centers of mass G, G' with respect to O, O' are located by the vectors $\boldsymbol{R}$ and $\boldsymbol{R}'$, where $\boldsymbol{R}'$ is defined by

$$\boldsymbol{R}' = \frac{1}{M}\sum_i m_i \boldsymbol{r}'_i. \tag{5.58}$$

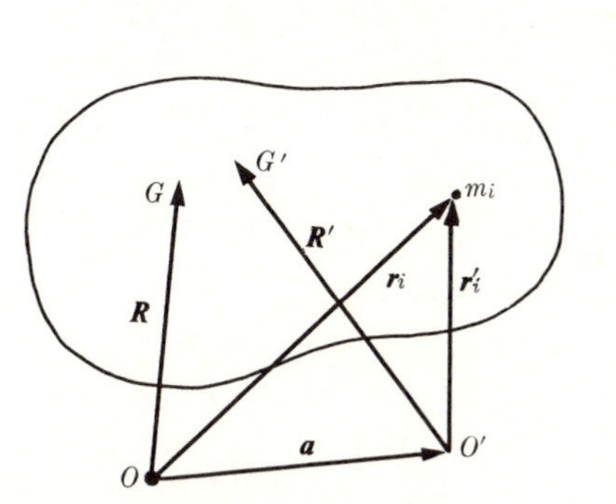

Fig. 5.7 Location of center of mass relative to two different origins.

Using Eq. (5.57), we can rewrite Eq. (5.52):

$$\begin{aligned}\boldsymbol{R} &= \frac{1}{M}\sum_i m_i(\boldsymbol{r}'_i+\boldsymbol{a}) \\ &= \frac{1}{M}\sum_i m_i\boldsymbol{r}'_i+\frac{1}{M}\boldsymbol{a}\sum_i m_i \\ &= \boldsymbol{R}'+\boldsymbol{a}.\end{aligned} \tag{5.59}$$

Thus $\boldsymbol{R}$ and $\boldsymbol{R}'$ are vectors locating the same point with respect to O and O' so that G and G' are the same point.

General theorems like the one above can be proved either for a system of particles or for a body described by a density ρ. Whichever point of view is adopted in any proof, a parallel proof can always be given from the other point of view.

Much of the labor involved in the calculation of the position of the center of mass from Eqs. (5.54), (5.55), (5.56) can often be avoided by the use of certain labor-saving theorems, including the theorem proved above which allows us a free choice of coordinate axes and origin. We have first the following theorem regarding symmetrical bodies:

Theorem *If a body is symmetrical with respect to a plane, its center of mass lies in that plane.* (5.60)

When we say a body is symmetrical with respect to a plane, we mean that for every particle on one side of the plane there is a particle of equal mass located at its mirror image in the plane. For a continuously distributed mass, we mean that the density at any point equals the density at its mirror image in the plane. Choose the origin in the plane of symmetry, and let the plane of symmetry be the xy-plane. Then in computing Z from Eq. (5.56) [or (5.52)], for each volume element (or particle) at a point (x, y, z) above the xy-plane, there is, by symmetry, a volume element of equal mass at the point $(x, y, -z)$ below the xy-plane, and the contributions of these two elements to the integral in Eq. (5.56) will cancel. Hence $Z = 0$, and the center of mass lies in the xy-plane. This proves Theorem (5.60). The theorem has a number of obvious corollaries:

If a body is symmetrical in two planes, its center of mass lies on their line of intersection. (5.61)

If a body is symmetrical about an axis, its center of mass lies on that axis. (5.62)

If a body is symmetrical in three planes with one common point, that point is its center of mass. (5.63)

If a body has spherical symmetry about a point (i.e., if the density depends only on the distance from that point), that point is its center of mass. (5.64)

These theorems enable us to locate the center of mass immediately in some cases, and to reduce the problem to a computation of only one or two coordinates of the center of mass in other cases. One should be on the lookout for symmetries, and use them to simplify the problem. Other cases not included in these theorems will occur (e.g., the parallelepiped), where it will be evident that certain integrals will be equal or will cancel, and the center of mass can be located without computing them.

Another theorem which often simplifies the location of the center of mass is that if a body is composed of two or more parts whose centers of mass are known, then the center of mass of the composite body can be computed by regarding its component parts as single particles located at their respective centers of mass. Let a body, or system of particles, be composed of n parts of masses $M_1, \ldots, M_n$. Let any part M_k be composed of N_k particles of masses $m_{k1}, \ldots, m_{kN_k}$, located at the points $\boldsymbol{r}_{k1}, \ldots, \boldsymbol{r}_{kN_k}$. Then the center of mass of the part M_k is located at the point

$$\boldsymbol{R}_k = \frac{1}{M_k} \sum_{l=1}^{N_k} m_{kl} \boldsymbol{r}_{kl}, \tag{5.65}$$

and

$$M_k = \sum_{l=1}^{N_k} m_{kl}. \tag{5.66}$$

The center of mass of the entire body is located at the point

$$\boldsymbol{R} = \frac{1}{M} \sum_{k=1}^{n} \sum_{l=1}^{N_k} m_{kl} \boldsymbol{r}_{kl}, \tag{5.67}$$

where

$$M = \sum_{k=1}^{n} \sum_{l=1}^{N_k} m_{kl}. \tag{5.68}$$

By Eq. (5.65), Eq. (5.67) becomes

$$\boldsymbol{R} = \frac{1}{M} \sum_{k=1}^{n} M_k \boldsymbol{R}_k, \tag{5.69}$$

and by Eq. (5.66), Eq. (5.68) becomes

$$M = \sum_{k=1}^{n} M_k. \tag{5.70}$$

Equations (5.69) and (5.70) are the mathematical statement of the theorem to be proved.

As an example, let us consider a uniform rectangular block with a cylindrical hole drilled out, as shown in Fig. 5.8. By the symmetry about the two vertical

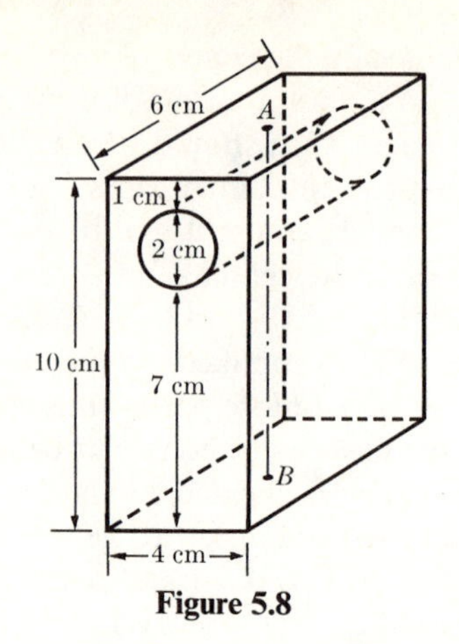

Figure 5.8

planes bisecting the block parallel to its sides, we conclude that the center of mass lies along the vertical line $\overline{AB}$ through the centers of the top and bottom faces. Let the center of mass of the block lie a distance Z below A, and let the density of the block be ρ. If the hole were not cut out, the mass of the block would be $6 \text{ cm} \times 4 \text{ cm} \times 10 \text{ cm} \times \rho$, and its center of mass would be at the midpoint of $\overline{AB}$, 5 cm from A. The mass of the material drilled out is $\pi \text{ cm}^2 \times 6 \text{ cm} \times \rho$, and its center of mass, before it was removed, was on $\overline{AB}$, 2 cm below A. Hence the theorem (5.69) above allows us to write

$$(6 \text{ cm} \times 4 \text{ cm} \times 10 \text{ cm} \times \rho) \times 5 \text{ cm} = (\pi \text{ cm}^2 \times 6 \text{ cm} \times \rho) \times 2 \text{ cm} + 6 \text{ cm} \times (4 \text{ cm} \times 10 \text{ cm} - \pi \text{ cm}^2) \times \rho \times Z.$$

The solution for Z is

$$Z = \frac{6 \times 4 \times 10 \times 5 - \pi \times 6 \times 2}{6 \times (4 \times 10 - \pi)} \text{ cm}.$$

As a second example, we locate the center of mass of a hemisphere of radius a. By symmetry, if the density is uniform, the center of mass lies on the axis of symmetry, which we take as the z-axis. We have then to compute only the integral in Eq. (5.56). The integral can be set up in rectangular, cylindrical, or spherical coordinates (Fig. 5.9):

$$\text{Rectangular: } Z = \frac{1}{M} \int_{z=0}^{a} \int_{y=-(a^2-z^2)^{1/2}}^{(a^2-z^2)^{1/2}} \int_{x=-(a^2-z^2-y^2)^{1/2}}^{(a^2-z^2-y^2)^{1/2}} \rho z \, dx \, dy \, dz.$$

$$\text{Cylindrical: } Z = \frac{1}{M} \int_{z=0}^{a} \int_{\varphi=0}^{2\pi} \int_{r=0}^{(a^2-z^2)^{1/2}} \rho z r \, dr \, d\varphi \, dz.$$

$$\text{Spherical: } Z = \frac{1}{M} \int_{r=0}^{a} \int_{\theta=0}^{\pi/2} \int_{\varphi=0}^{2\pi} (\rho r \cos\theta) r^2 \sin\theta \, dr \, d\theta \, d\varphi.$$

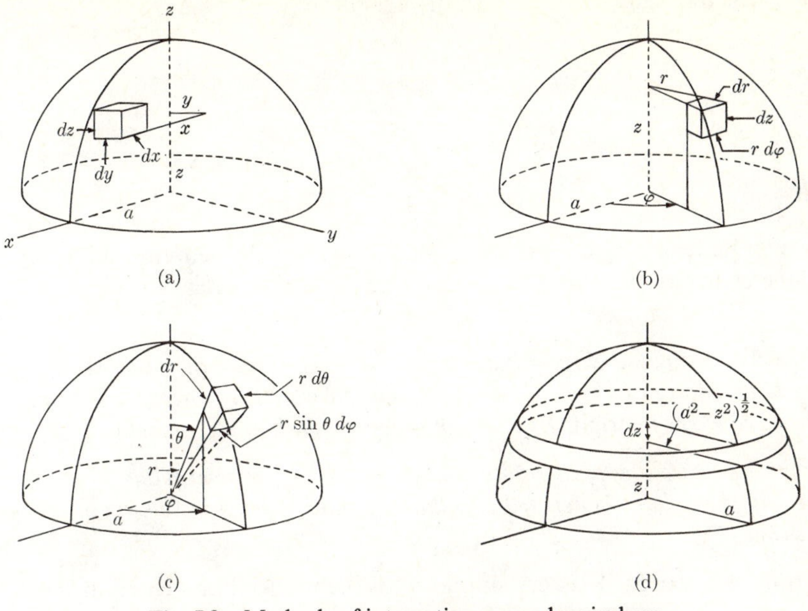

Fig. 5.9 Methods of integrating over a hemisphere.

Any one of these expressions can be used to evaluate Z for any density distribution. If ρ is uniform, we can also build up the hemisphere out of rings or disks and save one or two integrations. For example, building up the hemisphere out of disks perpendicular to the z-axis (this is equivalent to carrying out the integration over r and φ in cylindrical coordinates), we can write

$$\begin{aligned} Z &= \frac{1}{M}\int_{z=0}^{a} z\rho\pi(a^2-z^2)\,dz \\ &= \left(\frac{1}{\frac{2}{3}\pi a^3\rho}\right)\left(\frac{\pi a^4\rho}{4}\right) = \tfrac{3}{8}a, \end{aligned} \tag{5.71}$$

where the integrand is $z\rho$ times the volume of a disk of thickness dz, radius $(a^2-z^2)^{1/2}$.

When the density ρ is uniform, the center of mass of a body depends only on its geometrical shape, and is given by

$$\boldsymbol{R} = \frac{1}{V}\iiint_V \boldsymbol{r}\,dV. \tag{5.72}$$

The point G whose coordinate $\boldsymbol{R}$ is given by Eq. (5.72) is called the *centroid* of the volume V. If we replace the volume V by an area A or curve C in space, we obtain formulas for the centroid of an area or of a curve:

$$\boldsymbol{R} = \frac{1}{A}\iint_A \boldsymbol{r}\,dA, \tag{5.73}$$

$$\boldsymbol{R} = \frac{1}{s}\int_C \boldsymbol{r}\,ds, \tag{5.74}$$

where s is the length of the curve C. The following two theorems, due to Pappus, relate the centroid of an area or curve to the volume or area swept out by it when it is rotated about an axis:

Theorem 1. *If a plane curve rotates about an axis in its own plane which does not intersect it, the area of the surface of revolution which it generates is equal to the length of the curve multiplied by the length of the path of its centroid.* (5.75)

Theorem 2. *If a plane area rotates about an axis in its own plane which does not intersect it, the volume generated is equal to the area times the length of the path of its centroid.* (5.76)

The proof of Theorem 1 is very simple, with the notation indicated in Fig. 5.10:

$$A = \int_C 2\pi y\,ds = 2\pi \int_C y\,ds = 2\pi\,Ys, \tag{5.77}$$

where Y is the y-coordinate of the centroid of the curve C, and s is its length. The proof of Theorem 2 is similar and is left to the reader. These theorems may be used to determine areas and volumes of figures symmetrical about an axis when the centroids of the generating curves or areas are known, and conversely. We locate, for example, the position of the center of mass of a uniform semicircular disk of radius a, using Pappus' second theorem. If the disk is rotated about its diameter, the volume of the sphere generated, by Pappus' theorem (Fig. 5.11), is

$$\tfrac{4}{3}\pi a^3 = \left(\frac{\pi a^2}{2}\right)(2\pi Y),$$

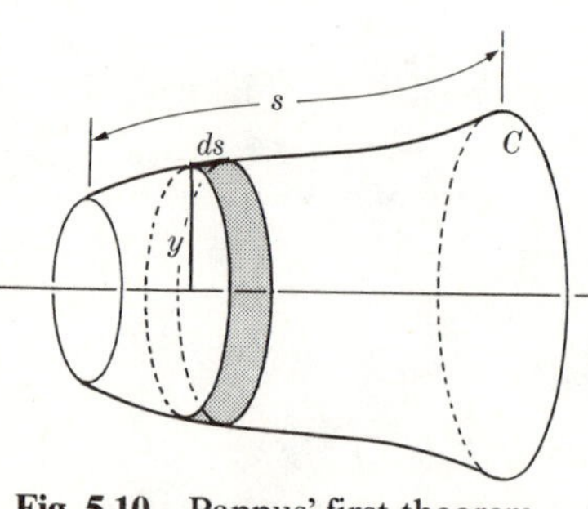

Fig. 5.10 Pappus' first theorem.

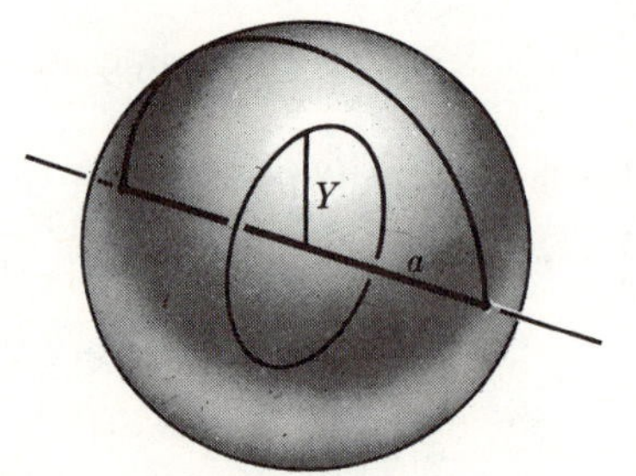

Fig. 5.11 Sphere formed by rotating a semicircle.

from which we obtain

$$Y = \frac{4a}{3\pi}. \tag{5.78}$$

The moment of inertia I of a body about an axis is defined by Eq. (5.10):

$$I = \sum_i m_i r_i^2, \tag{5.79}$$

or

$$I = \iiint \rho r^2 \, dV, \tag{5.80}$$

where r is the distance from each point or particle of the body to the given axis. We first prove several labor-saving theorems regarding moments of inertia:

Parallel axis theorem. *The moment of inertia of a body about any given axis is the moment of inertia about a parallel axis through the center of mass, plus the moment of inertia about the given axis if all the mass of the body were located at the center of mass.* (5.81)

To prove this theorem, let I_O be the moment of inertia about a z-axis through the point O, and let I_G be the moment of inertia about a parallel axis through the center of mass G. Let $\boldsymbol{r}$ and $\boldsymbol{r}'$ be the vectors to any point P in the body, from O and G, respectively, and let $\boldsymbol{R}$ be the vector from O to G. The components of these vectors will be designated by (x, y, z), (x', y', z'), and (X, Y, Z). Then, since (Fig. 5.12)

$$\boldsymbol{r} = \boldsymbol{r}' + \boldsymbol{R},$$

we see that

$$\begin{aligned} x^2 + y^2 &= (x' + X)^2 + (y' + Y)^2 \\ &= x'^2 + y'^2 + X^2 + Y^2 + 2Xx' + 2Yy', \end{aligned}$$

so that the moment of inertia I_O is

$$\begin{aligned} I_O &= \iiint (x^2 + y^2)\rho \, dV \\ &= \iiint (x'^2 + y'^2)\rho \, dV + (X^2 + Y^2) \iiint \rho \, dV + 2X \iiint x'\rho \, dV \\ &\quad + 2Y \iiint y'\rho \, dV. \end{aligned} \tag{5.82}$$

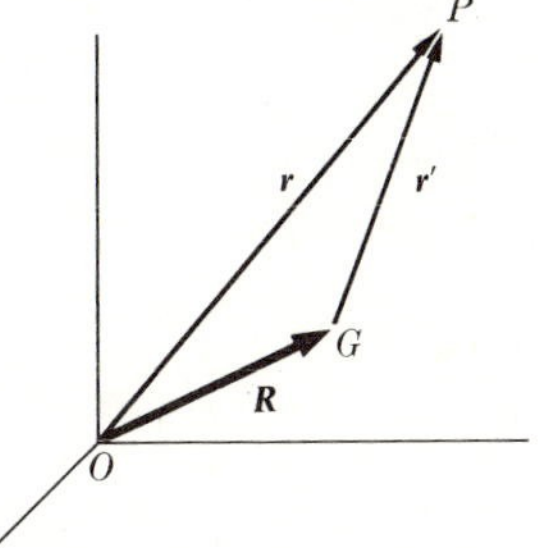

Fig. 5.12 Location of point P with respect to points O and G.

The first integral is I_G, and the integral in the second term is the total mass M of the body. The integrals in the last two terms are the same as the integrals occurring in Eqs. (5.54) and (5.55), and define the x- and y-coordinates of the center of mass relative to G. Since G is the center of mass, these integrals are zero, and we have

$$I_O = I_G + M(X^2 + Y^2). \tag{5.83}$$

This is the mathematical statement of the Parallel Axis Theorem. If we know the moment of inertia about any axis, and can locate the center of mass, we can use this theorem to determine the moment of inertia about any other parallel axis.

The moment of inertia of a composite body about any axis may be found by adding the moments of inertia of its parts about the same axis, a statement which is obvious from the definition of moment of inertia. This fact can be put to use in the same way as the analogous result for the center of mass of a composite body.

A body whose mass is concentrated in a single plane is called a *plane lamina*. We have the following theorem for a plane lamina:

Perpendicular axis theorem. *The sum of the moments of inertia of a plane lamina about any two perpendicular axes in the plane of the lamina is equal to the moment of inertia about an axis through their point of intersection perpendicular to the lamina.* (5.84)

The proof of this theorem is very simple. Consider any particle of mass m in the xy-plane. Its moments of inertia about the x- and y-axes are

$$I_x = my^2, \qquad I_y = mx^2. \tag{5.85}$$

Adding these, we have the moment of inertia of m about the z-axis:

$$I_x + I_y = m(x^2 + y^2) = I_z. \tag{5.86}$$

Since the moment of inertia of any lamina in the xy-plane is the sum of the moments of inertia of the particles of which it is composed, we have Theorem (5.84).

We illustrate these theorems by finding the moments of inertia of a uniform circular ring of radius a, mass M, lying in the xy-plane (Fig. 5.13). The moment of inertia about a z-axis perpendicular to the plane of the ring through its center is easily computed:

$$I_z = Ma^2. \tag{5.87}$$

The moments I_x and I_y are evidently equal, and we have, therefore, by Theorem (5.84),

$$I_x = \tfrac{1}{2}I_z = \tfrac{1}{2}Ma^2. \tag{5.88}$$

The moment of inertia about an axis A tangent to the ring is, by the Parallel Axis Theorem,

$$I_A = I_x + Ma^2 = \tfrac{3}{2}Ma^2. \tag{5.89}$$

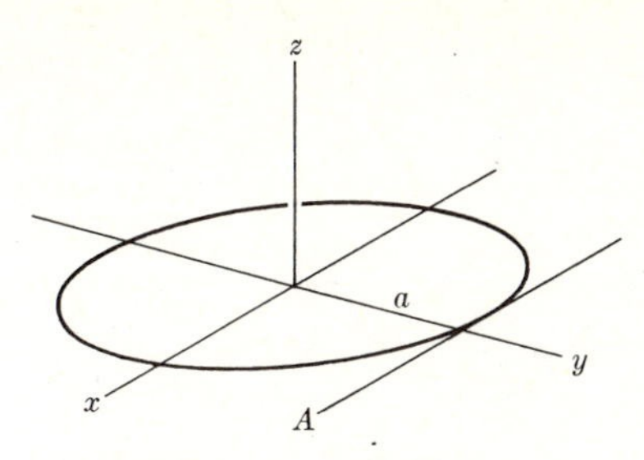

Fig. 5.13 A ring of radius *a*.

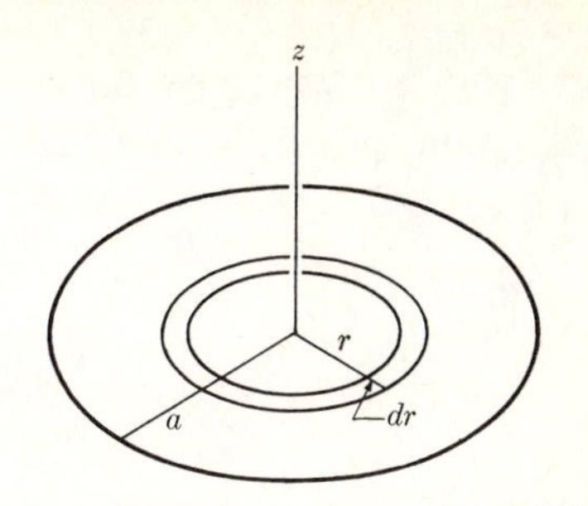

Fig. 5.14 Finding the moment of inertia of a disk.

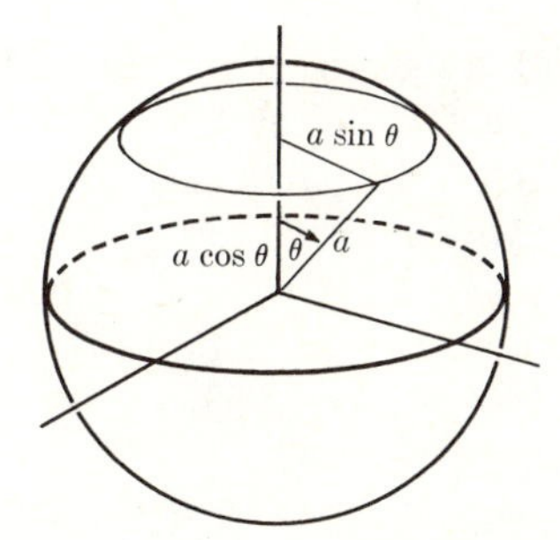

Fig. 5.15 Finding the moment of inertia of a solid sphere.

The moment of inertia of a solid body can be set up in whatever coordinate system may be convenient for the problem at hand. If the body is uniform and of simple shape, its moment of inertia can be computed by considering it as built up out of rods, rings, disks, etc. For example, the moment of inertia of a circular disk about an axis perpendicular to it through its center can be found by regarding the disk as made up of rings (Fig. 5.14) and using Eq. (5.87):

$$I_z = \int_0^a r^2 \rho 2\pi r \, dr = \frac{\pi a^4 \rho}{2} = \tfrac{1}{2} M a^2. \tag{5.90}$$

The moment of inertia of a solid sphere can be calculated from Eq. (5.90) by regarding the sphere as made up of disks (Fig. 5.15):

$$I = \int_\pi^0 \frac{a^2 \sin^2 \theta}{2} (\rho \pi a^2 \sin^2 \theta) \, d(a \cos \theta) = \frac{8\pi \rho a^5}{15} = \tfrac{2}{5} M a^2. \tag{5.91}$$

A body with a piece cut out can be treated by setting its moment of inertia equal to the moment of inertia of the original body minus the moment of inertia of the piece cut out, all moments being taken, of course, about the same axis.

5.6 STATICS OF RIGID BODIES

The equations of motion of a rigid body are Eqs. (5.4) and (5.5):

$$M\ddot{\boldsymbol{R}} = \sum_i \boldsymbol{F}_i^e, \tag{5.92}$$

$$\frac{d\boldsymbol{L}_O}{dt} = \sum_i \boldsymbol{N}_{iO}^e. \tag{5.93}$$

Equation (5.92) determines the motion of the center of mass, located by the vector $\boldsymbol{R}$, in terms of the sum of all external forces acting on the body. Equation (5.93) determines the rotational motion about a point O, which may be the center of mass or a point fixed in space, in terms of the total external torque about the point O. Thus if the total external force acting on a rigid body and the total external torque about a suitable point are given, its motion is determined. This would not be true if the body were not rigid, since then it would be deformed by the external forces in a manner depending on the particular points at which they are applied. Since we are concerned only with external forces throughout this section, we may omit the superscript e. It is only necessary to give the total torque about any one point O, since the torque about any other point O' can then be found from the following formula:

$$\sum_i \boldsymbol{N}_{iO'} = \sum_i \boldsymbol{N}_{iO} + (\boldsymbol{r}_O - \boldsymbol{r}_{O'}) \times \sum_i \boldsymbol{F}_i, \tag{5.94}$$

where $\boldsymbol{r}_O$, $\boldsymbol{r}_{O'}$ are vectors drawn to the points O, O' from any convenient origin. That is, the total torque about O' is the total torque about O plus the torque about O' if the total force were acting at O. The proof of Eq. (5.94) is very simple. Let $\boldsymbol{r}_i$ be the vector from the origin to the point at which $\boldsymbol{F}_i$ acts. Then

$$\begin{aligned}
\sum_i \boldsymbol{N}_{iO'} &= \sum_i (\boldsymbol{r}_i - \boldsymbol{r}_{O'}) \times \boldsymbol{F}_i \\
&= \sum_i (\boldsymbol{r}_i - \boldsymbol{r}_O + \boldsymbol{r}_O - \boldsymbol{r}_{O'}) \times \boldsymbol{F}_i \\
&= \sum_i (\boldsymbol{r}_i - \boldsymbol{r}_O) \times \boldsymbol{F}_i + \sum_i (\boldsymbol{r}_O - \boldsymbol{r}_{O'}) \times \boldsymbol{F}_i \\
&= \sum_i \boldsymbol{N}_{iO} + (\boldsymbol{r}_O - \boldsymbol{r}_{O'}) \times \sum_i \boldsymbol{F}_i.
\end{aligned}$$

If, in particular, a rigid body is at rest, the left members of Eqs. (5.92) and (5.93) are zero, and we have

$$\sum_i \boldsymbol{F}_i = \boldsymbol{0}, \tag{5.95}$$

$$\sum_i \boldsymbol{N}_i = \boldsymbol{0}. \tag{5.96}$$

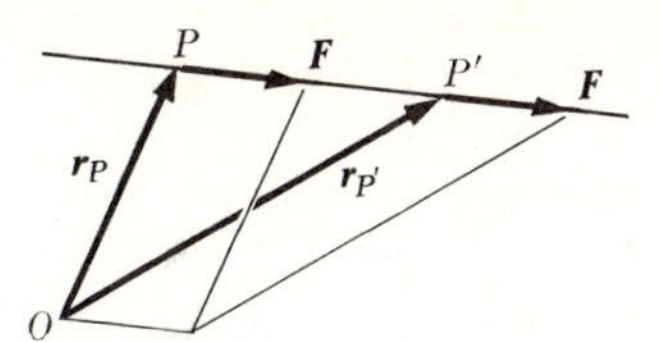

Fig. 5.16 The torque is independent of where along its line of action a force acts.

These are the conditions to be satisfied by the external forces and torques in order for a rigid body to be in equilibrium. They are not sufficient to guarantee that the body is at rest, for it might still be in uniform translational and rotational motion, but if the body is initially at rest, it will remain at rest when these conditions are satisfied. It is sufficient for the total torque in Eq. (5.96) to be zero about any point, since then, by Eq. (5.94), it will be zero also about every other point if Eq. (5.95) holds.

In computing the torque due to a force $\boldsymbol{F}$, it is necessary to know not only the vector $\boldsymbol{F}$ (magnitude and direction), but also the point P of the body at which the force acts. But if we draw a line through P in the direction of $\boldsymbol{F}$, then if $\boldsymbol{F}$ acts at any other point P' of this line, its torque will be the same, since, from the definition of the cross product, it can be seen (Fig. 5.16) that

$$\boldsymbol{r}_P \times \boldsymbol{F} = \boldsymbol{r}_{P'} \times \boldsymbol{F}. \tag{5.97}$$

(The areas of the parallelograms involved are equal.) The line through P in the direction of $\boldsymbol{F}$ is called the *line of action* of the force. It is often convenient in computing torques to remember that the force may be considered to act anywhere along its line of action. A distinction is sometimes made in this connection between "free" and "sliding" vectors, the force being a "sliding" vector. The terminology is likely to prove confusing, however, since as far as the motion of the center of mass is concerned [Eq. (5.92)], the force is a "free" vector, i.e., may act anywhere, whereas in computing torques, the force is a "sliding" vector, and for a nonrigid body, each force must be localized at the point where it acts. It is better to define vector, as we defined it in Section 3.1, as a quantity having magnitude and direction, without reference to any particular location in space. Then, in the case of force, we need for some purposes to specify not only the force vector $\boldsymbol{F}$ itself, but in addition the point or line on which the force acts.

A theorem due to Varignon states that if $\boldsymbol{C} = \boldsymbol{A}+\boldsymbol{B}$, then the moment of $\boldsymbol{C}$ about any point equals the sum of the moments of $\boldsymbol{A}$ and $\boldsymbol{B}$, provided $\boldsymbol{A}$, $\boldsymbol{B}$, and $\boldsymbol{C}$ act at the same point. The theorem is an immediate consequence of the vector identity given by Eq. (3.27):

$$\boldsymbol{r} \times \boldsymbol{C} = \boldsymbol{r} \times \boldsymbol{A} + \boldsymbol{r} \times \boldsymbol{B}, \qquad \text{if} \qquad \boldsymbol{C} = \boldsymbol{A}+\boldsymbol{B}. \tag{5.98}$$

This theorem allows us to compute the torque due to a force by adding the torques due to its components. Combining Varignon's theorem with the result of the

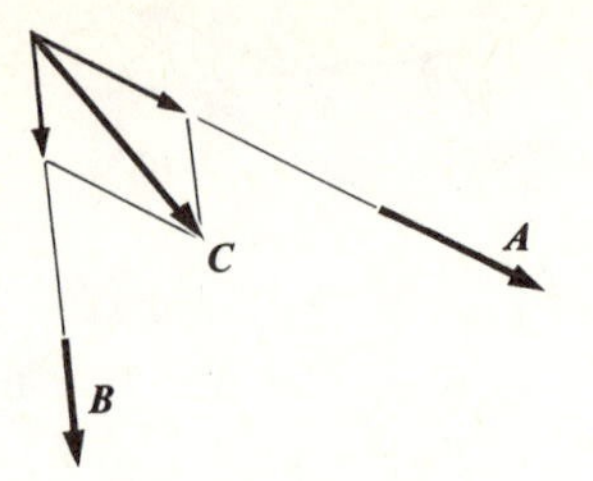

Fig. 5.17 A single force $\boldsymbol{C}$ whose torque is the sum of the torques of $\boldsymbol{A}$ and $\boldsymbol{B}$.

preceding paragraph, we may reduce the torque due to two forces $\boldsymbol{A}$, $\boldsymbol{B}$ acting in a plane, as shown in Fig. 5.17, to the torque due to the single force $\boldsymbol{C}$, since both $\boldsymbol{A}$ and $\boldsymbol{B}$ may be considered to act at the intersection of their lines of action, and Eq. (5.98) then allows us to add them. We could now add $\boldsymbol{C}$ similarly to any third force acting in the plane. This process can be continued so long as the lines of action of the forces being added are not parallel, and it is related to a more general theorem (5.111) regarding forces in a plane to be proved below.

Since, for a rigid body, the motion is determined by the total force and total torque, we shall call two systems of forces acting on a rigid body *equivalent* if they give the same total force, and the same total torque about every point. In view of Eq. (5.94), two systems of forces are then equivalent if they give the same total force, and the same total torque about any single point. It is of interest to know, for any system of forces, what is the simplest system of forces equivalent to it.

If a system of forces $\boldsymbol{F}_i$ acting at points $\boldsymbol{r}_i$ is equivalent to a single force $\boldsymbol{F}$ acting at a point $\boldsymbol{r}$, then the force $\boldsymbol{F}$ acting at $\boldsymbol{r}$ is said to be the *resultant* of the system of forces $\boldsymbol{F}_i$. If $\boldsymbol{F}$ is the resultant of the system of forces $\boldsymbol{F}_i$, then we must have

$$\boldsymbol{F} = \sum_i \boldsymbol{F}_i, \tag{5.99}$$

$$(\boldsymbol{r}-\boldsymbol{r}_O)\times\boldsymbol{F} = \sum_i (\boldsymbol{r}_i-\boldsymbol{r}_O)\times\boldsymbol{F}_i, \tag{5.100}$$

where $\boldsymbol{r}_O$ is any point about which moments are taken. By Eq. (5.94), if Eq. (5.99) holds, and Eq. (5.100) holds for any point $\boldsymbol{r}_O$, it holds for every point $\boldsymbol{r}_O$. The force $-\boldsymbol{F}$ acting at $\boldsymbol{r}$ is called the *equilibrant* of the system; if the equilibrant is added to the system of forces, the conditions for equilibrium are satisfied.

An example of a system of forces having a resultant is the system of gravitational forces acting on a body near the surface of the earth. We shall show that the resultant in this case acts at the center of mass. Let the acceleration of gravity be $\boldsymbol{g}$. Then the force acting on a particle m_i is

$$\boldsymbol{F}_i = m_i\boldsymbol{g}. \tag{5.101}$$

The total force is

$$\boldsymbol{F} = \sum_i m_i \boldsymbol{g} = M\boldsymbol{g}, \tag{5.102}$$

where M is the total mass. The total torque about any point O is, with O as origin,

$$\begin{aligned} \sum_i \boldsymbol{N}_{iO} &= \sum_i (\boldsymbol{r}_i \times m_i \boldsymbol{g}) \\ &= \sum_i (m_i \boldsymbol{r}_i \times \boldsymbol{g}) \\ &= \left(\sum_i m_i \boldsymbol{r}_i\right) \times \boldsymbol{g} \\ &= M\boldsymbol{R} \times \boldsymbol{g} \\ &= \boldsymbol{R} \times M\boldsymbol{g}, \end{aligned} \tag{5.103}$$

where $\boldsymbol{R}$ is the vector from O to the center of mass. Thus the total torque is given by the force $M\boldsymbol{g}$ acting at the center of mass. Because of this result, the center of mass is also called the *center of gravity*. We shall see in the next chapter that, in general, this result holds only in a uniform gravitational field, i.e., when $\boldsymbol{g}$ is the same at all points of the body. If the system of forces acting on a rigid body has a resultant, the forces may be replaced by this resultant in determining the motion of the body.

A system of forces whose sum is zero is called a *couple:*

$$\sum_i \boldsymbol{F}_i = \boldsymbol{0}. \tag{5.104}$$

A couple evidently has no resultant, except in the trivial case where the total torque is zero also, in which case the resultant force is zero. By Eqs. (5.94) and (5.104), a couple exerts the same total torque about every point:

$$\sum_i \boldsymbol{N}_{iO'} = \sum_i \boldsymbol{N}_{iO}. \tag{5.105}$$

Thus a couple is characterized by a single vector, the total torque, and all couples with the same total torque are equivalent. The simplest system equivalent to any given couple, if we exclude the trivial case where the total torque is zero, is a pair of equal and opposite forces $\boldsymbol{F}$, $-\boldsymbol{F}$, acting at points P, P' separated by a vector $\boldsymbol{r}$ (Fig. 5.18) such that

$$\sum_i \boldsymbol{N}_{iO} = \boldsymbol{r} \times \boldsymbol{F}. \tag{5.106}$$

Equation (5.106) states that the moment of the given couple about O equals the moment of the couple $(\boldsymbol{F}, -\boldsymbol{F})$ about P'; the two systems are therefore equivalent, since the point about which the moment of a couple is computed is immaterial. The force $\boldsymbol{F}$ and the points P and P' are by no means uniquely determined. Since only the cross product $\boldsymbol{r} \times \boldsymbol{F}$ is determined by Eq. (5.106), we can choose P

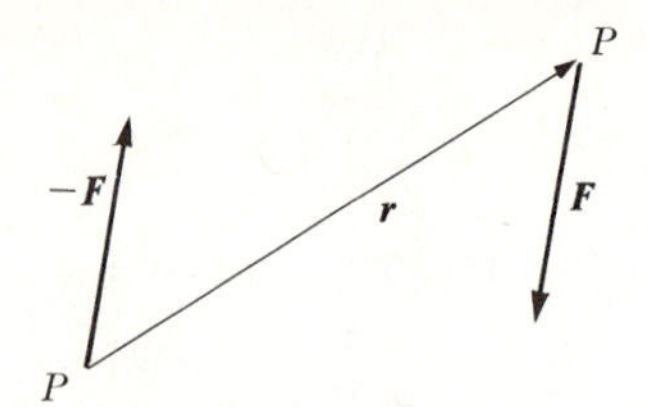

Fig. 5.18 A simple couple.

arbitrarily; we can choose the vector $\boldsymbol{F}$ arbitrarily except that it must lie in the plane perpendicular to the total torque; and we can then choose $\boldsymbol{r}$ as any vector lying in the same plane and determining with $\boldsymbol{F}$ a parallelogram whose area is the magnitude of the total torque.

The problem of finding the simplest system equivalent to any given system of forces is solved by the following theorems:

Theorem 1. *Every system of forces is equivalent to a single force through an arbitrary point, plus a couple (either or both of which may be zero).* (5.107)

To prove this, we show how to find the equivalent single force and couple. Let the arbitrary point P be chosen, let the sum of all the forces in the system be $\boldsymbol{F}$, and let their total torque about the point P be $\boldsymbol{N}$. Then, if we let the single force $\boldsymbol{F}$ act at P, and add a couple whose torque is $\boldsymbol{N}$, we have a system equivalent to the original system. Since the couple can be composed of two forces, one of which may be allowed to act at an arbitrary point, we may let one force of the couple act at the point P, and add it to $\boldsymbol{F}$ to get a single force acting at P plus the other force of the couple. This proves

Theorem 2. *Any system of forces can be reduced to an equivalent system which contains at most two forces.* (5.108)

The following theorem can be proved in two ways:

Theorem 3. *A single nonzero force and a couple in the same plane (i.e., such that the torque vector of the couple is perpendicular to the single force) have a resultant and, conversely, a single force is equivalent to an equal force through any arbitrary point, plus a couple.* (5.109)

Since a couple with torque $\boldsymbol{N}$ is equivalent to a pair of equal and opposite forces, $\boldsymbol{F}$, $-\boldsymbol{F}$, where $\boldsymbol{F}$ may be chosen arbitrarily in the plane perpendicular to $\boldsymbol{N}$, we may always choose $\boldsymbol{F}$ equal to the single force mentioned in the theorem. Furthermore, we may choose the point of action of $\boldsymbol{F}$ arbitrarily. Given a single nonzero force $\boldsymbol{F}$ acting at P, and a couple, we form a couple ($\boldsymbol{F}$, $-\boldsymbol{F}$) equivalent to the given couple, and let $-\boldsymbol{F}$ act at P; $\boldsymbol{F}$ and $-\boldsymbol{F}$ then cancel at P, and the remaining force $\boldsymbol{F}$ of the couple is the single resultant. The converse can be proved by a similar argument.

The other method of proof is as follows. Let the given force $\boldsymbol{F}$ act at a point P, and let the total torque of the couple be $\boldsymbol{N}$. Then the torque of the system about the point P is $\boldsymbol{N}$. We take any vector $\boldsymbol{r}$, in the plane perpendicular to $\boldsymbol{N}$, which forms with $\boldsymbol{F}$ a parallelogram of area N, and let P' be the point displaced from P by the vector $\boldsymbol{r}$. If the single force $\boldsymbol{F}$ acts at P', the torque about P will then be $\boldsymbol{N}$, and hence this single force is equivalent to the original force $\boldsymbol{F}$ acting at P plus the couple. We can combine Theorems 1 and 3 to obtain

Theorem 4. *Every system of forces is equivalent to a single force plus a couple whose torque is parallel to the single force.* (*Or, alternatively, every system of forces is equivalent to a couple plus a single force perpendicular to the plane of the couple.*) (5.110)

To prove this, we use Theorem 1 to reduce any system to a single force plus a couple, and use Theorem 3 to eliminate any component of the couple torque perpendicular to the single force. The point of application of the single force mentioned in Theorem 4 is no longer arbitrary, as its line of action will be fixed when we apply Theorem 3. Either the single force or the couple may vanish in special cases. For a system of forces in a plane, all torques about any point in the plane are perpendicular to the plane. Hence Theorem 4 reduces to

Theorem 5. *Any system of forces in a plane has a resultant, unless it is a couple.* (5.111)

In practice, the reduction of a complicated system of forces to a simpler system is a problem whose simplest solution is usually obtained by an ingenious application of the various theorems and techniques mentioned in this section. One method which always works, and which is often the simplest if the system of forces is very complicated, is to follow the procedure suggested by the proofs of the above theorems. Find the total force $\boldsymbol{F}$ by vector addition, and the total torque $\boldsymbol{N}$ about some conveniently chosen point P. Then $\boldsymbol{F}$ acting at P, plus a couple of torque $\boldsymbol{N}$, together form a system equivalent to the original system. If $\boldsymbol{F}$ is zero, the original system reduces to a couple. If $\boldsymbol{N}$ is perpendicular to $\boldsymbol{F}$, the system has a resultant, which can be found by either of the methods indicated in the proof of Theorem 3. If $\boldsymbol{N}$ is not perpendicular to $\boldsymbol{F}$, and neither is zero, then the system has no resultant, and can be reduced to a system of two forces, as in the derivation of Theorem 2, or to a single force and a couple whose torque is parallel to it, as in Theorem 4. It is a matter of taste, or of convenience for the purpose at hand, which of these latter reductions is regarded as the simplest. In fact, for determining the motion of a body, the most convenient reduction is certainly just the reduction given by Theorem 1, with the arbitrary point taken as the center of mass.

5.7 STATICS OF STRUCTURES

The determination of the forces acting at various points in a solid structure is a problem of utmost importance in all phases of mechanical engineering. There are two principal reasons for wanting to know these forces. First, the engineer must

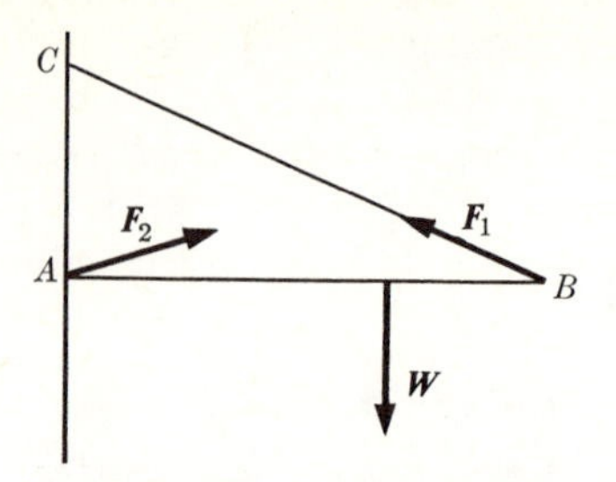

Fig. 5.19 The flagpole problem.

be sure that the materials and construction are such as will withstand the forces which will be acting, without breaking or crushing, and usually without suffering permanent deformation. Second, since no construction materials are really rigid, but deform elastically and sometimes plastically when subject to forces, it is necessary to calculate the amount of this deformation, and to take it into account, if it is significant, in designing the structure. When deformation or breaking of a structure is under consideration, the structure obviously cannot be regarded as a rigid body, and we are interested in the actual system of forces acting on and in the structure. Theorems regarding equivalent systems of forces are not of direct interest in such problems, but are often useful as tools in analyzing parts of the structure which may, to a sufficient approximation, be regarded as rigid, or in suggesting possible equivalent redistributions of forces which would subject the structure to less objectionable stresses while maintaining it in equilibrium.

If a structure is at rest, Eqs. (5.95) and (5.96) are applicable either to the structure as a whole, or to any part of it. It must be kept in mind that the forces and torques which are to be included in the sums are those which are external to and acting on whichever part of the structure is under consideration. If the structure is moving, the more general equations (5.92) and (5.93) are applicable. Either pair of vector equations represents, in general, six component equations, or three if all forces lie in a single plane. (Why three?) It may be that the structure is so constructed that when certain of the external forces and their points of application are given, all the internal forces and torques acting on each part of the structure can be determined by appropriate applications of Eqs. (5.95) and (5.96) (in the case of a structure at rest). Such a structure is said to be *statically determinate*. An elementary example is shown in Fig. 5.19, which shows a horizontal flagpole AB hinged at point A to a wall and supported by a cable BC. A force $\boldsymbol{W}$ acts on the pole as shown. When the force $\boldsymbol{W}$ and the dimensions of the structure are given, it is a simple matter to apply Eqs. (5.95) and (5.96) to the pole and to calculate the force $\boldsymbol{F}_1$ exerted by the cable and the force $\boldsymbol{F}_2$ acting through the hinge. Many examples of statically determinate structures are given in any elementary physics textbook.

Suppose now that the hinge at A in Fig. 5.19 were replaced by a welded joint, so that the flagpole would support the load even without the cable BC, provided the joint at A does not break. Then, given only the weight $\boldsymbol{W}$, it is evidently im-

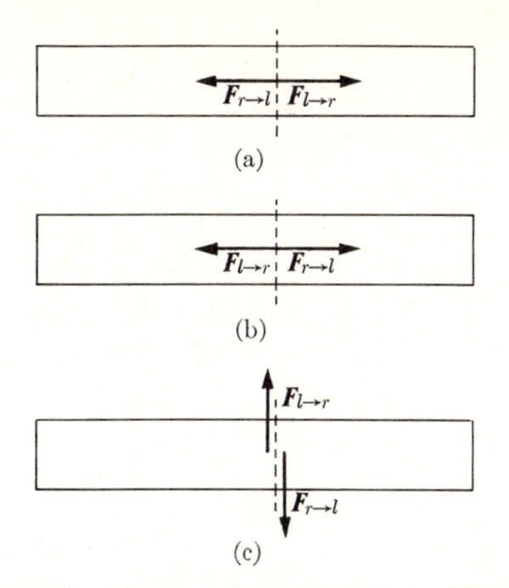

Fig. 5.20 Stresses in a beam: (a) compression; (b) tension; (c) shear.

possible to determine the force $\boldsymbol{F}_1$ exerted by the cable; F_1 may have any value from zero to a rather large value, depending on how tightly the cable is drawn up and on how much stress is applied to the joint at A. Such a structure is said to be *statically indeterminate.* A statically indeterminate structure is one in which the forces acting on its parts are not completely determined by the external forces, but depend also on the distribution of stresses within the structure. To find the internal forces in an indeterminate structure, we would need to know the elastic characteristics of its parts and the precise way in which these parts are distorted. Such problems are usually far more difficult than problems involving determinate structures. Many methods of calculating internal forces in mechanical structures have been developed for application to engineering problems, and some of these are useful in a wide variety of physical problems.

5.8 STRESS AND STRAIN

If an imaginary surface cuts through any part of a solid structure (a rod, string, cable, or beam), then, in general, the material on one side of this surface will be exerting a force on the material on the other side, and conversely according to Newton's third law. These internal forces which act across any surface within the solid are called *stresses.* The *stress* is defined as the force per unit area acting across any given surface in the material. If the material on each side of any surface pushes on the material on the other side with a force perpendicular to the surface, the stress is called a *compression.* If the stress is a pull perpendicular to the surface, it is called a *tension.* If the force exerted across the surface is parallel to the surface, it is called a *shearing stress.* Figure 5.20 illustrates these stresses in the case of a beam. The vector labeled $\boldsymbol{F}_{l\to r}$ represents the force exerted by the left half of the beam on the right half, and the equal and opposite force $\boldsymbol{F}_{r\to l}$ is exerted on the material on the left by the material on the right. A stress at an angle to a surface can be resolved into a shear component and a tension or compression component.

In the most general case, the stress may act in any direction relative to the surface, and may depend on the orientation of the surface. The description of the state of stress of a solid material in the most general case is rather complicated, and is best accomplished by using the mathematical techniques of tensor algebra to be developed in Chapter 10. We shall consider here only cases in which either the stress is a pure compression, independent of the orientation of the surface, or in which only one surface is of interest at any point, so that only a single stress vector is needed to specify the force per unit area across that surface.

If we consider a small volume ΔV of any shape in a stressed material, the material within this volume will be acted on by stress forces exerted across the surface by the material surrounding it. If the material is not perfectly rigid, it will be deformed so that the material in the volume ΔV may have a different shape and size from that which it would have if there were no stress. This deformation of a stressed material is called *strain.* The nature and amount of strain depend on the nature and magnitude of the stresses and on the nature of the material. A suitable definition of *strain*, stating how it is to be measured, will have to be made for each kind of strain. A tension, for example, produces an extension of the material, and the strain would be defined as the fractional increase in length.

If a wire of length l and cross-sectional area A is stretched to length $l+\Delta l$ by a force F, the definitions of stress and strain are

$$\text{stress} = F/A, \tag{5.112}$$

$$\text{strain} = \Delta l/l. \tag{5.113}$$

It is found experimentally that when the strain is not too large, the stress is proportional to the strain for solid materials. This is Hooke's law, and it is true for all kinds of stress and the corresponding strains. It is also plausible on theoretical grounds for the reasons suggested in the discussion of Eq. (2.57). The ratio of stress to strain is therefore constant for any given material if the strain is not too large. In the case of extension of a material in one direction due to tension, this ratio is called *Young's modulus*, and is

$$Y = \frac{\text{stress}}{\text{strain}} = \frac{Fl}{A\,\Delta l}. \tag{5.114}$$

If a substance is subjected to a pressure increment Δp, the resulting deformation will be a change in volume, and the strain will be defined by

$$\text{strain} = \frac{\Delta V}{V}. \tag{5.115}$$

The ratio of stress to strain in this case is called the *bulk modulus B:*

$$B = \frac{\text{stress}}{\text{strain}} = -\frac{\Delta p\, V}{\Delta V}, \tag{5.116}$$

where the negative sign is introduced in order to make B positive.

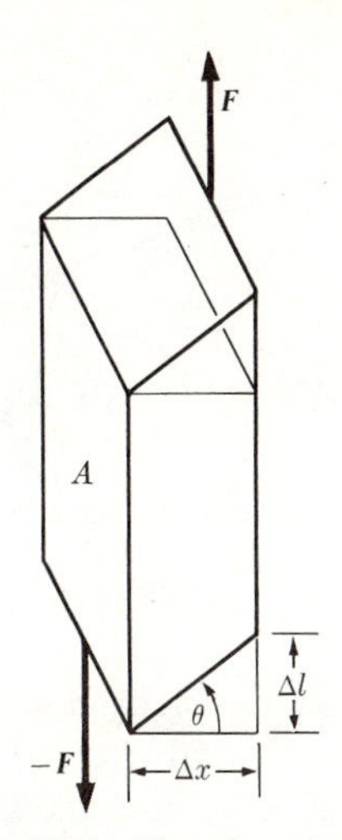

Fig. 5.21 Shearing strain.

In the case of a shearing stress, the stress is again defined by Eq. (5.112), where F is the force acting across and parallel to the area A. The resulting shearing strain consists in a motion of A parallel to itself through a distance Δl, relative to a plane parallel to A at a distance Δx from A (Fig. 5.21). The shearing strain is then defined by

$$\text{strain} = \frac{\Delta l}{\Delta x} = \tan \theta, \tag{5.117}$$

where θ is the angle through which a line perpendicular to A is turned as a result of the shearing strain. The ratio of stress to strain in this case is called the *shear modulus* n:

$$n = \frac{\text{stress}}{\text{strain}} = \frac{F}{A \tan \theta}. \tag{5.118}$$

An extensive study of methods of solving problems in statics is outside the scope of this text. We shall restrict ourselves in the next three sections to the study of three special types of problems which illustrate the analysis of a physical system, to determine the forces which act upon its parts and to determine the effect of these forces in deforming the system.

5.9 EQUILIBRIUM OF FLEXIBLE STRINGS AND CABLES

An ideal flexible string is one which will support no compression or shearing stress, nor any bending moment, so that the force exerted across any point in the string can only be a tension directed along the tangent to the string at that point. Chains and cables used in many structures can be regarded for most purposes as ideal flexible strings.

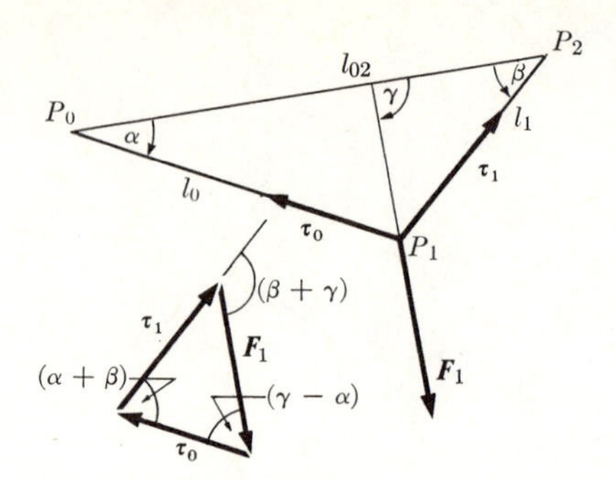

Fig. 5.22 A flexible string held at three points.

Let us first take a very simple problem in which a string of negligible weight is suspended between two points P_0 and P_2, and a force $\boldsymbol{F}_1$ acts at a point P_1 on the string (Fig. 5.22). Let τ_0 be the tension in the segment $\overline{P_0P_1}$, and τ_1 the tension in the segment $\overline{P_1P_2}$. Let l_0 and l_1 be the lengths of these segments of the string, and let l_{02} be the distance between P_0 and P_2. The angles α, β between the two segments of string and the line $\overline{P_0P_2}$ are determined by the cosine law:

$$\cos\alpha = \frac{l_{02}^2+l_0^2-l_1^2}{2l_0l_{02}}, \qquad \cos\beta = \frac{l_{02}^2+l_1^2-l_0^2}{2l_1l_{02}}, \tag{5.119}$$

so that the position of the point P_1 is independent of the force $\boldsymbol{F}_1$, provided the string does not stretch. Since the bit of string at the point P_1 is in equilibrium, the vector sum of the three forces $\boldsymbol{F}_1$, $\boldsymbol{\tau}_0$, and $\boldsymbol{\tau}_1$ acting on the string at P_1 must vanish, so that these forces form a closed triangle, as indicated in Fig. 5.22. The tensions are then determined in terms of the angle between the force $\boldsymbol{F}_1$ and the direction of the line $\overline{P_0P_2}$, by the sine law:

$$\tau_0 = F_1\frac{\sin(\beta+\gamma)}{\sin(\alpha+\beta)}, \qquad \tau_1 = F_1\frac{\sin(\gamma-\alpha)}{\sin(\alpha+\beta)}. \tag{5.120}$$

Now suppose that the string stretches according to Hooke's law, so that

$$l_0 = l_0'(1+k\tau_0), \qquad l_1 = l_1'(1+k\tau_1), \tag{5.121}$$

where l_0', l_1' are the unstretched lengths, and k is a constant [$1/k$ would be Young's modulus, Eq. (5.114), multiplied by the cross-sectional area of the string]. The unknown quantities τ_0, τ_1, l_0, and l_1 can be eliminated from Eqs. (5.119) by substitution from Eqs. (5.120) and (5.121). We then have two rather complicated equations to be solved for the angles α and β. The solution must be carried out by numerical methods when numerical values of l_0', l_1', k, l_{02}, F_1, and γ are given. When α and β are found, τ_0, τ_1, l_0, and l_1 can be found from Eqs. (5.120) and (5.121). One way of solving these equations by successive approximations is to assume first that the string does not stretch, so that $l_0 = l_0'$, $l_1 = l_1'$, and to calculate α and

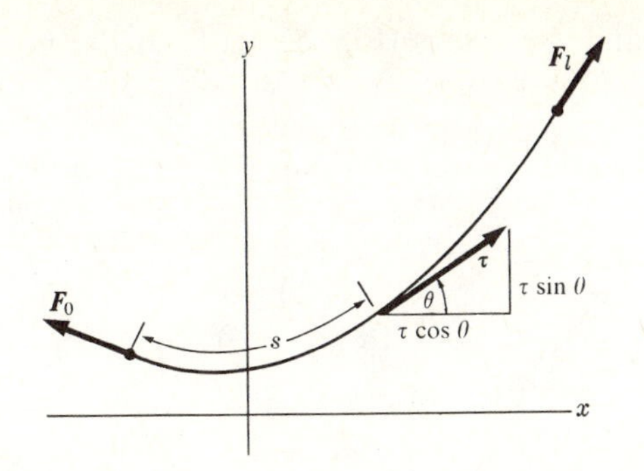

Fig. 5.23 A flexible string hanging under its own weight.

β from Eqs. (5.119), and τ_0, τ_1 from Eqs. (5.120). Using these values of τ_0, τ_1, we then calculate l_0, l_1 from Eqs. (5.121). The new values of l_0, l_1 can be used in Eqs. (5.119) to get better values for α, β from which better values of τ_0, τ_1 can be calculated. These can be used to get still better values for l_0, l_1 from Eqs. (5.121), and so on. As this process is repeated, the successive calculated values of α, β, τ_0, τ_1, l_0, l_1 will converge toward the true values. If the string stretches only very little, the first few repetitions will be sufficient to give very close values. The method suggested here is an example of a very general class of methods of solution of physical problems by successive approximations. It is an example of what are called *relaxation methods* of solving statics problems.

We next consider a string acted on by forces distributed continuously along the length of the string. A point on the string will be specified by its distance s from one end, measured along the string. Let $f(s)$ be the force per unit length at the point s, that is, the force on a small segment of length ds is $f\,ds$. Then the total force acting on the length of string between the end $s = 0$ and the point s is zero if the string is in equilibrium:

$$\boldsymbol{F}_0 + \int_0^s \boldsymbol{f}\, ds + \boldsymbol{\tau}(s) = \boldsymbol{0}, \tag{5.122}$$

where $\boldsymbol{F}_0$ is the supporting force at the end $s = 0$, and $\boldsymbol{\tau}(s)$ is a vector whose magnitude is the tension at the point s, oriented in the direction of increasing s. By differentiating Eq. (5.122) with respect to s, we obtain a differential equation for $\boldsymbol{\tau}(s)$:

$$\frac{d\boldsymbol{\tau}}{ds} = -\boldsymbol{f}. \tag{5.123}$$

The simplest and most important application of Eq. (5.123) is to the case of a string having a weight w per unit length. If the string is acted on by no other forces except at the ends, it will hang in a vertical plane, which we take to be the xy-plane, with the x-axis horizontal and the y-axis vertical. Let θ be the angle between the string and the x-axis (Fig. 5.23). Then the horizontal and vertical components of Eq. (5.123) become:

$$\frac{d}{ds}(\tau \sin \theta) = w, \tag{5.124}$$

$$\frac{d}{ds}(\tau \cos \theta) = 0. \tag{5.125}$$

Equation (5.125) implies that

$$\tau \cos \theta = C. \tag{5.126}$$

The horizontal component of tension is constant, as it should be since the external forces on the string are all vertical, except at the ends. By dividing Eq. (5.124) by C, and using Eq. (5.126), we eliminate the tension:

$$\frac{d \tan \theta}{ds} = \frac{w}{C}. \tag{5.127}$$

If we represent the string by specifying the function $y(x)$, we have the relations

$$\tan \theta = \frac{dy}{dx} = y', \tag{5.128}$$

$$ds = [(dx)^2 + (dy)^2]^{1/2} = dx(1+y'^2)^{1/2}, \tag{5.129}$$

so that Eq. (5.127) becomes

$$\frac{dy'}{dx} = \frac{w}{C}(1+y'^2)^{1/2}. \tag{5.130}$$

This can be integrated, if w is constant:

$$\int \frac{dy'}{(1+y'^2)^{1/2}} = \int \frac{w}{C}\, dx, \tag{5.131}$$

$$\sinh^{-1} y' = \frac{wx}{C} + \alpha, \tag{5.132}$$

where α is a constant. We solve for y':

$$y' = \frac{dy}{dx} = \sinh\left(\frac{wx}{C} + \alpha\right). \tag{5.133}$$

This can be integrated again, and we obtain

$$y = \beta + \frac{C}{w} \cosh\left(\frac{wx}{C} + \alpha\right). \tag{5.134}$$

The curve represented by Eq. (5.134) is called a *catenary*, and is the form in which a uniform string will hang if acted on by no force other than its own weight, except at the ends. The constants C, β, and α are to be chosen so that y has the proper value at the endpoints, and so that the total length of the string has the proper value. The total length is

$$l = \int ds = \int_{x_0}^{x_l} (1+y'^2)^{1/2}\, dx = \int_{x_0}^{x_l} \cosh\left(\frac{wx}{C}+\alpha\right) dx$$

$$= \frac{C}{w}\left[\sinh\left(\frac{wx_l}{C}+\alpha\right) - \sinh\left(\frac{wx_0}{C}+\alpha\right)\right]. \qquad (5.135)$$

5.10 EQUILIBRIUM OF SOLID BEAMS

A horizontal beam subject to vertical forces is one of the simplest examples of a structure subject to shearing forces and bending moments. To simplify the problem, we shall consider only the case when the beam is under no compression or tension, and we shall assume that the beam is so constructed and the forces so applied that the beam bends in only one vertical plane, without any torsion (twisting) about the axis of the beam. We find first the stresses within the beam from a knowledge of the external forces, and then determine the distortion of the beam due to these stresses.

Points along the beam will be located by a coordinate x measured horizontally from the left end of the beam (Fig. 5.24). Let vertical forces $F_1, \ldots, F_n$ act at the distances $x_1, \ldots, x_n$ from the left end. A force will be taken as positive if it is directed upward. Let AA' be a plane perpendicular to the beam at any distance x from the end. According to Theorem 1 (5.107), of Section 5.6, the system of forces exerted across the plane AA' by the material on the right against that on the left is equivalent to a single force $\boldsymbol{S}$ through any point in the plane, and a couple of torque $\boldsymbol{N}$. (Note that in applying Theorem 1, we are treating the plane AA' as a rigid body, that is, we are assuming that the cross-sectional plane AA' is not distorted by the forces acting on it.) In the case we are considering there is no compression or tension and all forces are vertical, so that $\boldsymbol{S}$ is directed vertically. We shall define the *shearing force* S as the vertical force acting across AA' from right to left; S will be taken as positive when this force is directed upward, negative

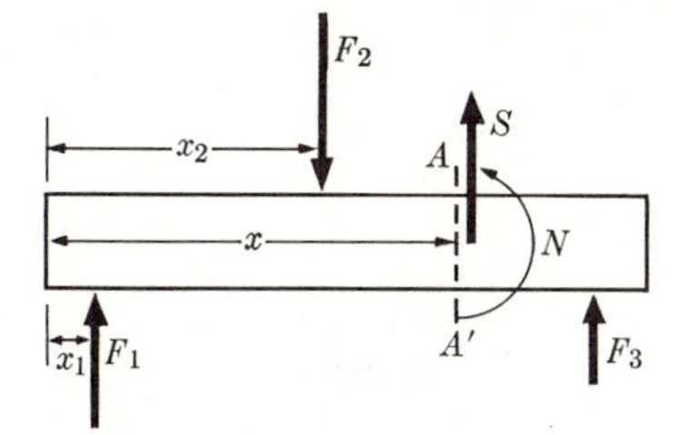

Fig. 5.24 Forces acting on a beam.

when it is downward.* By Newton's third law, the force acting across AA' from left to right is $-S$. Since we are assuming no torsion about the axis of the beam (x-axis), and since all the forces are vertical, the torque N will be directed horizontally and perpendicular to the beam. We shall define the *bending moment* N as the torque exerted from right to left across AA' about a horizontal axis in the plane AA'; N will be taken as positive when it tends to rotate the plane AA' in a counterclockwise direction. Since S is vertical, the torque will be the same about any horizontal axis in the plane AA'.

The shearing force S and bending moment N can be determined by applying the conditions of equilibrium [Eqs. (5.95) and (5.96)] to the part of the beam to the left of the plane AA'. The total force and total torque about a horizontal axis in the plane AA' are, if we neglect the weight of the beam,

$$\sum_{x_i<x} F_i + S = 0, \tag{5.136}$$

$$-N_0 - \sum_{x_i<x} (x-x_i)F_i + N = 0, \tag{5.137}$$

where the sums are taken over all forces acting to the left of AA' and N_0 is the bending moment, if any, exerted by the left end of the beam against its support. The torque N_0 will appear only if the beam is clamped or otherwise fastened at its left end. The force exerted by any clamp or other support at the end is to be included among the forces F_i. If the beam has a weight w per unit length, this should be included in the equilibrium equations:

$$\sum_{x_i<x} F_i - \int_0^x w\,dx + S = 0, \tag{5.138}$$

$$-N_0 - \sum_{x_i<x} (x-x_i)F_i + \int_0^x (x-x')w\,dx' + N = 0. \tag{5.139}$$

The shearing force and bending moment at a distance x from the end are therefore

$$S = -\sum_{x_i<x} F_i + \int_0^x w\,dx, \tag{5.140}$$

$$N = N_0 + \sum_{x_i<x} (x-x_i)F_i - \int_0^x (x-x')w\,dx'. \tag{5.141}$$

If there is any additional force distributed continuously along the beam, this can be included in w as an additional weight per unit length. If the beam is free at its ends, the shearing force and bending moment must be zero at the ends. If we set

*This sign convention for S is in agreement with sign conventions throughout this book, where the upward direction is taken as positive. Sign conventions for shearing force and bending moment are not uniform in physics and engineering texts, and one must be careful in reading the literature to note what sign convention is adopted by each author.

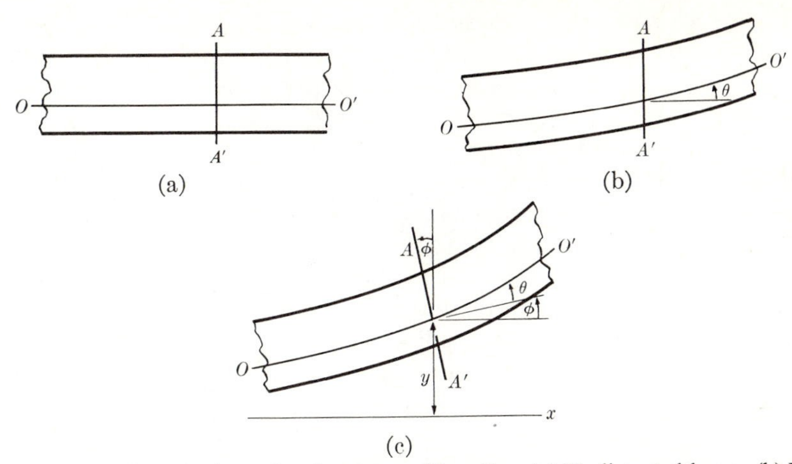

Fig. 5.25 Distortion of a beam by shearing and bending. (a) Undistorted beam. (b) Beam in shear. (c) Beam bent and in shear.

$S = N = 0$ at the right end of the beam, Eqs. (5.140) and (5.141) may be solved for two of the forces acting on the beam when the others are known. If the beam is fastened or clamped at either end, S and N may have any values there. Equations (5.140) and (5.141) determine S and N everywhere along the beam when all the forces are known, including the force and torque exerted through the clamp, if any, on the left end. The shearing force and bending moment may be plotted as functions of x whose slopes at any point are obtained by differentiating Eqs. (5.140) and (5.141):

$$\frac{dS}{dx} = w \text{ (except at } x_i), \tag{5.142}$$

$$\frac{dN}{dx} = \sum_{x_i < x} F_i - \int_0^x w \, dx' = -S. \tag{5.143}$$

The shearing force increases by $-F_i$ from left to right across a point x_i where a force F_i acts.

Let us now consider the distortion produced by the shearing forces and bending moments in a beam of uniform cross section throughout its length. In Fig. 5.25(a) is shown an undistorted horizontal beam through which are drawn a horizontal line OO' and a vertical plane AA'. In Fig. 5.25(b) the beam is under a shearing strain, the effect of which is to slide the various vertical planes relative to one another so that the line OO' makes an angle θ with the normal to the plane AA'. According to Eq. (5.118), the angle θ is given in terms of the shearing force S and the shear modulus n by:

$$\theta = \frac{S}{nA}, \tag{5.144}$$

where A is the cross-sectional area, and we have made the approximation $\tan\theta \doteq \theta$, since θ will be very small. In Fig. 5.25(c), we show the further effect of bending the beam. The plane AA' now makes an angle φ with the vertical. It is assumed that the cross-sectional surface AA' remains plane and retains its shape when the beam is under stress, although this may not be strictly true near the points where forces are applied. In order to determine φ, we consider two planes AA' and BB' initially vertical and a small distance l apart. When the beam is bent, AA' and BB' will make angles φ and $\varphi+\Delta\varphi$ with the vertical (Fig. 5.26). Due to the bending, the fibers on the outside of the curved beam will be stretched and those on the inside will be compressed. Somewhere within the beam will be a neutral layer of unstretched fibers, and we shall agree to draw the line OO' so that it lies in this neutral layer. A line between AA' and BB' parallel to OO' and a distance z above OO' will be compressed to a length $l-\Delta l$, where (see Fig. 5.26)

$$\Delta l = z\,\Delta\varphi. \tag{5.145}$$

The compressive force dF exerted across an element of area dA a distance z above the neutral layer OO' will be given by Eq. (5.114) in terms of Young's modulus:

$$\frac{dF}{dA} = Y\frac{\Delta l}{l} = Yz\frac{\Delta\varphi}{l}, \tag{5.146}$$

or, if we let $l = ds$, an infinitesimal element of length along the line OO',

$$\frac{dF}{dA} = Yz\frac{d\varphi}{ds}. \tag{5.147}$$

This equation is important in the design of beams, as it determines the stress of compression or tension at any distance z from the neutral layer. The total compressive force through the cross-sectional area A of the beam will be

$$F = \iint_A dF = Y\frac{d\varphi}{ds}\iint_A z\,dA. \tag{5.148}$$

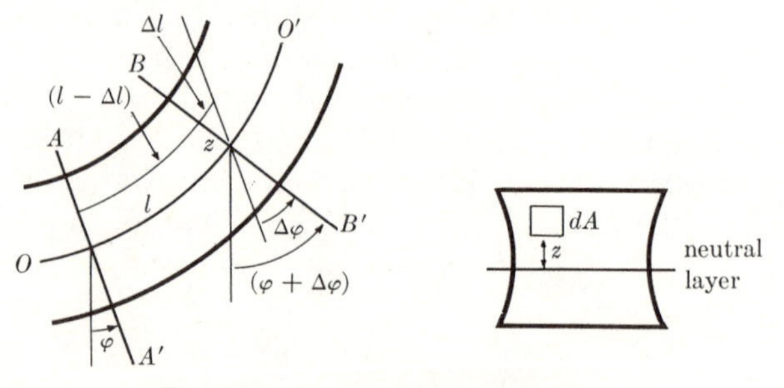

Fig. 5.26 Strains in a bent beam.

Since we are assuming no net tension or compression of the beam, $F = 0$, and

$$\iint_A z\,dA = 0. \tag{5.149}$$

This implies that the neutral layer contains the centroid of the area A of the beam, and we may require that OO' be drawn through the centroid of the cross-sectional area of the beam. The bending moment exerted by the forces dF is

$$\begin{aligned} N = \iint_A z\,dF &= Y\frac{d\varphi}{ds}\iint_A z^2\,dA \\ &= Yk^2A\frac{d\varphi}{ds}, \end{aligned} \tag{5.150}$$

where

$$k^2 = \frac{1}{A}\iint_A z^2\,dA, \tag{5.151}$$

so that k is the radius of gyration of the cross-sectional area of the beam about a horizontal axis through its centroid. The differential equation for φ is therefore

$$\frac{d\varphi}{ds} = \frac{N}{Yk^2A}. \tag{5.152}$$

Let the upward deflection of the beam from a horizontal x-axis be $y(x)$, measured to the line OO' (Fig. 5.25). Then $y(x)$ is to be determined by solving the equation

$$\frac{dy}{dx} = \tan(\theta+\varphi), \tag{5.153}$$

when θ and φ have been determined from Eqs. (5.144) and (5.152). If we assume that both θ and φ are very small angles, Eqs. (5.152) and (5.153) become

$$\frac{d\varphi}{dx} = \frac{N}{Yk^2A}, \tag{5.154}$$

$$\frac{dy}{dx} = \theta+\varphi. \tag{5.155}$$

When there are no concentrated forces F_i along the beam, we may differentiate Eq. (5.155) and make use of Eqs. (5.154), (5.144), (5.142), and (5.143) to obtain

$$\frac{d^2y}{dx^2} = \frac{w}{nA}+\frac{N}{Yk^2A}, \tag{5.156}$$

$$\frac{d^4y}{dx^4} = \frac{1}{nA}\frac{d^2w}{dx^2}-\frac{w}{Yk^2A}. \tag{5.157}$$

If bending can be neglected, as in a short, thick beam, Eq. (5.156) with $N = 0$ becomes a second-order differential equation to be solved for $y(x)$. For a longer beam, Eq. (5.157) must be used. These equations can also be used when concentrated loads F_i are present, by solving them for each segment of the beam between the points where the forces F_i are applied, and fitting the solutions together properly at these points. The solutions on either side of a point x_i where a force F_i is applied must be chosen so that y, φ, N are continuous across x_i, while S, dN/dx, dy/dx, d^3y/dx^3 increase across the point x_i by an amount determined by Eqs. (5.140), (5.143), (5.155), and (5.156). The solution of Eq. (5.156) will contain two arbitrary constants, and that of Eq. (5.157), four, which are to be determined by the conditions at the ends of the beam or segment of beam.

As an example, we consider a uniform beam of weight W, length L, clamped in a horizontal position (i.e., so that $\varphi = 0$)* at its left end ($x = 0$), and with a force $F_1 = -W'$ exerted on its right end ($x = L$). In this case, Eq. (5.157) becomes

$$\frac{d^4y}{dx^4} = -\frac{W}{Yk^2AL}. \tag{5.158}$$

The solution is

$$y = -\frac{Wx^4}{24Yk^2AL} + \tfrac{1}{6}C_3x^3 + \tfrac{1}{2}C_2x^2 + C_1x + C_0. \tag{5.159}$$

To determine the constants C_0, C_1, C_2, C_3, we have at the left end of the beam:

$$y = C_0 = 0, \tag{5.160}$$

$$\frac{dy}{dx} = C_1 = \theta = \frac{S}{nA} = -\frac{W+W'}{nA}, \tag{5.161}$$

where we have used Eqs. (5.155) and (5.144). We need two more conditions, which may be determined in a variety of ways. The easiest way in this case is to apply Eq. (5.156) and its derivative at the left end of the beam:

$$\frac{d^2y}{dx^2} = C_2 = \frac{W}{nAL} - \frac{W'L + \frac{1}{2}WL}{Yk^2A}, \tag{5.162}$$

$$\frac{d^3y}{dx^3} = C_3 = \frac{1}{Yk^2A}\frac{dN}{dx} = -\frac{S}{Yk^2A} = \frac{W'+W}{Yk^2A}, \tag{5.163}$$

where we have used Eq. (5.143). The deflection of the beam at any point x is then

$$y = -\frac{L^3}{Yk^2A}\left[\frac{Wx^2}{4L^2}\left(1 - \frac{2}{3}\frac{x}{L} + \frac{1}{6}\frac{x^2}{L^2}\right) + \frac{W'x^2}{2L^2}\left(1 - \frac{1}{3}\frac{x}{L}\right)\right] - \frac{L}{nA}\left[\frac{Wx}{L}\left(1 - \frac{1}{2}\frac{x}{L}\right) + \frac{W'x}{L}\right]. \tag{5.164}$$

*The condition $\varphi = 0$ means that the plane AA' is vertical; that is, the beam would be horizontal if there were no shearing strain.

The deflection at $x = L$ is

$$y = -\frac{L^3}{Yk^2A}[\tfrac{1}{8}W+\tfrac{1}{3}W'] - \frac{L}{nA}[\tfrac{1}{2}W+W']. \tag{5.165}$$

The first term in each equation is the deflection due to bending, and the second is that due to shear. The first term is proportional to L^3, and inversely proportional to k^2. The second term is proportional to L and independent of k. Hence bending is more important for long, thin beams, and shear is more important for short, thick beams. Our analysis here is probably not very accurate for short, thick beams, since, as pointed out above, some of our assumptions may not be valid near points of support or points where loads are applied (where "near" means relative to the cross-sectional dimensions of the beam).

5.11 EQUILIBRIUM OF FLUIDS

A *fluid* is defined as a substance which will support no shearing stress when in equilibrium. Liquids and gases fit this definition, and even very viscous substances like pitch, or tar, or the material in the interior of the earth, will eventually come to an equilibrium in which shearing stresses are absent, if they are left undisturbed for a sufficiently long time. The stress F/A across any small area A in a fluid in equilibrium must be normal to A, and in practically all cases it will be a compression rather than a tension.

We first prove that the stress F/A near any point in the fluid is independent of the orientation of the surface A. Let any two directions be given, and construct a small triangular prism with two equal faces $A_1 = A_2$ perpendicular to the two given directions. The third face A_3 is to form with A_1 and A_2 a cross section having the shape of an isosceles triangle (Fig. 5.27). Let $\boldsymbol{F}_1$, $\boldsymbol{F}_2$, $\boldsymbol{F}_3$ be the stress forces perpendicular to the faces A_1, A_2, A_3. If the fluid in the prism is in equilibrium,

$$\boldsymbol{F}_1 + \boldsymbol{F}_2 + \boldsymbol{F}_3 = \boldsymbol{0}. \tag{5.166}$$

The forces on the end faces of the prism need not be included here, since they are perpendicular to $\boldsymbol{F}_1$, $\boldsymbol{F}_2$, and $\boldsymbol{F}_3$, and must therefore separately add to zero. It follows from Eq. (5.166), and from the way the prism has been constructed, that

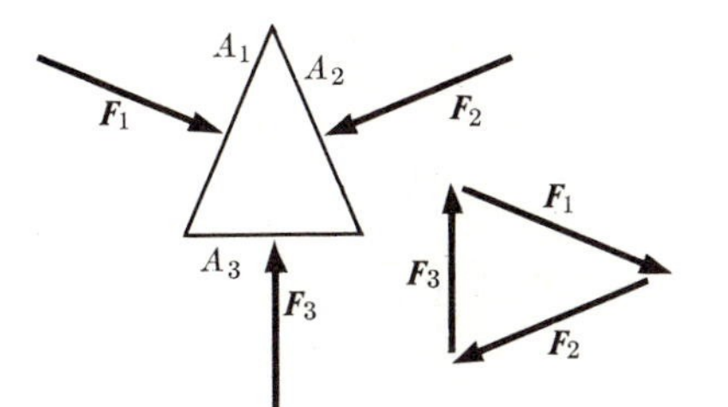

Fig. 5.27 Forces on a triangular prism in a fluid.

$\boldsymbol{F}_1$, $\boldsymbol{F}_2$, and $\boldsymbol{F}_3$ must form an isosceles triangle (Fig. 5.27), and therefore that

$$F_1 = F_2. \tag{5.167}$$

Since the directions of $\boldsymbol{F}_1$ and $\boldsymbol{F}_2$ are any two directions in the fluid, and since $A_1 = A_2$, the stress F/A is the same in all directions. The stress in a fluid is called the *pressure* p:

$$p = \frac{F_1}{A_1} = \frac{F_2}{A_2}. \tag{5.168}$$

Now suppose that in addition to the pressure the fluid is subject to an external force $\boldsymbol{f}$ per unit volume of fluid, that is, any small volume dV in the fluid is acted on by a force $\boldsymbol{f}\, dV$. Such a force is called a *body force;* $\boldsymbol{f}$ is the body force density. The most common example is the gravitational force, for which

$$\boldsymbol{f} = \rho \boldsymbol{g}, \tag{5.169}$$

where $\boldsymbol{g}$ is the acceleration due to gravity, and ρ is the density. In general, the body force density may differ in magnitude and direction at different points in the fluid. In the usual case, when the body force is given by Eq. (5.169), $\boldsymbol{g}$ will be constant and $\boldsymbol{f}$ will be constant in direction; if ρ is constant, $\boldsymbol{f}$ will also be constant in magnitude. Let us consider two nearby points P_1, P_2 in the fluid, separated by a vector $d\boldsymbol{r}$. We construct a cylinder of length $d\boldsymbol{r}$ and cross-sectional area dA, whose end faces contain the points P_1 and P_2. Then the total component of force in the direction of $d\boldsymbol{r}$ acting on the fluid in the cylinder, since the fluid is in equilibrium, will be

$$\boldsymbol{f} \cdot d\boldsymbol{r}\, dA + p_1\, dA - p_2\, dA = 0,$$

where p_1 and p_2 are the pressures at P_1 and P_2. The difference in pressure between two points a distance $d\boldsymbol{r}$ apart is therefore

$$dp = p_2 - p_1 = \boldsymbol{f} \cdot d\boldsymbol{r}. \tag{5.170}$$

The total difference in pressure between two points in the fluid located by vectors $\boldsymbol{r}_1$ and $\boldsymbol{r}_2$ will be

$$p_2 - p_1 = \int_{\boldsymbol{r}_1}^{\boldsymbol{r}_2} \boldsymbol{f} \cdot d\boldsymbol{r}, \tag{5.171}$$

where the line integral on the right is to be taken along some path lying entirely within the fluid from $\boldsymbol{r}_1$ to $\boldsymbol{r}_2$. Given the pressure p_1 at $\boldsymbol{r}_1$, Eq. (5.171) allows us to compute the pressure at any other point $\boldsymbol{r}_2$ which can be joined to $\boldsymbol{r}_1$ by a path lying within the fluid. The difference in pressure between any two points depends only on the body force. Hence any change in pressure at any point in a fluid in equilibrium must be accompanied by an equal change at all other points if the body force does not change. This is Pascal's law.

According to the geometrical definition (3.107) of the gradient, Eq. (5.170) implies that

$$\boldsymbol{f} = \nabla p. \tag{5.172}$$

The pressure gradient in a fluid in equilibrium must be equal to the body force density. This result shows that the net force per unit volume due to pressure is $-\nabla p$. The pressure p is like a potential energy per unit volume in the sense that its negative gradient represents a force per unit volume due to pressure. However, the integral of $p\,dV$ over a volume does not represent a potential energy except in very special cases. Equation (5.172) implies that the surfaces of constant pressure in the fluid are everywhere perpendicular to the body force. According to Eqs. (3.187) and (5.172), the force density $\boldsymbol{f}$ must satisfy the equation

$$\nabla \times \boldsymbol{f} = \mathbf{0}. \tag{5.173}$$

This is therefore a necessary condition on the body force in order for equilibrium to be possible. It is also a sufficient condition for the possibility of equilibrium. This follows from the discussion in Section 3.12, for if Eq. (5.173) holds, then it is permissible to define a function $p(\boldsymbol{r})$ by the equation

$$p(\boldsymbol{r}) = p_1 + \int_{\mathbf{r}_1}^{\mathbf{r}} \boldsymbol{f} \cdot d\boldsymbol{r}, \tag{5.174}$$

where p_1 is the pressure at some fixed point $\boldsymbol{r}_1$, and the integral may be evaluated along any path from $\boldsymbol{r}_1$ to $\boldsymbol{r}$ within the fluid. If the pressure in the fluid at every point $\boldsymbol{r}$ has the value $p(\boldsymbol{r})$ given by (5.174), then Eq. (5.172) will hold, and the body force $\boldsymbol{f}$ per unit volume will everywhere be balanced by the pressure force $-\nabla p$ per unit volume. Equation (5.174) therefore defines an equilibrium pressure distribution for any body force satisfying Eq. (5.173).

The problem of finding the pressure within a fluid in equilibrium, if the body force density $\boldsymbol{f}(\boldsymbol{r})$ is given, is evidently mathematically identical with the problem discussed in Section 3.12 of finding the potential energy for a given force function $\boldsymbol{F}(\boldsymbol{r})$. We first check that $\nabla \times \boldsymbol{f}$ is zero everywhere within the fluid, in order to be sure that an equilibrium is possible. We then take a point $\boldsymbol{r}_1$ at which the pressure is known, and use Eq. (5.174) to find the pressure at any other point, taking the integral along any convenient path.

The total body force acting on a volume V of the fluid is

$$\boldsymbol{F}_b = \iiint_V \boldsymbol{f}\, dV. \tag{5.175}$$

The total force due to the pressure on the surface A of V is

$$\boldsymbol{F}_p = -\iint_A \hat{\boldsymbol{n}} p\, dA, \tag{5.176}$$

where $\hat{\boldsymbol{n}}$ is the outward normal unit vector at any point on the surface. These two must be equal and opposite, since the fluid is in equilibrium:

$$\boldsymbol{F}_p = -\boldsymbol{F}_b. \tag{5.177}$$

Equation (5.176) gives the total force due to pressure on the surface of the volume V, whether or not V is occupied by fluid. Hence we conclude from Eq. (5.177) that a body immersed in a fluid in equilibrium is acted on by a force $\boldsymbol{F}_p$ due to pressure, equal and opposite to the body force $\boldsymbol{F}_b$ which would be exerted on the volume V if it were occupied by fluid in equilibrium. This is Archimedes' principle. Combining Eqs. (5.172), (5.175), (5.176), and (5.177), we have

$$\iint_A \hat{\boldsymbol{n}} p \, dA = \iiint_V \boldsymbol{\nabla} p \, dV. \tag{5.178}$$

This equation resembles Gauss' divergence theorem [Eq. (3.115)], except that the integrands are $\hat{\boldsymbol{n}}p$ and $\boldsymbol{\nabla}p$ instead of $\hat{\boldsymbol{n}} \cdot \boldsymbol{A}$ and $\boldsymbol{\nabla} \cdot \boldsymbol{A}$. Gauss' theorem can, in fact, be proved in a very useful general form which allows us to replace the factor $\hat{\boldsymbol{n}}$ in a surface integral by $\boldsymbol{\nabla}$ in the corresponding volume integral without any restrictions on the form of the integrand except that it must be so written that the differentiation symbol $\boldsymbol{\nabla}$ operates on the entire integrand.* Given this result, we could start with Eqs. (5.175), (5.176), and (5.177), and deduce Eq. (5.172):

$$\begin{aligned} \boldsymbol{F}_b + \boldsymbol{F}_p &= \iiint_V \boldsymbol{f} \, dV - \iint_A \hat{\boldsymbol{n}} p \, dA \\ &= \iiint_V (\boldsymbol{f} - \boldsymbol{\nabla} p) \, dV = \boldsymbol{0}. \end{aligned} \tag{5.179}$$

Since this must hold for any volume V, Eq. (5.172) follows.

So far we have been considering only the pressure, i.e., the stress, in a fluid. The strain produced by the pressure within a fluid is a change in volume per unit mass of the fluid or, equivalently, a change in density. If Hooke's law is satisfied, the change dV in a volume V produced by a small change dp in pressure can be calculated from Eq. (5.116), if the bulk modulus B is known:

$$\frac{dV}{V} = -\frac{dp}{B}. \tag{5.180}$$

If the mass of fluid in the volume V is M, then the density is

$$\rho = \frac{M}{V}, \tag{5.181}$$

and the change $d\rho$ in density corresponding to an infinitesimal change dV in volume is given by

$$\frac{d\rho}{\rho} = -\frac{dV}{V}, \tag{5.182}$$

so that the change in density produced by a small pressure change dp is

$$\frac{d\rho}{\rho} = \frac{dp}{B}. \tag{5.183}$$

*For the proof of this theorem, see Phillips, *Vector Analysis*. New York: John Wiley and Sons, 1933. (Chapter III, Section 34.)

After a finite change in pressure from p_0 to p, the density will be

$$\rho = \rho_0 \exp\left(\int_{p_0}^{p} \frac{dp}{B}\right). \tag{5.184}$$

In any case, the density of a fluid is determined by its equation of state in terms of the pressure and temperature. The equation of state for a perfect gas is

$$pV = RT, \tag{5.185}$$

where T is the absolute temperature, V is the volume per mole, and R is the universal gas constant:

$$R = 8.314 \times 10^7 \text{ erg-deg}^{-1} \text{ C-mole}^{-1}. \tag{5.186}$$

By substitution from Eq. (5.181), we obtain the density in terms of pressure and temperature:

$$\rho = \frac{Mp}{RT}, \tag{5.187}$$

where M is the molecular weight.

Let us apply these results to the most common case, in which the body force is the gravitational force on a fluid in a uniform vertical gravitational field [Eq. (5.169)]. If we apply Eq. (5.173) to this case, we have

$$\nabla \times \boldsymbol{f} = \nabla \times (\rho \boldsymbol{g}) = \boldsymbol{0}. \tag{5.188}$$

Since $\boldsymbol{g}$ is constant, the differentiation implied by the ∇ symbol operates only on ρ, and we can move the scalar ρ from one factor of the cross product to the other to obtain:

$$(\nabla \rho) \times \boldsymbol{g} = \boldsymbol{0}, \tag{5.189}$$

that is, the density gradient must be parallel to the gravitational field. The density must be constant on any horizontal plane *within* the fluid. Equation (5.189) may also be derived from Eq. (5.188) by writing out explicitly the components of the vectors $\nabla \times (\rho \boldsymbol{g})$ and $(\nabla \rho) \times \boldsymbol{g}$, and verifying that they are the same.* According to Eq. (5.172), the pressure is also constant in any horizontal plane *within* the fluid. Pressure and density are therefore functions only of the vertical height z within the fluid. From Eqs. (5.172) and (5.169) we obtain a differential equation for pressure as a function of z:

$$\frac{dp}{dz} = -\rho g. \tag{5.190}$$

If the fluid is incompressible, and ρ is uniform, the solution is

$$p = p_0 - \rho g z, \tag{5.191}$$

where p_0 is the pressure at $z = 0$. If the fluid is a perfect gas, either p or ρ may be

*Equation (5.189) holds also in a nonuniform gravitational field, since $\nabla \times \boldsymbol{g} = \boldsymbol{0}$, by Eq.(6.21).

eliminated from Eq. (5.190) by means of Eq. (5.187). If we eliminate the density, we have

$$\frac{dp}{dz} = -\frac{Mg}{RT}p. \tag{5.192}$$

As an example, if we assume that the atmosphere is uniform in temperature and composition, we can solve Eq. (5.192) for the atmospheric pressure as a function of altitude:

$$p = p_0 \exp\left(-\frac{Mg}{RT}z\right). \tag{5.193}$$

PROBLEMS

1. (a) Prove that the total kinetic energy of the system of particles making up a rigid body, as defined by Eq. (4.37), is correctly given by Eq. (5.16) when the body rotates about a fixed axis.

b) Prove that the potential energy given by Eq. (5.14) is the total work done against the external forces when the body is rotated from θ_s to θ, if N_z is the sum of the torques about the axis of rotation due to the external forces.

2. Using the scheme of analogy in Section 5.2, formulate a theorem analogous to that given by Eq. (2.8) and prove it, starting from Eq. (5.13).

3. Prove, starting with the equation of motion (5.13) for rotation, that if N_z is a function of θ alone, then $T+V$ is constant.

4. The balance wheel of a watch consists of a ring of mass M, radius a, with spokes of negligible mass. The hairspring exerts a restoring torque $N_z = -k\theta$. Find the motion if the balance wheel is rotated through an angle θ_0 and released.

5. A wheel of mass M, radius of gyration k, spins smoothly on a *fixed* horizontal axle of radius a which passes through a hole of slightly larger radius at the hub of the wheel. The coefficient of friction between the bearing surfaces is μ. If the wheel is initially spinning with angular velocity ω_0, find the time and the number of turns that it takes to stop.

6. A wheel of mass M, radius of gyration k is mounted on a horizontal axle. A coiled spring attached to the axle exerts a torque $N = -K\theta$ tending to restore the wheel to its equilibrium position $\theta = 0$. A mass m is located on the rim of the wheel at distance $2k$ from the axle at a point vertically above the axle when $\theta = 0$. Describe the kinds of motion which can occur, locate the positions of stable or unstable equilibrium of the wheel if any, and find the frequencies of small oscillations about the equilibrium points. Consider two cases: (a) $K > 2mgk$. (b) $K = 4mgk/\pi$. What if $K < 4mgk/5\pi$? [*Hint:* Solve the trigonometric equation graphically.]

7. An airplane propeller of moment of inertia I is subject to a driving torque

$$N = N_0(1+\alpha \cos \omega_0 t),$$

and to a frictional torque due to air resistance

$$N_f = -b\dot{\theta}.$$

Find its steady-state motion.

8. A motor armature weighing 2 kg has a radius of gyration of 5 cm. Its no-load speed is 1500 rpm. It is wound so that its torque is independent of its speed. At full speed, it draws a current of 2 amperes at 110 volts. Assume that the electrical efficiency is 80%, and that the friction is proportional to the square of the angular velocity. Find the time required for it to come up to a speed of 1200 rpm after being switched on without load.

9. Derive Eqs. (5.35) and (5.36).

10. Assume that a simple pendulum suffers a frictional torque $-mb_1\dot{\theta}$ due to friction at the point of support, and a frictional force $-b_2\boldsymbol{v}$ on the bob due to air resistance, where $\boldsymbol{v}$ is the velocity of the bob. The bob has a mass m, and is suspended by a string of length l. Find the time required for the amplitude to damp to $1/e$ of its initial (small) value. How should m, l be chosen if it is desired that the pendulum swing as long a time as possible? How should m, l be chosen if it is desired that the pendulum swing through as many cycles as possible?

11. A child of mass m sits in a swing of negligible mass suspended by a rope of length l. Assume that the dimensions of the child are negligible compared with l. His father pulls the child back until the rope makes an angle of one radian with the vertical, then pushes with a force $F = mg$ along the arc of a circle until the rope is vertical and releases the swing. (a) How high up will the swing go? (b) For what length of time did the father push on the swing? (Assume that it is permissible to write $\sin\theta \doteq \theta$ for $\theta < 1$.) Compare with the time required for the swing to reach the vertical if he simply releases the swing without pushing on it.

12. A baseball bat held horizontally at rest is struck at a point O' by a ball which delivers a horizontal impulse J' perpendicular to the bat. Let the bat be initially parallel to the x-axis, and let the baseball be traveling in the negative direction parallel to the y-axis. The center of mass G of the bat is initially at the origin, and the point O' is at a distance h' from G. Assuming that the bat is let go just as the ball strikes it, and neglecting the effect of gravity, calculate and sketch the motion $x(t)$, $y(t)$ of the center of mass, and also of the center of percussion, during the first few moments after the blow, say until the bat has rotated a quarter turn. Comment on the difference between the initial motion of the center of mass and that of the center of percussion.

13. A compound pendulum is arranged to swing about either of two parallel axes through two points O, O' located on a line through the center of mass. The distances h, h' from O, O' to the center of mass, and the periods τ, τ' of small amplitude vibrations about the axes through O and O' are measured. O and O' are arranged so that each is approximately the center of oscillation relative to the other. Given $\tau = \tau'$, find a formula for g in terms of measured quantities. Given that $\tau' = \tau(1+\delta)$, where $\delta \ll 1$, find a correction to be added to your previous formula so that it will be correct to terms of order δ.

14. Prove that if a body is composed of two or more parts whose centers of mass are known, then the center of mass of the composite body can be computed by regarding its component parts as single particles located at their respective centers of mass. Assume that each component part k is described by a density $\rho_k(\boldsymbol{r})$ of mass continuously distributed over the region occupied by part k.

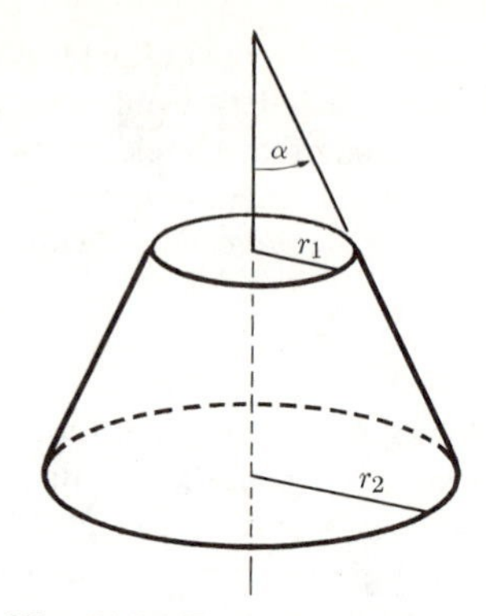

Fig. 5.28 Frustum of a cone.

15. A circular disk of radius a lies in the xy-plane with its center at the origin. The half of the disk above the x-axis has a density σ per unit area, and the half below the x-axis has a density 2σ. Find the center of mass G, and the moments of inertia about the x-, y-, and z-axes, and about parallel axes through G. Make as much use of labor-saving theorems as possible.

16. (a) Work out a formula for the moments of inertia of a cone of mass m, height h, and generating angle α, about its axis of symmetry, and about an axis through the apex perpendicular to the axis of symmetry. Find the center of mass of the cone.

b) Use these results to determine the center of mass of the frustum of a cone, shown in Fig. 5.28, and to calculate the moments of inertia about horizontal axes through each base and through the center of mass. The mass of the frustum is M.

17. Find the moments of inertia of the block shown in Fig. 5.8, about axes through its center of mass parallel to each of the three edges of the block.

18. Through a sphere of mass M, radius R, a plane saw cut is made at a distance $\frac{1}{2}R$ from the center. The smaller piece of the sphere is discarded. Find the center of mass of the remaining piece, and the moments of inertia about its axis of symmetry, and about a perpendicular axis through the center of mass.

19. How many yards of thread 0.03 inch in diameter can be wound on the spool shown in Fig. 5.29?

20. Given that the volume of a cone is one-third the area of the base times the height, locate by Pappus' theorem the centroid of a right triangle whose legs are of lengths a and b.

21. Prove that Pappus' second theorem holds even if the axis of revolution intersects the surface, provided that we take as volume the difference in the volumes generated by the two parts into which the surface is divided by the axis. What is the corresponding generalization of the first theorem?

22. Find the center of mass of a wire bent into a semicircle of radius a. Find the three radii of

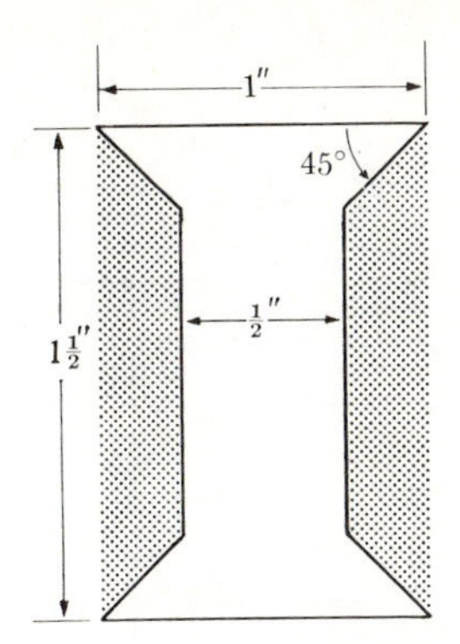

Fig. 5.29 How much thread can be wound on this spool?

gyration about x-, y-, and z-axes through the center of mass, where z is perpendicular to the plane of the semicircle and x bisects the semicircle. Use your ingenuity to reduce the number of calculations required to a minimum.

23. (a) Find a formula for the radius of gyration of a uniform rod of length l about an axis through one end making an angle α with the rod.

b) Using this result, find the moment of inertia of an equilateral triangular pyramid, constructed out of six uniform rods, about an axis through its centroid and one of its vertices.

24. Find the radii of gyration of a plane lamina in the shape of an ellipse of semimajor axis a, eccentricity ε, about its major and minor axes, and about a third axis through one focus perpendicular to the plane.

25. Forces 1 kg-wt, 2 kg-wt, 3 kg-wt, and 4 kg-wt act in sequence clockwise along the four sides of a square $0.5 \times 0.5\ \text{m}^2$. The forces are directed in a clockwise sense around the square. Find the equilibrant.

26. Forces 2 lb, 3 lb, and 5 lb act in sequence in a clockwise sense along the three sides of an equilateral triangle. The sides of the triangle have length 4 ft. Find the resultant.

27. (a) Reduce the system of forces acting on the cube shown in Fig. 5.30 to an equivalent single force acting at the center of the cube, plus a couple composed of two forces acting at two adjacent corners.

b) Reduce this system to a system of two forces, and state where these forces act.

c) Reduce this system to a single force plus a torque parallel to it.

28. A sphere weighing 500 g is held between thumb and forefinger at the opposite ends of a horizontal diameter. A string is attached to a point on the surface of the sphere at the end of a perpendicular horizontal diameter. The string is pulled with a force of 300 g in a direction parallel to the line from forefinger to thumb. Find the forces which must be exerted by forefinger and thumb to hold the sphere stationary. Is the answer unique? Does it correspond to your physical intuition about the problem?

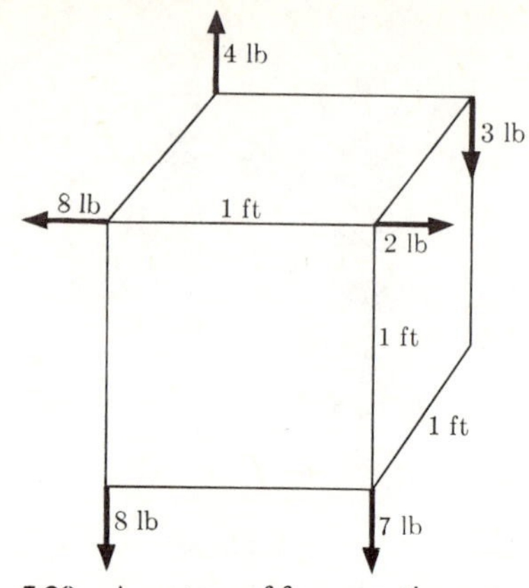

Fig. 5.30 A system of forces acting on a cube.

29. A sailboat sails in a direction making an angle α with the direction from which the wind is coming. Assume that the sail is a plane surface and that it is pulled in until it makes an angle β with the keel of the boat. (If the sail is to catch the wind, $\beta < \alpha$.) Assume that the force exerted on the sail by the wind is perpendicular to the sail and is proportional to the component of the wind velocity perpendicular to the sail. Find the optimum angle β at which to trim the sail in order that the component of force in the direction of travel will be a maximum. If the pilot wishes to reach a point directly upwind by tacking back and forth at an angle α from the wind, and if the friction which limits the velocity of the boat is proportional to its velocity, what is the optimum angle α (assuming that β is also optimized) which maximizes the component of velocity in a direction into the wind? How many inaccuracies can you find in the assumptions made, and what effect would each have on the optimum angles β and α?

30. A rope 10 m long is suspended between two points separated by 5 m horizontally. A mass of 50 kg is hung at the midpoint of the rope. The cross-sectional area of the rope is 1 cm^2 and $Y = 500$ kg/cm^2. Find the tension in the rope, taking its stretch into account to a first approximation.

31. A cable is to be especially designed to hang vertically and to support a load W at a distance l below the point of support. The cable is to be made of a material having a Young's modulus Y and a weight w per unit volume. Inasmuch as the length l of the cable is to be fairly great, it is desired to keep the weight of the cable to a minimum by making the cross-sectional area $A(z)$ of the cable, at a height z above the lower end, just great enough to support the load beneath it. The cable material can safely support a load just great enough to stretch it 1%. Determine the function $A(z)$ when the cable is supporting the given load.

32. (a) A cable is connected in a straight line between two fixed points. By exerting a sidewise force W at the center of the cable, a considerably greater force τ can be applied to the support points at each end of the cable. Find a formula for τ in terms of W, and the area A and Young's modulus Y of the cable, assuming that the angle through which the cable is pulled is small.

b) Show that this assumption is well satisfied if $W = 100$ lb, $A = 3$ in^2, and $Y = 60{,}000$ lb-in^{-2}. Find τ.

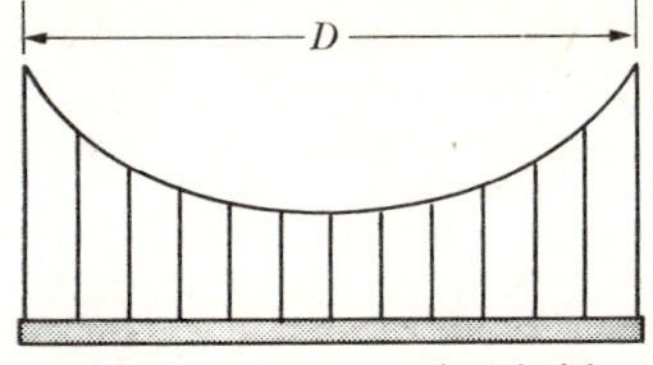

Fig. 5.31 A suspension bridge.

33. A cable 20 ft long is suspended between two points A and B, 15 ft apart. The line AB makes an angle of 30° with the horizontal (B higher). A weight of 2000 lb is hung from a point C 8 ft from the end of the cable at A.

a) Find the position of point C, and the tensions in the cable, if the cable does not stretch.

b) If the cable is $\frac{1}{2}$ inch in diameter and has a Young's modulus of 5×10^5 lb-in^{-2}, find the position of point C and the tensions, taking cable stretch into account. Carry out two successive approximations, and estimate the accuracy of your result.

34. (a) A cable of length l, weight w per unit length, is suspended from the points $x = \pm a$ on the x-axis. The y-axis is vertical. By requiring that $y = 0$ at $x = \pm a$, and that the total length of cable be l, show that $\alpha = 0$ in Eq. (5.134), and set up equations to be solved for β and C.

b) Show that the same results can be obtained for α and C by requiring that the cable be symmetrical about the y-axis, and that the forces at its ends balance the weight of the cable.

35. A bridge of weight w per unit length is to be hung from cables of negligible weight, as shown in Fig. 5.31. It is desired to determine the shape of the suspension cables so that the vertical cables, which are equally spaced, will support equal weights. Assume that the vertical cables are so closely spaced that we can regard the weight w per unit length as continuously distributed along the suspension cable. The problem then differs from that treated in the text, where the string had a weight w per unit length s along the string, in that here there is a weight w per unit horizontal distance x. Set up a differential equation for the shape $y(x)$ of the suspension cable, and solve for $y(x)$ if the ends are at the points $y = 0$, $x = \pm\frac{1}{2}D$, and if the maximum tension in the cable is to be τ_0.

36. A cable of length l, weight w per unit length, is suspended from points $x = \pm a$ on the x-axis. The y-axis is vertical. A weight W is hung from the midpoint of the cable. Set up the equations from which β, α, and C are to be determined.

37. A uniform beam of square cross section ($l \times l$) with ends cut off squarely is made of a material for which $Y = n$. It is held firmly in a square horizontal channel and hangs out a distance L from the end of the channel. For what value of L is the deflection at the end due equally to shear and bending? For that value of L, what angle does the end face make with a vertical plane? What angle does the top surface of the beam at the end make with the horizontal? Assume $W \ll Yl^2$.

38. A seesaw is made of a plank of wood of rectangular cross section 2×12 in^2 and 10 ft long, weighing 60 lb. Young's modulus is 1.5×10^6 lb-in^{-2}. The plank is balanced across a narrow support at its center. Two children weighing 100 lb each sit one foot from the ends. Find the shape of the plank when it is balanced in a stationary horizontal position. Neglect shear.

39. An empty pipe of inner radius a, outer radius b, is made of material with Young's modulus Y, shear modulus n, density ρ. A horizontal section of length L is clamped at both ends. Find the deflection at the center. Find the increase in deflection when the pipe is filled with a fluid of density ρ_0.

40. An I-beam has upper and lower flanges of width a, connected by a center web of height b. The web and flanges are of the same thickness c, assumed negligible with respect to a and b, and are made of a material with Young's modulus Y, shear modulus n. The beam has a weight W, length L, and rests on supports at each end. A load W' rests on the midpoint of the beam. Find the deflection of the beam at its midpoint. Separate the deflection into terms due to shear and to bending, and into terms due to the beam weight W and the load W'.

41. It was once proposed to construct a permanent space platform by erecting a pyramidal balloon 60 miles high and resting on the earth's surface, filled with a slight overpressure above atmospheric. A pressure of 1 lb/in^2 at the top would be adequate to support fairly heavy loads. Aside from other objections, show that this idea is impractical by calculating the pressure in the balloon at sea level, assuming it is filled with air at 0°C and that the pressure at 60 miles is 1 lb/in^2.

42. If the bulk modulus of water is B, and the atmospheric pressure at the surface of the ocean is p_0, find the pressure as a function of depth in the ocean, taking into account the compressibility of the water. Assume that B is constant. Look up B for water, and estimate the error that would be made at a depth of 5 miles if the compressibility were neglected.

43. Find the atmospheric pressure as a function of altitude on the assumption that the temperature decreases with altitude, the decrease being proportional to the altitude.

44. There is a slight flaw in the argument at the beginning of Section 5.11 due to the neglect of a possible body force in Eq. (5.166). Show that even when body forces are present, the stress at any point in a fluid in equilibrium is the same in all directions. [*Hint:* Let the size of the prism shrink to zero, and show that in this limit, the body force is negligible compared with the stresses across its surface.]

CHAPTER 6

GRAVITATION

6.1 CENTERS OF GRAVITY FOR EXTENDED BODIES

You will recall that we formulated the law of gravitation in Section 1.5. Any two particles of masses m_1 and m_2, a distance r apart, attract each other with a force whose magnitude is given by Eq. (1.11):

$$F = \frac{Gm_1m_2}{r^2}, \tag{6.1}$$

where

$$G = 6.67 \times 10^{-8} \text{ dyne-cm}^2\text{-g}^{-2}, \tag{6.2}$$

as determined by measurements of the forces between large lead spheres, carried out by means of a delicate torsion balance. Equation (6.1) can be written in a vector form which gives both the direction and magnitude of the attractive forces. Let $\boldsymbol{r}_1$ and $\boldsymbol{r}_2$ be the position vectors of the two particles. Then the gravitational force on m_2 due to m_1 is

$$\boldsymbol{F}_{1\to 2} = \frac{Gm_1m_2}{|\boldsymbol{r}_1 - \boldsymbol{r}_2|^3}(\boldsymbol{r}_1 - \boldsymbol{r}_2). \tag{6.3}$$

The vector $(\boldsymbol{r}_1 - \boldsymbol{r}_2)$ gives the force the correct direction, and its magnitude is divided out by the extra factor $|\boldsymbol{r}_1 - \boldsymbol{r}_2|$ in the denominator.

The law of gravitation as formulated in Eq. (6.3) is applicable only to particles or to bodies whose dimensions are negligible compared with the distance between them; otherwise the distance $|\boldsymbol{r}_1 - \boldsymbol{r}_2|$ is not precisely defined, nor is it immediately clear at what points and in what directions the forces act. For extended bodies, we must imagine each body divided into pieces or elements, small compared with the distances between the bodies, and compute the forces on each of the elements of one body due to each of the elements of the other bodies.

Consider now an extended body of mass M and a particle of mass m at a point P (Fig. 6.1). If the body of mass M is divided into small pieces of masses m_i, each piece is attracted toward m by a force which we shall call $\boldsymbol{F}_i$. Now the system of forces $\boldsymbol{F}_i$ can be resolved according to Theorem 1 of Section 5.6 (5.107) into a single force through an arbitrary point, plus a couple. Let this single force be $\boldsymbol{F}$:

$$\boldsymbol{F} = \sum_i \boldsymbol{F}_i, \tag{6.4}$$

and let the arbitrary point be taken as the point P. Since none of the forces $\boldsymbol{F}_i$ exerts any torque about P, the total torque about P is zero, and the couple vanishes.

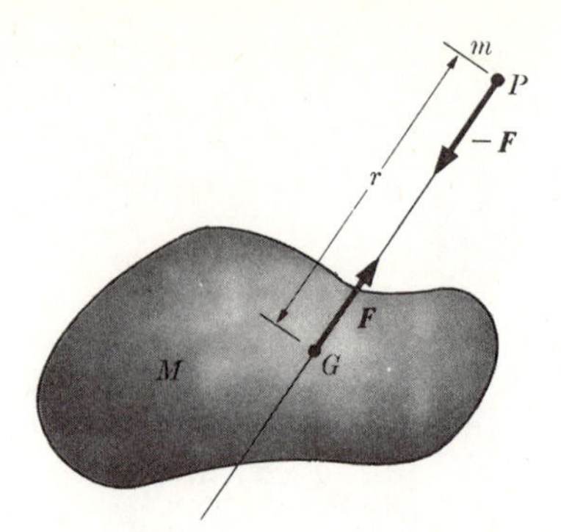

Fig. 6.1 Gravitational attraction between a particle and an extended body.

The system of forces therefore has a resultant $\boldsymbol{F}$ acting along a line through the mass m. The force acting on m is $-\boldsymbol{F}$, since Newton's third law applies to each of the forces $\boldsymbol{F}_i$ in Eq. (6.4). We locate on this line of action of $\boldsymbol{F}$ a point G a distance r from P such that

$$|\boldsymbol{F}| = \frac{GmM}{r^2}. \tag{6.5}$$

Then the system of gravitational forces between the body M and the particle m is equivalent to the single resultant forces $\boldsymbol{F}$ on M and $-\boldsymbol{F}$ on m which would act if all the mass of the body M were concentrated at G. The point G is called the *center of gravity* of the body *M relative to the point P;* G is not, in general, at the center of mass of body M, nor even on the line joining P with the center of mass. The parts of the body close to P are attracted more strongly than those farther away, whereas in finding the center of mass, all parts of the body are treated alike. Furthermore, the position of the point G will depend on the position of P. When P is far away compared with the dimensions of the body, the acceleration of gravity due to m will be nearly constant over the body and, in this case, we showed in Section 5.6 that G will coincide with the center of mass. Also, in the case of a uniform sphere or a spherically symmetrical distribution of mass, we shall show in the next section that the center of gravity always lies at the center of the sphere. The relative character of the concept of center of gravity makes it of little use except in the case of a sphere or of a body in a uniform gravitational field.

For two extended bodies, no unique centers of gravity can in general be defined, even relative to each other, except in special cases, as when the bodies are far apart, or when one of them is a sphere. The system of gravitational forces on either body due to the other may or may not have a resultant; if it does, the two resultants are equal and opposite and act along the same line. However, even in this case, we cannot define definite centers of gravity G_1, G_2 for the two bodies relative to each other, since Eq. (6.5) specifies only the distance $\overline{G_1G_2}$.

The general problem of determining the gravitational forces between bodies is usually best treated by means of the concepts of the field theory of gravitation discussed in the next section.

6.2 GRAVITATIONAL FIELD AND GRAVITATIONAL POTENTIAL

The gravitational force $\boldsymbol{F}_m$ acting on a particle of mass m at a point $\boldsymbol{r}$, due to other particles m_i at points $\boldsymbol{r}_i$, is the vector sum of the forces due to each of the other particles acting separately:

$$\boldsymbol{F}_m = \sum_i \frac{m m_i G(\boldsymbol{r}_i - \boldsymbol{r})}{|\boldsymbol{r}_i - \boldsymbol{r}|^3}. \tag{6.6}$$

If, instead of point masses m_i, we have mass continuously distributed in space with a density $\rho(\boldsymbol{r})$, the force on a point mass m at $\boldsymbol{r}$ is

$$\boldsymbol{F}_m = \iiint \frac{mG(\boldsymbol{r}' - \boldsymbol{r})\rho(\boldsymbol{r}')}{|\boldsymbol{r}' - \boldsymbol{r}|^3}\, dV'. \tag{6.7}$$

The integral may be taken over the region containing the mass whose attraction we are computing, or over all space if we let $\rho = 0$ outside this region. Now the force $\boldsymbol{F}_m$ is proportional to the mass m, and we define the *gravitational field intensity* (or simply *gravitational field*) $\boldsymbol{g}(\boldsymbol{r})$, at any point $\boldsymbol{r}$ in space, due to any distribution of mass, as the force per unit mass which would be exerted on any small mass m at that point:

$$\boldsymbol{g}(\boldsymbol{r}) = \frac{\boldsymbol{F}_m}{m}, \tag{6.8}$$

where $\boldsymbol{F}_m$ is the force that would be exerted on a point mass m at the point $\boldsymbol{r}$. We can write formulas for $\boldsymbol{g}(\boldsymbol{r})$ for point masses or continuously distributed mass:

$$\boldsymbol{g}(\boldsymbol{r}) = \sum_i \frac{m_i G(\boldsymbol{r}_i - \boldsymbol{r})}{|\boldsymbol{r}_i - \boldsymbol{r}|^3}, \tag{6.9}$$

$$\boldsymbol{g}(\boldsymbol{r}) = \iiint \frac{G(\boldsymbol{r}' - \boldsymbol{r})\rho(\boldsymbol{r}')}{|\boldsymbol{r}' - \boldsymbol{r}|^3}\, dV'. \tag{6.10}$$

The field $\boldsymbol{g}(\boldsymbol{r})$ has the dimensions of acceleration, and is in fact the acceleration experienced by a particle at the point $\boldsymbol{r}$, on which no forces act other than the gravitational force.

The calculation of the gravitational field $\boldsymbol{g}(\boldsymbol{r})$ from Eq. (6.9) or (6.10) is difficult except in a few simple cases, partly because the sum and integral call for the addition of a number of vectors. Since the gravitational forces between pairs of particles are central forces, they are conservative, as we showed in Section 3.12, and a potential energy can be defined for a particle of mass m subject to gravitational forces. For two particles m and m_i, the potential energy is given by Eqs. (3.229) and (3.230):

$$V_{mm_i} = \frac{-Gmm_i}{|\boldsymbol{r}-\boldsymbol{r}_i|}. \tag{6.11}$$

The potential energy of a particle of mass m at point r due to a system of particles m_i is then

$$V_m(\boldsymbol{r}) = \sum_i \frac{-Gmm_i}{|\boldsymbol{r}-\boldsymbol{r}_i|}. \tag{6.12}$$

We define the *gravitational potential* $\mathcal{G}(\boldsymbol{r})$ at point $\boldsymbol{r}$ as the negative of the potential energy per unit mass of a particle at point $\boldsymbol{r}$. [This choice of sign in $\mathcal{G}(\boldsymbol{r})$ is conventional in gravitational theory.]

$$\mathcal{G}(\boldsymbol{r}) = -\frac{V_m(\boldsymbol{r})}{m}. \tag{6.13}$$

For a system of particles,

$$\mathcal{G}(\boldsymbol{r}) = \sum_i \frac{m_i G}{|\boldsymbol{r}-\boldsymbol{r}_i|}. \tag{6.14}$$

If $\rho(\boldsymbol{r})$ represents a continuous distribution of mass, its gravitational potential is

$$\mathcal{G}(\boldsymbol{r}) = \iiint \frac{G\rho(\boldsymbol{r}')}{|\boldsymbol{r}-\boldsymbol{r}'|}\, dV'. \tag{6.15}$$

Because it is a scalar point function, the potential $\mathcal{G}(\boldsymbol{r})$ is easier to work with for many purposes than is the field $\boldsymbol{g}(\boldsymbol{r})$. In view of the relation (3.185) between force and potential energy, $\boldsymbol{g}$ may easily be calculated, when $\mathcal{G}$ is known, from the relation

$$\boldsymbol{g} = \nabla \mathcal{G}. \tag{6.16}$$

The inverse relation is

$$\mathcal{G}(\boldsymbol{r}) = \int_{\boldsymbol{r}_s}^{\boldsymbol{r}} \boldsymbol{g} \cdot d\boldsymbol{r}. \tag{6.17}$$

The definition of $\mathcal{G}(\boldsymbol{r})$, like that of potential energy $V(\boldsymbol{r})$, involves an arbitrary additive constant or, equivalently, an arbitrary point $\boldsymbol{r}_s$ at which $\mathcal{G} = 0$. Usually $\boldsymbol{r}_s$ is taken at an infinite distance from all masses, as in Eqs. (6.14) and (6.15).

The concepts of gravitational field and gravitational potential are mathematically identical to those of electric field intensity and electrostatic potential in electrostatics, except that the negative sign in Eq. (6.13) is conventional in gravitational theory, and except that all masses are positive and all gravitational forces are attractive, so that the force law has the opposite sign from that in

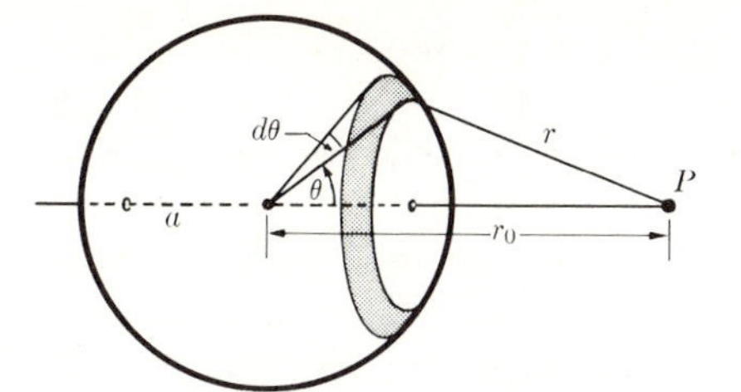

Fig. 6.2 Method of computing potential of a spherical shell.

electrostatics. The subject of potential theory is an extensive one, and we can give here only a very brief introductory treatment.

As an example of the use of the concept of potential, we calculate the potential due to a thin homogeneous spherical shell of matter of mass M, density σ per unit area, and radius a:

$$M = 4\pi a^2\sigma. \tag{6.18}$$

The potential at a point P is computed by integrating over a set of ring elements as in Fig. 6.2. The potential of a ring of radius $a \sin \theta$, width $a\, d\theta$, all of whose mass is at the same distance r from P, will be

$$d\mathcal{G} = \frac{G\sigma(2\pi a \sin \theta)a\, d\theta}{r},$$

and the total potential at P of the spherical shell is

$$\begin{aligned}\mathcal{G}(P) &= \int_0^\pi \frac{G\sigma(2\pi a \sin \theta)a\, d\theta}{r}\\ &= \frac{MG}{2}\int_0^\pi \frac{\sin \theta\, d\theta}{(r_0^2+a^2-2ar_0 \cos \theta)^{1/2}}\\ &= \frac{MG}{2ar_0}[(r_0+a)-|r_0-a|]. \end{aligned}\tag{6.19}$$

We have two cases, according to whether P is outside or inside the shell:

$$\mathcal{G}(P) = \frac{MG}{r_0}, \quad r_0 \geq a, \qquad \mathcal{G}(P) = \frac{MG}{a}, \quad r_0 \leq a. \tag{6.20}$$

Thus outside the shell the potential is the same as for a point mass M at the center of the shell. The gravitational field outside a spherical shell is then the same as if all the mass of the shell were at its center. The same statement then holds for the gravitational field outside any spherically symmetrical distribution of mass, since the total field is the sum of the fields due to the shells of which it is composed. This proves the statement made in the previous section; a spherically symmetrical distribution of mass attracts (and therefore is attracted by) any other mass outside it as if all its mass were at its center. Inside a spherical shell, the potential is constant, and it follows from Eq. (6.16) that the gravitational field is there zero. Hence a point inside a spherically symmetric distribution of mass at a distance r from the center is attracted as if the mass inside the sphere of radius r were at the center; the mass outside this sphere exerts no net force. These results would be somewhat more difficult to prove by computing the gravitational forces directly, as the reader can readily verify. Indeed, it took Newton twenty years! The calculation of the force of attraction on the moon by the earth described in the last section of Chapter 1 was made by Newton twenty years before he published his law of gravitation. It is likely that he waited until he could prove an assumption implicit in that calculation, namely, that the earth attracts any body outside it as if all the mass of the earth were concentrated at its center. The fact that today a college undergraduate can solve in twenty minutes a problem which stumped Newton for twenty years is not so much a measure of their relative mathematical talents as it is a measure of the enormous improvements that have been made in mathematical methods in the calculus since Newton invented it.

6.3 GRAVITATIONAL FIELD EQUATIONS

It is of interest to find differential equations satisfied by the functions $\boldsymbol{g}(\boldsymbol{r})$ and $\mathcal{G}(\boldsymbol{r})$. From Eq. (6.16) it follows that

$$\nabla \times \boldsymbol{g} = \mathbf{0}. \tag{6.21}$$

When written out in any coordinate system, this vector equation becomes a set of three partial differential equations connecting the components of the gravitational field. In rectangular coordinates,

$$\frac{\partial g_z}{\partial y} - \frac{\partial g_y}{\partial z} = 0, \qquad \frac{\partial g_x}{\partial z} - \frac{\partial g_z}{\partial x} = 0, \qquad \frac{\partial g_y}{\partial x} - \frac{\partial g_x}{\partial y} = 0. \tag{6.22}$$

These equations alone do not determine the gravitational field, for they are satisfied by every gravitational field. To determine the gravitational field, we need an equation connecting $\boldsymbol{g}$ with the distribution of matter.

Let us study the gravitational field $\boldsymbol{g}$ due to a point mass m. Consider any volume V containing the mass m, and let $\hat{\boldsymbol{n}}$ be the unit vector normal at each point to the surface S that bounds V (Fig. 6.3). Let us compute the surface integral

$$I = \iint_S \hat{\boldsymbol{n}} \cdot \boldsymbol{g} \, dS. \tag{6.23}$$

The physical or geometric meaning of this integral can be seen if we introduce the concept of lines of force, drawn everywhere in the direction of $\boldsymbol{g}$, and in such a manner that the number of lines per unit area at any point is equal to the gravitational field intensity. Then I is the number of lines passing out through the surface S, and is called the *flux* of $\boldsymbol{g}$ through S. The element of solid angle $d\Omega$

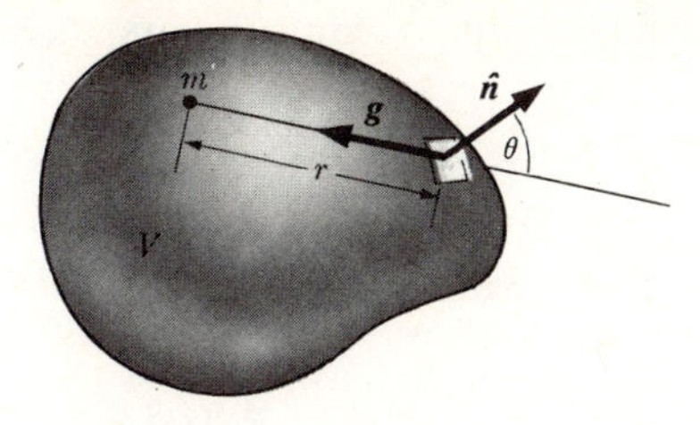

Fig. 6.3 A mass m enclosed in a volume V.

subtended at the position of m by an element of surface dS is defined as the area swept out on a sphere of unit radius by a radius from m which sweeps over the surface element dS. This area is

$$d\Omega = \frac{dS \cos \theta}{r^2}. \tag{6.24}$$

From Fig. 6.3, we have the relation

$$\hat{\boldsymbol{n}} \cdot \boldsymbol{g} = -\frac{mG \cos \theta}{r^2}. \tag{6.25}$$

When use is made of these two relations, the integral I [Eq. (6.23)] becomes

$$I = \iint_S -mG\, d\Omega = -4\pi mG. \tag{6.26}$$

The integral I is independent of the position of m within the surface S. This result is analogous to the corresponding result in electrostatics that there are 4π lines of force coming from every unit charge (in esu; there are ε_0^{-1} lines per unit charge in mks units.) Since the gravitational field of a number of masses is the sum of their individual fields, we have, for a surface S surrounding a set of masses m_i:

$$I = \iint_S \hat{\boldsymbol{n}} \cdot \boldsymbol{g}\, dS = -\sum_i 4\pi m_i G. \tag{6.27}$$

For a continuous distribution of mass within S, this equation becomes

$$\iint_S \hat{\boldsymbol{n}} \cdot \boldsymbol{g}\, dS = -\iiint_V 4\pi G\rho\, dV. \tag{6.28}$$

We now apply Gauss' divergence theorem [Eq. (3.115)] to the left side of this equation:

$$\iint_S \hat{\boldsymbol{n}} \cdot \boldsymbol{g}\, dS = \iiint_V \nabla \cdot \boldsymbol{g}\, dV. \tag{6.29}$$

Subtracting Eq. (6.28) from Eq. (6.29), we arrive at the result

$$\iiint_V (\nabla \cdot \boldsymbol{g} + 4\pi G\rho)\, dV = 0. \tag{6.30}$$

Now Eq. (6.30) must hold for any volume V, and this can only be true if the integrand vanishes:

$$\nabla \cdot \boldsymbol{g} = -4\pi G\rho. \tag{6.31}$$

This equation in cartesian coordinates has the form

$$\frac{\partial g_x}{\partial x} + \frac{\partial g_y}{\partial y} + \frac{\partial g_z}{\partial z} = -4\pi G\rho(x, y, z). \tag{6.32}$$

When $\rho(x, y, z)$ is given, the set of equations (6.22) and (6.32) can be shown to determine the gravitational field (g_x, g_y, g_z) uniquely, if we add the boundary condition that $\boldsymbol{g} \to 0$ as $|\boldsymbol{r}| \to \infty$. Substituting from Eq. (6.16), we get an equation satisfied by the potential:

$$\nabla^2 \mathcal{G} = -4\pi G\rho, \tag{6.33}$$

or

$$\frac{\partial^2 \mathcal{G}}{\partial x^2} + \frac{\partial^2 \mathcal{G}}{\partial y^2} + \frac{\partial^2 \mathcal{G}}{\partial z^2} = -4\pi G\rho. \tag{6.34}$$

This single equation determines $\mathcal{G}(x, y, z)$ uniquely if we add the condition that $\mathcal{G} \to 0$ as $|\boldsymbol{r}| \to \infty$. This result we quote from potential theory without proof. The solution of Eq. (6.33) is, in fact, Eq. (6.15). It is often easier to solve the partial differential equation (6.34) directly than to compute the integral in Eq. (6.15). Equations (6.33), (6.16), and (6.8) together constitute a complete summary of Newton's theory of gravitation, as likewise do Eqs. (6.31), (6.21), and (6.8); that is, all the results of the theory can be derived from either of these sets of equations.

Equation (6.33) is called *Poisson's equation.* Equations of this form turn up frequently in physical theories. For example, the electrostatic potential satisfies an equation of the same form, where ρ is the electric charge density. If $\rho = 0$, Eq. (6.33) takes the form

$$\nabla^2 \mathcal{G} = 0. \tag{6.35}$$

This is called *Laplace's equation.* An extensive mathematical theory of Eqs. (6.33) and (6.35) has been developed.* A discussion of potential theory is, however, outside the scope of this text.

---

*O. D. Kellogg, *Foundations of Potential Theory*. Berlin: J. Springer, 1929.

PROBLEMS

1. (a) Given Newton's laws of motion, and Kepler's first two laws of planetary motion (Section 3.15), show that the force acting on a planet is directed toward the sun and is inversely proportional to the square of the distance from the sun.

b) Use Kepler's third law to show that the forces on the planets are proportional to their masses.

c) If this suggests to you a universal law of attraction between any two masses, use Newton's third law to show that the force must be proportional to both masses.

2. Two equal masses m are separated by a distance a. Find the center of gravity of the two masses relative to a point P on the perpendicular bisector of the line joining them a distance y from the midpoint between them. Show that as $y \to \infty$, the center of gravity approaches the center of mass. What happens as $y \to 0$?

3. A mass αM is located at $x = a$, $y = 0$, and a second mass $(1-\alpha)M$ is located at $x = 0$, $y = b$, where $0 < \alpha < 1$. Find the coordinates x, y of the center of gravity of the two masses relative to the origin. Show that your formulas for x, y have the proper limits when $\alpha \to 0$ or $b \to \infty$.

4. (a) Find the gravitational field and gravitational potential at any point z on the symmetry axis of a uniform solid hemisphere of radius a, mass M. The center of the hemisphere is at $z = 0$.

b) Locate the center of gravity of the hemisphere relative to a point outside it on the z-axis, and show that as $z \to \pm\infty$, the center of gravity approaches the center of mass.

5. Assuming that the earth is a sphere of uniform density, with radius a, mass M, calculate the gravitational field intensity and the gravitational potential at all points inside and outside the earth, taking $\mathcal{G} = 0$ at an infinite distance.

6. Assume that the density of a star is a function only of the radius r measured from the center of the star, and is given by *

$$\rho = \frac{Ma^2}{2\pi r(r^2+a^2)^2}, \qquad 0 \leq r < \infty,$$

where M is the mass of the star, and a is a constant which determines the size of the star. Find the gravitational field intensity and the gravitational potential as functions of r.

7. Show that if the sun were surrounded by a spherical cloud of dust of uniform density ρ, the gravitational field within the dust cloud would be

$$\boldsymbol{g} = -\left(\frac{MG}{r^2}+\frac{4\pi}{3}\rho G r\right)\frac{\boldsymbol{r}}{r},$$

where M is the mass of the sun, and $\boldsymbol{r}$ is a vector from the sun to any point in the dust cloud.

---

*The expression for ρ is chosen to make the problem easy to solve, not because it has more than a remote resemblance to the density variations within any actual star.

Note: Problems 8 through 13 require an understanding of the material in Section 5.11.

8. Assuming that the interior of the earth can be treated as an incompressible fluid in equilibrium, (a) calculate the pressure within the earth as a function of distance from the center; (b) using appropriate values for the earth's mass and radius, calculate the pressure in tons per square inch at the center.

9. Set up the equations to be solved for the pressure as a function of radius in a spherically symmetric mass M of gas, assuming that the gas obeys the perfect gas laws and that the temperature is known as a function of radius.

10. (a) Assume that ordinary cold matter collapses, under a pressure greater than a certain critical pressure p_0, to a state of very high density ρ_1. A planet of mass M is constructed of matter of mean density ρ_0 in its normal state. Assuming uniform density and conditions of fluid equilibrium, at what mass M_0 and radius r_0 will the pressure at the center reach the critical value p_0?

b) If $M > M_0$, the planet will have a very dense core of density ρ_1 surrounded by a crust of density ρ_0. Calculate the resulting pressure distribution within the planet in terms of the radius r_1 of the core and the radius r_2 of the planet. Show that if M is somewhat larger than M_0, then the radius r_2 of the planet is less than r_0. (The planet Jupiter is said to have a mass very nearly equal to the critical mass M_0, so that if it were heavier it might be smaller.)

11. Find the pressure and temperature as functions of radius for the star of Problem 6 if the star is composed of a perfect gas of atomic weight A.

12. Find the density and gravitational field intensity as a function of radius inside a small spherically symmetric planet, to order $(1/B^2)$, assuming that the bulk modulus B is constant. The mass is M and the radius is a. [*Hint:* Calculate $g(r)$ assuming uniform density; then find the resulting pressure $p(r)$, and the density $\rho(r)$ to order $(1/B)$. Recalculate $g(r)$ using the new $\rho(r)$, and proceed by successive approximations to terms of order $(1/B^2)$.]

13. Consider a spherical mountain of radius a, mass M, floating in equilibrium in the earth, and whose density is half that of the earth. Assume that a is much less than the earth's radius, so that the earth's surface can be regarded as flat in the neighborhood of the mountain. If the mountain were not present, the gravitational field intensity near the earth's surface would be g_0.

a) Find the difference between g_0 and the actual value of g at the top of the mountain.

b) If the top of the mountain is eroded flat, level with the surrounding surface of the earth, and if this occurs in a short time compared with the time required for the mountain to float in equilibrium again, find the difference between g_0 and the actual value of g at the earth's surface at the center of the eroded mountain.

14. (a) Find the gravitational potential and the field intensity due to a thin rod of length l and mass M at a point a distance r from the center of the rod in a direction making an angle θ with the rod. Assume that $r \gg l$, and carry the calculations only to second order in l/r.

b) Locate the center of gravity of the rod relative to the specified point.

15. (a) Calculate the gravitational potential of a uniform circular ring of matter of radius a, mass M, at a distance r from the center of the ring in a direction making an angle θ with the axis of the ring. Assume that $r \gg a$, and calculate the potential only to second order in a/r.

b) Calculate to the same approximation the components of the gravitational field of the ring at the specified point.

16. A small body with cylindrical symmetry has a density $\rho(r, \theta)$ in spherical coordinates, which vanishes for $r > a$. The origin $r = 0$ lies at the center of mass. Approximate the gravitational potential at a point r, θ far from the body ($r \gg a$), by expanding in a power series in (a/r), and show that it has the form

$$\mathcal{G}(r, \theta) = \frac{MG}{r} + \frac{QG}{r^3} P_2(\cos\theta) + \frac{EG}{r^4} P_3(\cos\theta) + \cdots,$$

where $P_2(\cos\theta)$, $P_3(\cos\theta)$ are quadratic and cubic polynomials in $\cos\theta$ that do not depend on the body, and Q, E are constants which depend on the mass distribution. Find expressions for P_2, P_3, Q, and E, and show that Q is of the order of magnitude Ma^2, and E of the order Ma^3. It is conventional to normalize P_2 so that the constant term is $-\frac{1}{2}$, and P_3 so that the linear term is $-\frac{3}{2}\cos\theta$. The parameters Q, E are then called the *quadrupole moment* and the *octopole moment* of the body. (The polynomials $P_2, P_3, \ldots$ are the *Legendre polynomials* but a knowledge of Legendre polynomials is not needed to solve this problem.)

17. The earth has approximately the shape of an oblate ellipsoid of revolution whose polar diameter $2a(1-\eta)$ is slightly shorter than its equatorial diameter $2a$. ($\eta = 0.0034$.) To determine to first order in η, the effect of the earth's oblateness on its gravitational field, we may replace the ellipsoidal earth by a sphere of radius R so chosen as to have the same volume. The gravitational field of the earth is then the field of a uniform sphere of radius R with the mass of the earth, plus the field of a surface distribution of mass (positive or negative), representing the mass per unit area which would be added or subtracted to form the actual ellipsoid.

a) Show that the required surface density is, to first order in η,

$$\sigma = \tfrac{1}{3}\eta a\rho(1 - 3\cos^2\theta),$$

where θ is the colatitude, and ρ is the volume density of the earth (assumed uniform). Since the total mass thus added to the surface is zero, its gravitational field will represent the effect of the oblate shape of the earth.

b) Show that the resulting correction to the gravitational potential at a very great distance $r \gg a$ from the earth is, to order (a^3/r^3),

$$\delta\mathcal{G} \doteq \frac{1}{5}\eta \frac{MGa^2}{r^3}(1 - 3\cos^2\theta), \qquad (r \gg a).$$

18. (a) Use Gauss' theorem (6.26) to determine the gravitational field inside and outside a spherical shell of radius a, mass M, uniform density.

b) Calculate the resulting gravitational potential.

19. (a) Find the gravitational field at a distance x from an infinite plane sheet of density σ per unit area.

b) Compare this result with the field just outside a spherical shell of the same surface density.

What part of the field comes from the immediately adjacent matter and what part from more distant matter?

20. Show that the gravitational field equations (6.21), (6.31), and (6.33) are satisfied by the field intensity and potential which you calculated in Problem 5.

*__21.__ (a) Show that $\delta \mathcal{G}$ found in Problem 17(b) satisfies Laplace's equation (6.35). This, together with the fact that $\delta \mathcal{G}$ has the same angular dependence as the mass density which produces it, suggests that the formula given for $\delta \mathcal{G}$ may actually be valid everywhere outside the earth.

b) To show this, consider Poisson's equation (6.33) with $\rho = f(r)\,(1-3\cos^2\theta)$. Show that a solution $\mathcal{G} = h(r)\,(1-3\cos^2\theta)$ will satisfy Eq. (6.33) with this form of ρ, provided

$$\frac{d^2h}{dr^2}+\frac{2}{r}\frac{dh}{dr}-\frac{6h}{r^2} = -4\pi Gf.$$

c) Show that $h = r^{-3}$ satisfies this equation in the region where $f = 0$. Can you complete the proof that the formula for $\delta \mathcal{G}$ found in Problem 17(b) is in fact valid everywhere outside the earth?

CHAPTER 7

MOVING COORDINATE SYSTEMS

7.1 MOVING ORIGIN OF COORDINATES

Let a point in space be located by vectors $\boldsymbol{r}$, $\boldsymbol{r}^*$ with respect to two origins of coordinates O, O^*, and let O^* be located by a vector $\boldsymbol{h}$ with respect to O (Fig. 7.1). Then the relation between the coordinates $\boldsymbol{r}$ and $\boldsymbol{r}^*$ is given by

$$\boldsymbol{r} = \boldsymbol{r}^* + \boldsymbol{h}, \tag{7.1}$$

$$\boldsymbol{r}^* = \boldsymbol{r} - \boldsymbol{h}. \tag{7.2}$$

In terms of rectangular coordinates, with axes x^*, y^*, z^* parallel to axes x, y, z, respectively, these equations can be written:

$$x = x^* + h_x, \qquad y = y^* + h_y, \qquad z = z^* + h_z; \tag{7.3}$$

$$x^* = x - h_x, \qquad y^* = y - h_y, \qquad z^* = z - h_z. \tag{7.4}$$

Now if the origin O^* is moving with respect to the origin O, which we regard as fixed, the relation between the velocities relative to the two systems is obtained by differentiating Eq. (7.1):

$$\begin{aligned} \boldsymbol{v} = \frac{d\boldsymbol{r}}{dt} &= \frac{d\boldsymbol{r}^*}{dt} + \frac{d\boldsymbol{h}}{dt} \\ &= \boldsymbol{v}^* + \boldsymbol{v}_h, \end{aligned} \tag{7.5}$$

where $\boldsymbol{v}$ and $\boldsymbol{v}^*$ are the velocities of the moving point relative to O and O^*, and $\boldsymbol{v}_h$ is the velocity of O^* relative to O. We are supposing that the axes x^*, y^*, z^* remain parallel to x, y, z. This is called a *translation* of the starred coordinate system with respect to the unstarred system. Written out in cartesian components, Eq. (7.5) becomes the time derivative of Eq. (7.3). The relation between relative accelerations is

$$\begin{aligned} \boldsymbol{a} = \frac{d^2\boldsymbol{r}}{dt^2} &= \frac{d^2\boldsymbol{r}^*}{dt^2} + \frac{d^2\boldsymbol{h}}{dt^2} \\ &= \boldsymbol{a}^* + \boldsymbol{a}_h. \end{aligned} \tag{7.6}$$

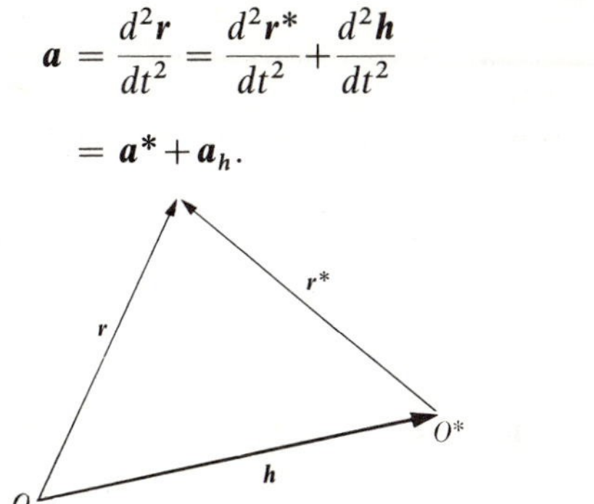

Fig. 7.1 Change of origin of coordinates.

Again these equations can easily be written out in terms of their rectangular components.

Newton's equations of motion hold in the fixed coordinate system, so that we have, for a particle of mass m subject to a force $\boldsymbol{F}$:

$$m\frac{d^2\boldsymbol{r}}{dt^2} = \boldsymbol{F}. \tag{7.7}$$

Using Eq. (7.6), we can write this equation in the starred coordinate system:

$$m\frac{d^2\boldsymbol{r}^*}{dt^2} + m\boldsymbol{a}_h = \boldsymbol{F}. \tag{7.8}$$

If O^* is moving at constant velocity relative to O, then $\boldsymbol{a}_h = \boldsymbol{0}$, and we have

$$m\frac{d^2\boldsymbol{r}^*}{dt^2} = \boldsymbol{F}. \tag{7.9}$$

Thus Newton's equations of motion, if they hold in any coordinate system, hold also in any other coordinate system moving with uniform velocity relative to the first. This is the Newtonian principle of relativity. It implies that, so far as mechanics is concerned, we cannot specify any unique fixed coordinate system or *frame of reference* to which Newton's laws are supposed to refer; if we specify one such system, any other system moving with constant velocity relative to it will do as well. This property of Eq. (7.7) is sometimes expressed by saying that Newton's equations of motion remain *invariant* in form, or that they are *covariant*, with respect to uniform translations of the coordinates. The concept of frame of reference is not quite the same as that of a coordinate system, in that if we make a change of coordinates that does not involve the time, we do not regard this as a change of frame of reference. A frame of reference includes all coordinate systems at rest with respect to any particular one. The principle of (special) relativity proposed by Einstein asserts that this relativity principle is not restricted to mechanics, but holds for all physical phenomena. The special theory of relativity is the result of the application of this principle to all types of phenomena, particularly electromagnetic phenomena. It turns out that this can only be done by modifying Newton's equations of motion slightly and, in fact, even Eqs. (7.5) and (7.6) require modification as we shall see in Chapter 13.

For any motion of O^*, we can write Eq. (7.8) in the form

$$m\frac{d^2\boldsymbol{r}^*}{dt^2} = \boldsymbol{F} - m\boldsymbol{a}_h. \tag{7.10}$$

This equation has the same form as the equation of motion (7.7) in a fixed coordinate system, except that in place of the force $\boldsymbol{F}$, we have $\boldsymbol{F} - m\boldsymbol{a}_h$. The term $-m\boldsymbol{a}_h$ we may call a fictitious force. We can treat the motion of a mass m relative to a moving coordinate system using Newton's equations of motion if we add this fictitious force to the actual force which acts. From the point of view of classical mechanics, it is not a force at all, but part of the mass times acceleration transposed to the other side of the equation. The essential distinction is that the real forces $\boldsymbol{F}$ acting on m depend on the positions and motions of other bodies, whereas the fictitious force depends on the acceleration of the starred coordinate system with respect to the fixed coordinate system. In the general theory of relativity, terms like $-m\boldsymbol{a}_h$ are regarded as legitimate forces in the starred coordinate system, on the same footing with the force $\boldsymbol{F}$, so that in all coordinate systems the same law of motion holds. This, of course, can only be done if it can be shown how to deduce the force $-m\boldsymbol{a}_h$ from the positions and motions of other bodies. The program is not so simple as it may seem from this brief outline, and modifications in the laws of motion are required to carry it through.†

7.2 ROTATING COORDINATE SYSTEMS

We now consider coordinate systems x, y, z and x^*, y^*, z^* whose axes are rotated relative to one another as in Fig. 7.2, where, for the present, the origins of the two sets of axes coincide. Introducing unit vectors $\hat{\boldsymbol{x}}, \hat{\boldsymbol{y}}, \hat{\boldsymbol{z}}$ associated with axes x, y, z, and unit vectors $\hat{\boldsymbol{x}}^*, \hat{\boldsymbol{y}}^*, \hat{\boldsymbol{z}}^*$ associated with axes x^*, y^*, z^*, we can express the position vector $\boldsymbol{r}$ in terms of its components along either set of axes:

$$\boldsymbol{r} = x\hat{\boldsymbol{x}} + y\hat{\boldsymbol{y}} + z\hat{\boldsymbol{z}}, \tag{7.11}$$

$$\boldsymbol{r} = x^*\hat{\boldsymbol{x}}^* + y^*\hat{\boldsymbol{y}}^* + z^*\hat{\boldsymbol{z}}^*. \tag{7.12}$$

Note that since the origins now coincide, a point is represented by the same vector $\boldsymbol{r}$ in both systems; only the components of $\boldsymbol{r}$ are different along the different axes. The relations between the coordinate systems can be obtained by taking the dot product of either the starred or the unstarred unit vectors with Eqs. (7.11)

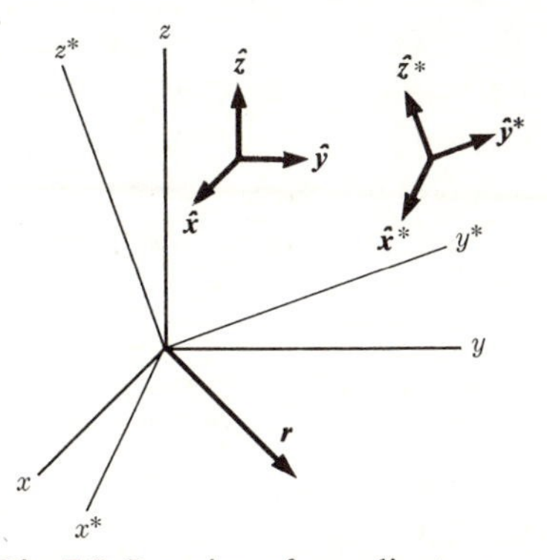

Fig. 7.2 Rotation of coordinate axes.

†P. G. Bergmann, *Introduction to the Theory of Relativity*. New York: Prentice-Hall, 1946. (Part 2.)

and (7.12). For example, if we compute $\hat{\boldsymbol{x}}\cdot\boldsymbol{r}$, $\hat{\boldsymbol{y}}\cdot\boldsymbol{r}$, $\hat{\boldsymbol{z}}\cdot\boldsymbol{r}$, from Eqs. (7.11) and (7.12) and equate the results, we obtain

$$\begin{aligned} x &= x^*(\hat{\boldsymbol{x}}^*\cdot\hat{\boldsymbol{x}})+y^*(\hat{\boldsymbol{y}}^*\cdot\hat{\boldsymbol{x}})+z^*(\hat{\boldsymbol{z}}^*\cdot\hat{\boldsymbol{x}}), \\ y &= x^*(\hat{\boldsymbol{x}}^*\cdot\hat{\boldsymbol{y}})+y^*(\hat{\boldsymbol{y}}^*\cdot\hat{\boldsymbol{y}})+z^*(\hat{\boldsymbol{z}}^*\cdot\hat{\boldsymbol{y}}), \\ z &= x^*(\hat{\boldsymbol{x}}^*\cdot\hat{\boldsymbol{z}})+y^*(\hat{\boldsymbol{y}}^*\cdot\hat{\boldsymbol{z}})+z^*(\hat{\boldsymbol{z}}^*\cdot\hat{\boldsymbol{z}}). \end{aligned} \tag{7.13}$$

The dot products $(\hat{\boldsymbol{x}}^*\cdot\hat{\boldsymbol{x}})$, etc., are the cosines of the angles between the corresponding axes. Similar formulas for x^*, y^*, z^* in terms of x, y, z can easily be obtained by the same process. Equations (7.11), (7.12), and (7.13) do not depend on the fact that the vector $\boldsymbol{r}$ is drawn from the origin. Analogous formulas apply in terms of the components of any vector $\boldsymbol{A}$ along the two sets of axes. If the starred axes are rotating, the cosines of the angles between starred and unstarred axes are functions of time.

The time derivative of any vector $\boldsymbol{A}$ was defined by Eq. (3.52):

$$\frac{d\boldsymbol{A}}{dt} = \lim_{\Delta t\to 0}\frac{\boldsymbol{A}(t+\Delta t)-\boldsymbol{A}(t)}{\Delta t}. \tag{7.14}$$

In attempting to apply this definition in the present case, we encounter a difficulty if the coordinate systems are rotating with respect to each other. A vector which is constant in one coordinate system is not constant in the other, but rotates. The definition requires us to subtract $\boldsymbol{A}(t)$ from $\boldsymbol{A}(t+\Delta t)$. During the time Δt, coordinate system x^*, y^*, z^* has rotated relative to x, y, z, so that at time $t+\Delta t$, the two systems will not agree as to which vector is (or was) $\boldsymbol{A}(t)$, i.e., which vector is in the same position that $\boldsymbol{A}$ was in at time t. The result is that the time derivative of a given vector will be different in the two coordinate systems. Let us use d/dt to denote the time derivative with respect to the unstarred coordinate system, which we regard as fixed, and d^*/dt to denote the time derivative with respect to the rotating starred coordinate system. We make this distinction with regard to vectors only; there is no ambiguity with regard to numerical quantities, and we denote their time derivatives by d/dt, or by a dot, which will have the same meaning in all coordinate systems. Let the vector $\boldsymbol{A}$ be given by

$$\boldsymbol{A} = A_x\hat{\boldsymbol{x}}+A_y\hat{\boldsymbol{y}}+A_z\hat{\boldsymbol{z}}, \tag{7.15}$$

$$\boldsymbol{A} = A_x^*\hat{\boldsymbol{x}}^*+A_y^*\hat{\boldsymbol{y}}^*+A_z^*\hat{\boldsymbol{z}}^*. \tag{7.16}$$

The unstarred time derivative of $\boldsymbol{A}$ may be obtained by differentiating Eq. (7.15), regarding $\hat{\boldsymbol{x}}$, $\hat{\boldsymbol{y}}$, $\hat{\boldsymbol{z}}$ as constant vectors in the fixed system:

$$\frac{d\boldsymbol{A}}{dt} = \dot{A}_x\hat{\boldsymbol{x}}+\dot{A}_y\hat{\boldsymbol{y}}+\dot{A}_z\hat{\boldsymbol{z}}. \tag{7.17}$$

Similarly, the starred derivative of $\boldsymbol{A}$ is given in terms of its starred components by

$$\frac{d^*\boldsymbol{A}}{dt} = \dot{A}_x^*\hat{\boldsymbol{x}}^*+\dot{A}_y^*\hat{\boldsymbol{y}}^*+\dot{A}_z^*\hat{\boldsymbol{z}}^*. \tag{7.18}$$

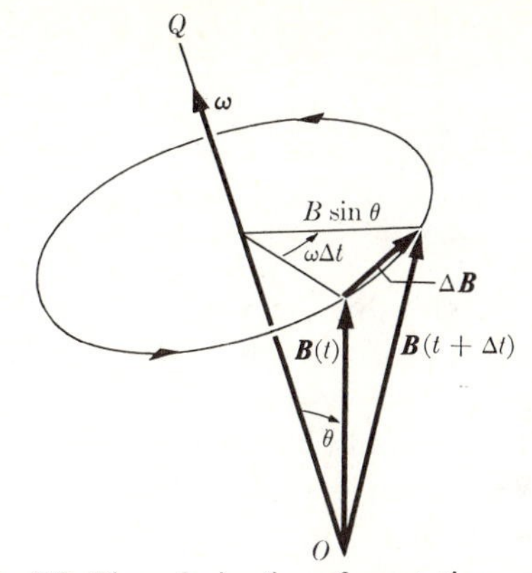

Fig. 7.3 Time derivative of a rotating vector.

We may regard Eqs. (7.17) and (7.18) as the definitions of unstarred and starred time derivatives of a vector. We can also obtain a formula for d/dt in starred components by taking the unstarred derivative of Eq. (7.16), remembering that the unit vectors $\hat{\boldsymbol{x}}^*$, $\hat{\boldsymbol{y}}^*$, $\hat{\boldsymbol{z}}^*$ are moving relative to the unstarred system, and have time derivatives:

$$\frac{d\boldsymbol{A}}{dt} = \dot{A}_x^*\hat{\boldsymbol{x}}^* + \dot{A}_y^*\hat{\boldsymbol{y}}^* + \dot{A}_z^*\hat{\boldsymbol{z}}^* + A_x^*\frac{d\hat{\boldsymbol{x}}^*}{dt} + A_y^*\frac{d\hat{\boldsymbol{y}}^*}{dt} + A_z^*\frac{d\hat{\boldsymbol{z}}^*}{dt}. \tag{7.19}$$

A similar formula could be obtained for $d^*\boldsymbol{A}/dt$ in terms of its unstarred components.

Let us now suppose that the starred coordinate system is rotating about some axis OQ through the origin, with an angular velocity ω (Fig. 7.3). We define the *vector angular velocity* $\boldsymbol{\omega}$ as a vector of magnitude ω directed along the axis OQ in the direction of advance of a right-hand screw rotating with the starred system. Consider a vector $\boldsymbol{B}$ at rest in the starred system. Its starred derivative is zero, and we now show that its unstarred derivative is

$$\frac{d\boldsymbol{B}}{dt} = \boldsymbol{\omega} \times \boldsymbol{B}. \tag{7.20}$$

In order to subtract $\boldsymbol{B}(t)$ from $\boldsymbol{B}(t+\Delta t)$, we draw these vectors with their tails together, and it will be convenient to place them with their tails on the axis of rotation. (The time derivative depends only on the components of $\boldsymbol{B}$ along the axes, and not on the position of $\boldsymbol{B}$ in space.) We first verify from Fig. 7.3 that the direction of $d\boldsymbol{B}/dt$ is given correctly by Eq. (7.20), recalling the definition [Eq. (3.24) and Fig. 3.11] of the cross product. The magnitude of $d\boldsymbol{B}/dt$ as given by Eq. (7.20) is

$$\left|\frac{d\boldsymbol{B}}{dt}\right| = |\boldsymbol{\omega} \times \boldsymbol{B}| = \omega B \sin\theta. \tag{7.21}$$

This is the correct formula, since it can be seen from Fig. 7.3 that, when Δt is small,

$$|\Delta\boldsymbol{B}| = (B \sin\theta)(\omega\,\Delta t).$$

When Eq. (7.20) is applied to the unit vectors $\hat{\boldsymbol{x}}^*$, $\hat{\boldsymbol{y}}^*$, $\hat{\boldsymbol{z}}^*$, Eq. (7.19) becomes, if we make use of Eqs. (7.18) and (7.16):

$$\begin{aligned}\frac{d\boldsymbol{A}}{dt} &= \frac{d^*\boldsymbol{A}}{dt} + A_x^*(\boldsymbol{\omega}\times\hat{\boldsymbol{x}}^*) + A_y^*(\boldsymbol{\omega}\times\hat{\boldsymbol{y}}^*) + A_z^*(\boldsymbol{\omega}\times\hat{\boldsymbol{z}}^*)\\ &= \frac{d^*\boldsymbol{A}}{dt} + \boldsymbol{\omega}\times\boldsymbol{A}.\end{aligned} \tag{7.22}$$

This is the fundamental relationship between time derivatives for rotating coordinate systems. It may be remembered by noting that the time derivative of any vector in the unstarred coordinate system is its derivative in the starred system plus the unstarred derivative it would have if it were at rest in the starred system. Equation (7.22) applies even when the angular velocity vector $\boldsymbol{\omega}$ is changing in magnitude and direction with time. Taking the derivative of right and left sides of Eq. (7.22), and applying Eq. (7.22) again to $\boldsymbol{A}$ and $d^*\boldsymbol{A}/dt$, we have for the second time derivative of any vector $\boldsymbol{A}$:

$$\begin{aligned}\frac{d^2\boldsymbol{A}}{dt^2} &= \frac{d}{dt}\left(\frac{d^*\boldsymbol{A}}{dt}\right) + \boldsymbol{\omega}\times\frac{d\boldsymbol{A}}{dt} + \frac{d\boldsymbol{\omega}}{dt}\times\boldsymbol{A}\\ &= \frac{d^{*2}\boldsymbol{A}}{dt^2} + \boldsymbol{\omega}\times\frac{d^*\boldsymbol{A}}{dt} + \boldsymbol{\omega}\times\left(\frac{d^*\boldsymbol{A}}{dt} + \boldsymbol{\omega}\times\boldsymbol{A}\right) + \frac{d\boldsymbol{\omega}}{dt}\times\boldsymbol{A}\\ &= \frac{d^{*2}\boldsymbol{A}}{dt^2} + 2\boldsymbol{\omega}\times\frac{d^*\boldsymbol{A}}{dt} + \boldsymbol{\omega}\times(\boldsymbol{\omega}\times\boldsymbol{A}) + \frac{d\boldsymbol{\omega}}{dt}\times\boldsymbol{A}.\end{aligned} \tag{7.23}$$

Since $\boldsymbol{\omega}\times\boldsymbol{A} = \boldsymbol{0}$ if $\boldsymbol{\omega}$ is parallel to $\boldsymbol{A}$, [Eq. (3.29)], the starred and unstarred derivatives of any vector parallel to the axis of rotation are the same, according to Eq. (7.22). In particular,

$$\frac{d\boldsymbol{\omega}}{dt} = \frac{d^*\boldsymbol{\omega}}{dt}.$$

It is to be noted that the vector $\boldsymbol{\omega}$ on both sides of this equation is the angular velocity of the starred system relative to the unstarred system, although its time derivative is calculated with respect to the unstarred system on the left side, and with respect to the starred system on the right. The angular velocity of the unstarred system relative to the starred system will be $-\boldsymbol{\omega}$.

We now show that the relations derived above for a rotating coordinate system are perfectly general, in that they apply to any motion of the starred axes relative to the unstarred axes. Let the unstarred rates of change of the starred unit vectors be given in terms of components along the starred axes by

$$\begin{aligned}\frac{d\hat{\boldsymbol{x}}^*}{dt} &= a_{11}\hat{\boldsymbol{x}}^* + a_{12}\hat{\boldsymbol{y}}^* + a_{13}\hat{\boldsymbol{z}}^*,\\ \frac{d\hat{\boldsymbol{y}}^*}{dt} &= a_{21}\hat{\boldsymbol{x}}^* + a_{22}\hat{\boldsymbol{y}}^* + a_{23}\hat{\boldsymbol{z}}^*,\\ \frac{d\hat{\boldsymbol{z}}^*}{dt} &= a_{31}\hat{\boldsymbol{x}}^* + a_{32}\hat{\boldsymbol{y}}^* + a_{33}\hat{\boldsymbol{z}}^*.\end{aligned} \tag{7.24}$$

By differentiating the equation

$$\hat{\mathbf{x}}^* \cdot \hat{\mathbf{x}}^* = 1, \tag{7.25}$$

we obtain

$$\frac{d\hat{\mathbf{x}}^*}{dt} \cdot \hat{\mathbf{x}}^* = 0. \tag{7.26}$$

From this and the corresponding equations for $\hat{\mathbf{y}}^*$ and $\hat{\mathbf{z}}^*$, we have

$$a_{11} = a_{22} = a_{33} = 0. \tag{7.27}$$

By differentiating the equation

$$\hat{\mathbf{x}}^* \cdot \hat{\mathbf{z}}^* = 0, \tag{7.28}$$

we obtain

$$\frac{d\hat{\mathbf{x}}^*}{dt} \cdot \hat{\mathbf{z}}^* = -\hat{\mathbf{x}}^* \cdot \frac{d\hat{\mathbf{z}}^*}{dt}. \tag{7.29}$$

From this and the other two analogous equations, we have

$$a_{31} = -a_{13}, \qquad a_{12} = -a_{21}, \qquad a_{23} = -a_{32}. \tag{7.30}$$

It is clear from Eqs. (7.27) and (7.30) that if the three coefficients a_{12}, a_{23}, a_{31} are specified, the other six are determined. Let us define a vector $\boldsymbol{\omega}$ whose components in the x^*, y^*, z^* coordinate system are

$$\omega_x^* = a_{23}, \qquad \omega_y^* = a_{31}, \qquad \omega_z^* = a_{12}. \tag{7.31}$$

At this point, this is simply an arbitrary definition. We may always define a vector by giving its components in some coordinate system. We call this vector $\boldsymbol{\omega}$ because we are going to show that it is in fact the angular velocity of the starred coordinate system. Equations (7.24) can now be rewritten, with the help of Eqs. (7.27), (7.30), and (7.31), in the form

$$\begin{aligned} \frac{d\hat{\mathbf{x}}^*}{dt} &= \boldsymbol{\omega} \times \hat{\mathbf{x}}^*, \\ \frac{d\hat{\mathbf{y}}^*}{dt} &= \boldsymbol{\omega} \times \hat{\mathbf{y}}^*, \\ \frac{d\hat{\mathbf{z}}^*}{dt} &= \boldsymbol{\omega} \times \hat{\mathbf{z}}^*. \end{aligned} \tag{7.32}$$

According to Eq. (7.20), these time derivatives of $\hat{\mathbf{x}}^*$, $\hat{\mathbf{y}}^*$, $\hat{\mathbf{z}}^*$ are just those to be expected if the starred unit vectors are rotating with an angular velocity $\boldsymbol{\omega}$. Thus no matter how the starred coordinate axes may be moving, we can define at any instant an angular velocity vector $\boldsymbol{\omega}$, given by Eq. (7.31), such that the time derivatives of any vector relative to the starred and unstarred coordinate systems are related by Eqs. (7.22) and (7.23).

Let us now suppose that the starred coordinate system is moving in such a way that its origin O^* remains fixed at the origin O of the fixed coordinate system. Then any point in space is located by the same position vector $\mathbf{r}$ in both coordinate systems [Eqs. (7.11) and (7.12)]. By applying Eqs. (7.22) and (7.23) to the position vector $\mathbf{r}$, we obtain formulas for the relation between velocities and accelerations in the two coordinate systems:

$$\frac{d\mathbf{r}}{dt} = \frac{d^*\mathbf{r}}{dt} + \boldsymbol{\omega} \times \mathbf{r}, \tag{7.33}$$

$$\frac{d^2\mathbf{r}}{dt^2} = \frac{d^{*2}\mathbf{r}}{dt^2} + \boldsymbol{\omega} \times (\boldsymbol{\omega} \times \mathbf{r}) + 2\boldsymbol{\omega} \times \frac{d^*\mathbf{r}}{dt} + \frac{d\boldsymbol{\omega}}{dt} \times \mathbf{r}. \tag{7.34}$$

Formula (7.34) is called *Coriolis' theorem.* The first term on the right is the acceleration relative to the starred system. The second term is called the *centripetal acceleration* of a point in rotation about an axis (*centripetal* means "toward the center"). Using the notation in Fig. 7.4, we readily verify that $\boldsymbol{\omega} \times (\boldsymbol{\omega} \times \mathbf{r})$ points directly toward and perpendicular to the axis of rotation, and that its magnitude is

$$\begin{aligned} |\boldsymbol{\omega} \times (\boldsymbol{\omega} \times \mathbf{r})| &= \omega^2 r \sin\theta \\ &= \frac{v^2}{r \sin\theta}, \end{aligned} \tag{7.35}$$

where $v = \omega r \sin\theta$ is the speed of circular motion and $(r \sin\theta)$ is the distance from the axis. The third term is present only when the point $\mathbf{r}$ is moving in the starred system, and is called the *coriolis acceleration.* The last term vanishes for a constant angular velocity of rotation about a fixed axis.

If we suppose that Newton's law of motion (7.7) holds in the unstarred coordinate system, we shall have in the starred system:

$$m\frac{d^{*2}\mathbf{r}}{dt^2} + m\boldsymbol{\omega} \times (\boldsymbol{\omega} \times \mathbf{r}) + 2m\boldsymbol{\omega} \times \frac{d^*\mathbf{r}}{dt} + m\frac{d\boldsymbol{\omega}}{dt} \times \mathbf{r} = \mathbf{F}. \tag{7.36}$$

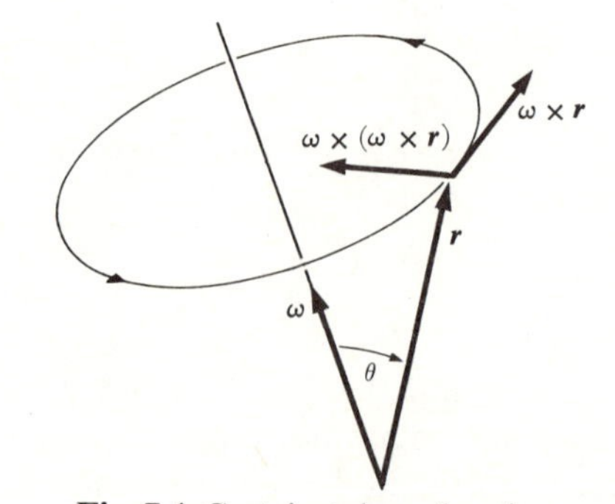

Fig. 7.4 Centripetal acceleration.

Transposing the second, third, and fourth terms to the right side, we obtain an equation of motion similar in form to Newton's equation of motion:

$$m\frac{d^{*2}\boldsymbol{r}}{dt^2} = \boldsymbol{F} - m\boldsymbol{\omega}\times(\boldsymbol{\omega}\times\boldsymbol{r}) - 2m\boldsymbol{\omega}\times\frac{d^*\boldsymbol{r}}{dt} - m\frac{d\boldsymbol{\omega}}{dt}\times\boldsymbol{r}. \tag{7.37}$$

The second term on the right is called the *centrifugal force* (*centrifugal* means "away from the center"); the third term is called the *coriolis force*. The last term has no special name, and appears only for the case of non-uniform rotation. If we introduce the fictitious centrifugal and coriolis forces, the laws of motion relative to a rotating coordinate system are the same as for fixed coordinates. A great deal of confusion has arisen regarding the term "centrifugal force." This force is not a real force, at least in classical mechanics, and is not present if we refer to a fixed coordinate system in space. We can, however, treat a rotating coordinate system as if it were fixed by introducing the centrifugal and coriolis forces. Thus a particle moving in a circle has no centrifugal force acting on it, but only a force toward the center which produces its centripetal acceleration. However, if we consider a coordinate system rotating with the particle, in this system the particle is at rest, and the force toward the center is balanced by the centrifugal force. It is very often useful to adopt a rotating coordinate system. In studying the action of a cream separator, for example, it is far more convenient to choose a coordinate system in which the liquid is at rest, and use the laws of diffusion to study the diffusion of cream toward the axis under the action of the centrifugal force field, than to try to study the motion from the point of view of a fixed observer watching the whirling liquid.

We can treat coordinate systems in simultaneous translation and rotation relative to each other by using Eq. (7.1) to represent the relation between the coordinate vectors $\boldsymbol{r}$ and $\boldsymbol{r}^*$ relative to origins O, O^* not necessarily coincident. In the derivation of Eqs. (7.32), no assumption was made about the origin of the starred coordinates, and therefore Eqs. (7.22) and (7.23) may still be used to express the time derivatives of any vector with respect to the unstarred coordinate system in terms of its time derivatives with respect to the starred system. Replacing $d\boldsymbol{r}^*/dt$, $d^2\boldsymbol{r}^*/dt$ in Eqs. (7.5) and (7.6) by their expressions in terms of the starred derivatives relative to the starred system as given by Eqs. (7.33) and (7.34), we obtain for the position, velocity, and acceleration of a point with respect to coordinate systems in relative translation and rotation:

$$\boldsymbol{r} = \boldsymbol{r}^* + \boldsymbol{h}, \tag{7.38}$$

$$\frac{d\boldsymbol{r}}{dt} = \frac{d^*\boldsymbol{r}^*}{dt} + \boldsymbol{\omega}\times\boldsymbol{r}^* + \frac{d\boldsymbol{h}}{dt}, \tag{7.39}$$

$$\frac{d^2\boldsymbol{r}}{dt^2} = \frac{d^{*2}\boldsymbol{r}^*}{dt^2} + \boldsymbol{\omega}\times(\boldsymbol{\omega}\times\boldsymbol{r}^*) + 2\boldsymbol{\omega}\times\frac{d^*\boldsymbol{r}^*}{dt} + \frac{d\boldsymbol{\omega}}{dt}\times\boldsymbol{r}^* + \frac{d^2\boldsymbol{h}}{dt^2}. \tag{7.40}$$

7.3 LAWS OF MOTION ON THE ROTATING EARTH

We write the equation of motion, relative to a coordinate system fixed in space, for a particle of mass m subject to a gravitational force $m\boldsymbol{g}$ and any other non-gravitational forces $\boldsymbol{F}$:

$$m\frac{d^2\boldsymbol{r}}{dt^2} = \boldsymbol{F} + m\boldsymbol{g}. \tag{7.41}$$

Now if we refer the motion of the particle to a coordinate system at rest relative to the earth, which rotates with constant angular velocity $\boldsymbol{\omega}$, and if we measure the position vector $\boldsymbol{r}$ from the center of the earth, we have, by Eq. (7.34):

$$\begin{aligned} m\frac{d^2\boldsymbol{r}}{dt^2} &= \boldsymbol{F} + m\boldsymbol{g} \\ &= m\frac{d^{*2}\boldsymbol{r}}{dt^2} + m\boldsymbol{\omega}\times(\boldsymbol{\omega}\times\boldsymbol{r}) + 2m\boldsymbol{\omega}\times\frac{d^*\boldsymbol{r}}{dt}, \end{aligned} \tag{7.42}$$

which can be rearranged in the form

$$m\frac{d^{*2}\boldsymbol{r}}{dt^2} = \boldsymbol{F} + m[\boldsymbol{g} - \boldsymbol{\omega}\times(\boldsymbol{\omega}\times\boldsymbol{r})] - 2m\boldsymbol{\omega}\times\frac{d^*\boldsymbol{r}}{dt}. \tag{7.43}$$

This equation has the same form as Newton's equation of motion. We have combined the gravitational and centrifugal force terms because both are proportional to the mass of the particle and both depend only on the position of the particle; in their mechanical effects these two forces are indistinguishable. We may define the effective gravitational acceleration $\boldsymbol{g}_e$ at any point on the earth's surface by:

$$\boldsymbol{g}_e(\boldsymbol{r}) = \boldsymbol{g}(\boldsymbol{r}) - \boldsymbol{\omega}\times(\boldsymbol{\omega}\times\boldsymbol{r}). \tag{7.44}$$

The gravitational force which we measure experimentally on a body of mass m at rest† on the earth's surface is $m\boldsymbol{g}_e$. Since $-\boldsymbol{\omega}\times(\boldsymbol{\omega}\times\boldsymbol{r})$ points radially outward from the earth's axis, $\boldsymbol{g}_e$ at every point north of the equator will point slightly to the south of the earth's center, as can be seen from Fig. 7.5. A body released near the earth's surface will begin to fall with acceleration $\boldsymbol{g}_e$, the direction determined by a plumb line is that of $\boldsymbol{g}_e$, and a liquid will come to equilibrium with its surface perpendicular to $\boldsymbol{g}_e$. This is why the earth has settled into equilibrium in the form of an oblate ellipsoid, flattened at the poles. The degree of flattening is just such as to make the earth's surface at every point perpendicular to $\boldsymbol{g}_e$ (ignoring local irregularities).

Equation (7.43) can now be written

$$m\frac{d^{*2}\boldsymbol{r}}{dt^2} = \boldsymbol{F} + m\boldsymbol{g}_e - 2m\boldsymbol{\omega}\times\frac{d^*\boldsymbol{r}}{dt}. \tag{7.45}$$

†A body in motion is subject also to the coriolis force.

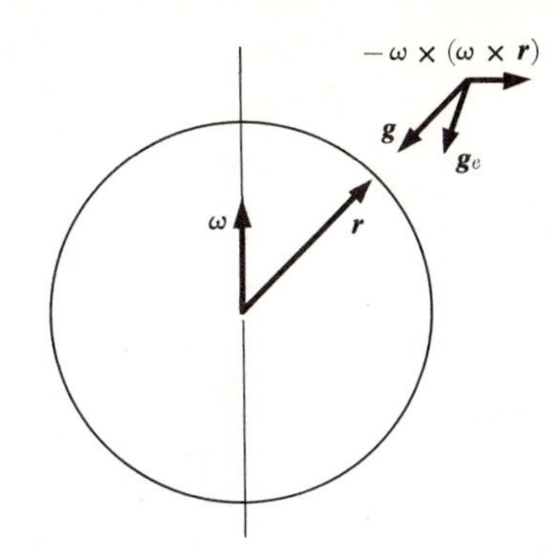

Fig. 7.5 Effective acceleration of gravity on the rotating earth.

The velocity and acceleration which appear in this equation are unaffected if we relocate our origin of coordinates at any convenient point at the surface of the earth; hence this equation applies to the motion of a particle of mass m at the surface of the earth relative to a local coordinate system at rest on the earth's surface. The only unfamiliar term is the coriolis force which acts on a moving particle. The reader can convince himself by a few calculations that this force is comparatively small at ordinary velocities $d^*\mathbf{r}/dt$. It will be instructive to try working out the direction of the coriolis force for various directions of motion at various places on the earth's surface. The coriolis force is of major importance in the motion of large air masses, and is responsible for the fact that in the northern hemisphere tornados and cyclones circle in the direction south to east to north to west. In the northern hemisphere, the coriolis force acts to deflect a moving object toward the right. As the winds blow toward a low pressure area, they are deflected to the right, so that they circle the low pressure area in a counterclockwise direction. An air mass circling in this way will have a low pressure on its left, and a higher pressure on its right. This is just what is needed to balance the coriolis force urging it to the right. An air mass can move steadily in one direction only if there is a high pressure to the right of it to balance the coriolis force. Conversely, a pressure gradient over the surface of the earth tends to develop winds moving at right angles to it. The prevailing westerly winds in the northern temperate zone indicate that the atmospheric pressure toward the equator is greater than toward the poles, at least near the earth's surface. The easterly trade winds in the equatorial zone are due to the fact that any air mass moving toward the equator will acquire a velocity toward the west due to the coriolis force acting on it. The trade winds are maintained by high pressure areas on either side of the equatorial zone.

7.4 THE FOUCAULT PENDULUM

An interesting application of the theory of rotating coordinate systems is the problem of the Foucault pendulum. The Foucault pendulum has a bob hanging from a string arranged to swing freely in any vertical plane. The pendulum is

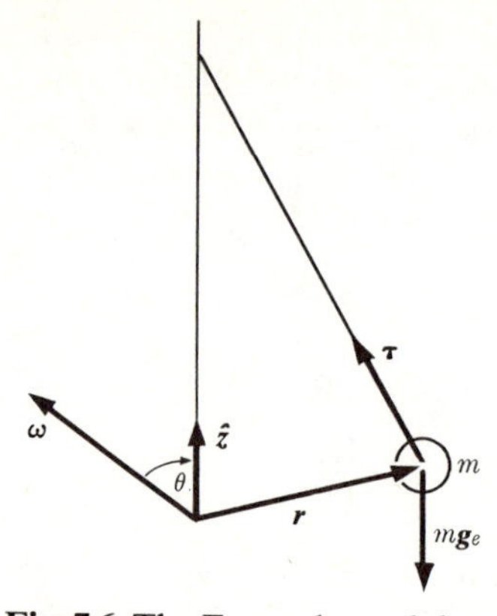

Fig. 7.6 The Foucault pendulum.

started swinging in a definite vertical plane and it is observed that the plane of swinging gradually precesses about the vertical axis during a period of several hours. The bob must be made heavy, the string very long, and the support nearly frictionless, in order that the pendulum can continue to swing freely for long periods of time. If we choose the origin of coordinates directly below the point of support, at the point of equilibrium of the pendulum bob of mass m, then the vector $\boldsymbol{r}$ will be nearly horizontal, for small amplitudes of oscillation of the pendulum. In the northern hemisphere, $\boldsymbol{\omega}$ makes an acute angle with the vertical, as in Fig. 7.6. Writing $\boldsymbol{\tau}$ for the tension in the string, we have as the equation of motion of the bob, according to Eq. (7.45):

$$m\frac{d^{*2}\boldsymbol{r}}{dt^2} = \boldsymbol{\tau} + m\boldsymbol{g}_e - 2m\boldsymbol{\omega} \times \frac{d^*\boldsymbol{r}}{dt}. \tag{7.46}$$

If the coriolis force were not present, this would be the equation for a simple pendulum on a nonrotating earth. The coriolis force is very small, less than 0.1% of the gravitational force if the velocity is 5 mph or less, and its vertical component is therefore negligible in comparison with the gravitational force. (It is the vertical force which determines the magnitude of the tension in the string.) However, the horizontal component of the coriolis force is perpendicular to the velocity $d^*\boldsymbol{r}/dt$, and as there are no other forces in this direction when the pendulum swings to and fro, it can change the nature of the motion. Any force with a horizontal component perpendicular to $d^*\boldsymbol{r}/dt$ will make it impossible for the pendulum to continue to swing in a fixed vertical plane. In order to solve the problem including the coriolis term, we use the experimental result as a clue, and try to find a new coordinate system rotating about the vertical axis through the point of support at such an angular velocity that in this system the coriolis terms, or at least their horizontal components, are missing. Let us introduce a new coordinate system rotating about the vertical axis with constant angular velocity $\hat{\boldsymbol{z}}\Omega$ (relative to the earth), where $\hat{\boldsymbol{z}}$ is a vertical unit vector. We shall call this precessing coordinate system the primed coordinate system, and denote the time derivative with respect to this system by d'/dt. Then we shall have, by Eqs. (7.33) and (7.34):

$$\frac{d^*\boldsymbol{r}}{dt} = \frac{d'\boldsymbol{r}}{dt} + \Omega\hat{\boldsymbol{z}} \times \boldsymbol{r}, \tag{7.47}$$

$$\frac{d^{*2}\boldsymbol{r}}{dt^2} = \frac{d'^2\boldsymbol{r}}{dt^2} + \Omega^2\hat{\boldsymbol{z}} \times (\hat{\boldsymbol{z}} \times \boldsymbol{r}) + 2\Omega\hat{\boldsymbol{z}} \times \frac{d'\boldsymbol{r}}{dt}. \tag{7.48}$$

Equation (7.46) becomes

$$\begin{aligned} m\frac{d'^2\boldsymbol{r}}{dt^2} &= \boldsymbol{\tau} + m\boldsymbol{g}_e - 2m\boldsymbol{\omega} \times \left(\frac{d'\boldsymbol{r}}{dt} + \Omega\hat{\boldsymbol{z}} \times \boldsymbol{r}\right) \\ &\quad - m\Omega^2\hat{\boldsymbol{z}} \times (\hat{\boldsymbol{z}} \times \boldsymbol{r}) - 2m\Omega\hat{\boldsymbol{z}} \times \frac{d'\boldsymbol{r}}{dt} \\ &= \boldsymbol{\tau} + m\boldsymbol{g}_e - 2m\Omega\boldsymbol{\omega} \times (\hat{\boldsymbol{z}} \times \boldsymbol{r}) - m\Omega^2\hat{\boldsymbol{z}} \times (\hat{\boldsymbol{z}} \times \boldsymbol{r}) \\ &\quad - 2m(\boldsymbol{\omega} + \hat{\boldsymbol{z}}\Omega) \times \frac{d'\boldsymbol{r}}{dt}. \end{aligned} \tag{7.49}$$

We expand the triple products by means of Eq. (3.35):

$$\begin{aligned} m\frac{d'^2\boldsymbol{r}}{dt^2} &= \boldsymbol{\tau} + m\boldsymbol{g}_e - m(2\Omega\boldsymbol{\omega}\cdot\boldsymbol{r} + \Omega^2\hat{\boldsymbol{z}}\cdot\boldsymbol{r})\hat{\boldsymbol{z}} \\ &\quad + m(2\Omega\hat{\boldsymbol{z}}\cdot\boldsymbol{\omega} + \Omega^2)\boldsymbol{r} - 2m(\boldsymbol{\omega} + \hat{\boldsymbol{z}}\Omega) \times \frac{d'\boldsymbol{r}}{dt}. \end{aligned} \tag{7.50}$$

Every vector on the right side of Eq. (7.50) lies in the vertical plane containing the pendulum, except the last term. Since, for small oscillations, $d'\boldsymbol{r}/dt$ is practically horizontal, we can make the last term lie in this vertical plane also by making $(\boldsymbol{\omega} + \hat{\boldsymbol{z}}\Omega)$ horizontal. We therefore require that

$$\hat{\boldsymbol{z}}\cdot(\boldsymbol{\omega} + \hat{\boldsymbol{z}}\Omega) = 0. \tag{7.51}$$

This determines Ω:

$$\Omega = -\omega\cos\theta, \tag{7.52}$$

where ω is the angular velocity of the rotating earth, Ω is the angular velocity of the precessing coordinate system relative to the earth, and θ is the angle between the vertical and the earth's axis, as indicated in Fig. 7.6. The vertical is along the direction of $-\boldsymbol{g}_e$, and since this is very nearly the same as the direction of $-\boldsymbol{g}$ (see Fig. 7.5), θ will be practically equal to the colatitude, that is, the angle between $\boldsymbol{r}$ and $\boldsymbol{\omega}$ in Fig. 7.5. For small oscillations, if Ω is determined by Eq. (7.52), the cross product in the last term of Eq. (7.50) is vertical. Since all terms on the right of Eq. (7.50) now lie in a vertical plane containing the pendulum, the acceleration $d'^2\boldsymbol{r}/dt^2$ of the bob in the precessing system is always toward the vertical axis, and if the pendulum is initially swinging to and fro, it will continue to swing to and fro in the same vertical plane in the precessing coordinate system. Relative to the earth, the plane of the motion precesses with angular velocity Ω of magnitude and sense given by Eq. (7.52). In the northern hemisphere, the precession is clockwise looking down.

Since the last three terms on the right in Eq. (7.50) are much smaller than the first two, the actual motion in the precessing coordinate system is practically the same as for a pendulum on a nonrotating earth. Even at large amplitudes, where the velocity $d'\mathbf{r}/dt$ has a vertical component, careful study will show that the last term in Eq. (7.50), when Ω is chosen according to Eq. (7.52), does not cause any additional precession relative to the precessing coordinate system, but merely causes the bob to swing in an arc which passes slightly east of the vertical through the point of support. At the equator, Ω is zero, and the Foucault pendulum does not precess; by thinking about it a moment, perhaps you can see physically why this is so. At the north or south pole, $\Omega = \pm\omega$, and the pendulum merely swings in a fixed vertical plane in space while the earth turns beneath it.

Note that we have been able to give a fairly complete discussion of the Foucault pendulum, by using Coriolis' theorem twice, without actually solving the equations of motion at all.

7.5 LARMOR'S THEOREM

The coriolis force in Eq. (7.37) is of the same form as the magnetic force acting on a charged particle (Eq. 3.281), in that both are given by the cross product of the velocity of the particle with a vector representing a force field. Indeed, in the general theory of relativity, the coriolis forces on a particle in a rotating system can be regarded as due to the relative motion of other masses in the universe in a way somewhat analogous to the magnetic force acting on a charged particle which is due to the relative motion of other charges. The similarity in form of the two forces suggests that the effect of a magnetic field on a system of charged particles may be canceled by introducing a suitable rotating coordinate system. This idea leads to Larmor's theorem, which we state first, and then prove:

Larmor's Theorem. *If a system of charged particles, all having the same ratio q/m of charge to mass, acted on by their mutual (central) forces, and by a central force toward a common center, is subject in addition to a weak uniform magnetic field* $\mathbf{B}$, *its possible motions will be the same as the motions it could perform without the magnetic field, superposed upon a slow precession of the entire system about the center of force with angular velocity*

$$\boldsymbol{\omega} = -\frac{q}{2mc}\mathbf{B} \text{ (gaussian units).*} \tag{7.53}$$

The definition of a *weak* magnetic field will appear as the proof is developed. We shall assume that all the particles have the same charge q and the same mass

*In mks units, omit the c here and in the following equations.

m, although it will be apparent that the only thing that needs to be assumed is that the ratio q/m is constant. Practically the only important applications of Larmor's theorem are to the behavior of an atom in a magnetic field. The particles here are electrons of mass m, charge $q = -e$, acted upon by their mutual electrostatic repulsions and by the electrostatic attraction of the nucleus.

Let the central force acting on the kth particle be $\boldsymbol{F}_k^c$, and let the sum of the forces due to the other particles be $\boldsymbol{F}_k^i$. Then the equations of motion of the system of particles, in the absence of a magnetic field, are

$$m\frac{d^2\boldsymbol{r}_k}{dt^2} = \boldsymbol{F}_k^c + \boldsymbol{F}_k^i, \qquad k = 1, \ldots, N, \tag{7.54}$$

where N is the total number of particles. The force $\boldsymbol{F}_k^c$ depends only on the distance of particle k from the center of force, which we shall take as origin, and the forces $\boldsymbol{F}_k^i$ depend only on the distances of the particles from one another. When the magnetic field is applied, the equations of motion become, by Eq. (3.281):

$$m\frac{d^2\boldsymbol{r}_k}{dt^2} = \boldsymbol{F}_k^c + \boldsymbol{F}_k^i + \frac{q}{c}\frac{d\boldsymbol{r}_k}{dt} \times \boldsymbol{B}, \qquad k = 1, \ldots, N. \tag{7.55}$$

In order to eliminate the last term, we introduce a starred coordinate system with the same origin, rotating about this origin with angular velocity $\boldsymbol{\omega}$. Making use of Eqs. (7.33) and (7.34), we can write the equations of motion in the starred coordinate system:

$$m\frac{d^{*2}\boldsymbol{r}_k}{dt^2} = \boldsymbol{F}_k^c + \boldsymbol{F}_k^i - m\boldsymbol{\omega} \times (\boldsymbol{\omega} \times \boldsymbol{r}_k) + \frac{q}{c}(\boldsymbol{\omega} \times \boldsymbol{r}_k) \times \boldsymbol{B}$$

$$+ \frac{d^*\boldsymbol{r}_k}{dt} \times \left(\frac{q\boldsymbol{B}}{c} + 2m\boldsymbol{\omega}\right). \tag{7.56}$$

We can make the last term vanish by setting

$$\boldsymbol{\omega} = -\frac{q}{2mc}\boldsymbol{B}. \tag{7.57}$$

Equation (7.56) then becomes

$$m\frac{d^{*2}\boldsymbol{r}_k}{dt^2} = \boldsymbol{F}_k^c + \boldsymbol{F}_k^i + \frac{q^2}{4mc^2}\boldsymbol{B} \times (\boldsymbol{B} \times \boldsymbol{r}_k), \qquad k = 1, \ldots, N. \tag{7.58}$$

The forces $\boldsymbol{F}_k^c$ and $\boldsymbol{F}_k^i$ depend only on the distances of the particles from the origin and on their distances from one another, and these distances will be the same in the starred and unstarred coordinate systems. Therefore, if we neglect the last term, Eqs. (7.58) have exactly the same form in terms of starred coordinates as Eqs. (7.54) have in unstarred coordinates. Consequently, their solutions will then be the same, and the motions of the system expressed in starred coordinates will be the same as the motions of the system expressed in unstarred coordinates in the absence of a magnetic field. This is Larmor's theorem.

The condition that the magnetic field be weak means that the last term in Eq. (7.58) must be negligible in comparison with the first two terms. Notice that the term we are neglecting is proportional to B^2, whereas the term in Eq. (7.55) which we have eliminated is proportional to B. Hence, for sufficiently weak fields, the former may be negligible even though the latter is not. The last term in Eq. (7.58) may be written in the form

$$\frac{q^2}{4mc^2}\boldsymbol{B}\times(\boldsymbol{B}\times\boldsymbol{r}_k) = m\boldsymbol{\omega}\times(\boldsymbol{\omega}\times\boldsymbol{r}_k). \tag{7.59}$$

Another way of formulating the condition for a weak magnetic field is to say that the Larmor frequency ω, given by Eq. (7.57), must be small compared with the frequencies of the motion in the absence of a magnetic field.

The reader who has understood clearly the above derivation should be able to answer the following two questions. The cyclotron frequency, given by Eq. (3.299), for the motion of a charged particle in a magnetic field is twice the Larmor frequency, given by Eq. (7.57). Why does not Larmor's theorem apply to the charged particles in a cyclotron? Equation (7.58) can be derived without any assumption as to the origin of coordinates in the starred system. Why is it necessary that the axis of rotation of the starred coordinate system pass through the center of force of the system of particles?

7.6 THE RESTRICTED THREE-BODY PROBLEM

We pointed out in Section 4.9 that the three-body problem, in which three masses move under their mutual gravitational forces, cannot be solved in any general way. In this section we will consider a simplified problem, the restricted problem of three bodies, which retains many features of the more general problem, among them the fact that there is no general method of solving it. In the restricted problem, we are given two bodies of masses M_1 and M_2 that revolve in circles under their mutual gravitational attraction and around their common center of mass. The third body of very small mass m moves in the gravitational field of M_1 and M_2. We are to assume that m is so small that the resulting disturbance of the motions of M_1 and M_2 can be neglected. We will further simplify the problem by assuming that m remains in the plane in which M_1 and M_2 revolve. The problem thus reduces to a one-body problem in which we must find the motion of m in the given (moving) gravitational field of the other two. An obvious example would be a rocket moving in the gravitational fields of the earth and the moon, which revolve very nearly in circles about their common center of mass.

If M_1 and M_2 are separated by a distance a, then according to the results of Section 4.7, their angular velocity is determined by equating the gravitational force to mass times acceleration in the reduced problem, in which M_1 is at rest and

M_2 has mass μ as given by Eq. (4.98):

$$\mu\omega^2 a = \frac{M_1 M_2 G}{a^2}, \tag{7.60}$$

so that

$$\omega^2 = \frac{(M_1+M_2)G}{a^3}. \tag{7.61}$$

The center of mass divides the distance a into segments that are proportional to the masses.

We now introduce a coordinate system rotating with angular velocity ω about the center of mass of M_1 and M_2. In this system, M_1 and M_2 are at rest, and we will take them to be on the x-axis at the points

$$x_1 = \frac{M_2}{M_1+M_2}a, \qquad x_2 = -\frac{M_1}{M_1+M_2}a. \tag{7.62}$$

The angular velocity $\boldsymbol{\omega}$ is taken to be along the z-axis. Then m moves in the xy-plane, and its equation of motion is

$$m\frac{d^{*2}\boldsymbol{r}}{dt^2} = \boldsymbol{F}_1 + \boldsymbol{F}_2 - m\boldsymbol{\omega}\times(\boldsymbol{\omega}\times\boldsymbol{r}) - 2m\boldsymbol{\omega}\times\frac{d^*\boldsymbol{r}}{dt}, \tag{7.63}$$

where $\boldsymbol{F}_1$ and $\boldsymbol{F}_2$ are the gravitational attraction of M_1 and M_2 on m. Written in terms of components, the two equations become

$$\ddot{x} = -\frac{M_1G(x-x_1)}{[(x-x_1)^2+y^2]^{3/2}} - \frac{M_2G(x-x_2)}{[(x-x_2)^2+y^2]^{3/2}} + \frac{(M_1+M_2)Gx}{a^3} + 2\omega\dot{y},$$

$$\ddot{y} = -\frac{M_1Gy}{[(x-x_1)^2+y^2]^{3/2}} - \frac{M_2Gy}{[(x-x_2)^2+y^2]^{3/2}} + \frac{(M_1+M_2)Gy}{a^3} - 2\omega\dot{x}. \tag{7.64}$$

Note that the mass m cancels in these equations.

Since the coriolis force is perpendicular to the velocity, it does no 'work' in this moving coordinate system. Moreover, the centrifugal force has zero curl and can be derived from the 'potential energy'

$$V_c = -\tfrac{1}{2}m\omega^2(x^2+y^2). \tag{7.65}$$

Therefore the total 'energy' in the moving coordinate system is a constant of the motion:

$$\text{`}E\text{'} = \tfrac{1}{2}m(\dot{x}^2+\dot{y}^2) + \text{`}V\text{'}, \tag{7.66}$$

where

$$\text{`}V\text{'} = -\frac{mM_1G}{[(x-x_1)^2+y^2]^{1/2}} - \frac{mM_2G}{[(x-x_2)^2+y^2]^{1/2}} - \frac{m(M_1+M_2)G(x^2+y^2)}{2a^3}. \tag{7.67}$$

The energy equation (7.66) enables us to make certain statements about the kinds of orbits that may be possible. In order to simplify the algebra, let us set

$$\xi = x/a, \qquad \eta = y/a, \tag{7.68}$$

$$\xi_1 = \frac{M_2}{M_1+M_2}, \qquad \xi_2 = -\frac{M_1}{M_1+M_2} = \xi_1 - 1. \tag{7.69}$$

Then Eq. (7.67) can be written as

$$\text{‘}V\text{’} = \frac{m(M_1+M_2)G}{a}\left\{\frac{\xi_2}{[(\xi-\xi_1)^2+\eta^2]^{1/2}} - \frac{\xi_1}{[(\xi-\xi_2)^2+\eta^2]^{1/2}} - \frac{1}{2}(\xi^2+\eta^2)\right\}. \tag{7.70}$$

In order to see the nature of this function, let us first look for its singular points, where $\partial\text{‘}V\text{’}/\partial\xi$ and $\partial\text{‘}V\text{’}/\partial\eta$ both vanish:

$$\begin{aligned} -\frac{\xi_2(\xi-\xi_1)}{[(\xi-\xi_1)^2+\eta^2]^{3/2}} + \frac{\xi_1(\xi-\xi_2)}{[(\xi-\xi_2)^2+\eta^2]^{3/2}} - \xi &= 0, \\ -\frac{\xi_2\eta}{[(\xi-\xi_1)^2+\eta^2]^{3/2}} + \frac{\xi_1\eta}{[(\xi-\xi_2)^2+\eta^2]^{3/2}} - \eta &= 0. \end{aligned} \tag{7.71}$$

A point (x, y) for which these equations are satisfied is an equilibrium point for the mass m (in the rotating coordinate system), since Eqs. (7.64) are evidently satisfied if m is at rest at this point. We first consider points on the $\eta = 0$ axis. The second equation is then satisfied, and the first becomes

$$-\frac{\xi_2(\xi-\xi_1)}{|\xi-\xi_1|^3} + \frac{\xi_1(\xi-\xi_2)}{|\xi-\xi_2|^3} - \xi = 0. \tag{7.72}$$

In Fig. 7.7, we plot the function ‘V’, as given by Eq. (7.70), along the $\eta = 0$ axis. The roots of Eq. (7.72) are the maxima of ‘V’ $(\xi, 0)$ in Fig. 7.7, where it can be seen that there are three such roots. Let us call them ξ_A, ξ_B, ξ_C as in the figure. Each is

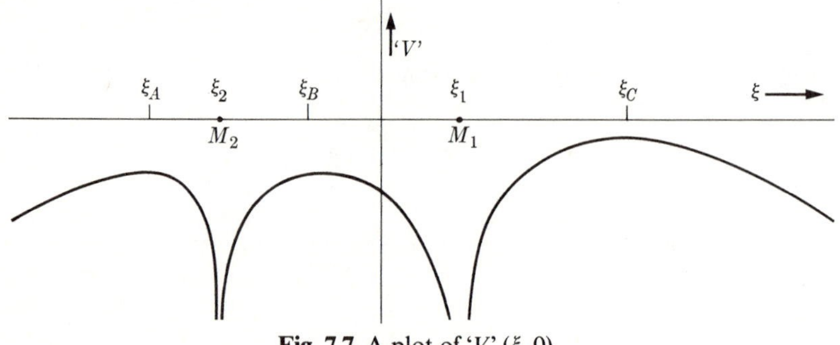

Fig. 7.7 A plot of ‘V’ $(\xi, 0)$.

the root of a quintic equation which may be derived from Eq. (7.72). It is possible to show that $\partial^2`V'/\partial\xi\,\partial\eta = 0$, $\partial^2`V'/\partial\xi^2 < 0$, and $\partial^2`V'/\partial\eta^2 > 0$ at these points A, B, and C. If we expand 'V' in a Taylor series about any one of these points, and consider only the quadratic terms, we see that the curves of constant 'V' are hyperbolas in the $\xi\eta$-plane in the neighborhood of points A, B, C, as shown in Fig. 7.8, where we plot the contours of constant 'V'. These points are *saddlepoints* of 'V'; that is, 'V' has a local maximum along the ξ-axis and a minimum along a line perpendicular to the ξ-axis at each of these points A, B, C. If $\eta \neq 0$, it can be factored from the second of Eqs. (7.71). We then multiply the second of Eqs. (7.71) by $(\xi - \xi_1)$ and subtract from the first of these equations. After some manipulation and using Eq. (7.69), we obtain

$$(\xi - \xi_2)^2 + \eta^2 = 1, \tag{7.73}$$

and, similarly,

$$(\xi - \xi_1)^2 + \eta^2 = 1. \tag{7.74}$$

These equations show that there are two singular points D, E, off the $\eta = 0$ axis,

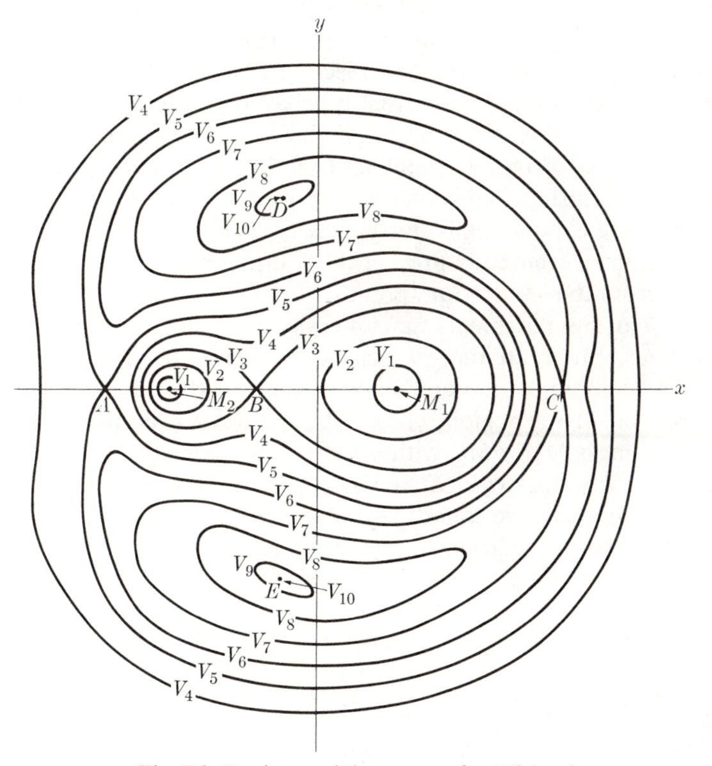

Fig. 7.8 Equipotential contours for 'V' (x, y).

which lie at unit distance from $(\xi_1, 0)$ and $(\xi_2, 0)$ which are themselves separated by a unit distance. By expanding 'V' in a Taylor series about point D or E, we can show that curves of constant 'V' are ellipses in the neighborhood of D or E, and that 'V' has a maximum at D and E. Knowing the behavior near the singular points, we can sketch the general appearance of the contours of constant 'V', as shown in Fig. 7.8. The curves are numbered in order of increasing 'V'.

If this were a fixed coordinate system, we could immediately conclude that equilibrium points A, B, C, D, E are all unstable, since the force $-\nabla$'V' is directed away from each equilibrium point when m is at some nearby points. However, this argument does not hold here because it neglects the coriolis force in Eqs. (7.64). If we expand the right members of Eqs. (7.64) in powers of the displacements (say $x-x_D$, $y-y_D$) from one of the equilibrium points (say D), and retain only linear terms, we may determine approximately the motion near the equilibrium point. If this is done near point D (or E), for example, we find that in linear approximation, the motion near D is stable if one of the masses M_1 or M_2 contains more than about 96% of the total mass (M_1+M_2). (See Problem 18.) For motions very near to point D, we may expect the linear approximation to yield a solution which is valid for very long times. Whether those motions which are stable in linear approximation are truly stable, in the sense that they remain near point D for all time, was until 1962 one of the unsolved problems of classical mechanics. This matter is discussed further at the end of Section 12.6.*

It is not difficult to show that, even in linear approximation, the equilibrium points A, B, C are unstable. If the motion in linear approximation is unstable, then the exact solution is certainly unstable. That is, regardless of how close m is initially to the equilibrium point (but not at it), it will not, in general, remain as close but will move exponentially away, at least at first. The neglected nonlinear terms may, of course, eventually prevent the solution from going more than some finite distance from the equilibrium point.

The only rigorous statements we can easily make about the motion of m, for very long periods of time, are those which can be derived from the energy equation (7.66). Given an initial position and velocity of m, we can calculate 'E'. The orbit then must remain in the region where 'V' $\leq$ 'E'. For example, motions which start near either mass M_1 or M_2, with 'E' $<$ V_3, must remain confined to a region near that mass. Motions with 'E' $>$ V_5 *may* go to arbitrarily large distances; whether they actually do, we cannot say from energy arguments.

If we could find another constant of the motion, say $F(x, y, \dot{x}, \dot{y})$, we could solve the problem by methods like those used in Chapter 3 for the central force

---

*A more complete discussion of the problem of three bodies, on a more advanced level than the present text, will be found in Aurel Wintner, *The Analytical Foundations of Celestial Mechanics*. Princeton: Princeton University Press, 1947. The proof that the motion near points D (or E) is often stable for all time was first given by A. M. Leontovitch in 1962 and is discussed in J. K. Moser, *Lectures on Hamiltonian Systems*, Memoirs of the Amer. Math. Soc., No. 81, 1968.

problem, where the angular momentum is also constant. Unfortunately, no other such constant is known, and it seems likely that none exists which could be used for this purpose. This problem has been studied very extensively.*

Faced with this situation, we may turn to the possibility of computing particular orbits from given initial conditions. This can be done either analytically, by approximation methods, or numerically, and in principle can be done to any desired accuracy and for any desired finite period of time.

In Chapter 12 we shall discuss a closely related special case of the three-body problem.

PROBLEMS

1. (a) Solve the problem of the freely falling body by introducing a translating coordinate system with an acceleration $\boldsymbol{g}$. Set up and solve the equations of motion in this accelerated coordinate system and transform the result back to a coordinate system fixed relative to the earth. (Neglect the earth's rotation.)

b) In the same accelerated coordinate system, set up the equations of motion for a falling body subject to an air resistance proportional to its velocity (relative to the fixed air).

2. A mass m is fastened by a spring (spring constant k) to a point of support which moves back and forth along the x-axis in simple harmonic motion at frequency ω, amplitude a. Assuming the mass moves only along the x-axis, set up and solve the equation of motion in a coordinate system whose origin is at the point of support.

3. Generalize Eq. (5.5) to the case when the origin of the coordinate system is moving, by adding fictitious torques due to the fictitious force on each particle. Express the fictitious torques in terms of the total mass M, the coordinate R^* of the center of mass, and a_h. Compare your result with Eq. (4.25).

4. Derive a formula for d^3A/dt^3 in terms of starred derivatives relative to a rotating coordinate system.

5. Westerly winds blow from west to east in the northern hemisphere with an average speed v. If the density of the air is ρ, what pressure gradient is required to maintain a steady flow of air from west to east with this speed? Make reasonable estimates of v and ρ, and estimate the pressure gradient in lb-in^{-2}-mile^{-1}. (You will need Eq. (5.172) from Chapter 5.)

6. (a) It has been suggested that birds may determine their latitude by sensing the coriolis force. Calculate the force a bird must exert in level flight at 30 mph against the sidewise component of coriolis force in order to fly in a straight line. Express your result in g's, that is, as a ratio of coriolis force to gravitational force, as a function of latitude and direction of flight.

b) If the bird's flight path is slightly circular, a centrifugal force will be present, which will add to the coriolis force and produce an error in estimated latitude. At 45° N latitude, how much may the flight path bend, in degrees per mile flown, if the latitude is to be determined within ± 100 miles? (Assume the sidewise force is measured as precisely as necessary!)

---

*See A. Wintner, *op. cit.*, and J. K. Moser, *op. cit.*

7. A body is dropped from rest at a height h above the surface of the earth.

a) Calculate the coriolis force as a function of time, assuming as a first approximation that it has a negligible effect on the motion, and using the velocity of a freely falling body with acceleration $\boldsymbol{g}_e$. Neglect air resistance, and assume h is small so that $\boldsymbol{g}_e$ can be taken as constant.

b) Now as a second approximation, calculate the net displacement of the point of impact due to the coriolis force calculated in part (a).

***8.** Find the answer to Problem 7(b) by solving for the motion in a nonrotating coordinate system. What approximations are needed to arrive at the same result?

9. An airplane flies across the North pole at 500 mph, following a meridian of longitude (which rotates with the earth). Find the angle between the direction of a plumb line hanging freely in the airplane as it passes over the pole and one hanging freely at the surface of the earth over the pole.

10. Assume the earth is a uniform sphere of mass M, radius R. Imagine a pipe extending vertically from the North pole to the center of the earth and out again at right angles to the equator. The pipe is filled with a fluid so that the fluid level at the North pole is at the earth's surface. Find the fluid level relative to the surface of the sphere at the equator. Does it change the answer much if the pipe runs near the surface of the earth? Does it change the answer much if you use the actual shape of the earth? (You need to know the material in Section 5.11 to answer this problem.)

11. A gyroscope consists of a wheel of radius r, all of whose mass is located on the rim. The gyroscope is rotating with angular velocity $\dot{\theta}$ about its axis, which is horizontal and is fixed relative to the earth's surface. We choose a coordinate system at rest relative to the earth whose z-axis coincides with the gyroscope axis and whose origin lies at the center of the wheel. The angular velocity $\boldsymbol{\omega}$ of the earth lies in the xz-plane, making an angle α with the gyroscope axis.

Find the x-, y-, and z-components of the torque $\boldsymbol{N}$ about the origin, due to the coriolis force in the xyz-coordinate system, acting on a mass m on the rim of the gyroscope wheel whose polar coordinates in the xy-plane are r, θ. Use this result to show that the total coriolis torque on the gyroscope, if the wheel has a mass M, is

$$\boldsymbol{N} = -\tfrac{1}{2}\boldsymbol{j}Mr^2\omega\dot{\theta}\sin\alpha.$$

This equation is the basis for the operation of the gyrocompass.

***12.** A mass m of a perfect gas of molecular weight M, at temperature T, is placed in a cylinder of radius a, height h, and whirled rapidly with an angular velocity ω about the axis of the cylinder. By introducing a coordinate system rotating with the gas, and applying the laws of static equilibrium, assuming that all other body forces are negligible compared with the centrifugal force, show that

$$p = \frac{RT}{M}\rho_0 \exp\left(\frac{M\omega^2 r^2}{2RT}\right),$$

where p is the pressure, r is the distance from the axis, and

$$\rho_0 = \frac{mM\omega^2}{2\pi hRT[\exp(M\omega^2 a^2/2RT) - 1]}.$$

***13.** A particle moves in the xy-plane under the action of a force

$$F = -kr,$$

directed toward the origin. Find its possible motions by introducing a coordinate system rotating about the z-axis with angular velocity ω chosen so that the centrifugal force just cancels the force F, and solving the equations of motion in this coordinate system. Describe the resulting motions, and show that your result agrees with that of Problem 45, Chapter 3.

14. A ball of mass m slides without friction on a horizontal plane at the surface of the earth. Show that it moves like the bob of a Foucault pendulum provided it remains near the point of tangency. Find its frequency of oscillation. Assume the earth is a sphere.

15. The bob of a pendulum is started so as to swing in a circle. By substituting in Eq. (7.46), find the angular velocity and show that the contribution due to the coriolis force is given very nearly by Eq. (7.52). Neglect the vertical component of the coriolis force, after showing that it is zero on the average for the assumed motion.

16. An electron revolves about a fixed proton in an ellipse of semimajor axis 10^{-8} cm. If the corresponding motion occurs in a magnetic field of 10,000 gausses, show that Larmor's theorem is applicable, and calculate the angular velocity of precession of the ellipse.

17. Write down a potential energy for the last term in Eq. (7.58). If the plane of the orbit in Problem 16 is perpendicular to $\boldsymbol{B}$, and if the orbit is very nearly circular, calculate (by the methods of Chapter 3) the rate of precession of the ellipse due to the last term in Eq. (7.58) in the rotating coordinate system. Is this precession to be added to or subtracted from that calculated in Problem 16?

18. Find the three second derivatives of 'V' with respect to ξ, η for the point D in Fig. 7.7. Expand the equations of motion (7.64), keeping terms linear in $\xi' = \xi - \xi_0$ and $\eta' = \eta - \eta_0$. Using the method of Section 4.10, find the condition on M_1, M_2 in order that the normal modes of oscillation be stable. If $M_1 > M_2$, what is the minimum value of $M_1/(M_1+M_2)$?

***19.** Prove the statements made in Section 7.6 regarding the second derivatives of 'V' at points A, B, and C in Fig. 7.7. Expand the equations of motion about points A and B, keeping terms linear in η and $\xi' = \xi - \xi_{A,B}$. Show by the method of Section 4.10 that some of the solutions are unstable for any values of the masses. (You cannot find the second derivatives explicitly, but the proof depends only on their signs.)

***20.** (a) Write out the quintic equation which must be solved for ξ_A in Fig. 7.7. Show that if $M_2 = 0$, the solution is $\xi_A = -1$.

b) Solve numerically for ξ_A to two decimal places for the earth-moon system.

c) Find the minimum launching velocity from the surface of the earth for which it is 'energetically' possible for a rocket to leave the earth-moon system. Compare with the escape velocity from the earth.

21. Two planets, each of mass M and radius R, revolve in circles about each other at a distance a apart, under their mutual gravitational attraction. Find the minimum (relative) velocity with which a rocket might leave one planet to arrive at the other. Show that the rocket must have a larger velocity than would be calculated if the motion of the planets were neglected.

22. (a) Locate all fixed points in the limiting case $M_2 \to 0$, and sketch Fig. 7.7 for this case. Show that the results in Section 7.6 applied to this case are consistent with the complete solution given in Section 3.14.

b) Show from this example for which the complete solution is known, that the minimum 'energetically' possible launching velocity for escape calculated as in Problem 18(c) is not necessarily the true minimum escape velocity.

CHAPTER 8

INTRODUCTION TO THE MECHANICS OF CONTINUOUS MEDIA

In this chapter we begin the study of the mechanics of continuous media, solids, fluids, strings, etc. In such problems, the number of particles is so large that it is not practical to study the motion of individual particles, and we instead regard matter as continuously distributed in space and characterized by its density. We are interested primarily in gaining an understanding of the concepts and methods of treatment which are useful, rather than in developing in detail methods of solving practical problems. In the first four sections, we shall treat the vibrating string, using concepts which are a direct generalization of particle mechanics. In the remainder of the chapter, the mechanics of fluids will be developed in a way less directly related to particle mechanics.

8.1 THE EQUATION OF MOTION FOR THE VIBRATING STRING

In this section we shall study the motion of a string of length l, stretched horizontally and fastened at each end, and set into vibration. In order to simplify the problem, we assume the string vibrates only in a vertical plane, and that the amplitude of vibration is small enough so that each point on the string moves only vertically, and so that the tension τ in the string does not change appreciably during the vibration.

We shall designate a point on the string by giving its horizontal distance x from the left-hand end (Fig. 8.1). The distance the point x has moved from the horizontal straight line representing the equilibrium position of the string will be designated by $u(x)$. Thus any position of the entire string is to be specified by specifying the function $u(x)$ for $0 \leq x \leq l$. This is precisely analogous, in the case of a system of N particles, to specifying the coordinates x_i, y_i, z_i, for $i = 1, \ldots, N$. In the case of the string, x is not a coordinate, but plays the same role as the subscript i; it designates a point on the string. Our idealized continuous string has infinitely many points, corresponding to the infinitely many values of x between 0 and l. For a given point x, it is $u(x)$ that plays the role of a coordinate locating

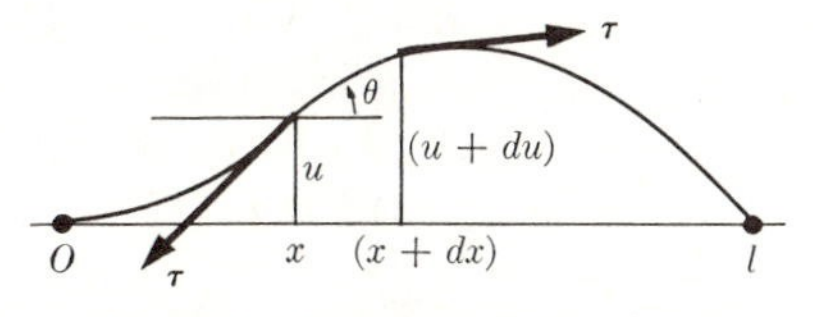

Fig. 8.1 The vibrating string.

that point, in analogy with the coordinates x_i, y_i, z_i of particle i. Just as a motion of the system of particles is to be described by functions $x_i(t)$, $y_i(t)$, $z_i(t)$, locating each particle at every instant of time, so a motion of the string is to be described by a function $u(x, t)$, locating each point x on the string at every instant of time.

In order to obtain an equation of motion for the string, we consider a segment of string of length dx between x and $x+dx$. If the density of the string per unit length is σ, then the mass of this segment is $\sigma\, dx$. The velocity of the string at any point is $\partial u/\partial t$, and its slope is $\partial u/\partial x$. The vertical component of tension exerted from right to left across any point in the string is

$$\tau_u = \tau \sin \theta, \tag{8.1}$$

where θ is the angle between the string and the horizontal (Fig. 8.1). We are assuming that θ is very small and, in this case,

$$\tau \sin \theta \doteq \tau \tan \theta = \tau \frac{\partial u}{\partial x}. \tag{8.2}$$

The net upward force dF due to the tension, on the segment dx of string, is the difference in the vertical component τ_u between the two ends of the segment:

$$\begin{aligned} dF &= [\tau_u]_{x+dx} - [\tau_u]_x \\ &\doteq \frac{\partial}{\partial x}\left(\tau \frac{\partial u}{\partial x}\right) dx. \end{aligned} \tag{8.3}$$

If we do not limit ourselves to very small slopes $\partial u/\partial x$, then a segment of string may also have a net horizontal component of force due to tension, and the segment will move horizontally as well as vertically, a possibility we wish to exclude. If there is, in addition, a vertical force f per unit length, acting along the string, the equation of motion of the segment dx will be

$$\sigma\, dx \frac{\partial^2 u}{\partial t^2} = \frac{\partial}{\partial x}\left(\tau \frac{\partial u}{\partial x}\right) dx + f\, dx. \tag{8.4}$$

For a horizontal string acted on by no horizontal forces except at its ends, and for small amplitudes of vibration, the tension is constant, and Eq. (8.4) can be rewritten:

$$\sigma \frac{\partial^2 u}{\partial t^2} = \tau \frac{\partial^2 u}{\partial x^2} + f. \tag{8.5}$$

The force f may be the gravitational force acting on the string, which is usually negligible unless the tension is very small. The force f may also represent an external force applied to the string to set it into vibration. We shall consider only the case $f = 0$, and we rewrite Eq. (8.5) in the form

$$\frac{\partial^2 u}{\partial x^2} - \frac{1}{c^2}\frac{\partial^2 u}{\partial t^2} = 0, \tag{8.6}$$

where

$$c = \left(\frac{\tau}{\sigma}\right)^{1/2}. \tag{8.7}$$

The constant c has the dimensions of a velocity, and we shall see in Section 8.3 that it is the velocity with which a wave travels along the string.

Equation (8.6) is a partial differential equation for the function $u(x, t)$; it is the mathematical expression of Newton's law of motion applied to the vibrating string. We shall want to find solutions $u(x, t)$ to Eq. (8.6), for any given initial position $u_0(x)$ of the string, and any given initial velocity $v_0(x)$ of each point along the string. If we take the initial instant at $t = 0$, this means that we want a solution $u(x, t)$ which satisfies the initial conditions:

$$\begin{aligned} u(x, 0) &= u_0(x), \\ \left[\frac{\partial u}{\partial t}\right]_{t=0} &= v_0(x). \end{aligned} \tag{8.8}$$

The solution must also satisfy the *boundary conditions:*

$$u(0, t) = u(l, t) = 0, \tag{8.9}$$

which express the fact that the string is tied at its ends. From the nature of the physical problem, we expect that there should be just one solution $u(x, t)$ of Eq. (8.6) which satisfies Eqs. (8.8) and (8.9), and this solution will represent the motion of the string with the given initial conditions. It is therefore reasonable to expect that the mathematical theory of partial differential equations will lead to the same conclusion regarding the number of solutions of Eq. (8.6), and indeed it does.

8.2 NORMAL MODES OF VIBRATION FOR THE VIBRATING STRING

We shall first try to find some solutions of Eq. (8.6) which satisfy the boundary conditions (8.9), without regard to the initial conditions (8.8). This is analogous to our treatment of the harmonic oscillator, in which we first looked for solutions of a certain type and later adjusted these solutions to fit the initial conditions of the problem. The method of finding solutions which we shall use is called the method of *separation of variables.* It is one of the few general methods so far devised for solving partial differential equations, and many important equations can be solved by this method. Unfortunately, it does not always work. In principle, any partial differential equation can be solved by numerical methods, but the labor involved in doing so is often prohibitive, sometimes even for the modern large-scale automatic computing machines.

The method of separation of variables consists in looking for solutions of the form

$$u(x, t) = X(x)\Theta(t), \tag{8.10}$$

that is, u is to be a product of a function X of x and a function Θ of t. The derivatives of u will then be

$$\frac{\partial^2 u}{\partial x^2} = \Theta \frac{d^2 X}{dx^2}, \qquad \frac{\partial^2 u}{\partial t^2} = X \frac{d^2 \Theta}{dt^2}. \tag{8.11}$$

If these expressions are substituted in Eq. (8.6), and if we divide through by ΘX, then Eq. (8.6) can be rewritten:

$$\frac{c^2}{X}\frac{d^2 X}{dx^2} = \frac{1}{\Theta}\frac{d^2\Theta}{dt^2}. \tag{8.12}$$

The left member of this equation is a function only of x, and the right member is a function only of t. If we hold t fixed and vary x, the right member remains constant, and the left member must therefore be independent of x. Similarly, the right member must actually be independent of t. We may set both members equal to a constant. It is clear on physical grounds that this constant must be negative, for the right member of Eq. (8.12) is the acceleration of the string divided by the displacement, and the acceleration must be opposite to the displacement or the string will not return to its equilibrium position. We shall call the constant $-\omega^2$:

$$\frac{1}{\Theta}\frac{d^2\Theta}{dt^2} = -\omega^2, \qquad \frac{c^2}{X}\frac{d^2 X}{dx^2} = -\omega^2. \tag{8.13}$$

The first of these equations can be rewritten as

$$\frac{d^2\Theta}{dt^2} + \omega^2\Theta = 0, \tag{8.14}$$

which we recognize as the equation for the harmonic oscillator, whose general solution, in the form most suitable for our present purpose, is

$$\Theta = A\cos\omega t + B\sin\omega t, \tag{8.15}$$

where A and B are arbitrary constants. The second of Eqs. (8.13) has a similar form:

$$\frac{d^2 X}{dx^2} + \frac{\omega^2}{c^2}X = 0, \tag{8.16}$$

and has a similar solution:

$$X = C\cos\frac{\omega x}{c} + D\sin\frac{\omega x}{c}. \tag{8.17}$$

The boundary condition (8.9) can hold for all times t only if X satisfies the conditions

$$\begin{aligned} X(0) &= C = 0, \\ X(l) &= C\cos\frac{\omega l}{c} + D\sin\frac{\omega l}{c} = 0. \end{aligned} \tag{8.18}$$

The first of these equations determines C, and the second then requires that

$$\sin \frac{\omega l}{c} = 0. \tag{8.19}$$

This will hold only if ω has one of the values

$$\omega_n = \frac{n\pi c}{l}, \qquad n = 1, 2, 3, \ldots. \tag{8.20}$$

Had we taken the separation constant in Eqs. (8.13) as positive, we would have obtained exponential solutions in place of Eq. (8.17), and it would have been impossible to satisfy the boundary conditions (8.18).

The frequencies $\nu_n = \omega_n/2\pi$ given by Eq. (8.20) are called the *normal frequencies of vibration* of the string. For a given n, we obtain a solution by substituting Eqs. (8.15) and (8.17) in Eq. (8.10), and making use of Eqs. (8.18), (8.20):

$$u(x, t) = A \sin \frac{n\pi x}{l} \cos \frac{n\pi ct}{l} + B \sin \frac{n\pi x}{l} \sin \frac{n\pi ct}{l}, \tag{8.21}$$

where we have set $D = 1$. This is called a *normal mode of vibration* of the string, and is entirely analogous to the normal modes of vibration which we found in Section 4.10 for coupled harmonic oscillators. Each point on the string vibrates at the same frequency ω_n with an amplitude which varies sinusoidally along the string. Instead of two coupled oscillators, we have an infinite number of oscillating points, and instead of two normal modes of vibration, we have an infinite number.

The initial position and velocity at $t = 0$ of the nth normal mode of vibration as given by Eq. (8.21) are

$$\begin{aligned} u_0(x) &= A \sin \frac{n\pi x}{l}, \\ v_0(x) &= \frac{n\pi cB}{l} \sin \frac{n\pi x}{l}. \end{aligned} \tag{8.22}$$

Only for these very special types of initial conditions will the string vibrate in one of its normal modes. However, we can build up more general solutions by adding solutions; for the vibrating string, like the harmonic oscillator, satisfies a principle of superposition. Let $u_1(x, t)$ and $u_2(x, t)$ be any two solutions of Eq. (8.6) which satisfy the boundary conditions (8.9). Then the function

$$u(x, t) = u_1(x, t) + u_2(x, t)$$

also satisfies the equation of motion and the boundary conditions. This is readily verified simply by substituting $u(x, t)$ in Eqs. (8.6) and (8.9), and making use of the fact that $u_1(x, t)$ and $u_2(x, t)$ satisfy these equations. A more general solution of Eqs. (8.6) and (8.9) is therefore to be obtained by adding solutions of the type (8.21), using different constants A and B for each normal frequency:

$$u(x, t) = \sum_{n=1}^{\infty} \left(A_n \sin \frac{n\pi x}{l} \cos \frac{n\pi ct}{l} + B_n \sin \frac{n\pi x}{l} \sin \frac{n\pi ct}{l} \right). \tag{8.23}$$

The initial position and velocity for this solution are

$$\begin{aligned} u_0(x) &= \sum_{n=1}^{\infty} A_n \sin \frac{n\pi x}{l}, \\ v_0(x) &= \sum_{n=1}^{\infty} \frac{n\pi c B_n}{l} \sin \frac{n\pi x}{l}. \end{aligned} \tag{8.24}$$

Whether or not Eq. (8.23) gives a general solution to our problem depends on whether, with suitable choices of the infinite set of constants A_n, B_n, we can make the functions $u_0(x)$ and $v_0(x)$ correspond to any possible initial position and velocity for the string. Our intuition is not very clear on this point, although it is clear that we now have a great variety of possible functions $u_0(x)$ and $v_0(x)$. The answer is provided by the Fourier series theorem, which states that any continuous function $u_0(x)$ for $(0 < x < l)$, which satisfies the boundary conditions (8.9), can be represented by the sum on the right in Eq. (8.24), if the constants A_n are properly chosen.* Similarly, with the proper choice of the constants B_n, any continuous function $v_0(x)$ for $(0 < x < l)$ can be represented. The expressions for A_n and B_n are, in this case,

$$\begin{aligned} A_n &= \frac{2}{l} \int_0^l u_0(x) \sin \frac{n\pi x}{l}\, dx, \\ B_n &= \frac{2}{n\pi c} \int_0^l v_0(x) \sin \frac{n\pi x}{l}\, dx. \end{aligned} \tag{8.25}$$

The most general motion of the vibrating string is therefore a superposition of normal modes of vibration at the fundamental frequency $\nu_1 = c/2l$ and its harmonics $\nu_n = nc/2l$.

---

*Dunham Jackson, *Fourier Series and Orthogonal Polynomials.* Menasha, Wisconsin: George Banta Publishing Co., 1941. (Chapter 1, Section 10.) Even functions with a finite number of discontinuities can be represented by Fourier series, but this point is not of great interest in the present application.

The Fourier series theorem was quoted in Section 2.11 in a slightly different form. The connection between Eqs. (8.24) and (2.205) is to be made by replacing t by x and T by $2l$ in Eq. (2.205). Both sine and cosine terms are then needed to represent an arbitrary function $u_0(x)$ in the interval $(0 < x < 2l)$, but only sine terms are needed if we want to represent $u_0(x)$ only in the interval $(0 < x < l)$. [Cosine terms alone would also do for this interval, but sine terms are appropriate if $u_0(x)$ vanishes at $x = 0$ and $x = l$.]

8.3 WAVE PROPAGATION ALONG A STRING

Equations (8.14) and (8.16) have also the complex solutions

$$\Theta = Ae^{\pm i\omega t}, \tag{8.26}$$

$$X = e^{\pm i(\omega/c)x}. \tag{8.27}$$

Hence Eq. (8.6) has complex solutions of the form

$$u(x, t) = Ae^{\pm i(\omega/c)(x \pm ct)}. \tag{8.28}$$

By taking the real part, or by adding complex conjugates and dividing by 2, we obtain the real solutions

$$u(x, t) = A \cos \frac{\omega}{c}(x - ct), \tag{8.29}$$

$$u(x, t) = A \cos \frac{\omega}{c}(x + ct). \tag{8.30}$$

By taking imaginary parts, or by subtracting complex conjugates and dividing by $2i$, we could obtain similar solutions with cosines replaced by sines. These solutions do not satisfy the boundary conditions (8.9), but they are of considerable interest in that they represent waves traveling down the string, as we now show.

A fixed point x on the string will oscillate harmonically in time, according to the solution (8.29) or (8.30), with amplitude A and angular frequency ω. At any given instant t, the string will be in the form of a sinusoidal curve with amplitude A and *wavelength* λ (distance between successive maxima):

$$\lambda = \frac{2\pi c}{\omega}. \tag{8.31}$$

We now show that this pattern moves along the string with velocity c, to the right in solution (8.29), and to the left in solution (8.30). Let

$$\xi = x - ct, \tag{8.32}$$

so that Eq. (8.29) becomes

$$u = A \cos \frac{\omega\xi}{c}, \tag{8.33}$$

where ξ is called the *phase* of the wave represented by the function u. For a fixed value of ξ, u has a fixed value. Let us consider a short time interval dt and find the increment dx required to maintain a constant value of ξ:

$$d\xi = dx - c\,dt = 0. \tag{8.34}$$

Now if dx and dt have the ratio given by Eq. (8.34),

$$\frac{dx}{dt} = c, \tag{8.35}$$

then the value of u at the point $x + dx$ at time $t + dt$ will be the same as its value at the point x at time t. Consequently, the pattern moves along the string with velocity c given by Eq. (8.7). The constant c is the *phase velocity* of the wave. Similarly, the velocity dx/dt for solution (8.30) is $-c$.

It is often convenient to introduce the *angular wave number* k defined by the equation

$$|k| = \frac{\omega}{c} = \frac{2\pi}{\lambda}, \tag{8.36}$$

where k is taken as positive for a wave traveling to the right, and negative for a wave traveling to the left. Then both solutions (8.29) and (8.30) can be written in the symmetrical form

$$u = A\cos(kx - \omega t). \tag{8.37}$$

The angular wave number k is measured in radians per unit length, just as the angular frequency ω is measured in radians per second. The expression for u in Eq. (8.37) is the real part of the complex function

$$u = Ae^{i(kx-\omega t)}. \tag{8.38}$$

This form is often used in the study of wave motion.

The possibility of superposing solutions of the form (8.29) and (8.30) with various amplitudes and frequencies, together with the Fourier series theorem, suggests a more general solution of the form

$$u(x, t) = f(x-ct) + g(x+ct), \tag{8.39}$$

where $f(\xi)$ and $g(\eta)$ are arbitrary functions of the variables $\xi = x-ct$, and $\eta = x+ct$. Equation (8.39) represents a wave of arbitrary shape traveling to the right with velocity c, and another traveling to the left. We can readily verify that Eq. (8.39) gives a solution of Eq. (8.6) by calculating the derivatives of u:

$$\frac{\partial u}{\partial x} = \frac{df}{d\xi}\frac{\partial \xi}{\partial x} + \frac{dg}{d\eta}\frac{\partial \eta}{\partial x} = \frac{df}{d\xi} + \frac{dg}{d\eta},$$

$$\frac{\partial^2 u}{\partial x^2} = \frac{d^2 f}{d\xi^2} + \frac{d^2 g}{d\eta^2},$$

$$\frac{\partial u}{\partial t} = \frac{df}{d\xi}\frac{\partial \xi}{\partial t} + \frac{dg}{d\eta}\frac{\partial \eta}{\partial t} = -c\frac{df}{d\xi} + c\frac{dg}{d\eta},$$

$$\frac{\partial^2 u}{\partial t^2} = c^2\frac{d^2 f}{d\xi^2} + c^2\frac{d^2 g}{d\eta^2}.$$

When these expressions are substituted in Eq. (8.6), it is satisfied identically, no matter what the functions $f(\xi)$ and $g(\eta)$ may be, provided, of course, that they have second derivatives. Equation (8.39) is, in fact, the most general solution of the equation (8.6); this follows from the theory of partial differential equations, according to which the general solution of a second-order partial differential equation contains two arbitrary functions. We can prove this without resorting to the theory of partial differential equations by assuming the string to be of infinite length, so that there are no boundary conditions to concern us, and by supposing that the initial position and velocity of all points on the string are given by the functions $u_0(x)$, $v_0(x)$. If the solution (8.39) is to meet these initial conditions, we must have, at $t = 0$:

$$u(x, 0) = f(x)+g(x) = u_0(x), \tag{8.40}$$

$$\left[\frac{\partial u}{\partial t}\right]_{t=0} = \left[-c\frac{df}{d\xi}+c\frac{dg}{d\eta}\right]_{t=0} = v_0(x). \tag{8.41}$$

At $t = 0$, $\xi = \eta = x$, so that Eq. (8.41) can be rewritten:

$$\frac{d}{dx}[-f(x)+g(x)] = \frac{v_0(x)}{c}, \tag{8.42}$$

which can be integrated to give

$$-f(x)+g(x) = \frac{1}{c}\int_0^x v_0(x)\,dx+C. \tag{8.43}$$

By adding and subtracting Eqs. (8.40) and (8.43), we obtain the functions f and g:

$$\begin{aligned} f(x) &= \tfrac{1}{2}\left(u_0(x)-\frac{1}{c}\int_0^x v_0(x)\,dx-C\right), \\ g(x) &= \tfrac{1}{2}\left(u_0(x)+\frac{1}{c}\int_0^x v_0(x)\,dx+C\right). \end{aligned} \tag{8.44}$$

The constant C can be omitted, since it will cancel out in $u = f+g$. These equations hold for all values of x, and we can replace x by ξ and η respectively:

$$\begin{aligned} f(\xi) &= \tfrac{1}{2}\left(u_0(\xi)-\frac{1}{c}\int_0^\xi v_0(\xi)\,d\xi\right), \\ g(\eta) &= \tfrac{1}{2}\left(u_0(\eta)+\frac{1}{c}\int_0^\eta v_0(\eta)\,d\eta\right). \end{aligned} \tag{8.45}$$

This gives a solution (8.39) to Eq. (8.6) for any initial position and velocity of the string.

Associated with a wave

$$u = f(x-ct), \tag{8.46}$$

there is a flow of energy down the string, as we show by computing the power delivered from left to right across any point x on the string. The power P is the product of the upward velocity of the point x and the upward force [see Eq. (8.2)] exerted by the left half of the string on the right half across the point x:

$$P = -\tau\frac{\partial u}{\partial x}\frac{\partial u}{\partial t}. \tag{8.47}$$

If u is given by Eq. (8.46), this is

$$P = c\tau\left(\frac{df}{d\xi}\right)^2, \tag{8.48}$$

which is always positive, indicating that the power flow is always from left to right for the wave (8.46). For a wave traveling to the left, P will be negative, indicating a flow of power from right to left. For a sinusoidal wave given by Eq. (8.37), the power is

$$P = k\omega\tau A^2 \sin^2(kx-\omega t), \tag{8.49}$$

or, averaged over a cycle,

$$\langle P\rangle_{\text{av}} = \tfrac{1}{2}k\omega\tau A^2. \tag{8.50}$$

We now consider a string tied at $x = 0$ and extending to the left from $x = 0$ to $x = -\infty$. The solution (8.39) must now satisfy the boundary condition

$$u(0,t) = f(-ct)+g(ct) = 0, \tag{8.51}$$

for all times t, so that the functions $f(\xi)$ and $g(\eta)$ must be such that

$$f(\xi) = -g(\eta) \qquad \text{whenever } \xi = -\eta. \tag{8.52}$$

The initial values $u_0(x)$ and $v_0(x)$ will now be given only for negative values of x, and Eqs. (8.45) will define $f(\xi)$ and $g(\eta)$ only for negative values of ξ and η. The values of $f(\xi)$ and $g(\eta)$ for positive values of ξ and η can then be found from Eq. (8.52):

$$f(\xi) = -g(-\xi), \qquad g(\eta) = -f(-\eta). \tag{8.53}$$

Let us consider a wave represented by $f(x-ct)$ traveling toward the end $x = 0$. A particular phase ξ_0, for which the wave amplitude is $f(\xi_0)$, will at time t_0 be at the point

$$x_0 = \xi_0+ct_0. \tag{8.54}$$

Let us suppose that ξ_0 and t_0 are so chosen that x_0 is negative. At a later time t_1, the phase ξ_0 will be at the point

$$x_1 = \xi_0+ct_1 = x_0+c(t_1-t_0). \tag{8.55}$$

At $t_1 = t_0-(x_0/c)$, $x_1 = 0$ and the phase ξ_0 reaches the end of the string. At later times x_1 will be positive, and $f(x_1-ct_1)$ will have no physical meaning, since the string does not extend to positive values of x. Now consider the phase η_0, of the leftward traveling wave $g(x+ct)$, defined by

$$\eta_0 = x+ct = -\xi_0. \tag{8.56}$$

The amplitude of the leftward wave $g(\eta_0)$ for the phase η_0 is related to the amplitude of the rightward wave $f(\xi_0)$ for the corresponding phase ξ_0 by Eq. (8.53):

$$g(\eta_0) = -f(\xi_0). \tag{8.57}$$

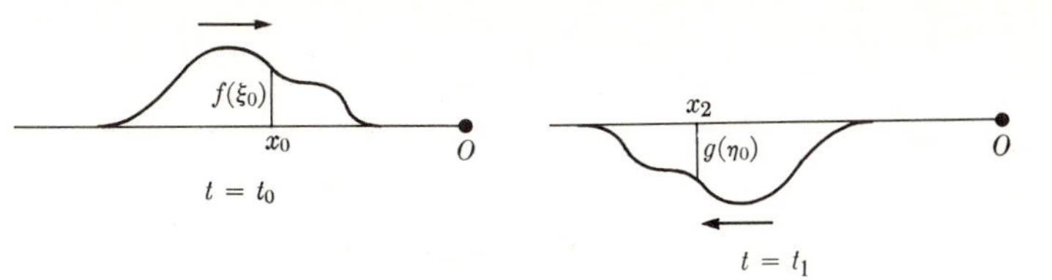

Fig. 8.2 A wave reflected at $x = 0$.

At time t_1, the phase η_0 will be at the point

$$x_2 = \eta_0 - ct_1 = -x_0 - c(t_1 - t_0). \tag{8.58}$$

If $t_1 > t_0 - (x_0/c)$, x_2 is negative and $g(\eta_0)$ represents a wave of equal and opposite amplitude to $f(\xi_0)$, traveling to the left. Thus the wave $f(x-ct)$ is reflected out of phase at $x = 0$ and becomes an equal and opposite wave traveling to the left. (See Fig. 8.2.) The total distance traveled by the wave during the time $(t_1 - t_0)$, from $x = x_0$ to $x = 0$ and back to $x = x_2$ is, by Eq. (8.58),

$$-x_0 - x_2 = c(t_1 - t_0), \tag{8.59}$$

as it should be.

The solution (8.39) can also be fitted to a string of finite length fastened at $x = 0$ and $x = l$. In this case, the initial position and velocity $u_0(x)$ and $v_0(x)$ are given only for $(0 \leq x \leq l)$. The functions $f(\xi)$ and $g(\eta)$ are then defined by Eq. (8.45) only for $(0 \leq \xi \leq l, 0 \leq \eta \leq l)$. If we define $f(\xi)$ and $g(\eta)$ for negative values of ξ and η by Eq. (8.53), in terms of their values for positive ξ and η, then the boundary condition (8.51) will be satisfied at $x = 0$. By an argument similar to that which led to Eq. (8.53), we can show that the boundary condition (8.9) for $x = l$ will be satisfied if, for all values of ξ and η,

$$\begin{aligned} f(\xi + l) &= -g(l - \xi), \\ g(\eta + l) &= -f(l - \eta). \end{aligned} \tag{8.60}$$

By means of Eqs. (8.53) and (8.60), we can find $f(\xi)$ and $g(\eta)$ for all values of ξ and η, once their values are given [by Eqs. (8.45)] for $0 \leq \xi \leq l$, $0 \leq \eta \leq l$. Thus we find a solution for the vibrating string of length l in terms of waves traveling in opposite directions and continuously being reflected at $x = 0$ and $x = l$. The solution is equivalent to the solution given by Eqs. (8.23) and (8.25) in terms of standing sinusoidal waves.

8.4 THE STRING AS A LIMITING CASE OF A SYSTEM OF PARTICLES

In the first three sections of this chapter, we have considered an idealized string characterized by a continuously distributed mass with density σ and tension τ. An actual string is made up of particles (atoms and molecules); our treatment of it as continuous is valid because of the enormously large number of particles in the string. A treatment of an actual string which takes into account the individual

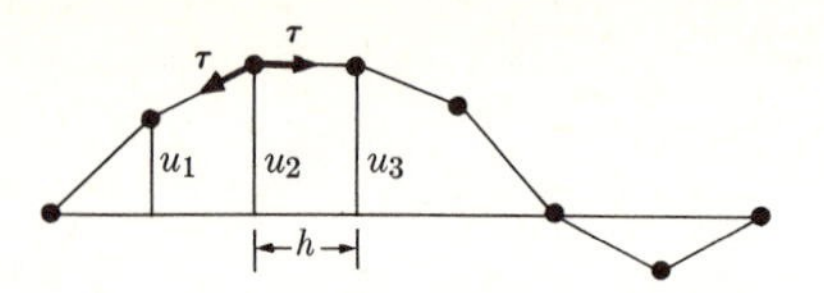

Fig. 8.3 A string made up of particles.

atoms would be hopelessly difficult, but we shall consider in this section an idealized model of a string made up of a finite number of particles, each of mass m. Figure 8.3 shows this idealized string, in which an attractive force τ acts between adjacent particles along the line joining them. The interparticle forces are such that in equilibrium the string is horizontal, with the particles equally spaced a distance h apart. The string is of length $(N+1)h$, with $N+2$ particles, the two end particles being fastened at the x-axis. The N particles which are free to move are numbered $1, 2, \ldots, N$, and the upward displacement of particle j from the horizontal axis will be called u_j. It will be assumed that the particles move only vertically and that only small vibrations are considered, so that the slope of the string is always small. Then the equations of motion of this system of particles are

$$m\frac{d^2u_j}{dt^2} = \tau\frac{u_{j+1}-u_j}{h} - \tau\frac{u_j-u_{j-1}}{h}, \qquad j = 1, \ldots, N, \tag{8.61}$$

where the expression on the right represents the vertical components of the forces τ between particle j and the two adjacent particles, and we are supposing that the forces τ are equal between all pairs of particles. Now let us assume that the number N of particles is very large, and that the displacement of the string is such that at any time t, a smooth curve $u(x, t)$ can be drawn through the particles, so that

$$u(jh, t) = u_j(t). \tag{8.62}$$

We can then represent the system of particles approximately as a continuous string of tension τ, and of linear density

$$\sigma = \frac{m}{h}. \tag{8.63}$$

The equations of motion (8.61) can be written in the form

$$\frac{d^2u_j}{dt^2} = \frac{\tau}{\sigma}\frac{1}{h}\left(\frac{u_{j+1}-u_j}{h} - \frac{u_j-u_{j-1}}{h}\right). \tag{8.64}$$

Now if the particles are sufficiently close together, we shall have, approximately,

$$\begin{aligned} \frac{u_{j+1}-u_j}{h} &\doteq \left[\frac{\partial u}{\partial x}\right]_{x=(j+1/2)h}, \\ \frac{u_j-u_{j-1}}{h} &\doteq \left[\frac{\partial u}{\partial x}\right]_{x=(j-1/2)h}, \end{aligned} \tag{8.65}$$

and hence

$$\frac{1}{h}\left(\frac{u_{j+1}-u_j}{h}-\frac{u_j-u_{j-1}}{h}\right) \doteq \left[\frac{\partial^2 u}{\partial x^2}\right]_{x=jh}. \tag{8.66}$$

The function $u(x, t)$ therefore, when h is very small, satisfies approximately the equation

$$\frac{\partial^2 u}{\partial t^2} = \frac{\tau}{\sigma}\frac{\partial^2 u}{\partial x^2}, \tag{8.67}$$

which is the same as Eq. (8.6) for the continuous string.

The solutions of Eqs. (8.61) when N is large will be expected to approximate the solutions of Eq. (8.6). If we were unable to solve Eq. (8.6) otherwise, one method of solving it numerically would be to carry out the above process in reverse, so as to reduce the partial differential equation (8.67) to the set of ordinary differential equations (8.61), which could then be solved by numerical methods. The solutions of Eqs. (8.61) are of some interest in their own right. Let us rewrite these equations in the form

$$m\frac{d^2 u_j}{dt^2}+\frac{2\tau}{h}u_j-\frac{\tau}{h}(u_{j+1}+u_{j-1}) = 0, \qquad j = 1, \ldots, N. \tag{8.68}$$

These are the equations for a set of harmonic oscillators, each coupled to the two adjacent oscillators. We are led, either by our method of treatment of the coupled oscillator problem or by considering our results for the continuous string, to try a solution of the form

$$u_j = a_j e^{\pm i\omega t}. \tag{8.69}$$

If we substitute this trial solution in Eqs. (8.68), the factor $e^{\pm i\omega t}$ cancels out, and we get a set of algebraic equations:

$$\left(\frac{2\tau}{h}-m\omega^2\right)a_j-\frac{\tau}{h}a_{j+1}-\frac{\tau}{h}a_{j-1} = 0, \qquad j = 1, \ldots, N. \tag{8.70}$$

This is a set of linear difference equations which could be solved for a_{j+1} in terms of a_j and a_{j-1}. Since $a_0 = 0$, if a_1 is given we can find the values of the remaining constants a_j by successive applications of these equations. A neater method of solution is to notice the analogy between the linear difference equations (8.70) and the linear differential equation (8.16), and to try the solution

$$a_j = Ae^{ipj}, \qquad j = 1, \ldots, N. \tag{8.71}$$

When this is substituted in Eqs. (8.70), we get, after canceling the factor Ae^{ipj}:

$$\left(\frac{2\tau}{h}-m\omega^2\right)-\frac{\tau}{h}(e^{ip}+e^{-ip}) = 0, \tag{8.72}$$

or

$$\cos p = 1-\frac{mh\omega^2}{2\tau}. \tag{8.73}$$

If ω is less than

$$\omega_c = \left(\frac{4\tau}{mh}\right)^{1/2}, \tag{8.74}$$

there will be real solutions for p. Let a solution be given by

$$p = kh, \qquad 0 \leq kh \leq \pi. \tag{8.75}$$

Then another solution is

$$p = -kh. \tag{8.76}$$

All other solutions for p differ from these by multiples of 2π, and in view of the form of Eq. (8.71), they lead to the same values of a_j, so we can restrict our attention to values of p given by Eqs. (8.75) and (8.76).

If we substitute Eq. (8.71) in Eq. (8.69), making use of Eq. (8.75), we have a solution of Eqs. (8.68) in the form

$$u_j = Ae^{\pm i(kjh - \omega t)}. \tag{8.77}$$

Since the horizontal distance of particle j from the left end of the string is

$$x_j = jh, \tag{8.78}$$

we see that the solution (8.77) corresponds to our previous solution (8.38) for the continuous string, and represents traveling sinusoidal waves. By combining the two complex conjugate solutions (8.77) and using Eq. (8.78), we obtain the real solution

$$u_j = A \cos (kx_j - \omega t), \tag{8.79}$$

which corresponds to Eq. (8.37). We thus have sinusoidal waves which may travel in either direction with the velocity [Eq. (8.36)]

$$c = \frac{\omega}{k} = \frac{h\omega}{|p|}, \tag{8.80}$$

where p is given by Eq. (8.73). If $\omega \ll \omega_c$ [Eq. (8.74)], then p will be nearly zero, and we can expand cos p in Eq. (8.73) in a power series:

$$\begin{aligned} 1 - \frac{p^2}{2} &\doteq 1 - \frac{mh\omega^2}{2\tau}, \\ |p| &\doteq \omega\left(\frac{mh}{\tau}\right)^{1/2}, \end{aligned} \tag{8.81}$$

and

$$c \doteq \left(\frac{h\tau}{m}\right)^{1/2}, \tag{8.82}$$

which agrees with Eq. (8.7) for the continuous string, in view of Eq. (8.63). However, for larger values of ω, the velocity c is smaller than for the continuous string, and approaches

$$c = \frac{h\omega_c}{\pi} = \frac{2}{\pi}\left(\frac{h\tau}{m}\right)^{1/2} \tag{8.83}$$

as $\omega \to \omega_c$. ($\omega_c = \infty$ for the continuous string for which $mh = \sigma h^2 = 0$.) Since the phase velocity given by Eq. (8.80) depends upon the frequency, we cannot superpose sinusoidal solutions to obtain a general solution of the form (8.39). If a wave of other than sinusoidal shape travels along the string, the sinusoidal components into which it may be resolved travel with different velocities, and consequently the shape of the wave changes as it moves along. This phenomenon is called *dispersion.*

When $\omega > \omega_c$, Eq. (8.73) has only complex solutions for p, of the form

$$p = \pi \pm i\gamma. \tag{8.84}$$

These lead to solutions u_j of the form

$$u_j = (-1)^j A e^{\pm \gamma j} \cos \omega t. \tag{8.85}$$

There is then no wave propagation, but only an exponential decline in amplitude of oscillation to the right or to the left from any point which may be set in oscillation. The minimum wavelength [Eq. (8.36)] which is allowed by Eq. (8.75) is

$$\lambda_c = \frac{2\pi}{k_c} = 2h. \tag{8.86}$$

It is evident that a wave of shorter wavelength than this would have no meaning, since there would not be enough particles in a distance less than λ_c to define the wavelength. The wavelength λ_c corresponds to the frequency ω_c, for which

$$u_j = Ae^{ij\pi}e^{\pm i\omega_c t} = (-1)^j A e^{\pm i\omega_c t}. \tag{8.87}$$

Adjacent particles simply oscillate out of phase with amplitude A.

We can build up solutions which satisfy the boundary conditions

$$u_0 = u_{N+1} = 0 \tag{8.88}$$

by adding and subtracting solutions of the form (8.77). We can, by suitably combining solutions of the form (8.77), obtain the solutions

$$\begin{aligned} u_j = A \sin pj \cos \omega t + B \sin pj \sin \omega t + C \cos pj \cos \omega t \\ + D \cos pj \sin \omega t. \end{aligned} \tag{8.89}$$

In order to satisfy the conditions (8.88), we must set

$$\begin{aligned} &C = D = 0, \\ &p = \frac{n\pi}{N+1}, \qquad n = 1, 2, \ldots, N, \end{aligned} \tag{8.90}$$

where the limitation $n \leq N$ arises from the limitation on p in Eq. (8.75). The normal frequencies of vibration are now given by Eq. (8.73):

$$\omega_n = \left[\frac{2\tau}{mh}\left(1 - \cos\frac{n\pi}{N+1}\right)\right]^{1/2}, \qquad n = 1, 2, \ldots, N. \tag{8.91}$$

If $n \ll N$, we can expand the cosine in a power series, to obtain

$$\begin{aligned} \omega_n &\doteq \left[\frac{n^2\pi^2\tau}{mh(N+1)^2}\right]^{1/2} \\ &= \frac{n\pi}{l}\left(\frac{\tau}{\sigma}\right)^{1/2}, \qquad [l = (N+1)h], \end{aligned} \tag{8.92}$$

which agrees with Eq. (8.20) for the continuous string.

A physical model which approximates fairly closely the string of particles treated in this section can be constructed by hanging weights m at intervals h along a stretched string. The mass m of each weight must be large in comparison with that of a length h of the string.

8.5 GENERAL REMARKS ON THE PROPAGATION OF WAVES

If we designate by F the upward component of force due to tension, exerted from left to right across any point in a stretched string, and by v the upward velocity of any point on the string, then, by Eq. (8.2), we have

$$F = -\tau\frac{\partial u}{\partial x}, \tag{8.93}$$

$$v = \frac{\partial u}{\partial t}. \tag{8.94}$$

By Eq. (8.4), if there is no other force on the string, we have

$$\frac{\partial v}{\partial t} = -\frac{1}{\sigma}\frac{\partial F}{\partial x}, \tag{8.95}$$

and by differentiating Eq. (8.93) with respect to t, assuming τ is constant in time, we obtain

$$\frac{\partial F}{\partial t} = -\tau\frac{\partial v}{\partial x}. \tag{8.96}$$

Equations (8.95) and (8.96) are easily understood physically. The acceleration of the string will be proportional to the difference in the upward force F at the ends of a small segment of string. Likewise, since F is proportional to the slope, the time rate of change of F will be proportional to the difference in upward velocities of the ends of a small segment of the string. The power delivered from left to right across any point in the string is

$$P = Fv. \tag{8.97}$$

Equations (8.95) and (8.96) are typical of many types of small amplitude wave propagation which occur in physics. There are two quantities, in this case F and v, such that the time rate of change of either is proportional to the space derivative of the other. For large amplitudes, the equations for wave propagation may become nonlinear, and new effects like the development of shock fronts may occur which are not described by the equations we have studied here. When there is dispersion, linear terms in v and F, or terms involving higher derivatives than the first, may appear. Whenever equations of the form (8.95) and (8.96) hold, a wave equation of the form (8.6) can be derived for either of the two quantities. For example, if we differentiate Eq. (8.95) with respect to t, and Eq. (8.96) with respect to x, assuming σ, τ to be constant, we obtain

$$\frac{\partial^2 F}{\partial x\,\partial t} = -\tau\frac{\partial v^2}{\partial x^2} = -\sigma\frac{\partial^2 v}{\partial t^2},$$

or

$$\frac{\partial^2 v}{\partial x^2} - \frac{1}{c^2}\frac{\partial^2 v}{\partial t^2} = 0, \tag{8.98}$$

where

$$c = \left(\frac{\tau}{\sigma}\right)^{1/2}. \tag{8.99}$$

Similarly, we can show that

$$\frac{\partial^2 F}{\partial x^2} - \frac{1}{c^2}\frac{\partial^2 F}{\partial t^2} = 0. \tag{8.100}$$

Usually one of these two quantities can be chosen so as to be analogous to a force (F), and the other to the corresponding velocity (v), and then the power transmitted will be given by an equation like Eq. (8.97). Likewise, all other quantities associated with the wave motion satisfy a wave equation, as, for example, u, which satisfies Eq. (8.6).

As a further example, the equations for a plane sound wave traveling in the x-direction, which will be derived in Section 8.10, can be written in the form

$$\frac{\partial v}{\partial t} = -\frac{1}{\rho}\frac{\partial p'}{\partial x}, \qquad \frac{\partial p'}{\partial t} = -B\frac{\partial v}{\partial x}, \tag{8.101}$$

where p' is the excess pressure (above atmospheric), v is the velocity, in the x-direction, of the air at any point, and where ρ is the density and B the bulk modulus. The physical meaning of these equations is clear almost without further discussion of the motion of gases. Both p' and v satisfy wave equations, easily derived from Eqs. (8.101):

$$\frac{\partial^2 p'}{\partial x^2} - \frac{1}{c^2}\frac{\partial^2 p'}{\partial t^2} = 0, \qquad \frac{\partial^2 v}{\partial x^2} - \frac{1}{c^2}\frac{\partial^2 v}{\partial t^2} = 0, \tag{8.102}$$

where

$$c = \left(\frac{B}{\rho}\right)^{1/2}, \tag{8.103}$$

and the power transmitted in the x-direction per unit area is

$$P = p'v. \tag{8.104}$$

In the case of a plane electromagnetic wave traveling in the x-direction and linearly polarized in the y-direction, the analogous equations can be shown to be (gaussian units)

$$\frac{\partial B_z}{\partial t} = -c\frac{\partial E_y}{\partial x}, \qquad \frac{\partial E_y}{\partial t} = -c\frac{\partial B_z}{\partial x}, \tag{8.105}$$

where E_y and B_z are the y- and z-components of electric and magnetic field intensities, and c is the speed of light. The components E_y and B_z satisfy wave equations with wave velocity c, and the power transmitted in the x-direction per unit area is

$$P = \frac{E_y B_z}{4\pi c}. \tag{8.106}$$

As a final example, on a two-wire electrical transmission line, the voltage E across the line and the current i through the line satisfy the equations

$$\frac{\partial E}{\partial t} = -\frac{1}{C}\frac{\partial i}{\partial x}, \qquad \frac{\partial i}{\partial t} = -\frac{1}{L}\frac{\partial E}{\partial x}, \tag{8.107}$$

where C is the shunt capacitance per unit length, and L is the series inductance per unit length. Again we can derive wave equations for i and E with the wave velocity

$$c = \left(\frac{1}{LC}\right)^{1/2}, \tag{8.108}$$

and again the power transmitted in the x-direction is

$$P = Ei. \tag{8.109}$$

Thus the study of wave propagation on a string is applicable to a wide variety of physical problems, many of them of greater practical and theoretical importance than the string itself. In many cases, our discussion of the string as made up of a number of discrete particles is also of interest. The electrical transmission line, for example, can be considered a limiting case of a series of low-pass filters. An electrical network made up of series inductances and shunt capacitances can be described by a set of equations of the same form as our Eqs. (8.61), with analogous results. In the case of sound waves, we are led by analogy to expect that at very high frequencies, when the wavelength becomes comparable to the distance between molecules, the wave velocity will begin to depend on the frequency, and that there will be a limiting frequency above which no wave propagation is possible.

8.6 KINEMATICS OF MOVING FLUIDS

In this section we shall develop the kinematic concepts useful in studying the motion of continuously distributed matter, with particular reference to moving fluids. One way of describing the motion of a fluid would be to attempt to follow the motion of each individual point in the fluid, by assigning coordinates x, y, z to each fluid particle and specifying these as functions of the time. We may, for example, specify a given fluid particle by its coordinates, x_0, y_0, z_0, at an initial instant $t = t_0$. We can then describe the motion of the fluid by means of functions $x(x_0, y_0, z_0, t)$, $y(x_0, y_0, z_0, t)$, $z(x_0, y_0, z_0, t)$ which determine the coordinates x, y, z at time t of the fluid particle which was at x_0, y_0, z_0 at time t_0. This would be an immediate generalization of the concepts of particle mechanics, and of the preceding treatment of the vibrating string. This program originally due to Euler leads to the so-called "Lagrangian equations" of fluid mechanics. A more convenient treatment for many purposes, due also to Euler, is to abandon the attempt to specify the history of each fluid particle, and to specify instead the density and velocity of the fluid at each point in space at each instant of time. This is the method which we shall follow here. It leads to the "Eulerian equations" of fluid mechanics. We describe the motion of the fluid by specifying the density $\rho(x, y, z, t)$ and the vector velocity $\boldsymbol{v}(x, y, z, t)$, at the point x, y, z at the time t. We thus focus our attention on what is happening at a particular point in space at a particular time, rather than on what is happening to a particular fluid particle.

Any quantity which is used in describing the state of the fluid, for example the pressure p, will be a function $[p(x, y, z, t)]$ of the space coordinates x, y, z and of the time t; that is, it will have a definite value at each point in space and at each instant of time. Although the mode of description we have adopted focuses attention on a point in space rather than on a fluid particle, we shall not be able to avoid following the fluid particles themselves, at least for short time intervals dt. For it is to the particles, and not to the space points, that the laws of mechanics apply. We shall be interested, therefore, in two time rates of change for any quantity, say p. The rate at which the pressure is changing with time at a fixed point in space will be the partial derivative with respect to time $(\partial p/\partial t)$; it is itself a function of x, y, z, and t. The rate at which the pressure is changing with respect to a point moving along with the fluid will be the total derivative

$$\frac{dp}{dt} = \frac{\partial p}{\partial t} + \frac{\partial p}{\partial x}\frac{dx}{dt} + \frac{\partial p}{\partial y}\frac{dy}{dt} + \frac{\partial p}{\partial z}\frac{dz}{dt}, \tag{8.110}$$

where dx/dt, dy/dt, dz/dt are the components of the fluid velocity $\boldsymbol{v}$. The change in pressure, dp, occurring during a time dt, at the position of a moving fluid particle which moves from x, y, z to $x+dx$, $y+dy$, $z+dz$ during this time, will be

$$\begin{aligned} dp &= p(x+dx, y+dy, z+dz, t+dt) - p(x, y, z, t) \\ &\doteq \frac{\partial p}{\partial x}dx + \frac{\partial p}{\partial y}dy + \frac{\partial p}{\partial z}dz + \frac{\partial p}{\partial t}dt, \end{aligned}$$

and if $dt \to 0$, this leads to Eq. (8.110). We can also write Eq. (8.110) in the forms:

$$\frac{dp}{dt} = \frac{\partial p}{\partial t} + v_x\frac{\partial p}{\partial x} + v_y\frac{\partial p}{\partial y} + v_z\frac{\partial p}{\partial z} \tag{8.111}$$

and

$$\frac{dp}{dt} = \frac{\partial p}{\partial t} + \boldsymbol{v}\cdot\nabla p, \tag{8.112}$$

where the second expression is a shorthand for the first, in accordance with the conventions for using the symbol ∇. The total derivative dp/dt is also a function of x, y, z, and t. A similar relation holds between partial and total derivatives of any quantity, and we may write, symbolically,

$$\frac{d}{dt} = \frac{\partial}{\partial t} + \boldsymbol{v}\cdot\nabla, \tag{8.113}$$

where total and partial derivatives have the meaning defined above.

Let us consider now a small volume δV of fluid, and we shall agree that δV always designates a volume element which moves with the fluid, so that it always contains the same fluid particles. In general, the volume δV will then change with time, and we wish to calculate this rate of change. Let us assume that δV is in the form of a rectangular box of dimensions δx, δy, δz (Fig. 8.4):

$$\delta V = \delta x\, \delta y\, \delta z. \tag{8.114}$$

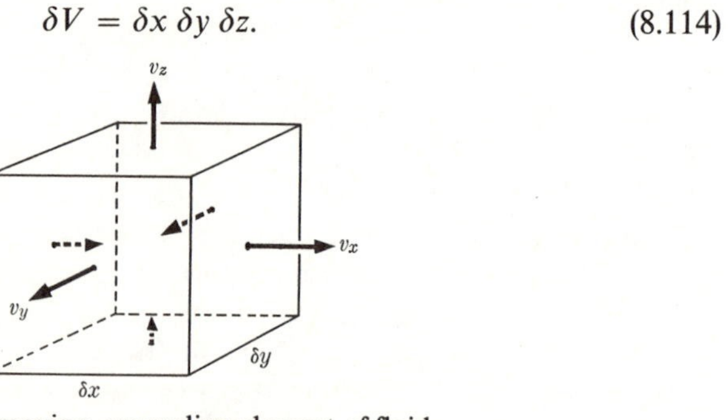

Fig. 8.4 A moving, expanding element of fluid.

The x-component of fluid velocity v_x may be different at the left and right faces of the box. If so, δx will change with time at a rate equal to the difference between these two velocities:

$$\frac{d}{dt}\delta x = \frac{\partial v_x}{\partial x}\delta x,$$

and, similarly,

$$\left.\begin{aligned} \frac{d}{dt}\delta y &= \frac{\partial v_y}{\partial y}\delta y, \\ \frac{d}{dt}\delta z &= \frac{\partial v_z}{\partial z}\delta z. \end{aligned}\right\} \tag{8.115}$$

The time rate of change of δV is then

$$\begin{aligned} \frac{d}{dt}\delta V &= \delta y\,\delta z\frac{d}{dt}\delta x + \delta x\,\delta z\frac{d}{dt}\delta y + \delta x\,\delta y\frac{d}{dt}\delta z \\ &= \left(\frac{\partial v_x}{\partial x} + \frac{\partial v_y}{\partial y} + \frac{\partial v_z}{\partial z}\right)\delta x\,\delta y\,\delta z, \end{aligned}$$

and finally,

$$\frac{d}{dt}\delta V = \nabla\cdot\boldsymbol{v}\,\delta V. \tag{8.116}$$

This derivation is not very rigorous, but it gives an insight into the meaning of the divergence $\nabla\cdot\boldsymbol{v}$. The derivation can be made rigorous by keeping careful track of quantities that were neglected here, like the dependence of v_x upon y and z, and showing that we arrive at Eq. (8.116) in the limit as $\delta V \to 0$. However, there is an easier way to give a more rigorous proof of Eq. (8.116). Let us consider a volume V of fluid which is composed of a number of elements δV:

$$V = \sum \delta V. \tag{8.117}$$

If we sum the left side of Eq. (8.116), we have

$$\sum \frac{d}{dt}\delta V = \frac{d}{dt}\sum \delta V = \frac{dV}{dt}. \tag{8.118}$$

The summation sign here really represents an integration, since we mean to pass to the limit $\delta V \to 0$, but the algebraic steps in Eq. (8.118) would look rather unfamiliar if the integral sign were used. Now let us sum the right side of Eq. (8.116), this time passing to the limit and using the integral sign, in order that we may apply Gauss' divergence theorem [Eq. (3.115)]:

$$\begin{aligned} \sum \nabla\cdot\boldsymbol{v}\,\delta V &= \iiint_V \nabla\cdot\boldsymbol{v}\,dV \\ &= \iint_S \hat{\boldsymbol{n}}\cdot\boldsymbol{v}\,dS, \end{aligned} \tag{8.119}$$

where S is the surface bounding the volume V, and $\hat{\boldsymbol{n}}$ is the outward normal unit vector. Since $\hat{\boldsymbol{n}}\cdot\boldsymbol{v}$ is the outward component of velocity of the surface element dS, the volume added to V by the motion of dS in a time dt will be $\hat{\boldsymbol{n}}\cdot\boldsymbol{v}\,dt\,dS$ (Fig. 8.5), and hence the last line in Eq. (8.119) is the proper expression for the rate of increase in volume:

$$\frac{dV}{dt} = \iint_S \hat{\boldsymbol{n}}\cdot\boldsymbol{v}\,dS. \tag{8.120}$$

Therefore Eq. (8.116) must be the correct expression for the rate of increase of a volume element, since it gives the correct expression for the rate of increase of any volume V when summed over V. Note that the proof is independent of the shape of δV. We have incidentally derived an expression for the time rate of change of a volume V of moving fluid:

$$\frac{dV}{dt} = \iiint_V \boldsymbol{\nabla}\cdot\boldsymbol{v}\,dV. \tag{8.121}$$

If the fluid is incompressible, then the volume of every element of fluid must remain constant:

$$\frac{d}{dt}\,\delta V = 0, \tag{8.122}$$

and consequently, by Eq. (8.116),

$$\boldsymbol{\nabla}\cdot\boldsymbol{v} = 0. \tag{8.123}$$

No fluid is absolutely incompressible, but for many purposes liquids may be regarded as practically so and, as we shall see, even the compressibility of gases may often be neglected.

Now the mass of an element of fluid is

$$\delta m = \rho\,\delta V, \tag{8.124}$$

and this will remain constant even though the volume and density may not:

$$\frac{d}{dt}\,\delta m = \frac{d}{dt}(\rho\,\delta V) = 0. \tag{8.125}$$

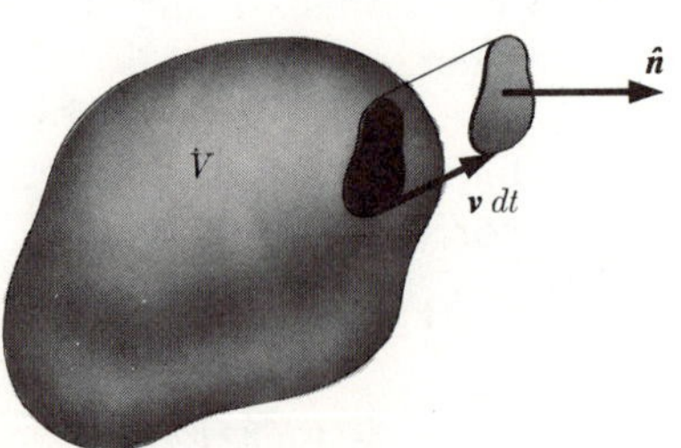

Fig. 8.5 Increase of volume due to motion of surface.

Let us carry out the differentiation, making use of Eq. (8.116):

$$\delta V \frac{d\rho}{dt} + \rho \frac{d\,\delta V}{dt} = \delta V \frac{d\rho}{dt} + \rho \nabla \cdot \boldsymbol{v}\, \delta V = 0,$$

or, when δV is divided out,

$$\frac{d\rho}{dt} + \rho \nabla \cdot \boldsymbol{v} = 0. \tag{8.126}$$

By utilizing Eq. (8.113), we can rewrite this in terms of the partial derivatives referred to a fixed point in space:

$$\frac{\partial \rho}{\partial t} + \boldsymbol{v} \cdot \nabla \rho + \rho \nabla \cdot \boldsymbol{v} = 0.$$

The last two terms can be combined, using the properties of ∇ as a symbol of differentiation:

$$\frac{\partial \rho}{\partial t} + \nabla \cdot (\rho \boldsymbol{v}) = 0. \tag{8.127}$$

This is the *equation of continuity* for the motion of continuous matter. It states essentially that matter is nowhere created or destroyed; the mass δm in any volume δV moving with the fluid remains constant.

We shall make frequent use in the remainder of this chapter of the properties of the symbol ∇, which were described briefly in Section 3.6. The operator ∇ has the algebraic properties of a vector and, in addition, when a product is involved, it behaves like a differentiation symbol. The simplest way to perform this sort of manipulation, when ∇ operates on a product, is first to write a sum of products in each of which only one factor is to be differentiated. The factor to be differentiated may be indicated by underlining it. Then each term may be manipulated according to the rules of vector algebra, except that the underlined factor must be kept behind the ∇ symbol. When the underlined factor is the only one behind the ∇ symbol, or when all other factors are separated out by parentheses, the underline may be omitted, as there is no ambiguity as to what factor is to be differentiated by the components of ∇. As an example, the relation between Eqs. (8.126) and (8.127) is made clear by the following computation:

$$\begin{aligned} \nabla \cdot (\rho \boldsymbol{v}) &= \nabla \cdot (\underline{\rho} \boldsymbol{v}) + \nabla \cdot (\rho \underline{\boldsymbol{v}}) \\ &= (\nabla \underline{\rho}) \cdot \boldsymbol{v} + \rho \nabla \cdot \underline{\boldsymbol{v}} \\ &= (\nabla \rho) \cdot \boldsymbol{v} + \rho \nabla \cdot \boldsymbol{v} \\ &= \boldsymbol{v} \cdot \nabla \rho + \rho \nabla \cdot \boldsymbol{v}. \end{aligned} \tag{8.128}$$

Any formulas arrived at in this way can always be verified by writing out both sides in terms of components, and the reader should do this a few times to convince himself. However, it is usually far less work to make use of the properties of the ∇ symbol.

We now wish to calculate the rate of flow of mass through a surface S fixed in space. Let dS be an element of surface, and let $\hat{\boldsymbol{n}}$ be a unit vector normal to dS.

If we construct a cylinder by moving dS through a distance $v\,dt$ in the direction of $-\boldsymbol{v}$, then in a time dt all the matter in this cylinder will pass through the surface dS (Fig. 8.6). The amount of mass in this cylinder is

$$\rho\hat{\boldsymbol{n}}\cdot\boldsymbol{v}\,dt\,dS,$$

where $\hat{\boldsymbol{n}}\cdot\boldsymbol{v}\,dt$ is the altitude perpendicular to the face dS. The rate of flow of mass through a surface S is therefore

$$\frac{dm}{dt} = \iint_S \rho\hat{\boldsymbol{n}}\cdot\boldsymbol{v}\,dS = \iint_S \hat{\boldsymbol{n}}\cdot(\rho\boldsymbol{v})\,dS. \tag{8.129}$$

If $\hat{\boldsymbol{n}}\cdot\boldsymbol{v}$ is positive, the mass flow across S is in the direction of $\hat{\boldsymbol{n}}$; if $\hat{\boldsymbol{n}}\cdot\boldsymbol{v}$ is negative, the mass flow is in the reverse direction. We see that $\rho\boldsymbol{v}$, the momentum density, is also the mass current, in the sense that its component in any direction gives the rate of mass flow per unit area in that direction. We can now give a further interpretation of Eq. (8.127) by integrating it over a fixed volume V bounded by a surface S with outward normal $\hat{\boldsymbol{n}}$:

$$\iiint_V \frac{\partial\rho}{\partial t}\,dV + \iiint_V \nabla\cdot(\rho\boldsymbol{v})\,dV = 0. \tag{8.130}$$

Since the volume V here is a fixed volume, we can take the time differentiation outside the integral in the first term. If we apply Gauss' divergence theorem to the second integral, we can rewrite this equation:

$$\frac{d}{dt}\iiint_V \rho\,dV = -\iint_S \hat{\boldsymbol{n}}\cdot(\rho\boldsymbol{v})\,dS. \tag{8.131}$$

This equation states that the rate of increase of mass inside the fixed volume V is equal to the negative of the rate of flow of mass outward across the surface. This

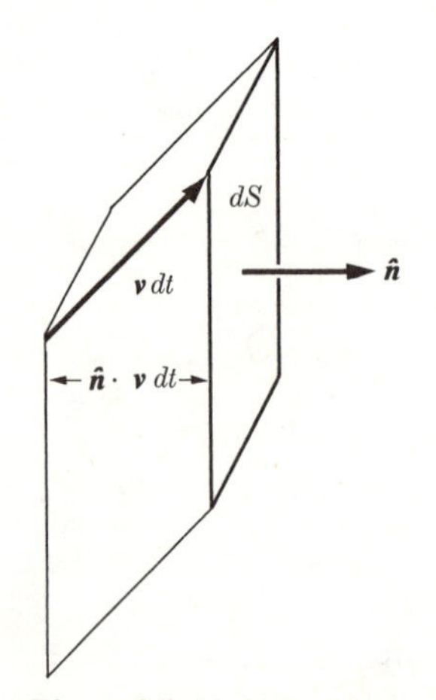

Fig. 8.6 Flow of fluid through a surface element.

result emphasizes the physical interpretation of each term in Eq. (8.127). In particular, the second term evidently represents the rate of flow of mass away from any point. Conversely, by starting with the self-evident equation (8.131) and working backwards, we have an independent derivation of Eq. (8.127).

Equations analogous to Eqs. (8.126), (8.127), (8.129), and (8.131) apply to the density, velocity, and rate of flow of any conserved physical quantity. An equation of the form (8.127) applies, for example, to the flow of electric charge, if ρ is the charge density and $\rho\boldsymbol{v}$ the electric current density.

The curl of velocity $\nabla \times \boldsymbol{v}$ is a concept which is useful in describing fluid flow. To understand its meaning, we compute the integral of the normal component of curl $\boldsymbol{v}$ across a surface S bounded by a curve C. By Stokes' theorem (3.117), this is

$$\iint_S \hat{\boldsymbol{n}}\cdot(\nabla \times \boldsymbol{v})\, dS = \int_C \boldsymbol{v}\cdot d\boldsymbol{r}, \tag{8.132}$$

where the line integral is taken around C in the positive sense relative to the normal $\hat{\boldsymbol{n}}$, as previously defined. If the curve C surrounds a vortex in the fluid, so that $\boldsymbol{v}$ is parallel to $d\boldsymbol{r}$ around C (Fig. 8.7), then the line integral on the right is positive and measures, in a sense, the rate at which the fluid is whirling around the vortex. Thus $\nabla \times \boldsymbol{v}$ is a sort of measure of the rate of rotation of the fluid per unit area; hence the name "curl $\boldsymbol{v}$." Curl $\boldsymbol{v}$ has a nonzero value in the neighborhood of a vortex in the fluid. Curl $\boldsymbol{v}$ may also be nonzero, however, in regions where there is no vortex, that is, where the fluid does not actually circle a point, provided there is a transverse velocity gradient. Figure 8.7 illustrates the two cases. In each case, the line integral of $\boldsymbol{v}$ counterclockwise around the circle C will have a positive value. If the curl of $\boldsymbol{v}$ is zero everywhere in a moving fluid, the flow is said to be *irrotational*. Irrotational flow is important chiefly because it presents fairly simple mathematical problems. If at any point $\nabla \times \boldsymbol{v} = \boldsymbol{0}$, then an element of fluid at that point will have no net angular velocity about that point, although its shape and size may be changing.

We arrive at a more precise meaning of curl $\boldsymbol{v}$ by introducing a coordinate system rotating with angular velocity $\boldsymbol{\omega}$. If $\boldsymbol{v}'$ designates the velocity of the fluid relative to the rotating system, then by Eq. (7.33),

$$\boldsymbol{v} = \boldsymbol{v}' + \boldsymbol{\omega} \times \boldsymbol{r},$$

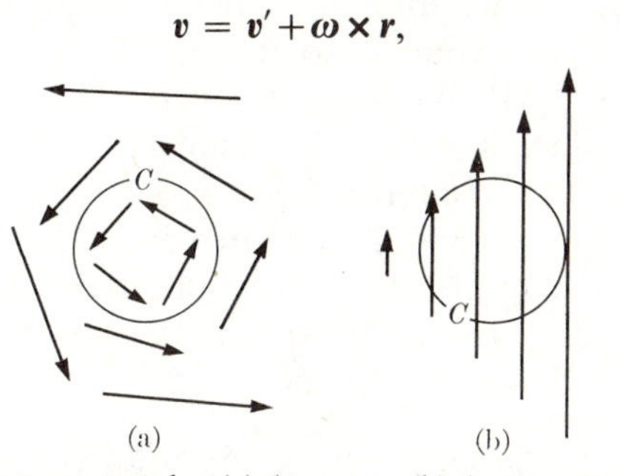

Fig. 8.7 Meaning of nonzero curl $\boldsymbol{v}$. (a) A vortex. (b) A transverse velocity gradient.

where $\boldsymbol{r}$ is a vector from the axis of rotation (whose location does not matter in this discussion) to a point in the fluid. Curl $\boldsymbol{v}$ is now

$$\begin{aligned}\nabla \times \boldsymbol{v} &= \nabla \times \boldsymbol{v}' + \nabla \times (\boldsymbol{\omega} \times \boldsymbol{r}) \\ &= \nabla \times \boldsymbol{v}' + \boldsymbol{\omega} \nabla \cdot \boldsymbol{r} - \boldsymbol{\omega} \cdot \nabla \boldsymbol{r} \\ &= \nabla \times \boldsymbol{v}' + 3\boldsymbol{\omega} - \boldsymbol{\omega} \\ &= \nabla \times \boldsymbol{v}' + 2\boldsymbol{\omega},\end{aligned}$$

where the second line follows from Eq. (3.35) for the triple cross product, and the third line by direct calculation of the components in the second and third terms. If at some point P in the fluid,

$$\boldsymbol{\omega} = \tfrac{1}{2}\nabla \times \boldsymbol{v}, \tag{8.133}$$

then at that point

$$\nabla \times \boldsymbol{v}' = \boldsymbol{0}. \tag{8.134}$$

Thus if $\nabla \times \boldsymbol{v} \neq \boldsymbol{0}$ at a point P, then in a coordinate system rotating with angular velocity $\boldsymbol{\omega} = \frac{1}{2}\nabla \times \boldsymbol{v}$, the fluid flow is irrotational at the point P. We may therefore interpret $\frac{1}{2}\nabla \times \boldsymbol{v}$ as the angular velocity of the fluid near any point. If $\nabla \times \boldsymbol{v}$ is constant, then it is possible to introduce a rotating coordinate system in which the flow is irrotational everywhere.

8.7 EQUATIONS OF MOTION FOR AN IDEAL FLUID

For the remainder of this chapter, except in the last section, we shall consider the motion of an ideal fluid, that is, one in which there are no shearing stresses, even when the fluid is in motion. The stress within an ideal fluid consists in a pressure p alone. This is a much greater restriction in the case of moving fluids than in the case of fluids in equilibrium (Section 5.11). A fluid, by definition, supports no shearing stress when in equilibrium, but all fluids have some viscosity and therefore there are always some shearing stresses between layers of fluid in relative motion. An ideal fluid would have no viscosity, and our results for ideal fluids will therefore apply only when the viscosity is negligible.

Let us suppose that, in addition to the pressure, the fluid is acted on by a body force of density $\boldsymbol{f}$ per unit volume, so that the body force acting on a volume element δV of fluid is $\boldsymbol{f}\,\delta V$. We need to calculate also the force density due to pressure. Let us consider a volume element $\delta V = \delta x\ \delta y\ \delta z$ in the form of a rectangular box (Fig. 8.8). The force due to pressure on the left face of the box is $p\ \delta y\ \delta z$, and acts in the x-direction. The force due to pressure on the right face of the box is also $p\ \delta y\ \delta z$, and acts in the opposite direction. Hence the net x-component of force δF_x on the box depends upon the difference in pressure between the left and right faces of the box:

$$\delta F_x = \left(-\frac{\partial p}{\partial x}\delta x\right)\delta y\ \delta z. \tag{8.135}$$

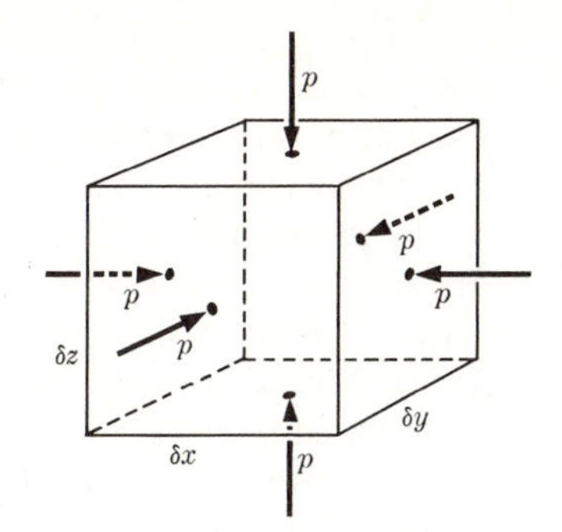

Fig. 8.8 Force on a volume element due to pressure.

Similar expressions may be derived for the components of force in the y- and z-directions. The total force on the fluid in the box due to pressure is then

$$\delta \boldsymbol{F} = \left(-\hat{\boldsymbol{x}}\frac{\partial p}{\partial x} - \hat{\boldsymbol{y}}\frac{\partial p}{\partial y} - \hat{\boldsymbol{z}}\frac{\partial p}{\partial z}\right)\delta V$$

$$= -\nabla p\, \delta V. \tag{8.136}$$

The force density per unit volume due to pressure is therefore $-\nabla p$. This result was also obtained in Section 5.11 [Eq. (5.172)].

We can now write the equation of motion for a volume element δV of fluid:

$$\rho\, \delta V \frac{d\boldsymbol{v}}{dt} = \boldsymbol{f}\, \delta V - \nabla p\, \delta V. \tag{8.137}$$

This equation is usually written in the form

$$\rho \frac{d\boldsymbol{v}}{dt} + \nabla p = \boldsymbol{f}. \tag{8.138}$$

By making use of the relation (8.113), we may rewrite this in terms of derivatives at a fixed point:

$$\frac{\partial \boldsymbol{v}}{\partial t} + \boldsymbol{v}\cdot\nabla\boldsymbol{v} + \frac{1}{\rho}\nabla p = \frac{\boldsymbol{f}}{\rho}, \tag{8.139}$$

where $\boldsymbol{f}/\rho$ is the body force per unit mass. This is Euler's equation of motion for a moving fluid.

If the density ρ depends only on the pressure p, we shall call the fluid *homogeneous*. This definition does not imply that the density is uniform. An incompressible fluid is homogeneous if its density is uniform. A compressible fluid of uniform chemical composition and uniform temperature throughout is homogeneous. When a fluid expands or contracts under the influence of pressure changes, work is done by or on the fluid, and part of this work may appear in the form of heat. If the changes in density occur sufficiently slowly so that there is adequate time for heat flow to maintain the temperature uniform throughout the fluid, the fluid may be considered homogeneous within the meaning of our definition. The relation between density and pressure is then determined by the equation of state of the fluid or by its isothermal bulk modulus (Section 5.11). In some cases, changes in density occur so rapidly that there is no time for any appreciable flow of heat. In such cases the fluid may also be considered homogeneous, and the adiabatic relation between density and pressure or the adiabatic bulk modulus should be used. In cases between these two extremes, the density will depend not only on pressure, but also on temperature, which, in turn, depends upon the rate of heat flow between parts of the fluid at different temperatures.

In a homogeneous fluid, there are four unknown functions to be determined at each point in space and time, the three components of velocity $\boldsymbol{v}$, and the pressure p. We have, correspondingly, four differential equations to solve, the three components of the vector equation of motion (8.139), and the equation of continuity (8.127). The only other quantities appearing in these equations are the body force, which is assumed to be given, and the density ρ, which can be expressed as a function of the pressure. Of course, Eqs. (8.139) and (8.127) have a tremendous variety of solutions. In a specific problem we would need to know the conditions at the boundary of the region in which the fluid is moving and the values of the functions $\boldsymbol{v}$ and p at some initial instant. In the following sections, we shall confine our attention to homogeneous fluids. In the intermediate case mentioned at the end of the last paragraph, where the fluid is inhomogeneous and the density depends on both pressure and temperature, we have an additional unknown function, the temperature, and we will need an additional equation determined by the law of heat flow. We shall not consider this case, although it is a very important one in many problems.

8.8 CONSERVATION LAWS FOR FLUID MOTION

Inasmuch as the laws of fluid motion are derived from Newton's laws of motion, we may expect that appropriate generalizations of the conservation laws of momentum, energy, and angular momentum also hold for fluid motion. We have already had an example of a conservation law for fluid motion, namely, the equation of continuity [Eq. (8.127) or (8.131)], which expresses the law of conservation of mass. Mass is conserved also in particle mechanics, but we did not find it necessary to write an equation expressing this fact.

A conservation law in fluid mechanics may be written in many equivalent forms. It will be instructive to study some of these in order to get a clearer idea of the physical meaning of the various mathematical expressions involved. Let ρ be the density of any physical quantity: mass, momentum, energy, or angular momentum. Then the simplest form of the conservation law for this quantity will be Eq. (8.125), which states that the amount of this quantity in an element δV of fluid remains constant. If the quantity in question is being produced at a rate Q per unit volume, then Eq. (8.125) should be generalized:

$$\frac{d}{dt}(\rho\,\delta V) = Q\,\delta V. \tag{8.140}$$

This is often called a conservation law for the quantity ρ, and we will use this terminology here. A conservation law gives us an accounting of where the quantity ρ is coming from and where it is going. It states that this quantity is appearing in the fluid at a rate Q per unit volume, or disappearing if Q is negative. In the sense in which we have used the term in Chapter 4, this would not be called a conservation law except when $Q = 0$. By a derivation exactly like that which led to Eq. (8.127), we can rewrite Eq. (8.140) as a partial differential equation:

$$\frac{\partial \rho}{\partial t} + \nabla\cdot(\rho \boldsymbol{v}) = Q. \tag{8.141}$$

This is probably the most useful form of conservation law. The meaning of the terms in Eq. (8.141) is brought out by integrating each term over a fixed volume V and using Gauss' theorem,* as in the derivation of Eq. (8.131):

$$\frac{d}{dt}\iiint_V \rho\,dV + \iint_S \hat{\boldsymbol{n}}\cdot\boldsymbol{v}\rho\,dS = \iiint_V Q\,dV. \tag{8.142}$$

According to the discussion preceding Eq. (8.129), this equation states that the rate of increase of the quantity within V, plus the rate of flow outward across the boundary S, equals the rate of appearance due to sources within V. Another form of the conservation law which is sometimes useful is obtained by summing equation (8.140) over a volume V moving with the fluid:

$$\sum \frac{d}{dt}(\rho\,\delta V) = \frac{d}{dt}\sum \rho\,\delta V = \sum Q\,\delta V. \tag{8.143}$$

If we pass to the limit $\delta V \to 0$, the summations become integrations:

$$\frac{d}{dt}\iiint_V \rho\,dV = \iiint_V Q\,dV. \tag{8.144}$$

The surface integral which appears in the left member of Eq. (8.142) does not appear in Eq. (8.144); since the volume V moves with the fluid, there is no flow across its boundary. Since Eqs. (8.140), (8.141), (8.142), and (8.144) are all equivalent, it is sufficient to derive a conservation law in any one of these forms. The others then follow. Usually it is easiest to derive an equation of the form (8.140), starting with the equation of motion in the form (8.138). We can also start with Eq. (8.139) and derive a conservation equation in the form (8.141), but a bit more manipulation is usually required.

*If ρ is a vector, as in the case of linear or angular momentum density, then a generalized form of Gauss' theorem [mentioned in Section 5.11 in connection with Eq. (5.178)] must be used.

In order to derive a conservation law for linear momentum, we first note that the momentum in a volume element δV is $\rho \boldsymbol{v}\, \delta V$. The momentum density per unit volume is therefore $\rho \boldsymbol{v}$, and this quantity will play the role played by ρ in the discussion of the preceding paragraph. In order to obtain an equation analogous to Eq. (8.140), we start with the equation of motion in the form (8.138), which refers to a point moving with the fluid, and multiply through by the volume δV of a small fluid element:

$$\rho\, \delta V \frac{d\boldsymbol{v}}{dt} + \nabla p\, \delta V = \boldsymbol{f}\, \delta V. \tag{8.145}$$

Since $\rho\, \delta V = \delta m$ is constant, we may include it in the time derivative:

$$\frac{d}{dt}(\rho \boldsymbol{v}\, \delta V) = (\boldsymbol{f} - \nabla p)\, \delta V. \tag{8.146}$$

The momentum of a fluid element, unlike its mass, is not, in general, constant. This equation states that the time rate of change of momentum of a moving fluid element is equal to the body force plus the force due to pressure acting upon it. The quantity $\boldsymbol{f} - \nabla p$ here plays the role of Q in the preceding general discussion. Equation (8.146) can be rewritten in any of the forms (8.141), (8.142), and (8.144). For example, we may write it in the form (8.144):

$$\frac{d}{dt} \iiint_V \rho \boldsymbol{v}\, dV = \iiint_V \boldsymbol{f}\, dV - \iiint_V \nabla p\, dV. \tag{8.147}$$

We can now apply the generalized form of Gauss' theorem [Eq. (5.178)] to the second term on the right, to obtain

$$\frac{d}{dt} \iiint_V \rho \boldsymbol{v}\, dV = \iiint_V \boldsymbol{f}\, dV + \iint_S -\hat{\boldsymbol{n}} p\, dS, \tag{8.148}$$

where S is the surface bounding V.

This equation states that the time rate of change of the total linear momentum in a volume V of moving fluid is equal to the total external force acting on it. This result is an immediate generalization of the linear momentum theorem (4.7) for a system of particles. The internal forces, in the case of a fluid, are represented by the pressure within the fluid. By the application of Gauss' theorem, we have eliminated the pressure within the volume V, leaving only the external pressure across the surface of V. It may be asked how we have managed to eliminate the internal forces without making explicit use of Newton's third law, since Eq. (8.138), from which we started, is an expression only of Newton's first two laws. The answer is that the concept of pressure itself contains Newton's third law implicitly, since the force due to pressure exerted from left to right across any surface element is equal and opposite to the force exerted from right to left across the same surface element. Furthermore, the points of application of these two forces are the same, namely, at the surface element. Both forces necessarily have the same line of action, and there is no distinction between the weak and strong forms of Newton's third law. The internal pressures will therefore also be expected to cancel out in the equation for the time rate of change of angular momentum. A similar remark applies to the forces due to any kind of stresses in a fluid or a solid; Newton's third law in strong form is implicitly contained in the concept of stress.

Equations representing the conservation of angular momentum, analogous term by term with Eqs. (8.140) through (8.144), can be derived by taking the cross product of the vector $\boldsymbol{r}$ with either Eq. (8.138) or (8.139), and suitably manipulating the terms. The vector $\boldsymbol{r}$ is here the vector from the origin about which moments are to be computed to any point in the moving fluid or in space. This development is left as an exercise. The law of conservation of angular momentum is responsible for the vortices formed when a liquid flows out through a small hole in the bottom of a tank. The only body force here is gravity, which exerts no torque about the hole, and it can be shown that if the pressure is constant, or depends only on vertical depth, there is no net vertical component of torque across any closed surface due to pressure. Therefore the angular momentum of any part of the fluid remains constant. If a fluid element has any angular momentum at all initially, when it is some distance from the hole, its angular velocity will have to increase in inverse proportion to the square of its distance from the hole in order for its angular momentum to remain constant as it approaches the hole.

In order to derive a conservation equation for the energy, we take the dot product of $\boldsymbol{v}$ with Eq. (8.146), to obtain

$$\frac{d}{dt}(\tfrac{1}{2}\rho v^2\,\delta V) = \boldsymbol{v}\cdot(\boldsymbol{f}-\nabla p)\,\delta V. \tag{8.149}$$

This is the energy theorem in the form (8.140). In place of the density ρ, we have here the kinetic energy density $\frac{1}{2}\rho v^2$. The rate of production of kinetic energy per unit volume is

$$Q = \boldsymbol{v}\cdot(\boldsymbol{f}-\nabla p). \tag{8.150}$$

In analogy with our procedure in particle mechanics, we shall now try to define additional forms of energy so as to include as much as possible of the right member of Eq. (8.149) under the time derivative on the left. We can see how to rewrite the second term on the right by making use of Eqs. (8.113) and (8.116):

$$\begin{aligned}\frac{d}{dt}(p\,\delta V) &= \frac{dp}{dt}\delta V + p\frac{d\,\delta V}{dt}\\ &= \frac{\partial p}{\partial t}\delta V + \boldsymbol{v}\cdot\nabla p\,\delta V + p\nabla\cdot\boldsymbol{v}\,\delta V,\end{aligned} \tag{8.151}$$

so that

$$-\boldsymbol{v}\cdot\nabla p\,\delta V = -\frac{d}{dt}(p\,\delta V) + \frac{\partial p}{\partial t}\delta V + p\nabla\cdot\boldsymbol{v}\,\delta V. \tag{8.152}$$

Let us now assume that the body force $\boldsymbol{f}$ is a gravitational force:

$$\boldsymbol{f} = \rho \boldsymbol{g} = \rho \nabla \mathcal{G}, \tag{8.153}$$

where $\mathcal{G}$ is the gravitational potential [Eq. (6.16)], i.e., the negative potential energy per unit mass due to gravitation. The first term on the right in Eq. (8.149) is then

$$\begin{aligned} \boldsymbol{v}\cdot\boldsymbol{f}\,\delta V &= (\boldsymbol{v}\cdot\nabla\mathcal{G})\rho\,\delta V = \left(\frac{d\mathcal{G}}{dt} - \frac{\partial\mathcal{G}}{\partial t}\right)\rho\,\delta V \\ &= \frac{d}{dt}(\rho\mathcal{G}\,\delta V) - \rho\frac{\partial\mathcal{G}}{\partial t}\,\delta V, \end{aligned} \tag{8.154}$$

since $\rho\,\delta V = \delta m$ is constant. With the help of Eqs. (8.152) and (8.154), Eq. (8.149) can be rewritten:

$$\frac{d}{dt}[(\tfrac{1}{2}\rho v^2 + p - \rho\mathcal{G})\,\delta V] = \left(\frac{\partial p}{\partial t} - \rho\frac{\partial\mathcal{G}}{\partial t}\right)\delta V + p\nabla\cdot\boldsymbol{v}\,\delta V. \tag{8.155}$$

The pressure p here plays the role of a potential energy density whose negative gradient gives the force density due to pressure [Eq. (8.136)]. The time rate of change of kinetic energy plus gravitational potential energy plus potential energy due to pressure is equal to the expression on the right.

Ordinarily, the gravitational field at a fixed point in space will not change with time (except perhaps in applications to motions of gas clouds in astronomical problems). If the pressure at a given point in space is constant also, then the first term on the right vanishes. What is the significance of the second term? For an incompressible fluid, $\nabla\cdot\boldsymbol{v} = 0$, and the second term would vanish also. We therefore suspect that it represents energy associated with compression and expansion of the fluid element δV. Let us check this hypothesis by calculating the work done in changing the volume of the element δV. The work dW done by the fluid element δV, through the pressure which it exerts on the surrounding fluid when it expands by an amount $d\,\delta V$, is

$$dW = p\,d\,\delta V. \tag{8.156}$$

The rate at which energy is supplied by the expansion of the fluid element is, by Eq. (8.116),

$$\frac{dW}{dt} = p\frac{d\,\delta V}{dt} = p\nabla\cdot\boldsymbol{v}\,\delta V, \tag{8.157}$$

which is just the last term in Eq. (8.155). So far, all our conservation equations are valid for any problem involving ideal fluids. If we restrict ourselves to homogeneous fluids, that is, fluids whose density depends only on the pressure, we can define a potential energy associated with the expansion and contraction of the fluid element δV. We shall define the potential energy $u\,\delta m$ of the fluid element

δV as the negative work done through its pressure on the surrounding fluid when the pressure changes from a standard pressure p_0 to any pressure p. The potential energy per unit mass u will then be a function of p:

$$u\,\delta m = -\int_{p_0}^{p} p\,d\,\delta V. \tag{8.158}$$

The volume $\delta V = \delta m/\rho$ is a function of pressure, and we may rewrite this in various forms:

$$\begin{aligned} u &= \int_{p_0}^{p} \frac{p\,d\rho}{\rho^2} \\ &= \int_{p_0}^{p} \frac{p}{\rho^2}\frac{d\rho}{dp}\,dp \\ &= \int_{p_0}^{p} \frac{p}{\rho B}\,dp, \end{aligned} \tag{8.159}$$

where the last step makes use of the definition of the bulk modulus [Eq. (5.116)]. The time rate of change of u is, by Eqs. (8.158) or (8.159) and (8.116),

$$\frac{d(u\,\delta m)}{dt} = \frac{-p\,d\,\delta V}{dt} = -p\boldsymbol{\nabla}\cdot\boldsymbol{v}\,\delta V. \tag{8.160}$$

We can now include the last term on the right in Eq. (8.155) under the time derivative on the left:

$$\frac{d}{dt}[(\tfrac{1}{2}\rho v^2 + p - \rho\mathcal{G} + \rho u)\,\delta V] = \left(\frac{\partial p}{\partial t} - \rho\frac{\partial \mathcal{G}}{\partial t}\right)\delta V. \tag{8.161}$$

The interpretation of this equation is clear from the preceding discussion. It can be rewritten in any of the forms (8.141), (8.142), and (8.144).

If p and $\mathcal{G}$ are constant at any fixed point in space, then the total kinetic plus potential energy of a fluid element remains constant as it moves along. It is convenient to divide by $\delta m = \rho\,\delta V$ in order to eliminate reference to the volume element:

$$\frac{d}{dt}\left(\frac{v^2}{2} + \frac{p}{\rho} - \mathcal{G} + u\right) = \frac{1}{\rho}\frac{\partial p}{\partial t} - \frac{\partial \mathcal{G}}{\partial t}. \tag{8.162}$$

This is Bernoulli's theorem. The term $\partial\mathcal{G}/\partial t$ is practically always zero; we have kept it merely to make clear the meaning of the term $(1/\rho)\,(\partial p/\partial t)$, which plays a similar role and is not always zero. When both terms on the right are zero, as in the case of steady flow, we have, for a point moving along with the fluid,

$$\frac{v^2}{2} + \frac{p}{\rho} - \mathcal{G} + u = \text{a constant}. \tag{8.163}$$

Other things being equal, that is if u, $\mathcal{G}$, and ρ are constant, the pressure of a moving fluid decreases as the velocity increases. For an incompressible fluid, ρ and u are necessarily constant.

The conservation laws of linear and angular momentum apply not only to ideal fluids, but also, when suitably formulated, to viscous fluids and even to solids, in view of the remarks made above regarding Newton's third law and the concept of stress. The law of conservation of energy (8.162) will not apply, however, to viscous fluids, unless conversion of mechanical to heat energy by viscous friction is included in the law, since the viscosity is due to an internal friction which results in a loss of kinetic and potential energies.

8.9 STEADY FLOW

By steady flow of a fluid we mean a motion of the fluid in which all quantities associated with the fluid, velocity, density, pressure, force density, etc., are constant in time at any given point in space. For steady flow, all partial derivatives with respect to time can be set equal to zero. The total time derivative, which designates the time rate of change of a quantity relative to a point moving with the fluid, will not in general be zero, but, by Eq. (8.113), will be

$$\frac{d}{dt} = \boldsymbol{v}\cdot\boldsymbol{\nabla}. \tag{8.164}$$

The path traced out by any fluid element as it moves along is called a *streamline*. A streamline is a line which is parallel at each point (x, y, z) to the velocity $\boldsymbol{v}(x, y, z)$ at that point. The entire space within which the fluid is flowing can be filled with streamlines such that through each point there passes one and only one streamline. If we introduce along any streamline a coordinate s which represents the distance measured along the streamline from some fixed point, we can regard any quantity associated with the fluid as a function of s along the streamline. The component of the symbol $\boldsymbol{\nabla}$ along the streamline at any point is d/ds, as we see if we choose a coordinate system whose x-axis is directed along the streamline at that point. Equation (8.164) can therefore be rewritten:

$$\frac{d}{dt} = v\frac{d}{ds}. \tag{8.165}$$

This equation is also evident from the fact that $v = ds/dt$. For example, Eq. (8.162), in the case of steady flow, can be written:

$$\frac{d}{ds}\left(\frac{v^2}{2}+\frac{p}{\rho}-\mathcal{G}+u\right) = 0. \tag{8.166}$$

The quantity in parentheses is therefore constant along a streamline.

The equation of continuity (8.127) in the case of steady flow becomes

$$\boldsymbol{\nabla}\cdot(\rho\boldsymbol{v}) = 0. \tag{8.167}$$

If we integrate this equation over a fixed volume V, and apply Gauss' theorem, we have

$$\iint_S \hat{\boldsymbol{n}}\cdot(\rho\boldsymbol{v})\,dS = 0, \tag{8.168}$$

where S is the closed surface bounding V. This equation simply states that the total mass flowing out of any closed surface is zero.

If we consider all the streamlines which pass through any (open) surface S, these streamlines form a tube, called a *tube of flow* (Fig. 8.9). The walls of a tube of flow are everywhere parallel to the streamlines, so that no fluid enters or leaves it. A surface S which is drawn everywhere perpendicular to the streamlines and through which passes each streamline in a tube of flow, will be called a *cross section* of the tube. If we apply Eq. (8.168) to the closed surface bounded by the walls of a tube of flow and two cross sections S_1 and S_2, then since $\hat{\boldsymbol{n}}$ is perpendicular to $\boldsymbol{v}$ over the walls of the tube, and $\hat{\boldsymbol{n}}$ is parallel or antiparallel to $\boldsymbol{v}$ over the cross sections, we have

$$\iint_{S_1} \rho v\,dS - \iint_{S_2} \rho v\,dS = 0, \tag{8.169}$$

or

$$\iint_S \rho v\,dS = I = \text{a constant}, \tag{8.170}$$

where S is any cross section along a given tube of flow. The constant I is called the *fluid current* through the tube.

The energy conservation equation (8.161), when rewritten in the form (8.141), becomes, in the case of steady flow,

$$\boldsymbol{\nabla}\cdot[(\tfrac{1}{2}\rho v^2 + p - \rho\mathcal{G} + \rho u)\boldsymbol{v}] = 0. \tag{8.171}$$

This equation has the same form as Eq. (8.167), and we can conclude in the same way that the energy current is the same through any cross section S of a tube of flow:

$$\iint_S (\tfrac{1}{2}\rho v^2 + p - \rho\mathcal{G} + \rho u)v\,dS = \text{a constant}. \tag{8.172}$$

This result is closely related to Eq. (8.166).

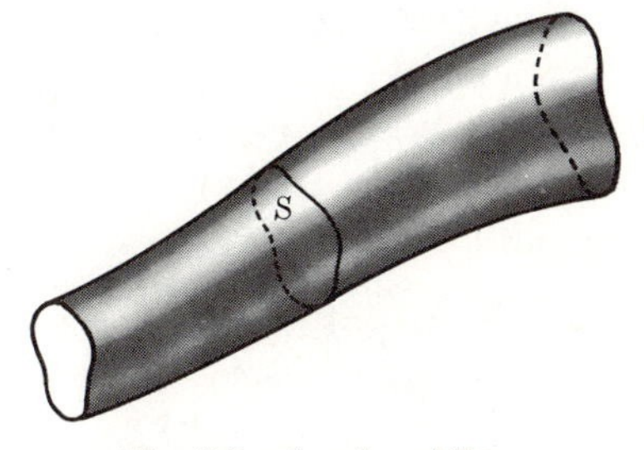

Fig. 8.9 A tube of flow.

If the flow is not only steady, but also irrotational, then

$$\nabla \times \boldsymbol{v} = \mathbf{0} \tag{8.173}$$

everywhere. This equation is analogous in form to Eq. (3.189) for a conservative force, and we can proceed as in Section 3.12 to show that if Eq. (8.173) holds, it is possible to define a *velocity potential function* $\phi(x, y, z)$ by the equation

$$\phi(\boldsymbol{r}) = \int_{\boldsymbol{r}_s}^{\boldsymbol{r}} \boldsymbol{v} \cdot d\boldsymbol{r}, \tag{8.174}$$

where $\boldsymbol{r}_s$ is any fixed point. The velocity at any point will then be

$$\boldsymbol{v} = \nabla \phi. \tag{8.175}$$

Substituting this in Eq. (8.167), we have an equation to be solved for ϕ:

$$\nabla \cdot (\rho \nabla \phi) = 0. \tag{8.176}$$

In the cases usually studied, the fluid can be considered incompressible, and this becomes

$$\nabla^2 \phi = 0. \tag{8.177}$$

This equation is identical in form with Laplace's equation (6.35) for the gravitational potential in empty space. Hence the techniques of potential theory may be used to solve problems involving irrotational flow of an incompressible fluid.

8.10 SOUND WAVES

Let us assume a fluid at rest with pressure p_0, density ρ_0, in equilibrium under the action of a body force $\boldsymbol{f}_0$, constant in time. Equation (8.139) then becomes

$$\frac{1}{\rho_0} \nabla p_0 = \frac{\boldsymbol{f}_0}{\rho_0}. \tag{8.178}$$

We may note that this equation agrees with Eq. (5.172) deduced in Section 5.11 for a fluid in equilibrium. Let us now suppose that the fluid is subject to a small disturbance, so that the pressure and density at any point become

$$p = p_0 + p', \tag{8.179}$$

$$\rho = \rho_0 + \rho', \tag{8.180}$$

where $p' \ll p$ and $\rho' \ll \rho$. We assume that the resulting velocity $\boldsymbol{v}$ and its space and time derivatives are everywhere very small. If we substitute Eqs. (8.179) and (8.180) in the equation of motion (8.139), and neglect higher powers than the first of p', ρ', $\boldsymbol{v}$ and their derivatives, making use of Eq. (8.178), we obtain

$$\frac{\partial \boldsymbol{v}}{\partial t} = -\frac{1}{\rho_0} \nabla p'. \tag{8.181}$$

Making a similar substitution in Eq. (8.127), we obtain

$$\frac{\partial \rho'}{\partial t} = -\rho_0 \nabla \cdot \boldsymbol{v} - \boldsymbol{v} \cdot \nabla \rho_0. \tag{8.182}$$

Let us assume that the equilibrium density ρ_0 is uniform, or nearly so, so that $\nabla\rho_0$ is zero or very small, and the second term can be neglected.

The pressure increment p' and density increment ρ' are related by the bulk modulus according to Eq. (5.183):

$$\frac{\rho'}{\rho_0} = \frac{p'}{B}. \tag{8.183}$$

This equation may be used to eliminate either ρ' *or* p' from Eqs. (8.181) and (8.182). Let us eliminate ρ' from Eq. (8.182):

$$\frac{\partial p'}{\partial t} = -B\nabla\cdot\boldsymbol{v}. \tag{8.184}$$

Equations (8.181) and (8.184) are the fundamental differential equations for sound waves. The analogy with the form (8.101) for one-dimensional waves is apparent. Here again we have two quantities, p' and $\boldsymbol{v}$, such that the time derivative of either is proportional to the space derivatives of the other. In fact, if $\boldsymbol{v} = \hat{\boldsymbol{x}}v_x$ and if v_x and p' are functions of x alone, then Eqs. (8.181) and (8.184) reduce to Eqs. (8.101).

We may proceed, in analogy with the discussion in Section 8.5, to eliminate either $\boldsymbol{v}$ or p' from these equations. In order to eliminate $\boldsymbol{v}$, we take the divergence of Eq. (8.181) and interchange the order of differentiation, again assuming ρ_0 nearly uniform:

$$\frac{\partial}{\partial t}(\nabla\cdot\boldsymbol{v}) = -\frac{1}{\rho_0}\nabla^2 p'. \tag{8.185}$$

We now differentiate Eq. (8.184) with respect to t, and substitute from Eq. (8.185):

$$\nabla^2 p' - \frac{1}{c^2}\frac{\partial^2 p'}{\partial t^2} = 0, \tag{8.186}$$

where

$$c = \left(\frac{B}{\rho_0}\right)^{1/2}. \tag{8.187}$$

This is the three-dimensional wave equation, as we shall show presently. Formula (8.187) for the speed of sound waves was first derived by Isaac Newton, and applies either to liquids or gases. For gases, Newton assumed that the isothermal bulk modulus $B = p$ should be used, but Eq. (8.187) does not then agree with the experimental values for the speed of sound. The sound vibrations are so rapid that they should be treated as adiabatic, and the adiabatic bulk modulus $B = \gamma p$ should be used, where γ is the ratio of specific heat at constant pressure to that at constant volume. Formula (8.187) then agrees with the experimental values of c.

If we eliminate p' by a similar process, we obtain a wave equation for $\boldsymbol{v}$:

$$\nabla^2\boldsymbol{v} - \frac{1}{c^2}\frac{\partial^2\boldsymbol{v}}{\partial t^2} = \boldsymbol{0}. \tag{8.188}$$

In deriving Eq. (8.188), it is necessary to use the fact that $\nabla \times (\nabla \times \boldsymbol{v}) = \mathbf{0}$. It follows from Eq. (8.181) that $\nabla \times \boldsymbol{v}$ is in any case independent of time, so that the time-dependent part of $\boldsymbol{v}$ which is present in a sound wave is irrotational. [We could add to the sound wave a small steady flow with $\nabla \times \boldsymbol{v} \neq \mathbf{0}$, without violating Eqs. (8.181) and (8.182). Equation (8.188) would then hold only for the time-dependent part of the velocity.]

In order to show that Eq. (8.186) leads to sound waves traveling with speed c, we note first that if p' is a function of x and t alone, Eq. (8.186) becomes

$$\frac{\partial^2 p'}{\partial x^2} - \frac{1}{c^2}\frac{\partial^2 p'}{\partial t^2} = 0. \tag{8.189}$$

This is of the same form as the one-dimensional wave equation (8.6), and therefore has solutions of the form

$$p' = f(x - ct). \tag{8.190}$$

This is called a *plane wave*, for at any time t the phase $x - ct$ and the pressure p' are constant along any plane (x = a constant) parallel to the yz-plane. A plane wave traveling in the direction of the unit vector $\hat{\boldsymbol{n}}$ will be given by

$$p' = f(\hat{\boldsymbol{n}} \cdot \boldsymbol{r} - ct), \tag{8.191}$$

where $\boldsymbol{r}$ is the position vector from the origin to any point in space. To see that this is a wave in the direction $\hat{\boldsymbol{n}}$, we rotate the coordinate system until the x-axis lies in this direction, in which case Eq. (8.191) reduces to Eq. (8.190). The planes f = a constant, at any time t, are now perpendicular to $\hat{\boldsymbol{n}}$, and travel in the direction of $\hat{\boldsymbol{n}}$ with velocity c. We can see from the argument just given that the solution (8.191) must satisfy Eq. (8.186), or we may verify this by direct computation, for any coordinate system:

$$\nabla p' = \frac{df}{d\xi}\nabla\xi = \frac{df}{d\xi}\hat{\boldsymbol{n}}, \tag{8.192}$$

where

$$\xi = \hat{\boldsymbol{n}} \cdot \boldsymbol{r} - ct, \tag{8.193}$$

and, similarly,

$$\nabla^2 p' = \frac{d^2 f}{d\xi^2}\hat{\boldsymbol{n}} \cdot \nabla\xi = \frac{d^2 f}{d\xi^2}\hat{\boldsymbol{n}} \cdot \hat{\boldsymbol{n}} = \frac{d^2 f}{d\xi^2}, \tag{8.194}$$

$$\frac{\partial^2 p'}{\partial t^2} = \frac{d^2 f}{d\xi^2}\left(\frac{\partial \xi}{\partial t}\right)^2 = c^2 \frac{d^2 f}{d\xi^2}, \tag{8.195}$$

so that Eq. (8.186) is satisfied, no matter what the function $f(\xi)$ may be.

Equation (8.188) will also have plane wave solutions:

$$\boldsymbol{v} = \boldsymbol{h}(\hat{\boldsymbol{n}}' \cdot \boldsymbol{r} - ct), \tag{8.196}$$

corresponding to waves traveling in the direction $\hat{\boldsymbol{n}}'$ with velocity c, where $\boldsymbol{h}$ is a vector function of $\xi' = \hat{\boldsymbol{n}}' \cdot \boldsymbol{r} - ct$. To any given pressure wave of the form (8.191) will correspond a velocity wave of the form (8.196), related to it by Eqs. (8.181) and (8.182). If we calculate $\partial \boldsymbol{v}/\partial t$ from Eq. (8.196), and $\nabla p'$ from Eq. (8.191), and substitute in Eq. (8.181), we will have

$$\frac{d\boldsymbol{h}}{d\xi'} = \frac{\hat{\boldsymbol{n}}}{(B\rho_0)^{1/2}} \frac{df}{d\xi}. \tag{8.197}$$

Equation (8.197) must hold at all points $\boldsymbol{r}$ at all times t. The right member of this equation is a function of ξ and is constant for a constant ξ. Consequently, the left member must be constant when ξ is constant, and must be a function only of ξ, which implies that $\xi' = \xi$ (or at least that ξ' is a function of ξ), and hence $\hat{\boldsymbol{n}}' = \hat{\boldsymbol{n}}$. This is obvious physically, that the velocity wave must travel in the same direction as the pressure wave. We can now set $\xi' = \xi$, and solve Eq. (8.197) for $\boldsymbol{h}$:

$$\boldsymbol{h} = \frac{\hat{\boldsymbol{n}}}{(B\rho_0)^{1/2}} f, \tag{8.198}$$

where the additive constant is zero, since both p' and $\boldsymbol{v}$ are zero in a region where there is no disturbance. Equations (8.198), (8.196), and (8.190) imply that for a plane sound wave traveling in the direction $\hat{\boldsymbol{n}}$, the pressure increment and velocity are related by the equation

$$\boldsymbol{v} = \frac{p'}{(B\rho_0)^{1/2}} \hat{\boldsymbol{n}}, \tag{8.199}$$

where $\boldsymbol{v}$, of course, is here the velocity of a fluid particle, not that of the wave, which is $c\hat{\boldsymbol{n}}$. The velocity of the fluid particles is along the direction of propagation of the sound wave, so that sound waves in a fluid are longitudinal. This is a consequence of the fact that the fluid will not support a shearing stress, and is not true of sound waves in a solid, which may be either longitudinal or transverse.

A plane wave oscillating harmonically in time with angular frequency ω may be written in the form

$$p' = A \cos(\boldsymbol{k} \cdot \boldsymbol{r} - \omega t) = \operatorname{Re} A e^{i(\boldsymbol{k} \cdot \boldsymbol{r} - \omega t)}, \tag{8.200}$$

where $\boldsymbol{k}$, the *wave vector*, is given by

$$\boldsymbol{k} = \frac{\omega}{c} \hat{\boldsymbol{n}}. \tag{8.201}$$

If we consider a surface perpendicular to $\hat{\boldsymbol{n}}$ which moves back and forth with the fluid as the wave goes by, the work done by the pressure across this surface in the direction of $\hat{\boldsymbol{n}}$ is, per unit area per unit time,

$$P = pv. \tag{8.202}$$

If v oscillates with average value zero, then since $p = p_0 + p'$, where p_0 is constant, the average power is

$$P_{\text{av}} = \langle p'v \rangle_{\text{av}} = \frac{\langle p'^2 \rangle_{\text{av}}}{(\rho_0 B)^{1/2}}, \tag{8.203}$$

where we have made use of Eq. (8.199). This gives the amount of energy per unit area per second traveling in the direction $\hat{\boldsymbol{n}}$.

The three-dimensional wave equation (8.186) has many other solutions corresponding to waves of various forms whose wave fronts (surfaces of constant phase) are of various shapes, and traveling in various directions. As an example, we consider a spherical wave traveling out from the origin. The rate of energy flow is proportional to p'^2 (a small portion of a spherical wave may be considered plane), and we expect that the energy flow per unit area must fall off inversely as the square of the distance, by the energy conservation law. Therefore p' should be inversely proportional to the distance r from the origin. We are hence led to try a wave of the form

$$p' = \frac{1}{r} f(r - ct). \tag{8.204}$$

This will represent a wave of arbitrary time-dependence, whose wave fronts, $\xi = r - ct =$ a constant, are spheres expanding with the velocity c. It can readily be verified by direct computation, using either rectangular coordinates, or using spherical coordinates with the help of Eq. (3.124), that the solution (8.204) satisfies the wave equation (8.186).

A slight difficulty is encountered with the above development if we attempt to apply to a sound wave the expressions for energy flow and mass flow developed in the two preceding sections. The rate of flow of mass per unit area per second, by Eqs. (8.199), (8.180), and (8.183), is

$$\rho \boldsymbol{v} = \rho_0 \left(1 + \frac{p'}{B}\right) \frac{p'}{(\rho_0 B)^{1/2}} \hat{\boldsymbol{n}}.$$

We should expect that $\rho \boldsymbol{v}$ would be an oscillating quantity whose average value is zero for a sound wave, since there should be no net flow of fluid. If we average the above expression, we have

$$\langle \rho \boldsymbol{v} \rangle_{\text{av}} = \frac{\rho_0^{1/2}}{B^{3/2}} (\langle p'^2 \rangle_{\text{av}} + B \langle p' \rangle_{\text{av}}) \hat{\boldsymbol{n}},$$

so that there is a small net flow of fluid in the direction of the wave, unless

$$\langle p' \rangle_{\text{av}} = -\frac{\langle p'^2 \rangle_{\text{av}}}{B}. \tag{8.205}$$

If Eq. (8.205) holds, so that there is no net flow of fluid, then it can be shown that, to second-order terms in p' and v, the energy current density given by Eq. (8.161) is, on the average, for a sound wave,

$$\langle(\tfrac{1}{2}\rho v^2+p-\rho\mathcal{G}+\rho u)\boldsymbol{v}\rangle_{\text{av}} = \frac{\langle p'^2\rangle_{\text{av}}}{(\rho_0 B)^{1/2}}\,\hat{\boldsymbol{n}}, \tag{8.206}$$

in agreement with Eq. (8.203). When approximations are made in the equations of motion, we may expect that the solutions will satisfy the conservation laws only to the same degree of approximation. By adding second-order (or higher) terms like (8.205) to a first-order solution, we can of course satisfy the conservation laws to second-order terms (or higher).

8.11 NORMAL VIBRATIONS OF FLUID IN A RECTANGULAR BOX

The problem of the vibrations of a fluid confined within a rigid box is of interest not only because of its applications to acoustical problems, but also because the methods used can be applied to problems in electromagnetic vibrations, vibrations of elastic solids, wave mechanics, and all phenomena in physics which are described by wave equations. In this section, we consider a fluid confined to a rectangular box of dimensions $L_xL_yL_z$.

We proceed as in the solution of the one-dimensional wave equation in Section 8.2. We first assume a solution of Eq. (8.186) of the form

$$p' = U(x, y, z)\Theta(t). \tag{8.207}$$

Substitution in Eq. (8.186) leads to the equation

$$\frac{1}{U}\nabla^2 U = \frac{1}{c^2\Theta}\frac{d^2\Theta}{dt^2}. \tag{8.208}$$

Again we argue that since the left side depends only on x, y, and z, and the right side only on t, both must be equal to a constant, which we shall call $-\omega^2/c^2$:

$$\frac{d^2\Theta}{dt^2}+\omega^2\Theta = 0, \tag{8.209}$$

$$\nabla^2 U+\frac{\omega^2}{c^2}U = 0. \tag{8.210}$$

The solution of Eq. (8.209) can be written:

$$\Theta = A\cos\omega t+B\sin\omega t, \tag{8.211}$$

or, alternatively,

$$\Theta = Ae^{-i\omega t}, \tag{8.212}$$

where A and B are constant. The form (8.212) leads to traveling waves of the form (8.200). We are concerned here with standing waves, and we therefore choose the form (8.211). In order to solve Eq. (8.210), we again use the method of separation of variables, and assume that

$$U(x, y, z) = X(x)Y(y)Z(z). \tag{8.213}$$

Substitution in Eq. (8.210) leads to the equation

$$\frac{1}{X}\frac{d^2X}{dx^2}+\frac{1}{Y}\frac{d^2Y}{dy^2}+\frac{1}{Z}\frac{d^2Z}{dz^2} = -\frac{\omega^2}{c^2}. \tag{8.214}$$

This can hold for all x, y, z only if each term on the left is constant. We shall call these constants $-k_x^2$, $-k_y^2$, $-k_z^2$, so that

$$\frac{d^2X}{dx^2}+k_x^2X = 0, \qquad \frac{d^2Y}{dy^2}+k_y^2Y = 0, \qquad \frac{d^2Z}{dz^2}+k_z^2Z = 0, \tag{8.215}$$

where

$$k_x^2+k_y^2+k_z^2 = \frac{\omega^2}{c^2}. \tag{8.216}$$

The solutions of Eqs. (8.215) in which we are interested are

$$\begin{aligned} X &= C_x \cos k_x x+D_x \sin k_x x, \\ Y &= C_y \cos k_y y+D_y \sin k_y y, \\ Z &= C_z \cos k_z z+D_z \sin k_z z. \end{aligned} \tag{8.217}$$

If we choose complex exponential solutions for X, Y, Z, and Θ, we arrive at the traveling wave solution (8.200), where k_x, k_y, k_z are the components of the wave vector $\boldsymbol{k}$.

We must now determine the appropriate boundary conditions to be applied at the walls of the box, which we shall take to be the six planes $x = 0$, $x = L_x$, $y = 0$, $y = L_y$, $z = 0$, $z = L_z$. The condition is evidently that the component of velocity perpendicular to the wall must vanish at the wall. At the wall $x = 0$, for example, v_x must vanish. According to Eq. (8.181),

$$\frac{\partial v_x}{\partial t} = -\frac{1}{\rho_0}\frac{\partial p'}{\partial x}. \tag{8.218}$$

We substitute for p' from Eqs. (8.207), (8.211), (8.213), and (8.217):

$$\frac{\partial v_x}{\partial t} = -\frac{k_xYZ}{\rho_0}(A \cos \omega t+B \sin \omega t)(-C_x \sin k_x x+D_x \cos k_x x). \tag{8.219}$$

Integrating, we have

$$v_x = -\frac{k_xYZ}{\omega\rho_0}(A \sin \omega t-B \cos \omega t)(-C_x \sin k_x x+D_x \cos k_x x) \tag{8.220}$$

plus a function of x, y, z, which vanishes, since we are looking for oscillating solutions. In order to ensure that v_x vanishes at $x = 0$, we must set $D_x = 0$, i.e., choose the cosine solution for X in Eq. (8.217). This means that the pressure p' must oscillate at maximum amplitude at the wall. This is perhaps obvious physically, and could have been used instead of the condition $v_x = 0$, which, however, seems more self-evident. The velocity component perpendicular to a wall must have a node at the wall, and the pressure must have an antinode. Similarly, the pressure must have an antinode (maximum amplitude of oscillation) at the wall $x = L_x$:

$$\cos k_x L_x = \pm 1, \tag{8.221}$$

so that

$$k_x = \frac{l\pi}{L_x}, \qquad l = 0, 1, 2, \ldots. \tag{8.222}$$

By applying similar considerations to the four remaining walls, we conclude that $D_y = D_z = 0$, and

$$\begin{aligned} k_y &= \frac{m\pi}{L_y}, \qquad m = 0, 1, 2, \ldots, \\ k_z &= \frac{n\pi}{L_z}, \qquad n = 0, 1, 2, \ldots. \end{aligned} \tag{8.223}$$

For each choice of three integers l, m, n, there is a normal mode of vibration of the fluid in the box. The frequencies of the normal modes of vibration are given by Eqs. (8.216), (8.222), and (8.223):

$$\omega_{lmn} = \pi c \left(\frac{l^2}{L_x^2} + \frac{m^2}{L_y^2} + \frac{n^2}{L_z^2} \right)^{1/2}. \tag{8.224}$$

The three integers l, m, n cannot all be zero, for this gives $\omega = 0$ and does not correspond to a vibration of the fluid. If we combine these results with Eqs. (8.217), (8.213), (8.211), and (8.207), we have for the normal mode of vibration characterized by the numbers l, m, n:

$$p' = (A \cos \omega_{lmn} t + B \sin \omega_{lmn} t) \cos \frac{l\pi x}{L_x} \cos \frac{m\pi y}{L_y} \cos \frac{n\pi z}{L_z}, \tag{8.225}$$

where we have suppressed the superfluous constant $C_x C_y C_z$. The corresponding velocities are [see Eq. (8.220)]

$$\begin{aligned} v_x &= \frac{l\pi}{L_x \rho_0 \omega_{lmn}} (A \sin \omega_{lmn} t - B \cos \omega_{lmn} t) \sin \frac{l\pi x}{L_x} \cos \frac{m\pi y}{L_y} \cos \frac{n\pi z}{L_z}, \\ v_y &= \frac{m\pi}{L_y \rho_0 \omega_{lmn}} (A \sin \omega_{lmn} t - B \cos \omega_{lmn} t) \cos \frac{l\pi x}{L_x} \sin \frac{m\pi y}{L_y} \cos \frac{n\pi z}{L_z}, \\ v_z &= \frac{n\pi}{L_z \rho_0 \omega_{lmn}} (A \sin \omega_{lmn} t - B \cos \omega_{lmn} t) \cos \frac{l\pi x}{L_x} \cos \frac{m\pi y}{L_y} \sin \frac{n\pi z}{L_z}. \end{aligned} \tag{8.226}$$

These four equations give a complete description of the motion of the fluid for a normal mode of vibration. The walls $x = 0$, $x = L_x$, and the $(l-1)$ equally spaced parallel planes between them are nodes for v_x and antinodes for p', v_y, and v_z. A similar remark applies to nodal planes parallel to the other walls.

It will be observed that the normal frequencies are not, in general, harmonically related to one another, as they were in the case of the vibrating string. If, however, one of the dimensions, say L_x, is much larger than the other two, so that the box becomes a long square pipe, then the lowest frequencies will correspond to the case where $m = n = 0$ and l is a small integer, and these frequencies are harmonically related. Thus, in a pipe, the first few normal frequencies above the lowest will be multiples of the lowest frequency. This explains why it is possible to get musical tones from an organ pipe, as well as from a vibrating string. Our treatment here applies only to a closed organ pipe, and a square one at that. The treatment of a closed circular pipe is not much more difficult than the above treatment and the general nature of the results is similar. The open-ended pipe is, however, much more difficult to treat exactly. The difficulty lies in the determination of the boundary condition at the open end; indeed, not the least of the difficulties is in deciding just where the boundary is. As a rough approximation, one may assume that the boundary is a plane surface across the end of the pipe, and that this surface is a pressure node. The results are then similar to those for the closed pipe, except that if one end of a long pipe is closed and one open, the first few frequencies above the lowest are all odd multiples of the lowest.

The general solution of the equations for sound vibrations in a rectangular cavity can be built up, as in the case of the vibrating string, by adding normal mode solutions of the form (8.225) for all normal modes of vibration. The constants A and B for each mode of vibration can again be chosen to fit the initial conditions, which in this case will be a specification of p' and $\partial p'/\partial t$ (or p' and $\mathbf{v}$) at all points in the cavity at some initial instant. We shall not carry out this development here. [In the above discussion, we have omitted the case $l = m = n = 0$, which corresponds to a constant pressure increment p'. Likewise, we omitted steady velocity solutions $\mathbf{v}(x, y, z)$ which do not oscillate in time. These solutions would have to be included in order to be able to fit all initial conditions.]

For cavities of other simple shapes, for example spheres and cylinders, the method of separation of variables used in the above example works, but in these cases instead of the variables x, y, z, coordinates appropriate to the shape of the boundary surface must be used, for example spherical or cylindrical coordinates. In most cases, except for a few simple shapes, the method of separation of variables cannot be made to work. Approximate methods can be used when the shape is very close to one of the simple shapes whose solution is known. Otherwise the only general methods of solution are numerical methods which may involve a prohibitive amount of labor. It can be shown, however, that the general features of our results for rectangular cavities hold for all shapes; that is, there are normal modes of vibration with characteristic frequencies, and the most general motion is a superposition of these.

8.12 SOUND WAVES IN PIPES

A problem of considerable interest is the problem of the propagation of sound waves in pipes. We shall consider a pipe whose axis is in the z-direction, and whose cross section is rectangular, of dimensions $L_x L_y$. This problem is the same as that of the preceding section except that there are no walls perpendicular to the z-axis.

We shall apply the same method of solution, the only difference being that the boundary conditions now apply only at the four walls $x = 0$, $x = L_x$, $y = 0$, $y = L_y$. Consequently, we are restricted in our choice of the functions $X(x)$ and $Y(y)$, just as in the preceding section, by Eqs. (8.217), (8.222), and (8.223). There are no restrictions on our choice of solution of the Z-equation (8.215). Since we are interested in solutions representing the propagation of waves down the pipe, we choose the exponential form of solution for Z:

$$Z = e^{ik_z z}, \tag{8.227}$$

and we choose the complex exponential solution (8.212) for Θ. Our solution for p', then, for a given choice of the integers l, m, is

$$\begin{aligned} p' &= \operatorname{Re} Ae^{i(k_z z - \omega t)} \cos \frac{l\pi x}{L_x} \cos \frac{m\pi y}{L_y} \\ &= A \cos \frac{l\pi x}{L_x} \cos \frac{m\pi y}{L_y} \cos (k_z z - \omega t). \end{aligned} \tag{8.228}$$

This represents a harmonic wave, traveling in the z-direction down the pipe, whose amplitude varies over the cross section of the pipe according to the first two cosine factors. Each choice of integers l, m corresponds to what is called a *mode of propagation* for the pipe. (The choice $l = 0$, $m = 0$ is an allowed choice here.) For a given l, m and a given frequency ω, the wave number k_z is determined by Eqs. (8.216), (8.222), and (8.223):

$$k_z = \pm \left[\frac{\omega^2}{c^2} - \left(\frac{l\pi}{L_x}\right)^2 - \left(\frac{m\pi}{L_y}\right)^2\right]^{1/2}. \tag{8.229}$$

The plus sign corresponds to a wave traveling in the $+z$-direction, and conversely. For $l = m = 0$, this is the same as the relation (8.201) for a wave traveling with velocity c in the z-direction in a fluid filling three-dimensional space. Otherwise, the wave travels with the velocity

$$c_{lm} = \frac{\omega}{|k_z|} = c\left[1 - \left(\frac{l\pi c}{\omega L_x}\right)^2 - \left(\frac{m\pi c}{\omega L_y}\right)^2\right]^{-1/2}, \tag{8.230}$$

which is greater than c and depends on ω. There is evidently a minimum frequency

$$\omega_{lm} = \left[\left(\frac{l\pi c}{L_x}\right)^2 + \left(\frac{m\pi c}{L_y}\right)^2\right]^{1/2} \tag{8.231}$$

below which no propagation is possible in the l, m mode; for k_z would be imaginary, and the exponent in Eq. (8.227) would be real, so that instead of a wave propagation we would have an exponential decline in amplitude of the wave in the z-direction. Note the similarity of these results to those obtained in Section 8.4 for the discrete string, where, however, there was an upper rather than a lower limit to the frequency. Since c_{lm} depends on ω, we again have the phenomenon of dispersion. A wave of arbitrary shape, which can be resolved into sinusoidally oscillating components of various frequencies ω, will be distorted as it travels along the pipe because each component will have a different velocity. We leave as an exercise the problem of calculating the fluid velocity $\mathbf{v}$, and the power flow, associated with the wave (8.228).

Similar results are obtained for pipes of other than rectangular cross section. Analogous methods and results apply to the problem of the propagation of electromagnetic waves down a wave guide. This is one reason for our interest in the present problem.

8.13 GROUP VELOCITY

We see from Formula (8.230) that the velocity c_{lm} becomes infinite when the frequency ω is equal to the cut-off frequency ω_{lm} for the l, m mode. This surprising result becomes less disturbing if we notice that c_{lm} is the velocity with which the wave pattern (8.228) travels down the pipe. The particles of fluid certainly do not travel with this velocity, as can readily be verified (see Problem 23). The fact that the pressure at each point in the fluid is oscillating at such a frequency and phase that the pressure pattern travels down the pipe with the velocity c_{lm} does not imply that there is anything physical that is traveling with this velocity. It is of interest to ask for the velocity with which sound waves can carry energy or momentum down the pipe, or for the velocity of some sort of a signal which could carry information down the pipe. It is clear that the harmonic wave (8.228), which extends uniformly in z from $-\infty$ to $+\infty$ cannot in itself carry any information since it is never changing. In order to carry information, the wave must change in some way. Let us imagine that a source of sound waves at one end of the pipe produces waves for a finite period of time, which then travel down the pipe to a receiver at the other end. The wave packet or wave group can carry information as well as energy and momentum from the sound source to the receiver. We now ask what is the velocity with which such a group of waves will travel.

In order to avoid inessential algebraic complications, let us ignore for the moment the x and y dependence of the pressure, and assume simply that we have waves

$$p'(z, t) = A \cos (kz - \omega t), \tag{8.232}$$

which can travel in the z-direction where $\omega(k)$ is some given function of k, given in the present case by Eqs. (8.216) or (8.229) where $k = k_z$. The x and y dependence in Eq. (8.228) is a specified function which is independent of z and t and can be inserted later. In order to find a solution in which the amplitude of the wave (8.232) is modulated in some way, let us try to find a solution in which at $t = 0$ the wave has the form

$$p'(z, 0) = F(z) \cos kz, \tag{8.233}$$

where $F(z)$ is a function which is confined to a finite region along the z-axis as shown in Fig. 8.10. We will assume furthermore that the function $F(z)$ is a slowly varying function of z as compared with the relatively rapid variation of the function $\cos kz$.

In order to make use of Fourier series, let us assume that the pipe has a finite length L, which however is sufficiently long so that during the period of interest we need not be concerned with the ends. We may then write the function $F(z)$ in the form of a Fourier series as in Eq. (8.24):

$$F(z) = \sum_n F_n \sin k_n z, \tag{8.234}$$

where

$$k_n = n\pi/L. \tag{8.235}$$

If the function $F(z)$ is slowly varying, as we have assumed, then the coefficients F_n will be zero or negligibly small unless $k_n \ll k$. The pressure wave (8.233) at $t = 0$ can now be written in the form

$$\begin{aligned} p'(z, 0) &= \sum_n F_n \sin k_n z \cos kz \\ &= \sum_n \tfrac{1}{2} F_n \{\sin [(k+k_n)z] - \sin [(k-k_n)z]\}. \end{aligned} \tag{8.236}$$

We now notice that by superposing waves of the form (8.232), we can write down a solution having the desired form (8.236) at $t = 0$:

$$p'(z, t) = \sum_n \tfrac{1}{2} F_n \{\sin [(k+k_n)z - \omega_{n+} t] - \sin [(k-k_n)z - \omega_{n-} t]\}, \tag{8.237}$$

where

$$\omega_{n+} = \omega(k+k_n), \qquad \omega_{n-} = \omega(k-k_n). \tag{8.238}$$

Let us expand these functions in a power series in k_n:

$$\omega_{n\pm} = \omega(k \pm k_n) = \omega \pm k_n v_g + \cdots, \tag{8.239}$$

where $\omega = \omega(k)$, and we have made the abbreviation

$$v_g = d\omega/dk. \tag{8.240}$$

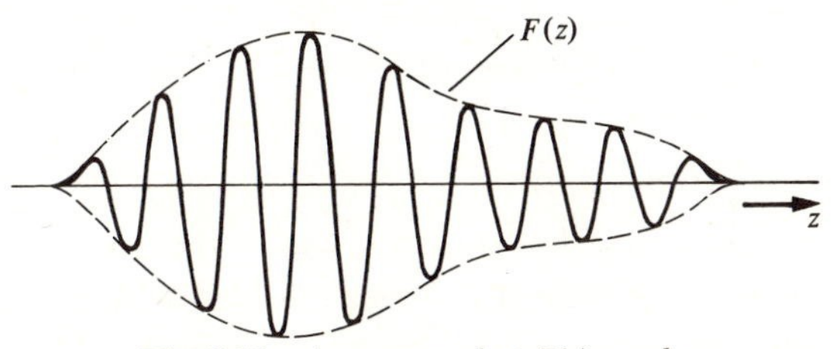

Fig. 8.10 A wave packet $F(z) \cos kz$.

The quantity v_g has the dimensions of a velocity. It will turn out to be the velocity with which the wave group travels down the pipe, and is therefore called the *group velocity.* We will assume that k_n is sufficiently small so that we can neglect all except the first two terms in Eq. (8.239). We may then write

$$\sin[(k \pm k_n)z - \omega_{n\pm}t] \doteq \sin(kz - \omega t)\cos[k_n(z - v_g t)] \pm \cos(kz - \omega t)\sin[k_n(z - v_g t)].$$

We insert this in Eq. (8.237) to obtain

$$\begin{aligned} p'(z, t) &\doteq \cos(kz - \omega t)\sum_n F_n \sin[k_n(z - v_g t)] \\ &= F(z - v_g t)\cos(kz - \omega t), \end{aligned} \tag{8.241}$$

where we have made use of Eq. (8.234). We have thus obtained an approximate solution consisting of a packet of waves of the form (8.232) modulated by an amplitude function $F(z - v_g t)$ which travels down the pipe with the group velocity v_g. Note that the wave pattern within the packet travels with the phase velocity $c_{lm} = \omega/k$. In the case of sound waves in a rectangular pipe the group velocity given by Eqs. (8.240), (8.216), (8.222), and (8.223) is

$$\begin{aligned} v_g &= ck\left[k^2 + \left(\frac{l\pi}{L_x}\right)^2 + \left(\frac{m\pi}{L_y}\right)^2\right]^{-1/2} \\ &= c^2 k/\omega = c^2/c_{lm}. \end{aligned} \tag{8.242}$$

We see that the group velocity in this case is always less than the phase velocity c_{lm} and becomes zero at the cutoff frequency ω_{lm}.

The approximation which we have made in neglecting the higher order terms in Eq. (8.239) is valid provided that the function $F(z)$ varies sufficiently slowly so that only long wavelength sine waves are needed in the Fourier series (8.234), so that the function $\omega(k)$ can be approximated by a straight line over a region within $\pm k_n$ of k for any value of k_n needed in the expansion. If higher order terms in Eq. (8.239) are included, then after a sufficient period of time the amplitude coefficient $F(z - v_g t)$ begins to change its shape.

Note that the above discussion will be of interest also in connection with the discrete string discussed in Section 8.4, where we also found wave solutions in which the frequency ω is a function of k, given by Eqs. (8.73) and (8.76). There are many physical phenomena in which waves occur whose velocity depends upon the frequency or wave number. In such cases wave packets can be formed which travel with the group velocity v_g given by Eq. (8.240). The group velocity will differ from the phase velocity $c = \omega/k$ unless c is independent of frequency or wavelength, as may easily be verified.

8.14 THE MACH NUMBER

Suppose we wish to consider two problems in fluid flow having geometrically similar boundaries, but in which the dimensions of the boundaries, or the fluid velocity, density, or compressibility are different. For example, we may wish to investigate the flow of a fluid in two pipes having the same shape but different sizes, or we may be concerned with the flow of a fluid at different velocities through pipes of the same shape, or with the flow of fluids of different densities. We might be concerned with the relation between the behavior of an airplane and the behavior of a scale model, or with the behavior of an airplane at different altitudes, where the density of the air is different. Two such problems involving boundaries of the same shape we shall call *similar* problems. Under what conditions will two similar problems have similar solutions?

In order to make this question more precise, let us assume that for each problem a characteristic distance s_0 is defined which determines the geometrical scale of the problem. In the case of similar pipes, s_0 might be a diameter of the pipe. In the case of an airplane, s_0 might be the wing span. We then define dimensionless coordinates x', y', z' by the equations

$$x' = x/s_0, \qquad y' = y/s_0, \qquad z' = z/s_0. \tag{8.243}$$

The boundaries for two similar problems will have identical descriptions in terms of the dimensionless coordinates x', y', z'; only the characteristic distance s_0 will be different. In a similar way, let us choose a characteristic speed v_0 associated with the problem. The speed v_0 might be the average speed of flow of fluid in a pipe, or the speed of the airplane relative to the stationary air at a distance from it, or v_0 might be the maximum speed of any part of the fluid relative to the pipe or the airplane. In any case, we suppose that v_0 is so chosen that the maximum speed of any part of the fluid is not very much larger than v_0. We now define a dimensionless velocity $\mathbf{v}'$, and a dimensionless time coordinate t':

$$\mathbf{v}' = \mathbf{v}/v_0, \tag{8.244}$$

$$t' = v_0 t/s_0. \tag{8.245}$$

We now say that two similar problems have similar solutions if the solutions are identical when expressed in terms of the dimensionless velocity $\mathbf{v}'$ as a function of x', y', z', and t'. The fluid flow pattern will then be the same in both problems, differing only in the distance and time scales determined by s_0 and v_0. We need also to assume a characteristic density ρ_0 and pressure p_0. In the case of the airplane, these would be the density and pressure of the undisturbed atmosphere; in the case of the pipe, they might be the average density and pressure, or the density and pressure at one end of the pipe. We shall define a dimensionless pressure increment p'' as follows:

$$p'' = \frac{p-p_0}{\rho_0 v_0^2}. \tag{8.246}$$

We shall now assume that the changes in density of the fluid are small enough so that we can write

$$\rho = \rho_0 + \frac{d\rho}{dp}(p-p_0), \tag{8.247}$$

where higher order terms in the Taylor series for ρ have been neglected. By making use of the definition (8.246) for p'', and of the bulk modulus B as given by Eq. (5.183), this can be written

$$\rho = \rho_0(1+M^2 p''), \tag{8.248}$$

where

$$M = v_0\left(\frac{B}{\rho_0}\right)^{-1/2} = \frac{v_0}{c}. \tag{8.249}$$

Here M is the ratio of the characteristic velocity v_0 to the velocity of sound c and is called the *Mach number* for the problem.

With the help of Eq. (8.248), we can rewrite the equation of continuity and the equation of motion in terms of the dimensionless variables introduced by Eqs. (8.243) to (8.246). The equation of continuity (8.127), when we divide through by the constant $\rho_0 v_0/s_0$ and collect separately the terms involving M, becomes

$$\boldsymbol{\nabla}'\cdot\boldsymbol{v}'+M^2\left[\frac{\partial p''}{\partial t'}+\boldsymbol{\nabla}'\cdot(p''\boldsymbol{v}')\right] = 0, \tag{8.250}$$

where

$$\boldsymbol{\nabla}' = \hat{\boldsymbol{x}}\frac{\partial}{\partial x'}+\hat{\boldsymbol{y}}\frac{\partial}{\partial y'}+\hat{\boldsymbol{z}}\frac{\partial}{\partial z'}. \tag{8.251}$$

The equation of motion (8.139), when we divide through by v_0^2/s_0, becomes, in the same way,

$$\frac{\partial \boldsymbol{v}'}{\partial t'}+\boldsymbol{v}'\cdot\boldsymbol{\nabla}'\boldsymbol{v}'+\frac{\boldsymbol{\nabla}'p''}{1+M^2p''} = \frac{s_0}{v_0^2}\frac{\boldsymbol{f}}{\rho}. \tag{8.252}$$

Equations (8.250) and (8.252) represent four differential equations to be solved for the four quantities p', $\boldsymbol{v}'$, subject to given initial and boundary conditions. If the body forces are zero, or if the body forces per unit mass $\boldsymbol{f}/\rho$ are made proportional to v_0^2/s_0, then the equations for two similar problems become identical if the Mach number M is the same for both. Hence, similar problems will have similar solutions if they have the same Mach number. Results of experiments on scale models in wind tunnels can be extrapolated to full-sized airplanes flying at speeds with corresponding Mach numbers. If the Mach number is much less than one, the terms in M^2 in Eqs. (8.250) and (8.252) can be neglected, and these equations then reduce to the equations for an incompressible fluid, as is obvious either from Eq. (8.250) or (8.248). Therefore at fluid velocities much less than the speed of

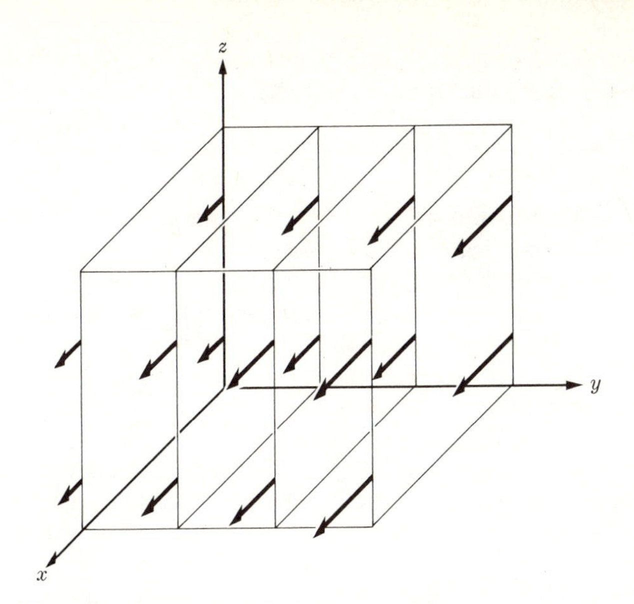

Fig. 8.11 Velocity distribution in the definition of viscosity.

sound, even air may be treated as an incompressible fluid. On the other hand, at Mach numbers near or greater than one, the compressibility becomes important, even in problems of liquid flow. Note that the Mach number involves only the characteristic velocity v_0, and the velocity of sound, which in turn depends on the characteristic density ρ_0 and the compressibility B. Changes in the distance scale factor s_0 have no effect on the nature of the solution, nor do changes in the characteristic pressure p_0 except insofar as they affect ρ_0 and B.

It must be emphasized that these results are applicable only to ideal fluids, i.e., when viscosity is unimportant, and to problems where the density of the fluid does not differ greatly at any point from the characteristic density ρ_0. The latter condition holds fairly well for liquids, except when there is cavitation (formation of vapor bubbles), and for gases except at very large Mach numbers.

8.15 VISCOSITY

In many practical applications of the theory of fluid flow, it is not permissible to neglect viscous friction, as has been done in the preceding sections. When adjacent layers of fluid are moving past one another, this motion is resisted by a shearing force which tends to reduce their relative velocity. Let us assume that in a given region the velocity of the fluid is in the x-direction, and that the fluid is flowing in layers parallel to the xz-plane, so that v_x is a function of y only (Fig. 8.11). Let the positive y-axis be directed toward the right. Then if $\partial v_x/\partial y$ is positive, the viscous friction will result in a positive shearing force F_x acting from right to left across an area A parallel to the xz-plane. The *coefficient of viscosity* η is defined as the ratio of the shearing stress to the velocity gradient:

$$\eta = \frac{F_x/A}{\partial v_x/\partial y}. \tag{8.253}$$

When the velocity distribution is not of this simple type, the stresses due to viscosity are more complicated. (See Section 10.6.)

We shall apply this definition to the important special case of steady flow of a fluid through a pipe of circular cross section, with radius a. We shall assume laminar flow; that is, we shall assume that the fluid flows in layers, as contemplated in the definition above. In this case, the layers are cylinders. The velocity is everywhere parallel to the axis of the pipe, which we take to be the z-axis, and the velocity v_z is a function only of r, the distance from the axis of the pipe. (See Fig. 8.12.) If we consider a cylinder of radius r and of length l, its area will be $A = 2\pi rl$, and according to the definition (8.253), the force exerted across this cylinder by the fluid outside on the fluid inside the cylinder is

$$F_z = \eta(2\pi rl)\frac{dv_z}{dr}. \tag{8.254}$$

Since the fluid within this cylinder is not accelerated, if there is no body force the viscous force must be balanced by a difference in pressure between the two ends of the cylinder:

$$\Delta p(\pi r^2) + F_z = 0, \tag{8.255}$$

where Δp is the difference in pressure between the two ends of the cylinder a

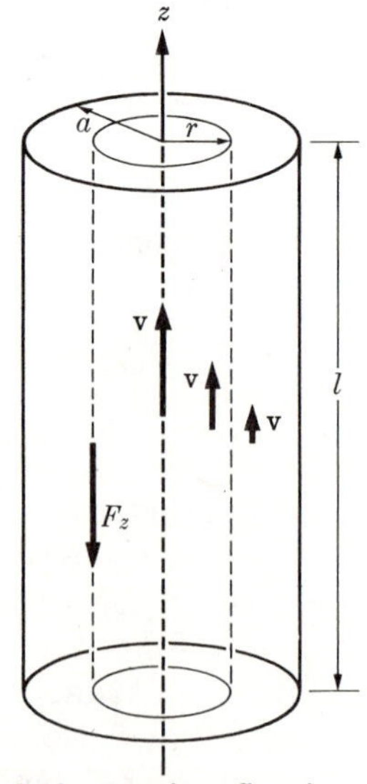

Fig. 8.12 Laminar flow in a pipe.

distance l apart, and we assume that the pressure is uniform over the cross section of the pipe. Equations (8.254) and (8.255) can be combined to give a differential equation for v_z:

$$\frac{dv_z}{dr} = -\frac{r\,\Delta p}{2\eta l}. \tag{8.256}$$

We integrate outward from the cylinder axis:

$$\begin{aligned} \int_{v_0}^{v_z} dv_z &= -\frac{\Delta p}{2\eta l}\int_0^r r\,dr, \\ v_z &= v_0 - \frac{r^2\,\Delta p}{4\eta l}, \end{aligned} \tag{8.257}$$

where v_0 is the velocity at the axis of the pipe. We shall assume that the fluid velocity is zero at the walls of the pipe:

$$[v_z]_{r=a} = v_0 - \frac{a^2\,\Delta p}{4\eta l} = 0, \tag{8.258}$$

although this assumption is open to question. Then

$$v_0 = \frac{a^2\,\Delta p}{4\eta l}, \tag{8.259}$$

and

$$v_z = \frac{\Delta p}{4\eta l}(a^2 - r^2). \tag{8.260}$$

The total fluid current through the pipe is

$$I = \int\int \rho v_z dS = 2\pi\rho \int_0^a v_z r\,dr. \tag{8.261}$$

We substitute from Eq. (8.260) and carry out the integration:

$$\frac{I}{\rho} = \frac{\pi a^4\,\Delta p}{8\eta l}. \tag{8.262}$$

This formula is called *Poiseuille's law*. It affords a convenient and simple way of measuring η.

Although we will not develop now the general equations of motion for viscous flow, we can arrive at a result analogous to that in Section 8.14, taking viscosity into account, without actually setting up the equations for viscous flow. Suppose that we are concerned, as in Section 8.14, with two similar problems in fluid flow, and let s_0, v_0, p_0, ρ_0 be a characteristic distance, velocity, pressure, and density, which again define the scale in any problem. However, let us suppose that in this case viscosity is to be taken into account, so that the equation of motion (8.139) is augmented by a term corresponding to the force of viscous friction. We do not at present know the precise form of this term, but at any rate it will consist of η multiplied by various derivatives of various velocity components, and divided by ρ [since Eq. (8.139) has already been divided through by ρ]. When we introduce the velocity $\boldsymbol{v}'$, and the dimensionless coordinates x', y', z', t', as in Section 8.14, and divide the equation of motion by v_0^2/s_0, we will obtain just Eq. (8.252), augmented by a term involving the coefficient of viscosity. Since all the terms in Eq. (8.252) are dimensionless, the viscosity term will be also, and will consist of derivatives of components of $\boldsymbol{v}'$ with respect to x', y', z', multiplied by numerical factors and by a dimensionless coefficient consisting of η times some combination of v_0 and s_0, and divided by $\rho = \rho_0(1+M^2p'')$ [Eq. (8.248)]. Now the dimensions of η, as determined by Eq. (8.253), are

$$[\eta] = \frac{\text{mass}}{\text{length} \times \text{time}}, \tag{8.263}$$

and the only combination of ρ_0, v_0, and s_0 having these dimensions is $\rho_0 v_0 s_0$. Therefore the viscosity term will be multiplied by the coefficient

$$\frac{1}{R(1+M^2p'')}, \tag{8.264}$$

where R is the *Reynolds number*, defined by

$$R = \frac{\rho_0 v_0 s_0}{\eta}. \tag{8.265}$$

We can now conclude that when viscosity is important, two similar problems will have the same equation of motion in dimensionless variables, and hence similar solutions, only if the Reynolds number R, as well as the Mach number M, is the same for both. If the Mach number is very small, then compressibility is unimportant. If the Reynolds number is very large, then viscosity may be neglected. It turns out that there is a critical value of Reynolds number for any given problem, such that the nature of the flow is very different for R larger than this critical value than for smaller values of R. For small Reynolds numbers, the flow is laminar, as the viscosity tends to damp out any vortices which might form. For large Reynolds numbers, the flow tends to be turbulent. This will be the case when the viscosity is small, or the density, velocity, or linear dimensions are large. Note that the Reynolds number depends on s_0, whereas the Mach number does not, so that the distance scale of a problem is important when the effects of viscosity are considered. Viscous effects are more important on a small scale than on a large scale.

It may be noted that the expression (8.265) for the Reynolds number, together with the fact that Eq. (8.139) is divided by v_0^2/s_0 to obtain the dimensionless equation of motion, implies that the viscosity term to be added to Eq. (8.139) has the dimensions of $(\eta v_0)/(\rho_0 s_0^2)$. This, in turn, implies that the viscous force density must be equal to η times a sum of second derivatives of velocity components with respect to x, y, and z. This is perhaps also evident from Eq. (8.253), since in calculating the total force on a fluid element, the differences in stresses on opposite faces of the element will be involved, and hence a second differentiation of velocities relative to x, y, and z will appear in the expression for the force. An expression for the viscous force density will be developed in Chapter 10.

PROBLEMS

1. A stretched string of length l is terminated at the end $x = l$ by a ring of negligible mass which slides without friction on a vertical rod.

a) Show that the boundary condition at this end of the string is

$$\left[\frac{\partial u}{\partial x}\right]_{x=l} = 0.$$

b) If the end $x = 0$ is tied, find the normal modes of vibration.

2. Find the boundary condition and the normal modes of vibration in Problem 1 if the ring at one end has a finite mass m. What is the significance of the limiting cases $m = 0$ and $m = \infty$? (Neglect the effect of gravity.)

3. The midpoint of a stretched string of length l is pulled a distance $u = l/10$ from its equilibrium position, so that the string forms two legs of an isosceles triangle. The string is then released. Find an expression for its motion by the Fourier series method.

4. A piano string of length l, tension τ, and density σ, tied at both ends, and initially at rest, is struck a blow at a distance a from one end by a hammer of mass m and velocity v_0. Assume that the hammer rebounds elastically with velocity $-v_0$, and that its momentum loss is transferred to a short length Δl of string centered around $x = a$. Find the motion of the string by the Fourier series method, assuming that Δl is negligibly small. If the finite length of Δl were taken into account, what sort of effect would this have on your result? If it is desired that no seventh harmonic of the fundamental frequency be present (it is said to be particularly unpleasant), at what points a may the string be struck?

5. A string of length l is tied at $x = l$. The end at $x = 0$ is forced to move sinusoidally so that

$$u(0, t) = A \sin \omega t.$$

a) Find the steady-state motion of the string; that is, find a solution in which all points on the string vibrate with the same angular frequency ω.

b) How would you find the actual motion if the string were initially at rest?

6. A force of linear density

$$f(x, t) = f_0 \sin \frac{n\pi x}{l} \cos \omega t,$$

where n is an integer, is applied along a stretched string of length l.

a) Find the steady-state motion of the string. [*Hint:* Assume a similar time and space dependence for $u(x, t)$, and substitute in the equation of motion.]

b) Indicate how one might solve the more general problem of a harmonic applied force

$$f(x, t) = f_0(x) \cos \omega t,$$

where $f_0(x)$ is any function which vanishes at the ends.

7. Assume that the friction of the air around a vibrating string can be represented as a force per unit length proportional to the velocity of the string. Set up the equation of motion for the string, and find the normal modes of vibration if the string is tied at both ends.

8. Find the motion of a horizontal stretched string of tension τ, density σ, and length l, tied at both ends, taking into account the weight of the string. The string is initially held straight and horizontal, and dropped. [*Hint:* Find the steady-state solution and add a suitable transient.]

9. A long string is terminated at its right end by a massless ring which slides on a vertical rod and is impeded by a frictional force proportional to its velocity. Set up a suitable boundary condition and discuss the reflection of a wave at the end. How does the reflected wave behave in the limiting cases of very large and very small friction? For what value of the friction constant is there no reflected wave?

10. Discuss the reflection of a wave traveling down a long string terminated by a massless ring, as in Problem 1.

11. Find a solution to Problem 3 by superposing waves $f(x-ct)$ and $g(x+ct)$ in such a way as to satisfy the initial and boundary conditions. Sketch the appearance of the string at times $t = 0, \frac{1}{4}l/c, \frac{1}{2}l/c$, and l/c.

12. a) A long stretched string of tension τ and density σ_1 is tied at $x = 0$ to a string of density σ_2. If the mass of the knot is negligible, show that u and $\partial u/\partial x$ must be the same on both sides of the knot.

b) A wave $A \cos (k_1 x - \omega t)$ traveling toward the right on the first string is incident on the junction. Show that in order to satisfy the boundary conditions at the knot, there must be a reflected wave traveling to the left in the first string and a transmitted wave traveling to the right in the second string, both of the same frequency as the incident wave. Find the amplitudes and phases of the transmitted and reflected waves.

c) Check your result in part (b) by calculating the power in the transmitted and reflected waves, and showing that the total is equal to the power in the incident wave.

13. Derive directly from Eq. (8.139) an equation expressing the conservation of angular momentum in a form analogous to Eq. (8.141).

14. Derive an equation expressing the law of conservation of angular momentum for a fluid in a form analogous to Eq. (8.140). From this, derive equations analogous to Eqs. (8.141), (8.142), and (8.144). Explain the physical meaning of each term in each equation. Show that the internal torques due to pressure can be eliminated from the integrated forms, and derive an equation analogous to Eq. (8.148).

15. Derive and interpret the following equation:

$$\frac{d}{dt}\iiint_V (\tfrac{1}{2}\rho v^2 - \rho\mathcal{G} + \rho u)\, dV + \iint_S \hat{\boldsymbol{n}}\cdot\boldsymbol{v}(\tfrac{1}{2}\rho v^2 - \rho\mathcal{G} + \rho u)\, dS = -\iint_S (\hat{\boldsymbol{n}}p)\cdot\boldsymbol{v}\, dS - \iiint_V \rho\frac{\partial\mathcal{G}}{\partial t}\, dV,$$

where V is a fixed volume bounded by a surface S with normal $\hat{\boldsymbol{n}}$, and the other variables have the same meanings as in Section 8.8.

16. a) A mass of initially stationary air at 45° N latitude flows inward toward a low-pressure spot at its center. Show that the coriolis torque about the low-pressure center depends only on the radial component of velocity. Hence, show that if frictional torques are neglected, the angular momentum per unit mass at radius r from the center depends only on r and on the initial radius r_0 at which the air is stationary, but does not depend on the details of the motion.
b) Calculate the azimuthal component of velocity around the low as a function of initial and final radius. If this were a reasonable model of a tornado, what would be the initial radius r_0 if the air at 264 ft from the center has a velocity of 300 mph?

17. Evaluate the potential energy u per unit mass as a function of p for a perfect gas of molecular weight M at temperature T. For the steady isothermal flow of this gas through a pipe of varying cross section and varying height above the earth, find expressions for the pressure, density, and velocity of the gas as functions of the cross section S of the pipe, the height h, and the pressure p_0 and velocity v_0 at a point in the pipe at height $h = 0$ where the cross section is S_0. Assume p, v, and ρ uniform over the cross section.

18. Work Problem 17 for an incompressible fluid of density ρ_0.

19. The function $\phi = a/r$, where a is a constant and r is the distance from a fixed point, satisfies Laplace's equation (8.177), except at $r = 0$, because it has the same form as the gravitational potential of a point mass. If this is a velocity potential, what is the nature of the fluid flow to which it leads?

20. a) Verify by direct computation that the spherical wave (8.204) satisfies the wave equation (8.186).
b) Write an analogous expression for a cylindrical wave of arbitrary time dependence, traveling out from the z-axis, independent of z and with cylindrical symmetry. Make the amplitude depend on the distance from the axis in such a way as to satisfy the requirement of conservation of energy. Show that such a wave cannot satisfy the wave equation. (It is a general property of cylindrical waves that they do not preserve their shape.)

***21.** Show that the normal mode of vibration given by Eqs. (8.225) and (8.226) can be represented as a superposition of harmonically oscillating plane waves traveling in appropriately chosen directions with appropriate phase relationships. Show that in the normal vibrations of a fluid in a box, the velocity oscillates 90° out of phase with the pressure at any point. How can this be reconciled with the fact that in a plane wave the velocity and pressure are in phase?

22. Find the normal modes of vibration of a square organ pipe with one end open and the other closed, on the assumption that the open end is a pressure node.

23. a) Calculate the fluid velocity $\boldsymbol{v}$ for the wave given by Eq. (8.228).
b) Calculate the mean rate of power flow through the pipe.

***24.** Show that the expression (8.228) for a sound wave in a pipe can be represented as a superposition of plane waves traveling with speed c in appropriate directions, and being reflected at the walls. Explain, in terms of this representation, why there is a minimum frequency for any given mode below which a wave cannot propagate through the pipe in this mode.

25. If the sound wave given by Eq. (8.228) is incident on a closed end of the pipe at $z = 0$, find the reflected wave, starting from the boundary condition $v_z = 0$ at $z = 0$.

26. Develop the theory of the propagation of sound waves in a circular pipe, using cylindrical coordinates and applying the method of separation of variables. Carry the solution as far as you can. You are not required to solve the equation for the radial part of the wave, but you should indicate the sort of solutions you would expect to find.

27. Calculate the group velocity for the wave (8.79) on the discrete string, and compare it with the phase velocity. Show that the two are nearly equal for wavelengths much greater than h.

28. Keep one more term in the expansion (8.239), and carry the development in Section 8.13 as far as you can, noting in what way it breaks down. Show that with this additional term, the modulation envelope $F(z\text{-}v_g t)$ cannot maintain its form as time progresses. Estimate the time after which a significant change in the modulation envelope may be expected.

29. A fluid of viscosity η flows steadily between two infinite parallel plane walls a distance l apart. The velocity of the fluid is everywhere in the same direction, and depends only on the distance from the walls. The total fluid current between the walls in any unit length measured along the walls perpendicular to the direction of flow is I. Find the velocity distribution and the pressure gradient parallel to the walls, assuming that the pressure varies only in the direction of flow.

30. Prove that the only combination of ρ_0, v_0, s_0 having the dimensions of viscosity is $\rho_0 v_0 s_0$.

CHAPTER 9

LAGRANGE'S EQUATIONS

9.1 GENERALIZED COORDINATES

Direct application of Newton's laws to a mechanical system results in a set of equations of motion in terms of the cartesian coordinates of each of the particles of which the system is composed. In many cases, these are not the most convenient coordinates in terms of which to solve the problem or to describe the motion of the system. For example, in the problem of the motion of a single particle acted on by a central force, which we treated in Section 3.13, we found it convenient to introduce polar coordinates in the plane of motion of the particle. The reason was that the force in this case can be expressed more simply in terms of polar coordinates. Again in the two-body problem, treated in Section 4.7, we found it convenient to replace the coordinates $\boldsymbol{r}_1$, $\boldsymbol{r}_2$ of the two particles by the coordinate vector $\boldsymbol{R}$ of the center of mass, and the relative coordinate vector $\boldsymbol{r}$ which locates particle 1 with respect to particle 2. We had two reasons for this choice of coordinates. First, the mutual forces which the particles exert on each other ordinarily depend on the relative coordinate. Second, in many cases we are interested in a description of the motion of one particle relative to the other, as in the case of planetary motion. In problems involving many particles, it is usually convenient to choose a set of coordinates which includes the coordinates of the center of mass, since the motion of the center of mass is determined by a relatively simple equation (4.18). In Chapter 7, we found the equations of motion of a particle in terms of moving coordinate systems, which are sometimes more convenient to use than the fixed coordinate systems contemplated in Newton's original equations of motion.

We shall include coordinate systems of the sort described above, together with cartesian coordinate systems, under the name *generalized coordinates.* A set of generalized coordinates is any set of coordinates by means of which the positions of the particles in a system may be specified. In a problem requiring generalized coordinates, we may set up Newton's equations of motion in terms of cartesian coordinates, and then change to the generalized coordinates, as in the problems studied in previous chapters. It would be very desirable and convenient, however, to have a general method for setting up equations of motion directly in terms of any convenient set of generalized coordinates. Furthermore it is desirable to have uniform methods of writing down, and perhaps of solving, the equations of motion in terms of any coordinate system. Such a method was invented by Lagrange and is the subject of this chapter.

In each of the cases mentioned in the first paragraph, the number of coordinates in the new system of coordinates introduced to simplify the problem was the same as the number of cartesian coordinates of all the particles involved. We may, for example, replace the two cartesian coordinates x, y of a particle moving in a plane by the two polar coordinates r, θ, or the three space coordinates x, y, z by three spherical or cylindrical coordinates. Or we may replace the six coordinates x_1, y_1, z_1, x_2, y_2, z_2 of a pair of particles by the three coordinates X, Y, Z of the center of mass plus the three coordinates x, y, z of one particle relative to the other. Or we may replace the three coordinates of a particle relative to a fixed system of axes by three coordinates relative to moving axes. (A vector counts as three coordinates.)

In our treatment of the rotation of a rigid body about an axis (Section 5.2), we described the position of the body in terms of the single angular coordinate θ. Here we have a case where we can replace a great many cartesian coordinates, three for each particle in the body, by a single coordinate θ. This is possible because the body is rigid and is allowed to rotate only about a fixed axis. As a result of these two facts, the position of the body is completely determined when we specify the angular position of some reference line in the body. The position of a free rigid body can be specified by six coordinates, three to locate its center of mass, and three to determine its orientation in space. This is a vast simplification compared with the $3N$ cartesian coordinates required to locate its N particles. A rigid body is an example of a system of particles subject to *constraints*, that is, conditions which restrict the possible sets of values of the coordinates. In the case of a rigid body, the constraint is that the distance between any two particles must remain fixed. If the body can rotate only about a fixed axis, then in addition the distance of each particle from the axis is fixed. This is the reason why specifying the value of the single coordinate θ is sufficient to determine the position of each particle in the body. We shall postpone the discussion of systems like this which involve constraints until Section 9.4. In this section, and the next, we shall set up the theory of generalized coordinates, assuming that there are as many generalized coordinates as cartesian coordinates. We shall then find, in Section 9.4, that this theory applies also to the motion of constrained systems.

When we want to speak about a physical system described by a set of generalized coordinates, without specifying for the moment just what the coordinates are, it is customary to designate each coordinate by the letter q with a numerical subscript. A set of n generalized coordinates would be written as q_1, $q_2, \ldots, q_n$. Thus a particle moving in a plane may be described by two coordinates q_1, q_2, which may in special cases be the cartesian coordinates x, y, or the polar coordinates r, θ, or any other suitable pair of coordinates. A particle moving in space is located by three coordinates, which may be cartesian coordinates x, y, z, or spherical coordinates r, θ, φ, or cylindrical coordinates ρ, z, φ, or, in general q_1, q_2, q_3.

The configuration of a system of N particles may be specified by the $3N$ cartesian coordinates $x_1, y_1, z_1, x_2, y_2, z_2, \ldots, x_N, y_N, z_N$ of its particles, or by any set of $3N$ generalized coordinates $q_1, q_2, \ldots, q_{3N}$. Since for each configuration of the system the generalized coordinates must have some definite set of values, the coordinates $q_1, \ldots, q_{3N}$ will be functions of the cartesian coordinates, and possibly also of the time in the case of moving coordinate systems:

$$\begin{aligned} q_1 &= q_1(x_1, y_1, z_1, x_2, y_2, \ldots, y_N, z_N; t), \\ q_2 &= q_2(x_1, y_1, \ldots\ldots\ldots\ldots, z_N; t), \\ &\vdots \\ q_{3N} &= q_{3N}(x_1, y_1, \ldots\ldots\ldots\ldots, z_N; t). \end{aligned} \tag{9.1}$$

Since the coordinates $q_1, \ldots, q_{3N}$ specify the configuration of the system, it must be possible also to express the cartesian coordinates in terms of the generalized coordinates:

$$\begin{aligned} x_1 &= x_1(q_1, q_2, \ldots, q_{3N}; t), \\ y_1 &= y_1(q_1, \ldots\ldots, q_{3N}; t), \\ &\vdots \\ z_N &= z_N(q_1, \ldots\ldots, q_{3N}; t). \end{aligned} \tag{9.2}$$

If Eqs. (9.1) are given, they may be solved for $x_1, y_1, \ldots, z_N$ to obtain Eqs. (9.2), and vice versa.

The mathematical condition that this solution be (theoretically) possible is that the Jacobian determinant of Eqs. (9.1) be different from zero at all points, or nearly all points:

$$\frac{\partial(q_1, \ldots, q_{3N})}{\partial(x_1, y_1, \ldots, z_N)} = \begin{vmatrix} \frac{\partial q_1}{\partial x_1} \frac{\partial q_2}{\partial x_1} \cdots \frac{\partial q_{3N}}{\partial x_1} \\ \frac{\partial q_1}{\partial y_1} \frac{\partial q_2}{\partial y_1} \cdots \frac{\partial q_{3N}}{\partial y_1} \\ \vdots \\ \frac{\partial q_1}{\partial z_N} \frac{\partial q_2}{\partial z_N} \cdots \frac{\partial q_{3N}}{\partial z_N} \end{vmatrix} \neq 0. \tag{9.3}$$

If this inequality does not hold, then Eqs. (9.1) do not define a legitimate set of generalized coordinates. In practically all cases of physical interest, it will be evident from the geometrical definitions of the generalized coordinates whether or not they are a legitimate set of coordinates. Thus we shall not have any occasion to apply the above test to our coordinate systems. [For a derivation of the condition (9.3), see W. F. Osgood, *Advanced Calculus*, New York: Macmillan, 1937, p. 129.]

As an example, we have the equations (3.72) and (3.73) connecting the polar coordinates r, θ of a single particle in a plane with its cartesian coordinates x, y. As an example of a moving coordinate system, we consider polar coordinates in which the reference axis from which θ is measured rotates counterclockwise with constant angular velocity ω (Fig. 9.1):

$$r = (x^2+y^2)^{1/2},$$
$$\theta = \tan^{-1}\frac{y}{x}-\omega t, \tag{9.4}$$

and conversely,

$$x = r\cos(\theta+\omega t),$$
$$y = r\sin(\theta+\omega t). \tag{9.5}$$

As an example of generalized coordinates for a system of particles, we have the center of mass coordinates X, Y, Z and relative coordinates x, y, z of two particles of masses m_1 and m_2, as defined by Eqs. (4.90) and (4.91), where X, Y, Z are the components of $\boldsymbol{R}$, and x, y, z are the components of $\boldsymbol{r}$. Because the transformation equations (4.90) and (4.91) do not contain the time explicitly, we regard this as a fixed coordinate system, even though x, y, z are the coordinates of m_1 referred to a moving origin located on m_2. The rule which defines the coordinates X, Y, Z, x, y, z is the same at all times.

If a system of particles is described by a set of generalized coordinates $q_1, \ldots, q_{3N}$, we shall call the time derivative $\dot{q}_k$, of any coordinate q_k, the *generalized velocity* associated with this coordinate. The generalized velocity associated with a cartesian coordinate x_i is just the corresponding component $\dot{x}_i$ of the velocity of the particle located by that coordinate. The generalized velocity associated with an angular coordinate θ is the corresponding angular velocity $\dot{\theta}$. The velocity associated with the coordinate X in the preceding example is $\dot{X}$, the x-component of velocity of the center of mass. The generalized velocities can be computed in terms of cartesian coordinates and velocities, and conversely, by differentiating Eqs. (9.1) or (9.2) with respect to t according to the rules for differentiating implicit functions. For example, the cartesian velocity components can be expressed in terms of the generalized coordinates and velocities by differentiating Eqs. (9.2):

$$\begin{aligned}\dot{x}_1 &= \sum_{k=1}^{3N}\frac{\partial x_1}{\partial q_k}\dot{q}_k+\frac{\partial x_1}{\partial t},\\ &\vdots\\ \dot{z}_N &= \sum_{k=1}^{3N}\frac{\partial z_N}{\partial q_k}\dot{q}_k+\frac{\partial z_N}{\partial t}.\end{aligned} \tag{9.6}$$

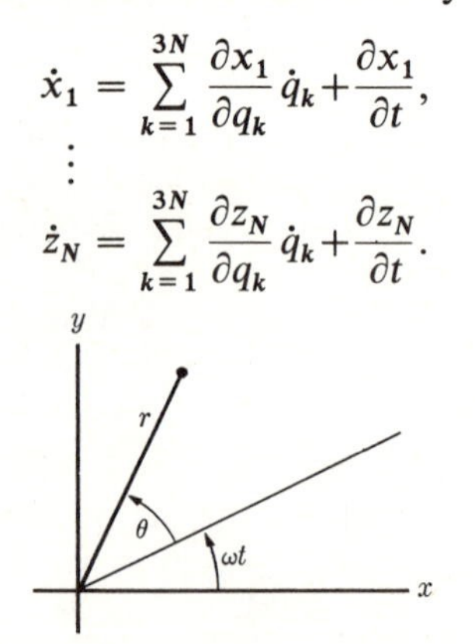

Fig. 9.1 A rotating polar coordinate system.

As an example, we have, from Eqs. (9.5):

$$\begin{aligned} \dot{x} &= \dot{r} \cos(\theta+\omega t) - r\dot{\theta} \sin(\theta+\omega t) - r\omega \sin(\theta+\omega t), \\ \dot{y} &= \dot{r} \sin(\theta+\omega t) + r\dot{\theta} \cos(\theta+\omega t) + r\omega \cos(\theta+\omega t). \end{aligned} \tag{9.7}$$

The kinetic energy of a system of N particles, in terms of cartesian coordinates, is

$$T = \sum_{i=1}^{N} \tfrac{1}{2} m_i(\dot{x}_i^2 + \dot{y}_i^2 + \dot{z}_i^2). \tag{9.8}$$

By substituting from Eqs. (9.6), we obtain the kinetic energy in terms of generalized coordinates. If we rearrange the order of summation, the result is

$$T = \sum_{k=1}^{3N} \sum_{l=1}^{3N} \tfrac{1}{2} A_{kl}\dot{q}_k\dot{q}_l + \sum_{k=1}^{3N} B_k\dot{q}_k + T_0, \tag{9.9}$$

where

$$A_{kl} = \sum_{i=1}^{N} m_i \left(\frac{\partial x_i}{\partial q_k}\frac{\partial x_i}{\partial q_l} + \frac{\partial y_i}{\partial q_k}\frac{\partial y_i}{\partial q_l} + \frac{\partial z_i}{\partial q_k}\frac{\partial z_i}{\partial q_l} \right), \tag{9.10}$$

$$B_k = \sum_{i=1}^{N} m_i \left(\frac{\partial x_i}{\partial q_k}\frac{\partial x_i}{\partial t} + \frac{\partial y_i}{\partial q_k}\frac{\partial y_i}{\partial t} + \frac{\partial z_i}{\partial q_k}\frac{\partial z_i}{\partial t} \right), \tag{9.11}$$

$$T_0 = \sum_{i=1}^{N} \tfrac{1}{2} m_i \left[\left(\frac{\partial x_i}{\partial t}\right)^2 + \left(\frac{\partial y_i}{\partial t}\right)^2 + \left(\frac{\partial z_i}{\partial t}\right)^2 \right]. \tag{9.12}$$

The coefficients A_{kl}, B_k, and T_0 are functions of the coordinates $q_1, \ldots, q_{3N}$, and also of t for a moving coordinate system. If A_{kl} is zero except when $k = l$, the coordinates are said to be *orthogonal*. The coefficients B_k and T_0 are zero when t does not occur explicitly in Eqs. (9.1), i.e., when the generalized coordinate system does not change with time. We see that the kinetic energy, in general, contains three sets of terms:

$$T = T_2 + T_1 + T_0, \tag{9.13}$$

where T_2 contains terms quadratic in the generalized velocities, T_1 contains linear terms, and T_0 is independent of the velocities. The terms T_1 and T_0 appear only in moving coordinate systems; for fixed coordinate systems, the kinetic energy is quadratic in the generalized velocities.

As an example, in plane polar coordinates [Eqs. (3.72)], the kinetic energy is

$$\begin{aligned} T &= \tfrac{1}{2} m(\dot{x}^2 + \dot{y}^2) \\ &= \tfrac{1}{2}(m\dot{r}^2 + mr^2\dot{\theta}^2), \end{aligned} \tag{9.14}$$

as may be obtained by direct substitution from Eqs. (3.72), or as a special case of

Eq. (9.9), where

$$\frac{\partial x}{\partial r} = \cos\theta, \qquad \frac{\partial x}{\partial \theta} = -r\sin\theta,$$
$$\frac{\partial y}{\partial r} = \sin\theta, \qquad \frac{\partial y}{\partial \theta} = r\cos\theta. \tag{9.15}$$

If we take the moving coordinate system defined by Eqs. (9.5), we find, by substituting from Eqs. (9.7), or by using Eq. (9.9),

$$\begin{aligned} T &= \tfrac{1}{2}m(\dot{x}^2+\dot{y}^2) \\ &= \tfrac{1}{2}(m\dot{r}^2+mr^2\dot{\theta}^2)+mr^2\omega\dot{\theta}+\tfrac{1}{2}mr^2\omega^2. \end{aligned} \tag{9.16}$$

In this case, a term linear in $\dot{\theta}$ and a term independent of $\dot{r}$ and $\dot{\theta}$ appear. The kinetic energy for the two-particle system can also easily be written down in terms of X, Y, Z, x, y, z, defined by Eqs. (4.90) and (4.91).

Instead of finding the kinetic energy first in cartesian coordinates and then translating into generalized coordinates, as in the examples above, it is often quicker to work out the kinetic energy directly in terms of generalized coordinates from a knowledge of their geometrical meaning. It may then be possible to start a problem from the beginning with a suitable set of generalized coordinates without writing out explicitly the transformation equations (9.1) and (9.2) at all. For example, we may obtain Eq. (9.14) immediately from the geometrical meaning of the coordinates r, θ (see Fig. 3.20) by noticing that the linear velocity associated with a change in r is $\dot{r}$ and that associated with a change in θ is $r\dot{\theta}$. Since the directions of the velocities associated with r and θ are perpendicular, the square of the total velocity is

$$v^2 = \dot{r}^2+r^2\dot{\theta}^2, \tag{9.17}$$

from which Eq. (9.14) follows immediately.

Care must be taken in applying this method if the velocities associated with changes of the various coordinates are not perpendicular. For example, let us consider a pair of coordinate axes u, w making an angle α less than 90° with each other, as in Fig. 9.2. Let u and w be the sides of a parallelogram formed by these axes and by lines parallel to the axes through the mass m as shown. Let $\hat{\boldsymbol{u}}$ and $\hat{\boldsymbol{w}}$ be unit vectors in the directions of increasing u and w. Using u and w as coordinates, the velocity of the mass m is

$$\boldsymbol{v} = \dot{u}\hat{\boldsymbol{u}}+\dot{w}\hat{\boldsymbol{w}}. \tag{9.18}$$

The kinetic energy is

$$T = \tfrac{1}{2}m\boldsymbol{v}\cdot\boldsymbol{v} = \tfrac{1}{2}m\dot{u}^2+\tfrac{1}{2}m\dot{w}^2+m\dot{u}\dot{w}\cos\alpha. \tag{9.19}$$

This is an example of a set of nonorthogonal coordinates in which a cross product

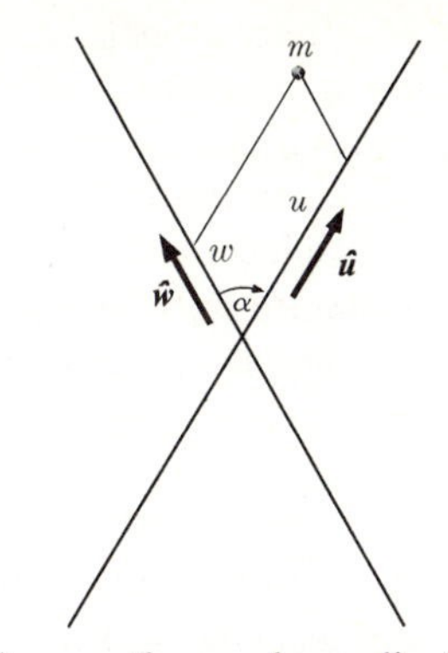

Fig. 9.2 A nonorthogonal coordinate system.

term in the velocities appears in the kinetic energy. The reason for using the term *orthogonal*, which means perpendicular, is clear from this example.

When systems of more than one particle are described in terms of generalized coordinates, it is usually safest to write out the kinetic energy first in cartesian coordinates and transform to generalized coordinates. However, in some cases, it is possible to write the kinetic energy directly in generalized coordinates. For example, if a rigid body rotates about an axis, we know that the kinetic energy is $\frac{1}{2}I\omega^2$, where ω is the angular velocity about that axis and I is the moment of inertia. Also, we can use the theorem proved in Section 4.9 that the total kinetic energy of a system of particles is the kinetic energy associated with the center of mass plus that associated with the internal coordinates. [See Eq. (4.127).] As an example, the kinetic energy of the two-particle system in terms of the coordinates X, Y, Z, x, y, z, defined by Eqs. (4.90) and (4.91), is

$$T = \tfrac{1}{2}M(\dot{X}^2+\dot{Y}^2+\dot{Z}^2)+\tfrac{1}{2}\mu(\dot{x}^2+\dot{y}^2+\dot{z}^2), \tag{9.20}$$

where M and μ are given by Eqs. (4.97) and (4.98). The result shows that this is an orthogonal coordinate system. If the linear velocity of each particle in a system can be written down directly in terms of the generalized coordinates and velocities, then the kinetic energy can immediately be written down.

We now note that the components of the linear momentum of particle i, according to Eq. (9.8), are

$$p_{ix} = m\dot{x}_i = \frac{\partial T}{\partial \dot{x}_i}, \qquad p_{iy} = m\dot{y}_i = \frac{\partial T}{\partial \dot{y}_i}, \qquad p_{iz} = m\dot{z}_i = \frac{\partial T}{\partial \dot{z}_i}. \tag{9.21}$$

In the case of a particle moving in a plane, the derivatives of T with respect to $\dot{r}$ and $\dot{\theta}$, as given by Eq. (9.14), are

$$p_r = m\dot{r} = \frac{\partial T}{\partial \dot{r}}, \qquad p_\theta = mr^2\dot{\theta} = \frac{\partial T}{\partial \dot{\theta}}, \tag{9.22}$$

where p_r is the component of linear momentum in the direction of increasing r, and p_θ is the angular momentum about the origin. Similar results will be found for spherical and cylindrical coordinates in three dimensions. In fact, it is not hard to show that for any coordinate q_k which measures the linear displacement of any particle or group of particles in a given direction, the linear momentum of that particle or group in the given direction is $\partial T/\partial \dot{q}_k$; and that for any coordinate q_k which measures the angular displacement of a particle or group of particles about an axis, their angular momentum about that axis is $\partial T/\partial \dot{q}_k$. This suggests that we define the *generalized momentum* p_k associated with the coordinate q_k by*

$$p_k = \frac{\partial T}{\partial \dot{q}_k}. \tag{9.23}$$

If q_k is a distance, p_k is the corresponding linear momentum. If q_k is an angle, p_k is the corresponding angular momentum. In other cases, p_k will have some other corresponding physical significance. According to Eq. (9.9), the generalized momentum p_k is

$$p_k = \sum_{l=1}^{3N} A_{kl}\dot{q}_l + B_k. \tag{9.24}$$

In the case of the coordinates X, Y, Z, x, y, z for the two-particle system, this definition gives

$$\begin{aligned} p_X &= M\dot{X}, \qquad p_Y = M\dot{Y}, \qquad p_Z = M\dot{Z}, \\ p_x &= \mu\dot{x}, \qquad p_y = \mu\dot{y}, \qquad p_z = \mu\dot{z}, \end{aligned} \tag{9.25}$$

where p_X, p_Y, p_Z are the components of the total linear momentum of the two particles, and p_x, p_y, p_z are the linear momentum components in the equivalent one-dimensional problem in x, y, z to which the two-body problem was reduced in Section 4.7. We shall see in the next section that the analogy between the generalized momenta p_k and the cartesian components of linear momentum can be extended to the equations of motion in generalized coordinates.

We now wish to define a generalized force. For this purpose it is convenient to define forces in terms of the work which they do when the particles move. Imagine a system of particles at positions specified by the coordinates $x_1, y_1, z_1, \ldots, z_N$ and acted on by forces $F_{1x}, F_{1y}, F_{1z}, \ldots, F_{Nz}$. If each particle in the system were to move to a nearby position, the new positions being specified by the coordinates $x_1+\delta x_1, y_1+\delta y_1, z_1+\delta z_1, \ldots, z_N+\delta z_N$, the work done would be

$$\delta W = \sum_{i=1}^{N} (F_{ix}\,\delta x_i + F_{iy}\,\delta y_i + F_{iz}\,\delta z_i). \tag{9.26}$$

*The kinetic energy T is defined by Eq. (9.9) as a function of $\dot{q}_1, \ldots, \dot{q}_{3N}; q_1, \ldots, q_{3N}$, and perhaps of t. The derivatives of this function T with respect to these variables will be denoted by the symbols for partial differentiation. Since $q_1, \ldots, q_{3N}; \dot{q}_1, \ldots, \dot{q}_{3N}$ are all functions of the time t for any given motion of the system, T is also a function of t alone for any given motion. The derivative of T with respect to time in this sense will be denoted by d/dt. The same remarks apply to any other quantity which may be written as a function of the coordinates and velocities and perhaps of t, and which is also a function of t alone for any given motion.

If we know the forces we can calculate δW for any set of small displacements δx_i, δy_i, δz_i. Conversely, if we knew either experimentally or theoretically the work δW for any displacement δx_i, δy_i, δz_i, then Eq. (9.26) would determine the forces. In order to solve for the forces we would need to write Eq. (9.26) for $3N$ independent sets of displacements. The set of increments δx_i, δy_i, δz_i is intended to represent any possible small displacements of the particles. We call this a *virtual displacement* of the system because it is not necessary that it represent any actual motion of the system. The increments δx_i, δy_i, δz_i may be expressed in terms of generalized coordinates as follows:

$$\begin{aligned} \delta x_i &= \sum_{k=1}^{3N} \frac{\partial x_i}{\partial q_k}\,\delta q_k, \\ \delta y_i &= \sum_{k=1}^{3N} \frac{\partial y_i}{\partial q_k}\,\delta q_k, \\ \delta z_i &= \sum_{k=1}^{3N} \frac{\partial z_i}{\partial q_k}\,\delta q_k, \end{aligned} \tag{9.27}$$

where $\delta q_1, \ldots, \delta q_{3N}$ are the differences in the generalized coordinates associated with the two sets of the positions of the particles. In the case of a moving coordinate system, we regard the time as fixed; that is, we express the changes in position in terms of the coordinate system at a particular time t. If we substitute Eqs. (9.27) in Eq. (9.26), we have, after rearranging terms:

$$\delta W = \sum_{k=1}^{3N} Q_k\,\delta q_k, \tag{9.28}$$

where

$$Q_k = \sum_{i=1}^{N} \left(F_{ix}\frac{\partial x_i}{\partial q_k} + F_{iy}\frac{\partial y_i}{\partial q_k} + F_{iz}\frac{\partial z_i}{\partial q_k} \right). \tag{9.29}$$

The coefficients Q_k depend on the forces acting on the particles, on the coordinates $q_1, \ldots, q_{3N}$, and possibly also on the time t. In view of the similarity in form between Eqs. (9.26) and (9.28), it is natural to call the quantity Q_k the *generalized force* associated with the coordinate q_k. We can define the generalized force Q_k directly, without reference to the cartesian coordinate system, as the coefficient which determines the work done in a virtual displacement in which q_k alone changes:

$$\delta W_k = Q_k\,\delta q_k, \tag{9.30}$$

where δW_k is the work done when the system moves in such a way that q_k increases by δq_k, all other coordinates remaining constant. Notice that the work in Eq. (9.26), and therefore also in Eq. (9.30), is to be computed from the values of the forces for the positions $x_1, \ldots, z_N$, or $q_1, \ldots, q_{3N}$; that is, we do not take account of any change in the forces during the virtual displacement.

If the forces $F_{1x}, \ldots, F_{Nz}$ are derivable from a potential energy $V(x_1, \ldots, z_N)$ [Eqs. (4.32)], then

$$\begin{aligned} \delta W &= -\delta V \\ &= -\sum_{i=1}^{N} \left(\frac{\partial V}{\partial x_i} \delta x_i + \frac{\partial V}{\partial y_i} \delta y_i + \frac{\partial V}{\partial z_i} \delta z_i \right). \end{aligned} \tag{9.31}$$

If V is expressed in terms of generalized coordinates, then

$$\begin{aligned} \delta W &= -\delta V \\ &= -\sum_{k=1}^{3N} \frac{\partial V}{\partial q_k} \delta q_k. \end{aligned} \tag{9.32}$$

By comparing this with Eq. (9.28), we see that

$$Q_k = -\frac{\partial V}{\partial q_k}, \tag{9.33}$$

which shows that in this sense also the definition of Q_k as a generalized force is a natural one. Equation (9.33) may also be verified by direct calculation of $\partial V/\partial q_k$:

$$\begin{aligned} \frac{\partial V}{\partial q_k} &= \sum_{i=1}^{N} \left(\frac{\partial V}{\partial x_i} \frac{\partial x_i}{\partial q_k} + \frac{\partial V}{\partial y_i} \frac{\partial y_i}{\partial q_k} + \frac{\partial V}{\partial z_i} \frac{\partial z_i}{\partial q_k} \right) \\ &= -\sum_{i=1}^{N} \left(F_{ix} \frac{\partial x_i}{\partial q_k} + F_{iy} \frac{\partial y_i}{\partial q_k} + F_{iz} \frac{\partial z_i}{\partial q_k} \right) \\ &= -Q_k. \end{aligned}$$

As an example, let us calculate the generalized forces associated with the polar coordinates r, θ, for a particle acted on by a force

$$\boldsymbol{F} = \hat{\boldsymbol{x}} F_x + \hat{\boldsymbol{y}} F_y = \hat{\boldsymbol{r}} F_r + \hat{\boldsymbol{\theta}} F_\theta. \tag{9.34}$$

If we use the definition (9.29), we have, using Eqs. (9.15):

$$\begin{aligned} Q_r &= F_x \frac{\partial x}{\partial r} + F_y \frac{\partial y}{\partial r} \\ &= F_x \cos\theta + F_y \sin\theta \\ &= F_r, \\ Q_\theta &= F_x \frac{\partial x}{\partial \theta} + F_y \frac{\partial y}{\partial \theta} \\ &= -rF_x \sin\theta + rF_y \cos\theta \\ &= rF_\theta. \end{aligned} \tag{9.35}$$

We see that Q_r is the component of force in the r-direction, and Q_θ is the torque acting to increase θ. It is usually quicker to use the definition (9.30), which enables us to bypass the cartesian coordinates altogether. If we consider a small displacement in which r changes to $r+\delta r$, with θ remaining constant, the work is

$$\delta W = F_r \, \delta r, \tag{9.36}$$

from which the first of Eqs. (9.35) follows. If we consider a displacement in which r is fixed and θ increases by $\delta\theta$, the work is

$$\delta W = F_\theta r \, \delta\theta, \tag{9.37}$$

from which the second of Eqs. (9.35) follows. In general, if q_k is a coordinate which measures the distance moved by some part of the mechanical system in a certain direction, and if F_k is the component in this direction of the total force acting on this part of the system, then the work done when q_k increases by δq_k, all other coordinates remaining constant, is

$$\delta W = F_k \, \delta q_k. \tag{9.38}$$

Comparing this with Eq. (9.30), we have

$$Q_k = F_k. \tag{9.39}$$

In this case, the generalized force Q_k is just the ordinary force F_k. If q_k measures the angular rotation of a certain part of the system about a certain axis, and if N_k is the total torque about that axis exerted on this part of the system, then the work done when q_k increases by δq_k is

$$\delta W = N_k \, \delta q_k. \tag{9.40}$$

Comparing this with Eq. (9.30), we have

$$Q_k = N_k. \tag{9.41}$$

The generalized force Q_k associated with an angular coordinate q_k is the corresponding torque.

9.2 LAGRANGE'S EQUATIONS

The analogy which led to the definitions of generalized momenta and generalized forces tempts us to suspect that the generalized equations of motion will equate the time rate of change of each momentum p_k to the corresponding force Q_k. To check this suspicion, let us calculate the time rate of change of p_k:

$$\frac{dp_k}{dt} = \frac{d}{dt}\left(\frac{\partial T}{\partial \dot{q}_k}\right). \tag{9.42}$$

We will need to start with Newton's equations of motion in cartesian form:

$$\begin{aligned} m_i\ddot{x}_i &= F_{ix}, \\ m_i\ddot{y}_i &= F_{iy}, \qquad [i = 1, \ldots, N] \\ m_i\ddot{z}_i &= F_{iz}. \end{aligned} \tag{9.43}$$

Therefore we express T in cartesian coordinates [Eq. (9.8)]. We then have

$$\frac{\partial T}{\partial \dot{q}_k} = \sum_{i=1}^{N} m_i \left(\dot{x}_i \frac{\partial \dot{x}_i}{\partial \dot{q}_k} + \dot{y}_i \frac{\partial \dot{y}_i}{\partial \dot{q}_k} + \dot{z}_i \frac{\partial \dot{z}_i}{\partial \dot{q}_k} \right), \tag{9.44}$$

where $\dot{x}_1, \dot{y}_1, \ldots, \dot{z}_N$ are given as functions of $q_1, \ldots, q_{3N}; \dot{q}_1, \ldots, \dot{q}_{3N}; t$ by Eqs. (9.6). Since $\partial x_i/\partial q_k$ and $\partial x_i/\partial t$ are functions only of $q_1, \ldots, q_{3N}; t$, we have by differentiating Eqs. (9.6):

$$\begin{aligned} \frac{\partial \dot{x}_i}{\partial \dot{q}_k} &= \frac{\partial x_i}{\partial q_k}, \\ \frac{\partial \dot{y}_i}{\partial \dot{q}_k} &= \frac{\partial y_i}{\partial q_k}, \qquad [i = 1, \ldots, N; k = 1, \ldots, 3N] \\ \frac{\partial \dot{z}_i}{\partial \dot{q}_k} &= \frac{\partial z_i}{\partial q_k}. \end{aligned} \tag{9.45}$$

By substituting from Eqs. (9.45) in Eq. (9.44), and differentiating again with respect to t, we obtain

$$\begin{aligned} \frac{dp_k}{dt} &= \sum_{i=1}^{N} m_i \left(\ddot{x}_i \frac{\partial x_i}{\partial q_k} + \ddot{y}_i \frac{\partial y_i}{\partial q_k} + \ddot{z}_i \frac{\partial z_i}{\partial q_k} \right) \\ &\quad + \sum_{i=1}^{N} m_i \left(\dot{x}_i \frac{d}{dt}\frac{\partial x_i}{\partial q_k} + \dot{y}_i \frac{d}{dt}\frac{\partial y_i}{\partial q_k} + \dot{z}_i \frac{d}{dt}\frac{\partial z_i}{\partial q_k} \right). \end{aligned} \tag{9.46}$$

According to Newton's equations of motion (9.43), and the definition (9.29), the first term in Eq. (9.46) is

$$\begin{aligned} \sum_{i=1}^{N} m_i \left(\ddot{x}_i \frac{\partial x_i}{\partial q_k} + \ddot{y}_i \frac{\partial y_i}{\partial q_k} + \ddot{z}_i \frac{\partial z_i}{\partial q_k} \right) &= \sum_{i=1}^{N} \left(F_{ix} \frac{\partial x_i}{\partial q_k} + F_{iy} \frac{\partial y_i}{\partial q_k} + F_{iz} \frac{\partial z_i}{\partial q_k} \right) \\ &= Q_k. \end{aligned} \tag{9.47}$$

The derivatives appearing in the last term in Eq. (9.46) are calculated as follows:

$$\frac{d}{dt}\frac{\partial x_i}{\partial q_k} = \sum_{l=1}^{3N} \frac{\partial^2 x_i}{\partial q_k \, \partial q_l} \dot{q}_l + \frac{\partial^2 x_i}{\partial q_k \, \partial t} = \frac{\partial}{\partial q_k} \left(\sum_{l=1}^{3N} \frac{\partial x_i}{\partial q_l} \dot{q}_l + \frac{\partial x_i}{\delta t} \right) = \frac{\partial \dot{x}_i}{\partial q_k}, \tag{9.48}$$

where we have made use of Eq. (9.6). Similar expressions hold for y and z. Thus the last sum in Eq. (9.46) is

$$\begin{aligned} &\sum_{i=1}^{N} m_i \left(\dot{x}_i \frac{d}{dt}\frac{\partial x_i}{\partial q_k} + \dot{y}_i \frac{d}{dt}\frac{\partial y_i}{\partial q_k} + \dot{z}_i \frac{d}{dt}\frac{\partial z_i}{\partial q_k} \right) \\ &\quad = \sum_{i=1}^{N} m_i \left(\dot{x}_i \frac{\partial \dot{x}_i}{\partial q_k} + \dot{y}_i \frac{\partial \dot{y}_i}{\partial q_k} + \dot{z}_i \frac{\partial \dot{z}_i}{\partial q_k} \right) = \frac{\partial}{\partial q_k} \sum_{i=1}^{N} \tfrac{1}{2} m_i (\dot{x}_i^2 + \dot{y}_i^2 + \dot{z}_i^2) \\ &\quad = \frac{\partial T}{\partial q_k}. \end{aligned} \tag{9.49}$$

We have finally:

$$\frac{dp_k}{dt} = Q_k + \frac{\partial T}{\partial q_k}, \qquad k = 1, \ldots, 3N. \tag{9.50}$$

Our original expectation was not quite correct, in that we must add to the generalized force Q_k another term $\partial T/\partial q_k$ in order to get the rate of change of momentum $\dot{p}_k$. To see its meaning, consider the kinetic energy of a particle in terms of plane polar coordinates, as given by Eq. (9.14). In this case,

$$\frac{\partial T}{\partial r} = mr\dot{\theta}^2, \tag{9.51}$$

and if we make use of Eqs. (9.22) and (9.35), the equation of motion (9.50) for $q_k = r$ is

$$m\ddot{r} = F_r + mr\dot{\theta}^2. \tag{9.52}$$

If we compare this with Eq. (3.207), which results from a direct application of Newton's law of motion, we see that the term $\partial T/\partial r$ is part of the mass times acceleration which appears here transposed to the right side of the equation. In fact, $\partial T/\partial r$ is the "centrifugal force" which must be added in order to write the equation of motion for r in the form of Newton's equation for motion in a straight line. Had we been a bit more clever originally, we should have expected that some such term might have to be included. We may call $\partial T/\partial q_k$ a "fictitious force" which appears if the kinetic energy depends on the coordinate q_k. This will be the case when the coordinate system involves "curved" coordinates, that is, if constant generalized velocities $\dot{q}_1, \ldots, \dot{q}_{3N}$ result in curved motions of some parts of the mechanical system. Equations (9.50) are usually written in the form

$$\frac{d}{dt}\left(\frac{\partial T}{\partial \dot{q}_k}\right) - \frac{\partial T}{\partial q_k} = Q_k, \qquad k = 1, \ldots, 3N. \tag{9.53}$$

These are Lagrange's equations.

If a potential energy exists, so that the forces Q_k are derivable from a potential energy function [Eq. (9.33)], we may introduce the *Lagrangian function*

$$L(q_1, \ldots, q_{3N}; \dot{q}_1, \ldots, \dot{q}_{3N}; t) = T - V, \tag{9.54}$$

where T depends on both $q_1, \ldots, q_{3N}$ and $\dot{q}_1, \ldots, \dot{q}_{3N}$, but V depends only on $q_1, \ldots, q_{3N}$ (and possibly t), so that

$$\frac{d}{dt}\frac{\partial L}{\partial \dot{q}_k} = \frac{d}{dt}\frac{\partial T}{\partial \dot{q}_k}, \tag{9.55}$$

$$\frac{\partial L}{\partial q_k} = \frac{\partial T}{\partial q_k} - \frac{\partial V}{\partial q_k} = \frac{\partial T}{\partial q_k} + Q_k. \tag{9.56}$$

Hence Eqs. (9.53) can be written in this case in the compact form

$$\frac{d}{dt}\left(\frac{\partial L}{\partial \dot{q}_k}\right) - \frac{\partial L}{\partial q_k} = 0, \qquad k = 1, \ldots, 3N. \tag{9.57}$$

The term *Lagrange's equations* is sometimes restricted to equations in the form (9.57). In nearly all cases of interest in physics (although not in engineering), the equations of motion can be written in the form (9.57). The most important exception is the case where frictional forces are involved, but such forces do not usually appear in atomic or astronomical problems.

Since Lagrange's equations have been derived from Newton's equations of motion, they do not represent a new physical theory, but merely a different but equivalent way of expressing the same laws of motion. As the example of Eqs. (9.52) and (3.207) illustrates, the equations we get by Lagrange's method can also be obtained by a direct application of Newton's law of motion. However, in complicated cases it is often easier to work out the kinetic energy and the forces or potential energy in generalized coordinates, and write the equations in Lagrangian form. Particularly in problems involving constraints, as we shall see in Section 9.4, the Lagrangian method is much easier to apply. The chief value of Lagrange's equations is, however, probably a theoretical one. From the manner in which they were derived, it is evident that Lagrange's equations (9.57) or (9.53) hold in the same form in any system of generalized coordinates. It can also be verified by direct computation (see Problem 24) that if Eqs. (9.57) hold in any coordinate system for any function $L(q_1, \ldots, q_{3N}; \dot{q}_1, \ldots, \dot{q}_{3N}; t)$, then equations of the same form hold in any other coordinate system. The Lagrangian function L has the same value, for any given set of positions and velocities of the particles, no matter in what coordinate system it may be expressed, but the form of the function L may be different in different coordinate systems. The fact that Lagrange's equations have the same form in all coordinate systems is largely responsible for their theoretical importance. Lagrange's equations represent a uniform way of writing the equations of motion of a system, which is independent of the kind of coordinate system used. They form a starting point for more advanced formulations of mechanics. In developing the general theory of relativity, in which cartesian coordinates may not even exist, Lagrange's equations are particularly important.

9.3 EXAMPLES

We first consider a system of particles $m_1, \ldots, m_N$, located by cartesian coordinates, and show that in this case Lagrange's equations become the Newtonian equations of motion. The kinetic energy is

$$T = \sum_{i=1}^{N} \tfrac{1}{2} m_i (\dot{x}_i^2 + \dot{y}_i^2 + \dot{z}_i^2), \tag{9.58}$$

and

$$\frac{\partial T}{\partial x_i} = \frac{\partial T}{\partial y_i} = \frac{\partial T}{\partial z_i} = 0, \tag{9.59}$$

$$\frac{\partial T}{\partial \dot{x}_i} = m_i \dot{x}_i, \qquad \frac{\partial T}{\partial \dot{y}_i} = m_i \dot{y}_i, \qquad \frac{\partial T}{\partial \dot{z}_i} = m_i \dot{z}_i. \tag{9.60}$$

The generalized force associated with each cartesian coordinate is just the ordinary force, as we see either from Eq. (9.29), or by comparing Eq. (9.28) with Eq. (9.26). Hence the equations of motion (9.53) are

$$\begin{aligned} \frac{d}{dt}\left(\frac{\partial T}{\partial \dot{x}_i}\right) - \frac{\partial T}{\partial x_i} &= m_i \ddot{x}_i = F_{ix}, \\ \frac{d}{dt}\left(\frac{\partial T}{\partial \dot{y}_i}\right) - \frac{\partial T}{\partial y_i} &= m_i \ddot{y}_i = F_{iy}, \qquad [i = 1, \ldots, N] \\ \frac{d}{dt}\left(\frac{\partial T}{\partial \dot{z}_i}\right) - \frac{\partial T}{\partial z_i} &= m_i \ddot{z}_i = F_{iz}. \end{aligned} \tag{9.61}$$

For a particle moving in a plane, the kinetic energy in polar coordinates is given by Eq. (9.14), and the forces Q_r and Q_θ by Eqs. (9.35). The Lagrange equations are

$$m\ddot{r} - mr\dot{\theta}^2 = F_r, \tag{9.62}$$

$$\frac{d}{dt}(mr^2\dot{\theta}) = rF_\theta. \tag{9.63}$$

These equations were obtained in Section 3.13 by elementary methods.

We now consider the rotating coordinate system defined by Eqs. (9.4) or (9.5). The kinetic energy is given by Eq. (9.16), and the generalized forces Q_r and Q_θ will be the same as in the previous example. Lagrange's equations in this case are

$$m\ddot{r} - mr\dot{\theta}^2 - 2m\omega r\dot{\theta} - m\omega^2 r = F_r, \tag{9.64}$$

$$\frac{d}{dt}(mr^2\dot{\theta}) + 2m\omega r\dot{r} = rF_\theta. \tag{9.65}$$

The reader should verify that the third term on the left in Eq. (9.64) is the negative of the coriolis force in the r-direction due to the rotation of the coordinate system, and that the fourth term is the negative of the centrifugal force. The second term in Eq. (9.65) is the negative of the coriolis torque in the θ-direction. Thus the necessary fictitious forces are automatically included when we write Lagrange's equations in a moving coordinate system. It must be noticed, however, that we use the actual kinetic energy [Eq. (9.16)] with respect to a coordinate system at rest, expressed in terms of the rotating coordinates, and not the kinetic energy as it would appear in the rotating system if we ignored the motion of the coordinate system.

9.4 SYSTEMS SUBJECT TO CONSTRAINTS

One important class of mechanical problems in which Lagrange's equations are particularly useful comprises systems which are subject to constraints.

A rigid body is a good example of a system of particles subject to constraints. A constraint is a restriction on the freedom of motion of a system of particles in the form of a condition which must be satisfied by their coordinates, or by the allowed changes in their coordinates. For example, a very simple hypothetical rigid body would be a pair of particles connected by a rigid weightless rod of length l. These particles are subject to a constraint which requires that they remain a distance l apart. In terms of their cartesian coordinates, the constraint is

$$[(x_2-x_1)^2+(y_2-y_1)^2+(z_2-z_1)^2]^{1/2} = l. \tag{9.66}$$

If we use the coordinates X, Y, Z of the center of mass and spherical coordinates r, θ, φ to locate particle 2 with respect to particle 1 as origin, the constraint takes the simple form:

$$r = l. \tag{9.67}$$

There are thus only five coordinates X, Y, Z, θ, φ left to determine. Each constraint which can be expressed in the form of an equation like (9.66) enables us to eliminate one of the coordinates by choosing coordinates in such a manner that one of them is held constant by the constraint. For a rigid body, the constraints require that the mutual distances of all pairs of particles remain constant. For a body containing N particles, there are $\frac{1}{2}N(N-1)$ pairs of particles. However, it is not hard to show that it is sufficient to specify the mutual distances of $3N-6$ pairs, if $N \geq 3$. Hence we can replace the $3N$ cartesian coordinates of the N particles by $3N-6$ mutual distances, 3 coordinates of the center of mass, and 3 coordinates describing the orientation of the body. Since the $3N-6$ mutual distances are all constant, the problem is reduced to one of finding the motion in terms of six coordinates. Another example of a system subject to a constraint is that of a bead sliding on a wire. The wire is situated along a certain curve in space, and the constraints require that the position of the bead lie on this curve. Since the coordinates of the points along a space curve satisfy two equations (e.g., the equations of two surfaces which intersect along the curve), there are two constraints, and we can locate the position of the bead by a single coordinate. (Can you suggest a suitable coordinate?) If the wire is moving, we have a moving constraint, and our single coordinate is relative to a moving system of reference. Constraints which can be expressed in the form of an equation relating the coordinates are called *holonomic*. All the above examples involve holonomic constraints.

Constraints may also be specified by a restriction on the velocities, rather than on the coordinates. For example, a cylinder of radius a, rolling and sliding

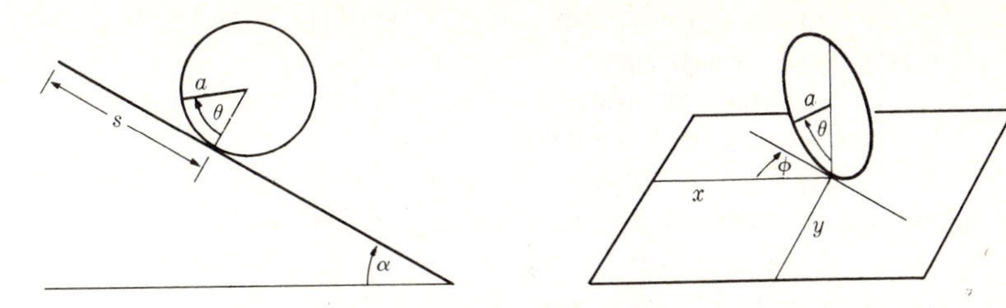

Fig. 9.3 A cylinder rolling down an incline.

Fig. 9.4 A disk rolling on a horizontal plane.

down an inclined plane, with its axis always horizontal, can be located by two coordinates s and θ, as in Fig. 9.3. The coordinate s measures the distance the cylinder has moved down the plane, and the coordinate θ is the angle that a fixed radius in the cylinder has rotated from the radius to the point of contact with the plane. Now suppose that the cylinder is rolling without slipping. Then the velocities $\dot{s}$ and $\dot{\theta}$ must be related by the equation

$$\dot{s} = a\dot{\theta}, \tag{9.68}$$

which may also be written

$$ds = a\, d\theta. \tag{9.69}$$

This equation can be integrated:

$$s - a\theta = C, \tag{9.70}$$

where C is a constant. This equation is of the same type as Eq. (9.66), and shows that the constraint is holonomic, although it was initially expressed in terms of velocities. If a constraint on the velocities, like Eq. (9.68), can be integrated to give a relation between the coordinates, like Eq. (9.70), then the constraint is holonomic. There are systems, however, in which such equations of constraint cannot be integrated. An example is a disk of radius a rolling on a horizontal table, as in Fig. 9.4. For simplicity, we assume that the disk cannot tip over, and that the diameter which touches the table is always vertical. Four coordinates are required to specify the position of the disk. The coordinates x and y locate the point of contact on the plane; the angle φ determines the orientation of the plane of the disk relative to the x-axis; and the angle θ is the angle between a radius fixed in the disk and the vertical. If we now require that the disk roll without slipping (it can also rotate about the vertical axis), this implies two equations of constraint. The velocity of the point of contact perpendicular to the plane of the disk must be zero:

$$\dot{x} \sin \varphi + \dot{y} \cos \varphi = 0, \tag{9.71}$$

and the velocity parallel to the plane of the disk must be

$$\dot{x} \cos \varphi - \dot{y} \sin \varphi = a\dot{\theta}. \tag{9.72}$$

It is not possible to integrate these equations to get two relations between the coordinates x, y, θ, φ. To see this, we note that by rolling the disk without slipping, and by rotating it about a vertical axis, we can bring the disk to any point x, y, with any angle φ between the plane of the disk and the x-axis, and with any point on the circumference of the disk in contact with the table, i.e., any angle θ. For if the disk is at any point x, y, and the desired point on the circumference is not in contact with the table, we may roll the disk around a circle whose circumference is of proper length, so that when it returns to x, y, the desired point will be in contact with the table. It may then be rotated to the desired angle φ. This shows that the four coordinates x, y, θ, φ are independent of one another, and there cannot be any relation between them. It must therefore be impossible to integrate Eqs. (9.71) and (9.72), and consequently this is an example of a nonholonomic constraint.

The number of independent ways in which a mechanical system can move without violating any constraints which may be imposed is called the number of *degrees of freedom* of the system. To be more precise, the number of degrees of freedom is the number of quantities which must be specified in order to determine the velocities of all particles in the system for any motion which does not violate the constraints. For example, a single particle moving in space has three degrees of freedom, but if it is constrained to move along a certain curve, it has only one. A system of N free particles has $3N$ degrees of freedom, a rigid body has 6 degrees of freedom (three translational and three rotational), and a rigid body constrained to rotate about an axis has one degree of freedom. The disk shown in Fig. 9.4 has four degrees of freedom if it is allowed to slip on the table, because we need then to specify $\dot{x}$, $\dot{y}$, $\dot{\theta}$, $\dot{\varphi}$. But if the disk is required to roll without slipping, there are only two degrees of freedom, because if $\dot{\varphi}$ and any one of the velocities $\dot{x}$, $\dot{y}$, $\dot{\theta}$ are given, the remaining two can be found from Eqs. (9.71) and (9.72). The disk is only free to roll, and to rotate about a vertical axis. For holonomic systems, the number of degrees of freedom is equal to the minimum number of coordinates required to specify the configuration of the system when coordinates held constant by the constraints are eliminated. Nonholonomic constraints occur in some problems in which bodies roll without slipping, but they are not of very great importance in physics. We shall therefore restrict our attention to holonomic systems.

For a holonomic system of N particles subject to c independent constraints, we can express the constraints as c relations which must hold between the $3N$ cartesian coordinates (including possibly the time if the constraints are changing with time):

$$\begin{aligned} h_1(x_1, y_1, \ldots, z_N; t) &= a_1, \\ h_2(x_1, y_1, \ldots, z_N; t) &= a_2, \\ &\vdots \\ h_c(x_1, y_1, \ldots, z_N; t) &= a_c, \end{aligned} \tag{9.73}$$

where $h_1, \ldots, h_c$ are c specified functions. The number of degrees of freedom will be

$$f = 3N - c. \tag{9.74}$$

As Eqs. (9.73) are independent, we may solve them for c of the $3N$ cartesian coordinates in terms of the other $3N-c$ coordinates and the constants $a_1, \ldots, a_c$. Thus only $3N-c$ coordinates need be specified, and the remainder can be found from Eqs. (9.73) if the constants $a_1, \ldots, a_c$ are known. We may take as generalized coordinates these $3N-c$ cartesian coordinates and the c quantities $a_1, \ldots, a_c$ defined by Eqs. (9.73), and held constant by the constraints. Or we may define $3N-c$ generalized coordinates $q_1, \ldots, q_f$, in any convenient way:

$$\begin{aligned} q_1 &= q_1(x_1, y_1, \ldots, z_N; t), \\ q_2 &= q_2(x_1, y_1, \ldots, z_N; t), \\ &\vdots \\ q_f &= q_f(x_1, y_1, \ldots, z_N; t). \end{aligned} \tag{9.75}$$

Equations (9.73) and (9.75) define a set of $3N$ coordinates $q_1, \ldots, q_f; a_1, \ldots, a_c$, and are analogous to Eqs. (9.1). They may be solved for the cartesian coordinates:

$$\begin{aligned} x_1 &= x_1(q_1, \ldots, q_f; a_1, \ldots, a_c; t), \\ y_1 &= y_1(q_1, \ldots, q_f; a_1, \ldots, a_c; t), \\ &\vdots \\ z_N &= z_N(q_1, \ldots, q_f; a_1, \ldots, a_c; t). \end{aligned} \tag{9.76}$$

Now let $Q_1, \ldots, Q_f, Q_{f+1}, \ldots, Q_{f+c}$ be the generalized forces corresponding to the coordinates $q_1, \ldots, q_f; a_1, \ldots, a_c$. We have then a set of Lagrange equations for the constrained coordinates and another for the unconstrained coordinates:

$$\frac{d}{dt}\frac{\partial T}{\partial \dot{q}_k} - \frac{\partial T}{\partial q_k} = Q_k, \qquad k = 1, \ldots, f, \tag{9.77}$$

$$\frac{d}{dt}\frac{\partial T}{\partial \dot{a}_j} - \frac{\partial T}{\partial a_j} = Q_{f+j}, \qquad j = 1, \ldots, c; \; c+f = 3N. \tag{9.78}$$

The importance of this separation of the problem into two groups of equations is that the forces of constraint can be so chosen that they do no work unless the constraints are violated, as we shall show in the next paragraph. If this is true, then according to the definition (9.30) of the generalized force, the forces of constraint do not contribute to the generalized force Q_k associated with an unconstrained coordinate q_k. Since the values of the constrained coordinates $a_1, \ldots, a_c$ are held constant, we can solve Eqs. (9.77) for the motion of the system in terms of the coordinates $q_1, \ldots, q_f$, treating $a_1, \ldots, a_c$ as given constants, without knowing the forces of constraint. This is a great advantage, for the forces of constraint depend upon how the system is moving, and cannot, in general, be determined until after the motion has been found. All we usually know about the constraining forces is that they have whatever values are required to maintain the constraints. Having solved Eqs. (9.77) for $q_1(t), \ldots, q_f(t)$, we may then, if we wish, substitute these functions in Eqs. (9.78) and calculate the forces of constraint. This may be a matter of considerable interest to the engineer who needs to verify that the constraining members are strong enough to withstand the constraining forces. Lagrange's equations thus reduce the problem of finding the motion of any holonomic system with f degrees of freedom to the problem of solving f second-order differential equations (9.77). When we speak of the generalized coordinates, the constrained coordinates $a_1, \ldots, a_c$ may or may not be included, as convenient.

If a bead slides on a frictionless wire, the wire can only exert constraining forces perpendicular to itself, so that no work is done on the bead so long as it stays on the wire.* If there is friction, we can separate the force on the bead into a component perpendicular to the wire which holds the bead on the wire without doing any work, and a frictional component along the wire which does work and will therefore have to be included in the generalized force associated with motion along the wire. If the frictional force depends on the normal force, as it does for dry sliding friction, then we cannot solve Eqs. (9.77) first, independently of Eqs. (9.78), and some of the advantage of the Lagrangian method is lost. In this case, we first use Eqs. (9.78) to find the normal forces in terms of $q_1, \ldots, q_f; \dot{q}_1, \ldots, \dot{q}_f$, and substitute these in the frictional force terms in Eqs. (9.77). If two particles are held a fixed distance apart by a rigid rod, then by Newton's third law, the force exerted by the rod on one particle is equal and opposite to that on the other. It was shown in Section 5.1 that no net work is done on the system by the rod so long as the constraint is not violated, that is, so long as the rod is not stretched or compressed. A similar situation will be found in all other cases; the constraints *could* always be maintained by forces which do no work.

If the forces $Q_1, \ldots, Q_f$ are derivable from a potential energy function, then we can define a Lagrangian function $L(q_1, \ldots, q_f; \dot{q}_1, \ldots, \dot{q}_f)$ which may in some cases depend on t, and which may also depend on the constants $a_1, \ldots, a_c$. The first f Lagrange equations (9.77) can then be written in the form

$$\frac{d}{dt}\frac{\partial L}{\partial \dot{q}_k} - \frac{\partial L}{\partial q_k} = 0, \qquad k = 1, \ldots, f. \tag{9.79}$$

9.5 EXAMPLES OF SYSTEMS SUBJECT TO CONSTRAINTS

A simple mechanical system involving constraints is the Atwood's machine shown in Fig. 9.5. Weights m_1, m_2 are connected by a rope of length l over a fixed pulley. We assume the weights move only vertically, so that we have only one degree of freedom. We take as coordinates the distance x of m_1 below the pulley axle, and l, the length of the rope. The coordinate l is constrained to have a constant value, and could be left out of consideration from the start if we wish only to find the

*If the wire is moving, the force exerted by the wire may do work on the bead, but the virtual displacements in terms of which the generalized forces have been defined are to be imagined as taking place at a fixed instant of time, and for such a displacement which does not violate the constraints, no work is done. Hence even in the case of moving constraints, the constraining forces do not appear in the generalized forces associated with the unconstrained coordinates.

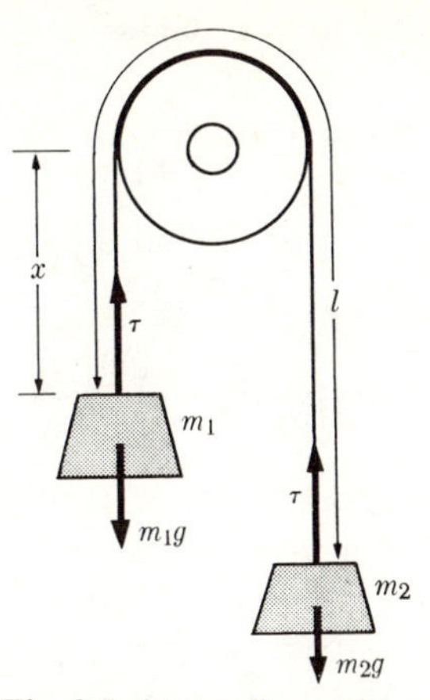

Fig. 9.5 Atwood's machine.

motion. If we also want to find the tension in the string, we must include l as a coordinate. The kinetic energy is

$$T = \tfrac{1}{2}m_1\dot{x}^2 + \tfrac{1}{2}m_2(\dot{l}-\dot{x})^2. \tag{9.80}$$

The only forces acting on m_1 and m_2 are the tension τ in the rope and the force of gravity. The work done when x increases by δx, l remaining constant, is

$$\begin{aligned}\delta W &= (m_1 g - \tau)\,\delta x - (m_2 g - \tau)\,\delta x \\ &= (m_1 - m_2)g\,\delta x = Q_x\,\delta x,\end{aligned} \tag{9.81}$$

so that

$$Q_x = (m_1 - m_2)g. \tag{9.82}$$

Note that Q_x is independent of τ. The work done when l increases by δl, x remaining constant, is

$$\delta W = (m_2 g - \tau)\,\delta l = Q_l\,\delta l, \tag{9.83}$$

so that

$$Q_l = m_2 g - \tau. \tag{9.84}$$

Notice that in order to obtain an equation involving the force of constraint τ, we must consider a motion which violates the constraint. This is also true if we wish to measure a force physically; we must allow at least a small motion in the direction of the force. The Lagrange equations of motion are (since $\dot{l} = \ddot{l} = 0$)

$$\frac{d}{dt}\left(\frac{\partial T}{\partial \dot{x}}\right) - \frac{\partial T}{\partial x} = (m_1 + m_2)\ddot{x} = (m_1 - m_2)g, \tag{9.85}$$

$$\frac{d}{dt}\left(\frac{\partial T}{\partial \dot{l}}\right) - \frac{\partial T}{\partial l} = -m_2\ddot{x} = m_2 g - \tau. \tag{9.86}$$

The first equation is to be solved to find the motion:

$$x = x_0 + v_0 t + \tfrac{1}{2}\frac{m_1 - m_2}{m_1 + m_2} g t^2. \tag{9.87}$$

The second equation can then be used to find the tension τ necessary to maintain the constraint:

$$\tau = m_2(g + \ddot{x}) = \frac{2m_1 m_2}{m_1 + m_2} g. \tag{9.88}$$

In this case the tension is independent of time and can be found from Eqs. (9.85) and (9.86) immediately, although in most cases the constraining forces depend on the motion and can be determined only after the motion is found. Equations (9.85) and (9.86) have an obvious physical interpretation and could be written down immediately from elementary considerations, as was done in Section 1.7.

A problem of little practical importance, but which is quite instructive, is that in which one cylinder rolls upon another, as shown in Fig. 9.6. The cylinder of radius a is fixed, and the cylinder of radius αa rolls around it under the action of gravity. Suppose we are given that the coefficient of static friction between the cylinders is μ, the coefficient of sliding friction is zero,* and that the moving cylinder starts from rest with its center vertically above the center of the fixed cylinder. We shall assume that the axis of the moving cylinder remains horizontal during the motion. It is advisable in all problems, and essential in this one, to think carefully about the motion before attempting to find the mathematical solution. It is clear that the moving cylinder cannot roll all the way around the fixed cylinder, for the normal force $\boldsymbol{F}$ which is exerted by the fixed cylinder on the moving one can only be directed outward, never inward. Therefore at some point, the

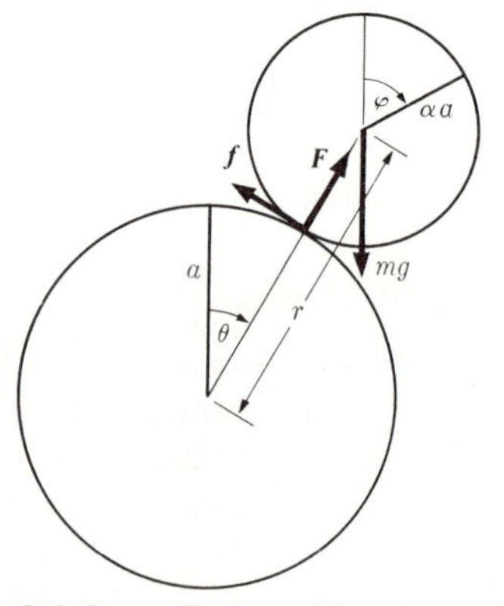

Fig. 9.6 One cylinder rolling on another.

*This implies that the moving cylinder either rolls without slipping, if the static friction is great enough, or slips without any friction at all. The latter assumption is made to simplify the problem.

moving cylinder will fly off the fixed one. The point at which it flies off is the point at which

$$F = 0. \tag{9.89}$$

Furthermore, the cylinder cannot continue to roll without slipping right up to the point at which it flies off, for the frictional force f which prevents slipping is limited by the condition

$$f \leq \mu F, \tag{9.90}$$

and μF will presumably become too small to prevent slipping before the point at which Eq. (9.89) holds. The motion therefore is divided into three parts. At first the cylinder rolls without slipping through an angle θ_1 determined by the condition

$$f = \mu F. \tag{9.91}$$

Beyond the angle θ_1, the cylinder slides without friction until it reaches the angle θ_2 determined by Eq. (9.89), after which it leaves the fixed cylinder and falls freely. We may anticipate some mathematical difficulties with the initial part of the motion due to the fact that the initial position of the moving cylinder is one of unstable equilibrium. Physically there is no difficulty, since the slightest disturbance will cause the cylinder to roll down, but mathematically there may be a difficulty which we must watch out for, inasmuch as the needed slight disturbance will not appear in the equations.

Let us find that part of the motion when the moving cylinder rolls without slipping. There is then only one degree of freedom, and we shall specify the position of the cylinder by the angle θ between the vertical and the line connecting the centers of the two cylinders. In order to compute the kinetic energy, we introduce the auxiliary angle φ through which the moving cylinder has rotated about its axis. The condition that the cylinder roll without slipping leads to the equation of constraint:

$$a\dot{\theta} = \alpha a(\dot{\varphi} - \dot{\theta}), \tag{9.92}$$

which can be integrated in the form

$$(1+\alpha)\theta = \alpha\varphi. \tag{9.93}$$

If we were concerned only with the rolling motion, we could now proceed to set up the Lagrange equation for θ, but inasmuch as we need to know the forces of constraint $\boldsymbol{F}$ and $\boldsymbol{f}$, it is necessary to introduce additional coordinates which are maintained constant by these constraining forces. The frictional force $\boldsymbol{f}$ maintains the constraint (9.93), and an appropriate coordinate is

$$\gamma = \theta - \frac{\alpha\varphi}{1+\alpha}. \tag{9.94}$$

So long as the cylinder rolls without slipping, $\gamma = 0$; γ measures the angle of slip around the fixed cylinder. The normal force $\boldsymbol{F}$ maintains the distance r between the centers of the cylinders:

$$r = a+\alpha a = (1+\alpha)a. \tag{9.95}$$

The kinetic energy of the rolling cylinder is the energy associated with the motion of its center of mass plus the rotational energy about the center of mass:

$$T = \tfrac{1}{2}m(\dot{r}^2+r^2\dot{\theta}^2)+\tfrac{1}{2}I\dot{\varphi}^2. \tag{9.96}$$

After substituting φ from Eq. (9.94), and since $I = \frac{1}{2}m\alpha^2a^2$, for a solid cylinder of radius αa, we have

$$T = \tfrac{1}{2}m\dot{r}^2+\tfrac{1}{2}mr^2\dot{\theta}^2+\tfrac{1}{4}m(1+\alpha)^2a^2(\dot{\theta}^2-2\dot{\gamma}\dot{\theta}+\dot{\gamma}^2). \tag{9.97}$$

The equations of constraint [Eq. (9.95) and $\gamma = 0$] *must not be used* until *after* the equations of motion are written down. The generalized forces are most easily determined with the help of Eq. (9.30); they are*

$$Q_\theta = mgr \sin \theta, \tag{9.98}$$

$$Q_\gamma = -fa(1+\alpha), \tag{9.99}$$

$$Q_r = F-mg \cos \theta. \tag{9.100}$$

The Lagrange equations for θ, γ, and r are now

$$m[r^2+\tfrac{1}{2}a^2(1+\alpha)^2]\ddot{\theta}+2mr\dot{r}\dot{\theta}-\tfrac{1}{2}ma^2(1+\alpha)^2\ddot{\gamma} = mgr \sin \theta, \tag{9.101}$$

$$-\tfrac{1}{2}ma^2(1+\alpha)^2\ddot{\theta}+\tfrac{1}{2}ma^2(1+\alpha)^2\ddot{\gamma} = -fa(1+\alpha), \tag{9.102}$$

$$m\ddot{r}-mr\dot{\theta}^2 = F-mg \cos \theta. \tag{9.103}$$

We can now insert the constraints $\gamma = 0$ and $r = (1+\alpha)a$, so that these equations become

$$\tfrac{3}{2}(1+\alpha)^2ma^2\ddot{\theta} = (1+\alpha)mga \sin \theta, \tag{9.104}$$

$$f = \tfrac{1}{2}(1+\alpha)ma\ddot{\theta}, \tag{9.105}$$

$$F = mg \cos \theta-(1+\alpha)ma\dot{\theta}^2. \tag{9.106}$$

Had we ignored the terms involving $\dot{\gamma}$ in the kinetic energy, the θ equation, which determines the motion, would have come out correctly, but the equation for the constraining force f would have been missing a term. This happens when the constrained coordinates are not orthogonal to the unconstrained coordinates, since a cross term ($\dot{\gamma}\dot{\theta}$) then appears in the kinetic energy. We had a similar situation in the Atwood machine problem as the reader may have noticed.

The equation of motion (9.104) can be solved by the energy method. The total energy, so long as the cylinder rolls without slipping, is

$$\tfrac{3}{4}(1+\alpha)^2ma^2\dot{\theta}^2+(1+\alpha)mga \cos \theta = E, \tag{9.107}$$

---

*The reader will find it an instructive exercise to verify these formulas.

and is constant, as can easily be shown from Eq. (9.104), and as we know anyway since the gravitational force is conservative and the forces of constraint do no work. Since the moving cylinder starts from rest at $\theta = 0$,

$$E = (1+\alpha)mga. \tag{9.108}$$

We substitute this in Eq. (9.107) and solve for $\dot{\theta}$:

$$\dot{\theta} = 2\left(\frac{\beta g}{a}\right)^{1/2} \sin\frac{\theta}{2}, \tag{9.109}$$

where

$$\beta = \frac{2}{3(1+\alpha)}. \tag{9.110}$$

We can now integrate to find $\theta(t)$:

$$\int_0^\theta \frac{\frac{1}{2}\,d\theta}{\sin\theta/2} = \left(\frac{\beta g}{a}\right)^{1/2} \int_0^t dt, \tag{9.111}$$

$$\left[\ln\tan\frac{\theta}{4}\right]_0^\theta = \left(\frac{\beta g}{a}\right)^{1/2} t. \tag{9.112}$$

When we substitute the lower limit $\theta = 0$, we run into a difficulty, for $\ln 0 = -\infty$! This is the expected difficulty due to the fact that $\theta = 0$ is a point of equilibrium, albeit unstable. If there is no disturbance whatever, it will take an infinite time for the cylinder to roll off the equilibrium point. Let us suppose, however, that it does roll off due to some slight disturbance, and let us take the time $t = 0$ as the time when the angle θ has some small value θ_0. There is now no difficulty, and we have

$$\tan\frac{\theta}{4} = \left(\tan\frac{\theta_0}{4}\right)\exp\left[\left(\frac{\beta g}{a}\right)^{1/2} t\right]. \tag{9.113}$$

As $t \to \infty$, $\theta \to 2\pi$, and the moving cylinder rolls all the way around the fixed one, *if* the constraints continue to hold. The rolling constraint holds, however, only so long as Eq. (9.90) holds. When we substitute from Eqs. (9.105), (9.106), and (9.109), Eq. (9.90) becomes

$$\tfrac{1}{3}mg\sin\theta \le \tfrac{1}{3}\mu mg(7\cos\theta - 4). \tag{9.114}$$

At $\theta = 0$, this certainly holds, so that the cylinder does initially roll, as we have supposed. At $\theta = \pi/2$, however, it certainly does not hold, since the left member is then positive and the right, negative. The angle θ_1 at which slipping begins is determined by the equation

$$\sin\theta_1 = \mu(7\cos\theta_1 - 4), \tag{9.115}$$

whose solution is

$$\cos\theta_1 = \frac{28\mu^2+[1+33\mu^2]^{1/2}}{1+49\mu^2}. \tag{9.116}$$

The second part of the motion, during which the moving cylinder slides without friction around the fixed one, can be found by solving Eqs. (9.101) and (9.102) for $\theta(t)$, $\gamma(t)$, with $f = 0$ and with only the single constraint $r = (1+\alpha)a$, and with the initial values $\theta = \theta_1, \dot{\theta} = \dot{\theta}_1$, determined from Eqs. (9.116) and (9.109). The solution can be found without essential difficulty, and the angle θ_2 at which the moving cylinder leaves the fixed one can then be determined from Eqs. (9.106) and (9.89). These calculations are left to the reader.

9.6 CONSTANTS OF THE MOTION AND IGNORABLE COORDINATES

We remarked in Chapter 3 that one general method for solving dynamical problems is to look for constants of the motion, that is, functions of the coordinates and velocities which are constant in time. One common case in which such constants can be found arises when the dynamical system is characterized by a Lagrangian function in which some coordinate q_k does not occur explicitly. The corresponding Lagrange equation (9.57) then reduces to

$$\frac{d}{dt}\left(\frac{\partial L}{\partial \dot{q}_k}\right) = 0. \tag{9.117}$$

This equation can be integrated immediately:

$$\frac{\partial L}{\partial \dot{q}_k} = p_k = \text{a constant}. \tag{9.118}$$

Thus, whenever a coordinate q_k does not occur explicitly in the Lagrangian function, the corresponding momentum p_k is a constant of the motion. Such a coordinate q_k is said to be *ignorable*. If q_k is ignorable, we can solve Eq. (9.118) for $\dot{q}_k$ in terms of the other coordinates and velocities, and of the constant momentum p_k, and substitute in the remaining Lagrange equations to eliminate $\dot{q}_k$ and reduce by one the number of variables in the problem; (q_k was already missing from the equations, since it was assumed ignorable.) When the remaining variables have been found, they can be substituted in Eq. (9.118), to give $\dot{q}_k$ as a function of t; q_k is then obtained by integration. If all but one of the coordinates are ignorable, the problem can thus be reduced to a one-dimensional problem and solved by the energy integral method, if L does not depend on the time t explicitly.

For example, in the case of central forces, the potential energy depends only on the distance r from the origin, so that if we use polar coordinates r, θ in a plane, V is independent of θ. Since T is also independent of θ according to Eq. (9.14) (T depends of course on $\dot{\theta}$), we will have

$$\frac{\partial L}{\partial \theta} = \frac{\partial}{\partial \theta}(T-V) = 0, \tag{9.119}$$

and hence

$$\frac{\partial L}{\partial \dot{\theta}} = mr^2\dot{\theta} = p_\theta = \text{a constant}, \tag{9.120}$$

a result which we obtained in Section 3.13 by a different argument. We see that the constancy of p_θ is a result of the fact that the system is symmetrical about the origin, so that L cannot depend on θ.

If a system of particles is acted on by no external forces, then if we displace the whole system in any direction, without changing the velocities and relative positions of the particles, there will be no change in T or V, or in L. If X, Y, and Z are rectangular coordinates of the center of mass, and if the remaining coordinates are relative to the center of mass, so that changing X corresponds to displacing the whole system, then

$$\frac{\partial L}{\partial X} = 0, \tag{9.121}$$

and therefore p_X, the total linear momentum in the x-direction, will be constant, a result we proved in Section 4.1 by a different method.

It is of interest to see how to show from Lagrange's equations that the total energy is a constant of the motion. In order to find an energy integral of the equations of motion in Lagrangian form, it is necessary to know how to express the total energy in terms of the Lagrangian function L. To this end, let us consider a system described in terms of a fixed system of coordinates, so that the kinetic energy T is a homogeneous quadratic function of the generalized velocities $\dot{q}_1, \ldots, \dot{q}_f$ [i.e., $T_1 = T_0 = 0$ in Eq. (9.13)]. By Euler's theorem,* we have

$$\sum_{k=1}^{f} \dot{q}_k \frac{\partial T}{\partial \dot{q}_k} = 2T. \tag{9.122}$$

Thus if

$$L = T_2 - V, \tag{9.123}$$

where V is a function of the coordinates $q_1, \ldots, q_f$ alone, then, by Eq. (9.122),

$$\sum_{k=1}^{f} \dot{q}_k \frac{\partial L}{\partial \dot{q}_k} - L = T + V = E. \tag{9.124}$$

We now consider the time derivative of the left member of Eq. (9.124). For greater generality, we shall at first allow L to depend explicitly on t. In the case we have considered, L does not depend explicitly on t. There are cases, however, when a system is subject to external forces that change with time and that can be

---

*W. Kaplan, *Advanced Calculus*. Reading, Mass.: Addison-Wesley, 1952. (Page 90, Problem 9.) The reader unfamiliar with Euler's theorem can readily verify Eq. (9.122) for himself by substituting for $T = T_2$ from Eq. (9.9).

derived from a potential V that varies with time. An example would be an atom subject to a varying external electric field. In such cases, the equations of motion can be written in the Lagrangian form (9.57) with the Lagrangian depending explicitly on the time t. In the case of moving coordinate systems also, the Lagrangian may depend on the time even though the forces are conservative. The time derivative of the left member of Eq. (9.124) is

$$\frac{d}{dt}\left(\sum_{k=1}^{f} \dot{q}_k \frac{\partial L}{\partial \dot{q}_k} - L\right) = \sum_{k=1}^{f}\left[\ddot{q}_k \frac{\partial L}{\partial \dot{q}_k} + \dot{q}_k \frac{d}{dt}\left(\frac{\partial L}{\partial \dot{q}_k}\right) - \frac{\partial L}{\partial q_k}\dot{q}_k - \frac{\partial L}{\partial \dot{q}_k}\ddot{q}_k\right] - \frac{\partial L}{\partial t}$$

$$= \sum_{k=1}^{f} \dot{q}_k\left[\frac{d}{dt}\left(\frac{\partial L}{\partial \dot{q}_k}\right) - \frac{\partial L}{\partial q_k}\right] - \frac{\partial L}{\partial t} = -\frac{\partial L}{\partial t}. \tag{9.125}$$

If L does not depend explicitly on t, the right side of Eq. (9.125) is zero, and

$$\sum_{k=1}^{f} \dot{q}_k \frac{\partial L}{\partial \dot{q}_k} - L = \text{a constant.} \tag{9.126}$$

When L has the form $(T_2 - V)$, as in a stationary coordinate system, this is the conservation of energy theorem. Regardless of the form of L, Eq. (9.126) represents an integral of Lagrange's equations (9.57), whenever L does not contain t explicitly, but the constant quantity on the left is not always the total energy. Note the analogy between the conservation of generalized momentum p_k when L is independent of q_k, and the conservation of energy when L is independent of t. There are many ways in which the relation between time and energy is analogous to the relation between a coordinate and the corresponding momentum.

We have seen that the familiar conservation laws of energy, momentum, and angular momentum can be regarded as consequences of symmetries exhibited by the mechanical systems to which they apply; that is, they are consequences of the fact that the Lagrangian function L, which determines the equations of motion, is independent of time and of the position and orientation of the entire system in space. This result, derived here for classical mechanics, holds generally throughout physics. In quantum mechanics and in relativity theory, even when we include electromagnetic and other kinds of force fields, conservation laws are associated with symmetries in the fundamental equations. We might, in fact, define energy as that quantity which is constant because the laws of physics do not change with time (if indeed they do not!).

9.7 FURTHER EXAMPLES

The spherical pendulum is a simple pendulum free to swing through the entire solid angle about a point. The pendulum bob is constrained to move on a spherical surface of radius R. We locate the bob by the spherical coordinates θ, φ (Fig. 9.7). We may include the length R of the pendulum as a coordinate if we wish to find the tension in the string, but we omit it here, as we are concerned only with finding the motion. If the bob swings above the horizontal, we will suppose that it still

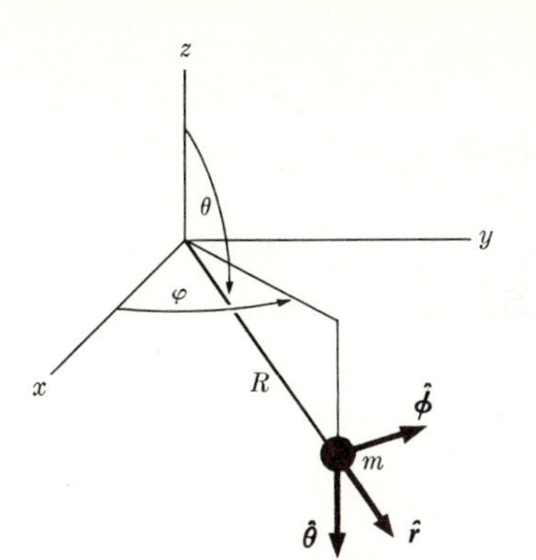

Fig. 9.7 A spherical pendulum.

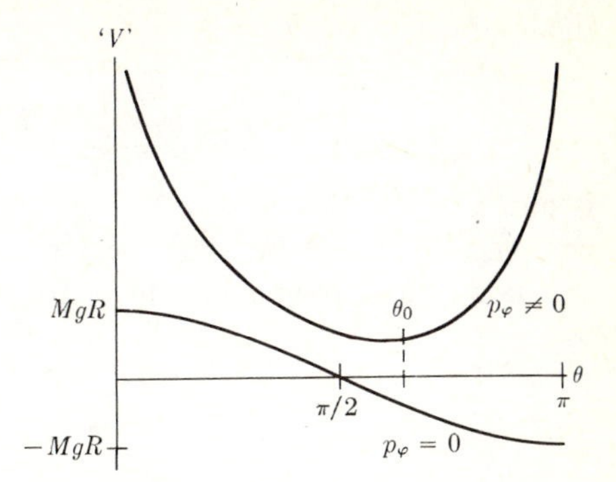

Fig. 9.8 Effective potential 'V'(θ) for spherical pendulum.

remains on the sphere, which would be true if the string were replaced by a rigid rod. Otherwise the constraint disappears whenever a compressional stress is required to maintain it, since a string will support only a tension and not a compression. The velocity of the bob is

$$\boldsymbol{v} = R\dot{\theta}\hat{\boldsymbol{\theta}} + R\sin\theta\,\dot{\varphi}\hat{\boldsymbol{\varphi}}. \tag{9.127}$$

Hence the kinetic energy is

$$T = \tfrac{1}{2}mv^2 = \tfrac{1}{2}mR^2\dot{\theta}^2 + \tfrac{1}{2}mR^2\sin^2\theta\,\dot{\varphi}^2. \tag{9.128}$$

The potential energy due to gravity, relative to the horizontal plane, is

$$V = mgR\cos\theta. \tag{9.129}$$

Hence the Lagrangian function is

$$L = T - V = \tfrac{1}{2}mR^2\dot{\theta}^2 + \tfrac{1}{2}mR^2\sin^2\theta\,\dot{\varphi}^2 - mgR\cos\theta. \tag{9.130}$$

The Lagrange equations are

$$\frac{d}{dt}(mR^2\dot{\theta}) - mR^2\dot{\varphi}^2\sin\theta\cos\theta - mgR\sin\theta = 0, \tag{9.131}$$

$$\frac{d}{dt}(mR^2\sin^2\theta\,\dot{\varphi}) = 0. \tag{9.132}$$

The coordinate φ is ignorable, and the second equation can be integrated immediately:

$$mR^2\sin^2\theta\,\dot{\varphi} = p_\varphi = \text{a constant}. \tag{9.133}$$

Also, since

$$\frac{\partial L}{\partial t} = 0, \tag{9.134}$$

the quantity

$$\dot{\theta}\frac{\partial L}{\partial \dot{\theta}}+\dot{\varphi}\frac{\partial L}{\partial \dot{\varphi}}-L = \tfrac{1}{2}mR^2\dot{\theta}^2+\tfrac{1}{2}mR^2 \sin^2\theta\, \dot{\varphi}^2+mgR \cos \theta \tag{9.135}$$

is constant, by Eq. (9.126). We recognize the quantity on the right as the total energy, as it should be, since we are using a fixed coordinate system. Calling this constant E, and substituting for $\dot{\varphi}$ from Eq. (9.133), we have

$$\tfrac{1}{2}mR^2\dot{\theta}^2+\frac{p_\varphi^2}{2mR^2 \sin^2 \theta}+mgR \cos \theta = E. \tag{9.136}$$

We may introduce an effective potential 'V'(θ) for the motion:

$$\text{`}V\text{'}(\theta) = mgR \cos \theta+\frac{p_\varphi^2}{2mR^2 \sin^2 \theta}, \tag{9.137}$$

so that

$$\tfrac{1}{2}mR^2\dot{\theta}^2 = E-\text{`}V\text{'}(\theta). \tag{9.138}$$

Since the left member cannot be negative, the motion is confined to those values of θ for which 'V'$(\theta) \leq E$. The effective potential 'V'(θ) is plotted in Fig. 9.8. We see that for $p_\varphi = 0$, 'V'(θ) is the potential curve for a simple pendulum, with a minimum at $\theta = \pi$ and a maximum at $\theta = 0$. For $E = -mgR$, the pendulum is at rest at $\theta = \pi$. For $mgR > E > -mgR$, the pendulum oscillates about $\theta = \pi$. For $E > mgR$, the pendulum swings in a circular motion through the top and bottom points $\theta = 0$ and π. When $p_\varphi \neq 0$, the motion is no longer that of a simple pendulum, and 'V'(θ) now has a minimum at a point θ_0 between $\pi/2$ and π, and rises to infinity at $\theta = 0$ and $\theta = \pi$. The larger p_φ, the larger the minimum value of 'V'(θ), and the closer θ_0 is to $\pi/2$. If $E =$ 'V'(θ_0), then θ is constant and equal to θ_0, and the pendulum swings in a circle about the vertical axis. As $p_\varphi \to \infty$, the pendulum swings more and more nearly in a horizontal plane. For $E >$ 'V'(θ_0), θ oscillates between a maximum and minimum value while the pendulum swings about the vertical axis. The reader should compare these results with his mechanical intuition or his experience regarding the motion of a spherical pendulum. The solution of Eq. (9.138) for $\theta(t)$ cannot be carried out in terms of elementary functions, but we can treat circular and nearly circular motions very easily. The relation between p_φ and θ_0 for uniform circular motion of the pendulum about the z-axis is

$$\left[\frac{d\text{`}V\text{'}}{d\theta}\right]_{\theta_0} = -mgR \sin \theta_0-\frac{p_\varphi^2 \cos \theta_0}{mR^2 \sin^3 \theta_0} = 0. \tag{9.139}$$

It is evident from this equation that $\theta_0 > \pi/2$, and that $\theta_0 \to \pi/2$ as $p_\varphi \to \infty$. By substituting from Eq. (9.133), we obtain a relation between $\dot{\varphi}$ and θ_0 for uniform circular motion:

$$\dot{\varphi}^2 = \frac{g}{R} \frac{1}{(-\cos \theta_0)}. \tag{9.140}$$

The energy for uniform circular motion at an angle θ_0, if we use Eqs. (9.136) and (9.139), and the fact that $\dot{\theta} = 0$, is

$$E_0 = \frac{mgR}{2} \left(\frac{2 - 3 \sin^2 \theta_0}{\cos \theta_0} \right). \tag{9.141}$$

For an energy slightly larger than E_0, and an angular momentum p_φ given by Eq. (9.139), the angle θ will perform simple harmonic oscillations about the value θ_0. For if we set

$$k = \left[\frac{d^2 \text{`}V\text{'}}{d\theta^2} \right]_{\theta_0} = \frac{mgR}{-\cos \theta_0} (1 + 3 \cos^2 \theta_0), \tag{9.142}$$

then, for small values of $\theta - \theta_0$, we can expand 'V'(θ) in a Taylor series:

$$\text{`}V\text{'}(\theta) \doteq E_0 + \tfrac{1}{2} k (\theta - \theta_0)^2. \tag{9.143}$$

The energy equation (9.138) now becomes

$$\tfrac{1}{2} m R^2 \dot{\theta}^2 + \tfrac{1}{2} k (\theta - \theta_0)^2 = E - E_0. \tag{9.144}$$

This is the energy for a harmonic oscillator with energy $E - E_0$, coordinate $\theta - \theta_0$, mass mR^2, spring constant k. The frequency of oscillation in θ is therefore given by

$$\omega^2 = \frac{k}{mR^2} = \frac{g}{R} \frac{1 + 3 \cos^2 \theta_0}{-\cos \theta_0}. \tag{9.145}$$

This oscillation in θ is superposed upon a circular motion around the z-axis with an angular velocity $\dot{\varphi}$ given by Eq. (9.133); $\dot{\varphi}$ will vary slightly as θ oscillates, but will remain very nearly equal to the constant value given by Eq. (9.140). It is of interest to compare $\dot{\varphi}$ and ω:

$$\frac{\dot{\varphi}^2}{\omega^2} = \frac{1}{1 + 3 \cos^2 \theta_0}. \tag{9.146}$$

Since $\theta_0 > \pi/2$, this ratio is less than 1, so that $\omega > \dot{\varphi}$, and the pendulum wobbles up and down as it goes around the circle. At $\theta_0 = \pi/2$, $\dot{\varphi} = \omega$, and the pendulum moves in a circle whose plane is tilted slightly from the horizontal; this case occurs only in the limit of very large values of p_φ. It is clear physically that when p_φ is so large that gravity may be neglected, the motion can be a circle in any plane through the origin. Can you show this mathematically? Near $\theta_0 = 0$, $\omega = 2\dot{\varphi}$, so that θ oscillates twice per revolution and the pendulum bob moves in an ellipse whose center is on the z-axis. This corresponds to the motion of the two-dimensional harmonic oscillator discussed in Section 3.10, with equal frequencies in the two perpendicular directions.

As a last example, we consider a system in which there are moving constraints. A bead of mass m slides without friction on a circular hoop of radius a. The hoop lies in a vertical plane which is constrained to rotate about a vertical diameter with constant angular velocity ω. There is just one degree of freedom, and inasmuch as we are not interested in the forces of constraint, we choose a single coordinate θ which measures the angle around the circle from the bottom of the vertical diameter to the bead (Fig. 9.9). The kinetic energy is then

$$T = \tfrac{1}{2}ma^2\dot{\theta}^2 + \tfrac{1}{2}ma^2\omega^2 \sin^2 \theta, \tag{9.147}$$

and the potential energy is

$$V = -mga \cos \theta. \tag{9.148}$$

The Lagrangian function is

$$L = \tfrac{1}{2}ma^2\dot{\theta}^2 + \tfrac{1}{2}ma^2\omega^2 \sin^2 \theta + mga \cos \theta. \tag{9.149}$$

The Lagrange equation of motion can easily be written out, but this is unnecessary, for we notice that

$$\frac{\partial L}{dt} = 0,$$

and therefore, by Eq. (9.126), the quantity

$$\dot{\theta}\frac{\partial L}{\partial \dot{\theta}} - L = \tfrac{1}{2}ma^2\dot{\theta}^2 - \tfrac{1}{2}ma^2\omega^2 \sin^2 \theta - mga \cos \theta = \text{‘}E\text{’} \tag{9.150}$$

is constant. The constant 'E' is not the total energy $T+V$, for the middle term has the wrong sign. The total energy is evidently not constant in this case. (What

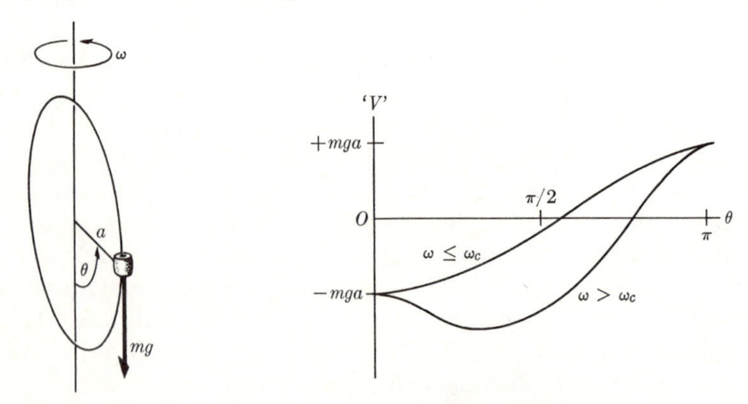

Fig. 9.9 A bead sliding on a rotating hoop.

Fig. 9.10 Effective potential energy for system shown in Fig. 9.9.

force does the work which produces changes in $T+V$?) We may note, however, that we can interpret Eq. (9.149) as a Lagrangian function in terms of a fixed coordinate system with the middle term regarded as part of an effective potential energy:

$$\text{`}V\text{'}(\theta) = -\tfrac{1}{2}ma^2\omega^2 \sin^2 \theta - mga \cos \theta. \tag{9.151}$$

The energy according to this interpretation is 'E'. The first term in 'V'(θ) is the potential energy associated with the centrifugal force which must be added if we regard the rotating system as fixed. The effective potential is plotted in Fig. 9.10. The shape of the potential curve depends on whether ω is greater or less than a critical angular velocity

$$\omega_c = (g/a)^{1/2}. \tag{9.152}$$

It is left to the reader to show this, and to discuss the nature of the motion of the bead in the two cases.

9.8 ELECTROMAGNETIC FORCES AND VELOCITY-DEPENDENT POTENTIALS

If the forces acting on a dynamical system depend upon the velocities, it may be possible to find a function $U(q_1, \ldots, q_f; \dot{q}_1, \ldots, \dot{q}_f; t)$ such that

$$Q_k = \frac{d}{dt}\frac{\partial U}{\partial \dot{q}_k} - \frac{\partial U}{\partial q_k}, \qquad k = 1, \ldots, f. \tag{9.153}$$

If such a function U can be found, then we can define a Lagrangian function

$$L = T - U, \tag{9.154}$$

so that the equations of motion (9.53) can be written in the form (9.57):

$$\frac{d}{dt}\frac{\partial L}{\partial \dot{q}_k} - \frac{\partial L}{\partial q_k} = 0, \qquad k = 1, \ldots, f. \tag{9.155}$$

The function U may be called a *velocity-dependent potential.* If there are also forces derivable from an ordinary potential energy $V(q_1, \ldots, q_f)$, V may be included in U, since Eq. (9.153) reduces to Eq. (9.33) for those terms which do not contain the velocities. The function U may depend explicitly on the time t. If it does not, and if the coordinate system is a fixed one, then L will be independent of t, and the quantity

$$E = \sum_{k=1}^{f} \dot{q}_k \frac{\partial L}{\partial \dot{q}_k} - L \tag{9.156}$$

will be a constant of the motion, according to Eq. (9.126). In this case, we may say that the forces are conservative even though they depend on the velocities. It is clear from this result that it cannot be possible to express frictional forces in the form (9.153), for the total energy is not constant when there is friction unless we include heat energy, and heat energy cannot be defined in terms of the coordinates and velocities $q_1, \ldots, q_f; \dot{q}_1, \ldots, \dot{q}_f$, and hence cannot be included in Eq. (9.156). It is not hard to show that if the velocity-dependent parts of U are linear in the velocities, as they are in all important examples, the energy E defined by Eq. (9.156) is just $T+V$, where V is the ordinary potential energy and contains the terms in U that are independent of the velocities.

As an example, a particle of charge q subject to a constant magnetic field $\boldsymbol{B}$ is acted on by a force (Gaussian units)

$$\boldsymbol{F} = \frac{q}{c}\boldsymbol{v}\times\boldsymbol{B}, \tag{9.157}$$

or

$$F_x = \frac{q}{c}(\dot{y}B_z - \dot{z}B_y),$$

$$F_y = \frac{q}{c}(\dot{z}B_x - \dot{x}B_z), \tag{9.158}$$

$$F_z = \frac{q}{c}(\dot{x}B_y - \dot{y}B_x).$$

Equations (9.158) have the form (9.153) if

$$U = \frac{q}{c}(z\dot{y}B_x + x\dot{z}B_y + y\dot{x}B_z). \tag{9.159}$$

It is, in fact, possible to express the electromagnetic force in the form (9.153) for any electric and magnetic field. The electromagnetic force on a particle of charge q is given by Eq. (3.283):

$$\boldsymbol{F} = q\boldsymbol{E} + \frac{q}{c}\boldsymbol{v}\times\boldsymbol{B}. \tag{9.160}$$

It is shown in electromagnetic theory* that for any electromagnetic field, it is possible to define a scalar function $\phi(x, y, z, t)$ and a vector function $\boldsymbol{A}(x, y, z, t)$ such that

$$\boldsymbol{E} = -\nabla\phi - \frac{1}{c}\frac{\partial \boldsymbol{A}}{\partial t}, \tag{9.161}$$

$$\boldsymbol{B} = \nabla\times\boldsymbol{A}. \tag{9.162}$$

The function ϕ is called the *scalar potential*, and $\boldsymbol{A}$ is called the *vector potential.* If these expressions are substituted in Eq. (9.160), we obtain

$$\boldsymbol{F} = -q\nabla\phi - \frac{q}{c}\frac{\partial \boldsymbol{A}}{\partial t} + \frac{q}{c}\boldsymbol{v}\times(\nabla\times\boldsymbol{A}). \tag{9.163}$$

---

*See, e.g., Slater and Frank, *Electromagnetism.* New York: McGraw-Hill Book Co., 1947. (Page 87.)

The last term can be rewritten using formula (3.35) for the triple cross product:

$$\boldsymbol{F} = -q\nabla\phi - \frac{q}{c}\frac{\partial \boldsymbol{A}}{\partial t} - \frac{q}{c}\boldsymbol{v}\cdot\nabla\boldsymbol{A} + \frac{q}{c}\nabla(\boldsymbol{v}\cdot\boldsymbol{A}). \tag{9.164}$$

[The components of $\boldsymbol{v}$ are $(\dot{x}, \dot{y}, \dot{z})$ and are independent of x, y, z, so that $\boldsymbol{v}$ is not differentiated by the operator ∇.] The two middle terms can be combined according to Eq. (8.113):

$$\boldsymbol{F} = -q\nabla\phi - \frac{q}{c}\frac{d\boldsymbol{A}}{dt} + \frac{q}{c}\nabla(\boldsymbol{v}\cdot\boldsymbol{A}), \tag{9.165}$$

where $d\boldsymbol{A}/dt$ is the time derivative of $\boldsymbol{A}$ evaluated at the position of the moving particle. It may now be verified by direct computation that the potential function

$$U = q\phi - \frac{q}{c}\boldsymbol{v}\cdot\boldsymbol{A}, \tag{9.166}$$

when substituted in Eqs. (9.153), with $q_1, q_2, q_3 = x, y, z$, yields the components of the force $\boldsymbol{F}$ given by Eq. (9.165). In view of the theorem stated in Problem 24 at the end of this chapter, the Lagrangian function $L = T - U$ will then also give the correct equations of motion when expressed in terms of any set of coordinates. It is also easy to show that the energy E defined by Eq. (9.156) with $L = T - U$ *is*

$$E = T + q\phi. \tag{9.167}$$

If $\boldsymbol{A}$ and ϕ are independent of t, then L is independent of t in a fixed coordinate system and the energy E is constant, a result derived by more elementary methods in Section 3.17 [Eq. (3.288)].

When there is a velocity-dependent potential, it is customary to define the momentum in terms of the Lagrangian function, rather than in terms of the kinetic energy:

$$p_k = \frac{\partial L}{\partial \dot{q}_k}. \tag{9.168}$$

If the potential is not velocity dependent, then this definition is equivalent to Eq. (9.23). In any case, it is $\partial L/\partial \dot{q}_k$ whose time derivative occurs in the Lagrange equation for q_k, and which is constant if q_k is ignorable. In the case of a particle subject to electromagnetic forces, the momentum components p_x, p_y, p_z will be, by Eqs. (9.168) and (9.166),

$$\begin{aligned} p_x &= m\dot{x} + \frac{q}{c}A_x, \\ p_y &= m\dot{y} + \frac{q}{c}A_y, \\ p_z &= m\dot{z} + \frac{q}{c}A_z. \end{aligned} \tag{9.169}$$

The second terms play the role of a potential momentum.

It appears that gravitational forces, electromagnetic forces, and indeed all the fundamental forces in physics can be expressed in the form (9.153), for a suitably chosen potential function U. [Frictional forces we do not regard as fundamental in this sense, because they are ultimately reducible to electromagnetic forces between atoms, and hence are in principle also expressible in the form (9.153) if we include all the coordinates of the atoms and molecules of which a physical system is composed.] Therefore the equations of motion of any system of particles can always be expressed in the Lagrangian form (9.155), even when velocity-dependent forces are present. It appears that there is something fundamental about the form of Eqs. (9.155). One important property of these equations, as we have already noted, is that they retain the same form if we substitute any new set of coordinates for $q_1, \ldots, q_f$. This can be verified by a straightforward, if somewhat tedious, calculation. Further insight into the fundamental character of the Lagrange equations must await the study of a more advanced formulation of mechanics utilizing the calculus of variations, which is beyond the scope of this book.*

9.9 LAGRANGE'S EQUATIONS FOR THE VIBRATING STRING

The Lagrange method can be extended also to the motion of continuous media. We shall consider only the simplest example, the vibrating string. Using the notation of Section 8.1, we could take $u(x)$ as a set of generalized coordinates analogous to q_k. In place of the subscript k denoting the various degrees of freedom, we have the position coordinate x denoting the various points on the string. The number of degrees of freedom is infinite for an ideal continuous string. The generalization of the Lagrange method to deal with a continuous index x denoting the various degrees of freedom introduces mathematical complications which we wish to avoid here.† Therefore we make use of the possibility of representing the function $u(x)$ as a Fourier series.

According to the Fourier series theorem quoted in Section 8.2, if the string is tied at the ends $x = 0, l$, we can represent its position $u(x)$ by the series (8.24):

$$u(x) = \sum_{k=1}^{\infty} q_k \sin \frac{k\pi x}{l}. \tag{9.170}$$

The coefficients q_k are given by Eq. (8.25):

$$q_k = \frac{2}{l} \int_0^l u(x) \sin \frac{k\pi x}{l} \, dx, \qquad k = 1, 2, 3, \ldots. \tag{9.171}$$

Since the coefficients q_k give a complete description of the position of the string, they represent a suitable set of generalized coordinates. When the string vibrates,

*See, e.g., H. Goldstein, *Classical Mechanics*. Reading, Mass.: Addison-Wesley, 1950. (Chapter 2.)

†For a treatment of this problem, see H. Goldstein, *op. cit.* (Chapter 11.)

the coordinates q_k become functions of t:

$$u(x, t) = \sum_{k=1}^{\infty} q_k(t) \sin \frac{k\pi x}{l}. \tag{9.172}$$

We have still an infinite number of coordinates q_k, but they depend on the discrete subscript k and can be treated exactly like the generalized coordinates considered earlier in this chapter. Since the string could in principle be treated as a system with a very large number of particles, and since we are allowed to describe the system by any suitable set of generalized coordinates, we need only express the Lagrangian function in terms of the coordinates q_k in order to write down the equations of motion.

We first need to calculate the kinetic energy, which is evidently

$$T = \int_0^l \tfrac{1}{2}\sigma \left(\frac{\partial u}{\partial t}\right)^2 dx. \tag{9.173}$$

If we differentiate Eq. (9.172) with respect to t and square, we obtain

$$\left(\frac{\partial u}{\partial t}\right)^2 = \sum_{k=1}^{\infty} \sum_{j=1}^{\infty} \dot{q}_k \dot{q}_j \sin \frac{k\pi x}{l} \sin \frac{j\pi x}{l}. \tag{9.174}$$

We now multiply by $\frac{1}{2}\sigma\, dx$ and integrate from 0 to l term by term.* Since

$$\int_0^l \sin \frac{k\pi x}{l} \sin \frac{j\pi x}{l}\, dx = \begin{cases} \frac{1}{2}l, & j = k, \\ 0, & j \neq k, \end{cases} \tag{9.175}$$

the result obtained is

$$T = \sum_{k=1}^{\infty} \tfrac{1}{4} l\sigma \dot{q}_k^2. \tag{9.176}$$

We next calculate the generalized force Q_k. If coordinate q_k increases by δq_k, while the rest are held fixed, a point x on the string moves up a distance given by Eq. (9.170):

$$\delta u = \delta q_k \sin \frac{k\pi x}{l}. \tag{9.177}$$

The upward force on an element dx of string is given by Eq. (8.3). The work done is therefore

$$\delta W = Q_k\, \delta q_k = \int_0^l \frac{\partial}{\partial x}\left(\tau \frac{\partial u}{\partial x}\right) \delta u\, dx. \tag{9.178}$$

---

*In order to differentiate and integrate infinite series term by term, and to rearrange orders of summation, as we shall do freely in this section, we must require that the series all converge uniformly. This will be the case if $u(x, t)$ and its derivatives are continuous functions. (For a precise statement and derivation of the conditions for manipulation of infinite series, see a text on advanced calculus, e.g., W. Kaplan, *Advanced Calculus.* Reading, Mass.: Addison-Wesley, 1952. Chapter 6.)

We substitute for $\partial u/\partial x$ from Eq. (9.170), and for δu from Eq. (9.177), and integrate term by term, to obtain (assuming τ constant):

$$Q_k = -\tfrac{1}{2}l\tau\left(\frac{\pi k}{l}\right)^2 q_k. \tag{9.179}$$

The forces Q_k are obviously derivable from the potential energy function,

$$V = \sum_{k=1}^{\infty} \tfrac{1}{4}l\tau\left(\frac{\pi k}{l}\right)^2 q_k^2. \tag{9.180}$$

It will be instructive to calculate V directly by calculating the work done against the tension τ in moving the string from its equilibrium position to the position $u(x)$. At the same time we shall verify that this work is independent of how we move the string to the position $u(x)$. Let $u(x, t)$ be the position of the string at any time t while the string is being moved to $u(x)$. [The function $u(x, t)$ is not necessarily a solution of the equation of motion, since we wish to consider an arbitrary manner of moving the string from $u = 0$ to $u = u(x)$.] At $t = 0$, the string is in its equilibrium position:

$$u(x, 0) = 0. \tag{9.181}$$

Let $t = t_1$ be the time the string arrives at its final position:

$$u(x, t_1) = u(x). \tag{9.182}$$

The work done against the vertical components of tension [Eq. (8.3)] during the interval dt is

$$dV = -\int_{x=0}^{l} \frac{\partial}{\partial x}\left(\tau\frac{\partial u}{\partial x}\right)\left(\frac{\partial u}{\partial t}\,dt\right)dx.$$

We integrate by parts, remembering that u and $\partial u/\partial t$ are 0 at $x = 0, l$:

$$\begin{aligned} dV &= \int_{x=0}^{l} \tau\frac{\partial u}{\partial x}\frac{\partial^2 u}{\partial t\,\partial x}\,dx\,dt \\ &= dt\frac{\partial}{\partial t}\int_{x=0}^{l} \tfrac{1}{2}\tau\left(\frac{\partial u}{\partial x}\right)^2 dx. \end{aligned} \tag{9.183}$$

The total work done is then

$$\begin{aligned} V &= \int_{t=0}^{t_1} dV \\ &= \left[\int_0^l \tfrac{1}{2}\tau\left(\frac{\partial u}{\partial x}\right)^2 dx\right]_{t=0}^{t_1} \\ &= \int_0^l \tfrac{1}{2}\tau\left(\frac{\partial u}{\partial x}\right)^2 dx, \end{aligned} \tag{9.184}$$

where in the last expression, $u = u(x)$ corresponds to the final position of the string. The result depends only on the final position of the string—an independent proof that the tension forces are conservative.

The work done against the tension is stored as potential energy in the stretched string. By substituting in Eq. (9.184) from Eq. (9.170), we again can obtain Eq. (9.180). In Eq. (8.61) for a string of particles, the right member contains two terms that represent the vertical components of force between adjacent pairs of particles. A third way of deriving the potential energy is to find the potential energy function between a pair of particles which yields this force. It must then be shown that, when this is summed over all pairs of adjacent particles, the result approaches Eq. (9.184) in the limit $h \to 0$.

The Lagrangian function for the vibrating string can now be written as

$$L = T - V = \sum_{k=1}^{\infty} \left[\tfrac{1}{4} l\sigma \dot{q}_k^2 - \tfrac{1}{4} l\tau \left(\frac{\pi k}{l}\right)^2 q_k^2 \right]. \tag{9.185}$$

The resulting Lagrange equation for q_k is

$$\tfrac{1}{2} l\sigma \ddot{q}_k + \tfrac{1}{2} l\tau \left(\frac{\pi k}{l}\right)^2 q_k = 0, \tag{9.186}$$

whose general solution is

$$q_k = A_k \cos \omega_k t + B_k \sin \omega_k t, \tag{9.187}$$

where

$$\omega_k = \frac{\pi k}{l} \left(\frac{\tau}{\sigma}\right)^{1/2} = \frac{\pi k c}{l}. \tag{9.188}$$

This result can be substituted in Eq. (9.172) to obtain the solution

$$u(x, t) = \sum_{k=1}^{\infty} \left(A_k \sin \frac{k\pi x}{l} \cos \omega_k t + B_k \sin \frac{k\pi x}{l} \sin \omega_k t \right), \tag{9.189}$$

which is in agreement with Eq. (8.23). If $u = u_0(x)$ and $\partial u/\partial t = v_0(x)$ are given at $t = 0$, we can find the constants A_k, B_k as in Eqs. (8.25):

$$\begin{aligned} A_k &= q_k(0) = \frac{2}{l} \int_0^l u_0(x) \sin \frac{k\pi x}{l}\, dx, \\ B_k &= \frac{\dot{q}_k(0)}{\omega_k} = \frac{2}{\omega_k l} \int_0^l v_0(x) \sin \frac{k\pi x}{l}\, dx. \end{aligned} \tag{9.190}$$

The coordinates q_k defined by Eqs. (9.170) and (9.171) are called the *normal coordinates* for the vibrating string. Each coordinate evidently represents one normal mode of vibration. The normal coordinates are also very useful in treating the case where a force $f(x, t)$ is applied along the string (see Problem 26 at the end of this chapter). Mathematically, the normal coordinates have the property that the Lagrangian L becomes a sum of terms, each term involving only one degree of freedom. Thus in normal coordinates the problem is subdivided into separate problems, one for each degree of freedom.

It was, of course, rather fortunate that the coordinates q_k which were chosen at the beginning of the problem turned out to be the normal coordinates. In general, this does not happen. For example, consider a string whose density varies along its length according to

$$\sigma = \sigma_0 + a \sin \frac{\pi x}{l}. \tag{9.191}$$

This string is heaviest near its center. We will use the same coordinates q_k as defined by Eqs. (9.170) and (9.171). We substitute Eqs. (9.191) and (9.172) in Eq. (9.173) and, instead of Eq. (9.176), we obtain (after some calculation),

$$T = \sum_{k=1}^{\infty} \sum_{j=1}^{\infty} \tfrac{1}{2} T_{kj} \dot{q}_k \dot{q}_j, \tag{9.192}$$

where

$$\begin{aligned} T_{kj} &= \tfrac{1}{2} l\sigma_0 + \frac{4la}{\pi} \frac{k^2}{4k^2-1}, && \text{if } k = j, \\ T_{kj} &= -\frac{4la}{\pi} \frac{kj}{[(k+j)^2-1][(k-j)^2-1]}, && \text{if } k \neq j, \end{aligned} \tag{9.193}$$

and k, j are both even or both odd; otherwise

$$T_{kj} = 0.$$

For this string, the q_k's are evidently not normal coordinates. In the Lagrange equations the q_k's, with k even, are all coupled together, as are those with k odd. The problem is then much more difficult, and a solution will not be attempted here.

9.10 HAMILTON'S EQUATIONS

The discussion in this section will be restricted to mechanical systems obeying Lagrange's equations in the form (9.57). The Lagrangian L is a function of the coordinates q_k, of the velocities $\dot{q}_k$, and perhaps of t. The state of the mechanical system at any time, that is, the positions and velocities of all its parts, is specified by giving the generalized coordinates and velocities q_k, $\dot{q}_k$. Lagrange's equations are second-order equations which relate the accelerations $\ddot{q}_k$ to the coordinates and velocities. The state of the system could equally well be specified by giving the coordinates q_k and the momenta p_k defined by Eq. (9.168):

$$p_k = \frac{\partial L}{\partial \dot{q}_k}, \qquad k = 1, 2, \ldots, f. \tag{9.194}$$

These equations specify p_k in terms of $q_1, \ldots, q_f; \dot{q}_1, \ldots, \dot{q}_f$. They can, in principle, be solved for $\dot{q}_k$ in terms of $q_1, \ldots, q_f; p_1, \ldots, p_f$.

It is an interesting exercise to try to write equations of motion in terms of the coordinates q_k and momenta p_k. Note first that, by use of the definition (9.194) and the equations of motion (9.57), we have

$$\begin{aligned} dL &= \sum_{k=1}^{f} \left(\frac{\partial L}{\partial \dot{q}_k} d\dot{q}_k + \frac{\partial L}{\partial q_k} dq_k \right) + \frac{\partial L}{\partial t} dt \\ &= \sum_{k=1}^{f} (p_k \, d\dot{q}_k + \dot{p}_k \, dq_k) + \frac{\partial L}{\partial t} dt. \end{aligned} \tag{9.195}$$

We next define a function $H(q_1, \ldots, q_f; p_1, \ldots, p_f; t)$ by

$$H = \sum_{k=1}^{f} p_k \dot{q}_k - L, \tag{9.196}$$

where for the velocities $\dot{q}_k$ we substitute their expressions in terms of coordinates and momenta. Then we have, from Eqs. (9.196) and (9.195),

$$dH = \sum_{k=1}^{f} (\dot{q}_k \, dp_k - \dot{p}_k \, dq_k) - \frac{\partial L}{\partial t} dt. \tag{9.197}$$

The definition (9.196) is chosen so that dH depends explicitly upon dp_k, dq_k, and dt. By inspection of Eq. (9.197), we see that

$$\dot{q}_k = \frac{\partial H}{\partial p_k}, \qquad \dot{p}_k = -\frac{\partial H}{\partial q_k}, \qquad k = 1, \ldots, f, \tag{9.198}$$

and

$$\frac{\partial H}{\partial t} = -\frac{\partial L}{\partial t}. \tag{9.199}$$

Equations (9.198) are the desired equations of motion that express $\dot{q}_k$ and $\dot{p}_k$ in terms of the coordinates and momenta.

Equations (9.198) are Hamilton's equations of motion for a mechanical system. The function H, defined by Eq. (9.196), is called the *Hamiltonian function.* We see from Eq. (9.124) that when V is a function only of the coordinates, for a stationary coordinate system, H is just the total energy expressed in terms of coordinates and momenta. For a moving coordinate system, where T is given by Eq. (9.13), the Hamiltonian is

$$H = T_2 + V - T_0, \tag{9.200}$$

with T_2 expressed in terms of coordinates and momenta. According to Section 9.8, H will also be the total energy in a stationary coordinate system when electromagnetic forces are present.

When L does not contain the time explicitly, neither does H according to Eq. (9.199), as is also obvious from the way in which H was defined. According to Eq. (9.125), H is a constant of the motion in this case. This can also be proved directly from Eqs. (9.198), since it is easy to show that

$$\frac{dH}{dt} = \frac{\partial H}{\partial t}, \tag{9.201}$$

as the reader may verify.

If any coordinate q_k does not appear explicitly in H, then Eqs. (9.198) give

$$p_k = \text{a constant}, \tag{9.202}$$

in agreement with Eq. (9.118). Since H does not contain q_k, we may take p_k as a given constant, and the $2(f-1)$ equations (9.198) for the other coordinates and momenta are then the Hamiltonian equations for a system of $f-1$ degrees of freedom. Thus degrees of freedom corresponding to coordinates that do not appear in H simply drop out of the problem. This is the origin of the term "ignorable coordinate." After the remaining equations of motion have been solved for the nonignorable coordinates and momenta, any ignorable coordinate is given by Eqs. (9.198) as an integral over t:

$$q_k(t) = q_k(0) + \int_0^t \frac{\partial H}{\partial p_k}\, dt. \tag{9.203}$$

Hamilton's equations are only a new formulation of Newton's laws of motion. In simple cases, they reduce to equations which could have been written immediately from Newton's laws. In the harmonic oscillator, for example, with coordinate x, the momentum is

$$p = m\dot{x}. \tag{9.204}$$

The Hamiltonian function is therefore

$$H = T + V = \frac{p^2}{2m} + \tfrac{1}{2}kx^2. \tag{9.205}$$

Equations (9.198) become

$$\dot{x} = \frac{p}{m}, \qquad \dot{p} = -kx. \tag{9.206}$$

The first of these is the definition of p, and the second is Newton's equation of motion.

Although they are of comparatively little value if used simply as a means of writing the equations of motion of a system, Hamilton's equations are important for two general reasons. First, they provide a useful starting point in setting up the laws of statistical mechanics and of quantum mechanics. Hamilton originally developed his equations by analogy with a similar mathematical formulation which he had found useful in optics. It is not surprising that Hamilton's equations should form the starting point for wave mechanics! Second, there are a number of methods of solution of mechanical problems based on Hamilton's formulation of the equations of motion. It is clear from the way in which they were derived that Hamilton's equations (9.198), like Lagrange's equations, are valid for any set of generalized coordinates $q_1, \ldots, q_f$ together with the corresponding momenta $p_1, \ldots, p_f$, defined by Eq. (9.194). In fact Hamilton's equations are valid for a much wider class of coordinate systems obtained by defining new coordinates and momenta as certain functions of the original coordinates and momenta. This is the basis for the utility of Hamilton's equations in the solution of mechanical problems. A further discussion of these topics is beyond the scope of this book.* We will, however, prove one general theorem in the next section which gives some insight into the importance of the variables p_k and q_k.

9.11 LIOUVILLE'S THEOREM

We may regard the coordinates $q_1, \ldots, q_f$ as the coordinates of a point in an f-dimensional space, the *configuration space* of the mechanical system. To each point in the configuration space there corresponds a configuration of the parts of the mechanical system. As the system moves, the point $q_1, \ldots, q_f$ traces a path in the configuration space. This path represents the history of the system. Of course many possible paths pass through each configuration $q_1, \ldots, q_f$ since the parts of the system may have any velocities $\dot{q}_1, \ldots, \dot{q}_f$.

If we wish to specify both the configuration and the motion of a system at any given instant, we must specify the coordinates and velocities, or equivalently, the coordinates and momenta. The $2f$-dimensional space whose points are specified by the coordinates and momenta $q_1, \ldots, q_f; p_1, \ldots, p_f$ is called the *phase space* of the mechanical system. As the system moves, the phase point $q_1, \ldots, q_f; p_1, \ldots, p_f$ traces out a path in the phase space. The velocity of the phase point is given by Hamilton's equations (9.198).

Each phase point represents a possible state of the mechanical system. Let us imagine that each phase point is occupied by a "particle" which moves according to the equations of motion (9.198). These particles trace out paths that represent all possible histories of the mechanical system. Through each phase point there is only one possible path, since if the positions and velocities, or positions and momenta, are given, the solution of the equations of motion is uniquely determined. The theorem of Liouville states that the phase "particles" move as an incompressible fluid. More precisely, the phase volume occupied by a set of "particles" is constant.

To prove Liouville's theorem, we make use of theorem (8.121) generalized to a space of $2f$ dimensions. We may either generalize the argument which led to Eq. (8.116), or we may use the generalization of Gauss' divergence theorem which

*See H. Goldstein, *op. cit.* (Chapters 7, 8, 9.)

is valid in any number of dimensions. In either case, we have for a volume V in phase space, moving with the "particles":

$$\frac{dV}{dt} = \int_V \cdots \int \sum_{k=1}^{f} \left(\frac{\partial \dot{q}_k}{\partial q_k} + \frac{\partial \dot{p}_k}{\partial p_k} \right) dq_1 \cdots dq_f \, dp_1 \cdots dp_f, \tag{9.207}$$

which is Eq. (8.121) written for the $2f$-dimensional phase space. We now substitute the velocities from Hamilton's equations (9.198):

$$\frac{dV}{dt} = \int_V \cdots \int \sum_{k=1}^{f} \left(\frac{\partial^2 H}{\partial q_k \, \partial p_k} - \frac{\partial^2 H}{\partial p_k \, \partial q_k} \right) dq_1 \cdots dq_f \, dp_1 \cdots dp_f = 0. \tag{9.208}$$

This is Liouville's theorem, and it should be noted that this theorem holds even when H depends explicitly on t.

In the case of a harmonic oscillator, the phase space is a plane with coordinate axes x and p. The phase points move around ellipses H = constant, given by Eq. (9.205), and with velocities as given by Eq. (9.206). According to Liouville's theorem, the motion is that of a two-dimensional incompressible fluid. In particular, a set of points that lie in a region of area A will at any later time lie in another region of area A.

Liouville's theorem makes the coordinates and momenta more useful for many purposes than coordinates and velocities. Because of this theorem, the concept of phase space is an important tool in statistical mechanics. Imagine a large number of mechanical systems identical to a given one, but with different initial conditions. Let each system be represented by a point in their common phase space, and let these points move according to Hamilton's equations. The statistical properties of this collection of systems may be specified at any time t by giving the density $\rho(q_1, \ldots, q_f; p_1, \ldots, p_f; t)$ in the phase space of system points per unit volume. Liouville's theorem implies that the density ρ in the immediate neighborhood of any system point must remain constant as that point moves through the phase space. (Why?) If we define statistical equilibrium as a distribution in which ρ is constant in time at each fixed point in the phase space, then clearly the necessary and sufficient condition for equilibrium is that ρ be uniform along the flow lines of the system points. (Why?)

We have been able in this section to give only a narrow glimpse of the power of the Hamiltonian methods.

PROBLEMS

1. Coordinates u, w are defined in terms of plane polar coordinates r, θ by the equations

$$u = \ln(r/a) - \theta \cot \zeta,$$

$$w = \ln(r/a) + \theta \tan \zeta,$$

where a and ζ are constants. Sketch the curves of constant u and of constant w. Find the kinetic energy for a particle of mass m in terms of u, w, $\dot{u}$, $\dot{w}$. Find expressions for Q_u, Q_w in terms of the polar force components F_r, F_θ. Find p_u, p_w. Find the forces Q_u, Q_w required to make the particle move with constant speed $\dot{s}$ along a spiral of constant $u = u_0$.

2. Two masses m_1 and m_2 move under their mutual gravitational attraction in a uniform external gravitational field whose acceleration is $\boldsymbol{g}$. Choose as coordinates the cartesian coordinates X, Y, Z of the center of mass (taking Z in the direction of $\boldsymbol{g}$), the distance r between m_1 and m_2, and the polar angles θ and φ which specify the direction of the line from m_1 to m_2. Write expressions for the kinetic energy, the six forces $Q_X, \ldots, Q_\varphi$, and the six momenta. Write out the six Lagrange equations of motion.

3. a) Set up the expression for the kinetic energy of a particle of mass m in terms of plane parabolic coordinates f, h, as defined in Problem 17 of Chapter 3. Find the momenta p_f and Ph.

b) Write out the Lagrange equations in these coordinates if the particle is not acted on by any force.

4. a) Find the forces Q_f and Q_h required to make the particle in Problem 3 move along a parabola $f = f_0$ = a constant, with constant generalized velocity $\dot{h} = \dot{h}_0$, starting from $h = 0$ at $t = 0$.

b) Find the corresponding forces F_x and F_y relative to a cartesian coordinate system.

5. a) Set up the Lagrange equations of motion in spherical coordinates r, θ, φ, for a particle of mass m subject to a force whose spherical components are F_r, F_θ, F_φ.

b) Set up Lagrange equations of motion for the same particle in a system of spherical coordinates rotating with angular velocity ω about the z-axis.

c) Identify the generalized centrifugal and coriolis forces 'Q_r', 'Q_θ', and 'Q_φ' by means of which the equations in the rotating system can be made to take the same form as in the fixed system. Calculate the spherical components 'F_r', 'F_θ', 'F_φ' of these centrifugal and coriolis forces, and show that your results agree with the expressions derived in Chapter 7.

6. Set up the Lagrangian function for the mechanical system shown in Fig. 4.16, using the coordinates x, x_1, x_2 as shown. Derive the equations of motion, and show that they are equivalent to the equations that would be written down directly from Newton's law of motion.

7. Choose suitable coordinates and write down the Lagrangian function for the restricted three-body problem. Show that it leads to the equations of motion obtained in Section 7.6.

8. Masses m and $2m$ are suspended from a string of length l_1 which passes over a pulley. Masses $3m$ and $4m$ are similarly suspended by a string of length l_2 over another pulley. These two pulleys hang from the ends of a string of length l_3 over a third fixed pulley. Set up Lagrange's equations, and find the accelerations and the tensions in the strings.

9. A massless tube is hinged at one end. A uniform rod of mass m, length l, slides freely in it. The axis about which the tube rotates is horizontal, so that the motion is confined to a plane. Choose a suitable set of generalized coordinates, one for each degree of freedom, and set up Lagrange's equations.

10. Set up Lagrange's equations for a uniform door whose axis is slightly out of plumb. What is the period of small vibrations?

11. A double pendulum is formed by suspending a mass m_2 by a string of length l_2 from a mass m_1 which in turn is suspended from a fixed support by a string of length l_1.

a) Choose a suitable set of coordinates, and write the Lagrangian function, assuming the double pendulum swings in a single vertical plane.

b) Write out Lagrange's equations, and show that they reduce to the equations for a pair of coupled oscillators if the strings remain nearly vertical.

c) Find the normal frequencies for small vibrations of the double pendulum. Describe the nature of the corresponding vibrations. Find the limiting values of these frequencies when $m_1 \gg m_2$, and when $m_2 \gg m_1$. Show that these limiting values are to be expected on physical grounds by considering the nature of the normal modes of vibration when either mass becomes vanishingly small.

12. A ladder rests against a smooth wall and slides without friction on wall and floor. Set up the equation of motion, assuming that the ladder maintains contact with the wall. If initially the ladder is at rest at an angle α with the floor, at what angle, if any, will it leave the wall?

13. One end of a uniform rod of mass M makes contact with a smooth vertical wall, the other with a smooth horizontal floor. A bead of mass m and negligible dimensions slides on the rod. Choose a suitable set of coordinates, set up the Lagrangian function, and write out the Lagrange equations. The rod moves in a single vertical plane perpendicular to the wall.

14. A ring of mass M rests on a smooth horizontal surface and is pinned at a point on its circumference so that it is free to swing about a vertical axis. A bug of mass m crawls around the ring with constant speed.

a) Set up the equations of motion, taking this as a system with two degrees of freedom, with the force exerted by the bug against the ring to be determined from the condition that he moves with constant speed.

b) Now set up the equation of motion, taking this as a system with one degree of freedom, the bug being constrained to be at a certain point on the ring at each instant of time. Show that the two formulations of the problem are equivalent.

15. A pendulum bob of mass m is suspended by a string of length l from a point of support. The point of support moves to and fro along a horizontal x-axis according to the equation

$$x = a \cos \omega t.$$

Assume that the pendulum swings only in a vertical plane containing the x-axis. Let the position of the pendulum be described by the angle θ which the string makes with a line vertically downward.

a) Set up the Lagrangian function and write out the Lagrange equation.

b) Show that for small values of θ, the equation reduces to that of a forced harmonic oscillator, and find the corresponding steady-state motion. How does the amplitude of the steady-state oscillation depend on m, l, a, and ω?

16. A pendulum bob of mass m is suspended by a string of length l from a car of mass M which moves without friction along a horizontal overhead rail. The pendulum swings in a vertical plane containing the rail.

a) Set up the Lagrange equations.

b) Show that there is an ignorable coordinate, eliminate it, and discuss the nature of the motion by the energy method.

17. Find the tension in the string for the spherical pendulum discussed in Section 9.7, as a function of E, p_φ, and θ. Determine, for a given E and p_φ, the angle θ_1 at which the string will collapse.

18. A particle of mass m slides over the inner surface of an inverted cone of half-angle α. The apex of the cone is at the origin, and the axis of the cone extends vertically upward. The only force acting on the particle, other than the force of constraint, is the force of gravity.

a) Set up the equations of motion, using as coordinates the horizontal distance ρ of the particle from the axis, and the angle φ measured in a horizontal circle around the cone. Show that φ is ignorable, and discuss the motion by the method of the effective potential.

b) For a given radius ρ_0, find the angular velocity $\dot{\varphi}_0$ of revolution in a horizontal circle, and the angular frequency ω of small oscillations about this circular motion. Show that the small oscillations are a wobbling or an up-and-down spiraling motion, depending on whether the angle α is greater than or less than the angle

$$\alpha_c = \sin^{-1}\frac{1}{\sqrt{3}}.$$

19. A flyball governor for a steam engine is shown in Fig. 9.11. Two balls, each of mass m, are attached by means of four hinged arms, each of length l, to sleeves which are located on a vertical rod. The upper sleeve is fastened to the rod; the lower sleeve has mass M and negligible moment of inertia, and is free to slide up and down the rod as the balls move out from or toward the rod. The rod-and-ball system rotates with constant angular velocity ω.

a) Set up the equation of motion, neglecting the weight of the arms and rod. Discuss the motion by the energy method.

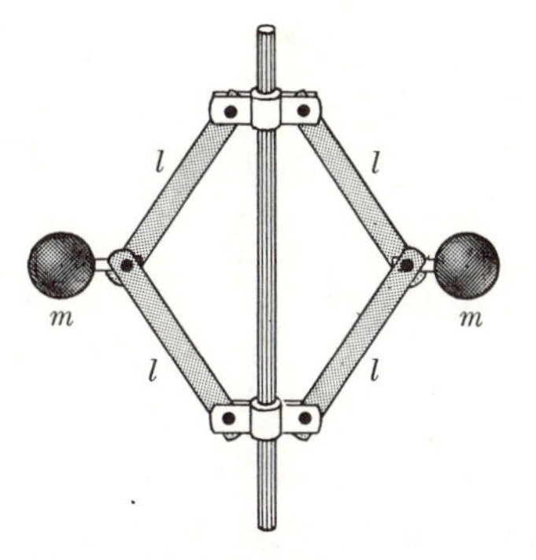

Fig. 9.11 A flyball governor.

b) Determine the value of the height z of the lower sleeve above its lowest point as a function of ω for steady rotation of the balls, and find the frequency of small oscillations of z about this steady value.

20. Discuss the motion of the governor described in Problem 19 if the shaft is not constrained to rotate at angular velocity ω, but is free to rotate, without any externally applied torque.

a) Find the angular velocity of steady rotation for a given height z of the sleeve.

b) Find the frequency of small vibrations about this steady motion.

c) How does this motion differ from that of Problem 19?

21. A rectangular coordinate system with axes x, y, z is rotating with uniform angular velocity ω about the z-axis. A particle of mass m moves under the action of a potential energy $V(x, y, z)$.

a) Set up the Lagrange equations of motion.

b) Show that these equations can be regarded as the equations of motion of a particle in a fixed coordinate system acted on by the force $-\nabla V$, and by a force derivable from a velocity-dependent potential U. Hence find a velocity-dependent potential for the centrifugal and coriolis forces.

c) Express U in spherical coordinates r, θ, φ, $\dot{r}$, $\dot{\theta}$, $\dot{\varphi}$, and verify that it gives rise to the forces 'Q_r', 'Q_θ', 'Q_φ' found in Problem 5.

22. Show that a uniform magnetic field B in the z-direction can be represented in cylindrical coordinates (Fig. 3.22) by the vector potential

$$\boldsymbol{A} = \tfrac{1}{2}B\rho\,\hat{\boldsymbol{\varphi}}.$$

Write out the Lagrangian function for a particle in such a field. Write down the equations of motion. Find three constants of the motion. Compare with Problem 75 of Chapter 3.

23. The kinetic part of the Lagrangian function for a particle of mass m in relativistic mechanics is

$$L_k = -mc^2[1-(v/c)^2]^{1/2}.$$

Show that this gives the proper formula (4.75) for the components of momentum. Show that if the potential function for electromagnetic forces Eq. (9.166) is subtracted, and if $\boldsymbol{A}$ and ϕ do not depend explicitly on t, then $T+q\phi$ is constant, with T given by formula (4.74).

***24.** Show by direct calculation that if Eqs. (9.155) hold for some function $L(q_1, \ldots, q_f; \dot{q}_1, \ldots, \dot{q}_f; t)$, and we introduce new coordinates $q_1^*, \ldots, q_f^*$, where

$$q_k = f_k(q_1^*, \ldots, q_f^*; t), \qquad k = 1, \ldots, f,$$

then

$$\frac{d}{dt}\frac{\partial L^*}{\partial \dot{q}_l^*} - \frac{\partial L^*}{\partial q_l^*} = 0, \qquad l = 1, \ldots, f,$$

where $L^*(q_1^*, \ldots, q_f^*; \dot{q}_1^*, \ldots, \dot{q}_f^*; t) = L(q_1, \ldots, q_f; \dot{q}_1, \ldots, \dot{q}_f; t)$ is obtained by substitution of $f_k(q_1^*, \ldots, q_f^*; t)$ for q_k.

25. Derive formula (9.184) by writing down a potential energy which gives the interparticle forces for the string of particles studied in Section 8.4, and passing to the limit $h \to 0$.

26. A stretched string is subject to an externally applied force of linear density $f(x, t)$. Introduce normal coordinates q_k, and find an expression for the generalized applied force $Q_k(t)$. Use the Lagrangian method to solve Problem 6(a), Chapter 8.

27. Solve Problem 7, Chapter 8, by using the coordinates q_k defined by Eqs. (9.170) and (9.171).

28. Write down the Hamiltonian function for the spherical pendulum. Write the Hamiltonian equations of motion, and derive from them Eq. (9.136).

***29.** Work out the relativistic Hamiltonian function for a particle subject to electromagnetic forces, using the Lagrangian function given in Problem 23. Write out the Hamiltonian equations of motion and show that they are equivalent to the Lagrange equations.

30. Work out the Hamiltonian function $H(q_k, p_k)$ for the vibrating string, starting from Eq. (9.185). Write down the equations which relate the momenta p_k to the function $u(x, t)$ which describes the motion of the string. Hence show that $H = T+V$, with T and V given by Eqs. (9.173) and (9.184).

31. Write down the Hamiltonian function for Problem 2. Write out Hamilton's equations. Identify the ignorable coordinates and show that there remain two separate one degree of freedom problems, each of which can be solved (in principle) by the energy method. What are the corresponding two potential-energy functions?

32. A beam of electrons is directed along the z-axis. The electrons are uniformly distributed over the beam cross section, which is a circle of radius a_0, and their transverse momentum components (p_x, p_y) are distributed uniformly in a circle (in momentum space) of radius p_0. Assuming that the electrons are focused by some lens system so as to form a spot of radius a_1, find the momentum distribution of electrons arriving at the spot. [*Hint:* use Liouville's theorem.]

33. A group of particles all of the same mass m, having initial heights and vertical momenta lying in the square $-a \leq z \leq a$, $-b \leq p \leq b$, fall freely in the earth's gravitational field for a time t. Find the region in the phase space within which they lie at time t, and show by direct calculation that its area is still $4ab$.

34. In an electron microscope, electrons scattered from an object of height z_O are focused by a lens at distance D_O from the object and form an image of height z_I at a distance D_I behind the lens. The aperture of the lens is A. Show by direct calculation that the phase area in the (z, p_z) phase plane occupied by electrons leaving the object (and destined to pass through the lens) is the same as the phase area occupied by electrons arriving at the image. Assume that $z_O \ll D_O$ and $z_I \ll D_I$.

CHAPTER 10

TENSOR ALGEBRA. INERTIA AND STRESS TENSORS

In this chapter we shall develop the algebra of linear vector functions, or *tensors*, as a mathematical tool which is useful in treating many problems. In particular, we shall need tensors in the study of the general motion of a rigid body and in the formulation of the concept of stress in a solid, or in a viscous fluid.

10.1 ANGULAR MOMENTUM OF A RIGID BODY

The equation of motion for the rotation of a rigid body is given by Eq. (5.5):

$$\frac{d\boldsymbol{L}}{dt} = \boldsymbol{N}, \tag{10.1}$$

where $\boldsymbol{L}$ is the angular momentum and $\boldsymbol{N}$ is the torque about a point P which may be either fixed or may be the center of mass of the body. In Section 5.2 we studied the rotation of a rigid body about a fixed axis. In order to treat the general problem of the rotation of a body about a point P, we must find the relation between the angular momentum vector $\boldsymbol{L}$ and the angular velocity vector $\boldsymbol{\omega}$.

Consider a body made up of point masses m_k situated at points $\boldsymbol{r}_k$ relative to an origin of coordinates at P. We have shown in Section 7.2 that the most general motion of the body about the point P is a rotation with angular velocity $\boldsymbol{\omega}$, and that the velocity $\boldsymbol{v}_k$ of each particle in the body is given by

$$\boldsymbol{v}_k = \boldsymbol{\omega} \times \boldsymbol{r}_k. \tag{10.2}$$

We sum the angular momentum given by Eq. (3.142) over all particles:

$$\begin{aligned} \boldsymbol{L} &= \sum_{k=1}^{N} m_k \boldsymbol{r}_k \times \boldsymbol{v}_k \\ &= \sum_{k=1}^{N} m_k \boldsymbol{r}_k \times (\boldsymbol{\omega} \times \boldsymbol{r}_k). \end{aligned} \tag{10.3}$$

Equation (10.3) expresses $\boldsymbol{L}$ as a function of $\boldsymbol{\omega}$, $\boldsymbol{L}(\boldsymbol{\omega})$. By substitution in Eq. (10.3) it is readily verified that the function $\boldsymbol{L}(\boldsymbol{\omega})$, for any two vectors $\boldsymbol{\omega}$, $\boldsymbol{\omega}'$, and any scalar c, satisfies the following relations:

$$\boldsymbol{L}(c\boldsymbol{\omega}) = c\boldsymbol{L}(\boldsymbol{\omega}), \tag{10.4}$$

$$\boldsymbol{L}(\boldsymbol{\omega}+\boldsymbol{\omega}') = \boldsymbol{L}(\boldsymbol{\omega})+\boldsymbol{L}(\boldsymbol{\omega}'). \tag{10.5}$$

A vector function $\boldsymbol{L}(\boldsymbol{\omega})$ with the properties (10.4), (10.5) is called a *linear vector function*. Linear vector functions are important because they occur frequently in physics, and because they have simple mathematical properties.

In order to develop an analogy between Eq. (10.3) and Eq. (5.9) for the case of rotation about an axis, we expand the triple cross product [Eq. (3.35)]:

$$\boldsymbol{L} = \sum_{k=1}^{N} [m_k r_k^2 \boldsymbol{\omega} - m_k \boldsymbol{r}_k(\boldsymbol{r}_k \cdot \boldsymbol{\omega})]. \tag{10.6}$$

The factor $\boldsymbol{\omega}$ is independent of k and can be factored from the sum over the first term. In a purely formal way, we may also factor $\boldsymbol{\omega}$ from the sum over the second term:

$$\boldsymbol{L} = \left(\sum_{k=1}^{N} m_k r_k^2\right)\boldsymbol{\omega} - \left(\sum_{k=1}^{N} m_k \boldsymbol{r}_k \boldsymbol{r}_k\right)\cdot\boldsymbol{\omega}. \tag{10.7}$$

The second term has no meaning, of course, since the juxtaposition $\boldsymbol{r}_k\boldsymbol{r}_k$ of two vectors has not yet been defined. We shall try to supply a meaning in the next section.

10.2 TENSOR ALGEBRA

The *dyad product* $\boldsymbol{AB}$ of two vectors is defined by the following equation, where $\boldsymbol{C}$ is any vector:

$$(\boldsymbol{AB})\cdot\boldsymbol{C} = \boldsymbol{A}(\boldsymbol{B}\cdot\boldsymbol{C}). \tag{10.8}$$

The right member of this equation is expressed in terms of products defined in Section 3.1. The left member is, by definition, the vector given by the right member. Note that the dyad $\boldsymbol{AB}$ is defined only in terms of its dot product with an arbitrary vector $\boldsymbol{C}$. We can readily show, from definition (10.8), that multiplication of a vector by a dyad is a linear operation in the sense that

$$(\boldsymbol{AB})\cdot(c\boldsymbol{C}) = c[(\boldsymbol{AB})\cdot\boldsymbol{C}], \tag{10.9}$$

$$(\boldsymbol{AB})\cdot(\boldsymbol{C}+\boldsymbol{D}) = (\boldsymbol{AB})\cdot\boldsymbol{C}+(\boldsymbol{AB})\cdot\boldsymbol{D}. \tag{10.10}$$

For fixed vectors $\boldsymbol{A}$, $\boldsymbol{B}$, the dyad $\boldsymbol{AB}$ therefore defines a linear vector function $\boldsymbol{F}(\boldsymbol{C})$:

$$\boldsymbol{F}(\boldsymbol{C}) = (\boldsymbol{AB})\cdot\boldsymbol{C}. \tag{10.11}$$

The dyad $\boldsymbol{AB}$ is an example of a *linear vector operator*, that is, it represents an operation which may be performed on any vector $\boldsymbol{C}$ to yield a new vector $(\boldsymbol{AB})\cdot\boldsymbol{C}$, which is a linear function of $\boldsymbol{C}$.

A linear vector operator is also called a *tensor*.* Tensors will be represented

---

*More precisely, a linear vector operator may be called a second-rank tensor, to distinguish it from third- and higher-rank tensors obtained as linear combinations of triads $\boldsymbol{ABC}$, etc. We shall be concerned in this book only with second-rank tensors, which we shall refer to simply as tensors.

by sans-serif boldface capitals, **A**, **B**, **C**, etc. We may for example let **T** be the tensor represented by the dyad $\boldsymbol{AB}$:

$$\mathsf{T} = \boldsymbol{AB}. \tag{10.12}$$

The meaning of the tensor **T** is specified by the definition,*

$$\mathsf{T}\cdot\boldsymbol{C} = \boldsymbol{A}(\boldsymbol{B}\cdot\boldsymbol{C}), \tag{10.13}$$

which gives the result of applying **T** to any vector $\boldsymbol{C}$. We can form more general linear vector operators by taking sums of dyads. The sum of two dyads, or tensors **S**, **T**, is defined as follows:

$$(\mathsf{S}+\mathsf{T})\cdot\boldsymbol{C} = \mathsf{S}\cdot\boldsymbol{C}+\mathsf{T}\cdot\boldsymbol{C}. \tag{10.14}$$

Note that all definitions of algebraic operations on tensors, like the above definition of (**S**+**T**), are formulated in terms of the application of the tensors to an arbitrary vector $\boldsymbol{C}$. The sum of one or more dyads is called a *dyadic*. According to the definition (10.14), the dyadic ($\boldsymbol{AB}+\boldsymbol{DE}$) operating on $\boldsymbol{C}$ yields the vector

$$(\boldsymbol{AB}+\boldsymbol{DE})\cdot\boldsymbol{C} = \boldsymbol{A}(\boldsymbol{B}\cdot\boldsymbol{C})+\boldsymbol{D}(\boldsymbol{E}\cdot\boldsymbol{C}). \tag{10.15}$$

We can readily show that the sum of two linear operators is a linear operator; therefore dyadics are also linear vector operators and we have for any dyadic or tensor **T**,

$$\mathsf{T}\cdot(c\boldsymbol{C}) = c(\mathsf{T}\cdot\boldsymbol{C}), \tag{10.16}$$

$$\mathsf{T}\cdot(\boldsymbol{C}+\boldsymbol{D}) = \mathsf{T}\cdot\boldsymbol{C}+\mathsf{T}\cdot\boldsymbol{D}. \tag{10.17}$$

The linearity relations (10.16), (10.17), together with the definition (10.14), guarantee that dyad products, sums of tensors, and dot products of tensors with vectors satisfy all the usual algebraic rules for sums and products. We can also define a dot product of a dyad with a vector on the left in the obvious way,

$$\boldsymbol{C}\cdot(\boldsymbol{AB}) = (\boldsymbol{C}\cdot\boldsymbol{A})\boldsymbol{B}, \tag{10.18}$$

and correspondingly for sums of dyads. Note that the dot product of a dyadic with a vector is not commutative;

$$\mathsf{T}\cdot\boldsymbol{C} = \boldsymbol{C}\cdot\mathsf{T} \tag{10.19}$$

does not hold in general. We can define, in an obvious way, a product c**T** of a tensor by a scalar, with the expected algebraic properties (see Problem 1).

A very simple tensor is given by the dyadic

$$\mathbf{1} = \hat{\boldsymbol{x}}\hat{\boldsymbol{x}}+\hat{\boldsymbol{y}}\hat{\boldsymbol{y}}+\hat{\boldsymbol{z}}\hat{\boldsymbol{z}}, \tag{10.20}$$

where $\hat{\boldsymbol{x}}$, $\hat{\boldsymbol{y}}$, $\hat{\boldsymbol{z}}$ are a set of perpendicular unit vectors along x-, y-, and z-axes. We calculate, using the definitions (10.14) and (10.8),

$$\mathbf{1}\cdot\boldsymbol{A} = \hat{\boldsymbol{x}}A_x+\hat{\boldsymbol{y}}A_y+\hat{\boldsymbol{z}}A_z = \boldsymbol{A}. \tag{10.21}$$

*The result of applying a tensor **T** to a vector $\boldsymbol{C}$ is often denoted by **T**$\boldsymbol{C}$, without the dot. We shall use the dot throughout this book.

The tensor $\mathbf{1}$ is called the *unit tensor*; it may be defined as the operator which, acting on any vector, yields that vector itself. Evidently $\mathbf{1}$ is one of the special cases for which

$$\mathbf{1}\cdot \boldsymbol{A} = \boldsymbol{A}\cdot\mathbf{1}. \tag{10.22}$$

If c is any scalar, the product $c\mathbf{1}$ is called a *constant tensor*, and has the property

$$(c\mathbf{1})\cdot \boldsymbol{A} = \boldsymbol{A}\cdot(c\mathbf{1}) = c\boldsymbol{A}. \tag{10.23}$$

Using the definitions above, we can now write Eq. (10.7) in the form

$$\boldsymbol{L} = \mathbf{I}\cdot\boldsymbol{\omega}, \tag{10.24}$$

where $\mathbf{I}$ is the *inertia tensor* of the rigid body, defined by

$$\mathbf{I} = \sum_{k=1}^{N} (m_k r_k^2 \mathbf{1} - m_k \boldsymbol{r}_k \boldsymbol{r}_k). \tag{10.25}$$

The inertia tensor $\mathbf{I}$ is the analog, for general rotations, of the moment of inertia for rotations about an axis. Note that $\boldsymbol{L}$ and $\boldsymbol{\omega}$ are not in general parallel. We will study the inertia tensor in more detail after we have developed the necessary properties of tensors.

If we write all vectors in terms of their components,

$$\boldsymbol{C} = C_x\hat{\boldsymbol{x}} + C_y\hat{\boldsymbol{y}} + C_z\hat{\boldsymbol{z}}, \tag{10.26}$$

then it is clear that by multiplying out dyad products and collecting terms, any dyadic can be written in the form

$$\begin{aligned}\mathbf{T} = {} & T_{xx}\hat{\boldsymbol{x}}\hat{\boldsymbol{x}} + T_{xy}\hat{\boldsymbol{x}}\hat{\boldsymbol{y}} + T_{xz}\hat{\boldsymbol{x}}\hat{\boldsymbol{z}} \\ & + T_{yx}\hat{\boldsymbol{y}}\hat{\boldsymbol{x}} + T_{yy}\hat{\boldsymbol{y}}\hat{\boldsymbol{y}} + T_{yz}\hat{\boldsymbol{y}}\hat{\boldsymbol{z}} \\ & + T_{zx}\hat{\boldsymbol{z}}\hat{\boldsymbol{x}} + T_{zy}\hat{\boldsymbol{z}}\hat{\boldsymbol{y}} + T_{zz}\hat{\boldsymbol{z}}\hat{\boldsymbol{z}}.\end{aligned} \tag{10.27}$$

Just as any vector $\boldsymbol{A}$ can be represented by its three components (A_x, A_y, A_z), so any dyadic can be specified by giving its nine components $T_{xx}, \ldots, T_{zz}$. These may conveniently be written in the form of a square array or matrix:

$$\mathbf{T} = \begin{pmatrix} T_{xx} & T_{xy} & T_{xz} \\ T_{yx} & T_{yy} & T_{yz} \\ T_{zx} & T_{zy} & T_{zz} \end{pmatrix}. \tag{10.28}$$

As an example, the reader may verify that the components of the inertia tensor (10.25) are

$$I_{xx} = \sum_{k=1}^{N} m_k(y_k^2 + z_k^2), \qquad I_{xy} = -\sum_{k=1}^{N} m_k x_k y_k, \qquad \text{etc.} \tag{10.29}$$

In order to simplify writing the tensor components, it often will be convenient to number the coordinate axes x_1, x_2, x_3 instead of using x, y, z:

$$x = x_1, \qquad y = x_2, \qquad z = x_3. \tag{10.30}$$

We shall write the corresponding unit vectors as $\hat{\boldsymbol{e}}_i$:

$$\hat{\boldsymbol{x}} = \hat{\boldsymbol{e}}_1, \qquad \hat{\boldsymbol{y}} = \hat{\boldsymbol{e}}_2, \qquad \hat{\boldsymbol{z}} = \hat{\boldsymbol{e}}_3. \tag{10.31}$$

Equations (10.26) and (10.27) can now be written as

$$\boldsymbol{C} = \sum_{i=1}^{3} C_i \hat{\boldsymbol{e}}_i, \tag{10.32}$$

and

$$\mathbf{T} = \sum_{i,j=1}^{3} T_{ij} \hat{\boldsymbol{e}}_i \hat{\boldsymbol{e}}_j. \tag{10.33}$$

Another advantage of this notation is that it allows the discussion to be generalized to vectors and tensors in a space of any number of dimensions simply by changing the summation limit.

By using the definitions of dyad products and sums, we can express the components of the vector $\mathbf{T}\cdot\boldsymbol{C}$ in terms of the components of $\mathbf{T}$ and $\boldsymbol{C}$:

$$(\mathbf{T}\cdot\boldsymbol{C})_i = \sum_{j=1}^{3} T_{ij} C_j, \tag{10.34}$$

as the reader should verify. Similarly,

$$(\boldsymbol{C}\cdot\mathbf{T})_i = \sum_{j=1}^{3} C_j T_{ji}. \tag{10.35}$$

We note that, by Eq. (10.33),

$$T_{ij} = \hat{\boldsymbol{e}}_i\cdot(\mathbf{T}\cdot\hat{\boldsymbol{e}}_j) = (\hat{\boldsymbol{e}}_i\cdot\mathbf{T})\cdot\hat{\boldsymbol{e}}_j. \tag{10.36}$$

We may omit the parentheses, since the order in which the multiplications are carried out does not matter.

We can now show that any linear vector function can be represented by a dyadic. Let $\boldsymbol{F}(\boldsymbol{C})$ be any linear function of $\boldsymbol{C}$. Consider first the case when $\boldsymbol{C}$ is a unit vector $\hat{\boldsymbol{e}}_j$, and let T_{ij} be the components of $\boldsymbol{F}(\boldsymbol{C})$ in that case:

$$\boldsymbol{F}(\hat{\boldsymbol{e}}_j) = \sum_{i=1}^{3} T_{ij} \hat{\boldsymbol{e}}_i. \tag{10.37}$$

Now any vector $\boldsymbol{C}$ can be written as

$$\boldsymbol{C} = \sum_{j=1}^{3} C_j \hat{\boldsymbol{e}}_j. \tag{10.38}$$

By use of the linear property of $\boldsymbol{F}(\boldsymbol{C})$, we have therefore

$$\begin{aligned}\boldsymbol{F}(\boldsymbol{C}) &= \sum_{j=1}^{3} \boldsymbol{F}(C_j\hat{\boldsymbol{e}}_j)\\ &= \sum_{j=1}^{3} C_j\boldsymbol{F}(\hat{\boldsymbol{e}}_j)\\ &= \sum_{i,j=1}^{3} C_jT_{ij}\hat{\boldsymbol{e}}_i.\end{aligned} \tag{10.39}$$

Thus the components of $\boldsymbol{F}(\boldsymbol{C})$ can be expressed in terms of the numbers T_{ij}:

$$[\boldsymbol{F}(\boldsymbol{C})]_i = \sum_{j=1}^{3} T_{ij}C_j. \tag{10.40}$$

If we define the dyadic

$$\mathsf{T} = \sum_{i,j=1}^{3} T_{ij}\hat{\boldsymbol{e}}_i\hat{\boldsymbol{e}}_j, \tag{10.41}$$

we see from Eqs. (10.40) and (10.34) that

$$\boldsymbol{F}(\boldsymbol{C}) = \mathsf{T}\cdot\boldsymbol{C}. \tag{10.42}$$

Thus the concepts of dyadic and linear vector operator or tensor are identical, and are equivalent to the concept of linear vector function in the sense that every linear vector function defines a certain tensor or dyadic, and conversely.

We can define a dot product of two tensors as follows:

$$(\mathsf{T}\cdot\mathsf{S})\cdot\boldsymbol{C} = \mathsf{T}\cdot(\mathsf{S}\cdot\boldsymbol{C}). \tag{10.43}$$

Application of the operator $\mathsf{T}\cdot\mathsf{S}$ to any vector means first applying S, and then T. We now calculate in terms of components, using the definition (10.43),

$$\begin{aligned}(\mathsf{T}\cdot\mathsf{S})\cdot\boldsymbol{C} &= \mathsf{T}\cdot\sum_{j,k=1}^{3} S_{jk}C_k\hat{\boldsymbol{e}}_j\\ &= \sum_{i=1}^{3}\sum_{j,k=1}^{3} T_{ij}S_{jk}C_k\hat{\boldsymbol{e}}_i\\ &= \sum_{i=1}^{3}\left[\sum_{k=1}^{3}\left(\sum_{j=1}^{3} T_{ij}S_{jk}\right)C_k\right]\hat{\boldsymbol{e}}_i.\end{aligned} \tag{10.44}$$

Comparing this result with Eq. (10.34), we see that

$$(\mathsf{T}\cdot\mathsf{S})_{ik} = \sum_{j=1}^{3} T_{ij}S_{jk}. \tag{10.45}$$

Equation (10.45) also results if we simply evaluate $\mathbf{T}\cdot\mathbf{S}$ in a formal way by writing the dot and dyad products and collecting terms:

$$\mathbf{T}\cdot\mathbf{S} = \sum_{ijkl=1}^{3} T_{ij}S_{kl}\hat{e}_i\hat{e}_j\cdot\hat{e}_k\hat{e}_l$$

$$= \sum_{ijl=1}^{3} T_{ij}S_{jl}\hat{e}_i\hat{e}_l, \tag{10.46}$$

and this shows that our definition (10.43) is consistent with the ordinary rules of algebra. If $\mathbf{T}$, $\mathbf{S}$ are written as matrices according to Eq. (10.28), then Eq. (10.45) is the usual mathematical rule for multiplying matrices. We can similarly show that the definition (10.14) implies that tensors are added by adding their component matrices according to the rule:

$$(\mathbf{T}+\mathbf{S})_{ij} = T_{ij}+S_{ij}. \tag{10.47}$$

Sums and products of tensors obey all the usual rules of algebra except that dot multiplication, in general, is not commutative:

$$\mathbf{T}+\mathbf{S} = \mathbf{S}+\mathbf{T}, \tag{10.48}$$

$$\mathbf{T}\cdot(\mathbf{S}+\mathbf{P}) = \mathbf{T}\cdot\mathbf{S}+\mathbf{T}\cdot\mathbf{P}, \tag{10.49}$$

$$\mathbf{T}\cdot(\mathbf{S}\cdot\mathbf{P}) = (\mathbf{T}\cdot\mathbf{S})\cdot\mathbf{P}, \tag{10.50}$$

$$\mathbf{1}\cdot\mathbf{T} = \mathbf{T}\cdot\mathbf{1} = \mathbf{T}, \tag{10.51}$$

and so on, but

$$\mathbf{T}\cdot\mathbf{S} \neq \mathbf{S}\cdot\mathbf{T}, \qquad \text{in general.} \tag{10.52}$$

It is useful to define the *transpose* $\mathbf{T}^t$ of a tensor $\mathbf{T}$ as follows:

$$\mathbf{T}^t\cdot C = C\cdot\mathbf{T}. \tag{10.53}$$

In terms of components,

$$T^t_{ij} = T_{ji}. \tag{10.54}$$

The transpose is often written $\tilde{\mathbf{T}}$, but the notation $\mathbf{T}^t$ is preferable for typographical reasons. The following properties are easily proved:

$$(\mathbf{T}+\mathbf{S})^t = \mathbf{T}^t+\mathbf{S}^t, \tag{10.55}$$

$$(\mathbf{T}\cdot\mathbf{S})^t = \mathbf{S}^t\cdot\mathbf{T}^t, \tag{10.56}$$

$$(\mathbf{T}^t)^t = \mathbf{T}. \tag{10.57}$$

A tensor is said to be *symmetric* if

$$\mathbf{T}^t = \mathbf{T}. \tag{10.58}$$

For example, the inertia tensor, given by Eq. (10.25), is symmetric. For a symmetric tensor,

$$T_{ji} = T_{ij}. \tag{10.59}$$

A symmetric tensor may be specified by six components; the remaining three are then determined by Eq. (10.59).

A tensor is said to be *antisymmetric* if

$$\mathbf{T}^t = -\mathbf{T}. \tag{10.60}$$

The components of an antisymmetric tensor satisfy the equation

$$T_{ji} = -T_{ij}. \tag{10.61}$$

Evidently the three diagonal components T_{ii} are all zero, and if three off-diagonal components are given, the three remaining components are determined by Eq. (10.61). An antisymmetric tensor has only three independent components (in three-dimensional space). An example is the linear operator defined by

$$\mathbf{T} \cdot \boldsymbol{C} = \boldsymbol{\omega} \times \boldsymbol{C}, \tag{10.62}$$

where $\boldsymbol{\omega}$ is a fixed vector. Comparing Eq. (10.62) with Eq. (7.20), we see that the operator $\mathbf{T}$ can be interpreted as giving the velocity of any vector $\boldsymbol{C}$ rotating with an angular velocity $\boldsymbol{\omega}$. Comparing Eq. (10.62) with Eq. (10.34), we see that the components of $\mathbf{T}$ are:

$$\begin{aligned} T_{11} &= T_{22} = T_{33} = 0, \\ T_{21} &= -T_{12} = \omega_3, \\ T_{32} &= -T_{23} = \omega_1, \\ T_{13} &= -T_{31} = \omega_2. \end{aligned} \tag{10.63}$$

Since an antisymmetric tensor, like a vector, has three independent components, we may associate with every antisymmetric tensor $\mathbf{T}$ a vector $\boldsymbol{\omega}$ (in three-dimensional space only!) whose components are related to those of $\mathbf{T}$ by Eq. (10.63). The operation $\mathbf{T}\cdot$ will then be equivalent to $\boldsymbol{\omega}\times$, according to Eq. (10.62).

Given any tensor $\mathbf{T}$, we can define a symmetric and an antisymmetric tensor by

$$\mathbf{T}_s = \tfrac{1}{2}(\mathbf{T} + \mathbf{T}^t), \tag{10.64}$$

$$\mathbf{T}_a = \tfrac{1}{2}(\mathbf{T} - \mathbf{T}^t), \tag{10.65}$$

such that

$$\mathbf{T} = \mathbf{T}_s + \mathbf{T}_a. \tag{10.66}$$

We saw, in the preceding paragraph, that an antisymmetric tensor could be represented geometrically by a certain vector $\boldsymbol{\omega}$. We will see in Section 10.4 how to represent a symmetric tensor. Since antisymmetric and symmetric tensors have rather different geometric properties, tensors which occur in physics are usually either symmetric or antisymmetric rather than a combination of the two. In three-dimensional space, the introduction of an antisymmetric tensor can always be avoided by the use of the associated vector. It is therefore not a coincidence that the two principal examples of tensors in this chapter, the inertia tensor and the stress tensor, are both symmetric.

10.3 COORDINATE TRANSFORMATIONS

We saw in the previous section that a tensor $\mathbf{T}$ may be defined geometrically as a linear vector operator by specifying the result of applying $\mathbf{T}$ to any vector $\boldsymbol{C}$. Alternatively, the tensor may be specified algebraically by giving its components T_{ij}. A discrepancy exists between the two definitions of a tensor, in that the algebraic definition appears to depend upon the choice of a particular coordinate system. A similar discrepancy in the case of a vector was noted in Section 3.1. We will now remove the discrepancy by learning how to transform the components of vectors and tensors when the coordinate system is changed. We will restrict the discussion to rectangular coordinates.

Let us consider two coordinate systems, x_1, x_2, x_3, and x'_1, x'_2, x'_3, having the same origin. The coordinates of a point in the two systems are related by Eqs. (7.13):

$$x'_i = \sum_{j=1}^{3} a_{ij}x_j, \tag{10.67}$$

where

$$a_{ij} = \hat{\boldsymbol{e}}'_i \cdot \hat{\boldsymbol{e}}_j \tag{10.68}$$

is the cosine of the angle between the x'_i- and x_j-axes. Likewise,

$$x_j = \sum_{i=1}^{3} a_{ij}x'_i. \tag{10.69}$$

The relations between the primed and unprimed components of any vector

$$\boldsymbol{C} = \sum_{i=1}^{3} C'_i\hat{\boldsymbol{e}}'_i = \sum_{j=1}^{3} C_j\hat{\boldsymbol{e}}_j \tag{10.70}$$

may be obtained in a similar manner by dotting $\hat{\boldsymbol{e}}'_i$ or $\hat{\boldsymbol{e}}_j$ into Eq. (10.70):

$$C'_i = \sum_{j=1}^{3} a_{ij}C_j, \tag{10.71}$$

$$C_j = \sum_{i=1}^{3} a_{ij}C'_i. \tag{10.72}$$

We can now define a vector algebraically as a set of three components (C_1, C_2, C_3) which transform like the coordinates (x_1, x_2, x_3) when the coordinate system is changed. By referring to all coordinate systems, this definition avoids giving preferential treatment to any particular coordinate system. In the same way, the primed and unprimed components of a tensor

$$\mathsf{T} = \sum_{i,k=1}^{3} T'_{ik}\hat{e}'_i\hat{e}'_k = \sum_{j,l=1}^{3} T_{jl}\hat{e}_j\hat{e}_l \tag{10.73}$$

are related by [see Eq. (10.36)]

$$T'_{ik} = \hat{e}'_i \cdot \mathsf{T} \cdot \hat{e}'_k = \sum_{j,l=1}^{3} a_{ij}a_{kl}T_{jl}, \tag{10.74}$$

$$T_{jl} = \hat{e}_j \cdot \mathsf{T} \cdot \hat{e}_l = \sum_{i,k=1}^{3} a_{ij}a_{kl}T'_{ik}. \tag{10.75}$$

A tensor may be defined algebraically as a set of nine components (T_{jl}) that transform according to the rule given in Eqs. (10.74) and (10.75). Note the distinction between a tensor and a matrix. The concept of a matrix is purely algebraic; matrices are arrays of numbers which may be added and multiplied according to the rules (10.45) and (10.47). The concept of a tensor is geometrical; a tensor may be represented in any particular coordinate system by a matrix, but the matrix must be transformed according to a definite rule if the coordinate system is changed.

The coefficients a_{ij} defined by Eq. (10.68) are the components of the unit vectors $\hat{e}'_i$ in the unprimed system, and conversely:

$$\hat{e}'_i = \sum_{j=1}^{3} a_{ij}\hat{e}_j, \tag{10.76}$$

and

$$\hat{e}_j = \sum_{i=1}^{3} a_{ij}\hat{e}'_i. \tag{10.77}$$

Since $\hat{e}'_1$, $\hat{e}'_2$, $\hat{e}'_3$ are a set of perpendicular unit vectors, we see that the numbers a_{ij} must satisfy the equations:

$$\hat{e}'_i \cdot \hat{e}'_k = \sum_{j=1}^{3} a_{ij}a_{kj} = \delta_{ik}, \tag{10.78}$$

where δ_{ik} is a shorthand notation for

$$\delta_{ik} = \begin{cases} 0 & \text{if } i \neq k, \\ 1 & \text{if } i = k. \end{cases} \tag{10.79}$$

There are six relations (10.78) among the nine coefficients a_{ik}. Hence, if three of the constants a_{ij} are specified, the rest may be determined from Eqs. (10.78). It is clear that three independent constants must be specified to locate the primed axes relative to the unprimed (or vice versa). For the x'_1-axis may point in any direction and two coordinates are therefore required to locate it. Once the position of the x'_1-axis is determined, the position of the x'_2-axis, which may be anywhere in a plane perpendicular to x'_1, may be specified by one coordinate. The position of the x'_3-axis is then determined (except for sign). We can write additional relations between the a_{ij}'s by the use of such relations as

$$\hat{e}_j \cdot \hat{e}_l = \delta_{jl}, \qquad \hat{e}'_1 \times \hat{e}'_2 = \pm \hat{e}'_3, \qquad \hat{e}_1 \cdot (\hat{e}_2 \times \hat{e}_3) = \pm 1, \qquad \text{etc.} \tag{10.80}$$

Since at least three of the a_{ij}'s must be independent, it is clear that the relations obtained from Eqs. (10.80) are not independent but could be obtained algebraically from Eqs. (10.78). An interesting relation is obtained from

$$\boldsymbol{e}_1 \cdot (\boldsymbol{e}_2 \times \boldsymbol{e}_3) = \begin{vmatrix} a_{11} & a_{21} & a_{31} \\ a_{12} & a_{22} & a_{32} \\ a_{13} & a_{23} & a_{33} \end{vmatrix} = \pm 1, \tag{10.81}$$

where the result is $+1$ if the primed and unprimed systems are both right- or both left-handed and is -1 if one is right-handed and the other left-handed. Hence the determinant $|a_{ij}|$ is $+1$ or -1 according to whether the handedness of the coordinate system is or is not changed.

In a left-handed system, the cross product is to be defined using the left in place of the right hand. [In Eq. (10.81), the triple product on the left is to be evaluated in the primed system.] The algebraic definition is then the same in either case:

$$(\boldsymbol{A} \times \boldsymbol{B}) = (A_2B_3 - A_3B_2,\ A_3B_1 - A_1B_3,\ A_1B_2 - A_2B_1). \tag{10.82}$$

This definition implies that the cross product $\boldsymbol{A} \times \boldsymbol{B}$ of two ordinary vectors is not itself an ordinary vector, since its direction reverses when we change the handedness of the coordinate system. An ordinary vector that has a direction independent of the coordinate system is called a *polar vector*. A vector whose sense depends upon the handedness of the coordinate system is called an *axial vector* or *pseudovector*. The angular velocity vector $\boldsymbol{\omega}$ is an axial vector, and so is any other vector whose sense is defined by a "right-hand rule." The vector associated with an (ordinary) antisymmetric tensor is an axial vector. The cross product $\boldsymbol{\omega} \times \boldsymbol{C}$ of an axial with a polar vector is itself a polar vector. The distinction between axial and polar vectors arises only if we wish to consider both right- and left-handed coordinate systems. In the applications in this book, we need only consider rotations of the coordinate system. Since rotations do not change the handedness of the system, we shall not be concerned with this distinction.

The transformation defined by Eqs. (10.67), (10.71), and (10.74), where the coefficients satisfy Eq. (10.78), is called *orthogonal*. As the name implies, an orthogonal transformation enables us to change from one set of perpendicular unit vectors to another.

The right member of Eq. (10.71) is formally similar to the right member of Eq. (10.34). This suggests an alternative interpretation of Eqs. (10.71). Let us define a tensor A with components

$$A_{ij} = a_{ij}, \tag{10.83}$$

and consider the vector

$$\boldsymbol{C}' = \mathsf{A}\cdot\boldsymbol{C}. \tag{10.84}$$

The components C_j' of $\boldsymbol{C}'$ are given by Eq. (10.71). Similarly, by Eq. (10.72),

$$\boldsymbol{C} = \mathsf{A}^t\cdot\boldsymbol{C}'. \tag{10.85}$$

Thus Eqs. (10.71) and (10.72) may be interpreted alternatively as representing the result of operating with the tensors A, A^t upon the vectors $\boldsymbol{C}$, $\boldsymbol{C}'$, respectively. In the original interpretation, C_j, C_i' are components of the same vector $\boldsymbol{C}$ in two different coordinate systems. In the alternative interpretation, C_j, C_i' are the components of two different vectors $\boldsymbol{C}$, $\boldsymbol{C}'$ in the same coordinate system. We are primarily interested in the first interpretation, in which these equations represent a coordinate transformation. However, the latter interpretation will often be useful in deriving certain algebraic properties of Eqs. (10.71), (10.72), which, of course, are independent of how we choose to interpret them. In case the primed axes are fixed in a rotating rigid body, either interpretation is useful. If the primed axes initially coincided with the unprimed axes, then at any later time we may interpret Eqs. (10.71), (10.72) as expressing the transformation from one coordinate system to the other. Alternatively, we may interpret A as the tensor which represents the operation of rotating the body from its initial to its present position, that is, a vector fixed in the body and initially coinciding with $\boldsymbol{C}$ will be rotated so as to coincide with $\boldsymbol{C}' = \mathsf{A}\cdot\boldsymbol{C}$. Making use of Eqs. (10.84), (10.85), and the definition (10.43) of the product of two tensors, we deduce that

$$\mathsf{A}^t\cdot(\mathsf{A}\cdot\boldsymbol{C}) = (\mathsf{A}^t\cdot\mathsf{A})\cdot\boldsymbol{C} = \boldsymbol{C}. \tag{10.86}$$

Hence, by Eq. (10.22),

$$\mathsf{A}^t\cdot\mathsf{A} = \mathsf{1}, \tag{10.87}$$

and similarly

$$\mathsf{A}\cdot\mathsf{A}^t = \mathsf{1}. \tag{10.88}$$

A tensor having this property is said to be *orthogonal*. Equation (10.87) is evidently equivalent to Eqs. (10.78) for the components a_{ij}. In the second interpretation, Eqs. (10.74) and (10.75) can be written as

$$\mathsf{T}' = \mathsf{A}\cdot\mathsf{T}\cdot\mathsf{A}^t, \tag{10.89}$$

$$\mathsf{T} = \mathsf{A}^t\cdot\mathsf{T}'\cdot\mathsf{A}. \tag{10.90}$$

The orthogonal tensor is the only example we shall have of a tensor with a definite geometrical significance, which is neither symmetric nor antisymmetric; it has, instead, the orthogonality property given by Eq. (10.87).

In view of the fact that the various vector operations were defined without reference to a coordinate system, it is clear that all algebraic rules for computing sums, products, transposes, etc., of vectors and tensors will be unaffected by an orthogonal transformation of coordinates. Thus, for example,

$$(\boldsymbol{B}+\boldsymbol{C})'_j = B'_j + C'_j, \tag{10.91}$$

$$(\mathbf{T}\cdot\boldsymbol{C})'_j = \sum_{l=1}^{3} T'_{jl} C'_l, \tag{10.92}$$

$$(\mathbf{T}^t)'_{ij} = T'_{ji}. \tag{10.93}$$

We can also verify directly the above equations, and others like them, by using the transformation equations and the rules of vector and tensor algebra. This is most easily done by taking advantage of the second interpretation of the transformation equations. For example, we can prove Eq. (10.93) by noting that

$$\begin{aligned}
(\mathbf{T}^t)' &= \mathbf{A}\cdot\mathbf{T}^t\cdot\mathbf{A}^t && \text{[by Eq. (10.89)]}\\
&= \mathbf{A}\cdot(\mathbf{A}\cdot\mathbf{T})^t && \text{[by Eqs. (10.56) and (10.57)]}\\
&= [(\mathbf{A}\cdot\mathbf{T})\cdot\mathbf{A}^t]^t && \text{[by Eqs. (10.56) and (10.57)]}\\
&= (\mathbf{T}')^t, \text{ Q.E.D.} && \text{[by Eq. (10.89)].}
\end{aligned}$$

Any property or relation between vectors and tensors which is expressed in the same algebraic form in all coordinate systems has a geometrical meaning independent of the coordinate system and is called an *invariant* property or relation.

Given a tensor **T**, we may define a scalar quantity called the *trace* of **T** as follows:

$$\operatorname{tr}(\mathbf{T}) = \sum_{i=1}^{3} T_{ii}. \tag{10.94}$$

Since this definition is in terms of components, we must show that the trace of **T** is the same in all coordinate systems:

$$\begin{aligned}
\operatorname{tr}(\mathbf{T}) &= \sum_{i=1}^{3} T_{ii}\\
&= \sum_{i=1}^{3}\sum_{j,l=1}^{3} a_{ji}a_{li}T'_{jl} && \text{[by Eq. (10.75)]}\\
&= \sum_{j,l=1}^{3}\left[\sum_{i=1}^{3} a_{ji}a_{li}\right]T'_{jl} && \text{[rearranging sums]}\\
&= \sum_{j,l=1}^{3} T'_{jl}\,\delta_{jl} && \text{[by Eq. (10.78)]}\\
&= \sum_{j=1}^{3} T'_{jj}, \text{ Q.E.D.} && \text{[by Eq. (10.79)].}
\end{aligned}$$

Another invariant scalar quantity associated with a tensor is the determinant

$$\det(\mathbf{T}) = \begin{vmatrix} T_{11} & T_{12} & T_{13} \\ T_{21} & T_{22} & T_{23} \\ T_{31} & T_{32} & T_{33} \end{vmatrix}, \tag{10.95}$$

as may be verified in a similar way by direct computation.

Let us now study the result of carrying out two coordinate transformations in succession. The primed coordinates are defined by Eq. (10.67), in terms of the unprimed coordinates. Let double-primed coordinates be defined by

$$x_k'' = \sum_{i=1}^{3} a_{ki}' x_i'. \tag{10.96}$$

We substitute for x_i' from Eq. (10.67) to obtain the double-primed coordinates in terms of the unprimed coordinates:

$$\begin{aligned} x_k'' &= \sum_{i=1}^{3} \sum_{j=1}^{3} a_{ki}' a_{ij} x_j \\ &= \sum_{j=1}^{3} \left[\sum_{i=1}^{3} a_{ki}' a_{ij} \right] x_j \\ &= \sum_{j=1}^{3} a_{kj}'' x_j, \end{aligned} \tag{10.97}$$

where the coefficients of the transformation $x \to x''$ are given by

$$a_{kj}'' = \sum_{i=1}^{3} a_{ki}' a_{ij}. \tag{10.98}$$

Thus the matrix of coefficients a_{kj}'' is obtained by multiplying the matrices a_{ki}', a_{ij} according to the rule for matrix multiplication. If we interpret the transformation coefficients as the components of tensors $\mathbf{A}$, $\mathbf{A}'$, $\mathbf{A}''$, we then see from Eqs. (10.98) and the definition (10.45) of the product of two tensors, that

$$\mathbf{A}'' = \mathbf{A}' \cdot \mathbf{A}. \tag{10.99}$$

This result also follows immediately from Eq. (10.84), applied twice, and we therefore have an alternative way to derive Eq. (10.98).

10.4 DIAGONALIZATION OF A SYMMETRIC TENSOR

The constant tensor, defined by Eq. (10.23), has in every coordinate system* the matrix:

$$c\mathbf{1} = \begin{pmatrix} c & 0 & 0 \\ 0 & c & 0 \\ 0 & 0 & c \end{pmatrix}. \tag{10.100}$$

---

*See Problem 10 at the end of this chapter.

A nonconstant tensor may, in a particular coordinate system, have the matrix:

$$\mathbf{T} = \begin{pmatrix} T_1 & 0 & 0 \\ 0 & T_2 & 0 \\ 0 & 0 & T_3 \end{pmatrix}. \tag{10.101}$$

The tensor $\mathbf{T}$ is then said to be in *diagonal form.* We do not call $\mathbf{T}$ a diagonal tensor, because the property (10.101) applies only to a particular coordinate system; after a change of coordinates [Eq. (10.74)], $\mathbf{T}$ will usually no longer be in diagonal form. If $\mathbf{T}$ is in diagonal form, then its effect on a vector is given simply by

$$(\mathbf{T}\cdot \boldsymbol{C})_i = T_i C_i, \qquad i = 1, 2, 3. \tag{10.102}$$

The importance of the diagonal form lies in the following fundamental theorem:

Any symmetric tensor can be brought into diagonal form by an orthogonal transformation. The diagonal elements are then unique except for their order, and the corresponding axes are unique except for degeneracy. (10.103)

Before proving this important theorem, let us try to understand its significance. The theorem states that, given any symmetric tensor $\mathbf{T}$, we can always choose the coordinate axes so that $\mathbf{T}$ is represented by a diagonal matrix. Furthermore, this can be done in essentially only one way; there is only one diagonal form (10.101) for a given tensor $\mathbf{T}$, except for the order in which the diagonal elements T_1, T_2, T_3 appear, and each element is associated with a unique axis in space, except for degeneracy, that is, except when two or three of the diagonal elements are equal. The axes $\hat{\boldsymbol{e}}_1$, $\hat{\boldsymbol{e}}_2$, $\hat{\boldsymbol{e}}_3$ in the coordinate system in which the tensor has a diagonal form are called its *principal axes.* The diagonal elements T_1, T_2, T_3 are called the *eigenvalues* or *characteristic values* of $\mathbf{T}$. Whenever we write a tensor element with a single subscript, we shall mean it to be an eigenvalue.

Any vector $\boldsymbol{C}$ parallel to a principal axis is called an *eigenvector* of $\mathbf{T}$. An eigenvector, according to Eq. (10.102), has the property that operation by $\mathbf{T}$ reduces to multiplication by the corresponding eigenvalue:

$$\mathbf{T}\cdot \boldsymbol{C} = T_i \boldsymbol{C}, \tag{10.104}$$

where T_i is the eigenvalue associated with the principal axis $\hat{\boldsymbol{e}}_i$ parallel to $\boldsymbol{C}$.

The theorem (10.103) allows us to picture a symmetric tensor $\mathbf{T}$ as a set of three numbers attached to three definite perpendicular directions in space. If we think of $\mathbf{T}$ as applied to every vector $\boldsymbol{C}$, then formula (10.102) shows that the effect is a stretching or a compression along each principal axis, together with a reflection if T_i is negative. In Section 10.2 we saw that a symmetric tensor may be specified by giving six components T_{ij} in any arbitrarily chosen coordinate system. We now see that we can alternatively specify $\mathbf{T}$ by specifying the principal axes (this requires three numbers, as we have seen), and the three associated eigenvalues.

If two or three of the eigenvalues are equal, we say the eigenvalue is *doubly* or *triply degenerate.* If the eigenvalue is triply degenerate, the tensor clearly is a constant tensor [Eq. (10.100)] and diagonal in every coordinate system. The principal axes are no longer unique; any axis is a principal axis. Every vector is an eigenvector of a constant tensor. If just two eigenvalues are equal, say $T_1 = T_2$, then if we consider rotating the coordinate axes in the $\hat{e}_1\hat{e}_2$-plane, we can see that the tensor will remain in diagonal form; the four elements referring to this plane behave like a constant tensor in that plane. Again the principal axes are not unique, since two of them may lie anywhere in the $\hat{e}_1\hat{e}_2$-plane. The third axis $\hat{e}_3$ associated with the nondegenerate eigenvalue T_3, however, is unique. We can prove that every axis in the $\hat{e}_1\hat{e}_2$-plane is a principal axis by considering the effect of $\mathbf{T}$ on any vector

$$\boldsymbol{C} = C_1\hat{e}_1 + C_2\hat{e}_2 \tag{10.105}$$

in this plane. In view of Eq. (10.102), if $T_1 = T_2$, we have

$$\begin{aligned} \mathbf{T}\cdot\boldsymbol{C} &= T_1C_1\hat{e}_1 + T_2C_2\hat{e}_2 \\ &= T_1\boldsymbol{C}, \end{aligned} \tag{10.106}$$

so that $\boldsymbol{C}$ is an eigenvector of $\mathbf{T}$. Every vector in the $\hat{e}_1\hat{e}_2$-plane is an eigenvector of $\mathbf{T}$ with eigenvalue T_1. If we like, we may say that there is a principal plane associated with a doubly degenerate eigenvalue.

We will now prove the theorem (10.103) by showing how the principal axes can be found. Let a symmetric tensor $\mathbf{T}$ be given in terms of its components T_{ij} in some coordinate system, which we will call the initial coordinate system. To find a principal axis, we must look for an eigenvector of $\mathbf{T}$. Let $\boldsymbol{C}$ be such an eigenvector, and T' the corresponding eigenvalue. We can rewrite Eq. (10.104) in the form

$$(\mathbf{T} - T'\mathbf{1})\cdot\boldsymbol{C} = \mathbf{0}. \tag{10.107}$$

If we write this equation in terms of components, we obtain

$$\begin{aligned} (T_{11} - T')C_1 + T_{12}C_2 + T_{13}C_3 &= 0, \\ T_{21}C_1 + (T_{22} - T')C_2 + T_{23}C_3 &= 0, \\ T_{31}C_1 + T_{32}C_2 + (T_{33} - T')C_3 &= 0. \end{aligned} \tag{10.108}$$

These equations for the unknown vector $\boldsymbol{C}$ have, of course, the trivial solution $\boldsymbol{C} = \mathbf{0}$. If we write the solution for C_i in terms of determinants, we see that $\boldsymbol{C} = \mathbf{0}$ is the only solution unless the determinant

$$\begin{vmatrix} T_{11} - T' & T_{12} & T_{13} \\ T_{21} & T_{22} - T' & T_{23} \\ T_{31} & T_{32} & T_{33} - T' \end{vmatrix} = 0, \tag{10.109}$$

in which case the solution for C_i is indeterminate. In this case it is shown in the theory of linear equations* that Eqs. (10.108) have also nontrivial solutions C_i. It is clear that Eqs. (10.108) cannot determine the numbers C_1, C_2, C_3 uniquely, but only their ratios to one another, $C_1:C_2:C_3$. This is also clear from Eq. (10.107), from which we began. Geometrically, only the direction of $\boldsymbol{C}$ is determined, not its magnitude (nor its sense). Equation (10.109), called the *secular equation*, represents a cubic equation to be solved for the eigenvalue T'. In general there will be three roots T'_1, T'_2, T'_3. Given any root T'_j, we can then substitute it in Eqs. (10.108) and solve for the ratios $C_1:C_2:C_3$. Any vector whose components are in the ratio $C_1:C_2:C_3$ is an eigenvector of $\mathbf{T}$ corresponding to the eigenvalue T'_j. For each eigenvalue T'_j, we can then take a unit vector $\hat{\boldsymbol{e}}'_j$ along the direction of the corresponding eigenvectors. The axes $\hat{\boldsymbol{e}}'_1$, $\hat{\boldsymbol{e}}'_2$, $\hat{\boldsymbol{e}}'_3$ are then the principal axes of $\mathbf{T}$. When we solve Eqs. (10.108) for the components of $\hat{\boldsymbol{e}}'_j$ for a given T'_j, we get three numbers a_{ji} (= C_i for $T' = T'_j$) which are the components of $\hat{\boldsymbol{e}}'_j$ along the axes $\hat{\boldsymbol{e}}_i$ of the initial coordinate system:

$$\hat{\boldsymbol{e}}'_j = \sum_{i=1}^{3} a_{ji}\hat{\boldsymbol{e}}_i. \tag{10.110}$$

This is just Eq. (10.76); hence the numbers a_{ji} are the coefficients of the orthogonal transformation from the initial coordinate system to the principal axes. We say that the transformation with coefficients a_{ji} *diagonalizes* $\mathbf{T}$.

In order to be sure we can carry out the above program, we must prove three lemmas, as the reader may have noted. First, we must prove that the roots T' of the secular equation (10.109) are real; otherwise we cannot find real solutions of Eqs. (10.108) for C_1, C_2, C_3. Second, we must prove that the vectors $\hat{\boldsymbol{e}}'_j$, obtained from Eqs. (10.108) for the different eigenvalues T'_j, are perpendicular; otherwise we do not obtain a set of perpendicular unit vectors. Third, we must show that in the degenerate case, two (or three) perpendicular unit vectors $\hat{\boldsymbol{e}}'_j$ can be found that correspond to a doubly (or triply) degenerate eigenvalue.

Lemma 1. *The roots of the secular equation* (10.109) *for a symmetric tensor are real.* (10.111)

Equation (10.109) is obtained from the eigenvalue equation (10.104):

$$\mathbf{T}\cdot\boldsymbol{C} = T'\boldsymbol{C}. \tag{10.112}$$

To prove the lemma, let us first allow T' to be complex. We will need also to allow the components C_i of the vector $\boldsymbol{C}$ to be complex. A vector $\boldsymbol{C}$ with complex components, has no geometric meaning in the usual sense, of course, but we can regard all the algebraic definitions of the various vector operations as applying also to vectors with complex components. The various theorems of vector algebra will hold also for vectors with complex components. [There is one exception to

*See any college algebra text, for example, Knebelman and Thomas, *Principles of College Algebra.* New York: Prentice-Hall, Inc., 1942. (Chapter IX, Theorem 10.)

these statements. The length of a complex vector cannot be defined by Eq. (3.13), but instead must be defined by

$$|A| = (A^* \cdot A)^{1/2}. \tag{10.113}$$

This definition will not be required here.] We will denote by $\boldsymbol{C}^*$ the vector whose components are the complex conjugates of those of $\boldsymbol{C}$. Let us multiply $\boldsymbol{C}^*$ into Eq. (10.112):

$$\boldsymbol{C}^* \cdot \mathbf{T} \cdot \boldsymbol{C} = T'(\boldsymbol{C}^* \cdot \boldsymbol{C}). \tag{10.114}$$

If we take the complex conjugate of this equation, we have

$$\boldsymbol{C} \cdot \mathbf{T} \cdot \boldsymbol{C}^* = T'^*(\boldsymbol{C}^* \cdot \boldsymbol{C}), \tag{10.115}$$

since $\mathbf{T}$ is real, and in view of Eq. (3.18). Now by definition (10.53),

$$\boldsymbol{C}^* \cdot \mathbf{T} = \mathbf{T}^t \cdot \boldsymbol{C}^*. \tag{10.116}$$

Hence

$$\begin{aligned} \boldsymbol{C}^* \cdot \mathbf{T} \cdot \boldsymbol{C} &= (\boldsymbol{C}^* \cdot \mathbf{T}) \cdot \boldsymbol{C} \\ &= (\mathbf{T}^t \cdot \boldsymbol{C}^*) \cdot \boldsymbol{C} \qquad \text{[by Eq. (10.53)]} \\ &= \boldsymbol{C} \cdot \mathbf{T}^t \cdot \boldsymbol{C}^* \qquad \text{[by Eq. (3.18)].} \end{aligned} \tag{10.117}$$

For a symmetric tensor, $\mathbf{T} = \mathbf{T}^t$, so that the left members of Eqs. (10.114) and (10.115) are equal, and

$$T' = T'^*, \tag{10.118}$$

so that T' is real.

***Lemma* 2.** *The eigenvectors of a symmetric tensor corresponding to different eigenvalues are perpendicular.* (10.119)

To prove this lemma, let us assume that T'_1, T'_2 are two eigenvalues of $\mathbf{T}$ corresponding to the eigenvectors $\boldsymbol{C}_1, \boldsymbol{C}_2$:

$$\mathbf{T} \cdot \boldsymbol{C}_1 = T'_1 \boldsymbol{C}_1, \tag{10.120}$$

$$\mathbf{T} \cdot \boldsymbol{C}_2 = T'_2 \boldsymbol{C}_2. \tag{10.121}$$

We multiply $\boldsymbol{C}_2$ into Eq. (10.120), and $\boldsymbol{C}_1$ into Eq. (10.121):

$$\boldsymbol{C}_2 \cdot \mathbf{T} \cdot \boldsymbol{C}_1 = T'_1(\boldsymbol{C}_2 \cdot \boldsymbol{C}_1), \tag{10.122}$$

$$\boldsymbol{C}_1 \cdot \mathbf{T} \cdot \boldsymbol{C}_2 = T'_2(\boldsymbol{C}_2 \cdot \boldsymbol{C}_1). \tag{10.123}$$

Since $\mathbf{T}$ is symmetric, the left members are equal [see the proof (10.117) above], and we have

$$(T'_1 - T'_2)(\boldsymbol{C}_2 \cdot \boldsymbol{C}_1) = 0. \tag{10.124}$$

If the eigenvalues T'_1, T'_2 are unequal, the eigenvectors $\mathbf{C}_2$, $\mathbf{C}_1$ are perpendicular.

Lemma 3. *In the case of double or triple degeneracy, Eqs.* (10.108) *have two or three mutually perpendicular solutions for the vector* $\mathbf{C}$. (10.125)

For the proof of (10.125), suppose that $T'_1 = T'_2$. If we substitute $T' = T'_1$ in Eqs. (10.108), then by the theorem referred to in the footnote on page 418, there is at least one nontrivial solution C_1, C_2, C_3. Let $\hat{e}'_1$ be a unit vector parallel to the vector (C_1, C_2, C_3). Then

$$\mathsf{T}\cdot\hat{e}'_1 = T'_1\hat{e}'_1. \tag{10.126}$$

Now, choose any pair of perpendicular unit vectors $\hat{e}''_2$, $\hat{e}''_3$ perpendicular to $\hat{e}'_1$, and use Eqs. (10.68) and (10.74) to transform the components of T into the double-primed coordinate system $\hat{e}'_1$, $\hat{e}''_2$, $\hat{e}''_3$. By a comparison of Eqs. (10.126) and (10.36), we see that we must get

$$T''_{11} = T'_1, \qquad T''_{21} = 0, \qquad T''_{31} = 0. \tag{10.127}$$

Since T is symmetric, its double-primed components must therefore be given by

$$\mathsf{T} = \begin{pmatrix} T'_1 & 0 & 0 \\ 0 & T''_{22} & T''_{23} \\ 0 & T''_{23} & T''_{33} \end{pmatrix}. \tag{10.128}$$

Furthermore, the secular equation

$$\begin{vmatrix} T'_1 - T' & 0 & 0 \\ 0 & T''_{22} - T' & T''_{23} \\ 0 & T''_{23} & T''_{33} - T' \end{vmatrix} = 0 \tag{10.129}$$

must have the same roots as Eq. (10.109). This is true since the left members of both equations are the determinants of the same tensor $(\mathsf{T} - T'\mathbf{1})$, expressed in the unprimed and double-primed coordinate systems, and we noted at the end of Section 10.3 that the determinant of a tensor has the same value in all coordinate systems. If we expand the determinant (10.129) by minors of the first row, we obtain

$$(T'_1 - T')\begin{vmatrix} T''_{22} - T' & T''_{23} \\ T''_{23} & T''_{33} - T' \end{vmatrix} = 0. \tag{10.130}$$

Since T'_1 is a double or triple root of this equation, it must be a root of the equation

$$\begin{vmatrix} T''_{22} - T' & T''_{23} \\ T''_{23} & T''_{33} - T' \end{vmatrix} = 0. \tag{10.131}$$

Therefore the equations

$$\begin{aligned} (T''_{22} - T'_1)C''_2 + T''_{23}C''_3 &= 0, \\ T''_{23}C''_2 + (T''_{33} - T'_1)C''_3 &= 0, \end{aligned} \tag{10.132}$$

have a nontrivial solution which defines an eigenvector $(0, C_2'', C_3'')$ in the $\hat{e}_2''\hat{e}_3''$-plane with the eigenvalue T_1'. We have therefore a second unit eigenvector $\hat{e}_2'$ parallel to $(0, C_2'', C_3'')$ and perpendicular to $\hat{e}_1'$. If we take a third unit vector $\hat{e}_3'$ perpendicular to $\hat{e}_1'$, $\hat{e}_2'$, then in this primed coordinate system, we must have

$$\begin{aligned} T'_{11} &= T'_1, \quad & T'_{21} &= 0, \quad & T'_{31} &= 0, \\ T'_{12} &= 0, \quad & T'_{22} &= T'_1, \quad & T'_{32} &= 0. \end{aligned} \tag{10.133}$$

Thus $\mathbf{T}$ must have the components

$$\mathbf{T} = \begin{pmatrix} T'_1 & 0 & 0 \\ 0 & T'_1 & 0 \\ 0 & 0 & T'_3 \end{pmatrix}, \tag{10.134}$$

and $\hat{e}_1'$, $\hat{e}_2'$, $\hat{e}_3'$ are principal axes. If T_1' happens to be a triple root of Eq. (10.109), it will also be a triple root of the secular equation

$$\begin{vmatrix} T'_1 - T' & 0 & 0 \\ 0 & T'_1 - T' & 0 \\ 0 & 0 & T'_3 - T' \end{vmatrix} = (T'_1 - T')(T'_1 - T')(T'_3 - T') = 0. \tag{10.135}$$

Therefore $T_3' = T_1'$, and we have three perpendicular eigenvectors corresponding to the triple root $T_1' = T_2' = T_3'$.

The above three lemmas complete the proof of the fundamental theorem (10.103).

The algebra in this section may be generalized to vector spaces of any number of dimensions, with analogous results regarding the existence of principal axes of a symmetric tensor.

At the end of Section 10.3 we noted that the trace and the determinant of a tensor $\mathbf{T}$ have the same value in all coordinate systems. We see from Eq. (10.101) that the trace is the sum of the eigenvalues of $\mathbf{T}$:

$$\operatorname{tr}(\mathbf{T}) = T_1 + T_2 + T_3, \tag{10.136}$$

and the determinant is the product of the eigenvalues:

$$\det(\mathbf{T}) = T_1 T_2 T_3. \tag{10.137}$$

We can form a third invariant scalar quantity associated with a symmetric tensor by summing the products of pairs of eigenvalues:

$$M(\mathbf{T}) = T_1 T_2 + T_2 T_3 + T_3 T_1. \tag{10.138}$$

We can evaluate $M(\mathbf{T})$ in any coordinate system by solving the secular equation (10.109) for the three roots T_1, T_2, T_3 and using Eq. (10.138). The solution of Eq. (10.109) can be avoided by noting that the sum (10.138) must be the coefficient of T' in Eq. (10.109), which is the sum of the diagonal minors of the determinant of $\mathbf{T}$:

$$M(\mathbf{T}) = \begin{vmatrix} T_{11} & T_{12} \\ T_{21} & T_{22} \end{vmatrix} + \begin{vmatrix} T_{22} & T_{23} \\ T_{32} & T_{33} \end{vmatrix} + \begin{vmatrix} T_{33} & T_{31} \\ T_{13} & T_{11} \end{vmatrix}. \tag{10.139}$$

We could also show by direct calculation that $M(\mathbf{T})$ as given by Eq. (10.139) has the same value after a coordinate transformation given by Eq. (10.74).

For any tensor $\mathbf{T}$, the determinant of $(\mathbf{T} - T'\mathbf{1})$ must have the same value in all coordinate systems. Therefore, in particular, the roots T' of Eq. (10.109) will be the same in all coordinate systems, even for a tensor $\mathbf{T}$ that is not symmetric. We still call the roots T' the eigenvalues of $\mathbf{T}$. If $\mathbf{T}$ is not symmetric, one eigenvalue will be real and the other two will usually be a conjugate complex pair. For the real eigenvalue we can find an eigenvector. For the complex eigenvalues we cannot always find eigenvectors. (That is, not unless we admit vectors with complex components, which have only algebraic significance. Even then, the eigenvectors are not always orthogonal.) In any case, the expressions given by Eqs. (10.136), (10.137), (10.138), and (10.139) are still real and independent of the coordinate system.

As an example of the diagonalization procedure, let us diagonalize the tensor

$$\mathbf{T} = \boldsymbol{AA} + \boldsymbol{BD} + \boldsymbol{DB},$$

which obviously is symmetric. We will take

$$\boldsymbol{A} = 4a\boldsymbol{e}_1,$$

$$\boldsymbol{B} = 7a\boldsymbol{e}_2 + a\boldsymbol{e}_3,$$

$$\boldsymbol{D} = a\boldsymbol{e}_2 - a\boldsymbol{e}_3.$$

The tensor $\mathbf{T}$ is then represented in this coordinate system by the matrix:

$$\mathbf{T} = \begin{pmatrix} 16a^2 & 0 & 0 \\ 0 & 14a^2 & -6a^2 \\ 0 & -6a^2 & -2a^2 \end{pmatrix}.$$

In this case, the secular equation (10.109) is

$$\begin{vmatrix} 16a^2 - T' & 0 & 0 \\ 0 & 14a^2 - T' & -6a^2 \\ 0 & -6a^2 & -2a^2 - T' \end{vmatrix} = (16a^2 - T') \times (T'^2 + 12a^2T' - 64a^4) = 0.$$

The roots (necessarily real) are

$$T'_1 = 16a^2, \qquad T'_2 = 16a^2, \qquad T'_3 = -4a^2.$$

Equations (10.108), for the doubly degenerate root $T' = 16a^2$, are

$$0 = 0,$$
$$-2a^2C_2 - 6a^2C_3 = 0,$$
$$-6a^2C_2 - 18a^2C_3 = 0.$$

Clearly, C_1 is arbitrary, and the last two equations are both satisfied if

$$C_2 = -3C_3.$$

Therefore any vector of the form

$$\boldsymbol{C} = C_1\hat{\boldsymbol{e}}_1 - 3C_3\hat{\boldsymbol{e}}_2 + C_3\hat{\boldsymbol{e}}_3$$

is an eigenvector for arbitrary C_1, C_3. Thus we have a two-parameter family of possible eigenvectors, from which we may select for $\hat{\boldsymbol{e}}'_1$, $\hat{\boldsymbol{e}}'_2$ any two perpendicular unit vectors. We will take

$$\hat{\boldsymbol{e}}'_1 = \hat{\boldsymbol{e}}_1, \text{ (that is, } C_1 = 1, C_3 = 0),$$

$$\hat{\boldsymbol{e}}'_2 = \frac{3}{\sqrt{10}}\hat{\boldsymbol{e}}_2 - \frac{1}{\sqrt{10}}\hat{\boldsymbol{e}}_3, \text{ (that is, } C_1 = 0, C_3 = 1/\sqrt{10}).$$

We could have guessed $\hat{\boldsymbol{e}}'_1$ from the form of $\mathbf{T}$. The reader should verify that $\hat{\boldsymbol{e}}'_1$ and $\hat{\boldsymbol{e}}'_2$ and, in fact, any vector in the $\hat{\boldsymbol{e}}'_1\hat{\boldsymbol{e}}'_2$-plane satisfy Eq. (10.104) with $T_i = 16a^2$. For $T' = -4a^2$, Eqs. (10.108) become

$$20a^2C_1 = 0,$$
$$18a^2C_2 - 6a^2C_3 = 0,$$
$$-6a^2C_2 + 2a^2C_3 = 0.$$

Now there is just a one-parameter family of solutions

$$C_1 = 0, \qquad C_3 = 3C_2,$$

of which there is one unit eigenvector (except for sign):

$$\hat{\boldsymbol{e}}'_3 = \frac{1}{\sqrt{10}}\hat{\boldsymbol{e}}_2 + \frac{3}{\sqrt{10}}\hat{\boldsymbol{e}}_3,$$

where positive signs were chosen so that $\hat{\boldsymbol{e}}'_1$, $\hat{\boldsymbol{e}}'_2$, $\hat{\boldsymbol{e}}'_3$ would form a righthanded system. The handedness may be determined either by visualizing the three vectors, or by checking that $\hat{\boldsymbol{e}}'_1 \cdot (\hat{\boldsymbol{e}}'_2 \times \hat{\boldsymbol{e}}'_3) = +1$. The vector $\hat{\boldsymbol{e}}'_3$ is perpendicular to the $\hat{\boldsymbol{e}}'_1\hat{\boldsymbol{e}}'_2$-plane as it must be according to Lemma 1. It may be verified that $\hat{\boldsymbol{e}}'_3$ is an eigenvector of $\mathbf{T}$ with the eigenvalue $-4a^2$.

By reference to Eq. (10.76), we may write the coefficients of the transformation to the principal axes of T:

$$\begin{pmatrix} a_{11} & a_{12} & a_{13} \\ a_{21} & a_{22} & a_{23} \\ a_{31} & a_{32} & a_{33} \end{pmatrix} = \begin{pmatrix} 1 & 0 & 0 \\ 0 & 3/\sqrt{10} & -1/\sqrt{10} \\ 0 & 1/\sqrt{10} & 3/\sqrt{10} \end{pmatrix}.$$

The reader should verify that these coefficients satisfy Eqs. (10.78); that the vectors $\hat{\boldsymbol{e}}'_j$ are properly transformed according to Eqs. (10.71) and (10.72), which in this case are

$$\delta_{jk} = \sum_{i=1}^{3} a_{ki} e'_{ji}, \qquad e'_{ji} = \sum_{k=1}^{3} a_{ki}\, \delta_{jk},$$

where e'_{ji} is the ith component of $\hat{\boldsymbol{e}}'_j$ in the unprimed coordinate system and δ_{jk} is the kth component of $\hat{\boldsymbol{e}}'_j$ in the primed coordinate system; and also that T is properly transformed according to Eq. (10.74) from its original form to its diagonal form.

10.5 THE INERTIA TENSOR

The inertia tensor of a rigid body is given by Eq. (10.25). For a body of density $\rho(x, y, z)$, we may rewrite the inertia tensor as

$$\mathsf{I}_o = \iiint \rho(r^2 \mathsf{1} - \boldsymbol{r}\boldsymbol{r})\, dV, \tag{10.140}$$

where we have used the subscript "o" to remind us that the inertia tensor is calculated with respect to a set of axes with origin at O. We will omit the subscript except when the discussion concerns more than one origin. The diagonal components of I are just the moments of inertia [Eq. (5.80)] about the three axes:

$$\begin{aligned} I_{xx} &= \iiint \rho(y^2+z^2)\, dV, \\ I_{yy} &= \iiint \rho(z^2+x^2)\, dV, \\ I_{zz} &= \iiint \rho(x^2+y^2)\, dV. \end{aligned} \tag{10.141}$$

The off-diagonal components, often called *products of inertia*, are

$$\begin{aligned} I_{xy} = I_{yx} &= -\iiint \rho xy\, dV, \\ I_{yz} = I_{zy} &= -\iiint \rho yz\, dV, \\ I_{zx} = I_{xz} &= -\iiint \rho zx\, dV. \end{aligned} \tag{10.142}$$

Since we may use Eq. (10.74) to calculate the components of the inertia tensor relative to any other set of axes through the same origin O, we see from Eqs. (10.74) and (10.141) that the moment of inertia about any axis through O, in a direction designated by the unit vector $\hat{\boldsymbol{n}}$, is

$$I_{\hat{n}} = \hat{\boldsymbol{n}} \cdot \mathsf{I} \cdot \hat{\boldsymbol{n}}. \tag{10.143}$$

It is often easier to calculate the components of the inertia tensor with respect to a conveniently chosen set of axes and then use Eq. (10.143), than to calculate $I_{\hat{n}}$ directly, if the axis $\hat{\boldsymbol{n}}$ is not an axis of symmetry of the body. (As an example, see Problem 25.)

We can obtain a useful analog to the Parallel Axis Theorem (5.81) for the moment of inertia by calculating the inertia tensor $\mathbf{I}_o$ relative to an arbitrary origin of coordinates O in terms of the inertia tensor $\mathbf{I}_G$ relative to the center of mass G. Let $\boldsymbol{r}$ and $\boldsymbol{r}'$ be position vectors of any point P in the body relative to O and G respectively, and let $\boldsymbol{R}$ be the coordinate of G relative to O (Fig. 5.12),

$$\boldsymbol{r} = \boldsymbol{r}' + \boldsymbol{R}. \tag{10.144}$$

Then we have, from the definition (10.140),

$$\begin{aligned}\mathbf{I}_o &= \iiint \rho[(\boldsymbol{r}'+\boldsymbol{R})\cdot(\boldsymbol{r}'+\boldsymbol{R})\,\mathbf{1} - (\boldsymbol{r}'+\boldsymbol{R})(\boldsymbol{r}'+\boldsymbol{R})]\,dV \\ &= \iiint \rho[(\boldsymbol{r}'\cdot\boldsymbol{r}')\,\mathbf{1} - \boldsymbol{r}'\boldsymbol{r}']\,dV + [(\boldsymbol{R}\cdot\boldsymbol{R})\,\mathbf{1} - \boldsymbol{R}\boldsymbol{R}]\iiint \rho\,dV \\ &\quad + 2\mathbf{1}\,[\boldsymbol{R}\cdot\iiint \rho\boldsymbol{r}'\,dV] - [\iiint \rho\boldsymbol{r}'\,dV]\,\boldsymbol{R} - \boldsymbol{R}\iiint \rho\boldsymbol{r}'\,dV.\end{aligned} \tag{10.145}$$

In view of the definition (5.53) of the center of mass, we have

$$\iiint \rho\boldsymbol{r}'\,dV = \mathbf{0}. \tag{10.146}$$

Equation (10.145) therefore reduces to

$$\mathbf{I}_o = \mathbf{I}_G + M(R^2\mathbf{1} - \boldsymbol{R}\boldsymbol{R}). \tag{10.147}$$

Note that both the statement and proof of this theorem are in precise analogy with the Parallel Axis Theorem (5.83) for the moment of inertia.

It is evident from the definition (10.140) that the inertia tensor of a composite body may be obtained by summing the inertia tensors of its parts, all relative to the same origin.

If a body rotates, the components of its inertia tensor, relative to stationary axes, will change with time. The components relative to axes fixed in the body, of course, will not change if the body is rigid. We may think of the inertia tensor $\mathbf{I}$ as rotating with the body. If the (constant) components along axes fixed in the body are given, the (changing) components along stationary axes are then given by Eq. (10.89), where $\mathbf{A}$ represents the transformation from body axes to space axes. The most convenient set of axes in the body for most purposes are the principal axes of the inertia tensor, also called the *principal axes of the body*. The eigenvalues of the inertia tensor are called the *principal moments of inertia*. We will learn more in the next chapter of the dynamical significance of the principal axes, but we may note here that, according to Eq. (10.24), if the body rotates about a principal axis, the angular momentum is parallel to the angular velocity. We may always choose arbitrary axes, compute $\mathbf{I}$, and then use the method of Section 10.4 to find the principal axes. It is often possible, however, to simplify the problem by choosing to begin with a coordinate system in which one or all of the axes are principal axes.

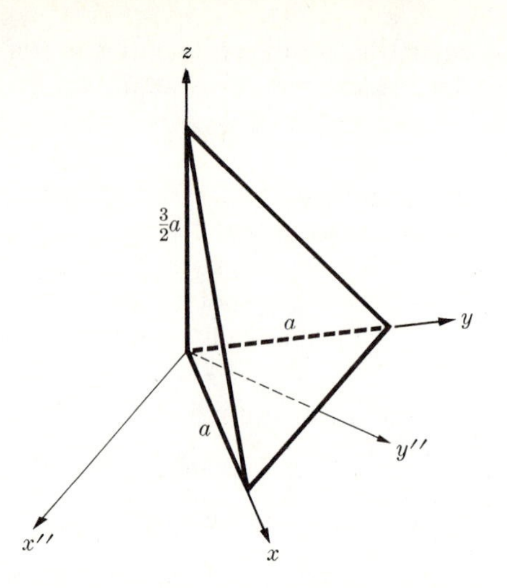

Fig. 10.1 A right triangular pyramid.

In many cases, a body will have some symmetry, so that we can see that certain of the products of inertia (10.142) will vanish if the axes are chosen in a certain way. For example, we can prove the following theorem:

Any plane of symmetry of a body is perpendicular to a principal axis. (10.148)

If we choose the yz-plane as the plane of symmetry, then

$$\rho(-x, y, z) = \rho(x, y, z). \tag{10.149}$$

It is easy to show that because of Eq. (10.149) the integrals (10.142) for I_{xy} and I_{zx} will vanish. Therefore the x-axis is a principal axis in this case. In a similar way, we can prove the theorem:

Any axis of symmetry of a body is a principal axis. The plane perpendicular to this axis is a principal plane corresponding to a degenerate principal moment of inertia. (10.150)

A sphere, or a body with spherical symmetry has, evidently, a constant inertia tensor.

As an example, consider the right triangular pyramid shown in Fig. 10.1. The components of the inertia tensor relative to the axes (x, y, z) are to be calculated from the formula:

$$\mathbf{I} = \int_{z=0}^{\frac{3}{2}a} \int_{y=0}^{a-\frac{2}{3}z} \int_{x=0}^{a-y-\frac{2}{3}z} \rho \begin{pmatrix} y^2+z^2 & -xy & -zx \\ -xy & z^2+x^2 & -yz \\ -zx & -yz & x^2+y^2 \end{pmatrix} dx\, dy\, dz,$$

where each component of $\mathbf{I}$ is to be obtained by evaluating the indicated integral over the corresponding component of the matrix. The density ρ is given in terms of the mass M by

$$M = \tfrac{1}{4}a^3\rho.$$

Because of the symmetry between x and y, it is necessary to evaluate only the four integrals

$$J_1 = \iiint \rho x^2\,dx\,dy\,dz = \iiint \rho y^2\,dx\,dy\,dz = \tfrac{1}{10}Ma^2,$$
$$J_2 = \iiint \rho z^2\,dx\,dy\,dz = \tfrac{9}{40}Ma^2,$$
$$J_3 = \iiint \rho xy\,dx\,dy\,dz = \tfrac{1}{20}Ma^2,$$
$$J_4 = \iiint \rho xz\,dx\,dy\,dz = \iiint \rho yz\,dx\,dy\,dz = \tfrac{3}{40}Ma^2.$$

The inertia tensor is then given by

$$\mathsf{I} = \begin{pmatrix} J_1+J_2 & -J_3 & -J_4 \\ -J_3 & J_1+J_2 & -J_4 \\ -J_4 & -J_4 & 2J_1 \end{pmatrix} = \begin{pmatrix} 13 & -2 & -3 \\ -2 & 13 & -3 \\ -3 & -3 & 8 \end{pmatrix}\frac{Ma^2}{40},$$

where the notation means that each element of the matrix is to be multiplied by $Ma^2/40$. Let us find the principal axes. By symmetry [theorem (10.148)] the axis x'' shown in Fig. 10.1 is a principal axis. Let us therefore first transform to the axes x'', y'', z. The coefficients of the transformation are, by Eq. (10.68),

$$\begin{pmatrix} a_{x''x} & a_{x''y} & a_{x''z} \\ a_{y''x} & a_{y''y} & a_{y''z} \\ a_{zx} & a_{zy} & a_{zz} \end{pmatrix} = \begin{pmatrix} 1/\sqrt{2} & -1/\sqrt{2} & 0 \\ 1/\sqrt{2} & 1/\sqrt{2} & 0 \\ 0 & 0 & 1 \end{pmatrix}.$$

Using Eq. (10.74), we now calculate the inertia tensor components along the x''-, y''-, and z-axes:

$$\mathsf{I} = \begin{pmatrix} 15 & 0 & 0 \\ 0 & 11 & -3\sqrt{2} \\ 0 & -3\sqrt{2} & 8 \end{pmatrix}\frac{Ma^2}{40}.$$

We see that the x''-axis is indeed a principal axis. The secular equation is

$$\begin{vmatrix} 15-\lambda & 0 & 0 \\ 0 & 11-\lambda & -3\sqrt{2} \\ 0 & -3\sqrt{2} & 8-\lambda \end{vmatrix} = 0, \qquad T' = \frac{Ma^2}{40}\lambda,$$

and the roots are

$$\lambda_{x'} = 15, \qquad \lambda_{y'} = 5, \qquad \lambda_{z'} = 14,$$

or

$$T_{x'} = \tfrac{3}{8}Ma^2, \qquad T_{y'} = \tfrac{1}{8}Ma^2, \qquad T_{z'} = \tfrac{7}{20}Ma^2.$$

Equations (10.108) can be solved for the components of the unit vectors $\hat{\mathbf{x}}'$, $\hat{\mathbf{y}}'$, $\hat{\mathbf{z}}'$ in terms of $\hat{\mathbf{x}}''$, $\hat{\mathbf{y}}''$, $\hat{\mathbf{z}}$:

$$\hat{\mathbf{x}}' = \hat{\mathbf{x}}'',$$

$$\hat{\mathbf{y}}' = \tfrac{1}{3}\sqrt{3}\hat{\mathbf{y}}'' + \tfrac{1}{3}\sqrt{6}\hat{\mathbf{z}},$$

$$\hat{\mathbf{z}}' = -\tfrac{1}{3}\sqrt{6}\hat{\mathbf{y}}'' + \tfrac{1}{3}\sqrt{3}\hat{\mathbf{z}}.$$

The choice of which root to call $\lambda_{y'}$ and which $\lambda_{z'}$ is of course arbitrary. The choice made here makes $\hat{\mathbf{x}}'$, $\hat{\mathbf{y}}'$, $\hat{\mathbf{z}}'$ a right-handed coordinate system.

As a second example, let us find the inertia tensor about the point O of the object shown in Fig. 10.2. The object is composed of three flat disks of mass M and radius a. By symmetry, the principal axes are the indicated axes x, y, z. We first calculate the inertia tensor of a single disk about its center, relative to its principal axes x', y', z', as shown in Fig. 10.3. The moment of inertia $I_{z'}$ is given by Eq. (5.90), and the moments of inertia $I_{x'}$, $I_{y'}$ are half $I_{z'}$, by the Perpendicular Axis Theorem (5.84). We can therefore write the inertia tensor of a disk, relative to its principal axes x', y', z', as

$$\mathbf{I}_G = \begin{pmatrix} 1 & 0 & 0 \\ 0 & 1 & 0 \\ 0 & 0 & 2 \end{pmatrix} \frac{Ma^2}{4}. \tag{10.151}$$

For the bottom disk, the principal axes are parallel to x, y, z, and we need only apply theorem (10.147) to obtain its inertia tensor relative to the x-, y-, z-axes with origin at O:

$$\mathbf{I}_o = \mathbf{I}_G + M(3a^2\mathbf{1} - 3a^2\hat{\mathbf{z}}\hat{\mathbf{z}})$$

$$= \begin{pmatrix} 13 & 0 & 0 \\ 0 & 13 & 0 \\ 0 & 0 & 2 \end{pmatrix} \frac{Ma^2}{4}.$$

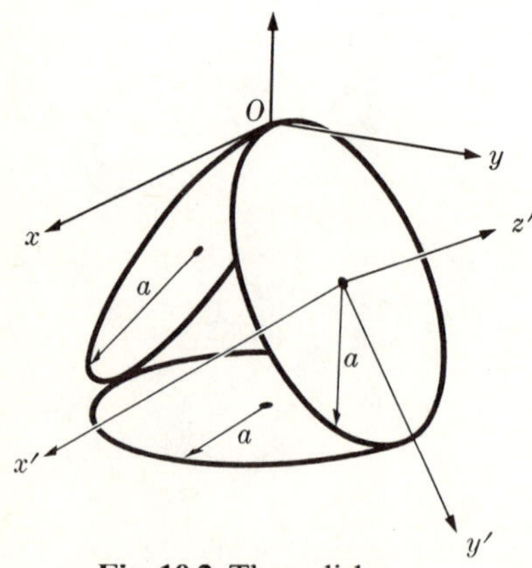

Fig. 10.2 Three disks.

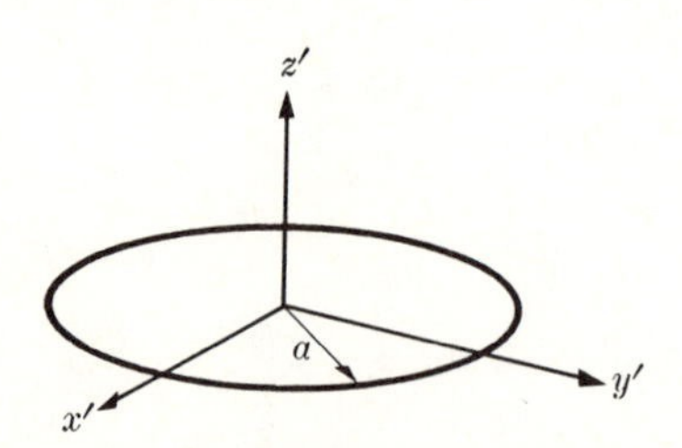

Fig. 10.3 A circular disk with its principal axes.

For the right-hand disk, with the axes x', y', z' oriented as shown, we first apply theorem (10.147) to obtain the inertia tensor about O, relative to axes parallel to x', y', z':

$$\mathbf{I}_{o(x'y'z')} = \begin{pmatrix} 5 & 0 & 0 \\ 0 & 1 & 0 \\ 0 & 0 & 6 \end{pmatrix} \frac{Ma^2}{4}.$$

The transformation from x'-, y'-, z'-axes to x-, y-, z-axes is given by

$$\mathbf{A} = \begin{pmatrix} a_{xx'} & a_{xy'} & a_{xz'} \\ a_{yx'} & a_{yy'} & a_{yz'} \\ a_{zx'} & a_{zy'} & a_{zz'} \end{pmatrix} = \begin{pmatrix} 1 & 0 & 0 \\ 0 & \frac{1}{2} & \frac{1}{2}\sqrt{3} \\ 0 & -\frac{1}{2}\sqrt{3} & \frac{1}{2} \end{pmatrix}.$$

We now use Eq. (10.74). It is perhaps easier to carry out the process in two steps, according to Eq. (10.89):*

$$\mathbf{A} \cdot \mathbf{I}_{o(x'y'z')} = \begin{pmatrix} 5 & 0 & 0 \\ 0 & \frac{1}{2} & 3\sqrt{3} \\ 0 & -\frac{1}{2}\sqrt{3} & 3 \end{pmatrix} \frac{Ma^2}{4}.$$

Now

$$(\mathbf{A} \cdot \mathbf{I}_{o(x'y'z')}) \cdot \mathbf{A}^t = \begin{pmatrix} 5 & 0 & 0 \\ 0 & \frac{1}{2} & 3\sqrt{3} \\ 0 & -\frac{1}{2}\sqrt{3} & 3 \end{pmatrix} \begin{pmatrix} 1 & 0 & 0 \\ 0 & \frac{1}{2} & -\frac{1}{2}\sqrt{3} \\ 0 & \frac{1}{2}\sqrt{3} & \frac{1}{2} \end{pmatrix} \frac{Ma^2}{4}$$

$$= \mathbf{I}_{o(xyz)} = \begin{pmatrix} 5 & 0 & 0 \\ 0 & \frac{19}{4} & \frac{5}{4}\sqrt{3} \\ 0 & \frac{5}{4}\sqrt{3} & \frac{9}{4} \end{pmatrix} \frac{Ma^2}{4}.$$

We may interpret this algebra as a computation of $\mathbf{I}_o$ relative to a new set of axes. Alternatively, we may interpret $\mathbf{A}$ as a tensor which rotates the disk through an angle of 60° about the x-axis; $\mathbf{I}_{o(x'y'z')}$ is then the moment of inertia of a disk whose principal axes are parallel to x, y, z, and the algebra is a computation of the effect of rotating the disk to its final position. The left-hand disk, correspondingly, has the inertia tensor

$$\mathbf{I}_{o(xyz)} = \begin{pmatrix} 5 & 0 & 0 \\ 0 & \frac{19}{4} & -\frac{5}{4}\sqrt{3} \\ 0 & -\frac{5}{4}\sqrt{3} & \frac{9}{4} \end{pmatrix} \frac{Ma^2}{4}.$$

---

*Matrices may be multiplied conveniently according to the rule (10.45) by noting that the element $(\mathbf{T} \cdot \mathbf{S})_{ik}$ is obtained by summing the products of pairs of elements across row i in $\mathbf{T}$ and down column k in $\mathbf{S}$.

The inertia tensors of the three disks may now be added and we obtain

$$\mathbf{I}_o = \begin{pmatrix} 23 & 0 & 0 \\ 0 & 22\frac{1}{2} & 0 \\ 0 & 0 & 6\frac{1}{2} \end{pmatrix} \frac{Ma^2}{4}.$$

Let us calculate the moment of inertia of the object shown in Fig. 10.2 about the y'-axis through O. By Eq. (10.143), we have

$$I_{\hat{y}'} = \hat{y}' \cdot \mathbf{I}_o \cdot \hat{y}' = 10\tfrac{1}{2} Ma^2.$$

We could use theorem (10.147) to obtain $\mathbf{I}$ about the center of gravity G, which is at the intersection of the z-and z'-axes. It is clear from symmetry that any axis perpendicular to one of the three disks through its center is a principal axis relative to G. This can only be true if the inertia tensor relative to G has a double degeneracy in the $yzy'z'$-plane. The reader should check this by carrying out the translation of $\mathbf{I}_o$ to the center of mass G. It should be noted that the principal axes of the inertia tensors of a body relative to two different points O and O', in general, will not be parallel, as experimentation with Eq. (10.147) will show. The parallel axis theorem does not yield parallel principal axes!

The kinetic energy T of a rotating body can also be expressed conveniently in terms of the inertia tensor. From Eqs. (10.2) and (10.3) and using the rules of vector algebra, we have

$$\begin{aligned} T &= \sum_{k=1}^{N} \tfrac{1}{2} m_k v_k^2 \\ &= \sum_{k=1}^{N} \tfrac{1}{2} m_k (\boldsymbol{\omega} \times \boldsymbol{r}_k) \cdot (\boldsymbol{\omega} \times \boldsymbol{r}_k) \\ &= \sum_{k=1}^{N} \tfrac{1}{2} m_k \boldsymbol{\omega} \cdot [\boldsymbol{r}_k \times (\boldsymbol{\omega} \times \boldsymbol{r}_k)] \\ &= \tfrac{1}{2} \boldsymbol{\omega} \cdot \boldsymbol{L}. \end{aligned} \tag{10.152}$$

Therefore T can be expressed in the form

$$T = \tfrac{1}{2} \boldsymbol{\omega} \cdot \mathbf{I} \cdot \boldsymbol{\omega}. \tag{10.153}$$

Equation (10.153) expressed in terms of components along any set of axes is then

$$\tfrac{1}{2} I_{xx} \omega_x^2 + \tfrac{1}{2} I_{yy} \omega_y^2 + \tfrac{1}{2} I_{zz} \omega_z^2 + I_{xy} \omega_x \omega_y + I_{yz} \omega_y \omega_z + I_{zx} \omega_z \omega_x = T. \tag{10.154}$$

This is the equation of a family of quadric surfaces in $\boldsymbol{\omega}$-space, each surface the locus of angular velocities for which the kinetic energy has a constant value T. If Eq. (10.153) is written in terms of components along principal axes x', y', z',

$$\tfrac{1}{2} I'_x \omega_x'^2 + \tfrac{1}{2} I'_y \omega_y'^2 + \tfrac{1}{2} I'_z \omega_z'^2 = T, \tag{10.155}$$

then we see that these surfaces are ellipsoids, since the moments of inertia are necessarily positive. If we define a vector

$$\boldsymbol{r} = \frac{a}{(2T)^{1/2}}\,\boldsymbol{\omega}, \tag{10.156}$$

where a is a constant, then Eq. (10.153) can be written as

$$\boldsymbol{r}\cdot\mathbf{I}\cdot\boldsymbol{r} = a^2. \tag{10.157}$$

This is the equation of the *inertia ellipsoid.* The scale factor a is introduced for dimensional reasons in order that $\boldsymbol{r}$ may be a coordinate vector in ordinary space measured in units of length. It is customary to set $a = 1$ in whatever units are being used, for example, $a = 1$ cm-sec-erg$^{\frac{1}{2}}$. In this case, we note that the size of the ellipsoid (but not its shape) depends on the units being used.

The inertia ellipsoid of a body, like its inertia tensor, is relative to a particular origin about which moments are computed. The six coefficients of the quadratic form on the left of Eq. (10.157) are the components of the inertia tensor:

$$I_{xx}x^2 + I_{yy}y^2 + I_{zz}z^2 + 2I_{xy}xy + 2I_{yz}yz + 2I_{zx}zx = a^2, \tag{10.158}$$

so that the inertia tensor is uniquely characterized by the corresponding inertia ellipsoid. This gives us another convenient geometrical way of picturing the inertia tensor.

By comparing Eq. (10.157) with Eq. (10.143), we see that the radius to any point on the inertia ellipsoid is

$$r = aI_{\mathrm{r}}^{-1/2}, \tag{10.159}$$

where I_{r} is the moment of inertia about an axis parallel to $\boldsymbol{r}$. In particular, the principal moments of inertia are related by Eq. (10.159) to the semiprincipal axes of the inertia ellipsoid. We see that if there is double degeneracy, the inertia ellipsoid is an ellipsoid of revolution. If the principal moments of inertia are all equal, the ellipsoid of inertia is a sphere.

For any symmetric tensor $\mathbf{T}$, we can form a quadratic equation of the form (10.157) which defines a quadric surface that uniquely characterizes $\mathbf{T}$. The principal axes of $\mathbf{T}$ are the principal axes of its associated quadric surface. If the eigenvalues of $\mathbf{T}$ are all positive, the surface is an ellipsoid. Otherwise, it will be a hyperboloid or a cylinder. If all the eigenvalues are negative, we would need to write $-a^2$ for the right member of the quadratic equation in order to define a real surface.

10.6 THE STRESS TENSOR

Let us represent any small surface element in a continuous medium by a vector $d\boldsymbol{S}$ whose magnitude dS is equal to the area of the surface element and whose direction is perpendicular to the surface element. To specify the sense of $d\boldsymbol{S}$, we will distinguish between the two sides of the surface element, calling one the back

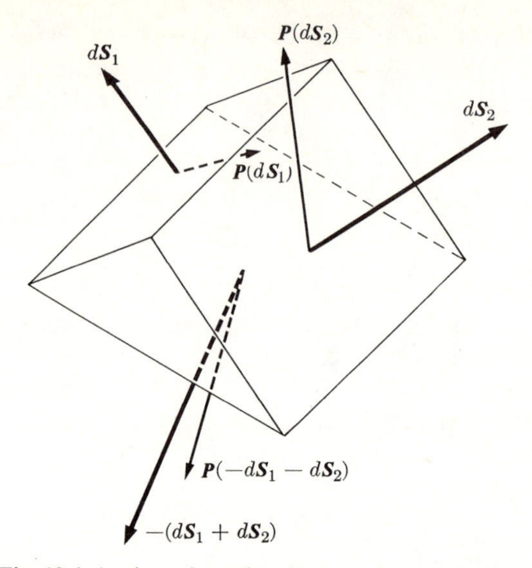

Fig. 10.4 A triangular prism in a continuous medium.

and the other the front. The sense of $d\boldsymbol{S}$ is then from the back to the front. We may then describe the state of stress of the medium at any point Q by specifying the force $\boldsymbol{P}(d\boldsymbol{S})$ exerted across any surface element $d\boldsymbol{S}$ at Q by the matter at the back on the matter at the front of $d\boldsymbol{S}$. We understand, of course, that the surface element $d\boldsymbol{S}$ is infinitesimal. That is, all statements we make are intended to be correct in the limit when all elements $d\boldsymbol{S} \to \boldsymbol{0}$. For a sufficiently small surface element, the force $\boldsymbol{P}$ may depend on the area and orientation of the surface element, but not on its shape. Thus $\boldsymbol{P}$ is indeed a function only of the vector $d\boldsymbol{S}$ at any particular point Q in the medium. We will show that $\boldsymbol{P}(d\boldsymbol{S})$ is a linear function of $d\boldsymbol{S}$. We may therefore represent the function $\boldsymbol{P}(d\boldsymbol{S})$ by a tensor P, the *stress tensor:**

$$\boldsymbol{P}(d\boldsymbol{S}) = \mathsf{P}\cdot d\boldsymbol{S}. \tag{10.160}$$

To show that $\boldsymbol{P}(d\boldsymbol{S})$ is a linear vector function, we note first that if $d\boldsymbol{S}$ is small enough so that the state of stress of the medium does not change over the surface element, then the force $\boldsymbol{P}$ will be proportional to the area dS so long as the orientation of the surface is kept fixed. Thus for a positive constant c,

$$\boldsymbol{P}(c\, d\boldsymbol{S}) = c\boldsymbol{P}(d\boldsymbol{S}). \tag{10.161}$$

If the direction of $d\boldsymbol{S}$ is reversed, the back and front of the surface element are interchanged, and therefore by Newton's third law, $\boldsymbol{P}(-d\boldsymbol{S}) = -\boldsymbol{P}(d\boldsymbol{S})$, so that Eq.

*The reader is cautioned that many authors define the stress tensor with the opposite sign from the definition adopted here, so that a tension is a positive stress and a pressure, a negative stress. The latter convention is almost universal in engineering practice, whereas the definition adopted here is more common in works on theoretical physics.

(10.161) holds also if c is negative. Now, given any two vectors $\boldsymbol{dS}_1$, $\boldsymbol{dS}_2$, let us imagine a triangular prism in the medium with two sides $\boldsymbol{dS}_1$, $\boldsymbol{dS}_2$, as in Fig. 10.4. If the end faces are perpendicular to the sides, then the third side is $-(\boldsymbol{dS}_1+\boldsymbol{dS}_2)$, as shown, because the three vectors $\boldsymbol{dS}$ must form a triangle similar to the end face of the prism. If the length of the prism is made much greater than the cross-sectional dimensions, we may neglect the forces on the end faces, and the total force on the prism is

$$\boldsymbol{dF} = \boldsymbol{P}(\boldsymbol{dS}_1)+\boldsymbol{P}(\boldsymbol{dS}_2)+\boldsymbol{P}(-\boldsymbol{dS}_1-\boldsymbol{dS}_2). \tag{10.162}$$

If the density is ρ, the acceleration of the prism is given by Newton's law of motion:

$$\rho\, dV\boldsymbol{a} = \boldsymbol{dF}. \tag{10.163}$$

Now if we reduce all linear dimensions of the prism by a factor α, the areas $\boldsymbol{dS}_i$ are multiplied by α^2; hence by Eq. (10.161), $\boldsymbol{dF}$ is multiplied by α^2, and dV is multiplied by α^3, so that

$$\alpha\rho\, dV\boldsymbol{a}(\alpha) = \boldsymbol{dF}, \tag{10.164}$$

where $\boldsymbol{a}(\alpha)$ is the acceleration of a prism α times smaller. Now as $\alpha \to 0$, the acceleration should not become infinite; hence we conclude that

$$\boldsymbol{dF} = \boldsymbol{0}, \tag{10.165}$$

from which, by Eqs. (10.162) and (10.161),

$$\boldsymbol{P}(\boldsymbol{dS}_1)+\boldsymbol{P}(\boldsymbol{dS}_2) = \boldsymbol{P}(\boldsymbol{dS}_1+\boldsymbol{dS}_2). \tag{10.166}$$

Equations (10.161) and (10.166) show that the function $\boldsymbol{P}(\boldsymbol{dS})$ is linear. Note that Eqs. (10.166) and (10.162) imply that there is no net force on the prism if the stress function $\boldsymbol{P}(\boldsymbol{dS})$ is the same function at all faces. Any net force can only result from differences in the stress function at different points of the medium; such differences reduce to zero as $\alpha \to 0$.

By considering small square prisms, and recognizing that the angular acceleration must not become infinite as the size shrinks to zero, we can show by a very similar argument (see Problem 32) that P must be a symmetric tensor. The stresses at each point Q in a medium are therefore given by specifying six components of the symmetric stress tensor P.

If the medium is an ideal fluid whose only stress is a pressure p in all directions, the stress tensor is evidently just

$$\mathsf{P} = p\mathsf{1}. \tag{10.167}$$

Note that we did not prove in Chapter 8 that in an ideal fluid, that is, one which can support no shearing stress, the pressure is the same in all directions. This was proved only in Chapter 5 for a fluid in equilibrium. This logical defect can now be remedied. (See Problem 33.)

According to the definition of $\mathbf{P}$, the total force due to the stress across any surface S is the vector sum of the forces on its elements:

$$\boldsymbol{F} = \iint_S \mathbf{P} \cdot d\boldsymbol{S}. \tag{10.168}$$

If S is the closed surface surrounding a volume V of the medium, and if we take $\hat{\boldsymbol{n}}$ to be the conventional outward normal unit vector, then the total force exerted on the volume V by the matter outside it is

$$\boldsymbol{F} = -\iint_S \hat{\boldsymbol{n}} \cdot \mathbf{P}\, dS, \tag{10.169}$$

and by the generalized Gauss' theorem, [see discussion below Eq. (5.178)],

$$\boldsymbol{F} = -\iiint_V \boldsymbol{\nabla} \cdot \mathbf{P}\, dV. \tag{10.170}$$

Since V is any volume in the medium, the force density due to stress is

$$\boldsymbol{f}_s = -\boldsymbol{\nabla} \cdot \mathbf{P}. \tag{10.171}$$

In agreement with an earlier discussion, we see that this force density arises only from differences in stress at different points in the medium. Equation (10.171) may also be derived by summing the forces on a small rectangular volume element.

The equation of motion (8.138) may now be generalized to apply to any continuous medium:

$$\rho \frac{d\boldsymbol{v}}{dt} + \boldsymbol{\nabla} \cdot \mathbf{P} = \boldsymbol{f}. \tag{10.172}$$

This equation may also be rewritten in the form (8.139):

$$\frac{\partial \boldsymbol{v}}{\partial t} + \boldsymbol{v} \cdot \boldsymbol{\nabla} \boldsymbol{v} + \frac{1}{\rho} \boldsymbol{\nabla} \cdot \mathbf{P} = \frac{\boldsymbol{f}}{\rho}. \tag{10.173}$$

This equation, together with the equation of continuity (8.127), determines the motion of the medium when the body force density $\boldsymbol{f}$ and the stress tensor $\mathbf{P}$ are given. The stress $\mathbf{P}$ at any point Q may be a function of the density and temperature, of the relative positions and velocities of the elements near Q, and perhaps also of the previous history of the medium, which may be a solid (elastic or plastic) or a fluid (ideal or viscous).

From Eqs. (10.172) and (10.173) we can derive conservation equations analogous to those derived in Section 8.8. The conservation equation for energy analogous to Eq. (8.149) is, for example,

$$\frac{d}{dt}(\tfrac{1}{2}\rho v^2\, \delta V) = \boldsymbol{v} \cdot (\boldsymbol{f} - \boldsymbol{\nabla} \cdot \mathbf{P})\, \delta V. \tag{10.174}$$

The further manipulations of the energy equation carried out in Section 8.8 cannot all be carried through in the same way for Eq. (10.174) because of the difference in form between the stress term here and the pressure term in Eq. (8.149), as the reader may verify. The energy changes associated with changes in volume and shape of an element in a continuous medium are in general more complicated than those associated with expansion and contraction of an ideal fluid.

In a viscous fluid, the stress tensor P will be expected to depend on the velocity gradients in the fluid. This is consistent with the dimensional arguments in Section 8.15, where we saw that the term $\nabla \cdot \mathsf{P}$ in Eq. (10.172) must consist of the coefficient of viscosity η multiplied by some combination of second derivatives of the velocity components with respect to x, y, and z. If the fluid is isotropic, as we shall assume, then the relation between P and the velocity gradients must not depend on the orientation of the coordinate system. We can guarantee that this will be so by expressing the relation in a vector form that does not refer explicitly to components. The dyad

$$\nabla \boldsymbol{v} = \begin{pmatrix} \dfrac{\partial v_x}{\partial x} & \dfrac{\partial v_y}{\partial x} & \dfrac{\partial v_z}{\partial x} \\ \dfrac{\partial v_x}{\partial y} & \dfrac{\partial v_y}{\partial y} & \dfrac{\partial v_z}{\partial y} \\ \dfrac{\partial v_x}{\partial z} & \dfrac{\partial v_y}{\partial z} & \dfrac{\partial v_z}{\partial z} \end{pmatrix} \tag{10.175}$$

has as its components the nine possible derivatives of the components of $\boldsymbol{v}$ with respect to x, y, and z. Hence we must try to relate P to $\nabla \boldsymbol{v}$. The dyad (10.175) is not symmetric, but we can separate it into a symmetric and an antisymmetric part, as in Eqs. (10.64) through (10.66):

$$\nabla \boldsymbol{v} = (\nabla \boldsymbol{v})_s + (\nabla \boldsymbol{v})_a, \tag{10.176}$$

$$(\nabla \boldsymbol{v})_s = \tfrac{1}{2}\nabla \boldsymbol{v} + \tfrac{1}{2}(\nabla \boldsymbol{v})^t, \tag{10.177}$$

$$(\nabla \boldsymbol{v})_a = \tfrac{1}{2}\nabla \boldsymbol{v} - \tfrac{1}{2}(\nabla \boldsymbol{v})^t. \tag{10.178}$$

The antisymmetric part is related, as in Eqs. (10.62) and (10.63), to a vector

$$\boldsymbol{\omega} = \tfrac{1}{2}\nabla \times \boldsymbol{v}, \tag{10.179}$$

such that for any vector $d\boldsymbol{r}$,

$$(\nabla \boldsymbol{v})_a \cdot d\boldsymbol{r} = \boldsymbol{\omega} \times d\boldsymbol{r}. \tag{10.180}$$

If $d\boldsymbol{r}$ is the vector from a given point Q to any nearby point Q', we see that the tensor $(\nabla \boldsymbol{v})_a$ selects out those parts of the velocity differences between Q and Q' which correspond to a (rigid) rotation of the fluid around Q with angular velocity $\boldsymbol{\omega}$. This is in agreement with the discussion of Eq. (8.133), which is identical with Eq. (8.179). Since no viscous forces will be associated with a pure rotation of the fluid, the viscous forces must be expressible in terms of the tensor $(\nabla \boldsymbol{v})_s$.

Since P is also symmetric, we are tempted to write simply

$$\mathsf{P} = C(\nabla \boldsymbol{v})_s, \tag{10.181}$$

where C is a constant. In the simple case depicted in Fig. 8.11, the only nonzero component of $\nabla \boldsymbol{v}$ is $\partial v_x/\partial y$, and Eqs. (10.181) and (10.177) then give

$$\mathbf{P} = \begin{pmatrix} 0 & \frac{1}{2}C\frac{\partial v_x}{\partial y} & 0 \\ \frac{1}{2}C\frac{\partial v_x}{\partial y} & 0 & 0 \\ 0 & 0 & 0 \end{pmatrix}, \tag{10.182}$$

and the viscous force across $d\boldsymbol{S} = \hat{\boldsymbol{y}}\, dS$ will be

$$d\boldsymbol{F} = \mathbf{P}\cdot d\boldsymbol{S} = \tfrac{1}{2}C\frac{\partial v_x}{\partial y}\, dS\, \hat{\boldsymbol{x}}, \tag{10.183}$$

in agreement with Eq. (8.254) if $C = -2\eta$. A negative sign is clearly needed, since the viscous force opposes the velocity gradient. However, Eq. (10.181) is not the most general linear relation between $\mathbf{P}$ and $\nabla \boldsymbol{v}$ that is independent of the coordinate system. For we can further decompose $(\nabla \boldsymbol{v})_s$ into a constant tensor and a traceless symmetric tensor in the following way:

$$(\nabla \boldsymbol{v})_s = (\nabla \boldsymbol{v})_c + (\nabla \boldsymbol{v})_{ts}, \tag{10.184}$$

$$(\nabla \boldsymbol{v})_c = \tfrac{1}{3}Tr(\nabla \boldsymbol{v})_s\mathbf{1} = \tfrac{1}{3}\nabla\cdot\boldsymbol{v}\mathbf{1}, \tag{10.185}$$

$$(\nabla \boldsymbol{v})_{ts} = (\nabla \boldsymbol{v})_s - \tfrac{1}{3}\nabla\cdot\boldsymbol{v}\mathbf{1}. \tag{10.186}$$

This decomposition is independent of the coordinate system, since we have shown that the trace is an invariant scalar quantity. We see by Eq. (8.116) that the tensor $(\nabla \boldsymbol{v})_c$ measures the rate of expansion or contraction of the fluid. The tensor $(\nabla \boldsymbol{v})_{ts}$, with five independent components, specifies the way in which the fluid is being sheared. We are therefore free to set

$$\mathbf{P} = -2\eta(\nabla \boldsymbol{v})_{ts} - \tfrac{1}{3}\eta'\nabla\cdot\boldsymbol{v}\mathbf{1}, \tag{10.187}$$

with a coefficient η which characterizes the viscous resistance to shear, and a coefficient η' which characterizes a viscous resistance, if any, to expansion and contraction. The last term corresponds to a uniform pressure (or tension) in all directions at the given point. The coefficient η' is small and not very well determined experimentally for actual fluids. According to the kinetic theory of gases, η' is zero for an ideal gas. To the viscous stress due to velocity gradients, given by formula (10.187), must be added a hydrostatic pressure which may also be present and which depends on the density, temperature, and composition of the fluid. If we lump the last term in Eq. (10.187) together with the hydrostatic pressure into a total pressure p, then the complete stress tensor is

$$\mathbf{P} = p\mathbf{1} - \eta[\nabla \boldsymbol{v} + (\nabla \boldsymbol{v})^t - \tfrac{2}{3}\nabla\cdot\boldsymbol{v}\mathbf{1}]. \tag{10.188}$$

The reader may readily write this out in terms of components.

Formula (10.188) is the most general expression for the stress in an isotropic fluid in which there is a hydrostatic pressure, plus viscous forces proportional to the velocity gradient. It is possible to imagine that the stress might also contain nonlinear terms in the velocity gradients, or even higher-order derivatives of the velocity, but such terms could be expected to be small in comparison with the linear terms. Experimentally, the viscous stresses in fluids are given very accurately in most cases by formula (10.188).

We have decomposed the tensor $\nabla \boldsymbol{v}$, with nine independent components, into a sum of three tensors with one, three, and five independent components, each a linear combination of the components of $\nabla \boldsymbol{v}$. A similar decomposition is clearly possible for any tensor. The reader may well ask whether any further decomposition is possible. This is a problem in group theory. We state without proof the result. Neither an antisymmetric tensor nor a symmetric traceless tensor can be further decomposed in a manner independent of the coordinate system. The reader can convince himself that this is plausible by a little experimentation.

Let us now consider an elastic solid. Let the solid be initially in an unstrained position, and let each point in the solid be designated by its position vector $\boldsymbol{r}$ relative to any convenient origin. Now let the solid be strained by moving each point $\boldsymbol{r}$ to a new position given by the vector $\boldsymbol{r}+\boldsymbol{\rho}(\boldsymbol{r})$ relative to the same origin. We will designate the components of $\boldsymbol{r}$ by (x, y, z) and of $\boldsymbol{\rho}$ by (ξ, η, ζ). If $\boldsymbol{\rho}$ were independent of $\boldsymbol{r}$, the motion would be a uniform displacement without deformation. Hence the strain at any point may be specified by the gradient dyad

$$\nabla \boldsymbol{\rho} = \begin{pmatrix} \frac{\partial \xi}{\partial x} & \frac{\partial \eta}{\partial x} & \frac{\partial \zeta}{\partial x} \\ \frac{\partial \xi}{\partial y} & \frac{\partial \eta}{\partial y} & \frac{\partial \zeta}{\partial y} \\ \frac{\partial \xi}{\partial z} & \frac{\partial \eta}{\partial z} & \frac{\partial \zeta}{\partial z} \end{pmatrix}. \tag{10.189}$$

Now $\nabla \boldsymbol{\rho}$ can again be decomposed into an antisymmetric part which corresponds to a rigid rotation about the point $\boldsymbol{r}+\boldsymbol{\rho}$ and a symmetric part which describes the deformation of the solid in the neighborhood of each point:

$$\mathbf{S} = \tfrac{1}{2}\nabla \boldsymbol{\rho} + \tfrac{1}{2}(\nabla \boldsymbol{\rho})^t. \tag{10.190}$$

The symmetric part can be further decomposed into a constant tensor describing a volume compression or expansion and a symmetric traceless tensor which describes the shear:

$$\mathbf{S}_c = \tfrac{1}{3}\nabla \cdot \boldsymbol{\rho}\mathbf{1} = \frac{1}{3}\frac{\Delta(\delta V)}{\delta V}\mathbf{1}, \tag{10.191}$$

$$\mathbf{S}_{st} = \tfrac{1}{2}\nabla \boldsymbol{\rho} + \tfrac{1}{2}(\nabla \boldsymbol{\rho})^t - \tfrac{1}{3}\nabla \cdot \boldsymbol{\rho}\mathbf{1}. \tag{10.192}$$

If the solid is isotropic, then this is the most general possible decomposition. Furthermore, if Hooke's law holds, the stress should be proportional to the strain:

$$\mathbf{P} = -\tfrac{1}{3}a\nabla\cdot\boldsymbol{\rho}\mathbf{1} - b\,\mathbf{S}_{st}. \tag{10.193}$$

The constants a and b are evidently related to the bulk modulus and the shear modulus.

In order to find a in terms of the bulk modulus, let us imagine a uniform expansion of the solid as in the definition (5.116) of the bulk modulus. Assume that every linear dimension of the solid increases by a factor α. The displacement of each point in the solid is then given by

$$\boldsymbol{\rho} = \alpha\mathbf{r},$$

where the origin of coordinates is chosen at a point which is not displaced. For this kind of a strain the stress is an isotropic pressure (or tension) Δp so that we can write, in view of Eq. (10.193),

$$\mathbf{P} = \Delta p\mathbf{1} = -a\alpha\mathbf{1}.$$

Since $\Delta V/V = 3\alpha$ in this case, we can read off the value of the constant a:

$$a = -3\Delta pV/\Delta V = 3B, \tag{10.194}$$

where we have used the definition (5.116) of the bulk modulus.

Let us now consider a shearing strain as shown in Fig. 5.21, used in the definition of the shear modulus. With the x-axis in the direction shown and the y-axis in the direction of the shear, the only nonzero component of $\nabla\rho$ is $\partial\eta/\partial x = \tan\theta$. The strain is therefore given by

$$\mathbf{S} = \mathbf{S}_{st} = \begin{pmatrix} 0 & \frac{1}{2}\tan\theta & 0 \\ \frac{1}{2}\tan\theta & 0 & 0 \\ 0 & 0 & 0 \end{pmatrix}. \tag{10.195}$$

We see from Fig. 5.21 that the stress tensor must contain the component $P_{yx} = -F/A$. Since the stress tensor must be symmetric we must also add the component $P_{xy} = -F/A$; it is also evident from Fig. 5.21 that we need to add this second component of stress if the solid in the figure is to be in equilibrium. Since no other stresses are associated with this particular strain, the stress tensor is

$$\mathbf{P} = \begin{pmatrix} 0 & -F/A & 0 \\ -F/A & 0 & 0 \\ 0 & 0 & 0 \end{pmatrix}. \tag{10.196}$$

If we compare Eqs. (10.195) and (10.196), making use of the definition (5.118) of the shear modulus, we see that

$$b = 2F/A\tan\theta = 2n. \tag{10.197}$$

If the solid is not isotropic, as for example, a crystal, then the relation between **P** and **S** may depend on the choice of axes, and must therefore be written:

$$P_{ij} = \sum_{k,l=1}^{3} c_{ijkl} S_{kl}. \tag{10.198}$$

Since **P** and **S** have six independent components each, there are thirty-six constants c_{ijkl}. By using the fact that there is an elastic potential energy which is a function of the strain, it can be shown that in the most general case there are twenty-one independent constants c_{ijkl}.

PROBLEMS

1. The product $c\mathbf{T}$ of a tensor by a scalar has been used in the text without formal definition. Remedy this defect by supplying a suitable definition and proving that this product has the expected algebraic properties.

2. Show that the centrifugal force in Eq. (7.37) is a linear function of the position vector $\boldsymbol{r}$ of the particle, and find an expression for the corresponding tensor in dyadic form. Write out the matrix of its coefficients.

3. Define time derivatives $d\mathbf{T}/dt$ and $d'\mathbf{T}/dt$ relative to fixed and rotating coordinate systems, as was done in Chapter 7 for derivatives of vectors. Prove that

$$\frac{d\mathbf{T}}{dt} = \frac{d'\mathbf{T}}{dt} + \omega \times \mathbf{T} - \mathbf{T} \times \omega,$$

where the cross product of a vector with a tensor is defined in the obvious way.

4. Write out the relations between the coefficients a_{ij} corresponding to the relations (10.80). Write down another relation between the unit vectors, involving a triple cross product, and write out the corresponding relations between the coefficients.

5. Transform the tensor

$$\mathbf{T} = \boldsymbol{AB} + \boldsymbol{BA},$$

where

$$\boldsymbol{A} = 5\hat{\boldsymbol{x}} - 3\hat{\boldsymbol{y}} + 2\hat{\boldsymbol{z}}, \qquad \boldsymbol{B} = 5\hat{\boldsymbol{y}} + 10\hat{\boldsymbol{z}},$$

into a coordinate system rotated 45° about the z-axis, using Eq. (10.74). Transform the vectors $\boldsymbol{A}$ and $\boldsymbol{B}$, using Eq. (10.71), and show that the results agree.

6. Write down and prove two additional relations like those in Eqs. (10.91) through (10.93), involving algebraic properties which are preserved by transformations of coordinates.

7. Prove Eqs. (10.91) and (10.92).

8. Write down the matrix for the orthogonal tensor **A** which produces a rotation by an angle α about the z-axis. Decompose **A** into a symmetric and an antisymmetric tensor as in Eq. (10.66). What is the geometrical interpretation of this decomposition?

9. Prove that Det (**T**) [Eq. (10.95)] is the same in all coordinate systems.

10. (a) Prove that the tensor given by formula (10.100) has the property given by Eq. (10.23).
b) Prove by direct calculation that this tensor is represented by the same matrix in all coordinate systems.

11. Prove by direct calculation that the quantity $M(\mathbf{T})$ defined by Eq. (10.139) has the same value after the coordinate transformation (10.74).

12. Diagonalize the tensor in Problem 5. [That is, find its eigenvalues and the corresponding principal axes.]

13. Diagonalize the tensor

$$\mathbf{T} = \begin{pmatrix} 7 & \sqrt{6} & -\sqrt{3} \\ \sqrt{6} & 2 & -5\sqrt{2} \\ -\sqrt{3} & -5\sqrt{2} & -3 \end{pmatrix}.$$

[*Hint:* The secular equation can be factored; the roots are all integers.]

14. What are the principal axes and corresponding eigenvalues of the tensor in Problem 2? Interpret physically.

15. Verify the statements made in the last paragraph of Section 10.4 regarding the principal axis transformation found in the worked-out example.

16. Prove that if two tensors **S** and **T** have a set of principal axes in common, then $\mathbf{S}\cdot\mathbf{T} = \mathbf{T}\cdot\mathbf{S}$. (The converse is also true.)

17. Prove that if a tensor **T** satisfies an algebraic equation

$$a_n\mathbf{T}^n + \cdots + a_2\mathbf{T}^2 + a_1\mathbf{T} + a_0\mathbf{1} = \mathbf{O},$$

where '$\mathbf{T}^n$' means $\mathbf{T}\cdot\mathbf{T}\cdots\mathbf{T}$ (n factors), then its eigenvalues must satisfy the same equation. The null tensor **O** is defined in the obvious way.

18. Use the result of Problem 17 to show that the eigenvalues of the tensor **A** representing a 180° rotation about some axis can only be ± 1. [*Hint:* Consider the result of applying **A** twice.] Show that the eigenvalues cannot all be $+1$. Then show that -1 must be a double root of the secular equation. [*Hint:* Use Eqs. (10.137) and (10.81).] Can you guess the corresponding eigenvectors? This entire problem is to be answered by using general arguments, without writing down the matrix for **A**.

19. Show that the eigenvalues of an orthogonal tensor [Eq. (10.87)] are complex (or real) numbers of unit magnitude. [*Hint:* Let $\boldsymbol{C}$ be an eigenvector (possibly complex) of $\mathbf{T}$, and consider the quantity $(\mathbf{T}\cdot\boldsymbol{C})\cdot(\mathbf{T}\cdot\boldsymbol{C}^*)$.] Hence show that one eigenvalue must be ± 1, and the other two are of the form $\exp(\pm i\alpha)$, for some angle α.

20. Write out the components of the orthogonal tensor $\mathbf{A}$ corresponding to a rotation by an angle θ about the z-axis. Find its eigenvalues. Find and interpret the eigenvectors corresponding to the real eigenvalue.

21. Find the components of the tensor corresponding to a rotation by an angle θ about the z-axis, followed by a rotation by an angle ψ about the y-axis. Find its eigenvalues. [*Hint:* According to Problem 19, one eigenvalue is ± 1; hence you can factor the secular equation.] Show that the result implies that this transformation is equivalent to a simple rotation about some axis. [You are not asked to find the axis.] Find the angle of rotation by comparing your result with the eigenvalues found in Problem 20.

22. Show that the eigenvalues of an antisymmetric tensor are pure imaginary (or zero). Hence show that an antisymmetric tensor must have one zero eigenvalue and two conjugate imaginary eigenvalues. Find the eigenvectors corresponding to the zero eigenvalue for the tensor (10.62).

23. Find the inertia tensor of a straight rod of length l, mass m, about its center. Use this result to find the inertia tensor about the centroid of an equilateral pyramid constructed out of six uniform rods. Show that this tensor can be written down immediately from symmetry considerations, given the result of Problem 23, Chapter 5.

24. Translate to the center of mass G the inertia tensor calculated about the origin for the three disks in Fig. 10.2. Verify the statement made in the text regarding the double degeneracy of $\mathbf{I}_G$.

25. Calculate the moment of inertia of a solid right-circular cone about a slant height. [*Hint:* Calculate the inertia tensor about the apex relative to principal axes, and use Eq. (10.143).]

26. Formulate and prove the most comprehensive theorem you can with regard to the inertia tensor of a plane lamina. What can you say about the principal axes and principal moments of inertia?

27. Find, by whatever method requires the least algebraic labor, the inertia tensor of a uniform rectangular block of mass M, dimensions $a\times b\times c$, about a set of axes through its center, of which the z-axis is parallel to side c, and the y-axis is a diagonal of the rectangle $a\times b$.

28. a) A uniform sphere of mass M, radius a, has two point masses $\frac{1}{4}M$, $\frac{1}{8}M$, located on its surface and separated by an angular distance of 45°. Find the principal axes and principal moments of inertia about the center of the sphere.

b) Find the inertia tensor about parallel axes through the center of mass. Are they still principal axes?

29. a) Find the inertia tensor of a plane rectangle of mass M, dimensions $a\times b$.

b) Use this result to find the inertia tensor about the center of mass of the house of cards

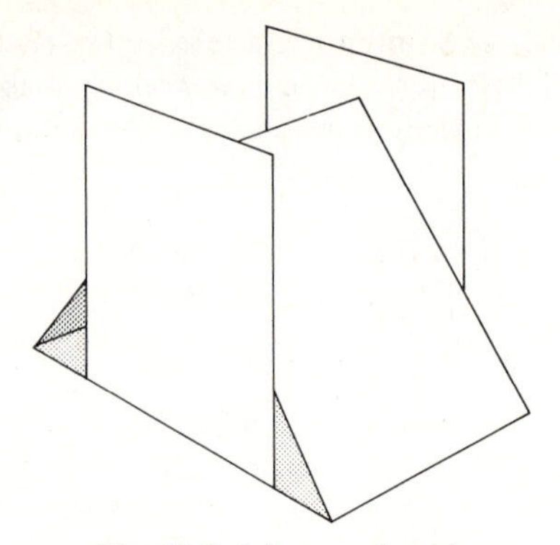

Fig. 10.5 A house of cards.

shown in Fig. 10.5. Each card has mass M, dimensions $a \times b (a < b)$. Use principal axes.

30. Find the equation for the ellipsoid of inertia of a uniform rectangular block of dimensions $l \times w \times h$.

31. Find the equation for the ellipsoid of inertia of an object in the shape of an ellipsoid whose equation is

$$\frac{x^2}{l^2} + \frac{y^2}{w^2} + \frac{z^2}{h^2} = 1.$$

32. Prove that the stress tensor $\mathbf{P}$ is symmetric.

33. Prove that if there is no shear on any surface element at some point, then the stress tensor $\mathbf{P}$ at that point is a constant tensor [Eq. (10.23)].

34. Derive Eq. (10.171) by calculating the net force on a rectangular volume element.

35. Derive from Eq. (10.173) the equation

$$\frac{\partial(\rho \boldsymbol{v})}{\partial t} + \nabla \cdot (\rho \boldsymbol{v}\boldsymbol{v} + \mathbf{P}) = \boldsymbol{f},$$

which expresses the conservation of linear momentum. Show from this equation that the momentum current tensor $(\rho \boldsymbol{v}\boldsymbol{v} + \mathbf{P})$ represents the flow of momentum, and interpret physically the two terms in this tensor.

36. Derive from Eq. (10.172) a law of conservation of angular momentum in a form analogous to Eq. (8.148).

37. Write the equations of motion (10.173) in cylindrical components for a moving viscous fluid. Use these equations, together with suitable assumptions, to derive Poiseuille's law (8.252) for steady viscous flow in a pipe. Write the stress tensor for this case, in cylindrical coordinates, as a function of r, z, φ.

38. Write out the components of the stress tensor $\mathbf{P}$ in a viscous fluid.

39. a) Show that the rate of production of kinetic energy per unit volume due to stresses in a moving medium is

$$Q = -\boldsymbol{v}\cdot(\nabla\cdot\mathbf{P}).$$

b) Show that the rate at which work is done by the stresses on the medium, per unit volume, is

$$\frac{dW}{dt} = -\nabla\cdot(\mathbf{P}\cdot\boldsymbol{v}).$$

[*Hint:* Calculate the work done across the surface of any volume V and use Gauss' theorem.]
c) Using these results, calculate the rate at which energy is dissipated per unit volume by viscous shearing stresses in a moving fluid. Write it out in terms of components.

40. An isotropic solid is subject to a stress consisting of a pure tension τ per unit area in one direction. Find the strain $\mathbf{S}$ in terms of τ, B, and n. Using this result, express Young's modulus Y [Eq. (5.114)] in terms of B and n. [*Hint:* Use symmetry to determine the form of $\mathbf{S}$.]

*__41.__ Find the most general linear relation between $\mathbf{S}$ and $\mathbf{P}$ for a nonisotropic elastic substance which possesses cylindrical symmetry relative to a specified direction.

*__42.__ a) Assume that in a nonisotropic elastic solid, there is an elastic potential energy V per unit volume, which is a quadratic function of the strain components. Show that there are 21 constants required to specify V.
b) Show that if the strain in a solid in equilibrium is increased by $\delta\mathbf{S}$, the work done per unit volume against the stresses (exclusive of any work done against body forces) is

$$\delta W = -\sum_{i,j=1}^{3} P_{ij}\,\delta S_{ij}.$$

[*Hint:* Calculate the work done on a volume element by the stresses on its surface and use Gauss' theorem.]
c) Combine results a) and b) to show that $\mathbf{P}$ is a linear function of $\mathbf{S}$ involving 21 independent constants, in general.

CHAPTER 11

THE ROTATION OF A RIGID BODY

11.1 MOTION OF A RIGID BODY IN SPACE

The motion of a rigid body in space is determined by Eqs. (5.4) and (5.5):

$$\frac{d\boldsymbol{P}}{dt} = \boldsymbol{F}, \tag{11.1}$$

$$\frac{d\boldsymbol{L}}{dt} = \boldsymbol{N}, \tag{11.2}$$

where

$$\boldsymbol{P} = M\boldsymbol{V}, \tag{11.3}$$

$$\boldsymbol{L} = \mathbf{I}\cdot\boldsymbol{\omega}, \tag{11.4}$$

$\boldsymbol{F}$ and $\boldsymbol{N}$ are the total force on the body and the total torque about a suitable point O, $\boldsymbol{V}$ is the velocity of the center of mass, and $\mathbf{I}$ and $\boldsymbol{\omega}$ are the inertia tensor and the angular velocity about the point O. For an unconstrained body moving in space, the point O is to be taken as the center of mass. If the body is constrained by external supports to rotate about a fixed point, that point is to be taken as the point O. If the point O is constrained to move in some fashion, the reader may supply the appropriate equation of motion. (See Chapter 7, Problem 3.)

Equations (11.2) and (11.4) for the rotation of a rigid body bear a formal analogy to Eqs. (11.1) and (11.3) for the motion of a point mass M. There are, however, three differences which spoil the analogy. In the first place, Eq. (11.4) involves a tensor $\mathbf{I}$, whereas Eq. (11.3) involves a scalar M; thus $\boldsymbol{P}$ is always parallel to $\boldsymbol{V}$, while $\boldsymbol{L}$ is not in general parallel to $\boldsymbol{\omega}$. A more serious difference is the fact that the inertia tensor $\mathbf{I}$ is not constant with reference to axes fixed in space, but changes as the body rotates, whereas M is constant (in Newtonian mechanics). Finally, and perhaps most serious, is the fact that no symmetrical set of three coordinates analogous to X, Y, Z exist with which to describe the orientation of a body in space. This point was made in Section 5.1, and it is suggested that the reader review the last paragraph in that section. For these reasons, we cannot proceed to solve the problem of rotation of a rigid body by analogy with the methods of Chapter 3.

There are two general approaches to the problem. We shall first, in Sections 11.2 and 11.3, try to obtain as much information as possible from the vector equations (11.2), (11.4) without introducing a set of coordinates to describe the orientation of the body. We shall then, in Sections 11.4 and 11.5, use Lagrange's equations to determine the motion in terms of a set of angular coordinates suggested by Euler.

11.2 EULER'S EQUATIONS OF MOTION FOR A RIGID BODY

The difficulty that $\mathbf{I}$ changes as the body rotates may be avoided by referring Eq. (11.2) to a set of axes fixed in the body. If we let "d'/dt" denote the time derivative with reference to axes fixed in the body, then by Eq. (7.22), Eq (11.2) becomes

$$\frac{d'\boldsymbol{L}}{dt}+\boldsymbol{\omega}\times\boldsymbol{L} = \boldsymbol{N}. \tag{11.5}$$

Since $\mathbf{I}$ is constant relative to body axes, we may substitute from Eq. (11.4) to obtain

$$\mathbf{I}\cdot\frac{d\boldsymbol{\omega}}{dt}+\boldsymbol{\omega}\times(\mathbf{I}\cdot\boldsymbol{\omega}) = \boldsymbol{N}. \tag{11.6}$$

(Recall that $d'\boldsymbol{\omega}/dt = d\boldsymbol{\omega}/dt$.) It is most convenient to choose as body axes the principal axes, $\hat{\boldsymbol{e}}_1, \hat{\boldsymbol{e}}_2, \hat{\boldsymbol{e}}_3$ of the body. Then Eq. (11.6) becomes

$$\begin{aligned} I_1\dot{\omega}_1+(I_3-I_2)\omega_3\omega_2 &= N_1, \\ I_2\dot{\omega}_2+(I_1-I_3)\omega_1\omega_3 &= N_2, \\ I_3\dot{\omega}_3+(I_2-I_1)\omega_2\omega_1 &= N_3. \end{aligned} \tag{11.7}$$

These are Euler's equations for the motion of a rigid body. If one point in the body is held fixed, that point is to be taken as the origin for the body axes, and the moments of inertia and torques are relative to that point. If the body is unconstrained, the center of mass is to be taken as origin for the body axes.

In order to derive the energy theorem from Euler's equations, we multiply Eq. (11.6) by $\boldsymbol{\omega}$:

$$\boldsymbol{\omega}\cdot\mathbf{I}\cdot\frac{d\boldsymbol{\omega}}{dt} = \boldsymbol{\omega}\cdot\boldsymbol{N}. \tag{11.8}$$

Since $\mathbf{I}$ is symmetric, the left member is

$$\boldsymbol{\omega}\cdot\mathbf{I}\cdot\frac{d\boldsymbol{\omega}}{dt} = \frac{d\boldsymbol{\omega}}{dt}\cdot\mathbf{I}\cdot\boldsymbol{\omega} = \frac{1}{2}\frac{d'}{dt}(\boldsymbol{\omega}\cdot\mathbf{I}\cdot\boldsymbol{\omega}) = \frac{dT}{dt}, \tag{11.9}$$

where T is given by Eq. (10.153). Here we have used the fact that d/dt, d'/dt have the same meaning when applied to a scalar quantity. Comparing Eqs. (11.8) and (11.9), we obtain the energy theorem:

$$\frac{dT}{dt} = \boldsymbol{\omega}\cdot\boldsymbol{N}, \tag{11.10}$$

in analogy with theorem (3.133) for the motion of a particle.

From Eqs. (11.7) we note immediately that a body cannot spin with constant angular velocity $\boldsymbol{\omega}$, except about a principal axis, unless external torques are applied. If $d\boldsymbol{\omega}/dt = \mathbf{0}$, Eq. (11.6) becomes

$$\boldsymbol{\omega}\times(\mathbf{I}\cdot\boldsymbol{\omega}) = \boldsymbol{N}. \tag{11.11}$$

The left member is zero only if $\mathbf{I}\cdot\boldsymbol{\omega}$ is parallel to $\boldsymbol{\omega}$, that is, if $\boldsymbol{\omega}$ is along a principal axis of the body. If a wheel is to spin freely without exerting forces and torques on its bearings, then it must be not only statically balanced, that is, with its center of mass on the axis of rotation, but also dynamically balanced, that is, the axis of rotation must be a principal axis of the inertia tensor, as any automobile mechanic knows.

In order to solve Eqs. (11.7) for $\boldsymbol{\omega}(t)$, we would need to know the components of torque along the (rotating) principal axes, an uncommon situation, except for the case $\boldsymbol{N} = \mathbf{0}$. We now consider a freely rotating symmetrical body, with no applied torque. Let the symmetry axis of the body be $\hat{\boldsymbol{e}}_3$, so that $I_1 = I_2$. Then the third of Eqs. (11.7) is

$$I_3\dot{\omega}_3 = 0, \tag{11.12}$$

and ω_3 is constant. The first two equations may be written

$$\dot{\omega}_1+\beta\omega_3\omega_2 = 0, \qquad \dot{\omega}_2-\beta\omega_3\omega_1 = 0, \tag{11.13}$$

where

$$\beta = \frac{I_3-I_1}{I_1}. \tag{11.14}$$

Equations (11.13) are a pair of coupled linear first-order equations in ω_1, ω_2. Let us look for a solution by setting

$$\omega_1 = A_1e^{pt}, \qquad \omega_2 = A_2e^{pt}. \tag{11.15}$$

We readily verify that formulas (11.15) satisfy Eqs. (11.13) provided that

$$p = \pm i\beta\omega_3, \tag{11.16}$$

and

$$A_2 = \mp iA_1. \tag{11.17}$$

We have found a complex conjugate pair of solutions,

$$\omega_1 = e^{\pm i\beta\omega_3t}, \qquad \omega_2 = \mp ie^{\pm i\beta\omega_3t}, \tag{11.18}$$

and these may be superposed with arbitrary constant multipliers to form the real solution:

$$\omega_1 = A\cos(\beta\omega_3t+\theta), \qquad \omega_2 = A\sin(\beta\omega_3t+\theta). \tag{11.19}$$

The angular velocity vector $\boldsymbol{\omega}$ therefore precesses in a circle of radius A about the $\hat{\boldsymbol{e}}_3$-axis, with angular velocity $\beta\omega_3$. The precession is in the same sense as ω_3 if $I_3 > I_1$, and in the opposite sense otherwise. The magnitude of $\boldsymbol{\omega}$ is

$$\omega = [\omega_3^2+A^2]^{1/2}, \tag{11.20}$$

and is constant, a result which can also be proved by direct calculation of $d(\omega^2)/dt$

from Eq. (11.7). The constants ω_3, A, θ are determined by the initial conditions. There are three arbitrary constants, since Euler's equations are three first-order differential equations. Since an unconstrained rotating rigid body has three rotational degrees of freedom, we should expect a total of six arbitrary constants to be determined by the initial conditions. What are the other three?

The instantaneous axis of rotation, determined by the vector $\boldsymbol{\omega}$, traces out a cone in the body (the *body cone*) as it precesses around the axis of symmetry. The half-angle α_b of the body cone is given by

$$\tan \alpha_b = \frac{A}{\omega_3}. \tag{11.21}$$

Alternatively, if the body is initially rotating with angular velocity ω about an axis making an angle α_b with the symmetry axis, then the constants ω_3 and A are given by

$$A = \omega \sin \alpha_b, \qquad \omega_3 = \omega \cos \alpha_b. \tag{11.22}$$

In order to find the motion in space, we need to locate the $\boldsymbol{\omega}$-axis with respect to a direction fixed in space. We could do this by tracing out step by step the motion relative to space axes, allowing the body to rotate with constant angular velocity about an axis in the body cone which precesses with angular velocity $\beta\omega_3$. It is easier to locate $\boldsymbol{\omega}$ relative to $\boldsymbol{L}$, since by Eq. (11.2) $\boldsymbol{L}$ is constant if $\boldsymbol{N} = \boldsymbol{0}$. The angle α_s between $\boldsymbol{\omega}$ and $\boldsymbol{L}$ is given by

$$\cos \alpha_s = \frac{\boldsymbol{\omega}\cdot \boldsymbol{L}}{\omega L} = \frac{\boldsymbol{\omega}\cdot \mathsf{I}\cdot\boldsymbol{\omega}}{\omega L} = \frac{2T}{\omega L}. \tag{11.23}$$

Since by Eq. (11.10) T is constant, the angle α_s is constant. The axis of rotation therefore traces out a cone in space, the *space cone*. The space cone has a half-angle α_s given by Eq. (11.23) and its axis is the direction of the angular momentum vector $\boldsymbol{L}$. The line of contact between the space cone and the body cone at any instant is the instantaneous axis of rotation. Since this axis in the body is instantaneously at rest, the body cone rolls without slipping around the space cone. This gives a complete description of the motion (see Fig. 11.1).

We can express α_s in terms of the constants ω, α_b. We have

$$\begin{aligned} \mathsf{I} &= (\hat{\boldsymbol{e}}_1\hat{\boldsymbol{e}}_1+\hat{\boldsymbol{e}}_2\hat{\boldsymbol{e}}_2)I_1+\hat{\boldsymbol{e}}_3\hat{\boldsymbol{e}}_3 I_3 \\ &= I_1\mathsf{1}+\hat{\boldsymbol{e}}_3\hat{\boldsymbol{e}}_3(I_3-I_1). \end{aligned} \tag{11.24}$$

By substituting from Eqs. (11.14), (11.22), and (11.24) into Eqs. (11.4), (10.153), and (11.23), and with $\boldsymbol{\omega} = \omega\hat{\boldsymbol{n}}$, we obtain:

$$2T = \omega^2 I_1[1+\beta \cos^2 \alpha_b], \tag{11.25}$$

$$\boldsymbol{L} = \omega I_1[\hat{\boldsymbol{n}}+\beta \cos \alpha_b \hat{\boldsymbol{e}}_3], \tag{11.26}$$

$$\cos \alpha_s = \frac{1+\beta \cos^2 \alpha_b}{[1+(2\beta+\beta^2) \cos^2 \alpha_b]^{1/2}}. \tag{11.27}$$

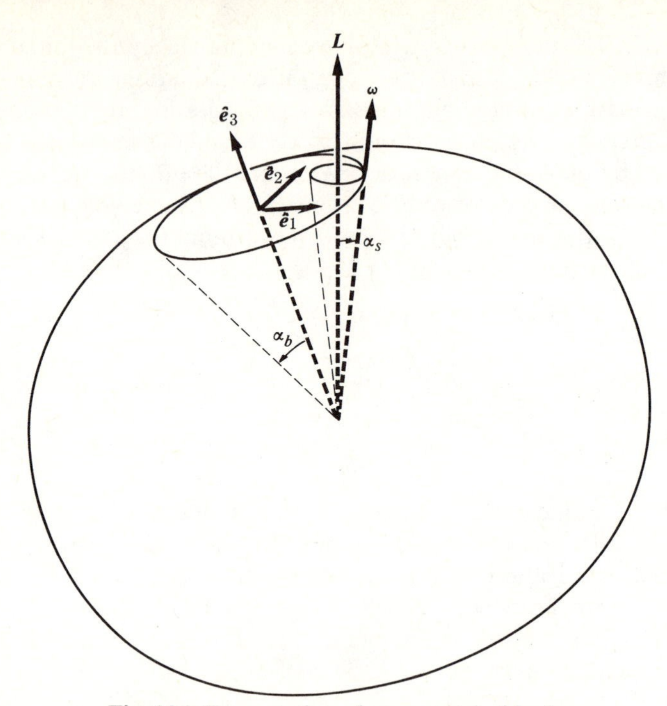

Fig. 11.1 Free rotation of a symmetrical body.

Note that α_s depends on only α_b and not on ω. It is clear from Eq. (11.26) that the space cone lies inside the body cone if $\beta > 0$ and outside if $\beta < 0$ (see Fig. 11.1). This is clear also if, when the body cone rolls on the space cone, the precession of the axis of rotation is to have the sense given by Eq. (11.19). [The reader should check this, remembering that Eq. (11.19) describes the motion of the axis relative to the body.]

Next, consider the case when the inertia tensor is nondegenerate. We shall number the principal axes so that $I_3 > I_2 > I_1$. It was shown above that a body may rotate freely about a principal axis. Let us study small deviations from this steady rotation. If $\boldsymbol{\omega}$ is not along a principal axis, then it cannot remain constant. Let us assume that $\boldsymbol{\omega}$ lies very close to a principal axis, say to $\hat{\mathbf{e}}_3$, so that $\omega_3 \gg \omega_1$ and $\omega_3 \gg \omega_2$. Then if $\mathbf{N} = \mathbf{0}$, we see from the third of Eqs. (11.7) that ω_3 is constant to first order in ω_1 and ω_2. The first two equations then become a pair of coupled linear equations in ω_1, ω_2, which we solve as in the preceding example to obtain

$$\begin{aligned} \omega_1 &= A[I_2(I_3-I_2)]^{1/2} \cos(\beta\omega_3 t+\theta), \\ \omega_2 &= A[I_1(I_3-I_1)]^{1/2} \sin(\beta\omega_3 t+\theta), \end{aligned} \tag{11.28}$$

where A and θ are arbitrary constants and

$$\beta = \left[\frac{(I_3-I_1)(I_3-I_2)}{I_1 I_2}\right]^{1/2}. \tag{11.29}$$

The vector ω therefore moves counterclockwise (looking down from the positive $\hat{e}_3$-axis) in a small ellipse about the $\hat{e}_3$-axis. In a similar manner, we can show that if ω is nearly parallel to the $\hat{e}_1$-axis, it moves clockwise in a small ellipse about that axis, and that if ω is nearly parallel to the $\hat{e}_2$-axis, the solution is of an exponential character. In the latter case, of course, the components ω_1 and ω_3 will not remain small, and the approximation that ω_2 is constant will hold only during the initial part of the motion. We conclude that rotation about the axes of maximum and minimum moments of inertia is stable, while rotation about the intermediate axis is unstable. This result is readily demonstrated by tossing a tennis racket in the air and attempting to make it spin about any principal axis. The general solution of Eqs. (11.7) for ω, when $\boldsymbol{N} = \mathbf{0}$, can also in principle be obtained. We shall solve the problem in the next section by a different method.

11.3 POINSOT'S SOLUTION FOR A FREELY ROTATING BODY

If there are no torques, $\boldsymbol{N} = \mathbf{0}$, then Eqs. (11.2) and (11.10) yield four integrals of the equations of motion:

$$\boldsymbol{L} = \mathbf{I} \cdot \boldsymbol{\omega} = \text{a constant}, \tag{11.30}$$

$$T = \tfrac{1}{2}\boldsymbol{\omega} \cdot \mathbf{I} \cdot \boldsymbol{\omega} = \text{a constant}. \tag{11.31}$$

Poinsot* has obtained a geometrical representation of the motion based on these constants, and utilizing the inertia ellipsoid. Let us imagine the inertia ellipsoid (10.157) rigidly fastened to the body and rotating with it. If we let $\boldsymbol{r}$ be the vector from the origin to the point where the axis of rotation intersects the inertia ellipsoid at any instant,

$$\boldsymbol{r} = \frac{r}{\omega}\boldsymbol{\omega}, \tag{11.32}$$

then comparison of Eqs. (11.31) and (10.157) shows that

$$T = \frac{a^2\omega^2}{2r^2}. \tag{11.33}$$

The normal to the ellipsoid at the point $\boldsymbol{r}$ is parallel to the vector

$$\nabla\,(\boldsymbol{r} \cdot \mathbf{I} \cdot \boldsymbol{r}) = 2I_1 x_1 \hat{e}_1 + 2I_2 x_2 \hat{e}_2 + 2I_3 x_3 \hat{e}_3 = 2\frac{r}{\omega}\boldsymbol{L}, \tag{11.34}$$

where x_1, x_2, x_3 are the components of $\boldsymbol{r}$ along the principal axes. The tangent plane to the ellipsoid at the point $\boldsymbol{r}$ is therefore perpendicular to the constant vector $\boldsymbol{L}$ (see Fig. 11.2). Let l be the perpendicular distance from the origin to this tangent plane:

$$l = \frac{\boldsymbol{r} \cdot \boldsymbol{L}}{L} = \frac{r}{\omega}\frac{\boldsymbol{\omega} \cdot \mathbf{I} \cdot \boldsymbol{\omega}}{L} = \frac{a(2T)^{1/2}}{L} = \text{a constant}. \tag{11.35}$$

---

*Poinsot, *Theorie Nouvelle de la Rotation des Corps*, 1834.

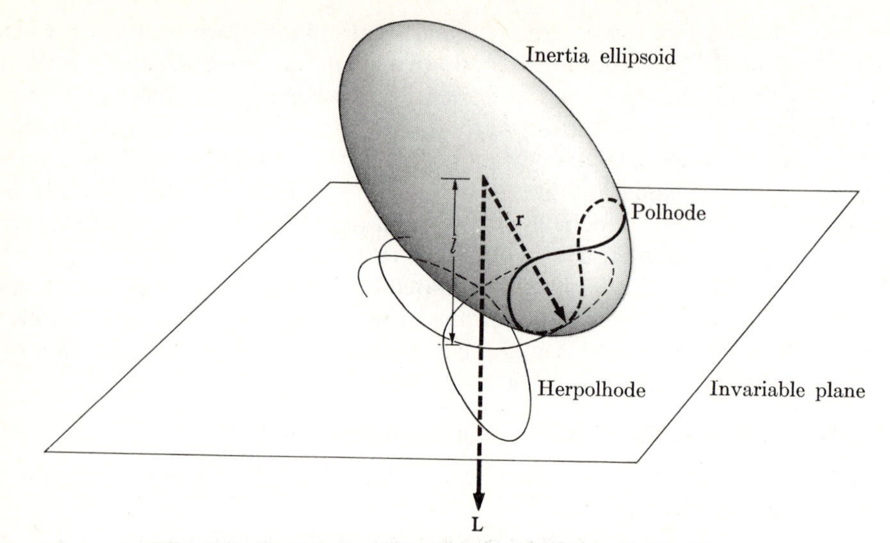

Fig. 11.2 The inertia ellipsoid rolls on the invariable plane.

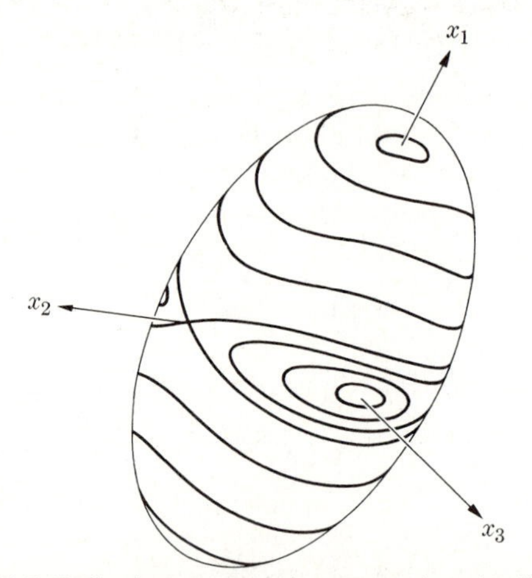

Fig. 11.3 Polhodes on a nondegenerate inertia ellipsoid.

The tangent plane is therefore fixed in space (relative to the origin O) and is called the *invariable plane*. Its position is determined by the initial conditions. Moreover, since the point of contact between the ellipsoid and the plane lies on the instantaneous axis of rotation, the ellipsoid rolls on the plane without slipping. The angular velocity at any instant has the magnitude

$$\omega = \frac{(2T)^{1/2}}{a} r. \tag{11.36}$$

This gives a complete description of the motion.

As the inertia ellipsoid rolls on the invariable plane, with its center fixed at the origin, the point of contact traces out a curve called the *polhode* on the inertia ellipsoid, and a curve called the *herpolhode* on the invariable plane. This is illustrated in Fig. 11.2. The polhode is a closed curve on the inertia ellipsoid, defined as the locus of points $\boldsymbol{r}$ where the tangent planes lie a fixed distance l from the center of the ellipsoid. In Fig. 11.3 are shown various polhodes on a nondegenerate inertia ellipsoid. Note that the topological features of the diagram are in agreement with the conclusions at the end of the preceding section. In general, the herpolhode is not closed but fills an annular ring in the invariable plane.

In the case of a symmetrical body it can be shown (Problem 7) that the polhodes are circles about the symmetry axis and the herpolhodes are circles in the invariable plane. In that case, r and therefore, by Eq. (11.36), ω (but not $\boldsymbol{\omega}$!) are constant during the motion. Poinsot's description of the motion in this case agrees with that in the preceding section. The polhode and herpolhode are the intersections of the body and the space cones with the inertia ellipsoid and the invariable plane, respectively.

11.4 EULER'S ANGLES

The results in Sections 11.2 and 11.3 regarding the motion of a rigid body were obtained without the use of any coordinates to describe the orientation of the body. In order to proceed further with the discussion, it is necessary to introduce a suitable set of coordinates. We choose a set of axes fixed in the body, which are most conveniently taken as the principal axes, with origin at the center of mass, or at the fixed point if one exists. These axes will be labeled with subscripts 1, 2, 3, as before. If there is an axis of symmetry, it will be numbered 3; otherwise, the axes may be numbered in any order. We need three coordinates to specify the orientation of the body axes with respect to a fixed set of space axes x, y, z. The relation between the two sets of axes could be specified by giving the coefficients of the transformation from coordinates x, y, z to x_1, x_2, x_3. There are nine coefficients but only three of them are independent, as we have seen, and to try to use three of the coefficients as coordinates is not convenient. As was pointed out in Section 5.1, there is no symmetric set of coordinates analogous to x, y, z with which to describe the orienta-

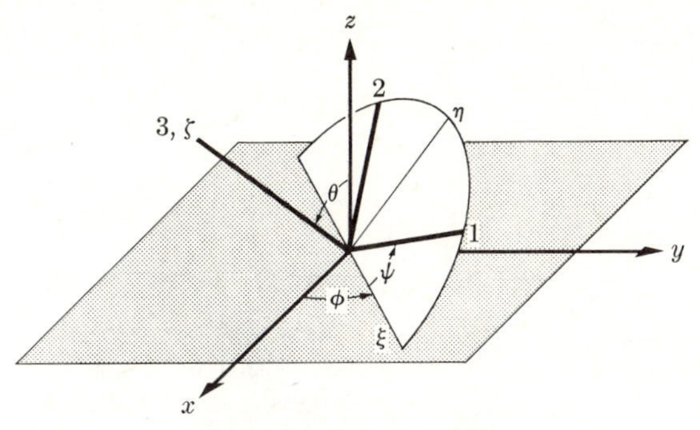

Fig. 11.4 Euler's angles.

tion of a body. Among the various coordinate systems that have been introduced for this purpose, one of the most useful is due to Euler.

In Fig. 11.4, the Euler angles θ, ϕ, ψ are shown. These are used to specify the position of the body axes 1, 2, 3 relative to the space axes x, y, z. The body axes 1, 2, 3 are shown as heavy lines; the space axes x, y, z are lighter. The angle θ is the angle between the 3-axis and the z-axis. Since the 3-axis is thus singled out for special treatment, if the body has an axis of symmetry, it should be taken as the 3-axis. Likewise, if the external torques possess an axis of symmetry in space, that axis should be taken as the z-axis. The intersection of the 1, 2-plane with the xy-plane is called the *line of nodes*, labeled ξ in the diagram. The angle ϕ is measured in the xy-plane from the x-axis to the line of nodes, as shown. The angle ψ is measured in the 1, 2-plane from the line of nodes to the 1-axis. We are assuming that both sets of axes x, y, z and 1, 2, 3 are right-handed. It will be convenient also to introduce a third (right-handed) set of axes, ξ, η, ζ, of which ξ is the line of nodes, ζ coincides with the body axis 3, and η is in the 1, 2-plane.*

In order to express the angular velocity vector ω in terms of Euler's angles, we first prove that angular velocities may be added like vectors, in the sense of the following theorem:

Given a primed coordinate system rotating with angular velocity $\boldsymbol{\omega}_1$ with respect to an unprimed system, and a starred coordinate system rotating with angular velocity $\boldsymbol{\omega}_2$ relative to the primed system, the angular velocity of the starred system relative to the unprimed system is $\boldsymbol{\omega}_1+\boldsymbol{\omega}_2$. (11.37)

To prove this theorem, let $\boldsymbol{A}$ be any vector at rest in the starred system:

$$\frac{d^*\boldsymbol{A}}{dt} = \mathbf{0}. \tag{11.38}$$

Then by theorem (7.22), its velocity relative to the primed system is

$$\frac{d'\boldsymbol{A}}{dt} = \boldsymbol{\omega}_2 \times \boldsymbol{A}. \tag{11.39}$$

Now applying theorem (7.22) again, we find the velocity of $\boldsymbol{A}$ relative to the unprimed system:

$$\frac{d\boldsymbol{A}}{dt} = \frac{d'\boldsymbol{A}}{dt} + \boldsymbol{\omega}_1 \times \boldsymbol{A} = (\boldsymbol{\omega}_1 + \boldsymbol{\omega}_2) \times \boldsymbol{A}. \tag{11.40}$$

A final comparison with theorem (7.22) shows that $(\boldsymbol{\omega}_1+\boldsymbol{\omega}_2)$ is the angular velocity of the starred system relative to the unprimed one.

Now consider Figure 11.4 and suppose that the body is moving so that θ, ϕ, ψ

*The reader is cautioned that the notation for Euler's angles, as well as the convention as to axes from which they are measured, and even the use of right-handed coordinate axes, are not standardized in the literature. It is therefore necessary to note carefully how each author defines the angles. The conventions adopted here are very common, but not universal.

are changing with time. If θ alone changes, while ϕ, ψ are fixed, the body rotates around the line of nodes with angular velocity $\dot{\theta}\hat{\xi}$. If ϕ alone changes, the body rotates around the z-axis with angular velocity $\dot{\phi}\hat{\mathbf{z}}$. If ψ alone changes, the body rotates around its 3-axis with angular velocity $\dot{\psi}\hat{\mathbf{e}}_3$. Now if we consider a primed coordinate system rotating with angular velocity $\dot{\phi}\hat{\mathbf{z}}$ about the z-axis, and let the ξ,η,ζ-system rotate with angular velocity $\dot{\theta}\hat{\xi}$ relative to this primed system, then by theorem (11.37), the angular velocity of the ξ,η,ζ-system is $\dot{\theta}\hat{\xi}+\dot{\phi}\hat{\mathbf{z}}$. The axes 1, 2, 3 rotate with angular velocity $\dot{\psi}\hat{\mathbf{e}}_3$ relative to ξ,η,ζ, hence the angular velocity of the body is

$$\boldsymbol{\omega} = \dot{\theta}\hat{\xi}+\dot{\phi}\hat{\mathbf{z}}+\dot{\psi}\hat{\mathbf{e}}_3. \tag{11.41}$$

We have, from Fig. 11.4, the relations

$$\begin{aligned} \hat{\xi} &= \hat{\mathbf{e}}_1 \cos\psi - \hat{\mathbf{e}}_2 \sin\psi, \\ \hat{\eta} &= \hat{\mathbf{e}}_1 \sin\psi + \hat{\mathbf{e}}_2 \cos\psi, \\ \hat{\zeta} &= \hat{\mathbf{e}}_3, \end{aligned} \tag{11.42}$$

and

$$\begin{aligned} \hat{\mathbf{z}} &= \hat{\zeta} \cos\theta + \hat{\eta} \sin\theta \\ &= \hat{\mathbf{e}}_1 \sin\theta \sin\psi + \hat{\mathbf{e}}_2 \sin\theta \cos\psi + \hat{\mathbf{e}}_3 \cos\theta. \end{aligned} \tag{11.43}$$

We may therefore express $\boldsymbol{\omega}$ in terms of its components along the principal axes:

$$\begin{aligned} \omega_1 &= \dot{\theta} \cos\psi + \dot{\phi} \sin\theta \sin\psi, \\ \omega_2 &= -\dot{\theta} \sin\psi + \dot{\phi} \sin\theta \cos\psi, \\ \omega_3 &= \dot{\psi} + \dot{\phi} \cos\theta. \end{aligned} \tag{11.44}$$

The kinetic energy is now given by Eq. (10.153):

$$T = \tfrac{1}{2}I_1\omega_1^2 + \tfrac{1}{2}I_2\omega_2^2 + \tfrac{1}{2}I_3\omega_3^2. \tag{11.45}$$

The kinetic energy is a rather complicated expression involving $\dot{\theta}$, $\dot{\phi}$, $\dot{\psi}$, θ, and ψ. Note that θ, ϕ, ψ are not orthogonal coordinates, that is, cross terms involving $\dot{\theta}\dot{\phi}$ and $\dot{\psi}\dot{\phi}$ appear in T. In the case of a symmetrical body ($I_1 = I_2$), the expression for T simplifies to the form:

$$T = \tfrac{1}{2}I_1\dot{\theta}^2 + \tfrac{1}{2}I_1\dot{\phi}^2 \sin^2\theta + \tfrac{1}{2}I_3(\dot{\psi} + \dot{\phi}\cos\theta)^2. \tag{11.46}$$

The generalized forces Q_θ, Q_ϕ, Q_ψ are easily shown to be the torques about the ξ-, z-, and 3-axes.

We are now in a position to write down Lagrange's equations for the rotation of a rigid body subject to given torques. If the torques are derivable from a potential energy $V(\theta,\phi,\psi)$, then there will be an energy integral. If V is independent of ϕ, then inspection of Eqs. (11.44) shows that ϕ will be an ignorable coordinate. Unfortunately, this is not enough to enable us to give a general solution of the problem. However, for a symmetrical body, if V is independent of ψ also, we see from Eq. (11.46) that both ϕ and ψ are ignorable. We have then three constants of the motion, enough to solve the problem. This case will be solved in the next section. A few other special cases are known for which the problem can be solved,* but for the general problem of the motion of an unsymmetrical body under the action of external torques, as for the many-body problem, there are no generally applicable methods of solution, except by numerical integration of the equations of motion.

11.5 THE SYMMETRICAL TOP

The symmetrical top, represented in Fig. 11.5, is a body for which $I_1 = I_2$. It pivots around a fixed point O that lies on the axis of symmetry a distance l from the center of mass G which also lies on the axis of symmetry. The only external forces are the forces of constraint at O and the force of gravity. Therefore, by Eq. (11.46), the Lagrangian function is

$$L = \tfrac{1}{2}I_1\dot{\theta}^2 + \tfrac{1}{2}I_1\dot{\phi}^2 \sin^2\theta + \tfrac{1}{2}I_3(\dot{\psi}+\dot{\phi}\cos\theta)^2 - mgl\cos\theta. \tag{11.47}$$

The coordinates ψ and ϕ are ignorable, and we have therefore three integrals of the motion:

$$\frac{dp_\psi}{dt} = \frac{\partial L}{\partial \psi} = 0, \tag{11.48}$$

$$\frac{dp_\phi}{dt} = \frac{\partial L}{\partial \phi} = 0, \tag{11.49}$$

$$\frac{dE}{dt} = -\frac{\partial L}{\partial t} = 0, \tag{11.50}$$

where

$$p_\psi = I_3(\dot{\psi}+\dot{\phi}\cos\theta), \tag{11.51}$$

$$p_\phi = I_1\dot{\phi}\sin^2\theta + I_3\cos\theta(\dot{\psi}+\dot{\phi}\cos\theta), \tag{11.52}$$

$$E = \tfrac{1}{2}I_1\dot{\theta}^2 + \tfrac{1}{2}I_1\dot{\phi}^2\sin^2\theta + \tfrac{1}{2}I_3(\dot{\psi}+\dot{\phi}\cos\theta)^2 + mgl\cos\theta. \tag{11.53}$$

We use Eqs. (11.51) and (11.52) to eliminate $\dot{\psi}$, $\dot{\phi}$ from Eq. (11.53):

$$E = \tfrac{1}{2}I_1\dot{\theta}^2 + \frac{(p_\phi - p_\psi\cos\theta)^2}{2I_1\sin^2\theta} + \frac{p_\psi^2}{2I_3} + mgl\cos\theta. \tag{11.54}$$

*See, for example, E. J. Routh, *The Advanced Part of a Treatise on the Dynamics of a System of Rigid Bodies*, 6th ed. London: Macmillan, 1905. (Also New York: Dover, 1955.)

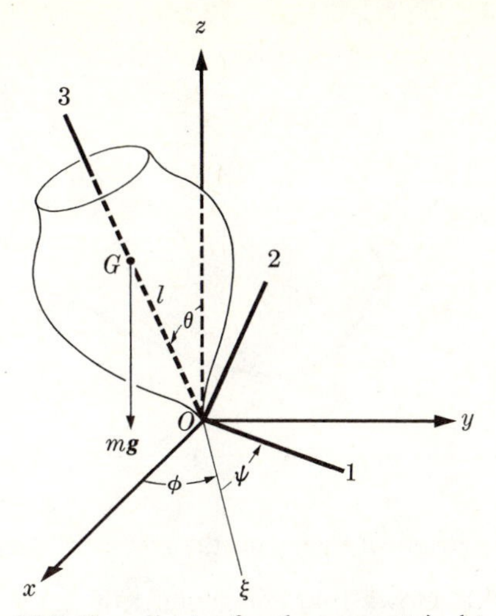

Fig. 11.5 Coordinates for the symmetrical top.

We can now solve the problem by the energy method. If we set

$$E' = E - \frac{p_\psi^2}{2I_3}, \tag{11.55}$$

$$`V' = \frac{(p_\phi - p_\psi \cos\theta)^2}{2I_1 \sin^2\theta} + mgl\cos\theta, \tag{11.56}$$

then

$$\dot{\theta} = \left\{ \frac{2}{I_1} [E' - `V'(\theta)] \right\}^{1/2}, \tag{11.57}$$

and θ is given, in principle, by computing the integral

$$\int_{\theta_i}^{\theta} \frac{d\theta}{[E' - `V'(\theta)]^{1/2}} = \left(\frac{I_1}{2}\right)^{1/2} t \tag{11.58}$$

and solving for $\theta(t)$. The constant θ_i is the initial value of θ. Once $\theta(t)$ is known, Eqs. (11.51) and (11.52) can be solved for $\dot{\psi}$ and $\dot{\phi}$ and integrated to give $\psi(t)$, $\phi(t)$.

Comparison of Eqs. (11.44) and (11.51) shows that

$$p_\psi = I_3\omega_3, \tag{11.59}$$

so that ω_3 is a constant of the motion. If $\omega_3 = 0$, then Eq. (11.56) reduces essentially to the formula (9.137) for a spherical pendulum, as it should. In Fig. 11.6, 'V'(θ) is plotted versus θ for $\omega_3 \neq 0$. The 'torque' associated with the 'potential energy' 'V'(θ) is

$$`N' = -\frac{\partial `V'}{\partial\theta} = mgl\sin\theta - \frac{(p_\phi - p_\psi\cos\theta)(p_\psi - p_\phi\cos\theta)}{I_1\sin^3\theta}. \tag{11.60}$$

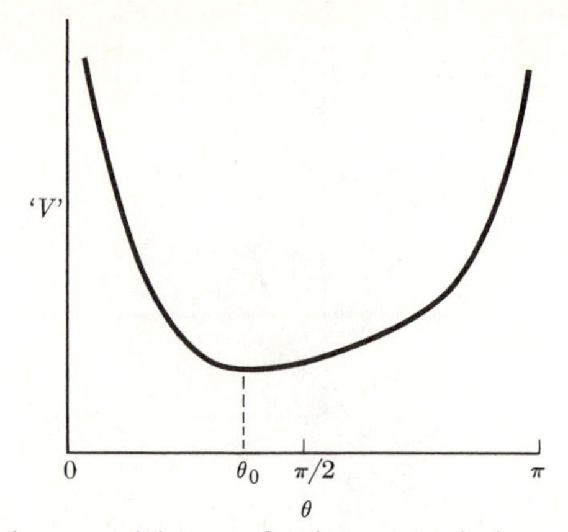

Fig. 11.6 Effective potential energy for the symmetrical top when $p_\phi \neq p_\psi$.

Inspection of Eq. (11.60) shows that, in general (if $p_\phi \neq p_\psi$), the 'torque' 'N' is positive for $\theta \doteq 0$ and negative for $\theta \doteq \pi$, and has one zero between 0 and π. Hence 'V' has one minimum, as shown in Fig. 11.6, at a point θ_0 satisfying the equation

$$mgl\, I_1 \sin^4 \theta_0 - (p_\phi - p_\psi \cos \theta_0)(p_\psi - p_\phi \cos \theta_0) = 0. \tag{11.61}$$

If $E' = $ 'V'(θ_0), the axis of the top precesses uniformly at an angle θ_0 with the vertical, and with angular velocity

$$\dot{\phi}_0 = \frac{p_\phi - p_\psi \cos \theta_0}{I_1 \sin^2 \theta_0}. \tag{11.62}$$

Solving Eq. (11.61) for $(p_\phi - p_\psi \cos \theta_0)$, and using Eq. (11.59), we obtain

$$(p_\phi - p_\psi \cos \theta_0) = \tfrac{1}{2} I_3 \omega_3 \frac{\sin^2 \theta_0}{\cos \theta_0} \left[1 \pm \left(1 - \frac{4mglI_1}{I_3^2 \omega_3^2} \cos \theta_0 \right)^{1/2} \right]. \tag{11.63}$$

We see that if $\theta_0 < \pi/2$, there is a minimum spin angular velocity below which the top cannot precess uniformly at the angle θ_0:

$$\omega_{\min} = \left(\frac{4mglI_1}{I_3^2} \cos \theta_0 \right)^{1/2}. \tag{11.64}$$

For $\omega_3 > \omega_{\min}$, there are two roots (11.63) and hence two possible values of $\dot{\phi}_0$, a slow and a fast precession, both in the same direction as the spin angular velocity ω_3. For $\omega_3 \gg \omega_{\min}$, the fast and slow precessions occur at angular velocities

$$\dot{\phi}_0 \doteq \frac{I_3}{I_1} \frac{\omega_3}{\cos \theta_0}, \tag{11.65}$$

and

$$\dot{\phi}_0 \doteq \frac{mgl}{I_3 \omega_3}. \tag{11.66}$$

It is the slow precession which is ordinarily observed with a rapidly spinning top. For $\theta_0 > \pi/2$ (top hanging with its axis below the horizontal), there is one positive and one negative value for $\dot{\phi}_0$. (To what do these motions of uniform precession reduce when $\omega_3 \to 0$?)

Study of Fig. 11.6 shows us that the more general motion involves a *nutation* or oscillation of the axis of the top in the θ-direction as it precesses. The axis oscillates between angles θ_1 and θ_2 which satisfy the equation

$$E' = \frac{(p_\phi - p_\psi \cos\theta)^2}{2I_1 \sin^2\theta} + mgl\cos\theta, \tag{11.67}$$

where p_ϕ, p_ψ, and E' are determined from the initial conditions. If we multiply Eq. (11.67) by $\sin^2\theta$, it becomes a cubic equation in $\cos\theta$. We see from Fig. 11.6 that there must be two real roots $\cos\theta_1$, $\cos\theta_2$ between -1 and $+1$. The third root for $\cos\theta$ must lie outside the physical range -1 to $+1$. In fact, inspection of Eq. (11.67) will show that the third root is greater than $+1$. (In the case of uniform precession discussed in the preceding paragraph, the two physical roots coincide, $\cos\theta_1 = \cos\theta_2 = \cos\theta_0$.) If initially $\dot{\theta} = 0$, then the initial value $\cos\theta_1$ of $\cos\theta$ satisfies Eq. (11.67); knowing one root of a cubic equation, we may factor the equation and find all three roots. During nutation, the precession velocity varies according to Eq. (11.52):

$$\dot{\phi} = \frac{p_\phi - p_\psi \cos\theta}{I_1 \sin^2\theta}. \tag{11.68}$$

If $|p_\phi| < |p_\psi|$, we can define an angle θ_3 as follows:

$$\cos\theta_3 = \frac{p_\phi}{p_\psi}. \tag{11.69}$$

For $\theta > \theta_3$, $\dot{\phi}$ has the same sign as ω_3, and for $\theta < \theta_3$, it has the opposite sign. The derivative with respect to θ of the right member of Eq. (11.67) is negative at $\theta = \theta_3$; hence we see from Fig. 11.6 that $\theta_3 < \theta_2$, where θ_2 is the largest angle satisfying Eq. (11.67). In fact, $\theta_3 < \theta_0$. If $\theta_3 < \theta_1$ (or if $|p_\phi| > |p_\psi|$ and p_ϕ, p_ψ have the same sign), then $\dot{\phi}$ has the same sign as ω_3 throughout the nutation, and the top axis traces out a curve like that shown in Fig. 11.7(a). If $\theta_3 > \theta_1$, $\dot{\phi}$ changes sign during the nutation and the top axis moves as in Fig. 11.7(b). It is clear that if the top is set in motion initially above the horizontal plane with $\dot{\phi}$ opposite in sign to ω_3, the motion necessarily will be like that shown in Fig. 11.7(b).

An important special case occurs when the top, spinning about its axis with angular velocity ω_3, is held with its axis initially at rest at an angle θ_1 and then released. Initially, we have

$$\theta = \theta_1, \qquad \dot{\theta} = 0, \qquad \dot{\phi} = 0, \qquad \dot{\psi} = \omega_3. \tag{11.70}$$

We substitute in Eqs. (11.51), (11.52), and (11.53) to find

$$p_\psi = I_3\omega_3, \qquad p_\phi = I_3\omega_3\cos\theta_1, \qquad E' = mgl\cos\theta_1. \tag{11.71}$$

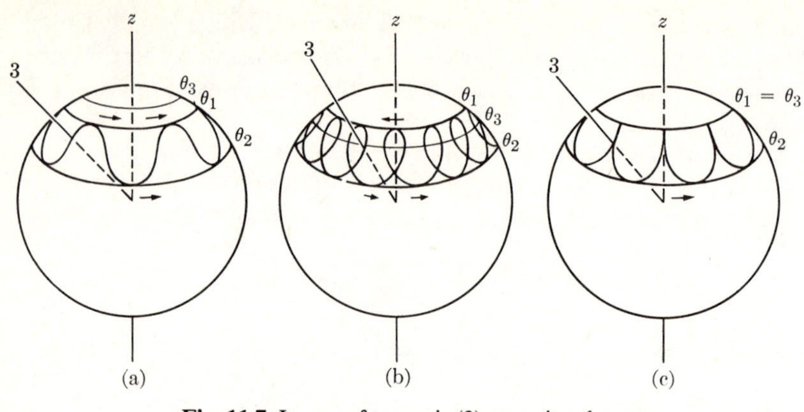

Fig. 11.7 Locus of top axis (3) on unit sphere.

In this case, we see that $\theta_3 = \theta_1$, and the motion is as shown in Fig. 11.7(c). An elementary discussion of this case, based on the conservation of angular momentum, was given in Section 4.2. Now Eq. (11.56) becomes

$$`V' = \frac{I_3^2\omega_3^2}{2I_1}\left[\frac{(\cos\theta_1 - \cos\theta)^2}{\sin^2\theta} + \alpha\cos\theta\right], \tag{11.72}$$

where

$$\alpha = \frac{2I_1 mgl}{I_3^2\omega_3^2}. \tag{11.73}$$

The turning points for the nutation are the roots of Eq. (11.67), which becomes in this case, if we multiply by $\sin^2\theta$,

$$(\cos\theta_1 - \cos\theta)^2 - \alpha(\cos\theta_1 - \cos\theta)(1 - \cos^2\theta) = 0. \tag{11.74}$$

The roots are

$$\begin{aligned} \cos\theta &= \cos\theta_1, \\ \cos\theta &= \frac{1}{2\alpha}\left[1 \pm (1 - 4\alpha\cos\theta_1 + 4\alpha^2)^{1/2}\right]. \end{aligned} \tag{11.75}$$

The angle θ_2 is given by the second formula, using the minus sign in the bracketed expression. The plus sign gives a root for $\cos\theta$ greater than $+1$. Let us consider the case of a rapidly spinning top, that is, when $\alpha \ll 1$. We then have

$$\cos\theta_2 \doteq \cos\theta_1 - \alpha\sin^2\theta_1. \tag{11.76}$$

The angle θ_2 is only slightly greater than θ_1, and the amplitude of nutation is proportional to α. If we set

$$\theta_2 = \theta_0 + a, \qquad \theta_1 = \theta_0 - a, \tag{11.77}$$

and substitute in Eq. (11.76), we find that, to first order in a and α,

$$a \doteq \tfrac{1}{2}\alpha \sin \theta_1. \tag{11.78}$$

We now set

$$\theta = \theta_0 + \delta \doteq \theta_1 + a + \delta, \tag{11.79}$$

and substitute in Eq. (11.72), which becomes, to second order in a and δ,

$$\text{`}V\text{'} \doteq V(\theta_0) + \tfrac{1}{2}\frac{I_3^2}{I_1}\omega_3^2\,\delta^2. \tag{11.80}$$

The first term is constant, and the second leads to harmonic oscillations in δ with a frequency

$$\omega_\theta = \frac{I_3}{I_1}\omega_3. \tag{11.81}$$

The nutation is given by

$$\theta \doteq \theta_1 + a - a\cos\omega_\theta t. \tag{11.82}$$

We substitute in Eq. (11.68) to obtain $\dot{\phi}$ to first order in a:

$$\dot{\phi} \doteq \frac{I_3\omega_3 a}{I_1 \sin\theta_1}[1-\cos\omega_\theta t]. \tag{11.83}$$

The average angular velocity of precession is

$$\langle\dot{\phi}\rangle_{\text{av}} \doteq \frac{I_3\omega_3 a}{I_1\sin\theta_1} = \frac{mgl}{I_3\omega_3}. \tag{11.84}$$

The top axis therefore precesses very slowly and nutates very rapidly with very small amplitude. In practice, the frictional torques which we have neglected usually damp out the nutation fairly quickly, leaving only the uniform precession.

As a final example, consider the case when the top is initially spinning with its symmetry axis vertical. In this case, so long as the 3- and z-axes coincide, the line of nodes is indeterminate. We see from Fig. 11.4 that the angle $\psi + \phi$ is determined as the angle between the x- and 1-axes, although ψ, ϕ separately are indeterminate. Hence we have initially

$$p_\psi = I_3(\dot{\psi}+\dot{\phi}) = I_3\omega_3, \tag{11.85}$$

$$p_\phi = I_3(\dot{\psi}+\dot{\phi}) = p_\psi. \tag{11.86}$$

Equation (11.56) in this case becomes

$$\text{`}V\text{'} = \frac{I_3^2\omega_3^2}{2I_1}\left[\frac{(1-\cos\theta)^2}{\sin^2\theta} + \alpha\cos\theta\right], \tag{11.87}$$

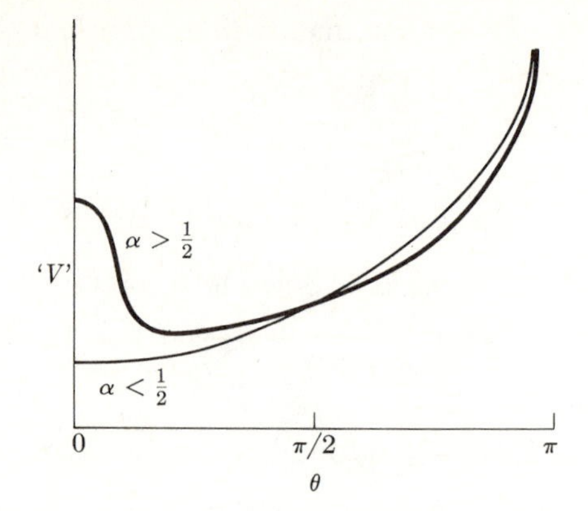

Fig. 11.8 Effective potential energy when $p_\phi = p_\psi$.

where α is given by Eq. (11.73). This, of course, is just a special case of Eq. (11.72). In Fig. 11.8 we plot 'V' for the case when $p_\phi = p_\psi$. The form of the curve depends upon the value of α. We see that a rapidly spinning top ($\alpha < \frac{1}{2}$) can spin stably about the vertical axis; if disturbed, it will exhibit a small nutation about the vertical axis. A slowly spinning top ($\alpha > \frac{1}{2}$) cannot spin stably about a vertical axis, but will execute a large nutation between $\theta_1 = 0$ and θ_2 given by Eq. (11.75). In this case,

$$\cos\theta_2 = \frac{1}{\alpha} - 1. \tag{11.88}$$

The minimum spin angular velocity below which the top cannot spin stably about a vertical axis occurs when $\alpha = \frac{1}{2}$, or, by Eq. (11.73),

$$\omega_{\min} = \left[\frac{4mglI_1}{I_3^2}\right]^{1/2}. \tag{11.89}$$

Note that this formula agrees with Eq. (11.64). If initially $\omega_3 > \omega_{\min}$, a top will spin with its axis vertical, but when friction reduces ω_3 below $\omega_{\min}$, it will begin to wobble.

All of the above conclusions about the behavior of a symmetrical top under various initial conditions can easily be verified experimentally with a top or with a gyroscope.

PROBLEMS

1. Use the result of Problem 3, Chapter 10, to derive Eq. (11.6) directly from the equation

$$\frac{d}{dt}(\mathbf{I}\cdot\boldsymbol{\omega}) = \boldsymbol{N}.$$

2. a) Assume that the earth is a uniform rigid ellipsoid of revolution, look up its equatorial and polar diameters, and calculate the angular velocity of precession of the North Pole on the earth's surface assuming that the polar axis (that is, the axis of rotation) deviates slightly from the axis of symmetry. (An irregular precession of roughly this sort is observed with an amplitude of a few feet, and a period of 427 days.)

b) Assume that the earth is a rigid sphere and that a mountain of mass 10^{-9} times the mass of the earth is added at a point 45° from the polar axis. Describe the resulting motion of the pole. How long does the pole take to move 1000 miles?

c) For a rigid ellipsoidal earth, as in part (a), how massive a "mountain" must be placed on the equator in order to make the polar precession unstable?

The earth is, of course, not of uniform density, but is more dense near its center. Even more important, the earth is not rigid, but behaves as an elastic spheroid for short times, and can deform plastically over long times. The results in this problem are therefore only suggestive and do not correspond to the actual motion of the earth. For example, the observed precession period of 427 days is longer than would be calculated for a rigid earth. When plastic deformation is taken into account, an appreciable wandering of the pole can result even for an ellipsoidal earth with a much smaller "mountain" than that calculated in part (c).*

3. Show that the axis of rotation of a freely rotating symmetrical rigid body precesses in space with an angular velocity

$$\omega_n = (2\beta + \beta^2 + \sec^2 \alpha_b)^{\frac{1}{2}}\omega_3,$$

where the notation is that used in Section 11.2.

4. Show that if the only torque on a symmetrical rigid body is about the axis of symmetry, then $(\omega_1^2 + \omega_2^2)$ is constant, where ω_1 and ω_2 are angular velocity components along axes perpendicular to the symmetry axis. If $N_3(t)$ is given, show how to solve for ω_1, ω_2, and ω_3.

5. A symmetrical rigid body moving freely in space is powered with jet engines symmetrically placed with respect to the 3-axis of the body, which supply a constant torque N_3 about the symmetry axis. Find the general solution for the angular velocity vector as a function of time, relative to body axes, and describe how the angular velocity vector moves relative to the body.

6. a) Consider a charged sphere whose mass m and charge e are both distributed in a spherically symmetrical way. That is, the mass and the charge densities are each functions of the radius r (but not necessarily the same function). Show that if this body rotates in a uniform magnetic field $\boldsymbol{B}$, the torque on it is

$$\boldsymbol{N} = \frac{eg}{2mc}\boldsymbol{L}\times\boldsymbol{B} \qquad \text{(Gaussian units)},$$

*An excellent short discussion of the rotation of the earth, treated as an elastic and plastic ellipsoid, will be found in an article by D. R. Inglis, *Reviews of Modern Physics*, vol. 29, p. 9 (1957).

where g is a numerical constant, and $g = 1$ if the mass density is everywhere proportional to the charge density.

b) Write an equation of motion for the body, and show that by introducing a suitably rotating coordinate system, you can eliminate the magnetic torque.

c) Compare this result with Larmor's theorem (Chapter 7). Why is no assumption needed here regarding the strength of the magnetic field?

d) Describe the motion. What points in the body are at rest in the rotating coordinate system?

7. Prove (without using the results of Section 11.2) that if two principal moments of inertia are equal, the polhode and the herpolhode are both circles.

8. a) Obtain equations, in terms of principal coordinates x_1, x_2, x_3, for two quadric surfaces whose intersection is the polhode. Your equations should contain the parameters I_1, I_2, I_3, l.

b) Find the equation for the projection of the polhode on any coordinate plane and show that the polhodes are closed curves around the major and minor poles of the ellipsoid, but that they are of the hyperbolic type near the intermediate axis, as shown in Fig. 11.3.

c) Find the radii of the circles on the invariable plane which bound the herpolhode.

9. Find the matrix (a_{ij}) which transforms the components of a vector from space axes to body axes. Express a_{ij} in terms of Euler's angles. [*Hint*: The transformation can be made up of three consecutive rotations by angles θ, ϕ, ψ, about suitable axes, and taken in proper order.]

10. Write out the Hamiltonian function in terms of θ, ψ, ϕ, p_θ, p_ψ, p_ϕ for a freely rotating unsymmetrical rigid body. Express the coefficients in terms of the parameters $I_1, I_3, (I_2 - I_1)$.

11. Use Lagrange's equations to treat the free rotation of an unsymmetrical rigid body near one of its principal axes, and show that your results agree with the last paragraph of Section 11.2.

12. Set up Lagrange's equations for a symmetrical top, the end of whose axis slides without friction on a smooth horizontal table. Discuss carefully the differences in the motions between this case and the case when the end of the top axis pivots about a fixed point.

13. A gyroscope is constructed of a disk of radius a, mass M, fastened rigidly at the center of an axle of length $(3a/2)$, mass $(2M/7)$, negligible cross section, and mounted inside two perpendicular rings, each of radius $(3a/2)$, mass $(M/3)$. The axle rotates in frictionless bearings at the intersection points of the rings. One of these intersection points pivots without friction about a fixed point O. Set up the Lagrangian function and discuss the kinds of motion which may occur (under the action of gravity).

14. Discuss the free rotation of a symmetrical rigid body, using the Lagrangian method. Find the angular velocity for uniform precession and the frequency of small nutations about this uniform precession. Describe the motion and show that your results agree with the solutions found in Section 11.2 and in Problem 3.

15. A top consists of a disk of mass M, radius r, mounted at the center of a cylindrical axle of length l, radius a, where $a \ll l$, and negligible mass. The end of the axle rests on a table, as shown in Fig. 11.9. The coefficient of sliding friction is μ. The top is set spinning about its

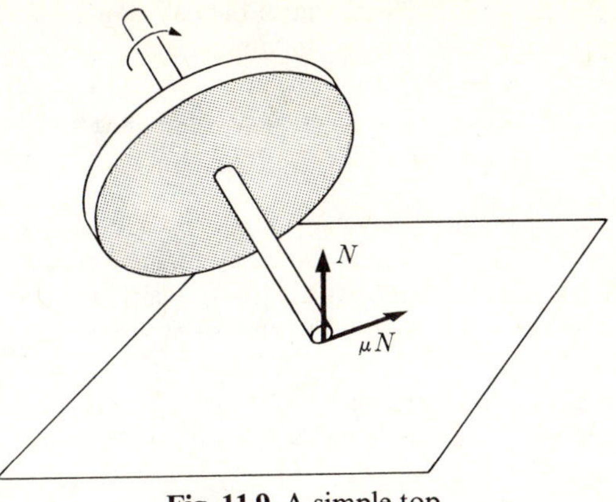

Fig. 11.9 A simple top.

symmetry axis with a very great angular velocity ω_{30}, and released with its axis at an angle θ_1 with the vertical. Assume that ω_3 is great enough compared with all other motions of the top so that the edge of the axle in contact with the table slides on the table in a direction perpendicular to the top axis, with the sense determined by ω_3. Write the equations of motion for the top. Assume that the nutation is small enough to be neglected, and that the friction is not too great, so that the top precesses slowly at an angle θ_0 which changes slowly due to the friction with the table. Show that the top axis will at first rise to a vertical position, and find approximately the time required and the number of complete revolutions of precession during this time. Describe the entire motion of the top relative to the table during this process. How long will it remain vertical before beginning to wobble?

16. Obtain a toy gyroscope, and make the necessary measurements in order to predict the rate at which it will precess, when spinning at its top speed, if its axis pivots about a fixed point at an angle of 45° with the vertical. Calculate the amplitude of nutation if the axis is held at an angle of 45° and released. Perform the experiment, and compare the measured rate of precession with the predicted rate.

17. A planet consists of a uniform sphere of radius a, mass M, girdled at its equator by a ring of mass m. The planet moves (in a plane) about a star of mass M'. Set up the Lagrangian function, using as coordinates the polar coordinates r, α in the plane of the orbit, and Euler's angles θ, ϕ, ψ, relative to space axes of which the z-axis is perpendicular to the plane of the orbit, and the x-axis is parallel to the axis from which α is measured. You may assume that $r \gg a$, and use the result of Problem 15, Chapter 6. Find the ignorable coordinates, and show that the period of rotation of the planet is constant.

18. Assume that the planet of Problem 17 revolves in a circle of radius r about the star, although this does not quite satisfy the equations of motion. Assume that the period of revolution is short in comparison with any precession of the axis of rotation, so that in studying the rotation it is permissible to average over the angle α. Show that uniform (slow) precession of the polar axis may occur if the axis is tilted at an angle θ_0 from the normal to the orbital plane, and find the angular velocity of precession in terms of the masses M, m, M', the radii a, r, the angle θ_0, and the angular velocity of rotation. Show that if the day is much shorter than the year, the above assumption regarding the period of revolution and the rate of precession is valid. Find the frequency of small nutations about this uniform precession and show that when the day is much shorter than the year, it corresponds to the free precession whose angular velocity is given in Problem 3.

19. Find the masses M, m required to give the planet in Problem 17 the same principal moments of inertia as a uniform ellipsoid of the same mass and shape as the earth. Show that, with the approximations made in Problem 18, if the sun and moon lie in the earth's orbital plane (they do very nearly), the effect of both sun and moon on the earth's rotation can be taken into account simply by adding the precession angular velocities that would be caused by each separately. The equator makes an angle of 23.5° with the orbital plane. Find the resulting total period of precession. [The measured value is 26,000 years.]

***20.** Write Lagrangian equations of motion for the rigid body in Problem 5. Carry the solution as far as you can. [Make use of the results of Problem 5 if you wish.] Show that you can obtain a second order differential equation involving θ alone. Can you find any particular solutions, or approximate solutions, of this equation for special cases? Describe the corresponding motions. [Note that this problem, to the extent that it can be solved, gives the motion of the body in space, in contrast to Problem 5, where we found the angular velocity relative to the body.]

21. An electron may for some purposes be regarded as a spinning charged sphere like that considered in Problem 6, with g very nearly equal to 2. Show that if g were exactly 2, and the electron spin angular momentum is initially parallel to its linear velocity, then as the electron moves through any magnetic field, its spin angular momentum would always remain parallel to its velocity.

22. An earth satellite consists of a spherical shell of mass 20 kg, diameter 1 m. It is directionally stabilized by a gyro consisting of a 4-kg disk, 20 cm in diameter, mounted on an axle of negligible mass whose frictionless bearings are fastened at the opposite ends of a diameter of the shell. The shell is initially not rotating, while the gyro rotates at angular velocity ω_0. A 1-mg dust grain traveling perpendicular to the gyro axis with a velocity of 3×10^4 m/sec buries itself in the shell at one end of the axis. What must be the rotation frequency of the gyro in order that the gyro axis shall thereafter remain within 0.1 degree of its initial position? An accuracy of two significant figures in the result will be satisfactory.

23. A gyrocompass is a symmetrical rigid body mounted so that its axis is constrained to move in a horizontal plane at the earth's surface. Choose a suitable pair of coordinate angles and set up the Lagrangian function if the gyrocompass is at a fixed point on the earth's surface of colatitude θ_0. Neglect friction. Show that the angular velocity component ω_3 along the symmetry axis remains constant, and that if $\omega_3 > (I_1/I_3)\omega_0 \sin \theta_0$, where ω_0 is the angular velocity of the earth, then the symmetry axis oscillates in the horizontal plane about a north-south axis. Find the frequency of small oscillations. In an actual gyrocompass, the rotor must be driven to make up for frictional torques about the symmetry axis, while frictional torques in the horizontal plane damp the oscillations of the symmetry axis, which comes to rest in a north-south line.

CHAPTER 12

THEORY OF SMALL VIBRATIONS

An important and frequently recurring problem is to determine whether a given motion of a dynamical system is stable, and if it is, to determine the character of small vibrations about the given motion. The simplest problem of this kind is that of the stability of a point of equilibrium, which we shall discuss first. In this case, we can use the machinery of tensor algebra developed in Chapter 10 to give an elegant method of solution for the small oscillations. A more general problem occurs when we are given any particular solution to the equations of motion. We may then ask whether that solution is stable, in the sense that every solution which starts from initial conditions near enough to those of the given solution will remain near that solution. This problem will be discussed in Section 12.6. Methods of solution will be given for the special case of steady motion.

12.1 CONDITION FOR STABILITY NEAR AN EQUILIBRIUM CONFIGURATION

Let us consider a mechanical system described by generalized coordinates $x_1, \ldots, x_f$, and subject to forces derivable from a potential energy $V(x_1, \ldots, x_f)$ independent of time. If the system is subject to constraints, we will suppose the coordinates are so chosen that $x_1, \ldots, x_f$ are unconstrained. The coordinate system is to be fixed in time; therefore the kinetic energy has the form

$$T = \sum_{k,l=1}^{f} \frac{1}{2} M_{lk}\dot{x}_l\dot{x}_k. \tag{12.1}$$

Lagrange's equations then become

$$\sum_{l=1}^{f} \frac{d}{dt}(M_{lk}\dot{x}_l) - \sum_{l,m=1}^{f} \frac{1}{2}\frac{\partial M_{lm}}{\partial x_k}\dot{x}_l\dot{x}_m + \frac{\partial V}{\partial x_k} = 0, \qquad k = 1, \ldots, f. \tag{12.2}$$

These equations have a solution corresponding to an equilibrium configuration for which the coordinates all remain constant if they can be solved when all velocity-dependent terms are set equal to zero. The system can therefore be in equilibrium in any configuration for which the generalized forces vanish:

$$\frac{\partial V}{\partial x_k} = 0, \qquad k = 1, \ldots, f. \tag{12.3}$$

These f equations are to be solved for the equilibrium points, if any, of the system.

The question of stability is easily answered in this case. If $V(x_1, \ldots, x_f)$ is a minimum for an equilibrium configuration $x_1^0, \ldots, x_f^0$ relative to all nearby configurations $x_1^0 + \delta x_1, \ldots, x_f^0 + \delta x_f$, then this is a stable configuration. The total energy

$$E = T + V \tag{12.4}$$

is constant. Let

$$E = V(x_1^0, \ldots, x_f^0) + \delta E \tag{12.5}$$

be the energy corresponding to any initial conditions $x_1^0 + \delta x_1^0, \ldots, x_f^0 + \delta x_f^0$; $\dot{x}_1^0, \ldots, \dot{x}_f^0$ near equilibrium. Then if $\delta x_1^0, \ldots, \delta x_f^0$; $\dot{x}_1^0, \ldots, \dot{x}_f^0$ are small enough, we can make δE as small as we please. Since T is never negative, the motion is restricted by Eq. (12.4) to a region in the configuration space for which

$$V(x_1, \ldots, x_f) \leq V(x_1^0, \ldots, x_f^0) + \delta E. \tag{12.6}$$

Since V is a minimum at $(x_1^0, \ldots, x_f^0)$, if δE is sufficiently small, the motion is restricted to a small region near $x_1^0, \ldots, x_f^0$. Furthermore, since

$$T \leq \delta E, \tag{12.7}$$

the velocities $\dot{x}_1, \ldots, \dot{x}_f$ are limited to small values. Therefore the equilibrium is stable in the sense that motions at small velocities near the equilibrium configuration remain near the equilibrium configuration.

Conversely, if V is not a minimum near $x_1^0, \ldots, x_f^0$, then it is plausible that the equilibrium is unstable, because in some direction away from $x_1^0, \ldots, x_f^0$, V will decrease. If we can choose the coordinates so that x_1, say, corresponds to that direction, and so that x_1 is orthogonal to the other coordinates, then Eq. (12.2) for x_1 is

$$\frac{d}{dt}(M_{11}\dot{x}_1) - \sum_{l,m=1}^{f} \frac{1}{2} \frac{\partial M_{lm}}{\partial x_1} \dot{x}_l \dot{x}_m = -\frac{\partial V}{\partial x_1}. \tag{12.8}$$

For small enough velocities such that quadratic terms in the velocities are negligible, this becomes

$$M_{11}\ddot{x}_1 = -\frac{\partial V}{\partial x_1}. \tag{12.9}$$

But as we move away from equilibrium in the x_1-direction, $\partial V/\partial x_1$ becomes negative, and x_1 has a positive acceleration away from the equilibrium point. In Section 12.3 we shall present a more rigorous proof that the equilibrium is unstable if V is not a minimum there.

There is a test for a minimum point which is sometimes useful. If $x_1^0, \ldots, x_f^0$ is an equilibrium configuration for which Eq. (12.3) holds, then it is a minimum of $V(x_1, \ldots, x_f)$ relative to nearby configurations, provided that all the determinants in the following sequence are positive:

$$\frac{\partial^2 V}{\partial x_1^2} > 0, \quad \begin{vmatrix} \dfrac{\partial^2 V}{\partial x_1^2} & \dfrac{\partial^2 V}{\partial x_1 \partial x_2} \\ \dfrac{\partial^2 V}{\partial x_2 \partial x_1} & \dfrac{\partial^2 V}{\partial x_2^2} \end{vmatrix} > 0, \ldots, \quad \begin{vmatrix} \dfrac{\partial^2 V}{\partial x_1^2} & \cdots & \dfrac{\partial^2 V}{\partial x_1 \partial x_f} \\ \vdots & \vdots & \vdots \\ \dfrac{\partial^2 V}{\partial x_f \partial x_1} & \cdots & \dfrac{\partial^2 V}{\partial x_f^2} \end{vmatrix} > 0, \tag{12.10}$$

where the derivatives are evaluated at $x_1^0, \ldots, x_f^0$.*

12.2 LINEARIZED EQUATIONS OF MOTION NEAR AN EQUILIBRIUM CONFIGURATION

We wish now to study the motion of a system in the neighborhood of an equilibrium configuration. The coordinates will be chosen so that the equilibrium configuration lies at the origin $x_1 = \cdots = x_f = 0$. The potential energy V is to be expanded in a Taylor series in $x_1, \ldots, x_f$. The constant term $V(0, \ldots, 0)$ may be omitted as it does not enter into the equations of motion. The linear terms are absent, in view of Eqs. (12.3). If our study is restricted to small values of $x_1, \ldots, x_f$, we may neglect cubic and higher-order terms in $x_1, \ldots, x_f$, so that

$$V = \sum_{k,l=1}^{f} \tfrac{1}{2} K_{kl} x_k x_l, \tag{12.11}$$

where

$$K_{kl} = \left(\frac{\partial^2 V}{\partial x_k \partial x_l} \right)_{x_1 = \cdots = x_f = 0}. \tag{12.12}$$

Since the coordinate system is stationary, the kinetic energy is

$$T = \sum_{k,l=1}^{f} \tfrac{1}{2} M_{kl} \dot{x}_k \dot{x}_l. \tag{12.13}$$

In general, the coefficients M_{kl} may be functions of the coordinates, but since the velocities are to be small, to second order in $x_1, \ldots, x_f$; $\dot{x}_1, \ldots, \dot{x}_f$ we may take M_{kl} to be the values of the coefficients at $x_1 = \cdots = x_f = 0$.

Equations (12.11) and (12.13) can be written in a suggestive way by introducing in the f-dimensional configuration space a configuration vector $\boldsymbol{x}$ with components $x_1, \ldots, x_f$:

$$\boldsymbol{x} = (x_1, \ldots, x_f). \tag{12.14}$$

The coefficients K_{kl} and M_{kl} become the components of tensors

$$\mathsf{K} = \begin{pmatrix} K_{11} & \cdots & K_{1f} \\ \vdots & \vdots\vdots\vdots & \vdots \\ K_{f1} & \cdots & K_{ff} \end{pmatrix}, \qquad \mathsf{M} = \begin{pmatrix} M_{11} & \cdots & M_{1f} \\ \vdots & \vdots\vdots\vdots & \vdots \\ M_{f1} & \cdots & M_{ff} \end{pmatrix}. \tag{12.15}$$

These tensors are symmetric, or can be taken as such, since by Eq. (12.12)

$$K_{kl} = K_{lk}, \tag{12.16}$$

*W. F. Osgood, *Advanced Calculus*, New York: Macmillan, 1925, p. 179.

and in the defining equation (12.13) only the sum $\frac{1}{2}(M_{kl} + M_{lk})$ is defined as the coefficient of $\dot{x}_k\dot{x}_l = \dot{x}_l\dot{x}_k$. Therefore, we may require that

$$M_{kl} = M_{lk}. \tag{12.17}$$

The kinetic and potential energies may now be written as

$$T = \tfrac{1}{2}\dot{\boldsymbol{x}} \cdot \mathbf{M} \cdot \dot{\boldsymbol{x}}, \tag{12.18}$$

$$V = \tfrac{1}{2}\boldsymbol{x} \cdot \mathbf{K} \cdot \boldsymbol{x}. \tag{12.19}$$

The Lagrange equations (9.79) may be written as

$$\mathbf{M} \cdot \ddot{\boldsymbol{x}} + \mathbf{K} \cdot \boldsymbol{x} = \mathbf{0}. \tag{12.20}$$

This equation bears a formal resemblance to Eq. (2.84) for the simple harmonic oscillator. If we write Eq. (12.20) in terms of components, we obtain a direct generalization of Eqs. (4.135) and (4.136) for two coupled harmonic oscillators.

We may solve Eq. (12.20) by the same method used to solve Eqs. (4.135) and (4.136). We try

$$\boldsymbol{x} = \boldsymbol{C}e^{pt}, \tag{12.21}$$

where $\boldsymbol{C} = (C_1, \ldots, C_f)$ is a constant vector whose components $C_1, \ldots, C_f$ may be complex. We substitute in Eq. (12.20) and divide by e^{pt}:

$$p^2\mathbf{M} \cdot \boldsymbol{C} + \mathbf{K} \cdot \boldsymbol{C} = \mathbf{0}. \tag{12.22}$$

If we write this in terms of components, we obtain

$$\sum_{l=1}^{f} (p^2 M_{kl} + K_{kl})C_l = 0, \qquad k = 1, \ldots, f. \tag{12.23}$$

If $C_1, \ldots, C_f$ are not all zero, the determinant of the coefficients must vanish:

$$\begin{vmatrix} p^2M_{11}+K_{11} & \cdots & p^2M_{1f}+K_{1f} \\ \vdots & \vdots\vdots & \vdots \\ p^2M_{f1}+K_{f1} & \cdots & p^2M_{ff}+K_{ff} \end{vmatrix} = 0. \tag{12.24}$$

This is an equation of order f in p^2 whose f roots p_j^2 are always real as we shall show in the next section. If any solution for p^2 is positive, then at least some of the roots p_j will be positive and some of the solutions (12.21) will grow exponentially in time. If the equilibrium configuration is at a minimum of the potential energy, we have shown that the solutions are all stable. In that case, the f solutions, $p_j^2 = -\omega_j^2$, give the f normal frequencies of oscillation. We may then substitute any p_j^2 in Eqs. (12.23) and solve for the components C_{lj} of the vector $\boldsymbol{C}_j$ (except for an arbitrary factor). The solution may then be obtained as a superposition of normal vibrations, just as in Section 4.10 for two coupled oscillators. In the next section we shall consider an alternative way, utilizing the methods of tensor algebra developed in Chapter 10, of determining the same solution.

12.3 NORMAL MODES OF VIBRATION

If the coordinates $x_1, \ldots, x_f$ are orthogonal, the tensor M will be in diagonal form:

$$M_{kl} = M_k \, \delta_{kl}. \tag{12.25}$$

[The symbol δ_{kl} is defined in Eq. (10.79).] If the coordinates are not orthogonal, we can diagonalize M by the method of Section 10.4, generalized to f dimensions. (We will use the same method below to diagonalize the potential energy.) Let us suppose that this has been done, and that the coordinates $x_1, \ldots, x_f$ are the components of $\boldsymbol{x}$ along the principal axes of M, so that Eq. (12.25) holds. (If $x_1, \ldots, x_f$ are rectangular coordinates of a set of particles, M_k is the mass of the particle whose coordinate is x_k.)

We now define a new vector $\boldsymbol{y}$ with coordinates $y_1, \ldots, y_f$ given by

$$y_k = M_k^{\frac{1}{2}} \, x_k, \qquad k = 1, \ldots, f. \tag{12.26}$$

Note that the configuration of the system is specified now by a vector $\boldsymbol{y}$ in a new vector space related to the x-space by a stretch or compression along each axis, as given by Eq. (12.26). The kinetic energy in terms of $\boldsymbol{y}$ is

$$T = \tfrac{1}{2}\dot{\boldsymbol{y}} \cdot \dot{\boldsymbol{y}} = \sum_{k=1}^{f} \tfrac{1}{2}\dot{y}_k^2. \tag{12.27}$$

Clearly, the expression for the kinetic energy does not change if we rotate the y-coordinate system, which is our reason for introducing the vector $\boldsymbol{y}$.

The potential energy is given by

$$V = \tfrac{1}{2}\boldsymbol{y} \cdot \mathsf{W} \cdot \boldsymbol{y} = \sum_{k,l=1}^{f} \tfrac{1}{2} W_{kl} y_k y_l, \tag{12.28}$$

where

$$W_{kl} = \frac{K_{kl}}{M_k^{1/2} M_l^{1/2}}. \tag{12.29}$$

The equations of motion are

$$\ddot{\boldsymbol{y}} + \mathsf{W} \cdot \boldsymbol{y} = \mathbf{0}. \tag{12.30}$$

The tensor W is symmetric, and can therefore be diagonalized by the method given in Section 10.4. Let $\hat{\boldsymbol{e}}_j$ be an eigenvector of W corresponding to the eigenvalue W_j:

$$\mathsf{W} \cdot \hat{\boldsymbol{e}}_j = W_j \hat{\boldsymbol{e}}_j. \tag{12.31}$$

Let a_{lj} be the components of $\hat{\boldsymbol{e}}_j$ in the y-coordinate system:

$$\hat{\boldsymbol{e}}_j = (a_{1j}, \ldots, a_{fj}), \qquad j = 1, \ldots, f. \tag{12.32}$$

Then we may write Eq. (12.31) in terms of components in a form corresponding to Eqs. (10.108):

$$\sum_{l=1}^{f} (W_{kl} - W_j \delta_{kl}) a_{lj} = 0, \qquad k = 1, \ldots, f. \tag{12.33}$$

Again, the condition for a nonzero solution is

$$\begin{vmatrix} W_{11} - W_j & W_{12} & \cdots & W_{1f} \\ W_{21} & W_{22} - W_j & \cdots & W_{2f} \\ \vdots & \vdots & \cdots & \vdots \\ W_{f1} & W_{f2} & \cdots & W_{ff} - W_j \end{vmatrix} = 0. \tag{12.34}$$

This is an algebraic equation of order f to be solved for the f roots W_j. Note that it is the same as Eq. (12.24) if $p^2 = -W_j$ and we divide the left side of Eq. (12.24) by $M_1 \cdot M_2 \ldots M_f$, remembering that M_{kl} is now given by Eq. (12.25). Each root W_j is to be substituted in Eq. (12.33), which may then be solved for the ratios $a_{1j}:a_{2j}: \cdots :a_{fj}$. The a_{lj} can then be determined so that $\hat{\boldsymbol{e}}_j$ is a unit vector:

$$\sum_{l=1}^{f} a_{lj}^2 = 1. \tag{12.35}$$

The proofs given in Section 10.4 can be extended to spaces of any number of dimensions, so we know that the roots W_j are real, and therefore the coefficients a_{lj} are also real. Moreover, the unit vectors $\hat{\boldsymbol{e}}_j$, $\hat{\boldsymbol{e}}_l$ are orthogonal* for $W_j \neq W_l$. We have therefore

$$\hat{\boldsymbol{e}}_j \cdot \hat{\boldsymbol{e}}_r = \delta_{jr}, \tag{12.36}$$

or

$$\sum_{l=1}^{f} a_{lj} a_{lr} = \delta_{jr}. \tag{12.37}$$

In the case of degeneracy, when two or more roots W_j are equal, we can still choose the a_{lj} so that the corresponding $\hat{\boldsymbol{e}}_j$ are orthogonal. The situation is precisely analogous to that described in Section 10.4, except that for $f > 3$ it cannot be visualized geometrically. The proof of lemma (10.125) can be generalized to multiple degeneracies in spaces of any number of dimensions.

Now let the components of the configuration vector $\boldsymbol{y}$ along $\hat{\boldsymbol{e}}_1, \ldots, \hat{\boldsymbol{e}}_f$ be $q_1, \ldots, q_f$:

$$\boldsymbol{y} = \sum_{j=1}^{f} q_j \hat{\boldsymbol{e}}_j. \tag{12.38}$$

---

*It is customary to use the term "orthogonal" rather than "perpendicular" in abstract vector algebra when the vectors have only an algebraic, and not necessarily a geometric, significance.

In terms of components in the original y-coordinate system,

$$y_k = \sum_{j=1}^{f} a_{kj}q_j. \tag{12.39}$$

Conversely, by dotting $\hat{\mathbf{e}}_r$ into Eq. (12.38) and using Eqs. (12.32) and (12.36), we obtain:

$$q_r = \sum_{k=1}^{f} a_{kr}y_k. \tag{12.40}$$

These equations are analogous to Eqs. (10.67) and (10.69).

The potential energy in the coordinate system $q_1, \ldots, q_f$, which diagonalizes **W**, is

$$V = \sum_{j=1}^{f} \tfrac{1}{2}W_j q_j^2. \tag{12.41}$$

Let us consider the case when V is a minimum at the origin $\mathbf{y} = \mathbf{0}$. The eigenvalues $W_1, \ldots, W_f$ must all be positive; otherwise for some values of $q_1, \ldots, q_f$, V would be negative. If V were not a minimum, some of the eigenvalues W_j would be negative. (The special case $W_j = 0$ may or may not correspond to a minimum, depending on higher-order terms which we have neglected.) Let us set

$$W_j = \omega_j^2. \tag{12.42}$$

The kinetic energy (12.27) in this case is

$$T = \sum_{j=1}^{f} \tfrac{1}{2}\dot{q}_j^2. \tag{12.43}$$

In view of Eqs. (12.41) and (12.43), the Lagrange equations separate into equations for each coordinate q_j:

$$\ddot{q}_j + \omega_j^2 q_j = 0, \qquad j = 1, \ldots, f. \tag{12.44}$$

The coordinates q_j are called the *normal coordinates*. The solution is

$$q_j = A_j \cos \omega_j t + B_j \sin \omega_j t, \qquad j = 1, \ldots, f, \tag{12.45}$$

where A_j, B_j are arbitrary constants. We may write the solution in terms of the original coordinates, using Eqs. (12.26) and (12.39):

$$x_k = M_k^{-1/2} \sum_{j=1}^{f} a_{kj}(A_j \cos \omega_j t + B_j \sin \omega_j t). \tag{12.46}$$

The coefficients are

$$A_j = q_j(0) = \sum_{k=1}^{f} a_{kj}M_k^{1/2}x_k(0) \tag{12.47}$$

and

$$B_j = \omega_j^{-1}\dot{q}_j(0) = \sum_{k=1}^{f} \omega_j^{-1} a_{kj} M_k^{1/2} \dot{x}_k(0). \tag{12.48}$$

We therefore have the complete solution for small vibrations about a point of stable equilibrium.

When the number of degrees of freedom is large, solving Eq. (12.34) may be a formidable job which, in general, can be done only numerically for numerical values of the coefficients. However, in some cases we may know some of the roots beforehand (often we know that certain normal frequencies are zero), or from symmetry considerations we may know that certain roots are equal. Any such information helps in factoring Eq. (12.34).

If V is not a minimum at $x_1 = \cdots = x_f = 0$, and some of the coefficients W_j are negative, then we obtain exponential-type solutions. This proves that the motion is unstable in this case, since the solution (except for very special initial conditions) will contain terms which increase exponentially with time, at least until the linear approximation we have made in the equations of motion is no longer valid. The case when some W_j is zero will not be discussed in detail here. In the linear approximation we are making, the corresponding $\dot{q}_j$ is constant in that case, and this corresponds to what was called neutral equilibrium in Chapter 2. The motion will proceed at constant $\dot{q}_j$ until q_j is large enough so that nonlinear terms in q_j must be considered.

It may be noted that in finding the normal coordinates, we have found a transformation from coordinates $x_1, \ldots, x_f$ to $q_1, \ldots, q_f$ which simultaneously diagonalizes two tensors $\mathbf{M}$ and $\mathbf{K}$, or more correctly, which simultaneously diagonalizes two quadratic forms, T and V. Unless two tensors have the same principal axes, it is of course impossible simultaneously to diagonalize them by a rotation of the coordinate system. However, if the coordinate system is allowed to stretch or compress along chosen axes, as in the transformation (12.26), then we can bring two quadratic expressions to diagonal form simultaneously (provided that at least one is positive or negative definite). We first find the principal axes of the first tensor. By stretching and compressing along the principal axes, we can reduce this to a constant tensor (provided that the eigenvalues are all positive or all negative). In the case above, we reduced $\mathbf{M}$ to $\mathbf{1}$ with the transformation (12.26). Since all axes are principal axes for a constant tensor, the principal axes of the second tensor, as modified by the stretching of coordinates, will reduce both tensors to diagonal form. The reader will find it instructive to give a geometrical interpretation of this procedure, in the case of tensors in two or three dimensions, by representing each tensor by its associated quadric curve or surface, just as the inertia tensor was represented in Section 10.5 by the inertia ellipsoid. When we are dealing with vectors and tensors in physical space, we ordinarily do not consider nonuniform stretching of axes because this distorts the geometry of the space. When we deal with an abstract vector space, we may consider any transformation which is convenient for the algebraic purpose at hand.

12.4 FORCED VIBRATIONS, DAMPING

We now wish to determine the motion of the system considered in the preceding section when it is subject to prescribed external forces $F_1(t), \ldots, F_f(t)$ acting on the coordinates $x_1, \ldots, x_f$. We will again restrict our consideration to motions which remain close enough to a stable equilibrium configuration so that only linear terms in $x_1, \ldots, x_f$ need to be included in the equations of motion. If we introduce the vector

$$\boldsymbol{F}(t) = (F_1, \ldots, F_f), \tag{12.49}$$

we may write the equations of motion in the abbreviated form,

$$\mathsf{M} \cdot \ddot{\boldsymbol{x}} + \mathsf{K} \cdot \boldsymbol{x} = \boldsymbol{F}(t), \tag{12.50}$$

where we have simply added the forces $\boldsymbol{F}(t)$ to Eq. (12.20). Note that Eq. (12.50) may be obtained from the Lagrangian function

$$L = T - V - V', \tag{12.51}$$

where T and V are given by Eqs. (12.11) and (12.13), and

$$V' = -\sum_{k=1}^{f} x_k F_k(t). \tag{12.52}$$

Again suppose that the coordinates $x_1, \ldots, x_f$ are chosen to be orthogonal so that M is diagonal. If the coordinates x_k are not initially orthogonal, and a rotation of the coordinate system is performed to principal axes of M, then the components $F_k(t)$ must be subject to the same transformation as the coordinates x_k. Since we shall follow this process through in the case where we diagonalize the tensor K, we shall not follow it through in detail for M, but simply assume that, if necessary, it has been carried out and that M is diagonal.

We transform now to the normal coordinates found in the preceding section [Eqs. (12.26), (12.39), and (12.40)]:

$$x_k = \sum_{j=1}^{f} M_k^{-1/2} a_{kj} q_j, \tag{12.53}$$

$$q_j = \sum_{k=1}^{f} M_k^{1/2} a_{kj} x_k. \tag{12.54}$$

The generalized forces $Q_j(t)$ associated with $F_k(t)$ are obtained by using Eq. (9.30).

$$Q_j(t) = \sum_{k=1}^{f} M_k^{-1/2} a_{kj} F_k(t). \tag{12.55}$$

The inverse transformation is

$$F_k(t) = \sum_{j=1}^{f} M_k^{1/2} a_{kj} Q_j(t). \tag{12.56}$$

The reader may also verify Eqs. (12.55) by substituting Eqs. (12.53) in Eq. (12.52) and calculating

$$Q_j = -\frac{\partial V'}{\partial q_j}. \tag{12.57}$$

In normal coordinates,

$$V' = -\sum_{j=1}^{f} q_j Q_j(t), \tag{12.58}$$

and the equations of motion are

$$\ddot{q}_j + \omega_j^2 q_j = Q_j(t), \qquad j = 1, \ldots, f. \tag{12.59}$$

Each of these equations is identical in form with Eq. (2.86) for the undamped forced harmonic oscillator ($b = 0$). Therefore the normal modes behave like independent forced oscillators, and the solution can be obtained by the methods described in Chapter 2.

It is tempting to try to generalize our results to the case when linear damping forces are also present. We can easily write down the appropriate equations. In the general case when the coordinates are not orthogonal and there is frictional coupling between coordinates, the equations of motion will be

$$\sum_{l=1}^{f} (M_{kl}\ddot{x}_l + B_{kl}\dot{x}_l + K_{kl}x_l) = 0, \qquad k = 1, \ldots, f, \tag{12.60}$$

or, in vector form,

$$\mathbf{M}\cdot\ddot{x} + \mathbf{B}\cdot\dot{x} + \mathbf{K}\cdot x = \mathbf{0}. \tag{12.61}$$

The tensor $\mathbf{B}$ is symmetric if the frictional forces obey Newton's third law. This is easy to show if the coordinates x are rectangular coordinates of individual particles or bodies. It can also be shown if the coordinates are orthogonal and if separate coordinates are assigned to each body. A linear transformation to generalized coordinates will then preserve the symmetry of $\mathbf{B}$.

Unfortunately, as the reader may perhaps convince himself with some experimentation, it is generally not possible simultaneously to diagonalize three tensors $\mathbf{M}, \mathbf{B}, \mathbf{K}$ with any linear transformation of coordinates, even if stretching is allowed. The method of the preceding section therefore fails in this case, and there are no ormal coordinates. The situation is not improved by assuming that $x_1, \ldots, x_f$ e orthogonal so that $\mathbf{M}$ is diagonal, or even by assuming that there is no frictional coupling so that $\mathbf{B}$ is diagonal. If we apply the transformations (12.53) and (12.54) which diagonalize T and V, the coordinates q_j are in general still coupled by frictional forces:

$$\ddot{q}_j + \sum_{r=1}^{f} b_{jr}\dot{q}_r + \omega_j^2 q_j = 0, \tag{12.62}$$

where

$$b_{jr} = \sum_{k,l=1}^{f} M_k^{-1/2} M_l^{-1/2} a_{kj} a_{lr} B_{kl}. \tag{12.63}$$

Note that the matrix b_{jr} is not diagonal even if B_{kl} is. There is a special case which sometimes occurs when the frictional forces are proportional to the masses, so that $\mathbf{B} = 2\gamma\mathbf{M}$. The method of Section 12.3 then works, since in the y-coordinate system in which $\mathbf{M} \to \mathbf{1}$, we have $B \to 2\gamma\mathbf{1}$ and the normal coordinates q_j along the principal axes of $\mathbf{W}$ satisfy the separated equations:

$$\ddot{q}_j + 2\gamma\dot{q}_j + \omega_j^2 q_j = 0. \tag{12.64}$$

It may, of course, also happen that in the y-space in which $\mathbf{M}$ becomes $\mathbf{1}$ the transformed tensors $\mathbf{B}$ and $\mathbf{K}$ have the same principal axes, but this would be an unlikely accident. When the damping forces are very small, a perturbation method similar to that which will be developed in the next section can be applied to find an approximate solution in terms of damped normal modes.

Except in these special cases, the problem of damped vibrations can be handled only by direct substitution of a trial solution like (12.21) in the equations of motion (12.60). The secular equation analogous to Eq. (12.24) is then of order $2f$ in p. Each root allows a solution for the vector $\boldsymbol{C}$. If there are complex roots, they occur in conjugate pairs, p, p^*, with corresponding conjugate vectors $\boldsymbol{C}$, $\boldsymbol{C}^*$. The two solutions (12.21) can then be combined to yield a real solution which in the stable case will be damped and oscillatory, and can be called a normal mode. If all $2f$ solutions are combined with appropriate arbitrary constants, the general solution to Eqs. (12.60) can be written. It is clear on physical grounds, since the frictional forces reduce the energy of the system, that the real parts of all roots p must be negative if $V(\boldsymbol{x})$ has a minimum at $\boldsymbol{x} = \mathbf{0}$; the mathematical proof of this statement is a difficult exercise in algebra.

When magnetic forces are present, and a velocity-dependent potential appears in the Lagrangian function, the above treatment is not applicable. Such cases can be treated by the method discussed in Section 12.6.

12.5 PERTURBATION THEORY

It may happen that the potential energy is given by

$$V = V^0 + V', \tag{12.65}$$

where $V^0(x_1, \ldots, x_f)$ is a potential energy for which we can solve the problem of small vibrations about a minimum point $x_1 = \cdots = x_f = 0$, and where $V'(x_1, \ldots, x_f)$ is very small for small values of $x_1, \ldots, x_f$. We will call V^0 the unperturbed potential energy, and V' the perturbation. We expect that the solutions for the potential energy V will approximate those for the unperturbed problem. In this section we shall develop an approximate method of solution based on this idea.

We shall assume that V' is also stationary at $x_1 = \cdots = x_f = 0$, so that

$$\left(\frac{\partial V'}{\partial x_k}\right)_{x_1=\cdots=x_f=0} = 0. \tag{12.66}$$

If this is not the case, it is not difficult to find approximately the values of $x_1^0, \ldots, x_f^0$ for which V is stationary. If we expand about the point $x_1^0, \ldots, x_f^0$, the linear terms in V^0 cancel the linear terms in V'. We leave this as an exercise. The origin of coordinates should then be shifted slightly to $x_1^0, \ldots, x_f^0$. This will alter slightly the quadratic terms in V^0, but these small changes can be included in V'. In any case, we will therefore have an expansion of V around the equilibrium point of the form (12.65), with

$$V^0 = \sum_{k,l} \tfrac{1}{2} K_{kl}^0 x_k x_l, \tag{12.67}$$

$$V' = \sum_{k,l} \tfrac{1}{2} K'_{kl} x_k x_l, \tag{12.68}$$

where the coefficients K'_{kl} are small. The precise criteria that must be satisfied in order that K'_{kl} can be considered small will be developed as we proceed.

We first transform to the normal coordinates $q_1^0, \ldots, q_f^0$ for the unperturbed problem. We then have

$$V^0 = \sum_{j=1}^{f} \tfrac{1}{2} W_j^0 (q_j^0)^2, \tag{12.69}$$

$$V' = \sum_{j,r} \tfrac{1}{2} W'_{jr} q_j^0 q_r^0, \tag{12.70}$$

$$W'_{jr} = \sum_{k,l} M_k^{-1/2} M_l^{-1/2} a_{kj} a_{lr} K'_{kl}, \tag{12.71}$$

where W_j^0 are the roots of the secular determinant (12.34) for the unperturbed problem, and where again we assume, for simplicity, that $x_1, \ldots, x_f$ are orthogonal coordinates. The coefficients W'_{jr} are to be treated as small. The superscript "o" will remind us that the variables q_j^0 are normal coordinates for the unperturbed problem.

The equations of motion for $q_1^0, \ldots, q_j^0$ are

$$\ddot{q}_j^0 + W_j^0 q_j^0 + \sum_{r=1}^{f} W'_{jr} q_r^0 = 0, \qquad j = 1, \ldots, f. \tag{12.72}$$

We see that the diagonal element of $\mathbf{W}'$ adds to the coefficient of q_j^0, while the off-diagonal elements couple the unperturbed normal modes. We expect that if $\mathbf{W}'$ is small, there will be a normal mode of the perturbed problem close to each normal mode of the unperturbed problem, that is, a solution with frequency ω_j near $\omega_j^0 = (W_j^0)^{1/2}$ and for which q_j^0 is large while the remaining q_r^0, $r \neq j$, are small. However, if the tensor $\mathbf{W}^0$ has degenerate eigenvalues, so that two or more of the unperturbed frequencies are equal (or perhaps nearly equal), then we expect that even a small amount of coupling can radically change the motion, as in the case of two coupled oscillators that we worked out in Chapter 4. This insight will help in developing a perturbation method.

If we try to find a normal mode of oscillation by substituting

$$q_j^0 = C_j e^{pt}, \qquad j = 1, \ldots, f, \tag{12.73}$$

in Eqs. (12.72), we obtain

$$(p^2 + W_j^0)C_j + \sum_{r=1}^{f} W'_{jr} C_r = 0. \tag{12.74}$$

Now let us assume that the mode which we seek is close to some unperturbed mode, say $j = 1$. We then set

$$\begin{aligned} p^2 &= -W_1^0 - W'_1, \\ C_1 &= 1 + C'_1, \\ C_j &= C'_j, \qquad\qquad j = 2, \ldots, f. \end{aligned} \tag{12.75}$$

If $W'_1, C'_1, \ldots, C'_f$ are zero, Eq. (12.73) then represents a solution of the unperturbed problem. Hence for the perturbed problem, we assume that W'_1, $C'_1, \ldots, C'_f$ are small. We substitute Eqs. (12.75) in Eqs. (12.74), and collect second-order terms on the right-hand side:

$$-W'_1 + W'_{11} = -\sum_{r=1}^{f} W'_{1r} C'_r + W'_1 C'_1, \tag{12.76}$$

$$(W_j^0 - W_1^0)C'_j + W'_{j1} = -\sum_{r=1}^{f} W'_{jr} C'_r + W'_1 C'_j, \qquad j = 2, \ldots, f. \tag{12.77}$$

When we neglect second-order terms, the first equation gives W'_1:

$$W'_1 \doteq W'_{11}. \tag{12.78}$$

Therefore

$$\omega_1^2 = -p^2 \doteq W_1^0 + W'_{11}. \tag{12.79}$$

Equations (12.77), if we neglect the right members, yield the coefficients C'_j:

$$C'_j \doteq \frac{W'_{j1}}{W_1^0 - W_j^0}, \qquad j = 2, \ldots, f. \tag{12.80}$$

The coefficient C'_1 is not determined; this corresponds to the fact that the normal mode (12.73) may have an arbitrary amplitude (and phase), although it must, of course, be near the amplitude (and phase) $C_1^0 = 1$ which was chosen in Eqs. (12.75) for the unperturbed solution. It will be convenient to require that $C_1, \ldots, C_f$ be the coefficients of a unit vector:

$$\sum_{j=1}^{f} C_j^2 = 1. \tag{12.81}$$

We then obtain the following equation for C'_1:

$$C'_1 = -\tfrac{1}{2} \sum_{j=1}^{f} (C'_j)^2. \tag{12.82}$$

To first order in small quantities,

$$C'_1 \doteq 0. \tag{12.83}$$

By substituting in Eqs. (12.73), multiplying by an arbitrary constant $\frac{1}{2}Ae^{i\theta}$, and superposing the complex conjugate solution, we obtain the first-order approximations to the perturbed normal mode:

$$\begin{aligned} q_1^0 &\doteq A \cos(\omega_1 t + \theta), \\ q_j^0 &\doteq \frac{AW'_{j1}}{W_1^0 - W_j^0} \cos(\omega_1 t + \theta), \qquad j = 2, \ldots, f, \end{aligned} \tag{12.84}$$

where ω_1 is given by Eq. (12.79). We see that the first-order effect of the perturbation is to shift ω_1^2 by the diagonal perturbation coefficient W'_{11} and to excite the other unperturbed modes weakly with an amplitude that is proportional to the perturbation coupling coefficients W'_{j1} and inversely proportional to the differences in unperturbed normal frequencies (squared). This is a physically reasonable result.

We can now formulate more precisely the requirement that $\mathbf{W}'$ be small. In our derivation, we have assumed that

$$W'_1 \ll |W_j^0 - W_1^0|, \qquad j = 2, \ldots, f, \tag{12.85}$$

$$C'_j \ll 1. \tag{12.86}$$

Equations (12.78) and (12.80) show that this is justified if

$$\begin{aligned} W'_{j1} &\ll |W_j^0 - W_1^0|, \qquad j = 2, \ldots, f, \\ W'_{11} &\ll |W_j^0 - W_1^0|, \qquad j = 2, \ldots, f. \end{aligned} \tag{12.87}$$

This is the condition for the validity of formulas (12.79) and (12.84).

First-order approximations to the remaining modes are obtained from these formulas by interchanging the subscript "1" with any other.

The astute reader will note that the condition (12.87) in the diagonal coefficient W'_{11} is necessary only because we have neglected the last term in Eqs. (12.77). Equations (12.77) are easily solved for C'_j even if the last term on the right is included. This allows us to remove the restriction on the size of the diagonal coefficients if we wish. This is also obvious because we can always include any diagonal coefficient in V^0 [Eq. (12.69)]. The normal coordinates for the unperturbed problem are still the same; only the (squared) frequencies are altered by adding additional diagonal terms. However, in the transformation to normal coordinates, diagonal and off-diagonal terms become intermixed, so that unless all terms in $V'(x_1, \ldots, x_f)$ are small the off-diagonal terms of $V'(q_1^0, \ldots, q_f^0)$ are unlikely all to be small.

Conditions (12.87) clearly cannot be satisfied if there is a degeneracy—if, for example, $W_1^0 = W_2^0 = W_3^0$. In that case, as mentioned earlier, we expect that even with very small coupling of the unperturbed modes, any perturbed mode with ω^2 near W_1^0 will show appreciable excitation of all three unperturbed modes. We therefore set

$$p^2 = -W_1^0 - W', \tag{12.88}$$

and assume that only $C_4, \ldots, C_f$ are small, while C_1, C_2, C_3 may all be of order 1. We substitute in Eqs. (12.74) and transpose second-order terms to the right members:

$$\begin{aligned}
(W'_{11} - W')C_1 + W'_{12}C_2 + W'_{13}C_3 &= -\sum_{r=4}^{f} W'_{1r}C_r, \\
W'_{21}C_1 + (W'_{22} - W')C_2 + W'_{23}C_3 &= -\sum_{r=4}^{f} W'_{2r}C_r, \\
W'_{31}C_1 + W'_{32}C_2 + (W'_{33} - W')C_3 &= -\sum_{r=4}^{f} W'_{3r}C_r,
\end{aligned} \tag{12.89}$$

and

$$(W_j^0 - W_1^0)C_j + \sum_{r=1}^{3} W'_{jr}C_r = -\sum_{r=4}^{f} W'_{jr}C_r + W'C_j, \qquad j = 4, \ldots, f. \tag{12.90}$$

If we neglect the right members, then Eqs. (12.89) become a standard three-dimensional eigenvalue problem for the eigenvalue W' and the associated eigenvector (C_1, C_2, C_3). There will be three solutions corresponding to three perturbed normal modes with frequencies $\omega^2 = W_1^0 + W'$ near the degenerate unperturbed frequency. In general, the three roots W' will be different, and so the perturbed modes will no longer be degenerate. The remaining coefficients $C_4, \ldots, C_f$ can be found to a first-order approximation from Eqs. (12.90) by neglecting the right members. We may also require that $\mathbf{C}$ be a unit vector [Eq. (12.81)]. In analogy with Eq. (12.83), this will mean that to first order the three-dimensional vector (C_1, C_2, C_3) should be a unit vector. The case of double or multiple degeneracy of any order must be treated in the same way. Clearly, if the degeneracy is of high order, the first-order perturbation equations [Eqs. (12.89) with right members zero] may be almost as difficult to solve as the exact equations (12.74). When $f \geq 4$, we may have more than one degenerate normal frequency; in that case, the above method can be applied separately to each group of degenerate unperturbed modes to find the perturbed modes.

In cases of approximate degeneracy ($W_1^0 \doteq W_2^0 \doteq W_3^0$) when conditions (12.87) fail for a group of neighboring unperturbed modes, the method of the previous paragraph can also be applied. Equations (12.89) are slightly modified by the addition of small terms, like $W_2^0 - W_1^0$, in the diagonal coefficients. The reader can readily formulate the procedure for himself.

When the first-order approximate solution has been found, the approximate values of the coefficients W'_1, C'_j may be substituted in the right members of Eqs. (12.76), (12.77) [or Eqs. (12.89), (12.90)]. The resulting equations are then solved to find a second-order approximation. If, for example, we substitute Eqs. (12.80) in Eq. (12.76), we obtain the second-order approximation to the frequency correction:

$$W'_1 \doteq W'_{11} + \sum_{r=2}^{f} \frac{(W'_{1r})^2}{W_1^0 - W_r^0}, \tag{12.91}$$

where we have used the fact that $\mathbf{W}'$ is a symmetric tensor. We see that the second-order frequency shift in mode 1 contains a contribution due to coupling with each of the other modes. The modes tend to repel one another in second order; that is, each higher frequency mode ($W_r^0 > W_1^0$) reduces the frequency of mode 1, and each lower frequency mode increases it. The same result was observed in the solution to the problem of two coupled oscillators in Chapter 4. The procedure can be carried out in a straightforward way to successively higher-order approximations, but the labor involved rapidly increases.

To any order of approximation, we may introduce normal coordinates $q_1, \ldots, q_f$ for the perturbed problem by setting

$$q_j^0 = \sum_{r=1}^{f} C_{jr} q_r, \tag{12.92}$$

where C_{jr}, $j = 1, \ldots, f$, are the coefficients for the rth perturbed normal mode found to any order of approximation by the perturbation theory. The vectors $\boldsymbol{C}_r = (C_{1r}, \ldots, C_{fr})$ are, to the given order of approximation, orthogonal unit vectors (or can be made so), as we shall see presently. We may therefore solve Eqs. (12.92) for (to this order of approximation)

$$q_r = \sum_{j=1}^{f} C_{jr} q_j^0. \tag{12.93}$$

From Eqs. (12.92) and (12.73), if $p^2 = -W_r^0 - W'_r = -\omega_r^2$ is the approximate value for the frequency, we see that the approximate solution for q_r must have the time dependence e^{pt}, and hence the general real solution is

$$q_r \doteq A_r \cos \omega_r t + B_r \sin \omega_r t. \tag{12.94}$$

Comparison with Eqs. (12.45) shows that the q_r are (approximate) normal coordinates. The Lagrangian, to the given order of approximation, must therefore be

$$L \doteq \sum_{r=1}^{f} (\tfrac{1}{2}\dot{q}_r^2 - \tfrac{1}{2}\omega_r^2 q_r^2), \tag{12.95}$$

as may also be verified by straightforward substitution of Eqs. (12.92) in Eqs. (12.69), (12.70), and (12.43), to any order of approximation in C_{jr}. Alternatively, we may note that Eqs. (12.74) are just the equations we would obtain if we were to look for an eigenvector $\boldsymbol{C}$ of $\mathbf{W} = \mathbf{W}^0 + \mathbf{W}'$ corresponding to the eigenvalue $-p^2$. Hence the approximate solutions we have obtained for Eqs. (12.74) are also approximate solutions to the problem of diagonalizing $\mathbf{W}$. Equations (12.92) must therefore define approximate normal coordinates for the perturbed motion.

12.6 SMALL VIBRATIONS ABOUT STEADY MOTION

Let a mechanical system be described by coordinates $x_1, \ldots, x_f$, and by a Lagrangian function $L(x_1, \ldots, x_f; \dot{x}_1, \ldots, \dot{x}_f; t)$. If a solution $x_1^0(t), \ldots, x_f^0(t)$ is known, we may look for solutions close to the known solution by defining new coordinates $y_1, \ldots, y_f$:

$$x_k = x_k^0(t) + y_k, \qquad k = 1, \ldots, f. \tag{12.96}$$

We substitute in the Lagrangian L, and expand in powers of $y_1, \ldots y_f; \dot{y}_1, \ldots, \dot{y}_f$. Since $x_1^0(t), \ldots, x_f^0(t)$ satisfy the equations of motion, the reader can readily show that no linear terms in $y_1, \ldots, y_f; \dot{y}_1, \ldots, \dot{y}_f$ occur in L. Terms in L independent of $y_1, \ldots, y_f; \dot{y}_1, \ldots, \dot{y}_f$ do not affect the equations of motion and may be ignored. If we assume that $y_1, \ldots, y_f; \dot{y}_1, \ldots, \dot{y}_f$ are small, and that we may neglect cubic and higher powers of small quantities, L becomes a quadratic function of the new variables. The equations of motion will then be linear in $y_1, \ldots, y_f; \dot{y}_1, \ldots, \dot{y}_f$; $\ddot{y}_1, \ldots, \ddot{y}_f$. However, the coefficients in the equations will, in general, be functions of the time t, and the methods we have so far developed will not suffice to solve them. To develop methods for solving equations with time-varying coefficients is beyond the scope of this book. We will therefore consider only cases when the coefficients in the linearized equations turn out to be constant.

We can guarantee that the coefficients will be constant by restricting ourselves to *steady motions*. Suppose that some of the coordinates $x_1, \ldots, x_f$ are ignorable, that is, do not appear in the Lagrangian function. We will also assume that L does not depend explicitly on t. We define a steady motion as one in which all of the nonignorable coordinates are constant. This definition evidently depends upon the system of coordinates chosen. We should perhaps define steady motion as motion for which, in some coordinate system, the nonignorable coordinates are all constant.

We have seen in Section 9.10 that ignorable coordinates are particularly easy to handle in terms of the Hamiltonian equations of motion. Let us therefore introduce coordinates $x_1, \ldots, x_f$, and corresponding momenta $p_1, \ldots, p_f$, of which $x_{N+1}, \ldots, x_f$ are ignorable. The Hamiltonian function is

$$H = H(x_1, \ldots, x_N; p_1, \ldots, p_N, p_{N+1}, \ldots, p_f). \tag{12.97}$$

In view of Eqs. (9.198), the momenta $p_{N+1}, \ldots, p_f$ are all constant. We have therefore to deal only with $2N$ equations (9.198), which for steady motion reduce to

$$\frac{\partial H}{\partial p_k} = 0, \qquad \frac{\partial H}{\partial x_k} = 0, \qquad k = 1, \ldots, N. \tag{12.98}$$

For given values $p_{N+1}^0, \ldots, p_f^0$, we are to find the solutions, if any, of these equations for $x_1, \ldots, x_N; p_1, \ldots, p_N$. Any such solution $x_1^0, \ldots, x_N^0; p_1^0, \ldots, p_N^0$ defines a steady motion. The ignorable coordinates will all have constant velocities given by

$$\dot{x}_j^0 = \left(\frac{\partial H}{\partial p_j}\right)_0, \qquad j = N+1, \ldots, f, \tag{12.99}$$

where the subscript „0" implies that the derivative is to be evaluated at $x_1^0, \ldots, x_N^0$; $p_1^0, \ldots, p_f^0$.

Given a steady motion, let us choose the origin of the coordinate system so that $x_1^0 = \cdots = x_N^0 = p_1^0 = \cdots = p_N^0 = 0$. In order to look for motions near this steady motion we hold $p_{N+1}, \ldots, p_f$ fixed and expand H in powers of $x_1, \ldots, x_N$, $p_1, \ldots, p_N$, which we regard as small. We may omit any terms which do not depend on $x_1, \ldots, p_N$. Linear terms are absent because of Eqs. (12.98). If we neglect cubic terms in small quantities, H becomes a quadratic function of $x_1, \ldots, x_N$, $p_1, \ldots, p_N$, with constant coefficients. It may be that H separates into a positive definite "kinetic energy," 'T'$(p_1, \ldots, p_N)$, and a "potential energy," 'V'$(x_1, \ldots, x_N)$. In that case, the methods of the preceding sections are applicable. In order to apply these methods, we must express the "kinetic energy," 'T', in terms of $\dot{x}_1, \ldots, \dot{x}_N$, which may be done by solving for $p_1, \ldots, p_N$ the linear equations

$$\dot{x}_k = \frac{\partial \text{`}T\text{'}}{\partial p_k}, \qquad k = 1, \ldots, N. \tag{12.100}$$

The problem is thus reconverted to Lagrangian form, with 'L' $(x_1, \ldots, x_N; \dot{x}_1, \ldots, \dot{x}_N) =$ 'T'$-$'V'. Note, however, that we cannot obtain the correct 'L' simply by substituting $\dot{x}_j^0$ from Eq. (12.99) in the original L. The transition to Hamiltonian form is necessary in order to be able to eliminate the ignorable coordinates from the problem by regarding $p_{N+1}, \ldots, p_f$ as given constants. The "potential energy" 'V' will contain terms involving $p_{N+1}, \ldots, p_f$ from the original kinetic energy, and these will appear with opposite sign in 'L'.

If 'V'$(x_1, \ldots, x_N)$ has a minimum at $x_1 = \cdots = x_N = 0$, then $x_1, \ldots, x_N$, if they are small enough, undergo stable oscillations. We may then say that the given steady motion is stable in the sense that for nearby motions (with the same $p_{N+1}, \ldots, p_f$), the coordinates oscillate about their steady values. These oscillations can be described by normal coordinates, which may be found by the method of Section 12.3.

If the coordinate system we use is a moving one, or if magnetic forces are present and must be described by a velocity-dependent potential (9.166), or if, as often happens, the ignorable coordinates are not orthogonal to the nonignorable coordinates, then cross product terms $x_k p_l$ $(k,l \leq N)$ may appear in H. Thus, in general, the quadratic terms in H have the form

$$H = \sum_{k,l=1}^{N} \left(\tfrac{1}{2}a_{kl}p_k p_l + b_{kl}x_k p_l + \tfrac{1}{2}c_{kl}x_k x_l\right), \tag{12.101}$$

where we may as well assume that

$$a_{lk} = a_{kl}, \qquad c_{lk} = c_{kl}. \tag{12.102}$$

The coefficients a_{kl}, b_{kl}, c_{kl} are functions of the constants $p_{N+1}, \ldots, p_f$ and depend on the particular steady motion whose stability is in question. We no longer have a separation into kinetic and potential energies, and the methods of the preceding sections can no longer be applied. It may be that H as given by Eq. (12.101) is positive (or negative) definite in $x_1, \ldots, x_N, p_1, \ldots, p_N$, in which case we may be sure that the steady motion is stable. For $x_1, \ldots, x_N, p_1, \ldots, p_N$ must remain on a surface of constant H, and if H is positive definite, this surface will be an "ellipsoid" in the $2N$-dimensional phase space.

In any case, we may study the small vibrations about steady motion by solving the linearized equations given by the Hamiltonian function (12.101):

$$\begin{aligned} \dot{x}_k &= \sum_{l=1}^{N} (a_{kl}p_l + b_{lk}x_l), \\ \dot{p}_k &= -\sum_{l=1}^{N} (b_{kl}p_l + c_{kl}x_l), \qquad k = 1, \ldots, N. \end{aligned} \tag{12.103}$$

We could return to a Lagrangian formulation involving N second-order equations in $x_1, \ldots, x_N$, but it is just as easy to deal directly with Eqs. (12.103). Let us look for a normal mode in which all quantities have the same time dependence:

$$x_k = X_k e^{pt}, \qquad p_k = P_k e^{pt}. \tag{12.104}$$

We substitute in Eqs. (12.103) and obtain the $2N$ linear equations

$$\begin{aligned} &\sum_{l=1}^{N} [(b_{lk} - p\,\delta_{kl})X_l + a_{kl}P_l] = 0, \\ &\sum_{l=1}^{N} [c_{kl}X_l + (b_{kl} + p\,\delta_{kl})P_l] = 0, \qquad k = 1, \ldots, N. \end{aligned} \tag{12.105}$$

The determinant of the coefficients must vanish:

$$\begin{vmatrix} b_{11}-p & b_{21} & \cdots & a_{11} & \cdots & a_{1N} \\ b_{12} & b_{22}-p & \cdots & a_{21} & \cdots & a_{2N} \\ \vdots & \vdots & \vdots & \vdots & & \vdots \\ c_{11} & c_{12} & \cdots & b_{11}+p & \cdots & b_{1N} \\ c_{21} & c_{22} & \cdots & b_{21} & \cdots & b_{2N} \\ \vdots & \vdots & \vdots & \vdots & \vdots & \vdots \\ c_{N1} & c_{N2} & \cdots & b_{N1} & \cdots & b_{NN}+p \end{vmatrix} = 0. \tag{12.106}$$

We now note that if p is any root of this equation, so is $-p$. First, let us set $p = -p'$ in the determinant. Now, interchange the upper N rows and the lower N rows. Next, interchange the left N columns with the right N columns. Finally, interchange rows and columns, that is, rotate about the main diagonal. None of these operations changes the value of the determinant (except possibly its sign, which does not matter). We now have the same equation for p' that we originally had for p, so that if p is a root, so is p'. We see therefore that when we expand the determinant (12.106), only even powers of p appear, and we have an algebraic equation of degree N in p^2. If the roots are all negative, as they will be if H in Eq. (12.101) is positive definite, then the normal modes are all stable. Each root $p^2 = -\omega_j^2$ gives two values $p = \pm i\omega_j$. We substitute $p = i\omega_j$ in Eqs. (12.105) and solve for X_{lj}, P_{lj}, which will, in general, be complex. There is, of course, an arbitrary constant which may be chosen in any convenient way. The solutions for $p = -i\omega_j$ will be X_{lj}^*, P_{lj}^*. We substitute in Eqs. (12.104), multiply by an arbitrary constant $A_j e^{i\theta j}$ and superpose the two complex conjugate solutions to obtain the real solution for the normal mode j:

$$\begin{aligned} x_k &= A_j C_{kj} \cos(\omega_j t + \beta_{kj} + \theta_j), \\ p_k &= A_j D_{kj} \cos(\omega_j t + \gamma_{kj} + \theta_j), \end{aligned} \tag{12.107}$$

where

$$\begin{aligned} X_{kj} &= \tfrac{1}{2} C_{kj} e^{i\beta_{kj}}, \\ P_{kj} &= \tfrac{1}{2} D_{kj} e^{i\gamma_{kj}}, \end{aligned} \tag{12.108}$$

and A_j and θ_j are an arbitrary amplitude and phase. The general solution is now a superposition of normal modes:

$$\begin{aligned} x_k &= \sum_{j=1}^{N} A_j C_{kj} \cos(\omega_j t + \beta_{kj} + \theta_j), \\ p_k &= \sum_{j=1}^{N} A_j D_{kj} \cos(\omega_j t + \gamma_{kj} + \theta_j). \end{aligned} \tag{12.109}$$

Note that we cannot represent the above result in terms of normal coordinates $q_1, \ldots, q_N$ linearly related to $x_1, \ldots, x_N$ on account of the phase differences β_{kj}, γ_{kj} which arise because of the cross terms in coordinates and momenta. If the frequencies are all different, it is possible to find a linear transformation connecting the $2N$ variables $x_1, \ldots, x_N; p_1, \ldots, p_N$ with a set of normal coordinates and momenta $\underline{q}_1, \ldots, \underline{q}_N; \underline{p}_1, \ldots, \underline{p}_N$, each of which oscillates at the corresponding normal frequency. Such transformations belong to the theory of canonical transformations of Hamiltonian dynamics, and are beyond the scope of this book.

If any root p^2 of Eq. (12.106) is positive or complex, the corresponding normal mode is unstable, and the coordinates and momenta move exponentially away from their steady values. A root $p^2 = 0$ would correspond to neutral stability. One common case in which roots $p^2 = 0$ arise occurs when an ignorable coordinate has been included among the $x_1, \ldots, x_N$. If x_j is ignorable, and is included in $x_1, \ldots, x_N$, then p_j is constant and may take any value. If we take p_j slightly different from p_j^0 for the initially given steady motion, there is a new steady motion with $\dot{x}_j$ constant and slightly different from $\dot{x}_j^0$. This new motion is given by

$$x_j = A_j + B_j \dot{x}_j t, \qquad \text{all other } x_k = A_k, \tag{12.110}$$

where A_k may be slightly different from x_k^0. This motion corresponds to a normal mode with $p^2 = 0$. It may be that there is an ignorable coordinate x_j, but we have chosen a coordinate system in which x_j does not appear. A root $p^2 = 0$ of Eq. (12.106) will still occur. Since the coordinates actually used will be functions of the ignorable one (among others), in the corresponding normal mode, several or all of the coordinates may exhibit constant velocities. These may be found by substituting in the equations of motion (12.103). In some cases, the algebra required to use explicitly some ignorable coordinate x_j is too formidable, and we may prefer to use a coordinate system in which x_j does not appear. This increases the degree of the secular equation (12.106) by one, but since the extra root is $p^2 = 0$, we know that p^2 will factor out and the remaining equation will have the same degree as if we had ignored x_j.

The case of degeneracy, when a multiple root for p^2 occurs, is more complicated for Eqs. (12.103) than for Eqs. (12.20), where the force is derivable from a potential energy depending only on $x_1, \ldots, x_N$. We can no longer make use of the diagonalization theory for a symmetric tensor to show that for a multiple root p^2, Eqs. (12.105) have a corresponding multiplicity of independent solutions, as we did for the analogous equations (12.23) or (12.33). Sometimes Eqs. (12.105) may have only one independent solution, even when p^2 is a multiple root of Eq. (12.106), and we must look for other forms of solution than (12.104). We will not carry out the algebraic details here,* but the result is that when Eqs. (12.105) do not yield enough independent solutions for X_k, P_k for a multiple root p^2, X_k, P_k should be replaced in Eqs. (12.104) by polynomials in t of degree $(n - 1)$, where n is the multiplicity of the root p^2. The resulting expressions must be substituted in Eqs. (12.103), which then give $2Nn$ relations between the $2Nn$ coefficients in the $2N$ polynomials. These relations can be shown to leave just n arbitrary coefficients, so that the correct number of arbitrary constants are available. Alternatively, we can slightly alter the coefficients in the Hamiltonian (12.101) so that the degeneracy in p^2 is removed, find the solution, and then find its limiting form as the coefficients approach their original values. (See Problem 37, Chapter 2.) When powers of t appear in the solution, it is clear that the solution is not stable even when p^2 is real and negative, but represents an oscillation whose amplitude after a long time will increase as some power of t. Hence degeneracy generally implies instability in the case of Eqs. (12.103). It will be found that multiple roots of Eq. (12.106) ordinarily mark the boundary between real and complex solutions for p^2 in the sense that a small change in some coefficient a_{kl}, b_{kl}, or c_{kl} will split the degeneracy and lead on the one hand to two real, or on the other hand to two complex, roots for p^2, depending on the sense of the change. The situation is closely analogous mathematically to the problem of the damped harmonic oscillator, where a double root for p in

*For a further discussion of problems of small vibrations, see E. J. Routh, *Dynamics of a System of Rigid Bodies*, Advanced Part. New York: Dover Publications, 1955. Chapter 6. A less complete but more elegant treatment using matrix methods is given in R. Bellman, *Stability Theory of Differential Equations*, New York: McGraw-Hill, 1953.

Eq. (2.125) marks the dividing line between the overdamped and underdamped cases and leads to solutions linear in t. In the present case, there is no damping; Eq. (12.106) contains only even powers of p; and if complex conjugate roots for p^2 occur, then the corresponding four roots p have the form $\pm\gamma \pm i\omega$, and some of the solutions grow exponentially. Thus multiple roots can mark the boundary between stable and unstable cases. In the case of Eqs. (12.20), where the forces are not velocity-dependent, the boundary between stability and instability occurs only when some root for p^2 is zero; degenerate negative roots for p^2 always correspond to stable solutions.

We should further remark that even when the roots p^2 are all negative and distinct, so that the solutions of Eqs. (12.103) are all stable, we cannot guarantee that the exact solutions of the nonlinear equations arising from the complete Hamiltonian (12.97) are stable. For vibrations about an equilibrium point when the forces are derivable from a potential energy, we were able to prove that stable solutions of the linearized equations are obtained only around a potential minimum, and that in that case, the solutions are absolutely stable if the amplitude is small enough. We saw above that if the Hamiltonian H is positive (or negative) definite near the steady motion, then we can also show that the solutions are stable. But the solutions of Eqs. (12.103) may all be stable even when H is not positive or negative definite. In that case, all we can say is that, for as long a time as we please, if we start with sufficiently small amplitudes, the solutions of the exact problem given by the Hamiltonian (12.97) will approximate those of the linearized problem given by the Hamiltonian (12.101). This is true because the nonlinear terms can be made as small as we like by making the amplitude sufficiently small, and then their effect on the motion can be appreciable, if at all, only if they are integrated over a very long time. Nevertheless, if we neglect the nonlinear terms, we are prevented from asserting complete stability for all time. Cases are indeed known where the linearized solutions are stable, and yet no matter how near the steady-state motion we begin, the exact solution eventually deviates from the steady-state motion by a large amount. To find the criteria that determine ultimate stability in the general case is perhaps the outstanding unsolved problem of classical mechanics.*

12.7 BETATRON OSCILLATIONS IN AN ACCELERATOR

In a circular particle accelerator, for example a cyclotron, betatron, or synchrotron, charged particles revolve in a magnetic guide field which holds them within a circular vacuum chamber as they are accelerated. Since the particles revolve many times while they are being accelerated, it is essential that the orbits be stable. Since the particle gains only a small energy increment at each revolution, it is permissible to study first the stability of the orbits at constant energy, and then to consider

*It has been shown by V. I. Arnold, *Soviet Math.* **2**, 247, 1961, that, for $N \leq 2$, in a certain sense, in almost all cases when the linear equations (12.103) are stable the exact nonlinear equations also have stable solutions for motions sufficiently close to the steady motion.

separately the acceleration process itself. We will be concerned here only with the stability problem at constant energy E. Let us assume that the magnetic field is symmetrical about a vertical axis, so that we may write, using cylindrical polar coordinates (Fig. 3.22),

$$\boldsymbol{B}(\rho, \varphi, z) = B_z(\rho, z)\hat{\boldsymbol{z}} + B_\rho(\rho, z)\hat{\boldsymbol{\rho}}. \tag{12.111}$$

The magnetic field in a synchrotron or betatron is also a function of time, increasing as the energy increases, but since we are treating E as constant, we also take B as constant. We will suppose that in the median plane $z = 0$, the field is entirely vertical:

$$\boldsymbol{B}(\rho, \varphi, 0) = B_{z0}(\rho)\hat{\boldsymbol{z}}. \tag{12.112}$$

A particle of appropriate energy E may travel in a circle of constant radius $\rho = a(E)$. We call this orbit the *equilibrium orbit*. We are interested in the stability of this orbit; that is, we want to know whether particles near this orbit execute small vibrations about it. Such vibrations are called *betatron oscillations* because the theory was first worked out for the betatron.

The vector potential (Section 9.8) for a magnetic field with symmetry about the z-axis can be taken to be entirely in the φ-direction:

$$\boldsymbol{A} = A_\varphi(\rho, z)\hat{\boldsymbol{\varphi}}, \tag{12.113}$$

so that

$$\begin{aligned}\boldsymbol{B} = \nabla \times \boldsymbol{A} &= \left(\hat{\boldsymbol{z}}\frac{\partial}{\partial z} + \hat{\boldsymbol{\rho}}\frac{\partial}{\partial \rho} + \frac{\hat{\boldsymbol{\varphi}}}{\rho}\frac{\partial}{\partial \varphi}\right) \times \hat{\boldsymbol{\varphi}} A_\varphi(\rho, z) \\ &= \frac{\hat{\boldsymbol{z}}}{\rho}\frac{\partial}{\partial \rho}(\rho A_\varphi) - \hat{\boldsymbol{\rho}}\frac{\partial A_\varphi}{\partial z}.\end{aligned} \tag{12.114}$$

We see from Eq. (12.114) that $\boldsymbol{A}$ is given by

$$A_\varphi(\rho, z) = \rho^{-1}\int_0^\rho \rho B_z(\rho, z)\, d\rho, \tag{12.115}$$

since A_φ must vanish at $\rho = 0$ because of the ambiguity in the direction of $\hat{\boldsymbol{\varphi}}$. Note that $2\pi\rho A_\varphi$ is the magnetic flux through a circle of radius ρ.

The Lagrangian function is given by Eqs. (9.154) and (9.166):

$$L = \tfrac{1}{2}m(\dot{\rho}^2 + \rho^2\dot{\varphi}^2 + \dot{z}^2) + \frac{e}{c}\rho\dot{\varphi}A_\varphi(\rho, z), \tag{12.116}$$

where e, m are the charge and mass of the particle to be accelerated. If the velocity of the particle is comparable with the speed of light, that is, if the kinetic energy is comparable with or greater than mc^2, the relativistic form for the Lagrangian should be used (Problem 23, Chapter 9). The momenta are

$$p_\rho = \frac{\partial L}{\partial \dot{\rho}} = m\dot{\rho},$$

$$p_\varphi = m\rho^2\dot{\varphi} + \frac{e}{c}\rho A_\varphi, \tag{12.117}$$

$$p_z = m\dot{z}.$$

The Hamiltonian function is given by Eq. (9.196) or (9.200):

$$H = \frac{p_\rho^2}{2m} + \frac{p_z^2}{2m} + \frac{[p_\varphi - (e/c)\rho A_\varphi]^2}{2m\rho^2}. \tag{12.118}$$

We see that φ is ignorable and p_φ may be taken to be a given constant. The Hamiltonian function then has the form

$$H = \text{`}T\text{'} + \text{`}V\text{'}, \tag{12.119}$$

with

$$\text{`}T\text{'} = \frac{p_\rho^2 + p_z^2}{2m} = \frac{1}{2}m(\dot{\rho}^2 + \dot{z}^2), \tag{12.120}$$

$$\text{`}V\text{'} = \frac{[p_\varphi - (e/c)\rho A_\varphi]^2}{2m\rho^2}. \tag{12.121}$$

The problem reduces to an equivalent problem of static equilibrium. The steady motions are the equilibrium orbits given by the solutions for ρ, z of the equations

$$\frac{\partial \text{`}V\text{'}}{\partial z} = \frac{e}{c}\rho\dot{\varphi}B_\rho = 0, \tag{12.122}$$

$$\frac{\partial \text{`}V\text{'}}{\partial \rho} = -\rho\dot{\varphi}\left(m\dot{\varphi} + \frac{e}{c}B_z\right) = 0, \tag{12.123}$$

where we have used Eqs. (12.117) and (12.114). The first equation above is satisfied in the median plane $z = 0$, and usually nowhere else ($\rho\dot{\varphi} \neq 0$). The second equation gives

$$\dot{\varphi} = -\frac{e}{mc}B_{z0}(\rho), \tag{12.124}$$

which is equivalent to Eq. (3.299). We may solve Eq. (12.124) for ρ, given $\dot{\varphi}$, or alternatively, for $\dot{\varphi}$ with $\rho = a$, the radius of the equilibrium orbit. Note that the energy $\frac{1}{2}ma^2\dot{\varphi}^2$ of a particle executing the steady motion is less than the total energy H of the particle whose motion we are studying, but the difference is of second order in small quantities for vibrations near the steady motion. According to the development in the preceding section, the two particles should be chosen to have the same p_φ.

We now set

$$\rho = a + x, \tag{12.125}$$

where a is the radius of the orbit for the steady motion. Next, by expanding in powers of x, z, $\dot{x}$, $\dot{z}$, we obtain

$$\text{`}T\text{'} = \tfrac{1}{2}m(\dot{x}^2 + \dot{z}^2), \tag{12.126}$$

$$\text{`}V\text{'} = \tfrac{1}{2}m\omega^2(1-n)x^2 + \tfrac{1}{2}m\omega^2 n z^2, \tag{12.127}$$

where we have set

$$\omega = \dot{\varphi} = -\frac{eB_{z0}(a)}{mc}, \tag{12.128}$$

$$n = -\left(\frac{a}{B_z}\frac{\partial B_z}{\partial \rho}\right)_{z=0,\rho=a}, \tag{12.129}$$

and have used Eqs. (12.114), (12.117), (12.124), and

$$\frac{\partial B_z}{\partial \rho} = \frac{\partial B_\rho}{\partial z}, \tag{12.130}$$

which follows from the fact that

$$\nabla \times \boldsymbol{B} = \boldsymbol{0}, \tag{12.131}$$

if there are no currents in the vacuum chamber. The quantity n is called the *field index*. We see immediately that the motion is stable only if

$$0 < n < 1. \tag{12.132}$$

In a cyclotron, the field is nearly constant at the center, so that $n \ll 1$, and then falls rapidly near the outside edge of the magnet. In a betatron or synchrotron, the magnetic field has a constant value of n and $\boldsymbol{B}$ increases in magnitude as the particles are accelerated so as to keep a constant. We see from Eq. (12.129) that the value of n does not change as B_z is increased provided the shape of the magnetic field as a function of radius does not change, that is, provided $\partial B_z/\partial z$ increases in proportion to B_z.

Since the variables x, z are separated in 'T' and 'V', we can immediately write down the betatron oscillation frequencies. It is convenient to express them in terms of the numbers of betatron oscillations per revolution, ν_x and ν_z:

$$\begin{aligned} \nu_x &= \frac{\omega_x}{\omega} = (1-n)^{1/2}, \\ \nu_z &= \frac{\omega_z}{\omega} = n^{1/2}. \end{aligned} \tag{12.133}$$

If there are imperfections in the accelerator, so that B_z is not independent of φ, the difference between B_z and its average value gives rise to a periodic force acting on the coordinate x. The resulting perturbation of the orbit can be treated by solving the corresponding forced harmonic oscillator equations. If B_ρ is not zero everywhere in the median plane ($z = 0$), vertical forces act which drive the vertical betatron oscillations. In general, such imperfections also lead to variations in the field index n, so that $n = n(\varphi)$, and n for the steady motion becomes a periodic function of time. In alternating gradient accelerators, the field index n is deliberately made to vary periodically in azimuth φ. The solution of this problem is too complex for inclusion here.

12.8 STABILITY OF LAGRANGE'S THREE BODIES

A particular solution of the problem of three bodies moving under their mutual gravitational attractions was discovered by Lagrange. This solution is a steady motion in which the three masses remain at the corners of an equilateral triangle as they revolve around their common center of mass. We wish to investigate the stability of this steady motion. This problem is an example of a rather general class of problems in celestial mechanics concerned with the stability of particular solutions of the equations of motion. When the particular solution is a steady motion, the problem can be treated by the method of Section 12.6.

We will simplify the problem by considering only motions confined to a single plane. There are then six coordinates, two for each particle. A little study shows that there are three ignorable coordinates. Two of them represent rigid translations of the three particles, and may be taken as the cartesian coordinates of the center of mass. The corresponding constant momenta are the components of the total linear momentum. The third ignorable coordinate will represent a rigid rotation of the three particles about the center of mass. The corresponding constant momentum is the total angular momentum. The remaining three nonignorable coordinates will specify the relative positions of the three particles with respect to each other.

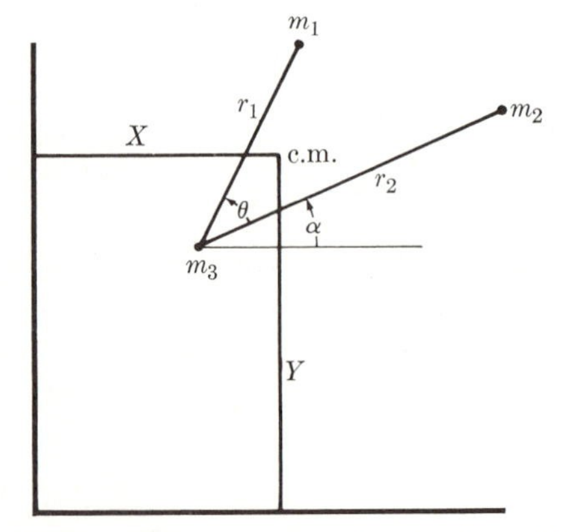

Fig. 12.1 Coordinates for the three-body problem.

These must be constant in a steady motion; therefore any steady motion must be a rigid translation and rotation of the system of three bodies. We may, for example, choose the coordinates as in Fig. 12.1. Here, X and Y are coordinates of the center of mass, variation of α with the remaining coordinates held fixed represents a rotation of the entire system about the center of mass, and r_1, r_2, θ determine the shape and size of the triangle formed by the masses m_1, m_2, m_3. We expect to find three normal modes of vibration of r_1, r_2, θ about their steady values. If we should happen to overlook any ignorable coordinate, we will find only those steady motions in which that coordinate is constant. The ignorable coordinate will then reveal itself as a zero root ($p^2 = 0$) of the secular equation (12.106).

According to Eq. (4.127), the kinetic energy will separate into a part depending on X and Y, and a part depending on r_1, r_2, θ, and α. The potential energy depends only on r_1, r_2, and θ. The coordinates X, Y are therefore orthogonal to r_1, r_2, θ, and α, and the center-of-mass motion separates out of the problem. The center-of-mass energy

$$T_{\text{c.m.}} = \tfrac{1}{2}M(\dot{X}^2+\dot{Y}^2) = \frac{1}{2M}(p_X^2+p_Y^2), \qquad M = m_1+m_2+m_3, \quad (12.134)$$

is constant, and may be omitted from the Hamiltonian. The center of mass moves with constant velocity, and we may study separately the motion relative to the center of mass. The ignorable coordinate α is evidently not orthogonal to θ, since the angular velocity of m_1 involves $(\dot{\theta} + \dot{\alpha})$ and this appears squared in the kinetic energy. This is not an accidental result of our choice of coordinates, but an inherent consequence of the fact that rotation of the system as a whole influences the "internal" motion described by r_1, r_2, θ.

It would be a straightforward algebraic exercise to set up the kinetic energy in terms of r_1, r_2, θ, α, find the momenta $p_{r1}p_{r2}$, p_θ, p_α, set up the Hamiltonian, find the steady motions, and carry through the procedure of Section 12.6 to find the secular equation (12.106), which would be a third-order equation in p^2 whose roots determine the character of small deviations from steady motion. This procedure, however, is extremely tedious, as the reader may verify. It turns out, interestingly enough, that a less laborious way of finding the actual solution is to abandon the Hamilton-Lagrange formalism, and to set up the equations of motion from first

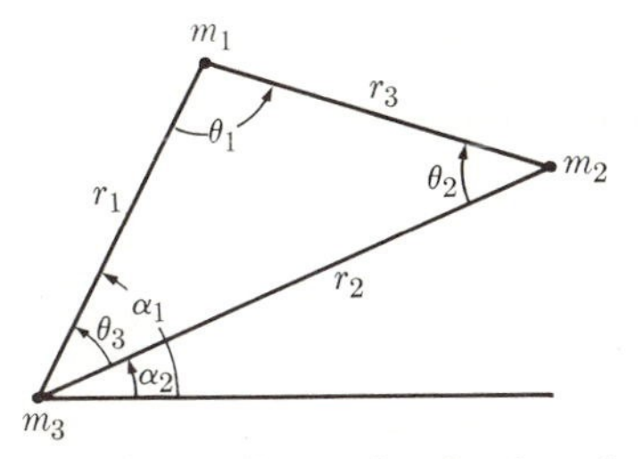

Fig. 12.2 Alternative coordinates for the three-body problem.

principles. We will still need the results of the above general considerations as a guide to the solution. The algebra is still sufficiently involved so that there is a rather high probability of algebraic mistakes. It is therefore desirable to replace θ, α by the coordinates α_1, α_2 shown in Fig. 12.2, so as to introduce an algebraic symmetry between particles m_1 and m_2. This reduces the amount of algebra needed and provides a check on the results, in that our formulas must exhibit the proper symmetry between subscripts "1" and "2." Neither α_1 nor α_2 is ignorable now, and we see that the ignorable coordinate no longer appears explicitly. We could not make use of the ignorable property anyway, since we are not going to write the equations in Hamiltonian form. Our secular equation will turn out to be of fourth order in p^2, but we know that one root will be $p^2 = 0$, and can be factored out. We show also in Fig. 12.2 several auxiliary variables $r_3, \theta_1, \theta_2, \theta_3$ which will be needed.

We will write the equations of motion of m_1 in terms of components directed radially away from m_3 and perpendicular to the radius r_1. In applying Newton's laws of motion directly, we must refer all accelerations to a coordinate system at rest. The acceleration of m_1 is its acceleration relative to m_3 plus the acceleration of m_3. The latter acceleration can be found by applying Newton's law of motion to m_3, which is attracted by m_1 and m_2. The force on m_1 is the gravitational attraction of m_2 and m_3. We have therefore, in the radial direction,

$$m_1\left(\ddot{r}_1 - r_1\dot{\alpha}_1^2 + \frac{m_1 G}{r_1^2} + \frac{m_2 G}{r_2^2}\cos\theta_3\right) = -\frac{m_1 m_3 G}{r_1^2} - \frac{m_1 m_2 G}{r_3^2}\cos\theta_1. \tag{12.135}$$

The corresponding equation for the motion of m_1, perpendicular to r_1, is

$$m_1\left(r_1\ddot{\alpha}_1 + 2\dot{r}_1\dot{\alpha}_1 - \frac{m_2 G}{r_2^2}\sin\theta_3\right) = -\frac{m_1 m_2 G}{r_3^2}\sin\theta_1. \tag{12.136}$$

Two similar equations can be written for the motion of m_2. The four equations can be conveniently rewritten in the form:

$$\begin{aligned}
\ddot{r}_1 - r_1\dot{\alpha}_1^2 + \frac{(m_1+m_3)G}{r_1^2} + \frac{m_2 G}{r_2^2}\cos\theta_3 + \frac{m_2 G}{r_3^2}\cos\theta_1 &= 0,\\
\ddot{r}_2 - r_2\dot{\alpha}_2^2 + \frac{(m_2+m_3)G}{r_2^2} + \frac{m_1 G}{r_1^2}\cos\theta_3 + \frac{m_1 G}{r_3^2}\cos\theta_2 &= 0,\\
r_1\ddot{\alpha}_1 + 2\dot{r}_1\dot{\alpha}_1 - \frac{m_2 G}{r_2^2}\sin\theta_3 + \frac{m_2 G}{r_3^2}\sin\theta_1 &= 0,\\
r_2\ddot{\alpha}_2 + 2\dot{r}_2\dot{\alpha}_2 + \frac{m_1 G}{r_1^2}\sin\theta_3 - \frac{m_1 G}{r_3^2}\sin\theta_2 &= 0.
\end{aligned} \tag{12.137}$$

The algebraic symmetry between subscripts "1" and "2" is exhibited in the above equations. (Note that $\theta_3 = \alpha_1 - \alpha_2$ and changes sign if we interchange the two particles.) The auxiliary variables $\theta_1, \theta_2, \theta_3, r_3$ may be expressed in terms of r_1, r_2,

α_1, α_2 by using the sine and cosine laws for the triangle.

We now see why it is easier to use Newton's laws directly here. We can express the acceleration of m_3 very simply in terms of the gravitational forces on m_3. In the Lagrangian formulation of the corresponding equations for r_1, r_2, α_1, α_2 (or α, θ), the terms which represent the acceleration of m_3 have to be expressed kinematically, that is, in terms of the coordinates, velocities, and accelerations of m_1 and m_2, because they are derived by differentiation of the kinetic energy T in generalized coordinates. This is very complicated, and involves explicitly the position of the center of mass relative to m_1, m_2, m_3, which we do not need in the present formulation. The resulting equations are equivalent to Eqs. (12.137), but considerably more complicated in form. One reason for the simplicity of Eqs. (12.137) is, of course, the use of the auxiliary variables $\theta_1, \theta_2, \theta_3, r_3$; in the Lagrangian formulation, any such auxiliary variables would have to be differentiated with respect to r_1, r_2, α_1, α_2 in order to write down the equations of motion.

We first look for steady motions. We know from our preliminary discussion that a steady motion can only be a rigid rotation about the center of mass (plus a uniform translation). We therefore take $r_1, r_2, r_3, \theta_1, \theta_2, \theta_3$ to be constant, and set

$$\alpha_2 = \omega t, \qquad \alpha_1 = \omega t + \theta_3. \tag{12.138}$$

If we substitute into the last of Eqs. (12.137), we obtain

$$\frac{1}{r_1^2} \sin \theta_3 = \frac{1}{r_3^2} \sin \theta_2. \tag{12.139}$$

From the law of sines,

$$\frac{r_1}{\sin \theta_2} = \frac{r_3}{\sin \theta_3}, \tag{12.140}$$

we then have, unless $\sin \theta_3 = \sin \theta_2 = 0$,

$$r_1^3 = r_3^3. \tag{12.141}$$

In the same way, from the third of Eqs. (12.137), we find $r_2 = r_3$, unless $\sin \theta_3 = \sin \theta_1 = 0$. We may therefore set

$$\begin{aligned} r_1 = r_2 = r_3 &= a, \\ \theta_1 = \theta_2 = \theta_3 &= \frac{\pi}{3}. \end{aligned} \tag{12.142}$$

The only possible steady motion, unless the masses lie in a straight line, is one in which the three masses lie at the corners of an equilateral triangle. We must still verify that the first two of Eqs. (12.137) are satisfied. This is the case if

$$\omega^2 = \frac{MG}{a^3}, \tag{12.143}$$

where M is the total mass. This is the particular solution of the three-body problem found by Lagrange. The case when the three masses lie in a straight line is left as an exercise.

We now seek solutions for motions near the steady motion. Let us set

$$\begin{aligned} r_1 &= a+x_1, & r_2 &= a+x_2, \\ \alpha_1 &= \omega t+\tfrac{1}{3}\pi+\varepsilon_1, & \alpha_2 &= \omega t+\varepsilon_2, \end{aligned} \tag{12.144}$$

where x_1, x_2, ε_1, ε_2 are four new variables which we will regard as small. We substitute in Eqs. (12.137), retaining only linear terms. We first calculate

$$\theta_3 = \tfrac{1}{3}\pi+\varepsilon_1-\varepsilon_2, \tag{12.145}$$

and, to first order, from the law of cosines,

$$r_3^2 = a^2\left[1+\frac{x_1+x_2}{a}+\sqrt{3}\,(\varepsilon_1-\varepsilon_2)\right]. \tag{12.146}$$

Now, from the law of sines,

$$\sin\theta_1 = \frac{r_2}{r_3}\sin\theta_3; \tag{12.147}$$

hence, to first order,

$$\theta_1 = \frac{\pi}{3}-\frac{1}{2}(\varepsilon_1-\varepsilon_2)-\frac{1}{2}\sqrt{3}\,\frac{x_1-x_2}{a}, \tag{12.148}$$

and, similarly,

$$\theta_2 = \frac{\pi}{3}-\frac{1}{2}(\varepsilon_1-\varepsilon_2)+\frac{1}{2}\sqrt{3}\,\frac{x_1-x_2}{a}. \tag{12.149}$$

We note as a check that $\theta_1+\theta_2+\theta_3 = \pi$. We are ready to substitute in Eqs. (12.137), which, to first order, become

$$\begin{aligned} \ddot{x}_1-2a\omega\dot{\varepsilon}_1-\left(\omega^2+\frac{2m_1+2m_3-\frac{1}{4}m_2}{a^3}G\right)x_1-\frac{9m_2G}{4a^3}x_2-\frac{3\sqrt{3}\,m_2G}{4a^2}(\varepsilon_1-\varepsilon_2) &= 0, \\ \ddot{x}_2-2a\omega\dot{\varepsilon}_2-\left(\omega^2+\frac{2m_2+2m_3-\frac{1}{4}m_1}{a^3}G\right)x_2-\frac{9m_1G}{4a^3}x_1-\frac{3\sqrt{3}\,m_1G}{4a^2}(\varepsilon_1-\varepsilon_2) &= 0, \\ a\ddot{\varepsilon}_1+2\omega\dot{x}_1-\frac{3\sqrt{3}\,m_2G}{4a^3}(x_1-x_2)-\frac{9m_2G}{4a^2}(\varepsilon_1-\varepsilon_2) &= 0, \\ a\ddot{\varepsilon}_2+2\omega\dot{x}_2-\frac{3\sqrt{3}\,m_1G}{4a^3}(x_1-x_2)+\frac{9m_1G}{4a^2}(\varepsilon_1-\varepsilon_2) &= 0. \end{aligned} \tag{12.150}$$

Note that the second and fourth equations may be obtained from the first and third by interchanging the subscripts "1" and "2" and reversing the signs of α_1, α_2, and ω; this symmetry follows from the choice of coordinates in Fig. 12.2.

A normal mode is found by setting

$$\begin{aligned} x_1 &= X_1 e^{pt}, & x_2 &= X_2 e^{pt}, \\ \varepsilon_1 &= E_1 e^{pt}, & \varepsilon_2 &= E_2 e^{pt}. \end{aligned} \tag{12.151}$$

For convenience we set

$$p = (G/a^3)^{1/2} P. \tag{12.152}$$

Substituting Eqs. (12.151) in Eqs. (12.150) and using Eq. (12.143), we obtain

$$\begin{aligned}
&\left(P^2 - 3M + \frac{9}{4} m_2\right)\frac{X_1}{a} - \frac{9}{4} m_2 \frac{X_2}{a} - \left(\frac{3\sqrt{3}}{4} m_2 + 2M^{1/2}P\right) E_1 + \frac{3\sqrt{3}}{4} m_2 E_2 = 0, \\
&-\frac{9}{4} m_1 \frac{X_1}{a} + \left(P^2 - 3M + \frac{9}{4} m_1\right)\frac{X_2}{a} - \frac{3\sqrt{3}}{4} m_1 E_1 + \left(\frac{3\sqrt{3}}{4} m_1 - 2M^{1/2}P\right) E_2 = 0, \\
&\left(2M^{1/2}P - \frac{3\sqrt{3}}{4} m_2\right)\frac{X_1}{a} + \frac{3\sqrt{3}}{4} m_2 \frac{X_2}{a} + \left(P^2 - \frac{9}{4} m_2\right) E_1 + \frac{9}{4} m_2 E_2 = 0, \\
&-\frac{3\sqrt{3}}{4} m_1 \frac{X_1}{a} + \left(2M^{1/2}P + \frac{3\sqrt{3}}{4} m_1\right)\frac{X_2}{a} + \frac{9}{4} m_1 E_1 + \left(P^2 - \frac{9}{4} m_1\right) E_2 = 0.
\end{aligned} \tag{12.153}$$

The secular determinant is

$$\begin{vmatrix}
\left(P^2 - 3M + \frac{9}{4} m_2\right) & \left(-\frac{9}{4} m_2\right) & -\left(2M^{1/2}P + \frac{3\sqrt{3}}{4} m_2\right) & \left(\frac{3\sqrt{3}}{4} m_2\right) \\
\left(-\frac{9}{4} m_1\right) & \left(P^2 - 3M + \frac{9}{4} m_1\right) & -\left(\frac{3\sqrt{3}}{4} m_1\right) & -\left(2M^{1/2}P - \frac{3\sqrt{3}}{4} m_1\right) \\
\left(2M^{1/2}P - \frac{3\sqrt{3}}{4} m_2\right) & \left(\frac{3\sqrt{3}}{4} m_2\right) & \left(P^2 - \frac{9}{4} m_2\right) & \left(\frac{9}{4} m_2\right) \\
-\left(\frac{3\sqrt{3}}{4} m_1\right) & \left(2M^{1/2}P + \frac{3\sqrt{3}}{4} m_1\right) & \left(\frac{9}{4} m_1\right) & \left(P^2 - \frac{9}{4} m_1\right)
\end{vmatrix} = 0. \tag{12.154}$$

We know from previous considerations that this must be a fourth-degree equation in P^2, one root of which is $P^2 = 0$. This fact, together with the symmetries in the array of coefficients, encourages us to try to manipulate the above determinant to simplify its expansion and to bring out explicitly the factor P^2. We add the second column to the first, and the third to the fourth, then subtract the first row from the second, and the third from the fourth:

$$\begin{vmatrix} (P^2-3M) & \left(-\frac{9}{4}m_2\right) & -\left(2M^{1/2}P+\frac{3\sqrt{3}}{4}m_2\right) & -(2M^{1/2}P) \\ 0 & \left[P^2-3M+\frac{9}{4}(m_1+m_2)\right] & \left[2M^{1/2}P-\frac{3\sqrt{3}}{4}(m_1-m_2)\right] & 0 \\ (2M^{1/2}P) & \left(\frac{3\sqrt{3}}{4}m_2\right) & \left(P^2-\frac{9}{4}m_2\right) & P^2 \\ 0 & \left[2M^{1/2}P+\frac{3\sqrt{3}}{4}(m_1-m_2)\right] & -\left[P^2-\frac{9}{4}(m_1+m_2)\right] & 0 \end{vmatrix} = 0. \tag{12.155}$$

We factor P from the last column, then multiply the last column by $3M^{1/2}/2$, and subtract from the first column. We can then factor P from the first column, to obtain

$$P^2\begin{vmatrix} P & \left(-\frac{9}{4}m_2\right) & -\left(2M^{1/2}P+\frac{3\sqrt{3}}{4}m_2\right) & -(2M^{1/2}) \\ 0 & \left[P^2-3M+\frac{9}{4}(m_1+m_2)\right] & \left[2M^{1/2}P-\frac{3\sqrt{3}}{4}(m_1-m_2)\right] & 0 \\ \left(\frac{1}{2}M^{1/2}\right) & \left(\frac{3\sqrt{3}}{4}m_2\right) & \left(P^2-\frac{9}{4}m_2\right) & P \\ 0 & \left[2M^{1/2}P+\frac{3\sqrt{3}}{4}(m_1-m_2)\right] & -\left[P^2-\frac{9}{4}(m_1+m_2)\right] & 0 \end{vmatrix} = 0. \tag{12.156}$$

The factor P^2 is now in evidence. To simplify the expansion of the determinant, we multiply the third row by $2PM^{-1/2}$ and subtract from the first row:

$$P^2\begin{vmatrix} 0 & * & * & -(2M^{1/2}+2P^2M^{-1/2}) \\ 0 & \left[P^2-3M+\frac{9}{4}(m_1+m_2)\right] & \left[2M^{1/2}P-\frac{3\sqrt{3}}{4}(m_1-m_2)\right] & 0 \\ \frac{1}{2}M^{1/2} & * & * & * \\ 0 & \left[2M^{1/2}P+\frac{3\sqrt{3}}{4}(m_1-m_2)\right] & -\left[P^2-\frac{9}{4}(m_1+m_2)\right] & 0 \end{vmatrix} = 0. \tag{12.157}$$

The stars indicate terms that we have not bothered to calculate, since they will not appear in the result. We can now expand in minors of the first column, and expand the resulting three-rowed determinant in minors of the last column, with the final result:

$$P^2(M+P^2)[P^4+MP^2+\tfrac{27}{4}(m_1m_2+m_2m_3+m_3m_1)] = 0. \tag{12.158}$$

Note, as a check on the algebra, that the three masses enter symmetrically in this equation, as they must. It is a fortunate accident that an additional factor appears explicitly, so that we have only to solve a quadratic equation for P^2. We have, finally, four roots:

$$P^2 = 0, \qquad P^2 = -M,$$
$$P^2 = -\tfrac{1}{2}M \pm \tfrac{1}{2}[M^2 - 27(m_1m_2 + m_2m_3 + m_3m_1)]^{1/2}. \tag{12.159}$$

As we know, the zero root results from the fact that there is an additional ignorable coordinate. The root $P^2 = -M$ yields a stable oscillatory mode. The last two roots for P^2 will both be real and negative provided that

$$(m_1 + m_2 + m_3)^2 > 27(m_1m_2 + m_2m_3 + m_3m_1). \tag{12.160}$$

If this inequality is reversed, the last two roots are complex, and we have four complex values of P. Of these roots two give rise to damped and two to antidamped oscillatory solutions. In the intermediate case when the two members of the inequality (12.160) are equal, it can be shown that the amplitude of oscillation grows linearly in time. Hence the Lagrangian motion of three bodies is unstable when condition (12.160) is not satisfied. If one of the bodies, say m_1, is much smaller than the other two, we have the restricted problem of three bodies studied in Section 7.6, and the condition for stability reduces to

$$(m_2 + m_3)^2 > 27m_2m_3, \tag{12.161}$$

or, if m_3 is the largest,

$$m_3 > 24.96m_2. \tag{12.162}$$

If m_3 is the sun and m_2 is the planet Jupiter, this condition is satisfied; hence, if we neglect the effects of all other planets, there are stable steady motions in which a small body revolves around the sun with the same period as Jupiter and at the corner of an equilateral triangle relative to the sun and Jupiter. (There are two such positions.) The Trojan asteroids are a group of bodies with the same period as Jupiter which appear to be in this position. Since Eq. (12.161) is also satisfied by the earth-moon system, the corresponding steady motion of an artificial satellite in the earth-moon system is stable. Consideration of motions perpendicular to the plane of steady motion does not alter these conclusions. However, in view of the remarks at the end of Section 12.6, our conclusions about the stability may be valid only for limited periods of time.

We leave as an exercise the solution of Eqs. (12.153), to determine the ratios of the variables and hence the oscillation pattern for each normal mode (see Problem 30). In solving Eqs. (12.153), it may be helpful to subject them to the same series of manipulations which led from the determinant (12.154) to the determinant (12.157). Note that adding one row of the determinant to another corresponds to adding the corresponding equations. Adding two columns corresponds to grouping the corresponding variables, that is, to introducing a new variable which is the sum of the two original ones.*

---

*A more extensive and more advanced treatment of the three-body problem can be found in C. L. Siegel, *Vorlesungen über Himmelsmechanik*, Berlin: Springer-Verlag, 1956.

PROBLEMS

1. Find the transformation to normal coordinates for the two coupled oscillators shown in Fig. 4.10.

2. Solve Problem 40, Chapter 4, by transforming from x_1, x_2 to normal coordinates by the method of Section 12.3.

3. A mass m moving in space is subject to a force whose potential energy is

$$V = V_0 \exp\,[(5x^2+5y^2+8z^2-8yz-26ya-8za)/a^2],$$

where the constants V_0 and a are positive. Show that V has one minimum point. Find the normal frequencies of vibration about the minimum.

4. A mass m is hung from a fixed support by a spring of constant k whose relaxed length is $l = 2mg/k$. A second equal mass is hung from the first mass by an identical spring. Find the *six* normal coordinates and the corresponding frequencies for small vibrations of this system from its equilibrium position. Each spring exerts a force only along the line joining its two ends, but may pivot freely in any direction at its ends. Neglect the mass of the springs.

5. An ion of mass m, charge q, is held by a linear attractive force $F = -kr$ to a point A, where r is the distance from the ion to the point A. An identical ion is similarly bound to a second point B a distance l from A. The two ions move (in three-dimensional space) under the action of these forces and their mutual electrostatic repulsion. Find the normal modes of vibration, and write down the most general solution for small vibrations about the equilibrium point.

6. The mass m_2 in Fig. 4.10 is subject to a force $F_2 = B \sin \omega t$. The system is at rest at $t = 0$. Find the motion by the method of normal coordinates, using the result of Problem 1.

7. The pair of ions in Problem 5 is subject to a plane polarized electromagnetic wave incident perpendicular to the line $\overline{AB}$, whose electric field $E_0 \cos \omega t$ is directed at 45° to the line $\overline{AB}$. Find the steady-state motion.

8. The mass in Problem 3 is subject to a force

$$F_x = F_y = F_z = Be^{-at}.$$

Find a particular solution.

9. Set up the tensors **M**, **B**, **K** for Problem 39, Chapter 4, show that all three can be simultaneously diagonalized, and solve the problem by the method of normal coordinates.

10. The masses m_1 and m_2 in Fig. 4.10 are subject to frictional forces $-\gamma m_1 \dot{x}_1$, $-\gamma m_2 \dot{x}_2$, respectively. Find the general solution.

11. Assume that $V^0(x_1, \ldots, x_f)$ has a minimum at $x_1 = \cdots = x_f = 0$, and that $V'(x_1, \ldots, x_f)$ is small, but that Eq. (12.66) does not necessarily hold. Find approximate expressions to first order in V' and its derivatives at $x_1 = \cdots = x_f = 0$, for the coordinates $x_1^0, \ldots, x_f^0$ of the new equilibrium point for $V = V^0 + V'$. Assuming the expansion of V^0 about $x_1 = \cdots = x_f = 0$ is given by Eq. (12.67) (plus higher-order terms) and that the quadratic terms in the expansion of V are to be

$$V = \sum_{k,l} \tfrac{1}{2}(K^0_{kl} + K'_{kl})y_k y_l,$$

where $y_k = x_k - x_k^0$, find approximate first order expressions for the coefficients K'_{kl}.

12. Find the second-order approximations for the coefficients C'_j, C'_1, which are given to first order by Eqs. (12.80) and (12.83).

13. Find the third-order approximation to the frequency correction given to second order by Eq. (12.91).

14. Formulate the equations to be solved to obtain a first-order approximation in the case when conditions (12.87) fail for a group of four nearby modes, that is, the case of approximate degeneracy.

15. A triple pendulum is formed by suspending a mass M by a string of length l from a fixed support. A mass m is hung from M by a string of length l, and from this second mass a third mass m is hung by a third string of length l. The masses swing in a single vertical plane. Set up the equations for small vibrations of the system, using as coordinates the angles $\theta_1, \theta_2, \theta_3$ made by each string with the vertical. Show that if $M \ggg m$, the normal coordinates can be found if terms of order $(m/M)^{1/2}$ are neglected. Find the approximate normal frequencies to order m/M. [*Hint:* Transform $\mathbf{K}$ to a constant tensor, and diagonalize $\mathbf{M}$.]

16. In Fig. 12.3, the four masses move only along a horizontal straight line under the action of four identical springs of constant k, and a weak spring of constant $k' \ll k$. Find, to first order in k', an approximate solution for the normal modes of vibration.

17. Find the approximate solution to Problem 16 for the case when the masses are all equal. Does the approximate result suggest a way to solve the problem exactly?

18. A uniform ellipsoid of revolution of mass M, whose axis of symmetry is two thirds as long as its equatorial diameter, is modified by placing masses m, $2m$, $3m$, m, $2m$, $3m$ in sequence around its equator at points 60° apart. Two masses, each $4m$, are placed at opposite ends of a diameter, making an angle of 45° with the axis and at the longitude of the masses m. If $m \ll M$, find, to first order in m/M, the new principal axes. [*Hint*: The perturbation procedure developed in Section 12.5 for diagonalizing the tensor $\mathbf{W}$ may be applied to diagonalize approximately any symmetric tensor $\mathbf{I}^\circ + \mathbf{I}'$ if the eigenvectors of $\mathbf{I}^\circ$ are known and $\mathbf{I}'$ is small.]

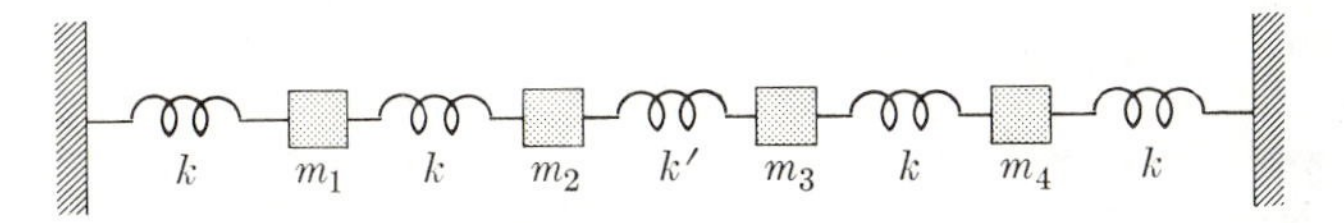

Fig. 12.3 Four coupled harmonic oscillators.

19. Apply the perturbation method to the problem of the string with variable density considered in the last paragraph of Section 9.9, assuming that $a \ll \sigma_0$. Find the lowest normal frequency to second order in a, and write out the corresponding solution $u(x, t)$ to first order in a.

20. Formulate a first-order perturbation method of solving Eqs. (12.60), treating the friction as a small perturbation, and assuming the solution without friction is already known. Show why, even in first order, one cannot introduce normal coordinates which include the effects of friction.

21. Two charges $+Ze$ are located at fixed points a distance $2a$ apart. An electron of mass m, charge $-e$, moves in the field of these charges. Find the steady motions and the small vibrations about the steady motions.

***22.** Two charges $+Ze$ and $-Ze$ are located at the fixed points $z = a$ and $z = -a$. An electron of mass m, charge $-e$, moves in the field of these charges. Sketch a graph of z vs. r, where r is the distance from the z-axis, showing the values of z, r for which there are steady motions. Investigate the stability of these steady motions.

23. A mass m slides without friction on a smooth horizontal table. It is tied to a weightless string of total length l which passes through a hole in the table and is tied at its lower end to a mass M which hangs below the table. Set up the Hamiltonian function using as coordinates the polar coordinates r, α of the mass m relative to the hole, and the spherical angles θ, φ of the mass M relative to the hole. Find the steady motions and the normal frequencies of small vibrations about a steady motion.

24. Assume that the three masses in Fig. 4.16 are free to move in a plane but are constrained to remain in a straight line relative to one another. Choose your coordinates so that as many as possible will be ignorable, find the steady motions, and find the normal modes of vibration about them.

25. A symmetrical rigid body is mounted in weightless, frictionless gimbal rings. A hairspring is attached to one of the rings so as to exert a restoring torque $-k\phi$ about the z-axis, where ϕ is the Euler angle. Find the steady motions and investigate the character of small vibrations about them.

26. In Problem 13, Chapter 11, a hairspring is connected between the disk axle and the rings which exerts a restoring torque $-k\psi'$, where ψ' is the relative angle of rotation between disk and rings. The "gyroscope" moves freely in space with no external forces. Find the steady motions and investigate the small vibrations about them.

27. Two masses m are connected by a rigid weightless rod of length $2l$. One mass is connected with the origin by a spring of constant k, the other by a spring of constant $2k$. The relaxed length of both springs is zero. The masses move in a single plane. Choose as coordinates the polar coordinates r, θ of the center of mass relative to the origin, and the angle α which the rod makes with the radius from the origin to the center of mass, taking $\alpha = 0$ when the stronger spring is stretched least. Find the steady motions and the conditions under which they are stable.

***28.** In Problem 23, a second mass m slides without friction on the table, and is connected to the first mass by a rigid weightless rod of length a. (Assume that the arrangement is ingeniously contrived so that the rod and string do not become entangled.) Assume also that $M = 2m$. Use as an additional coordinate the angle β between the rod and the string. Find the steady motions, and determine which are stable. Find the normal vibrations about the stable steady motions.

29. The vector potential due to a magnetic dipole of magnetic moment μ is, in spherical coordinates relative to the dipole axis,

$$A = \frac{\mu \sin \theta}{2\pi r^2} \hat{\varphi}.$$

Find the steady motions for a charged particle moving in such a field, and show that they are unstable.

30. Find the solution of Eqs. (12.153) for X_1, X_2, E_1, E_2, when $P^2 = -M$, and describe the corresponding oscillation. [See the hint in the last paragraph of Section 12.8.]

***31.** Analyze the case which was omitted in Section 12.8 when the three bodies m_1, m_2, m_3 lie in a straight line. Show that there are three possible steady motions, one for each mass lying between the other two. [*Hint*: You will need Descartes' rule of signs.] Show that motions near each of these steady motions are unstable. Compare your results with those of Section 7.6 and Problem 19 of Chapter 7.

32. Find the solution of Eqs. (12.153) for the double root $P^2 = -\frac{1}{2}M$, when the inequality (12.160) becomes an equality. Show that in this case, Eqs. (12.150) have a second solution in which X_1, X_2, E_1, E_2 are certain linear functions of t, say $X_1 = X_1' + X_1''t$, etc. [You can simplify the algebra a little by assuming that one of the additive constants, say X_1', is zero. This is allowable, since X_1' can always be made zero by subtracting from the second solution a suitable multiple of the first solution you found in which X_1 is constant. The linearity of the equations permits linear superposition of solutions.]

33. Find the solution of Eqs. (12.153) for X_1, X_2, E_1, E_2, for the root $P^2 = 0$, and show that it corresponds to a new steady motion near the chosen one. Since your solution has only one arbitrary constant, there must be another solution of Eqs. (12.150) corresponding to $P^2 = 0$. Guess its form, and verify by substitution.

***34.** Show that if motions of Lagrange's three bodies out of the plane of the steady motion are considered, at least one of the three additional coordinates is ignorable. Choose as two non-ignorable coordinates the distances $q_1 = z_1 - z_3$ and $q_2 = z_2 - z_3$, where z_i is the perpendicular distance of m_i from the plane of steady motion. Set up the linearized equations of motion by the method used in Section 12.8. Solve for the corresponding normal vibrations and show that the result can be interpreted as corresponding simply to a small change in the orientation of the plane of steady motion.

CHAPTER 13

BASIC POSTULATES OF THE SPECIAL THEORY OF RELATIVITY

13.1 THE POSTULATES OF THE SPECIAL THEORY OF RELATIVITY*

We have seen in Section 7.1 that Newton's laws of motion have the property that if they hold in one coordinate system, then they hold also in any other coordinate system moving at a constant velocity with respect to the first. The equations of motion have the same form in both coordinate systems (with the same forces on the right-hand side). Einstein's postulate of special relativity asserts that all the laws of physics must have this same property:

Postulate of Special Relativity. *Every law of physics must be such that if it holds in any coordinate system, it holds also in any other coordinate system moving at a constant velocity with respect to the first.* (13.1)

This postulate implies that there is a special set of coordinate systems, moving uniformly with respect to one another, in any one of which the laws of physics are valid. Such systems are called *inertial coordinate systems.* In a coordinate system accelerated relative to an inertial system, the laws of physics do not necessarily hold, at least in the same form. Postulate (13.1) leaves it to experiment to determine which coordinate systems are inertial. Given a suitable physical law, for example Newton's first law of motion, the law of inertia, we are to determine experimentally in which coordinate system the law holds.

To most people the postulate of relativity is intuitively appealing. From the operational point of view, physical quantities like position and velocity must be defined by specifying the way they are measured. We measure the position and velocity of a body with respect to some other body or with respect to a coordinate system whose origin and axes are located with reference to certain bodies. Then position and velocity have no meaning except relative to some other body or relative to some coordinate system. From this point of view, all coordinate systems should be equivalent. While this argument makes the postulate of relativity plausible, it does not prove its validity. Indeed if some law of physics were to hold in one inertial coordinate system and not in the others, that very fact would give us an operational way of distinguishing between coordinate systems, just as the law of inertia allows us to distinguish inertial coordinate systems from accelerated systems. We could then define an absolute velocity as the velocity with respect to the coordinate system in which that particular law of physics holds. It is also clear that, to the extent that the postulate of special relativity is intuitively plausible, a more general

*The reader may wish to review Sections 1.4, 7.1, and 7.2 in preparation for reading this chapter.

postulate is equally plausible, namely that the laws of physics should be the same in all coordinate systems no matter how they move. This idea leads to the general theory of relativity, which is considerably more difficult mathematically and which is less well established experimentally. We will confine ourselves in this chapter and the next to the special theory of relativity, except for some remarks in the last section.

Note that the postulate of relativity, like the laws of thermodynamics, is a general principle applying to all the laws of physics. We must examine each presently accepted or proposed physical law to determine whether it agrees with the postulate, and if it does not, we must find a suitable modification which is in agreement with the postulate. Einstein in his first paper* succeeded in doing this with the laws of mechanics and the laws of electromagnetism. He was led, as we shall see, to a number of startling conclusions, all of which have turned out to be in agreement with many different kinds of experiments. The special theory of relativity is now on a very sound experimental basis.

The equations of electricity and magnetism as formulated by Maxwell seem at first glance to violate the postulate of relativity. In particular, Maxwell's equations predict the existence of electromagnetic waves traveling with the speed c of light. If it is a law of physics that light travels with a speed c in some coordinate system, then Eq. (7.5) seems to show that this law would not be the same in a coordinate system moving with a constant velocity relative to the first. One might try to modify the law of propagation of light so that the velocity c is not with respect to the coordinate system but with respect to the source of the light, or with respect to some medium within which the light propagates and which fills all space. Either of these suggestions would require some modification in Maxwell's equations, and furthermore they are not in agreement with experiment. Indeed it turns out experimentally that within the precision with which the experiments can be performed, the law that light travels (in vacuum) with a universal speed c is indeed true in all coordinate systems. Einstein in 1905 proposed his special theory of relativity based on the postulate of relativity (13.1) and on the following postulate:

The speed of light is a fixed universal constant c *relative to every coordinate system.* (13.2)

Among the experiments which justify this postulate are those of Michelson and Morley,† on the velocity of light relative to the moving earth, observations of the aberration in the direction of light from stars, and observations of light from revolving double star systems (see Problem 1). Unfortunately, space does not permit a discussion here of these experiments. For a discussion of the experimental basis of the theory of relativity the reader should consult one of the excellent popular books on the subject or a text on the theory of relativity. (See the Bibliography.) One of the most direct verifications of postulate (13.2) is the measurement by

*A. Einstein, *Annalen der Physik*, **17** (1905).

†A. A. Michelson and E. W. Morley, *Amer. J. Science*, **34**, 333 (1887).

Kennedy and Thorndike* of the difference in the speed of light relative to the Earth during the course of a year. It showed that the speed is the same within 2 m-sec^{-1} in two coordinate systems moving with a relative velocity of 60,000 m-sec^{-1}.

Postulate (13.2) is essentially the statement of one particular law of physics regarding the propagation of light, a law moreover which is itself just one consequence of the more general laws of electricity and magnetism formulated by Maxwell. Einstein selected this particular law for his postulate because the existence of such a law is crucial as we shall see in the setting up of a coordinate system and in determining the relations between relatively moving coordinate systems. The speed c plays a central role in the theory of relativity which goes beyond its role as the speed of propagation of light. We could have formulated the second postulate in the more general form:

The laws of physics involve a universal constant speed c [*which by postulate* (13.1) *must be the same in all coordinate systems*]. (13.2′)

The fact that the laws of physics can contain a universal speed which is the same in every coordinate system was discovered by Einstein. There can only be one such speed (see Problem 12), and hence any law of physics which refers to a universal constant speed must contain the speed c.

The first step in the development of the theory of relativity is to set up suitable definitions of the various physical quantities which appear in the laws of physics—positions, velocities, times, energies, etc. Many of these quantities, for example position, are defined relative to a particular coordinate system. Clearly if Postulate (13.1) is to apply, the various physical quantities must be defined in the same way in each coordinate system. The statement that the speed of light has the same value in all coordinate systems would be wrong or at any rate meaningless unless the units in which velocity is measured are defined in the same way in each coordinate system. Once it is specified how each coordinate system is to be set up and how various quantities are to be measured relative to it, the next step is to determine the relationships between corresponding quantities in different inertial coordinate systems. In classical physics, the transformation between two coordinate systems moving relative to one another is given by Eqs. (7.1), (7.5), and (7.6) for the positions, velocities, and accelerations. As we noted above, these transformation equations are incompatible with Postulate (13.2). Instead we will find in Section 13.5 a different set of transformation equations, the Lorentz transformation equations, which relate different inertial systems.

When we know how to transform the various physical quantities from one coordinate system to another, we can check each law of physics to see whether it has the same form in different inertial systems. It turns out that this is the case for Maxwell's equations of electrodynamics, a result which is perhaps not surprising in view of our having adopted Postulate (13.2) from the beginning. Of Newton's three laws of motion, only the first is compatible with the Lorentz transformation; the other two require modification.

*R. J. Kennedy and E. M. Thorndike, *Phys. Rev.*, **42**, 400 (1932).

In discovering how to modify Newton's laws of motion, Einstein was guided by the observation that these laws have been very well verified experimentally for all cases involving velocities that are slow in comparison with the speed of light. The Lorentz transformation equations connecting two inertial coordinate systems reduce to the corresponding classical equations whenever the relative velocity of the two coordinate systems is negligible compared with the speed of light, as indeed they must. Einstein therefore required that the new relativistic laws of mechanics must agree with the old classical mechanics whenever all the velocities involved are much less than c. A similar kind of reasoning was used by Neils Bohr in his early work on the quantum theory. Bohr used a principle which he called the *correspondence principle*, which we may formulate in a somewhat more general way as follows:

Correspondence Principle. *Any new theory which is proposed must agree with an older theory in its predictions of those phenomena for which the older theory gave the correct predictions.* (13.3)

This seemingly trivial observation has proved to be a very powerful tool in the development of modern physical theories. It guided Einstein to the development of the correct relativistic equations of motion.

In seeking the relativistic laws of mechanics, we will first ask the question whether it is possible to formulate the conservations laws, in particular the law of conservation of momentum, in a way that is compatible with the postulates of relativity. We shall find that this is the case, provided that we change the definition of the momentum of a particle in accordance with Eq. (4.74). It turns out that the law of conservation of momentum is compatible with the postulates of relativity only if we include with it the law of conservation of energy. The energy of a particle is defined to be the kinetic energy given by Eq. (4.73), plus the rest energy mc^2. If a body has a mass m, then the simple formula mc^2, according to the theory of relativity, includes within it all of the various kinds of energy—heat energy, potential energy, nuclear energy, etc.—contained within the body. When we know the correct relativistic definition of momentum, we may attempt to generalize Newton's law of motion by specifying that the force acting on a particle is to be equal to the time rate of change of its relativistic momentum. This gives a relativistic law of motion. The law of motion implies a particular transformation for the force when we change from one coordinate system to another. It is then necessary to examine the various laws of force and to modify them if necessary so that they agree with this transformation law. These developments are the subject matter for the rest of this chapter and the next.

13.2 THE APPARENT PARADOX CONCERNING THE VELOCITY OF LIGHT

Although the postulate of relativity (13.1) is intuitively appealing, when combined with the second postulate (13.2), it seems to lead to a number of paradoxical results which make it difficult, at first, to accept the theory based on these two postulates. Many of these paradoxes are related to a very simple one which follows directly

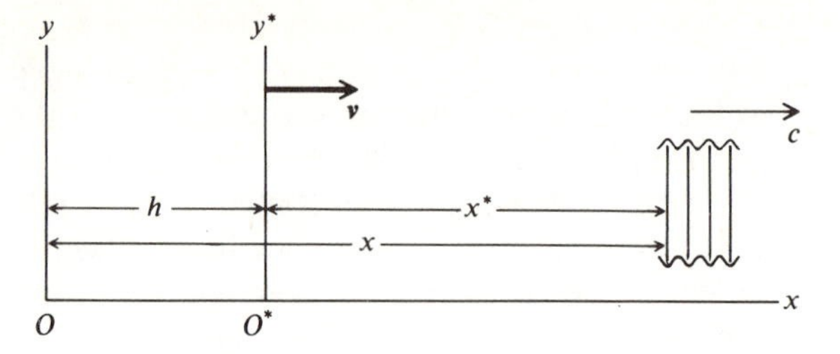

Fig. 13.1 A light pulse traveling with velocity c in the x-direction, and its relationship to a fixed origin O and a moving origin O^*.

from the second postulate and which we shall study in this section.

Consider a light pulse which leaves the origin O at $t = 0$ and which at a time t has traveled a distance x (Fig. 13.1). Suppose now that we set up a second starred coordinate system whose origin O^* moves along the x-axis with velocity v. Suppose that at $t = 0$ the origin O^* coincides with O, and that at time t it has moved a distance h as shown in Fig. 13.1. We see from the figure that

$$x = x^* + h. \tag{13.4}$$

Now if the velocity of light is c relative to O,

$$x = ct, \tag{13.5}$$

and if

$$h = vt, \tag{13.6}$$

then the velocity of light relative to O^* is evidently

$$c^* = x^*/t = c - v. \tag{13.7}$$

Hence $c^* \neq c$, and we seem to have a conflict with Postulate (13.2). Since the argument leading to Eq. (13.7) depends only upon simple mathematical and geometrical relationships, it is difficult to see how it could be wrong. *There is, in fact, nothing wrong with this argument.* Equation (13.7) is certainly valid, in relativity theory as well as in classical physics, provided that we understand clearly what we mean by the symbols appearing in this equation. Note that Eq. (13.4) assumes implicitly that the three distances x, h, and x^* are all measured in the same set of units, or if not, that corrections are made for the differences of the units before the addition is performed. If these three quantities were not measured in the same units, and if no allowance for this fact were made in writing down the equation, then clearly the equation would not hold. Likewise, we have assumed that the same time interval t, measured in the same units, is used in all three equations (13.5), (13.6), and (13.7). Our conclusion, which must certainly be valid in any theory, is that

Eq. (13.7) gives the relationship between the velocity c relative to O and the velocity c^* relative to O^* which moves with velocity v with respect to O, provided that the three velocities are obtained by dividing the appropriate distances by the same time t, where all distances are measured in the same way and in the same units. We see that in order to resolve the paradox between Eq. (13.7) and Postulate (13.2), we must suppose that when one coordinate system moves relative to another, either the units of time or length, or both, are different in the two coordinate systems. If for example, the distance x^* is measured in smaller units than the distance x, then the velocity c^* will come out bigger than $c-v$. Likewise, if the time t^* since the light pulse left the origin is measured in longer units than the time t, then again $c^* = x^*/t^*$ will come out bigger than $c-v$. Either or both of these possibilities might allow c^* to be equal to c.

It is assumed in classical physics, although the assumption is not usually stated explicitly, that the results of measurements are not affected by the motion of the measuring instruments, so long as we exclude obvious effects such as distortions produced by violent accelerations. If we accept the postulates of the theory of relativity, then we will have to abandon this assumption and allow for the possibility that moving meter sticks are shorter than stationary meter sticks, or that a moving clock may run slower than a fixed clock. If that is so, then the operational definition of any physical quantity must include not only a specification of the way in which it is to be measured, but also a specification of the velocity with which the measuring instruments involved are moving. It is customary to adopt the convention that unless otherwise explicitly stated, all quantities referred to a particular coordinate system are understood to be measured with measuring instruments at rest with respect to that coordinate system. If now we understand by x^* the distance of the light pulse from O^* as measured in the starred coordinate system, and if t^* is the time measured in the starred coordinate system, then it follows from Postulate (13.2) that

$$c = x/t = x^*/t^*, \tag{13.8}$$

where x and t are measured in the unstarred system. In Section 13.4, we will draw conclusions regarding the behavior of moving clocks and meter sticks from this assertion.

There is another more subtle assumption involved in the argument which leads to Eq. (13.7). Since the light pulse and the origins O and O^* are all moving relative to each other, the three distances h, x, and x^* in Fig. 13.1 must all be measured at the same time t. We will see in the next section that we can define what we mean by the *same time* at two different places only relative to a particular coordinate system. Two events which are simultaneous relative to one coordinate system are not necessarily simultaneous relative to another system. The distance x^* as measured in the starred coordinate system may therefore be measured not only in different units, but even between different points than it would be if all measurements were carried out in the unstarred system as assumed in the derivation of Eq. (13.7).

13.3 COORDINATE SYSTEMS, FRAMES OF REFERENCE

In setting up a coordinate system, we first specify its origin O, whose location is to be specified in some way, presumably with reference to some identifiable object or events. We then specify three mutually perpendicular axes x, y, z, whose directions again are to be specified with reference to some identifiable objects. We will assume that the coordinate system has been chosen in such a way that it is an inertial system. If we assume that the law of inertia is valid, then one way of guaranteeing that our coordinate system is inertial is to locate the origin and the axes with respect to bodies upon which no force is acting. [Because of the equality of gravitational and inertial mass, it may be difficult or even in principle impossible to determine when no forces act on a body. That is a difficulty which we must ignore if we are to make progress with the special theory of relativity at this point. We will return to it in the last section of this chapter.]

Having chosen an origin and a set of axes for our coordinate system, we must next specify the way in which the various physical quantities are to be measured. Since our postulates imply that the behavior of measuring instruments may depend on the way in which they are moving, it will be necessary to specify the way in which the instruments used to measure any particular quantity are moving. This specification must refer only to the coordinate system and perhaps to the bodies on which measurements are being made. Unless otherwise stated, we will specify that all measuring instruments are to be at rest relative to the coordinate system. In the case of certain measurements of the properties of a specific physical object, as for example its mass, it may be convenient to specify that the measuring instruments are to be at rest relative to the body whose properties are being measured.

Any definition of a physical quantity involves assumptions regarding the physical laws governing the behavior of the measuring instruments used. We will for example make use of the law (13.2) to define the way in which the time is to be measured at any particular point in a coordinate system. When new physical phenomena are discovered which are in conflict with previous physical laws, it is necessary not only to change the laws, but to examine the definitions of the various physical concepts and quantities in order to determine whether they depend upon assumptions which are no longer valid. It is not possible practically or even theoretically to specify explicitly all of the assumptions that are being made in any particular operational definition. We specify those assumptions which seem relevant at any particular stage in the development of physics, but unstated assumptions may turn out to be important in later analysis. Many of these assumptions have to do with those factors which can affect the results of a particular measurement. It is assumed in classical physics that with certain obvious exceptions the results of a measurement do not depend (among other things) upon the color of the measuring instrument, the past history of the measuring instrument, the phase of the moon at the time the measurement is made, or the velocity with which the instrument is moving. The latter assumption is no longer made in the theory of relativity, but we will continue to assume that with certain obvious exceptions the other three factors, and indeed many others, do not affect the results of physical measurements.

Another kind of assumption is that different ways of measuring the same quantity will give the same result. There are for example many different ways of measuring a length, some of which are appropriate for certain cases and some of which are not. Measurements of length involve in some way a comparison of the unknown length to be measured with some known length which is a part of the measuring apparatus and which according to our assumptions must be at rest relative to the coordinate system with respect to which the measurement is being made. This known length in turn is calibrated against some specified standard unit length. We will continue to assume, as in classical physics, that when different methods can be used to measure the same quantity, they will give the same result, provided of course that the measuring instruments are at rest relative to one another. In discussing a particular measurable quantity, we may therefore choose whatever method of measuring it is convenient for our purpose. In particular, we will suppose in the analysis to follow that the distance between any two fixed points in our coordinate system is to be measured in the usual way by means of a meter stick. The length of a moving object requires a more careful definition which will be given subsequently. We assume in the same way that suitable operational definitions have been given or will be given of all the physical quantities with which we will have to deal.

Each point in our coordinate system is to be located by coordinates x, y, z, which we imagine to be measured by means of meter sticks at rest relative to the coordinate system. Any physical event will be located in space by the coordinates x, y, z, at the point at which it occurs, and will also be identified by a time t at which it occurs, which is to be measured in the following way. We assume that at the origin of the coordinate system is located a stationary clock which reads the time t measured from some chosen reference time $t = 0$. Just as the origin in space may be located by some identifiable object upon which no forces act, so the time origin may be located by some identifiable event occurring at the space origin. This event may be simply the zero reading of the clock itself, if that is convenient. Any event which occurs at the space origin of the coordinate system is to be assigned the time t read on the clock at the moment when the event occurred.

There are a number of ways of specifying the time of an event which occurs elsewhere than at the space origin. One way is to assume that at each point in space is located a clock at rest relative to the coordinate system, which is identical to the reference clock located at the origin, or at least runs at the same speed. It is then necessary to specify how these clocks are to be synchronized. Since we have at the moment only one physical law (13.2), we will have to make use of that law for this purpose. We will therefore agree to synchronize the clock at the point (x, y, z) by reading the clock at the origin from the point (x, y, z), that is, by receiving at the point (x, y, z) a light signal sent out by the clock at the origin at a specified time. We will correct the reading from the origin clock by adding the time $(x^2+y^2+z^2)^{\frac{1}{2}}/c$ required by the light signal to go from the origin to the point (x, y, z), and will set the local clock at the corresponding corrected time. An equivalent alternative way of defining the time t of an event occurring at the point (x, y, z), which does not require so many clocks, can be given as follows. We observe at the origin a light signal sent out from the given event and note the time at which this signal arrives, as measured by the clock at the origin. We then subtract from this time the time required for light to travel at velocity c from the point (x, y, z) to the origin. The time t of the given event is then this corrected time.

Any moving body can now be assigned coordinates $x(t)$, $y(t)$, $z(t)$, which can in principle be measured in accordance with these definitions. The velocity and acceleration components of the body can now be defined in the usual way as derivatives of these functions. The mass of a body may be measured either by using a balance or by comparing its acceleration with that of a standard mass, as explained in Section 1.3. In either case, the measuring apparatus and the standard mass will be at rest or nearly so, relative to the unknown mass.

In order to ensure that physical quantities are defined in the same way in different inertial coordinate systems, we must have some way of comparing the basic units in the two coordinate systems. There are two ways in which the fundamental units of measurement may be defined. In the case of length for example, we may in the first case choose some particular physical object whose length is then defined to be the standard unit of length. Thus, for a long time the meter was defined to be the distance between two marks on a bar kept in the International Bureau of Weights and Measures, Sevres, France. If we define the unit of length in this way, we are faced with the problem of what we mean by a meter in a (starred) coordinate system that is moving at some velocity with respect to the (unstarred) laboratory in France. We may do this by arranging for the laboratory to calibrate a second standard meter bar, which of course can only be done while the second standard meter bar is at rest with respect to the original meter. The second standard bar is then accelerated up to the appropriate velocity so that it is at rest relative to the moving coordinate system. We suppose that this acceleration is carried out sufficiently slowly and carefully so that no obvious distortions of the second standard meter bar are produced and we then define it as the standard meter in the starred coordinate system.

A somewhat more satisfactory way of comparing standards of length in two different inertial systems is based upon the observation that lengths perpendicular to the direction of relative motion can be compared directly. Let us take two points at rest in the unstarred coordinate system, and one meter apart, measured in a direction perpendicular to the direction in which the starred system is moving. As these two points move through the starred coordinate system, they will trace out a pair of parallel lines. The distance between these two lines is then defined as the standard meter in the starred coordinate system. Notice that we are assuming that the length of a meter stick is unaffected by a rotation of the meter stick in space, so that if a stick is calibrated in one orientation, it can be used to measure lengths in any direction.

A second way of defining the unit of length is to choose some universally identifiable physical phenomenon and to define a meter with reference to some measureable length involved in this phenomenon. The meter is now defined as 1,650,763.73 times the wavelength of a particular red line in the spectrum of a particular isotope of krypton, Kr^{86}. The krypton atoms emitting the radiation are to be at rest in the coordinate system in which it is measured, and the wavelength is to be measured in a vacuum. This definition of the unit of length can then be applied in any coordinate system and the problem of comparing units in the two systems does not arise.

We will assume that any of these ways of identifying units in two inertial systems would give the same results. In particular, an important relationship between measurements of lengths relative to two inertial systems follows from our discussion of the comparison of standard meter bars. Lengths perpendicular to the direction of relative motion of the two systems can be compared directly, as we saw above. Therefore *lengths and distances perpendicular to the direction of motion have the same values in both coordinate systems.*

A similar discussion can be given for definitions of units in which other physical quantities are to be measured. One way of specifying the unit of time, when the unit of distance has been specified (or vice versa), would be to specify a precise value for the speed of light and to let the unit of time be determined by comparing the measured speed of light with the specified standard value. At the present time (1970) it is technically possible to measure both time and distance intervals more accurately than the speed of light can be measured, and for that reason the unit of time is presently defined in terms of the period of revolution of the earth around the sun in the year 1900. It is anticipated that when the appropriate measurement techniques are sufficiently improved, the unit of time will be defined in terms of the frequency of a certain spectral line of a particular atom. If the same spectral line were chosen as that used to define the standard of length, this would amount to specifying the speed of light.

As we have seen, the length of a moving body may be measured directly in a straightforward way if the length to be measured is perpendicular to the direction of motion. There are several plausible ways to measure a length parallel to its direction of motion. Suppose two marks A and B on a moving body indicate the ends of the length to be measured. At some time t, we mark the points A', B', fixed in our coordinate system, occupied at that moment (simultaneously) by the moving marks A and B. The distance $l = \overline{A'B'}$ between the fixed points A' and B', which can be measured with a meter stick at rest, is now defined as the length in our coordinate system between the moving marks A and B. This method can be used to measure a length $\overline{AB}$ in any orientation relative to its velocity.

A second way of measuring a length parallel to its motion if its velocity $\boldsymbol{v}$ has been measured (as previously defined), is to measure the time interval t between the passage of A and B past a fixed point. The length $\overline{AB}$ is then $l = vt$. A third way is suggested in Problem 4. We will assume that all these ways of measuring a length relative to a given coordinate system give the same result. This assumption is consistent with the postulates of relativity. (See Problems 4 and 6.)

It will turn out that moving bodies, including moving measuring instruments, behave in unexpected ways if their velocity is comparable with the velocity of light. However, kinematical considerations having to do with the positions and motions of bodies relative to a single coordinate system, or relative to each other, can be carried out in the usual way so long as all distances, times, velocities, etc., are measured by instruments at rest relative to a single coordinate system. For example, as noted above, the arguments leading to Eq. (13.7) are perfectly valid provided that all lengths and times are measured relative to a single coordinate system. When we wish to relate quantities measured in different coordinate systems in rapid relative motion, relativistic effects need to be considered.

One remarkable result which follows immediately from the above definitions is that two events E_1 and E_2 which occur at the same time $t_1 = t_2$ in two different places in one inertial coordinate system may occur at different times $t_1^* \neq t_2^*$ in an inertial system moving with respect to the first. This result could perhaps have been anticipated, if we remember that the synchronization of clocks in each of the two coordinate systems is to be carried out using light signals and assuming in each coordinate system that light propagates with the same speed c. It is even possible, and more disturbing, to find situations in which there are two events E_1 and E_2 such that E_1 occurs before E_2 in one inertial coordinate system and occurs after E_2 in another.

As an example consider the situation shown in Fig. 13.2. We have again two coordinate systems oriented with their axes parallel, and where the starred coordi-

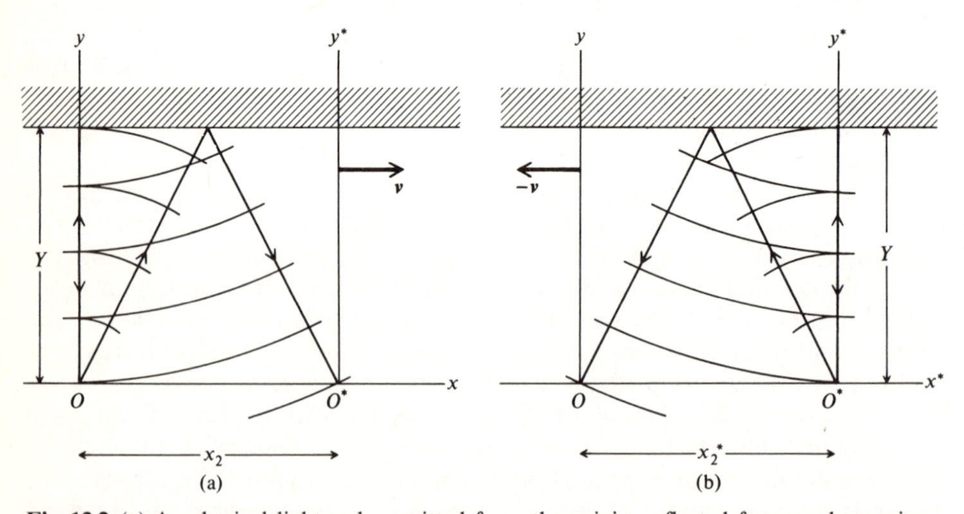

Fig. 13.2 (a) A spherical light pulse emitted from the origin, reflected from a plane mirror at $y = Y$, and returning to O and O^*, as seen in the unstarred coordinate system. (b) The same events as seen in the starred coordinate system.

nate system is moving with a velocity v along the x-axis. The time origin in both coordinate systems is chosen as the time when the space origins O and O^* coincide. At this instant a spherical pulse of light is sent out from the coincident origins, propagates up to a plane mirror parallel to the xz-plane located at $y = Y$, and is reflected back. A part of the reflected spherical wave returns to the origin O, an event which we will call E_1, and another part of the wave returns to the moving origin O^* which has now traveled to the point $x = x_2$. The reception of the reflected pulse at O^* we will call the event E_2. The event represented by the original emission of the light pulse occurs at the space-time origin $x = y = z = t = 0$ in the unstarred system and also at the origin $x^* = y^* = z^* = t^* = 0$ in the starred system. Let us first look at this sequence of events from the point of view of the unstarred coordinate system (Fig. 13.2a), remembering that so long as we stick to one coordinate system, we can use ordinary kinematical considerations based on the propagation of light at speed c. The portion of the wave which travels up to the mirror and returns to O evidently arrives at the time

$$t_1 = 2Y/c, \tag{13.9}$$

which is the time assigned to the event E_1 in the unstarred coordinate system. The event E_2 occurs at the point

$$x_2 = vt_2, \qquad y_2 = z_2 = 0.$$

The light arriving at O^* has traveled a distance

$$ct_2 = 2[\tfrac{1}{4}x_2^2 + Y^2]^{1/2}.$$

We substitute for x_2 and solve:

$$t_2 = \frac{2Y}{c}\left(1 - \frac{v^2}{c^2}\right)^{-1/2}. \tag{13.10}$$

Thus, in the unstarred coordinate system, the event E_2 occurs after the event E_1, that is $t_2 > t_1$.

Let us now consider the same sequence of events from the point of view of the starred coordinate system as shown in Fig. 13.2(b). The unstarred origin O is now traveling to the left and we will argue that its speed v as measured in O^* must be the same as the speed of O^* measured in O. If we have two inertial systems, and if in each we measure the speed v with which the other is moving, we must get the same speed in both cases. If this were not the case, we would have an experimental way of distinguishing between the two coordinate systems. This would be in conflict with our postulate (13.1). The direction of the relative velocity of each system with respect to the other will of course depend upon the relative orientations of the starred and unstarred axes. If as in the present case the two sets of axes are parallel, and if O^* moves in the positive direction relative to O along the x-axis, then the velocity of O^* relative to O is $v_x = v$, $v_y = v_z = 0$; and the velocity of O relative to O^* is $v_x^* = -v$, $v_y^* = v_z^* = 0$. Also, as we have shown earlier, the coordinate

$y^* = Y$ of the mirror will be the same in the starred coordinate system as the y-coordinate of the mirror in the unstarred system, since both are measured perpendicular to the motion. We now see, as above, that the event E_2 occurs at the origin $x_2^* = y_2^* = z_2^* = 0$ at the time

$$t_2^* = 2Y/c, \tag{13.11}$$

and the event E_1 occurs at the point $x_1^* = -vt_1^*$, $y_1^* = z_1^* = 0$, at the time

$$t_1^* = \frac{2Y}{c}\left(1-\frac{v^2}{c^2}\right)^{-1/2}! \tag{13.12}$$

The times assigned to the two events in the starred coordinate system are just interchanged with the times assigned in the unstarred system, and the event E_2 now occurs before the event E_1 in the starred system!

This paradoxical result is so contrary to ordinary common sense as to make us feel that the theory which predicts it must be self-contradictory, or at least must lead to physically absurd results. If we examine carefully the reason for this opinion, we will see that it is due to our concept of causality. If one event is the cause of a second event, then the first event must surely precede the second event in any coordinate system. In the present case, the event E_0, the emission of the light pulse, is in part the cause of the other two events E_1 and E_2 and must therefore precede them, as indeed it does in both coordinate systems. If there could be any causal relationship between the events E_1 and E_2, then in one or the other of the two coordinate systems we would have a contradiction with the principle of causality, that the cause must precede the effect. If for example a signal of any kind could be sent out from the origin O at the time t_1, signaling the occurrence of the event E_1, and if this signal could arrive at the point O^* at the time t_2, then we would have a contradiction, because in the starred coordinate system the signal would be received before it was sent out, and we would learn of the occurrence of the event E_1 before it had occurred in the starred coordinate system. We conclude that if we are not to abandon the principle of causality, then it must be impossible in a situation like this for the events E_1 and E_2 to have any direct causal relationship. In particular it must be impossible for any kind of signal to leave one event and arrive at the other. Signals may be carried by some kind of a wave, a light wave or a sound wave for example, or may be carried by some physical object which travels from one event to the other, a message in a bottle for example. The time between the two events E_1 and E_2 is, in both coordinate systems,

$$\begin{aligned} t_2-t_1 = t_1^*-t_2^* &= \frac{2Y}{c}\left[\left(1-\frac{v^2}{c^2}\right)^{-1/2}-1\right] \\ &= (cx_2-2vY)/c^2 = (-cx_1^*-2vY)/c^2. \end{aligned} \tag{13.13}$$

This is less than the time required for light to travel the distance $x_2 = -x_1^*$ between the two events. A similar result is found in all cases where the time ordering of two events can be different in two different inertial coordinate systems; the time difference between the two events is always less than the time required for light to travel a corresponding distance in either coordinate system. We therefore conclude that if the theory is to be consistent with the principle of causality, it must be impossible for any signal, or any material object which could carry a signal, to travel faster than the universal speed c in any coordinate system.

Since the laws of physics will surely allow an object to be at rest at the origin of a coordinate system, we conclude in particular that the relative velocity v of two inertial coordinate systems can never exceed c. In fact we could already have noticed from Eqs. (13.10) that v must be less than c or the theory becomes self contradictory. For it is certainly true with respect to the starred coordinate system that the light pulse travels up to the mirror and back in a finite time $t_2^* = 2Y/c$, so that the event E_2 certainly does occur. Because of peculiarities in the behavior of clocks, the time t_2 assigned to this event may be different in the unstarred coordinate system, but the event certainly does occur and therefore must be assigned some real time t_2. We conclude that the speed v must be less than c.

This conclusion of the theory of relativity, that material objects cannot travel faster than the speed of light is well verified experimentally. Engineers must routinely take this and other conclusions of the theory into account in designing particle accelerators which accelerate subatomic particles to high energies. An electron in a high-voltage x-ray tube with an energy of 500,000 eV travels at a speed about 87% of the speed of light. If classical mechanics were correct an electron with 10 times this energy would travel with a speed $\sqrt{10}$ times as great or over twice the speed of light. However, a 5,000,000 eV electron travels at a speed only 99.88% of the speed of light. The fact that electrons with energies over about a million volts all travel at nearly the same speed simplifies considerably the design of electron linear accelerators. In the two-mile electron linear accelerator at Stanford University, electrons can be accelerated to 2000 times this energy or 10^{10} eV, at which energy they are traveling only 10 cm sec^{-1} slower than light.

The difficulty with simultaneity occurs only when we are trying to compare the time of occurrence of two events which occur at different places in space. If two events occur at the same place, or near enough together so that the time required for light to propagate from one to the other is negligible in comparison with the times of interest for any particular discussion, then the difficulty with simultaneity does not occur. Thus, events which occur at the same place, or nearly the same place in this sense, and at the same time, or nearly the same time, in one coordinate system, are also simultaneous or nearly so in any other coordinate system. In particular, we may assume that if a moving clock passes near a fixed clock, we may compare their readings in an unambiguous way at the moment of passage, and we may if we choose synchronize the clocks so that they have the same reading at that moment. Likewise, if two relatively moving observers pass by one another at a certain instant, and if at that instant they observe the time read on a distant clock (by observing light or radio signals from it), they will not disagree as to the reading which they observe at that moment. They will however disagree as to how to correct for the time required for the light signal to travel from the distant clock, since each assumes that the light signal travels with speed c relative to him. This pictorial language referring to relatively moving observers is often convenient in bringing out the differences between physical quantities relative to different coordinate systems. The reader will recognize that such language does not imply any real disagreement about the physical situation. Anyone, however he may be moving, is free to make use of measurements made by any instruments, and is free to use in his calculations quantities measured relative to any coordinate system, provided he does so correctly. A given quantity may have different values relative to two different coordinate systems, but all observers should agree as to what those values are. In the above example, either observer may calculate the correction for the travel time of light relative to a coordinate system in which the other is at rest, and they will of course agree about that correction, unless one of them makes a mistake.

If two events E_1 and E_2 are simultaneous in an unstarred coordinate system, then they are also simultaneous in any coordinate system moving at right angles to the line connecting the two events. To show this, let us take the origin O at the midpoint of the line joining the two events, and let us assume that the origin O^* moves along the perpendicular bisector of this line as shown in Fig. 13.3. A fixed clock at O and a moving clock at O^* record the arrivals of light signals from the two events. We see from the symmetry of the situation that at any given moment and as viewed from either coordinate system, each clock is equidistant from the points at which the events E_1 and E_2 occurred. If the light signals propagating from the two events arrive simultaneously at one of the clocks, they must also arrive simultaneously at the other clock. Therefore the same correction will be deducted from the arrival time at the origin in calculating the time of occurrence of both events E_1

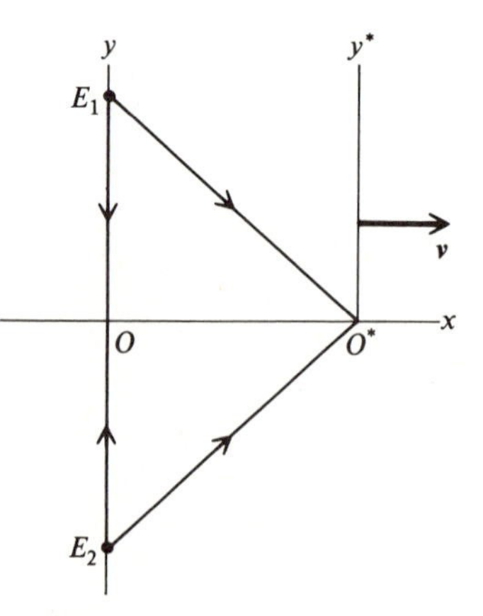

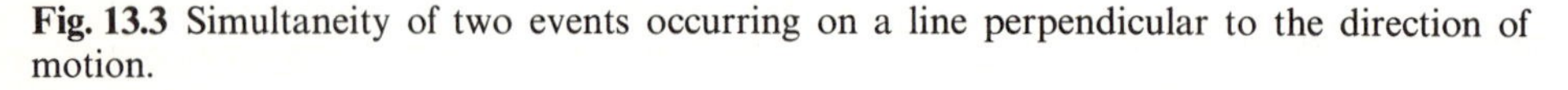
Fig. 13.3 Simultaneity of two events occurring on a line perpendicular to the direction of motion.

and E_2, and this statement will be true in both coordinate systems. Hence, if $t_1 = t_2$, we will also have $t_1^* = t_2^*$, although t_1^* and t_1 will not in general be equal.

To speak of the position of a moving point in space has a meaning only if we specify the time to which we refer. If we intend to transform from one set of coordinates to another coordinate system moving relative to the first, we can make the transformation only if we specify, in addition to the coordinates x, y, z of a point, also the time t for which the transformation is to be made. Conversely, events which are assigned the same time in one coordinate are assigned different times in another coordinate system. This implies that if we are to transform the time variable from one coordinate system to another, we must not only specify the time t but also the space point x, y, z to which that time refers. As long as we refer only to a single coordinate system, we may speak separately of a point x, y, z in space and the time t. If however we wish to refer to two or more relatively moving coordinate systems, or if we wish to make statements without reference to any particular coordinate system, then we must speak not of a point in space or a time separately, but rather of an *event* occurring at a particular point at a particular time and which in a particular coordinate system has the coordinates (x, y, z, t). It is often convenient to introduce geometric language and to refer to the four variables (x, y, z, t) as designating a "point" in a four-dimensional "space." In order to distinguish whether we are speaking of ordinary three-dimensional space or of this four-dimensional space, we will refer to the four-dimensional "point" (x, y, z, t) as an *event*, and to the corresponding space as *space-time*. This use of geometrical terminology is convenient, but it must not be misunderstood to imply that there is no distinction between space and time. Lengths are to be measured with meter sticks, and time intervals with a clock. They are different but related physical concepts, measured in different but related ways. We are required to think of the four variables (x, y, z, t) together because of the fact that two events which occur at the same place at different times relative to one coordinate system will occur at different places in a moving coordinate system and conversely, two events which occur at the same time in different places in one coordinate system occur at different times in a moving coordinate system.

Let us now consider an event E_0 with coordinates (x_0, y_0, z_0, t_0) in some coordinate system. Let us assume that a spherical light signal leaves this event and travels throughout space. This light signal will arrive at any other point (x, y, z) in space at a time t which is evidently given by

$$(x-x_0)^2+(y-y_0)^2+(z-z_0)^2-c^2(t-t_0)^2 = 0. \tag{13.14}$$

The set of events (x, y, z, t) whose coordinates satisfy Eq. (13.14) are said to lie on the *light cone* relative to the event E_0. Equation (13.14) is also satisfied by events with $t < t_0$ which are so located that a light signal starting out from the event (x, y, z, t) will arrive at the event E_0. Because the law of propagation of light is the same in all coordinate systems, if we consider a starred coordinate system in which the event E_0 has the coordinates $(x_0^*, y_0^*, z_0^*, t_0^*)$, then in this coordinate system the

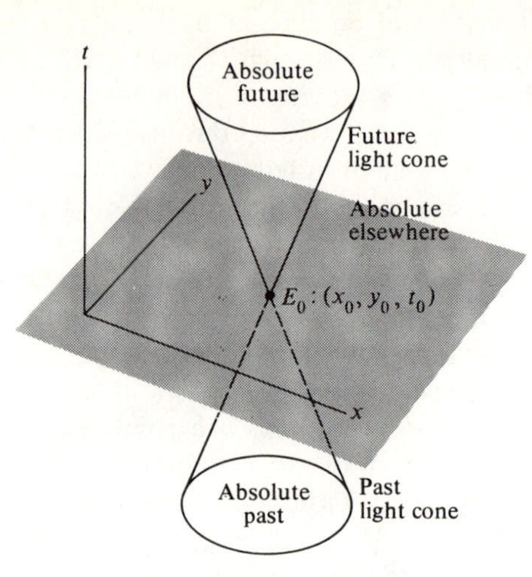

Fig. 13.4 The light cone.

same light cone (13.14) consists of the events whose coordinates satisfy the equation

$$(x^* - x_0^*)^2 + (y^* - y_0^*)^2 + (z_0 - z_0^*)^2 - c^2(t^* - t_0^*)^2 = 0. \tag{13.15}$$

We see that the equation for the light cone has the same form in all coordinate systems, as it should, since we may define the light cone relative to a particular event E_0 without reference to any coordinate system. It contains two parts, a *future light cone* containing all those events which may be reached by a light signal leaving the event E_0, and a *past light cone* consisting of all those events from which a light signal can reach the event E_0.

We can more easily visualize the situation by restricting our consideration to the xy-plane. A light signal starting from the event E_0 travels out in an expanding circle in this plane. In Fig. 13.4, we plot the two coordinates x and y horizontally and vertically we plot the time t to some suitable scale. In such a plot, the events in the plane $z = z_0$ whose coordinates satisfy Eq. (13.14) lie on a pair of cones as shown.

We have seen that the light cone itself is defined in a manner independent of the coordinate system and contains the same events in all coordinate systems. Likewise, the interior of the light cone consists of the same set of events in all coordinate systems, although these events will have different coordinates in different coordinate systems. Any event E in the interior of the future light cone will have in all coordinate systems the property that it can be reached from E_0 by an object traveling at a velocity less than c, although the particular velocity required will be different in different coordinate systems. There is a coordinate system, for example one whose origin travels from E_0 to E, in which E and E_0 occur at the same place but at different times. The interior of the light cone is often called the *absolute future* relative to the event E_0; the word "absolute" implies that this set of events is the same in all coordinate systems, and the word "future" is used because for each of these events E, there is a coordinate system in which E and E_0 occur at the same place with E occurring at a later time than E_0. It is left to the reader to supply the corresponding discussion of the interior of the past light cone which is called the *absolute past* relative to the event E_0.

An event E outside the light cone relative to E_0 could only be reached by a point leaving E_0 and traveling with a speed greater than c if $t > t_0$, or conversely if $t < t_0$. Two such events cannot communicate with one another or affect one another causally; no signal or physical object can leave one of them and arrive at the other. The collection of events outside the light cone is sometimes called the *absolute elsewhere* relative to E_0. We have seen that an event E which is in the absolute elsewhere relative to E_0 has the property that in some coordinate systems it may occur before E_0, and in other coordinate systems it may occur after E_0. By continuity, we may argue that there is some coordinate system in which these two events occur simultaneously but at different places. Later when we have developed the equations for the Lorentz transformation, the reader will be able to find that coordinate system explicitly for any given pair of events E_0 and E. (See Problem 10.)

There are thus five different possible relationships between two events E and E_0, which have a meaning independent of any coordinate system. The event E may be absolutely elsewhere from E_0, it may be in the absolute future or in the absolute past relative to E_0, or it may lie on the future light cone or on the past light cone relative to E_0. If in some coordinate system E and E_0 have the coordinates (x, y, z, t) and (x_0, y_0, z_0, t_0), we can test which of these relationships holds according to whether the quantity

$$S = (x-x_0)^2+(y-y_0)^2+(z-z_0)^2-c^2(t-t_0)^2 \tag{13.16}$$

is greater than, less than, or equal to zero, and in the latter two cases, whether the quantity $(t-t_0)$ is greater than or less than zero. The quantity S given by Eq. (13.16) is itself independent of the coordinate system in which it is calculated (Problem 11).

13.4 BEHAVIOR OF CLOCKS AND METER STICKS

Preparatory to deriving the transformation equations between two relatively moving coordinate systems, we will first make use of postulate (13.2) in order to determine the behavior of moving clocks and meter sticks. In order to derive the new law for moving clocks, as always in physics, we choose the simplest possible situation to study. We consider two clocks, one at rest at the origin O of the unstarred coordinate system, and the other at the origin O^* of a coordinate system moving with speed v along the x-axis. We will assume that the two clocks are synchronized so that both read zero at the moment the moving clock passes the fixed one. Since we will have to use the law of propagation of light in both coordinate systems, we will need to compare distances relative to the two coordinate systems. The only distances that we presently know how to compare relative to two coordinate systems are distances measured perpendicular to the direction of motion. We therefore choose the arrangement shown in Fig. 13.2 (p. 512). We have already derived Eqs. (13.10) and (13.11) for the time of arrival of the light pulse at the point O^* as measured by the two clocks, from which we conclude

$$t_2^* = t_2[1-(v^2/c^2)]^{1/2}. \tag{13.17}$$

The moving clock therefore runs more slowly than the fixed clock by a factor $[1-(v^2/c^2)]^{1/2}$, as measured in the unstarred coordinate system.

This effect, called *time dilation*, is well verified experimentally. The measured half-lives of radioactive particles, for example muons, are considerably longer when the particles move at speeds close to the speed of light than when the same particles are at rest,* and the ratio of life times is just given by $[1-(v^2/c^2)]^{1/2}$. The Mossbauer effect† gives a way of measuring frequencies of light (gamma rays) emitted by radioactive nuclei to extremely high precision, better than one part in 10^{14}. This is adequate to measure the change in frequency due to a velocity of 30 m $\sec^{-1}$. Thus, we may check formula (13.17) for clocks moving at quite ordinary velocities.

If a moving clock changes its velocity, and if we *assume* that the acceleration is small enough so that it does not affect the rate of the clock, then we expect from the above argument that the moving clock at any instant will run slowly by the factor $[1-(v^2/c^2)]^{1/2}$, where v is the instantaneous speed of the clock. If t^* is the time shown by the moving clock, and t is the time relative to an inertial coordinate system in which the clock is moving with speed v, then

$$dt^* = dt[1-(v^2/c^2)]^{1/2}. \tag{13.18}$$

The relation between t^* and t is to be found by integrating this equation. We are required by the postulates of relativity to conclude that the rate of a clock depends upon its velocity. We are not required to assume that it depends upon the acceleration. Experimentally, Formula (13.18) is rather well verified, even for accelerating clocks. No effect of acceleration has been observed, except for the obvious effects of violent accelerations, which depend upon the particular clock mechanism involved. On the theoretical side, the special theory of relativity gives a consistent set of physical laws if we assume that clocks behave according to Formula (13.18).

Let us return now to the two clocks, one at rest in the unstarred coordinate system, and the other at rest in the starred coordinate system moving with velocity $\mathbf{v}$ relative to the unstarred system. It follows from the postulate of relativity that the unstarred clock must likewise appear to be running more slowly than the starred clock, as measured from the starred coordinate system. In the starred coordinate system, the sequence of events is shown in Fig. 13.2(b). The corresponding equations are (13.9) and (13.12), which lead to the conclusion that, as viewed from the starred

*D. H. Frisch and J. H. Smith, "Measurement of the Relativistic Time Dilation Using μ-Mesons," *Amer. J. Phys.*, **31**, 342 (May, 1963).

†S. DeBenedetti, "The Mossbauer Effect", *Scientific American*, April, 1960, p. 72. Available as Offprint No. 271 from W. H. Freeman and Co., San Francisco.

coordinate system, the unstarred clock runs slow by the same factor $[1-(v^2/c^2)]^{1/2}$. When one clock moves at a constant velocity $\boldsymbol{v}$ with respect to the other, the two clocks can be near the same point in space only once. Therefore only at one time is it possible to compare their readings directly. In the present case, we have chosen to set both clocks so that they read zero at this time. Since at any other time the two clocks will be separated by some distance, in order to compare their readings at any later time it is necessary to define what we mean by the same time at two different places. As we have seen, it is impossible to do this in a way that is independent of the coordinate system. In the unstarred coordinate system, the clock at the origin reads the time t_2 when the light pulse arrives at O^*. However, in the starred coordinate system, the event at O when the clock at O reads t_2 occurs after the light pulse has arrived at O^*. It is for this reason that it is possible for clocks in each coordinate system to run slowly as compared to clocks in the other coordinate system.

If the starred clock were at some time to reverse its velocity and to return past O, we could then again directly compare the two readings, and our conclusion from Eq. (13.18) is that the starred clock at that moment would read a shorter time than the unstarred clock by a factor which could be calculated by integrating Eq. (13.18) during the time since the clocks last met. The fact that the starred clock shows a shorter time interval than the unstarred clock between their two encounters is a statement which is independent of any coordinate system, since the two clocks can be directly compared at both encounters, but this does not violate the postulate of relativity because the starred clock has been accelerated and therefore cannot serve as the origin of any inertial coordinate system. Although this remark removes any logical inconsistency with Postulate (13.1), it leaves a somewhat unsatisfactory situation conceptually. We will study this situation more carefully in Section 13.6.

Let us use the above result to determine the behavior of a moving meter stick, or of any other moving rigid body, where in this case by a *rigid body* we simply mean a body all of whose parts move with the same velocity. Let the meter stick be oriented parallel to the x-axis and moving with a constant velocity $\boldsymbol{v}$ parallel to the x-axis. We introduce a starred coordinate system as shown in Fig. 13.5 moving with velocity $\boldsymbol{v}$. Two points O^* and A are fixed on the moving meter stick. We wish

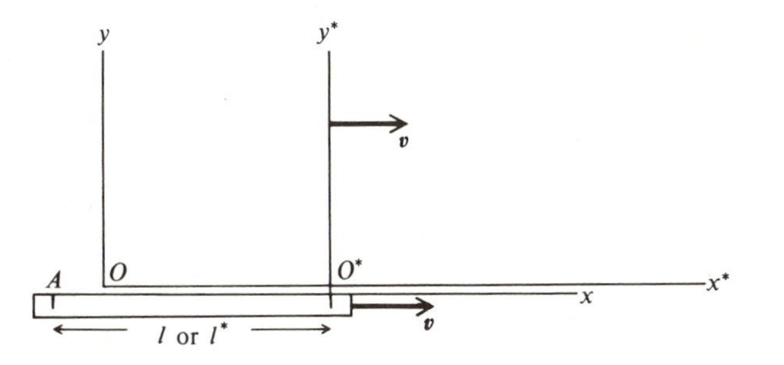

Fig. 13.5 Measuring the length of a moving body.

to determine the distance l^* between the points O^* and A as measured in the moving coordinate system in terms of the distance l between the same two points as measured in the fixed coordinate system. We will measure, in both coordinate systems, the time interval between the passage of O^* and A past the clock at the origin O. Since the speed of each coordinate system relative to the other is v, we will have

$$l = vt, \tag{13.19}$$

$$l^* = vt^*, \tag{13.20}$$

where we suppose as usual that the clocks at O and O^* are synchronized to read zero at the moment when the two origins pass one another, and where t and t^* are the measured times when the point A passes the origin O. Since both events whose time interval is to be measured occur at the position O, there is no ambiguity in the measured time t. However, these two events occur at different places in the starred coordinate system, and the time t^* therefore involves a correction using the law of propagation of light in the starred coordinate system. We must therefore view the situation from the viewpoint of the starred coordinate system, since it is in that coordinate system that the time t^* is measured. In that coordinate system, the clock at O is running slowly by a factor $[1-(v^2/c^2)]^{1/2}$. We therefore conclude from our previous discussion that

$$t = t^*[1-(v^2/c^2)]^{1/2}, \tag{13.21}$$

and therefore that

$$l = l^*[1-(v^2/c^2)]^{1/2}. \tag{13.22}$$

The measured length l of the moving meter stick is shorter by the factor $[1-(v^2/c^2)]^{1/2}$ than the length l^* measured in the coordinate system in which the meter stick is at rest. The length of any moving object as measured in a coordinate system in which that object is at rest is called the *rest-length* of the object. The rest-length of any object is an absolute concept, because it is defined in terms of the object itself. If we designate the rest-length of an object by l_0, we have for the length of any object moving with a velocity $\boldsymbol{v}$ in a direction parallel to that length

$$l = l_0[1-(v^2/c^2)]^{1/2}. \tag{13.23}$$

The contraction in the length of a moving rigid body given by Eq. (13.23) is called the *Lorentz contraction.* It was first proposed by Fitzgerald in 1893 and used in 1895 by H. A. Lorentz to explain the null result of the Michelson-Morley experiment. Lorentz showed that if Michelson's apparatus were to contract in the direction in which it is moving according to Eq. (13.23), then this contraction would just cancel the effect of its motion, so that light would appear to have the same velocity in all directions even relative to the moving instrument. Einstein pointed out that if nature conspires in this way to prevent us from determining experimentally the absolute motion of our instruments, then the notion of an absolute velocity has no meaning. He therefore proposed to adopt the postulate of relativity, from which as we have seen the Lorentz contraction formula can be deduced. Lorentz actually went further and showed that if an instrument is constructed in some way from particles upon which only electrical forces act according to Maxwell's equations, then as a consequence of these equations it would indeed contract according to Formula (13.23). We see that if the postulates of relativity are valid, it will contract according to this formula in any case, no matter what the law of force between its parts may be.

Since these quantities occur frequently in the theory of relativity, it is convenient to introduce the following abbreviations:

$$\beta = v/c, \tag{13.24}$$

$$\gamma = (1-\beta^2)^{-1/2}. \tag{13.25}$$

The speed v here may be the speed of any moving body, or it may be the speed of some coordinate system relative to another coordinate system. Equation (13.23) may now be written

$$l = l_0/\gamma. \tag{13.26}$$

If $d\tau$ is a time interval recorded by a clock moving with the speed v relative to some coordinate system and if dt is the corresponding time interval as measured in that coordinate system, we have from Eq. (13.18) the following relation

$$dt = \gamma d\tau. \tag{13.27}$$

The time which would be recorded by a clock located on and moving along with any moving body is called the *proper time* relative to that body. This definition is also independent of any coordinate system. The relation between the proper time relative to any moving body and the coordinate time in a particular coordinate system is obtained by integrating Eq. (13.27), evaluating γ from the speed $v(t)$ of the body relative to the coordinate system.

13.5 THE LORENTZ TRANSFORMATION

We are now in a position to derive the relations between the coordinates of an event in two different inertial coordinate systems. To avoid any unnecessary complication, we assume for convenience that the starred and unstarred coordinate systems have their axes parallel, that the x and x^* axes coincide, that the origins O and O^* coincide at $t = t^* = 0$, and that the direction of motion is parallel to the x and x^* axes as shown in Fig. 13.6. Consider an event E which occurs at a point (x, y, z) at a time t in the unstarred coordinate system. Let the starred coordinates of this event be (x^*, y^*, z^*, t^*). We have already shown that the coordinates measured perpendicular to the direction of motion are the same in both systems:

$$y^* = y, \qquad z^* = z. \tag{13.28}$$

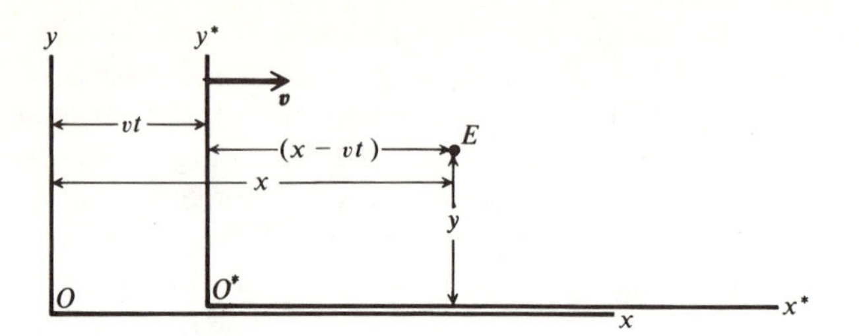

Fig. 13.6 Two coordinate systems in relative motion.

Figure 13.6 shows the situation as viewed in the unstarred coordinate system. We see that at the time t the distance from the y^*z^*-plane to the event E as measured in the unstarred coordinate system is

$$l = x - vt. \tag{13.29}$$

Note that l is also the constant distance between the y^*z^*-plane and a point which is moving with the velocity $\boldsymbol{v}$, and is therefore at rest in the starred coordinate system, and which passes through the event E. We have therefore, by Eq. (13.26),

$$x^* = l_0 = \gamma(x - vt). \tag{13.30}$$

In order to determine the time t^* assigned to the event E in the starred coordinate system, let us assume that a light signal emitted from the event E arrives at a point $(0, y^*, z^*)$ on the y^*z^*-plane. Since the light signal is moving to the left with speed c, and the y^*z^*-plane is moving to the right with the speed v, their relative velocity *in the unstarred coordinate system* is $v + c$, and hence the time of arrival of the signal at the y^*z^*-plane in the unstarred coordinate system is

$$t_1 = t + \frac{x - vt}{v + c}. \tag{13.31}$$

Since the moving clocks at the points $(0, y^*, z^*)$ and O^* lie on a line perpendicular to the direction of relative motion, if these two clocks are synchronized in the starred coordinate system, then they are also synchronized in the unstarred coordinate system, although of course they are running slow and therefore do not read the correct unstarred time. Therefore the time t_1^* recorded by the starred clock when the light signal arrives at the point $(0, y^*, z^*)$ is the same as that recorded simultaneously by a clock at the origin O^*, which is by Eq. (13.17),

$$t_1^* = t_1/\gamma. \tag{13.32}$$

The time t^* is determined by correcting for the travel time of light in the starred coordinate system:

$$t^* = t_1^* - x^*/c. \tag{13.33}$$

We substitute from the preceding three equations, and after some straightforward algebra, arrive at the result

$$t^* = \gamma[t-(vx/c^2)]. \tag{13.34}$$

We record for future reference the Lorentz transformation equations (13.28), (13.30), and (13.34) connecting two coordinate systems arranged as in Fig. 13.6:

$$\begin{aligned} x^* &= \gamma(x-vt), \\ y^* &= y, \\ z^* &= z, \\ t^* &= \gamma[t-(vx/c^2)]. \end{aligned} \tag{13.35}$$

These important equations should be committed to memory, a task which is relatively easy if we note the symmetry between the first and last equations, and remember that the first equation is just the classical equation corrected by the Lorentz contraction factor γ.

Given Eqs. (13.35), we can construct in a straightforward way the transformation equations between any two inertial coordinate systems. If the origins O and O^* do not coincide, then we simply add to the right-hand sides of the four Eqs. (13.35) the coordinates x_0^*, y_0^*, z_0^*, t_0^* in the starred coordinate system of the event $x = y = z = t = 0$ at the origin O. If the axes of the starred and unstarred systems are not parallel, we introduce a third coordinate system, call it the primed system, whose origin coincides with the origin O, and whose axes are parallel to the starred axes. Since the primed and unprimed coordinate systems are not moving with respect to one another, the relation between them is given by Eqs. (7.13):

$$\begin{aligned} x' &= x(\hat{x}\cdot\hat{x}')+y(\hat{y}\cdot\hat{x}')+z(\hat{z}\cdot\hat{x}'), \\ y' &= x(\hat{x}\cdot\hat{y}')+y(\hat{y}\cdot\hat{y}')+z(\hat{z}\cdot\hat{y}'), \\ z' &= x(\hat{x}\cdot\hat{z}')+y(\hat{y}\cdot\hat{z}')+z(\hat{z}\cdot\hat{z}'), \\ t' &= t. \end{aligned} \tag{13.36}$$

The starred coordinates are now given by Eqs. (13.35) with x, y, z, t replaced by x', y', z', t' on the right-hand side, with the latter variables given by Eqs. (13.36), and with the addition of x_0^*, y_0^*, z_0^*, t_0^* if the origins O^* and O do not coincide.

As we have seen, the velocity of the unstarred coordinate system relative to the starred coordinate system is $-\boldsymbol{v}$. The equations for the inverse transformation, giving x, y, z, t in terms of x^*, y^*, z^*, t^*, are obtained from Eqs. (13.35) simply by removing the stars on the left, starring the variables on the right, and changing the sign of v. (See Problem 9.)

We have deduced the Lorentz transformation equations (13.35) from Postulates (13.1) and (13.2). Conversely, given Eqs. (13.35) and our conventions regarding the way in which the coordinates and time are to be defined in any coordinate system, we can show that the law of propagation of light is the same in all inertial systems. Let us assume that we are given that light propagates with speed c in the unstarred coordinate system, and that we are also given the transformation equations (13.35). In order to be completely general, let us consider a light pulse propagating at an angle α with the x-axis. With an appropriate choice of axes, the equations describing the motion of the light pulse are

$$\begin{aligned} x &= ct \cos \alpha, \\ y &= ct \sin \alpha, \\ z &= 0. \end{aligned} \tag{13.37}$$

If we substitute these equations into Eqs. (13.35), we obtain

$$\begin{aligned} x^* &= \gamma ct(\cos \alpha - \beta), \\ y^* &= ct \sin \alpha, \\ z^* &= 0, \\ t^* &= \gamma t(1 - \beta \cos \alpha). \end{aligned} \tag{13.38}$$

(Recall that $\beta = v/c$.) We solve the last of these equations for t and substitute into the first three, to obtain the equations for the motion of the light pulse in the starred system, which we write in the following form:

$$\begin{aligned} x^* &= ct^* \cos \alpha^*, \\ y^* &= ct^* \sin \alpha^*, \\ z^* &= 0, \end{aligned} \tag{13.39}$$

where we have made the abbreviations

$$\begin{aligned} \cos \alpha^* &= (\cos \alpha - \beta)/(1 - \beta \cos \alpha), \\ \sin \alpha^* &= \sin \alpha/\gamma(1 - \beta \cos \alpha). \end{aligned} \tag{13.40}$$

Equations (13.40) define a real angle α^* provided that

$$\cos^2 \alpha^* + \sin^2 \alpha^* = 1. \tag{13.41}$$

Equation (13.41) is readily verified by substitution from Eqs. (13.40). We have therefore shown that the light pulse also propagates with the velocity c in the starred coordinate system but at a different angle α^* from the x^*-axis. The Lorentz transformation equations guarantee that the law of propagation of light holds in all coordinate systems if it holds in one. It follows that if we start from Eqs. (13.35), we can deduce the results obtained in Section 13.4.

We can now reformulate the two postulates (13.1) and (13.2) more explicitly in the following form:

The laws of physics must be such that if the coordinates and time are transformed according to the Lorentz transformation equations (13.35), *they take the same form in the new coordinate system.* (13.42)

Our program is to check the various physical laws against this postulate and to find suitable modifications of those which do not satisfy it. Of course the laws of physics contain other physical quantities besides the coordinates and the time. In order to check their form when the coordinate system is changed, it is necessary to know how these other physical quantities change when the coordinate system is changed. This may be done in two ways. We may examine the operational definition of each quantity and attempt to determine how the quantity will transform by requiring that it have the same operational definition in each coordinate system. When the equations for the transformations of all the physical quantities involved have been determined, one can then test any supposed law to see whether it satisfies the postulate of relativity. Alternatively one may ask the question: what transformation laws must we assume for quantities other than the coordinates and the time in order that a proposed set of laws of physics may be consistent with the postulate of relativity? We then have to show of course that these assumed transformation laws are compatible with the same operational definition of each physical quantity in every inertial coordinate system.

13.6 SOME APPLICATIONS OF THE LORENTZ TRANSFORMATION

To illustrate the use of the Lorentz transformation equations, we will derive a formula for the relativistic Doppler effect due to the motion of the light source. A source traveling with velocity $\boldsymbol{v}$ emits an electromagnetic signal at (angular) frequency ω^* relative to the source. We wish to find the frequency ω of the signal as observed at some stationary point. This problem can readily be solved directly in a single fixed coordinate system by making use of the law (13.18) for the rate of a moving clock (see Problem 13). However to illustrate the use of the Lorentz transformation, we will first set-up an equation for the light wave in a starred coordinate system in which the light source is at rest, and then transform this equation to the unstarred system.

A plane sinusoidal wave with frequency ω^* propagating in the x^*-direction with velocity c will have the form

$$W = A \cos [(\omega^*/c)(x^* - ct^*) + \theta],$$

as shown in the second paragraph of Section 8.3. We may generalize this to a light wave traveling in the direction of a unit vector $\hat{\boldsymbol{n}}^*$ by defining the *wave vector*

$$\boldsymbol{k}^* = \hat{\boldsymbol{n}}^*\omega^*/c, \tag{13.43}$$

and putting

$$W = A \cos (\boldsymbol{k}^* \cdot \boldsymbol{r}^* - \omega^* t^* + \theta), \tag{13.44}$$

where W is any component of the electromagnetic field, and $\boldsymbol{r}^* = (x^*, y^*, z^*)$. We are interested in the phase of the wave, that is, in the argument of the cosine, which contains information about the frequency, velocity, and propagation direction of the wave. The phase can be written

$$\boldsymbol{k}^* \cdot \boldsymbol{r}^* - \omega^* t^* + \theta = k_x^* x^* + k_y^* y^* + k_z^* z^* - \omega^* t^* + \theta. \tag{13.45}$$

We substitute from the Lorentz transformation equations (13.35) and collect terms separately containing x, y, z, and t to obtain the phase in the unstarred coordinate system in the form

$$\boldsymbol{k}^* \cdot \boldsymbol{r}^* - \omega^* t^* + \theta = k_x x + k_y y + k_z z - \omega t + \theta, \tag{13.46}$$

where we have made the abbreviations

$$\begin{aligned} k_x &= \gamma[k_x^* + (v\omega^*/c^2)], \\ k_y &= k_y^*, \\ k_z &= k_z^*, \\ \omega &= \gamma(\omega^* + v k_x^*). \end{aligned} \tag{13.47}$$

We see that the phase which describes the propagation of the wave has the same form in both coordinate systems. We will see below that the propagation vector $\boldsymbol{k}$ is related to the frequency ω by an equation similar to (13.43) and with the same propagation velocity c as it must. The reader will note the similarity between the transformation equations (13.47) for the quantities $\boldsymbol{k}$ and ω and the transformation equations (13.35); we will return to this point in the next chapter.

Let us now assume that the wave propagates in the starred coordinate system in the x^*y^*-plane at an angle α^* with the x^*-axis so that

$$\begin{aligned} k_x^* &= (\omega^*/c) \cos \alpha^*, \\ k_y^* &= (\omega^*/c) \sin \alpha^*, \\ k_z^* &= 0. \end{aligned} \tag{13.48}$$

We substitute these expressions in Eqs. (13.47) to obtain

$$\begin{aligned} k_x &= (\gamma\omega^*/c)(\cos \alpha^* + \beta), \\ k_y &= (\omega^*/c) \sin \alpha^*, \\ k_z &= 0, \\ \omega &= \gamma\omega^*(1 + \beta \cos \alpha^*). \end{aligned} \tag{13.49}$$

We see that the wave propagates in the unstarred coordinate system at an angle α with the x-axis given by

$$\begin{aligned} \cos \alpha &= k_x c/\omega = (\cos \alpha^* + \beta)/(1 + \beta \cos \alpha^*), \\ \sin \alpha &= k_y c/\omega = \sin \alpha^*/\gamma(1 + \beta \cos \alpha^*). \end{aligned} \tag{13.50}$$

It is left to the reader to check that $\sin^2 \alpha + \cos^2 \alpha = 1$, so that Eqs. (13.50) define a real angle α. We solve the first of these equations for

$$\cos \alpha^* = (\cos \alpha - \beta)/(1 - \beta \cos \alpha). \tag{13.51}$$

Equation (13.51) can also be obtained from the first of Eqs. (13.50) by simply reversing the sign of β and interchanging α and α^*. We substitute Eq. (13.51) in the last of Eqs. (13.49) to obtain after one algebraic step

$$\omega = \omega^*/\gamma(1 - \beta \cos \alpha). \tag{13.52}$$

This is the formula for the relativistic Doppler effect. If we put $\gamma = 1$, then this is just the classical formula for the Doppler shift when the source moves with a velocity $v = \beta c$ at an angle α with the direction of observation. The factor γ represents the time dilation effect due to the motion of the light source. Note also the reciprocal relation between Eq. (13.52) and the last of Eqs. (13.49), solved for ω^*.

Let us now derive the Lorentz transformation law for velocities. We consider a particle moving with a velocity $\boldsymbol{u}$ and we choose the origin at the position of the particle at $t = 0$. The particle then moves according to the equations

$$\begin{aligned} x &= u_x t, \\ y &= u_y t, \\ z &= u_z t. \end{aligned} \tag{13.53}$$

We substitute in Eqs. (13.35) to obtain

$$\begin{aligned} x^* &= \gamma(u_x - v)t, \\ y^* &= u_y t, \\ z^* &= u_z t, \\ t^* &= \gamma[1 - (vu_x/c^2)]t. \end{aligned} \tag{13.54}$$

We substitute from the last of these equations in the first three to obtain

$$\begin{aligned} x^* &= \frac{u_x - v}{1 - (vu_x/c^2)} t^*, \\ y^* &= \frac{u_y}{\gamma[1 - (vu_x/c^2)]} t^*, \\ z^* &= \frac{u_z}{\gamma[1 - (vu_x/c^2)]} t^*, \end{aligned} \tag{13.55}$$

from which we can read off the velocity components in the starred coordinate system:

$$\begin{aligned} u_x^* &= (u_x - v)/[1 - (vu_x/c^2)], \\ u_y^* &= u_y/\gamma[1 - (vu_x/c^2)], \\ u_z^* &= u_z/\gamma[1 - (vu_x/c^2)]. \end{aligned} \tag{13.56}$$

In the nonrelativistic limit, these equations reduce to the classical formulas. It can be shown (see Problem 17) that for a given speed u less than c, the speed u^* is a maximum when $\boldsymbol{u}$ is directed along the x-axis in the opposite direction to $\boldsymbol{v}$, and that u^* is always less than c.

We remarked in Section 13.4 that if a clock at rest in an inertial coordinate system is passed twice by a moving clock, the time interval recorded on the moving clock between the two passages will be less than that on the clock at rest. This result is sometimes called the *twin paradox* and may be formulated pictorially as follows. An astronaut sets out on a high-speed rocket journey, leaving his identical twin at rest in an inertial system. After traveling at a high speed for a long time, the astronaut reaches a distant planet, reverses his course, and returns home again at high speed. According to the theory of relativity, he will find that his twin who has remained at home has aged more than he during the time he was on his trip. There is no logical inconsistency here, in that the two twins are no longer identical, because they have not had identical histories. The astronaut was accelerated during his trip and did not remain at rest in an inertial coordinate system. Nevertheless it is instructive to study this situation in more detail, both from the point of view of the astronaut, and from the point of view of the twin who stays at home. (See Fig. 13.7.)

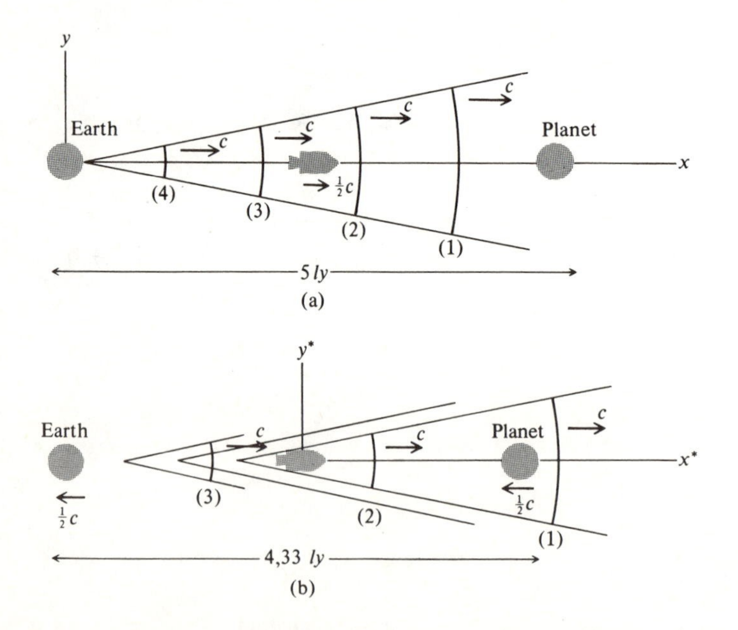

Fig. 13.7 (a) A voyage to a distant planet, in the Earth coordinate system, at the moment when the rocket is half-way. (b) Same voyage, in the rocket coordinate system, at the moment when the voyage is half over. Shown in both cases are light signals sent out once a year from Earth, labeled in parentheses with the year of emission. (Rocket and planets not drawn to scale.)

Let us take the sedentary twin to be at rest at the origin of an unstarred coordinate system. The distant planet is also at rest in this coordinate system. To simplify the arithmetic, let us assume that the astronaut travels at a speed $\frac{1}{2}c$ along the x-axis for a time of 10 years, arriving at the planet located at $x_1 = 5$ light years. He then decelerates and comes to rest in a time which we will assume is short in comparison with 10 years, and accelerates rapidly to a speed $\frac{1}{2}c$ in the opposite direction, returning home 20 years after leaving, according to the clock at rest at the origin. For $\beta = 1/2$, $\gamma = 2/\sqrt{3}$. The astronaut's clock will read $20\sqrt{3}/2$ years when he returns, assuming that the clock was not damaged in the deceleration process, and assuming that the deceleration occurred in a short enough time so that we need not be concerned about the change in reading of the astronaut's clock during this period. If radio signals are sent out every year from the origin according to the time on the origin clock, the signals will overtake the astronaut during his journey out with a velocity of $\frac{1}{2}c$ relative to the astronaut as measured from the unstarred system. These yearly signals will be received by the astronaut at two year intervals as measured in the unstarred coordinate system, and he will receive the fifth signal at the point where he turns around. On the return journey, he will receive the remaining fifteen signals at regular intervals.

Let us now describe the sequence of events from the point of view of the astronaut. Let us introduce a starred coordinate system whose origin coincides with the astronaut during his outward journey. In this coordinate system, due to the Lorentz contraction, the distance between the Earth and the planet is only $5\sqrt{3}/2$ light years. The earth and the planet move with a speed $\frac{1}{2}c$, so that if the Earth leaves the astronaut at $t^* = 0$, the planet will arrive at $t_1^* = 10\sqrt{3}/2$ years, in agreement with our previous discussion, from the point of view of the unstarred system, of the reading of the astronaut's clock. The unstarred clock left on Earth will be running slow from the point of view of the starred system and will, when the distant planet arrives at the astronaut, be reading a time $(10\sqrt{3}/2)(\sqrt{3}/2) = 7.5$ years. The signal arriving at the planet from the Earth clock indicates five years; if the astronaut corrects for the fact that the Earth's clock is running slowly, he concludes that this signal was sent out at $t^* = 5(2/\sqrt{3})$ years. Since the Earth is traveling at half the speed of light, it sent out the five-year signal when it was $5/\sqrt{3}$ light years from the astronaut's rocket. If we add the time $5/\sqrt{3}$ years for the signal to reach the astronaut to the time $10/\sqrt{3}$ years at which it was sent out, we arrive at the present time $5\sqrt{3}$ read on the astronaut's clock. All is therefore consistent.

Having made these calculations, the astronaut rapidly decelerates and lands on the planet. At the location of the astronaut on the planet, nothing has changed, except that he is now at rest in the unstarred coordinate system. His clock still reads $5\sqrt{3}$ years (plus the negligible time added while he was decelerating). The local unstarred clocks on the planet read 10 years, as they did just before the astronaut stopped to land. The 5 year radio signal from the Earth has just passed. If however, the astronaut now calculates the present time which must be shown by the clock on the Earth, his definition of simultaneity has changed. He will now add the five-year light-travel time from the Earth to the five-year signal which he receives and will conclude that the Earth clock, like the unstarred clock on the planet, is now reading 10 years. This does not mean that he concludes that the Earth clock suddenly jumped from 7.5 to 10 years. Everything has been running smoothly back on Earth. It merely means that his definition of simultaneity has changed, so that he now regards a different moment in the history of the twin back home as being simultaneous with his arrival at the planet.

It is left to the reader to supply a similar discussion for the trip home and to show that in a double-starred coordinate system whose origin coincides with the astronaut during the trip home, the Earth clock also runs slowly by the factor $\sqrt{3}/2$, but that again because of a changed definition of simultaneity, the earth clock is ahead of the astronaut's clock from the point of view of the double-starred system when he leaves the planet, so that, in spite of running more slowly, it still reads 20 years when he arrives home with his clock reading $20/\sqrt{3}$ years.

PROBLEMS

1. A pair of stars A and B revolve about one another in circular orbits with a period T. The Earth lies in the orbital plane at a distance x from the pair of stars, so that their orbits are seen from Earth edge-on. Make the hypothesis that light travels with speed c relative to the source which emits the light. Show that if this were the case, each star as seen from Earth would appear to traverse the semicircle on the far side of its orbit more rapidly than it traverses the near semicircle. Show that if

$$4vx/T = c^2 - v^2,$$

where v is the speed of star A in its orbit, then A would be seen simultaneously at both ends of the diameter of its orbit. See if you can pick reasonable values of v, T, and x for which this condition holds. No such effect has ever been seen.

2. Assume that light travels with speed c relative to a fixed coordinate system through which the Earth travels with velocity $\mathbf{v}$. Show that a light pulse traveling back and forth along a line parallel to $\mathbf{v}$ has an average speed, relative to the Earth, given by

$$\langle v_\parallel \rangle = c[1-(v^2/c^2)],$$

in the sense that the time required for a round trip between two points a distance l apart on Earth is $2l/\langle v_\parallel \rangle$. Find the average speed $\langle v_\perp \rangle$ relative to the earth for a round trip along a line perpendicular to $\mathbf{v}$. If $v = 30$ km-sec^{-1} (speed of the earth in its orbit), find the ratio $\langle v_\perp \rangle / \langle v_\parallel \rangle$. This is the basis of the Michelson-Morley experiment, which showed that $\langle v_\perp \rangle / \langle v_\parallel \rangle = 1$ to within about one part in 10^9.

3. Three rocket ships A, C, and B fly in formation in a straight line at a velocity of 45 km-sec^{-1} relative to the Earth. Ships A and B are 20,000 km apart (as measured in a coordinate system in which the Earth is at rest), and C is midway between them. They all fly in the direction of the line from B to A. To synchronize their clocks, ships A and B send radio signals which are received by ship C. A particular signal a from A is received by C simultaneously with receipt of a signal b from B. Relative to a starred coordinate system in which the rocket ships are at rest, the signals a, b were therefore sent out simultaneously from A and B. (By symmetry, C is also equidistant from A and B in the starred system.)

Show that relative to a coordinate system in which the Earth is at rest, the signals a, b were not sent out simultaneously from A and B, although an Earth observer would agree that they were received simultaneously at C. Which signal was sent out first? How much later was the other signal sent out? (Note that you do not need to know how to transform from one coordinate system to another to do this problem, since all calculations can be carried out relative to the Earth.)

4. A meter stick with ends O^*, P^* moves with velocity $\boldsymbol{v}$ parallel to its length. An observer moving with the stick measures its length l^* in the following way. As the end O^* passes the origin O of a fixed coordinate system, a light pulse is sent out from O^* and travels to the other end P^* where it is reflected from a mirror. The pulse returns to the end O^* at a time t^* as recorded on a moving clock, so that $l^* = 2\,ct^*$. If the same light pulse is used to measure the length l of the stick relative to the fixed coordinate system, write a formula for l in terms of the measured time t of arrival of the pulse back at O^*. Using the relation (13.21), show that this method of measurement will give the same relation (13.22) between l and l^*.

5. Equation (13.21) was derived by an argument carried out relative to the starred coordinate system, in order to simplify the argument. Show that the same equation can be derived by an argument carried out relative to the unstarred coordinate system. Use the fact that the starred clocks run slowly by the factor $[1-(v^2/c^2)]^{1/2}$ relative to the unstarred clocks, and take into account the way in which the starred observer synchronizes his clocks, but make all calculations relative to the unstarred coordinate system.

6. An object A is at rest in an unstarred coordinate system and has a length l_A in that system. An object B is at rest in a starred system moving relative to A with velocity $\boldsymbol{v}$ parallel to the length l_A. The length of B in the starred system is $l_B^* = l_A$. The measured length of B in the unstarred system is $l_B = l_B^*/\gamma$, and likewise $l_A^* = l_A/\gamma$. Show in detail how to reconcile the apparent paradox that each object is shorter than the other, in a coordinate system in which the other is at rest. Assume that each length is measured by the method of noting the simultaneous positions of its two ends at a fixed time, and measuring the distance between the noted points. [*Hint*: Show that the difference in simultaneity between the two coordinate systems just accounts for the discrepancy between the measured lengths.]

7. A small box of volume dV, moving with velocity $\boldsymbol{v}$ contains a clock moving with it. The clock reads a time interval $d\tau$ between two successive events occurring within the box. Let dV_0 be the volume of the box as measured by instruments at rest within the box. Show, using the properties of moving clocks and meter sticks, that

$$dV_0 d\tau = dV dt,$$

where dt is the time interval (in our coordinate system) between the same two events.

8. Derive the Lorentz transformation equations (13.35) in the following way.
a) Assume the relation between starred and unstarred variables is linear:

$$x^* = a_{11}x + a_{12}y + a_{13}z + a_{10}t,$$

$$y^* = a_{21}x + a_{22}y + a_{23}z + a_{20}t,$$

$$z^* = a_{31}x + a_{32}y + a_{33}z + a_{30}t,$$

$$t^* = a_{01}x + a_{02}y + a_{03}z + a_{00}t.$$

Show that this assumption is equivalent to assuming that space and time are homogeneous, that is, show that a nonlinear relation would imply that different regions in space-time are to be transformed differently.

b) Find the coefficients a_{ij} by assuming (1) that the axes of the two systems are parallel (and therefore coincide at $t = 0$), (2) that the origin O^* moves along the x-axis with velocity v, (3) that a light wave travels with speed c in both systems, that is, Eq. (13.15) must go over into Eq. (13.14) when the above substitutions are made, (4) that increasing t increases t^* (that is, $a_{00} > 0$), (5) that the coefficients in the inverse transformation, from x^* back to x, must be obtained from a_{ij} by replacing v by $-v$, and (6) that any scalar quantity appearing in the transformation equations can depend only on the magnitude and not on the sign of v. [*Hints and cautions*: If you follow the most expeditious algebraic route, you should find that you need the assumptions in the order that they are given. Note that assumption (1) does not mean that at $t = 0$ a point on the y-axis has the same coordinates y and y^*, but only that it lies on the y^*-axis, and that increasing y increases y^*. If you find you do not need assumptions (5) and (6), you have made some other, unwarranted, assumption.]

9. Solve Eqs. (13.35) for x, y, z, t and show that the result has the same form with v replaced by $-v$.

10. Given two events E_0 and E for which the quantity S given by Eq. (13.16) is positive, find a Lorentz transformation to a starred coordinate system in which E_0 and E occur simultaneously. To simplify the algebra, choose the original coordinate system so that E_0 is at the origin, and E is at the point $(x, 0, 0, t)$. Find the velocity βc of the starred coordinate system, and verify that S is the square of the distance between the two events in the starred system.

11. Given two events E, E_0 with coordinates (x, y, z, t) and (x_0, y_0, z_0, t_0), make a Lorentz transformation according to Eqs. (13.35) and verify that the quantity S given by Eq. (13.16) is given by the same formula in the starred coordinate system. [You may use the result of Problem 9.]

12. Show that there can be no more than one speed which is the same in all coordinate systems, that is, show that if we tried to postulate some phenomenon traveling at a universal constant speed v different from c, we could not find any reasonable transformation which would make both v and c the same in all coordinate systems. [There are many ways to show this, some quite simple, some more involved. Try to find an argument which convinces you.]

13. Derive formula (13.52) for the relativistic Doppler effect by taking ω^* to be the frequency of a moving light source as read by a clock moving with the source. Assume that the moving clock runs slowly according to Eq. (13.18) relative to a fixed coordinate system, and that the emitted light waves travel with speed c in the fixed system. Calculate the frequency ω with which these waves pass a fixed point.

14. A straight meter stick at rest in an unstarred coordinate system makes an angle α with the x-axis. A starred coordinate system with parallel axes moves with velocity v along the x-axis. Show that the meter stick is also straight in the starred system and find the angle α^* it makes with the x^*-axis.

15. Given the Lorentz transformation equations (13.35), derive formula (13.26) for the Lorentz contraction. [*Suggestion*: Write down the coordinates of the ends of a rod of length l in a coordinate system in which it is at rest, and transform to a moving coordinate system.]

16. Carry out a second Lorentz transformation of the type (13.35) to a double-starred system moving with velocity v^* along the x^*-axis, relative to the starred system. Express x^{**}, y^{**}, z^{**}, t^{**} in terms of x, y, z, t and show that the result has the form (13.35), with a relative velocity given by Eq. (13.56).

17. Show that the speed u^* of a particle relative to a starred coordinate system is given in terms of the velocity $\mathbf{v}$ of the coordinate system and the velocity $\mathbf{u}$ of the particle relative to an unstarred coordinate system by the equation

$$(u^*)^2 = c^2\left\{1 - \frac{[1-(v^2/c^2)][1-(u^2/c^2)]}{[1-(\mathbf{u}\cdot\mathbf{v})/c^2]^2}\right\}.$$

Show that for fixed u and v, u^* is a maximum when $\mathbf{u}$ is in the direction of $-\mathbf{v}$, and that $u^* < c$ if $u < c$.

18. Discuss the homeward journey of the astronaut described at the end of Section 13.6. Relative to a coordinate system moving with the astronaut on the return journey, find the time shown by the Earth clock when the astronaut begins the return journey. Add the time interval recorded by the Earth clock during the return journey, corrected for the Earth's motion, and show that the result is 20 years.

Show that the five yearly signals received from Earth on the outward journey and the fifteen signals received on the return journey are consistent with Formula (13.52) for the relativistic Doppler shift for a signal frequency of one per year relative to the Earth.

19. A small rectangular box having dimensions dx, dy, dz moves with velocity $\mathbf{u}$ during a time dt. Show by using Eqs. (13.35) that

$$dx\,dy\,dz\,dt = dx^*dy^*dz^*dt^*,$$

where dx^*, dy^*, dz^* are the dimensions of the same box and dt^* is the corresponding time interval, in a starred coordinate system. Take the dimensions small enough so that the higher order terms in the differentials may be neglected. You may assume that a rectangular box remains a rectangular box if one of its edges is parallel to the direction of motion. [This problem must be done very carefully. Let one corner of the box be at the origin at $t = 0$, and let E_1 be an event occurring at the same corner at time dt later. Find the time dt^* of E_1 in the starred system. (Do not forget the box is moving.) (If dx, dy, dz are small, you would get the same result for dt^*/dt at any corner of the box, to first order.) To find dx^*, remember that it is the distance between two events $(dx^*, 0, 0, 0)$ and $(0, 0, 0, 0)$ which are simultaneous in the starred coordinate system.]

Can you prove the above result by a simple argument using the result of Problem 7?

20. An alternative way to synchronize clocks would be to bring them to a common point where they can be synchronized unambiguously, and then move them very slowly (so that time-dilation is negligible) to their permanent positions in the coordinate system. Show that this works by analyzing, in an unstarred coordinate system, the application of this method to synchronize clocks in a starred coordinate system moving in the x-direction with speed v. A clock is brought to the origin O^* and synchronized with a standard starred clock so that it reads zero at $t^* = 0$. It is then moved with a small velocity $\boldsymbol{u}^*$ to a new position in the starred coordinate system. Express the proper time interval $d\tau$ for the moving clock in terms of the unstarred time dt, the speed v of the starred system, and the velocity $\boldsymbol{u}^*$ of the clock relative to the starred system. Expand $d\tau$ in a power series in $\boldsymbol{u}^*$. Assume that $\boldsymbol{u}^*$ (but not v) is small but is otherwise an arbitrary function $\boldsymbol{u}^*(t)$. Show that if terms involving higher than the first power of $\boldsymbol{u}^*$ can be neglected, then the moving clock at any time t reads the correct starred time t^* corresponding to its location at that time. Note that the term linear in $\boldsymbol{u}^*$ just gives the proper phase shift (relative to the unstarred system) between starred clocks at different locations. What would be the effect of keeping quadratic terms in $\boldsymbol{u}^*$? Show that the linear term must be kept even in the limit $u^* \to 0$ if the clock is to be moved a finite distance away from O^*, but that the higher terms do vanish in this limit.

CHAPTER 14

RELATIVISTIC DYNAMICS

14.1 SPACE-TIME VECTOR ALGEBRA

In Chapter 3 we developed the algebra of vectors in three-dimensional space as a useful mathematical tool for writing physical equations and doing calculations in a form independent of any particular three-dimensional coordinate system. We now wish to formulate an analogous algebra in the four-dimensional space-time whose "points" are events labeled by coordinates (x, y, z, t), where these variables transform according to the formulas for the Lorentz transformation when we go to a moving coordinate system. We will define a *4-scalar* or *world scalar* as a quantity whose value does not change when we transform the coordinates according to the equations for the Lorentz transformation. Likewise we will define a *4-vector* as a set of four quantities which transform like the differences of the four coordinates of two events under a Lorentz transformation. This 4-vector algebra greatly simplifies the problem of discovering the proper form for the laws of physics so that they have the same form in all coordinate systems. Thus for example, if we formulate a law by asserting that two 4-scalars are equal, or that two 4-vectors are equal, it will be evident from the form of the equation that it is the same in all coordinate systems.

It is convenient to define a time variable which is measured in the same units as the space variables. We will therefore define the four coordinates of an event in space-time as follows:

$$\begin{aligned} x_0 &= ct, \\ x_1 &= x, \\ x_2 &= y, \\ x_3 &= z. \end{aligned} \tag{14.1}$$

We will use a Greek letter to indicate any one of the four subscripts 0, 1, 2, 3. Thus 'x_μ' designates any one of the four variables x_0, x_1, x_2, x_3; we may also sometimes use 'x_μ' to represent the set of four variables. A Latin letter will designate any one of the values 1, 2, 3. Thus 'x_i' will designate any one of the three space coordinates.

Given any two events* E^1 and E^2, we saw in Section 13.3 that the quantity

$$S^{21} = (x_1^2-x_1^1)^2+(x_2^2-x_2^1)^2+(x_3^2-x_3^1)^2-(x_0^2-x_0^1)^2, \tag{14.2}$$

*When several events are under discussion we will use superscripts to designate a particular event, in order to reserve subscripts for designating one of the four coordinates.

given by Eq. (13.16), is zero on the light cone, is negative when one of the events can be reached from the other by a body traveling at a speed less than that of light, and is positive when that is not possible. If the quantity S^{21} is zero (or negative or positive) in one inertial system, then it is also zero (or negative or positive) in any other inertial system. It is a matter of straightforward algebra to verify that if we substitute from the Lorentz transformation equations (13.35) the quantity S^{21} has in all cases the same value in the new coordinate system. (See Problem 11, Chapter 13.) We can see from Eq. (14.2) that S^{21} is unchanged in value if we change the space or time origin of the coordinate system. It is also evident that if we simply rotate the space coordinates to a new orientation, keeping the origin fixed, S^{21} retains its value. In fact if we simply transform to a new set of space axes, then the last term remains separately constant, and likewise the sum of the first three terms remains constant. When we go to a moving coordinate system all four terms may change, but the sum of the four terms remains constant. We conclude that the quantity S^{21} has the same value in any inertial coordinate system. A quantity with this property is called a *4-scalar* or sometimes a *world scalar*. It is convenient to introduce the following abbreviations:

$$g_0 = -1, g_1 = g_2 = g_3 = 1. \tag{14.3}$$

We can now write Eq. (14.2) in the following abbreviated form:

$$S^{21} = \Sigma_\mu g_\mu (x_\mu^2 - x_\mu^1)^2. \tag{14.4}$$

The reader will note the analogy between the quantity (14.2), which has the same value in all inertial coordinate systems, and the quantity

$$(x_1^2 - x_1^1)^2 + (x_2^2 - x_2^1)^2 + (x_3^2 - x_3^1)^2,$$

which has the same value in all three-dimensional rectangular coordinate systems which are stationary with respect to one another. We will base our development of the algebra of space-time on this analogy.

Some authors define a time variable

$$x_4 = ict, \tag{14.5}$$

instead of our variable x_0. Since $(x_4)^2 = -(x_0)^2$, this makes the quantity S^{21} look formally exactly like the corresponding quantity in three-dimensional space. This formal identity however tends to obscure the physical difference between the time coordinate and the three space coordinates, which in the case of the definition (14.5) is preserved by requiring that the fourth coordinate must always be pure imaginary while the other three coordinates are real. We prefer here to stick to the definition (14.1), which has the advantage that all four coordinates are real and that the difference between the space and time coordinates is kept clearly before us by the minus sign in the form (14.2). Although the space and time coordinates are not physically the same, they are related, and they are intermixed under a Lorentz transformation (13.35).

The change of variables from any unstarred inertial coordinate system to any other starred inertial coordinate system can be written in the form

$$x_\mu^* = x_\mu^{*O} + \Sigma_\nu a_{\mu\nu} x_\nu, \tag{14.6}$$

where the four quantities x_μ^{*O} are the coordinates of the unstarred origin O in the starred coordinate system, and the coefficients $a_{\mu\nu}$ depend upon the orientation of the starred axes relative to the unstarred axes and upon the relative motion of the two coordinate systems. If the starred and unstarred axes are parallel and if the origin O^* is moving with a velocity v along the x-axis, then these coefficients are given by Eqs. (13.35):

$$a_{\mu\nu} = \begin{array}{c} \mu\backslash^\nu \\ 0 \\ 1 \\ 2 \\ 3 \end{array} \begin{array}{c} \begin{array}{cccc} 0 & 1 & 2 & 3 \end{array} \\ \begin{pmatrix} \gamma & -\beta\gamma & 0 & 0 \\ -\beta\gamma & \gamma & 0 & 0 \\ 0 & 0 & 1 & 0 \\ 0 & 0 & 0 & 1 \end{pmatrix} \end{array}. \tag{14.7}$$

If the starred coordinate system is at rest relative to the unstarred coordinate system but the starred axes are not parallel to the unstarred axes, then the coefficients $a_{\mu\nu}$ are given by Eqs. (7.13):

$$a_{\mu\nu} = \begin{pmatrix} 1 & 0 & 0 & 0 \\ 0 & \hat{x}_1\cdot\hat{x}_1 & \hat{x}_1\cdot\hat{x}_2 & \hat{x}_1\cdot\hat{x}_3 \\ 0 & \hat{x}_2\cdot\hat{x}_1 & \hat{x}_2\cdot\hat{x}_2 & \hat{x}_2\cdot\hat{x}_3 \\ 0 & \hat{x}_3\cdot\hat{x}_1 & \hat{x}_3\cdot\hat{x}_2 & \hat{x}_3\cdot\hat{x}_3 \end{pmatrix}. \tag{14.8}$$

If we now make a transformation to a third (primed) inertial system according to the equations

$$x'_\lambda = x'^{O*}_\lambda + \Sigma_\mu a^*_{\lambda\mu} x^*_\mu, \tag{14.9}$$

then it is a simple matter to verify that if we substitute from Eqs. (14.6), we obtain the relation between the primed and unstarred coordinate systems as follows

$$x'_\lambda = x'^{O}_\lambda + \Sigma_\nu a'_{\lambda\nu} x_\nu, \tag{14.10}$$

where

$$x'^{O}_\lambda = x'^{O*}_\lambda + \Sigma_\mu a^*_{\lambda\mu} x^{*O}_\mu, \tag{14.11}$$

and

$$a'_{\lambda\mu} = \Sigma_{\ \mu} a^*_{\lambda\mu} a_{\mu\nu}. \tag{14.12}$$

The reader may notice that Eq. (14.12) is just the rule for multiplying two 4×4 matrices. By a succession of transformations of the special forms (14.7) and (14.8), we can obtain the relation (14.6) between any two inertial coordinate systems, regardless of the locations of their origins or the orientations of their axes, or their relative velocities with respect to one another. It will be convenient to refer to the transformation (14.6) as a "Lorentz transformation" and thus to define a Lorentz transformation as any transformation between any two inertial coordinate systems.* We will however restrict ourselves to rectangular spatial coordinate systems, at least when we wish to deal with transformations between moving reference systems. In solving a particular problem in one single reference system, we are of course free to introduce curvilinear space coordinate systems if that is convenient.

The coefficients $a_{\mu\nu}$ depend upon the velocity of the starred coordinate system relative to the unstarred system (which may be specified by giving its three components) and upon the orientation of the starred coordinate axes relative to the unstarred axes (which requires three angles for its specification). Since they depend upon six parameters, the sixteen coefficients $a_{\mu\nu}$ cannot all be independent, and we expect that there should be ten relations between them. It is not hard to show (see Problem 2) that if the scalar S^{21} [Eq. (14.4)] is to have the same value in the starred and unstarred coordinate systems, for any two events E^1, E^2, then the following ten relations must hold:

$$\Sigma_\mu\, g_\mu a_{\mu\nu} a_{\mu\lambda} = g_\nu \delta_{\nu\lambda} = \begin{cases} g_\nu & \text{if } \nu = \lambda, \\ 0 & \text{if } \nu \neq \lambda. \end{cases} \tag{14.13}$$

The reader who has studied Chapter 10 will note the analogy between Eqs. (14.13) and Eqs. (10.78), which hold between the coefficients a_{ij} in the transformation (10.69) between two three-dimensional coordinate systems.

The coefficients $a_{\mu\nu}^{-1}$ for the inverse of the transformation (14.6),

$$x_\nu = x_\nu^{O*} + \Sigma_\mu a_{\nu\mu}^{-1} x_\mu^*, \tag{14.14}$$

are given by

$$a_{\nu\mu}^{-1} = g_\mu g_\nu a_{\mu\nu}. \tag{14.15}$$

To see this, note that it follows from Eqs. (14.13), since $(g_\nu)^2 = 1$, that

$$\Sigma_\lambda a_{\mu\lambda}^{-1} a_{\lambda\nu} = \delta_{\mu\nu}.$$

By comparing this equation with Eq. (14.12), we see that the coefficients $a_{\mu\nu}^{-1}$ are indeed those of the inverse transformation (14.14).

We have defined a 4-scalar as any physical quantity whose value is unchanged

---

*The name *Lorentz transformation* is often restricted to transformations in which the origins O, O^* coincide. The larger group of transformations which allow also a new choice of origin is called the Poincaré group. Although for general statements we want to consider any two inertial coordinate systems, in specific examples we will usually use only the special case (13.35) which is the one given explicitly by Lorentz although in a somewhat less transparent form. (H. A. Lorentz, "Electromagnetic phenomena in a system moving with any velocity less than that of light", *Proc. Acad. Sci., Amsterdam*, **6**, 1904.) The algebra of 4-vectors was introduced by Poincaré in 1905 and developed later by Minkowski in 1908.

by a Lorentz transformation. A typical 4-scalar is the quantity S^{21} defined by Eq. (14.4). We now define a *4-vector* as any set of four quantities A_μ which transform under a Lorentz transformation (14.6) according to the rule

$$A_\mu^* = \Sigma_\nu \, a_{\mu\nu} A_\nu. \tag{14.16}$$

A typical 4-vector is the set of coordinate differences $(x_\mu^2 - x_\mu^1)$ between two events E^1 and E^2. It is evident from Eq. (14.6) that this 4-vector, which we may call the *4-vector displacement* between the two events, transforms according to Eq. (14.16). In the special case when the starred and unstarred axes are parallel, and when the velocity is along the x-axis, the transformation rule for the components of a 4-vector is

$$\begin{aligned} A_0^* &= \gamma(A_0 - \beta A_1), \\ A_1^* &= \gamma(A_1 - \beta A_0), \\ A_2^* &= A_2, \\ A_3^* &= A_3. \end{aligned} \tag{14.17}$$

In a transformation between two coordinate systems which are at rest relative to one another, the component A_0 does not change (it is a 3-scalar) and the components A_1, A_2, A_3 transform like an ordinary 3-vector. Thus we see that, so long as we do not change to a moving frame of reference, a 4-vector consists of a scalar plus the three components of a vector in ordinary three-dimensional space. When we transform to a moving coordinate system however, the 3-scalar part A_0 gets mixed with the 3-vector part, and hence we have to consider the four quantities A_μ as components of a single entity, the 4-vector.

The reader will observe that if A_μ, B_μ are 4-vectors, then so are the sets of four quantities obtained by multiplying the components of one vector by a scalar or by adding the two component by component:

$$cA_\mu \textit{ is a 4-vector if } A_\mu \textit{ is, and if } c \textit{ is a 4-scalar,} \tag{14.18}$$

and

$$A_\mu + B_\mu \textit{ is a 4-vector if } A_\mu, B_\mu \textit{ are.} \tag{14.19}$$

Given two 4-vectors A_μ, B_μ, it follows from the transformation equations (14.16) and (14.13) that the quantity

$$\begin{aligned} (A_\mu, B_\mu) &= \Sigma_\mu \, g_\mu A_\mu B_\mu, \\ &= A_1 B_1 + A_2 B_2 + A_3 B_3 - A_0 B_0, \end{aligned} \tag{14.20}$$

is a 4-scalar. Its value is not changed under a Lorentz transformation. We leave this to the reader to show (see Problem 3). The converse statement is also true (see Problem 4).

Given four quantities A_μ transforming according to some rule upon a change in coordinate system, if the product (A_μ, B_μ) is a 4-scalar for any arbitrarily chosen 4-vector B_μ, then A_μ is a 4-vector. (14.21)

The quantity (14.20) will evidently play a role analogous to the scalar product of two vectors in the vector algebra of three-dimensions. We may call it the 4-scalar product of two 4-vectors. Note however that the self product (A_μ, A_μ) is not positive definite as is the corresponding product in three-dimensional vector algebra.

It is not possible to define a 4-vector analog of the cross product, as the reader will readily verify with a little experimentation. In fact, the analog of the cross product (3.32) does not even have the right number of components to be a 4-vector.*

It is not customary to introduce a special typeface for a 4-vector as we did for vectors in three-dimensional space. We will simply designate a 4-vector by writing its general component as 'A_μ'.

If for two given events E^1, E^2, the scalar S^{21} defined by Eq. (14.2) is positive, we say that the interval between the two events is *space-like*, and we define a *proper distance* between the two events by the equation

$$\sigma^{21} = (S^{21})^{1/2}. \tag{14.22}$$

The proper distance σ^{21} is evidently a 4-scalar in this case, and can be defined in an absolute way as the distance between the two events as measured by a meter stick at rest in a coordinate system which is so moving that in it the two events occur simultaneously.

If for two events E^1, E^2, the scalar S^{21} is negative, then we say that the interval between the two events is *time-like*. In this case we can define a 4-scalar as follows:

$$\tau^{21} = (-S^{21})^{1/2}/c. \tag{14.23}$$

The quantity τ^{21} we will call the *proper time interval* between the two events, and we will take it to be positive if event 2 follows event 1 and negative if it precedes event 1. We may define the scalar τ^{21} in a manner independent of the coordinate system as the time interval read on a clock traveling at the proper constant velocity so as to leave the event E^1 and arrive at the event E^2 (or vice versa).

As we saw in the preceding chapter, the velocity components dx_i/dt of a moving particle transform under a Lorentz transformation according to the rather complicated Eqs. (13.56). It is evident from these equations that there is no way in which we can add a 3-scalar u_0 to the 3-velocity u_i in order to form a 4-vector. We can however construct a 4-vector which is closely related to the velocity in the following way. Let two nearby events in the history of a moving particle have the coordinates x_μ and $x_\mu + dx_\mu$. The coordinate differences dx_μ evidently are the

*The reader who has studied Section 10.2 will recognize that the correct analog of the cross product is the antisymmetric 4-tensor $A_\mu B_\nu - A_\nu B_\mu$. Since it has six independent components, it cannot be correlated with any 4-vector. It is a numerical accident that in three-dimensional space the antisymmetric tensor and the vector each have just three components.

components of an (infinitesimal) 4-vector. The proper time interval between the two events

$$d\tau = [-(dx_\mu, dx_\mu)]^{1/2}/c \tag{14.24}$$

is a 4-scalar. We can therefore [by rule (14.18)] form a 4-vector, which we will call the *4-velocity*, as follows:

$$U_\mu = dx_\mu/d\tau. \tag{14.25}$$

Because the 4-velocity U_μ is a 4-vector, we will find it useful when we come to attempt to write equations of motion in a form which will be invariant under a Lorentz transformation. The 4-velocity is evidently related to the 3-velocity by the equations

$$\begin{aligned} U_0 &= c\,dt/d\tau = \gamma c, \\ U_i &= dx_i/d\tau = \gamma u_i, \end{aligned} \tag{14.26}$$

where

$$\gamma = [1-(u^2/c^2)]^{-1/2}.$$

We can readily calculate the 4-scalar product

$$\Sigma_\mu g_\mu (U_\mu)^2 = -c^2. \tag{14.27}$$

The 4-vector U_μ is an example of a 4-vector which has actually only three independent components, because its self-scalar product must satisfy the condition (14.27). If we are given the three velocity components u_i, then the four components of the 4-velocity U_μ are given by Eqs. (14.26). Conversely of course, if we know the 4-velocity U_μ, or even its space part U_i, we can find the 3-velocity u_i. Note also that in the nonrelativistic limit $u \ll c$, $U_i \doteq u_i$, $U_0 \doteq c$.

If a physical law is formulated in such a form that it asserts that two 4-scalars are equal, or that two 4-vectors have equal components, then these equations will clearly have the same form in all inertial coordinate systems. We may say that they are *manifestly covariant*. It is not necessary that the laws of physics be written in terms of 4-scalars and 4-vectors. In fact since 4-vector algebra had not yet been invented, Einstein in his first paper did not have this tool available. He found a covariant way of writing the equations of mechanics and electrodynamics, that is, a form which was the same in all inertial coordinate systems. In order to prove this, however, it was necessary to discover the special transformation laws for each of the physical quantities appearing in the equations (for example the rather complicated law (13.56) for the velocity), and then to show by explicit calculation that when the appropriate Lorentz transformation is made on all quantities in the equations, the equations take on the same form in the new starred coordinate system.

Given a set of four quantities A_0, A_1, A_2, A_3, each perhaps determined by some kind of a measurement with reference to some unstarred coordinate system, we could always define in a formal way a 4-vector, by specifying that its components in that coordinate system are to be the four quantities A_μ and that its components in any other coordinate system are to be calculated from the transformation Eqs. (14.16). Such a formal 4-vector, however, would not in general be a physical 4-vector. That is, the operational definition specifying the way in which the quantities A_ν^* are to be measured in the starred coordinate system will not in general be the same as the operational definition specifying how the quantities A_μ are measured in the unstarred coordinate system. In order to define a physical 4-vector, we must define the four quantities A_μ by specifying how they are to be measured in any particular coordinate system, using the same definition in each coordinate system, and then we must show that the relation between the four quantities in any two coordinate systems is given by Eqs. (14.16). Alternatively, we may give an operational definition of the quantities A_μ in one coordinate system, define the components A_ν^* in any other coordinate system by Eqs. (14.16), and then we must prove that the quantities A_ν^* may in fact be found in the starred coordinate system by using the same operational definition as was used in the unstarred coordinate system. As an example, the 4-velocity U_μ was constructed above in such a way as to make it clear from the beginning that it was a 4-vector. Equations (14.26) then make it clear that the 4-velocity U_μ has the same operational definition in each coordinate system; we measure the 3-velocity u_i in each coordinate system in the usual way, and then calculate the components U_μ from Eqs. (14.26). Conversely, we could start with the operational definition (14.26) and proceed to prove that the four quantities U_μ do indeed transform like a 4-vector. It is evident that if a law of physics which asserts the equality of two 4-vectors is to satisfy Postulate (13.1) [or (13.42)], then the 4-vectors which appear in this equation must be physical 4-vectors in the above sense. Postulate (13.1) requires that the laws of physics have the same form and the same meaning in every inertial coordinate system. The same sort of remarks evidently apply also to 4-scalars.

In analogy with the corresponding terminology in the case of the 4-vector interval between two events, we sometimes say that a 4-vector A_μ is *space-like* if the self scalar $(A_\mu, A_\mu) > 0$, and is *time-like* if $(A_\mu, A_\mu) < 0$. According to Eq. (14.27), the 4-velocity is a time-like 4-vector. In a particular coordinate system, we may write the self-scalar product of a 4-vector in terms of its time component A_0 and its space part $\boldsymbol{A} = (A_1, A_2, A_3)$:

$$(A_\mu, A_\mu) = \boldsymbol{A}\cdot\boldsymbol{A} - (A_0)^2. \tag{14.28}$$

Each of the two terms $(A_0)^2$ and $\boldsymbol{A}\cdot\boldsymbol{A}$ is a 3-scalar; only the combination (14.28) is a 4-scalar.

As an application of the algebra of 4-vectors, let us consider a plane wave with angular frequency ω and wave number $\boldsymbol{k}$ given by the equation

$$W = Ae^{i(\boldsymbol{k}\cdot\boldsymbol{r} - \omega t)}. \tag{14.29}$$

If we define the four quantities

$$k_0 = \omega/c,\ k_1 = k_x,\ k_2 = k_y,\ k_3' = k_z, \tag{14.30}$$

we can write the wave W in the form

$$W = Ae^{i(k_1x_1+k_2x_2+k_3x_3-k_0x_0)}. \tag{14.31}$$

Since the phase $(\boldsymbol{k}\cdot\boldsymbol{r}-\omega t) = (k_\mu,\ x_\mu)$ at any particular point on the wave is determined by its location either in space or time relative to a crest or a trough of the wave, and since a crest or a trough can be identified unambiguously in any coordinate system, the phase must be a scalar, having the same value (up to an additive multiple of π) in all coordinate systems. We conclude, using theorem (14.21), that the four quantities k_μ are the components of a 4-vector, since their product with x_μ is a scalar. We will call the 4-vector k_μ the *wave-vector*. The fact that its components transform like a 4-vector allows us to find the relationship between the frequency ω and wave number $\boldsymbol{k}$ in one coordinate system and the frequency ω^* and wave number $\boldsymbol{k}^*$ in any other coordinate system. In particular, we can now give a more elegant derivation of Eq. (13.52) for the relativistic Doppler effect (see Problem 5). From the wave vector k_μ we can form the invariant scalar

$$S = c^2\,(k_\mu,\ k_\mu) = c^2k^2-\omega^2. \tag{14.32}$$

For a light wave $S = 0$. For a wave whose phase velocity (ω/k) is less than c, S is positive. For a given value of S, Eq. (14.32), which gives the relationship between ω and k for any particular wave, is often called the *dispersion relation* for the wave. The quantity S may itself be given as a function of ω, or of k. Since ω and k are not themselves 4-scalars, if S is a function of ω or of k, it will be a different function in different coordinate systems. A sound wave for example which travels with a velocity v in air has the dispersion relation $\omega = kv$. If we substitute in Eq. (14.32), we find

$$S = k^2\,(v^2-c^2), \qquad \text{if } \omega = kv. \tag{14.33}$$

Formula (14.33) holds of course only in a coordinate system at rest with respect to the air in which the sound is propagating. In every coordinate system the scalar S given in terms of the wave vector k_μ by Eq. (14.32) will have the same value, but in a coordinate system in which the air is moving, its functional expression in terms of k will be different from Eq. (14.33). The law $\omega = kv$, which is equivalent to the assertion that sound waves propagate with phase velocity v, is true only in the coordinate system in which the air is at rest. The correct relativistic formulation of the law of propagation of a sound wave would be to say that it propagates with speed v relative to the air. Stated that way, the law is independent of any arbitrary choice of coordinate system; it does of course depend upon the way in which the air is moving. Contrast this situation with the case of a light wave, for which $v = c$ and the dispersion relation $S = 0$ holds in the same form in all inertial systems.

Let us now try to define a 4-vector differential operator which will be the analog of the 3-vector operator ∇ defined in Section 3.6. We first inquire whether the differentiation symbols $\partial/\partial x_\mu$ transform like the components of a 4-vector. To find out, let us consider a scalar function $f(x_0, x_1, x_2, x_3)$. Let us assume that f is a 4-scalar, that is, it has the same value at the same space-time point no matter what inertial coordinate system is being used. If we then express f as a function of a new set of coordinates x_μ^*, its derivatives with respect to the new variables are given according to the usual rules of calculus by the formula

$$\frac{\partial f}{\partial x_\mu^*} = \Sigma_\nu \frac{\partial f}{\partial x_\nu} \frac{\partial x_\nu}{\partial x_\mu^*}.$$

From Eqs. (14.14) and (14.15) we obtain

$$\frac{\partial x_\nu}{\partial x_\mu^*} = a_{\nu\mu}^{-1} = g_\mu g_\nu a_{\mu\nu}. \tag{14.34}$$

From the above equations we see that the transformation law for the coordinate derivatives is

$$\frac{\partial}{\partial x_\mu^*} = \Sigma_\nu g_\mu g_\nu a_{\mu\nu} \frac{\partial}{\partial x_\nu}. \tag{14.35}$$

We see that the coordinate derivatives do not quite transform according to the law (14.16) for the components of a 4-vector, because some of the terms on the right in Eq. (14.35) involve a negative sign.* Noting that $(g_\mu)^2 = 1$, we can obtain a set of differential operators which do transform like the components of a 4-vector by multiplying Eq. (14.35) by g_μ:

$$g_\mu \frac{\partial}{\partial x_\mu^*} = \Sigma_\nu a_{\mu\nu} g_\nu \frac{\partial}{\partial x_\nu}. \tag{14.36}$$

We will define a 4-vector differential operator $\Box_\mu$, which we may call the *4-gradient*, as follows

$$\Box_\mu = g_\mu \frac{\partial}{\partial x_\mu},$$

that is,

$$\Box_0 = -\frac{1}{c}\frac{\partial}{\partial t}, \qquad \Box_1 = \frac{\partial}{\partial x}, \qquad \Box_2 = \frac{\partial}{\partial y}, \qquad \Box_3 = \frac{\partial}{\partial z}. \tag{14.37}$$

---

*The reader who is familiar with the extension of vector algebra to non-orthogonal systems will recognize the distinction between the transformation laws (14.35) and (14.16) as corresponding to the distinction between covariant and contravariant vectors. It arises because of the sign differences in the invariant scalar product (14.20). Because in the present case only a sign difference is involved, we have chosen to avoid introducing the somewhat more elaborate formalism that is required for dealing with general nonorthogonal coordinate systems. We introduce here only one kind of 4-vector, transforming according to the (contravariant) transformation rule (14.16).

The space components of the 4-vector $\Box_\mu$ are just the components of the gradient operator in ordinary space, while the time component involves differentiation with respect to t.

By forming its self-scalar product, we obtain a 4-scalar differential operator which is familiar from the theory of waves (see for example Eq. 8.186),

$$(\Box_\mu, \Box_\mu) = \Box^2 = \nabla^2 - \frac{1}{c^2}\frac{\partial^2}{\partial t^2}. \tag{14.38}$$

The operator $\Box_\mu$ plays an important role in the theory of waves. For example, if we assume that the quantity W in the wave (14.29) is a scalar, then we may apply the operator $\Box_\mu$ to obtain the 4-vector

$$\Box_\mu W = ik_\mu W. \tag{14.39}$$

For a light wave ($\omega = kc$), each quantity W characterizing some component of the wave satisfies (in empty space) the equation

$$\Box^2 W = 0. \tag{14.40}$$

Provided the set of quantities W is properly transformed, the invariance in form of this equation under a Lorentz transformation is guaranteed by the fact that $\Box^2$ is a 4-scalar. We may use our machinery to construct a Lorentz-covariant equation for a wave which travels with a velocity different from that of light in the following way:

$$\Box^2 W - \kappa^2 W = 0, \tag{14.41}$$

where κ is some universal constant having the dimensions of an inverse length. Since $\Box^2$ and κ^2 are both scalars, the required covariance is manifest. If we substitute from Eq. (14.29), we obtain the dispersion relation

$$-k^2 + (\omega^2/c^2) - \kappa^2 = 0,$$

or in the notation of Eq. (14.32),

$$S = c^2k^2 - \omega^2 = -c^2\kappa^2. \tag{14.42}$$

The reader who has studied Chapter 8 will recognize that this is a wave with a phase velocity

$$v_p = \omega/k = (\pm)\, c\,(1 + \kappa^2 k^{-2})^{1/2}, \tag{14.43}$$

which is greater than c (!), and a group velocity

$$v_g = d\omega/dk = (\pm)\, c\,(1 + \kappa^2 k^{-2})^{-1/2}, \tag{14.44}$$

which is less than c. The phase velocity is the velocity with which a pure sinusoidal pattern appears to move; if no physical object travels with that speed, and since a perfect sinusoidal pattern carries no information, the phase velocity of a wave may exceed c without violating the postulates of relativity. A low-frequency modulation of the sinusoidal pattern travels with approximately the group velocity. Since the wave therefore can carry signals and transport energy at the group velocity, v_g must be less than c to satisfy the postulate of relativity.

14.2 THE RELATIVISTIC CONSERVATION LAWS

We now seek to formulate the laws of mechanics in such a way that they satisfy the postulate of relativity (13.42). We will be guided by the correspondence principle; the correct relativistic laws of mechanics must not only satisfy Postulate (13.42), but they must also reduce to Newton's laws whenever all velocities involved are negligible compared with c. From a strictly logical point of view, the postulate of relativity and the correspondence principle are not sufficient to define the new theory uniquely. However the laws of relativistic mechanics proposed by Einstein in his first paper,* and independently also by Poincaré,† are the only plausible set of laws satisfying these requirements which have yet been suggested.

The simplest approach is to look first for a relativistic generalization of the law of conservation of momentum. We therefore make the hypothesis that the new theory will contain a conservation law which will reduce in the limit of small velocities to the classical law of conservation of momentum. A conservation law asserts that a particular physical quantity remains constant, or alternatively that it has the same value at any two different times. Classically the momentum of a system of particles is a vector quantity formed by adding the vector momenta of the individual particles as defined by Eq. (4.4). It is pretty clear that a generalization of the law of conservation of momentum will have to involve a 4-vector P_μ whose space part P_i reduces to the classical momentum vector in the limit of small velocities. A suitable definition can be found by making use of the 4-velocity Eq. (14.25). We therefore define a 4-vector, which we will call the (4-vector) *momentum* of a single particle, as follows:

$$p_\mu = mU_\mu, \tag{14.45}$$

where m is the *rest mass* of the particle, often called just the *mass*. It is defined as the mass measured in the usual way in a coordinate system in which the particle is at rest or nearly so. The mass m defined in this way is evidently a scalar, that is, it has the same value in all coordinate systems. In every coordinate system we are to use the same mass m, namely the mass measured by instruments moving with the same velocity as the particle; these instruments are the same instruments, no matter in what coordinate system we happen to be writing the equations. Because the rest mass is a scalar, the momentum P_μ of a particle is a 4-vector.

*A. Einstein, *Annalen der Physik*, **17** (1905). The definition of force chosen by Einstein was less convenient than the one adopted later which we will present in the next section.

†H. Poincaré, *Compt. Rend.*, **140**, 1504 (1905).

The time and space parts of the momentum are given by

$$p_0 = \gamma mc,$$
$$\boldsymbol{p} = \gamma m\boldsymbol{u}. \tag{14.46}$$

Since $\gamma \doteq 1$ if $u \ll c$, the space part of the 4-vector p_μ reduces to the classical momentum vector for small velocities.*

We now define the total momentum of a system of particles as the 4-vector obtained by summing the momenta of the individual particles:

$$P_\mu = \Sigma_j p_{j\mu} = \Sigma_j m_j U_{j\mu}. \tag{14.47}$$

The space part of the vector P_μ will reduce to the classical total momentum if the velocities of all of the particles are small in comparison with c.

We now propose that the relativistic conservation law asserts that the total 4-momentum remains constant,

$$(P_\mu)_{t=t_1} = (P_\mu)_{t=t_2}, \tag{14.48}$$

for any two times t_1 and t_2, for a system of particles whose total 4-momentum is P_μ, and which does not interact with any other bodies outside the system. The law (14.48) contains four equations corresponding to the four values of the subscript μ. The three equations for $\mu = 1, 2, 3$ assert that the 3-momentum,

$$\boldsymbol{P} = (P_1, P_2, P_3) = \Sigma_j m_j \gamma_j \boldsymbol{u}_j, \tag{14.49}$$

remains constant. In the limiting case of small velocities, this is just the classical law of conservation of momentum. We may therefore call the 3-vector $\boldsymbol{P}$ defined by Eq. (14.49) the *3-momentum* in general, regardless of the velocities. There is an additional equation corresponding to $\mu = 0$ in Eqs. (14.48) which asserts that the quantity

$$P_0 = \Sigma_j m_j c \gamma_j \tag{14.50}$$

is constant, where

$$\gamma_j = [1-(u_j^2/c^2)]^{-1/2}. \tag{14.51}$$

In order to see the significance of this equation, let us expand P_0 in a power series in u^2/c^2:

$$P_0 = c\Sigma_j m_j + c^{-1}\Sigma_j \tfrac{1}{2} m_j u_j^2 + c^{-3}\Sigma_j \tfrac{3}{4} m_j u_j^4 + \cdots. \tag{14.52}$$

The first term is just the sum of the rest masses of the particles multiplied by c, and would be constant according to the laws of classical physics. The second

*In relativity, the correct definition of the 3-momentum $\boldsymbol{p}$ of a particle turns out to be simply the space part p_i of the 4-momentum p_μ. We therefore use the same symbol (p) for both. Note that the 3-velocity $\boldsymbol{u}$ is not the space part of the 4-velocity U_μ, so we have used different symbols (u, U) to distinguish them.

term is the classical expression for the kinetic energy, divided by c. If we assume that the velocities are small enough so that the higher order terms are negligible, then the $\mu = 0$ component of the conservation law seems to be telling us that the kinetic energy is constant. This is an unexpected and slightly embarrassing dividend, since we started out to write simply a law of conservation of momentum. It is clear that in relativity theory the law of conservation of momentum may entail some additional conservation law, since it must assert the conservation of a 4-vector with four components. Thus energy and momentum conservation go together in relativity theory; we cannot have one without the other. This is at first a bit puzzling, for in classical mechanics the conservation laws of momentum and energy are on a rather different footing. The conservation law of momentum depends only on Newton's third law, or on some other equivalent feature of the theory; it is generally true for all interactions between particles, at least provided that we take account of all the momentum involved. The conservation of kinetic energy, which would correspond to the second term in P_0, holds only for the restricted class of completely elastic interactions among the particles, in classical mechanics. There is a general law of conservation of energy, but it requires us to take into account all forms of energy, including heat energy and the potential energies associated with various force fields. Therefore the total energy in classical mechanics cannot be expressed in purely mechanical concepts, that is, in terms of the masses and velocities of the particles involved. Since we want the conservation of the momentum P_i, given by Eq. (14.49) to hold in general so long as there are no exchanges of momentum between the given system and the outside world, we are forced to the conclusion that the quantity P_0 given by Eq. (14.50) is also conserved under the same circumstances, that is, provided there is no exchange of energy with the outside world. Since we know that in the case of inelastic collisions, and for nonrelativistic velocities so that the higher order terms in Eq. (14.52) are negligible, the second term does not remain constant, we are forced to conclude that the first term must compensate for changes in the second term. *All changes in the internal energy of a body in whatever form must therefore be reflected in its rest mass.*

We will define the relativistic energy of a body in terms of the mass and velocity by the formula

$$E = cp_0 = \gamma mc^2 = mc^2 + \tfrac{1}{2}mu^2 + \tfrac{3}{4}mu^4/c^2 + \cdots. \tag{14.53}$$

The first term in the expansion, mc^2, is called the *rest energy*; this term is present even when the velocity of the body is zero. As we saw in the last paragraph, this term must contain all forms of internal energy of the body, including heat energy, internal potential energy of various kinds, and also rotational energy if any. In an inelastic collision, kinetic energy may be converted into some form of internal energy or vice versa, so that the rest energy of a body and hence also its mass may change. In classical physics, mass and energy are separately conserved. In relativistic mechanics, these two conservation laws become a single conservation law, the conservation of total relativistic energy E. The second term in the expansion in Eq. (14.53) is the classical expression for the kinetic energy. The succeeding terms are higher-order corrections which vanish in the limit $u \ll c$. We may define the relativistic kinetic energy as the energy of motion of a body in the following way:

$$T = E - mc^2 = \tfrac{1}{2}mu^2 + \tfrac{3}{4}mu^4/c^2 + \cdots. \tag{14.54}$$

It is of interest to calculate the changes in mass associated with changes in internal energy. A gram of water has a total rest energy $mc^2 = 9 \times 10^{13}$ J. If the water is heated from 0°C to 100°C, the total heat energy added is 418 J. The corresponding increase in mass is $418\ \mathrm{J}/c^2 = 4.64 \times 10^{-12}$ g. This increase in mass is too small to be measured on even the most sensitive balance. Let us next consider an ordinary chemical reaction. If one gram of charcoal is burned with 2.7 g of oxygen to make 3.7 g of CO_2, the energy released is 34,000 J. The total mass of CO_2 produced is therefore less than the total mass of charcoal plus oxygen by 3.8×10^{-10} g. This is somewhat beyond the precision which can presently (1970) be achieved with the best chemical balances. The energy which can be released in nuclear reactions is very much greater than that which can be released in chemical reactions per gram of reacting products. If a gram of uranium undergoes fission for example, the energy released is 7.6×10^{10} J! If we divide by c^2, we conclude that the sum of the masses of all the fission products is less than the one gram mass of the original uranium by about 0.8×10^{-6} g. This is a small fraction of a gram but it is readily measureable. Thus the energy released in any hypothetical nuclear reaction can be determined by weighing the constituents and the products of the reaction and multiplying the mass difference by c^2.

If p_μ is the 4-momentum of a particle of mass m, we can form a scalar by taking the self product as follows:

$$(p_\mu, p_\mu) = p_1^2 + p_2^2 + p_3^2 - p_0^2 = -m^2c^2. \tag{14.55}$$

The fact that this scalar has the value $-m^2c^2$ can be verified in any coordinate system by a direct calculation, or more readily by noticing that since it is a scalar, we can evaluate it in a coordinate system in which $\boldsymbol{u} = \mathbf{0}$, and $\gamma = 1$. The 4-momentum of a particle of mass m, like the 4-velocity, is a time-like 4-vector.

Einstein noticed that the theory of relativity allows the possibility of a particle traveling with the velocity of light and having zero rest mass. If in Eq. (14.46) we let $m \to 0$, $u \to c$, keeping the product $\gamma mc^2 = E$ constant, we arrive at the relations

$$p_0 = E/c, \qquad \boldsymbol{p} = \boldsymbol{c}E/c^2, \tag{14.56}$$

where $\boldsymbol{c}$ is the velocity of the particle and has the magnitude c. Such a particle is the *photon* associated with electromagnetic waves. It has no rest mass, and its energy is not determined by its speed, which is always c, but is given rather according to Einstein's hypothesis by the formula

$$E = h\nu = \hbar\omega, (\hbar = h/2\pi), \tag{14.57}$$

where h is Planck's constant and $\nu = \omega/2\pi$ is the frequency of the electromagnetic wave. The energy-momentum 4-vector of a photon has the self-scalar product 0; we may appropriately call it a *light-like* 4-vector.

In the quantum theory, there is an equivalence between waves and particles. Any wave phenomenon can alternatively be described in terms of particles or quanta. Conversely, phenomena involving particles can alternatively be described in terms of associated waves. The wave associated with a particle whose 4-momentum is p_μ has the wave 4-vector k_μ given by the formula

$$p_\mu = \hbar k_\mu. \tag{14.58}$$

This relationship is the same in all inertial coordinate systems, so that the quantum hypothesis is compatible with the postulate of relativity. If we substitute Eq. (14.58) in Eq. (14.55) and make use of Eq. (14.30), we obtain the dispersion relation between the frequency and wave number of the wave associated with the particle of mass m:

$$\omega^2 - k^2c^2 = m^2c^4/\hbar^2. \tag{14.59}$$

These relationships form the basis for the development of wave mechanics.

14.3 COLLISION THEORY

The relativistic expressions for the momentum and energy of a particle as given by Eqs. (14.46), (14.53), and (14.54) have already been given in Chapter 4 (Eqs. 4.73, 4.74, and 4.75), where they were introduced for the purpose of using the conservation laws of energy and momentum to obtain relations between initial and final momenta, energies, and angles of scattering in collisions between high energy particles. As pointed out in Chapter 4, the conservation equations (4.76), (4.77), and (4.78) can be used for high-energy particles if the relativistic expressions for momentum and energy are used. We may now add to these three equations the following equation for the energy Q released in a collision process involving two initial particles of masses m_1, m_2, and two final particles of masses m_3, m_4:

$$Q = (m_1 + m_2 - m_3 - m_4)\,c^2. \tag{14.60}$$

Consider now a collection of N particles, including perhaps some photons, whose total energy-momentum is

$$P_\mu = \sum_{j=1}^{N} p_{j\mu}, \tag{14.61}$$

Since the 4-vectors $p_{j\mu}$ are all time-like or light-like, we can show (see Problem 9) that the total 4-momentum P_μ is time-like. The only exception is the case when all N particles are photons all traveling in the same direction, in which case P_μ is also light-like. We may therefore define a scalar M by the equation

$$(P_\mu, P_\mu) = P_1^2 + P_2^2 + P_3^2 - P_0^2 = -M^2c^2. \tag{14.62}$$

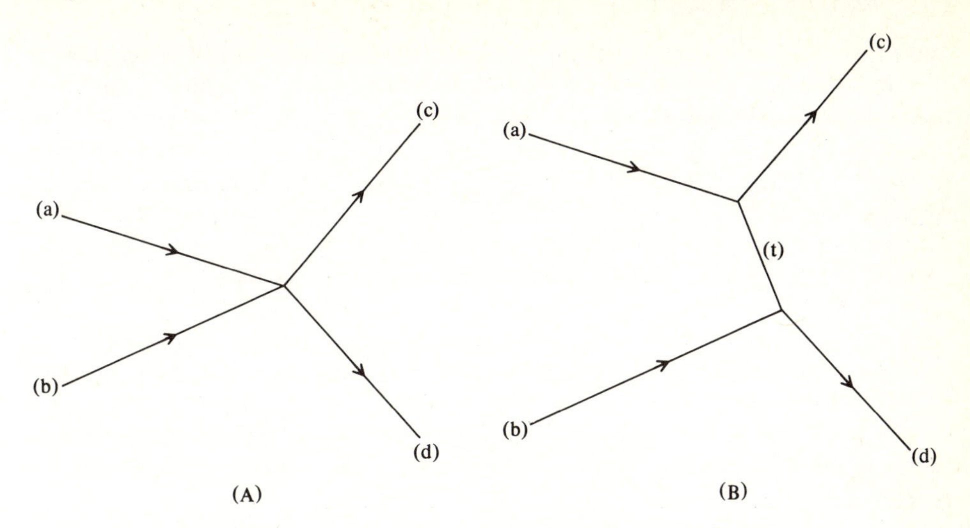

Fig. 14.1 A two-particle reaction producing two final particles.

The scalar M is called the *invariant mass* of the system. "Invariant" refers to the fact that the expression (14.62) is invariant under a Lorentz transformation. If there are no external forces on a system of particles, then by the conservation law, the momentum P_μ and therefore also M will remain constant.

The invariant mass is often useful in studying reactions between high-energy particles. Suppose for example that a particular reaction which may be observed in a bubble chamber produces a large number of pions which can be seen to leave the reaction. The mass of the pion is known, and the momentum of any (charged) pion can be measured by observing its direction of motion and the radius of curvature of its path in a magnetic field; hence the 4-momentum p_μ of each pion observed to come from the reaction can be measured. Suppose it is suspected that there may be a particle with a lifetime too short to be observed which decays into, say 3 pions. The mass of such a particle would be the invariant mass M calculated from the sum of the 4-momenta of the 3 pions into which it decays. We may therefore check our hypothesis by grouping the observed pions from each reaction into groups of three in all possible ways and calculating for each group the invariant mass M. We do this for a large number of observed reactions and study the distribution of calculated values of M. If our hypothesis is correct, many of the values of M will be found to cluster around some particular value which is then the mass of the hypothetical particle for which we are searching. If our hypothesis is wrong, then we will expect to find a more or less random distribution of values of M.

Let us now consider a reaction produced by a collision of two incident particles (a) and (b), producing two final particles (c) and (d), as shown schematically in Fig. 14.1(A). We may characterize the reaction by giving the 4-momenta $p_{a\mu}$

and $p_{b\mu}$ of the incident particles and $p_{c\mu}$, $p_{d\mu}$ of the final particles. The reaction is therefore specified by the sixteen components of these four 4-vectors. If the same reaction is described in a starred coordinate system, the 4-momenta will be transformed according to the Lorentz transformation equation (14.16). The Lorentz transformation coefficients $a_{\mu\nu}$ depend upon six parameters, namely, three components which specify the velocity of the starred coordinate system relative to the unstarred coordinate system, and three angles which specify the orientation of the starred axes relative to the unstarred axes. We therefore expect that the sixteen components of the four 4-momenta can be specified by ten invariant parameters which characterize the reaction in a manner independent of the coordinate system, and six further parameters which depend upon the particular coordinate system in which the 4-momenta are calculated. We can in fact readily construct ten scalars from the four 4-momenta by forming the four self-products $(p_{a\mu}, p_{a\mu})$, etc., and the six pairwise products $(p_{a\mu}, p_{b\mu})$, etc. The conservation law of energy-momentum

$$p_{a\mu}+p_{b\mu} = p_{c\mu}+p_{d\mu} \tag{14.63}$$

gives four equations between the momentum components, so that of the ten scalar invariants, only six can be independent. Four of these may be taken to be the masses of the four particles involved in the reaction, which are given by the four self-scalar products:

$$\begin{aligned} m_a^2c^2 &= (E_a^2/c^2)-|\boldsymbol{p}_a|^2, \\ m_b^2c^2 &= (E_b^2/c^2)-|\boldsymbol{p}_b|^2, \\ m_c^2c^2 &= (E_c^2/c^2)-|\boldsymbol{p}_c|^2, \\ m_d^2c^2 &= (E_d^2/c^2)-|\boldsymbol{p}_d|^2. \end{aligned} \tag{14.64}$$

A fifth convenient parameter is the invariant mass M of the initial pair of particles, which by the conservation law is also equal to the invariant mass of the final pair of particles and which is given by

$$\begin{aligned} M^2c^2 &= -(p_{a\mu}+p_{b\mu}, p_{a\mu}+p_{b\mu}) = -(p_{c\mu}+p_{d\mu}, p_{c\mu}+p_{d\mu}) \\ &= (E_a+E_b)^2/c^2-|\boldsymbol{p}_a+\boldsymbol{p}_b|^2. \end{aligned} \tag{14.65}$$

The sixth invariant parameter may be conveniently taken to be the self-scalar product

$$S = (p_{a\mu}-p_{c\mu}, p_{a\mu}-p_{c\mu}) = (p_{d\mu}-p_{b\mu}, p_{d\mu}-p_{b\mu}), \tag{14.66}$$

where the second equality follows from the conservation law. There is a seventh similar scalar product

$$T = (p_{a\mu}-p_{d\mu}, p_{a\mu}-p_{d\mu}) = (p_{c\mu}-p_{b\mu}, p_{c\mu}-p_{b\mu}), \tag{14.67}$$

which, however, according to our argument above, cannot be independent of the preceding six. (See Problem 12.)

If the scalar S (or T) is negative or zero, we can provide the following interesting interpretation of S. Let us imagine that the reaction in question proceeds in two steps as shown in Fig. 14.1(B). The particle (a) first decays into two particles (c) and (t). Particle (t) then reacts with particle (b) to form the final particle (d). It is then evident from the conservation law that the quantity $p_{a\mu}-p_{c\mu} = p_{t\mu}$ is the 4-momentum of particle (t). The invariant $S = -M_t^2c^2$ then gives the mass of particle (t). The 4-vector $p_{a\mu}-p_{c\mu}$ is often called the *momentum transfer* in the reaction. The scalar S is sometimes called the *invariant momentum transfer squared.*

We have seen that, in addition to the masses of the four particles, there are just two invariants M and S associated with the reaction. Any invariant quantity associated with the reaction, for example the cross section to be defined below, must therefore be expressible as a function of M and S.

We saw in Chapter 4 that in classical mechanics it is convenient for theoretical analyses to use the center-of-mass coordinate system. The same is true in studying relativistic processes, where we define the *center-of-mass coordinate system* as one in which the total 3-momentum $\boldsymbol{P}$ is zero. Because P_μ is time-like, it is always possible to find a coordinate system in which $\boldsymbol{P}$ vanishes. (See Problem 13.) In the center-of-mass system, the time-component P_0 will be Mc, where M is the invariant mass. The reader will recognize that because the same coefficients $a_{\mu\nu}$ appear in the transformation equations (14.16) for 4-vectors and (14.6) for coordinates, the problem of finding the center-of-mass system is identical algebraically with the problem of finding a coordinate system in which two events E^1, E^2 (separated by a time-like interval) occur at the same space point at different times. Experiments are usually done in a coordinate system in which one of the initial particles, say (b), is at rest. Such a coordinate system is called the *laboratory coordinate system.*

The quantity Mc^2, where M is the invariant mass of a group of particles, is the total energy in the center-of-mass coordinate system, and is therefore called the *center-of-mass energy*. If a group of particles is produced in a reaction, the sum of their rest energies ($\Sigma_i m_i c^2$) is always less than Mc^2, except when all the particles are at rest, and then the sum of their rest energies just equals the center-of-mass energy. By calculating Mc^2 for the initial particles in the laboratory coordinate system, we can determine the *threshold* for any particular reaction, when the incident particle energy is just large enough so that the center-of-mass energy equals the sum of the rest energies of the particles to be produced. (See Problem 14.) For incident particle energies less than threshold, that particular reaction cannot take place.

Let us now try to define a cross-section for a reaction in a relativistically invariant way. We assume there are initially two groups of particles (a) and (b) of momenta $p_{a\mu}$, $p_{b\mu}$ which enter into some reaction whose cross-section we wish to define. We will not at the moment assume anything about the nature of the reaction, except that it is some identifiable reaction or set of reactions and that if we consider transformations of coordinates, we refer to the same reaction in each coordinate system. The same reaction will in general look different in different coordinate systems—scattering angles for example may change when we transform the coordinates—so that any parameters needed to specify the particular reaction we are considering will have to be properly transformed when we make a change of coordinate systems. For simplicity we will consider only coordinate systems in which the velocities of particles (*a*) and (*b*) are parallel or in which one of the two velocities is zero. Let us then assume that particles (*a*) are traveling parallel to the x-axis with the velocity u_a and particles (*b*), with the velocity u_b. Let us call particles (*b*) the target particles, and imagine as we did in Section 3.16 that each target particle carries along with it a cross-sectional area σ perpendicular to the direction of its motion. We define σ by requiring that the probability that the reaction will occur is equal to the probability that a point particle (*a*) will pass through the cross-section σ associated with the target particle (*b*). We note that since the area σ is perpendicular to the direction of motion of particle (*b*), it will be independent of the velocity of particle (*b*) and will therefore be the same in all coordinate systems of the type we are considering. Let the densities of particles of each type per unit volume be n_a and n_b respectively. If we imagine a surface parallel to the yz-plane moving along with velocity u_b (the sign of u_b indicates the direction along the x-axis), the number of particles (*a*) per unit area per unit time which pass through this surface is evidently $n_a\,|u_a - u_b|$. The number of reactions dN occurring in a time dt in a volume dV is therefore given by

$$dN/dt\,dV = |u_a - u_b|\,n_a n_b \sigma. \tag{14.68}$$

The density n_a is given by

$$n_a = dN_a/dV, \tag{14.69}$$

where dN_a is the number of particles in a volume dV. If we now change to another coordinate system, keeping our attention on a fixed group of particles of type (*a*), then the number dN_a will be the same in the new coordinate system but the volume dV will change. If we consider a coordinate system in which particles of type (*a*) are at rest, and if we let the volume occupied by dN_a of these particles in this rest coordinate system be dV_a, then the volume which they occupy in any other coordinate system is given by

$$dV = dV_a/\gamma_a, \tag{14.70}$$

as we see if we note that the dimensions of dV in the direction of motion of particles (*a*) are reduced by a factor γ_a while dimensions perpendicular to this direction are not affected by the motion. We may therefore write the density n_a in terms of the density n_{0a} in a coordinate system in which particles (*a*) are at rest:

$$n_a = \gamma_a n_{0a}. \tag{14.71}$$

The proper density n_{0a} is evidently a scalar, since it is defined (in any coordinate system) as the density calculated in a coordinate system in which particles (*a*) are at rest.

We may now write Eq. (14.68) in the form

$$dN/dt\, dV = \gamma_a \gamma_b \left| u_a - u_b \right| n_{0a} n_{0b} \sigma. \tag{14.72}$$

It can be shown (see Problem 20) that the 4-volume element $dV\,dt$ is an invariant under Lorentz transformations. Since dN is simply the number of reactions which occur in that 4-volume element, dN is also a scalar, and hence the left-hand side of Eq. (14.72) is a scalar. Since as we have argued above, the last three factors in Eq. (14.72) are scalars, the quantity $\gamma_a \gamma_b |u_a - u_b|$ must also be invariant under special Lorentz transformations of the type (14.7). It is left to the reader to show by direct calculation that this is indeed the case (see Problem 21). Equation (14.72) provides a definition of the invariant cross-section σ in any coordinate system in which $\boldsymbol{u}_a$ and $\boldsymbol{u}_b$ are parallel (or antiparallel). The definition can be generalized to arbitrary coordinate systems but we will not carry out this development here. (See Problem 22.)

Let us see how much we can learn about the cross-section just from its transformation properties. We are often interested in a reaction in which some final particle, say (c), appears with momentum $\boldsymbol{p}_c$, energy E_c. Any other final particles (d), (e), etc., may have any energies allowed by the conservation laws. (If there are only two final particles (c) and (d), then $p_{d\mu}$ is determined by the conservation laws when $p_{c\mu}$ is given.) Since ordinarily $\boldsymbol{p}_a$ and E_a may have any values within some range, we are interested in the differential cross-section $d\sigma$ for producing particle (c) with energy and momentum within some small range of values. If the mass m_c can also vary over a range, an uncommon case,* we could write

$$d\sigma = F(M, \text{etc.})\, dp_{cx} dp_{cy} dp_{cz} dE_c \tag{14.73}$$

for the cross-section for producing particle (c) with momentum $\boldsymbol{p}_c$ lying within a small range $dp_{cx}dp_{cy}dp_{cz}$, and energy between E_c and $E_c + dE_c$. Now we can show (Problem 20) that the volume element $dp_{cx}dp_{cy}dp_{cz}dE_c$ in energy-momentum space is a 4-scalar. Since $d\sigma$ is also a scalar, it follows that the function F is also, and can therefore depend only on the invariants associated with the process (M and S in the case of a two-body final state). To find the function $F(M, S, \ldots)$, we would need to work out the theory of the particular reaction. For most examples of interest, this would require quantum mechanics.

In the usual case when the mass m_c is fixed, the energy and momentum are related by Eq. (14.55):

$$p_c^2 c^2 + m_c^2 c^4 = E_c^2. \tag{14.74}$$

We may then specify only the final momentum $\boldsymbol{p}_c$, as E_c is then determined. We are interested in the cross-section $d\sigma$ for producing $\boldsymbol{p}_c$ in a range $dp_{cx}dp_{cy}dp_{cz}$. The momentum-space element $dp_{cx}dp_{cy}dp_{cz}$ is not invariant. In order to find its transformation properties, replace E_c by the mass m_c related to it by Eq. (14.74), and allow m_c for the moment to have a range of values. Then we find, either by

---

*Except for very short lived particles, according to quantum mechanics.

differentiating Eq. (14.74) holding p_{cx}, p_{cy}, p_{cz} constant, or by taking the appropriate Jacobian determinant, that

$$dp_{cx}dp_{cy}dp_{cz}dE = dp_{cx}dp_{cy}dp_{cz}\, c^4 m_c dm_c/E_c. \tag{14.75}$$

Now the rest mass m_c, as well as dm_c, is a 4-scalar, as is the left member of Eq. (14.75), so we conclude that

$$dp_{cx}dp_{cy}dp_{cz}/E_c \text{ is invariant under a Lorentz transformation.} \tag{14.76}$$

Although we have derived this result by considering a range of values of m_c, the final result is independent of dm_c and holds no matter how small a range of values we allow, from which we conclude that it holds for a fixed m_c. We therefore write

$$d\sigma = F(M, \text{etc.})\, dp_{cx}dp_{cy}dp_{cz}/E_c\,, \tag{14.77}$$

where again F must be a scalar. Since

$$dp_{cx}dp_{cy}dp_{cz} = p_c^2 dp_c d\Omega_c = [E_c^2 - m_c^2 c^4]^{\frac{1}{2}}\, E_c dE_c d\Omega_c/c^2, \tag{14.78}$$

we may also write the cross-section for sending particle (c) into a solid angle $d\Omega_c$ with momentum p_c to $p_c + dp_c$ or with energy E_c to $E_c + dE_c$..

The astute reader will perhaps observe that in the case when there are only two final particles of fixed masses m_c, m_d, there are only eight components of the 4-momenta $p_{c\mu}$, $p_{d\mu}$, the conservation laws give four relations between them, and we have two equations like Eq. (14.74), so that only two variables remain whose differentials can appear in $d\sigma$. If we give the angular range $d\Omega_c$ into which particle (c) goes, its momentum p_c and energy E_c must be determined. We could in principle handle this case by a method similar to that in the preceding paragraph by finding the transformation properties of $d\Omega_c$ under these conditions. The algebra is rather complicated, and we will not carry it through.

14.4 THE RELATIVISTIC EQUATIONS OF MOTION

Let us define the *(3-)force* in relativistic mechanics as the time rate of change of the relativistic momentum so that the equation of motion is

$$d\boldsymbol{p}/dt = \boldsymbol{F}. \tag{14.79}$$

If we take Eq. (14.79) as the definition of the force $\boldsymbol{F}$, then we must find the correct transformation equations for the force so that Eq. (14.79) will hold in all coordinate systems. To do this, let us first try to write an equivalent equation in terms of 4-vectors. If we take as the independent variable the proper time τ as measured on a clock moving with the particle, then the equation of motion can be written in the 4-vector form

$$dp_\mu/d\tau = \mathscr{F}_\mu, \tag{14.80}$$

where the four quantities $\mathscr{F}_\mu$ must be a 4-vector if this equation is to have the same form in all systems. We will call $\mathscr{F}_\mu$ the *4-force*. By Eqs. (14.46), (13.27), (14.53), and (13.25), we see that Eqs. (14.79) and (14.80) imply the following relation between the 4-force and the ordinary or 3-force

$$\mathscr{F}_0 = \frac{\gamma}{c}\frac{dE}{dt} = \frac{\gamma}{c}\boldsymbol{u}\cdot\boldsymbol{F},$$
$$\mathscr{F}_1 = \gamma F_x, \qquad \mathscr{F}_2 = \gamma F_y, \qquad \mathscr{F}_3 = \gamma F_z. \tag{14.81}$$

If we differentiate Eq. (14.55) with respect to τ, we see that $\mathscr{F}_\mu$ must satisfy the following equation:

$$(p_\mu, \mathscr{F}_\mu) = (U_\mu, \mathscr{F}_\mu) = 0. \tag{14.82}$$

In deriving Eq. (14.82), we have assumed that the rest mass m is constant; we assumed this also in obtaining the first of Eqs. (14.81). In inelastic collisions, as we have seen, the rest mass does not stay constant. In such cases, when the forces change the internal energy, and hence the mass, of a body, Eq. (14.82) no longer holds, $dE/dt \neq \boldsymbol{u}\cdot\boldsymbol{F}$, and the first of Eqs. (14.80) may be independent of the other three. Equations (14.79) would then have to be supplemented by an equation giving dE/dt or dm/dt. Any force which does not change the rest mass of the particle on which it acts, and in particular the electromagnetic force, satisfies Eq. (14.82). We will assume Eq. (14.82) holds except during inelastic collisions.

The reader may verify that the components $\mathscr{F}_\mu$ given by Eqs. (14.81) do indeed satisfy Eq. (14.82). We note that the 4-vector form (14.80) of the equations of motion contains four equations, of which the fourth is (usually) redundant since it can be deduced from the first three by making use of Eq. (14.82). Although the form (14.80) for the equations of motion is more convenient if we are concerned with examining relativistic invariance, the form (14.79) is most convenient if we are interested in writing down and solving the equations of motion in one coordinate system, because they are expressed in terms of the coordinate time t. The proper time τ is not a very convenient variable, particularly if we are dealing with more than one particle, since each particle carries along its own individual proper time.

Since we know how to transform the 4-vector $\mathscr{F}_\mu$ from one inertial coordinate system to another, we can transform the force $\boldsymbol{F}$ by simply calculating the components $\mathscr{F}_\mu$ from Eqs. (14.81), calculating $\mathscr{F}_\mu^*$ in the new coordinate system, and then using Eqs. (14.81) to find $\boldsymbol{F}^*$ in the new coordinate system. If we carry out this procedure for the special case when the axes are parallel and the relative motion is along the x-axis, we find

$$F_x^* = F_x - \beta(1-\beta u_x/c)^{-1}(u_yF_y + u_zF_z)/c,$$
$$F_y^* = \gamma^{-1}(1-\beta u_x/c)^{-1} F_y, \tag{14.83}$$
$$F_z^* = \gamma^{-1}(1-\beta u_x/c)^{-1} F_z,$$

where βc is the velocity of the starred coordinate system relative to the unstarred system, $\gamma = (1-\beta^2)^{-1/2}$, and u_x, u_y, u_z are the components of the velocity of the particle or point on which the force $\boldsymbol{F}$ acts. Note that the transformation equations for the 3-force involve not only the velocity of the coordinate system, but also the velocity of the particle on which the force is acting.

14.5 SOLUTIONS OF THE EQUATIONS OF MOTION

We will postpone until the next section the rather difficult problem of finding force laws which transform according to Eqs. (14.83). Let us assume that we know the force $\boldsymbol{F}$ acting on a particle as a function of the coordinates, and let us ask under what circumstances we can solve the problem to find the motion of the particle. If the force is given as a function of t alone, $\boldsymbol{F}(t)$, then Eqs. (14.79) can be integrated directly to find the momentum $\boldsymbol{p}(t)$. Once $\boldsymbol{p}$ is known, we may solve Eqs. (14.46) for the velocity $\boldsymbol{u}$. This is usually most easily done by solving first for the factor γ by making use of the following relation which follows from Eq. (14.46):

$$p^2/m^2c^2 = \beta^2\gamma^2 = \gamma^2-1. \tag{14.84}$$

When $\gamma(t)$ is known, the velocity $\boldsymbol{u}$ is then immediately given by

$$\boldsymbol{u} = d\boldsymbol{r}/dt = \boldsymbol{p}/m\gamma. \tag{14.85}$$

We may now integrate to find the position $\boldsymbol{r}(t)$.

As an example, let us consider a particle acted on by a constant force $\boldsymbol{F}$ which we will take to be in the x-direction. If the particle starts from rest at $t = 0$, its momentum will at any later time be given by

$$p = Ft. \tag{14.86}$$

We substitute in Eq. (14.84) and solve for

$$\gamma = [1+(Ft/mc)^2]^{1/2}. \tag{14.87}$$

We now have to integrate the equation

$$dx/dt = c(Ft/mc)\,[1+(Ft/mc)^2]^{-1/2}, \tag{14.88}$$

to obtain

$$x = x_0+(mc^2/F)\,[1+(Ft/mc)^2]^{1/2}-mc^2/F. \tag{14.89}$$

We see that $u \to c$ as $t \to \infty$, as we would expect.

If the force $\boldsymbol{F}$ is derivable from a potential energy function $V(\boldsymbol{r})$, then the first of Eqs. (14.81), which is also derivable from Eqs. (14.79), can be written in the form

$$\frac{d}{dt}(mc^2\gamma) = -\boldsymbol{u}\cdot\nabla V, \tag{14.90}$$

where we have written $mc^2\gamma$ instead of E, in order to reserve the latter symbol for the total energy. Since $\boldsymbol{u} = d\boldsymbol{r}/dt$, and in view of the definition (3.107) of the gradient,

Eq. (14.90) becomes

$$\frac{d}{dt}(mc^2\gamma + V) = 0, \tag{14.91}$$

or

$$mc^2\gamma + V = E, \tag{14.92}$$

where E is a constant which we may call the *total energy*.

We may solve Eq. (14.92) for the magnitude of the velocity,

$$u = c[1 - m^2c^4(E-V)^{-2}]^{1/2}. \tag{14.93}$$

This equation gives the speed as a function of the position. If the motion is one-dimensional so that $u = dx/dt$, then we can in principle integrate Eq. (14.93) to obtain the motion $x(t)$. Unfortunately, there are hardly any potential functions $V(x)$ for which the necessary algebra can easily be carried out. In any case Eq. (14.92) or (14.93) can be used to give a qualitative description of the motion.

When the motion is 3-dimensional, we may, as in Chapter 3, define the angular momentum about the origin as the vector

$$\boldsymbol{L} = \boldsymbol{r} \times \boldsymbol{p}, \tag{14.94}$$

and the torque as

$$\boldsymbol{N} = \boldsymbol{r} \times \boldsymbol{F}. \tag{14.95}$$

We may now readily show from Eq. (14.79) that

$$\frac{d\boldsymbol{L}}{dt} = \boldsymbol{N}. \tag{14.96}$$

For a central force, we can again prove that the angular momentum $\boldsymbol{L}$ is a constant of the motion. The problem of a central force can then again in principle be solved along the lines outlined in Section 3.13 for the classical case. Again the actual solution in most cases involves prohibitive algebraic difficulties because of the more complicated form of Eq. (14.91).

Let us now consider a particle subject to a conservative force F derivable from a potential energy $V(\boldsymbol{r})$, and let us use as independent variable the time τ measured by a clock traveling with the particle. With this independent variable, the "velocity" is just the space part of the 4-velocity

$$\frac{dx_i}{d\tau} = U_i. \tag{14.97}$$

The "acceleration" is given, according to Eqs. (14.45) and (14.79), by

$$m\frac{d^2x_i}{d\tau^2} = \gamma F_i = -\gamma\frac{\partial V}{\partial x_i}. \tag{14.98}$$

If we substitute for γ from Eq. (14.92), we see that this can be written in the form

$$m\frac{d^2\boldsymbol{r}}{d\tau^2} = \boldsymbol{\nabla}\frac{(E-V)^2}{2mc^2}. \tag{14.99}$$

This has the same form as Newton's equation of motion for a particle subject to a potential energy $-(E-V)^2/2mc^2$. We can bring out more clearly the connection with the classical law if we set

$$E = E_c + mc^2, \tag{14.100}$$

so that E_c is the total kinetic plus potential energy, with the constant rest energy omitted. We now note that

$$\boldsymbol{\nabla}\frac{(E-V)^2}{2mc^2} = -\boldsymbol{\nabla}\left[V - \frac{(E_c-V)^2}{2mc^2}\right],$$

so that if we put

$$V_{\text{rel}} = V - \frac{(E_c-V)^2}{2mc^2}, \tag{14.101}$$

then Eq. (14.99) takes the classical form

$$m\frac{d^2\boldsymbol{r}}{d\tau^2} = -\boldsymbol{\nabla}V_{\text{rel}}. \tag{14.102}$$

This proves the theorem quoted in Problem 57, Chapter 3, that the orbit given by the special theory of relativity for a particle moving under a potential energy $V(\boldsymbol{r})$ is the same as the orbit which it would follow according to Newtonian mechanics if the potential energy were V_{rel}. The time in the equivalent classical problem is, however, not the coordinate time, but rather the proper time relative to the particle. Note also that the equivalent potential energy V_{rel} depends upon the energy E_c.

It will turn out as we will see later that the electromagnetic force on a charged particle in the theory of relativity is also given by the familiar classical Lorentz formula for the force (3.283). The relativistic equation of motion for a charged particle in an electromagnetic field is therefore

$$\frac{d\boldsymbol{p}}{dt} = q\boldsymbol{E} + \frac{q}{c}\boldsymbol{u}\times\boldsymbol{B}. \tag{14.103}$$

Since the magnetic force is perpendicular to the velocity $\boldsymbol{u}$, it again does no work, and the rate of change of the mechanical energy of the particle is again given by the formula

$$\frac{d}{dt}(\gamma mc^2) = q\boldsymbol{u}\cdot\boldsymbol{E}. \tag{14.104}$$

In the case of a static electromagnetic field, the electric field is derivable from a potential as in Eq. (3.286):

$$\boldsymbol{E} = -\boldsymbol{\nabla}\phi. \tag{14.105}$$

In this case the energy

$$\gamma mc^2 + q\phi = E \tag{14.106}$$

is a constant of the motion. If the electric field is zero, then by Eq. (14.104) the mechanical energy is constant and the factor $m\gamma$ in $\boldsymbol{p}$ can be removed from the differentiation in Eq. (14.103), which then becomes

$$m\gamma \frac{d\boldsymbol{u}}{dt} = \frac{q}{c} \boldsymbol{u} \times \boldsymbol{B}. \tag{14.107}$$

This is the same as the classical equation of motion for a particle in a magnetic field, except that the rest mass m is replaced by the constant $m\gamma$ sometimes called the "transverse mass." The relativistic motion of a charged particle in a magnetic field is therefore the same as its classical motion, except that its mass is increased by the factor γ. In particular, according to the results obtained in Section 3.17, a charged particle moving in a uniform constant magnetic field travels in a circle of radius

$$r = \frac{cm\gamma u}{qB} = \frac{cp}{qB}, \tag{14.108}$$

with frequency

$$\nu = \frac{qB}{2\pi m\gamma c}, \tag{14.109}$$

according to Eqs. (3.295) and (3.299).

As a final example, let us generalize the derivation in Section 4.5 so as to obtain the equation of motion for a rocket in which either the rocket or the exhaust gases or both may be traveling with relativistic velocities. We proceed as in Section 4.5, by writing down an equation expressing the conservation of momentum of the system consisting of the rocket plus the gases which are exhausted during a short time dt. The momentum of the rocket is

$$\boldsymbol{p} = M\gamma\boldsymbol{u}, \tag{14.110}$$

where $\boldsymbol{u}$ is the velocity of the rocket, $\gamma = (1-u^2/c^2)^{-1/2}$, and M is the rest mass of the rocket, which changes with time as gases are exhausted from the rocket engine. If during the time dt the material exhausted has a rest mass dM_e, and a velocity $\boldsymbol{u}_e$ relative to whatever coordinate system we are using, then the momentum of these exhaust gases is $dM_e\gamma_e\boldsymbol{u}_e$. If in addition some external force $\boldsymbol{F}$ acts on the rocket, then the momentum balance equation during the time dt reads

$$d\boldsymbol{p} + dM_e\gamma_e\boldsymbol{u}_e = \boldsymbol{F}\,dt.$$

The time rate of change of momentum of the rocket is therefore given by

$$\frac{d\boldsymbol{p}}{dt} = \boldsymbol{F} - \frac{dM_e}{dt}\gamma_e \boldsymbol{u}_e. \tag{14.111}$$

Since the exhaust velocity and the rate of exhaust of mass are usually known relative to the rocket itself, it will be convenient to express the second term in terms of quantities measured in the coordinate system in which the rocket is momentarily at rest. This is probably most conveniently done by writing the equation in 4-vector notation. We therefore multiply the equation through by γ so that it now reads

$$\frac{dp_i}{d\tau} = \mathscr{F}_i - \frac{dM_e}{d\tau} U_{ei}, \tag{14.112}$$

where $d\tau$ is the proper time relative to the rocket and $\mathscr{F}_i$ is given by Eq. (14.81). Since the rest mass dM_e is a scalar, the corresponding 4-vector equation is evidently

$$\frac{dp_\mu}{d\tau} = \mathscr{F}_\mu - \frac{dM_e}{d\tau} U_{e\mu}. \tag{14.113}$$

The extra equation for $\mu = 0$ is

$$\frac{dp_0}{d\tau} = \mathscr{F}_0 - \frac{dM_e}{d\tau}\gamma_e c. \tag{14.114}$$

If we multiply by c/γ, we obtain the energy equation

$$\frac{dE}{dt} + \gamma_e \frac{dM_e}{dt} c^2 = \boldsymbol{u}\cdot\boldsymbol{F}, \tag{14.115}$$

which states correctly that the rate of increase of energy of the rocket plus the rate at which energy is appearing in the rocket exhaust must be equal to the rate at which energy is being supplied by the external force if any.

If we regard the last term on the right in Eq. (14.113) as a force acting on the rocket, we note that it does not satisfy Eq. (14.82), but that instead

$$\left(p_\mu, -\frac{dM_e}{d\tau} U_{e\mu}\right) = -M \frac{dM_e}{d\tau}(U_\mu, U_{e\mu}), \tag{14.116}$$

which is not in general zero. The rest mass of the rocket is not constant, and this is reflected in the fact that the right-hand member of Eq. (14.113) does not satisfy Eq. (14.82). If we differentiate Eq. (14.55) for the present case, we obtain the useful relation

$$\left(p_\mu, \frac{dp_\mu}{d\tau}\right) = -Mc^2 \frac{dM}{d\tau} = -M \frac{dM_e}{d\tau}(U_\mu, U_{e\mu}), \tag{14.117}$$

where the last member follows from Eqs. (14.113) and (14.116), if we assume that the external 4-force $\mathscr{F}_\mu$ does satisfy Eq. (14.82). The scalar product $(U_\mu, U_{e\mu})$ has the same value in all coordinate systems; it is most conveniently evaluated in a coordinate system in which the rocket is momentarily at rest so $U_0 = c$, $U_i = 0$. The result is

$$(U_\mu, U_{e\mu}) = -\Gamma c^2, \tag{14.118}$$

where

$$\Gamma = (1 - V^2/c^2)^{-1/2}, \tag{14.119}$$

and where V is the exhaust velocity relative to the rocket itself. We thus obtain the following relation between the rate of change of the rest mass of the rocket and the rate at which rest mass appears in the exhaust:

$$\frac{dM}{d\tau} = -\Gamma \frac{dM_e}{d\tau}. \tag{14.120}$$

If we multiply through by c^2, we see that this equation simply asserts that in the coordinate system in which the rocket is momentarily at rest, the total kinetic plus rest energy appearing in the exhaust is equal to the rate at which the total rest energy of the rocket is decreasing.

In analogy with our procedure for the nonrelativistic rocket, let us put $p_\mu = MU_\mu$ in Eq. (14.113) and transpose the term involving $dM/d\tau$ to the right-hand side:

$$M\frac{dU_\mu}{d\tau} = \mathscr{F}_\mu + \frac{dM}{d\tau}\left(\frac{U_{e\mu}}{\Gamma} - U_\mu\right), \tag{14.121}$$

where we have made use of Eq. (14.120). Let us now express the second term on the right in terms of velocities relative to the coordinate system in which the rocket is momentarily at rest. We use a superscript '0' to denote quantities in the rest coordinate system of the rocket, so that $U_i^0 = 0$, $U_0^0 = c$, $U_{ei}^0 = \Gamma V_i$, $U_{e0}^0 = \Gamma c$. If $a_{\mu\nu}$ are the transformation coefficients from the rest system to the system in which the rocket 4-velocity is U_μ,

$$U_{e\mu} = \Sigma_\nu a_{\mu\nu} U_{e\nu}^0 = \Gamma a_{\mu 0} c + \Gamma \Sigma_i a_{\mu i} V_i. \tag{14.122}$$

The coefficients $a_{\mu 0}$ can be calculated from the equation

$$U_\mu = \Sigma_\nu a_{\mu\nu} U_\nu^0 = a_{\mu 0} c. \tag{14.123}$$

We substitute in the preceding equation to obtain

$$U_{e\mu} = \Gamma U_\mu + \Gamma \Sigma_i a_{\mu i} V_i. \tag{14.124}$$

Since Γ is a scalar (it is the same constant no matter what coordinate system we are using), the first term on the right in Eq. (14.124) is a 4-vector, and therefore the second term must be also since the left-hand side is a 4-vector. We conclude that the four quantities $\Sigma_i a_{\mu i} V_i$ for $\mu = 0, 1, 2, 3$ are the components of a 4-vector. We substitute in Eq. (14.121) to obtain the equation of motion for the rocket in the form

$$M \frac{dU_\mu}{d\tau} = \mathscr{F}_\mu + \mathscr{T}_\mu, \tag{14.125}$$

where the 4-vector $\mathscr{T}_\mu$, which we may call the *4-thrust* of the rocket engine, is given by

$$\mathscr{T}_\mu = \frac{dM}{d\tau} \Sigma_i a_{\mu i} V_i. \tag{14.126}$$

If we take the x-axis in the direction of motion of the rocket so that the coefficients $a_{\mu i}$ are given by Eq. (14.7), then the components of the 4-thrust are

$$\mathscr{T}_0 = \beta\gamma V_x \frac{dM}{d\tau}, \quad \mathscr{T}_1 = \gamma V_x \frac{dM}{d\tau}, \quad \mathscr{T}_2 = V_y \frac{dM}{d\tau}, \quad \mathscr{T}_3 = V_z \frac{dM}{d\tau}. \tag{14.127}$$

It is interesting to note that the 4-thrust $\mathscr{T}_\mu$ does satisfy Eq. (14.82), as we see most easily by calculating the scalar product $(U_\mu, \mathscr{T}_\mu)$ in the rest-frame of the rocket.

Because $\mathscr{T}_\mu$ satisfies Eq. (14.82), we may also write Eq. (14.125) in terms of the coordinate time in the form

$$M \frac{d(\gamma \boldsymbol{u})}{dt} = \boldsymbol{F} + \boldsymbol{T}, \tag{14.128}$$

where the *3-thrust* $\boldsymbol{T}$ is given by

$$T_x = V_x \frac{dM}{d\tau}, \quad T_y = \frac{V_y}{\gamma} \frac{dM}{d\tau}, \quad T_z = \frac{V_z}{\gamma} \frac{dM}{d\tau}. \tag{14.129}$$

If the speed u of the rocket is much less than c, this reduces to the nonrelativistic expression given by the first term on the right in Eq. (4.54), when we note that $\boldsymbol{u}$ in Eq. (4.54) corresponds to $\boldsymbol{V}$ in the present equation. If the rocket engine is held at rest in a test stand, the left-hand side of Eq. (14.128) vanishes, and we see that the thrust $\boldsymbol{T}$ is just the negative of the force required to hold the rocket engine at rest. The thrust in a coordinate system in which the rocket engine is moving is then simply obtained by transforming from a coordinate system in which the engine is at rest, using the usual force-transformation law (14.83) for the thrust. [Note that the transformation law (14.83) for the force is based on the relation (14.81) between the 3-force and the 4-force which implies that the 4-force satisfies Eq. (14.82).]

Let us now solve Eq. (14.128) for the special case of a rocket accelerated from rest with no other force acting ($\boldsymbol{F} = \boldsymbol{0}$). We then take the x-axis in the direction of the acceleration so that the x-component of Eq. (14.128) reads

$$M\frac{d(\gamma u)}{dt} = -\gamma V\frac{dM}{dt},$$

where we have put $dM/d\tau = \gamma dM/dt$ and $V_x = -V$. Using the relation (14.84), we can rewrite this equation in the form

$$M\frac{d(\beta\gamma)}{dt} = -[1+(\beta\gamma)^2]^{1/2}B\frac{dM}{dt}, \tag{14.130}$$

where

$$B = V/c.$$

Equation (14.130) is readily integrated, starting from $\beta = 0$, $M = M_0$, at $t = 0$, to obtain

$$\beta\gamma+[1+(\beta\gamma)^2]^{1/2} = (M_0/M)^B,$$

whose solution is

$$\beta\gamma = \tfrac{1}{2}[(M_0/M)^B-(M_0/M)^{-B}]. \tag{14.131}$$

In the nonrelativistic limit, we may write

$$(M_0/M)^B = e^{B\ln(M_0/M)} \doteq 1+B\ln(M_0/M)+\cdots,$$

so that Eq. (14.131) becomes in the limit

$$u \doteq V\ln(M_0/M), \qquad \text{if } V\ln(M_0/M) \ll c,\ u \ll c, \tag{14.132}$$

in agreement with the classical result (4.56). In the opposite extreme relativistic case $(M_0/M)^B \gg 1$, we have

$$\gamma \doteq \tfrac{1}{2}(M_0/M)^{V/c}, \qquad \text{if } (M_0/M)^{V/c} \gg 1. \tag{14.133}$$

To reach a relativistic velocity, say $\gamma = 5$, even with exhaust velocity $V = c$, requires exhausting 90% of the initial mass of the rocket in photons or high energy particles.

14.6 RELATIVISTIC FORCE LAWS. ELECTRODYNAMICS

Difficulties arise in the theory of relativity when we try to formulate a law of force for the interaction between two particles at a distance from one another. If we try to formulate Newton's law of gravitation, for example, in a relativistically invariant way, we are faced with the problem that the distance between the two interacting bodies has a different value in different coordinate systems. If the distance between the two bodies is changing, a more fundamental difficulty arises due to the relativity of simultaneity. The force on particle (a) at time t is supposed to depend upon the relative positions of particles (a) and (b) at the same time t. If the two particles are located some distance apart, then the moment or event at

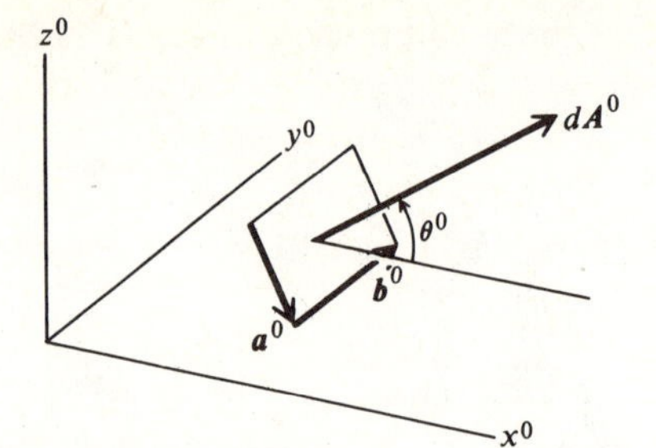

Fig. 14.2 An element of an area in a fluid.

particle (*b*) which is simultaneous with the event at particle (*a*) at time t will be different in different coordinate systems, yet the force is supposed to depend upon the distance between these two events! There is even difficulty with Newton's third law, which asserts that the force $\boldsymbol{F}_{a\to b}$ exerted by particle (*a*) on particle (*b*) is equal and opposite to the force $\boldsymbol{F}_{b\to a}$ exerted by particle (*b*) on particle (*a*). If the forces are changing with time, then the third law presumably asserts that these forces are equal and opposite if we measure them at the same time. However, if the two bodies are far apart, then the time at particle (*b*) which is simultaneous with a given time at particle (*a*) will depend upon the coordinate system chosen, and the third law will not satisfy the postulate of relativity.

This difficulty does not arise in the case of forces of contact, where the two interacting particles or bodies are located at the same point in space. In this case, the action and reaction forces occur at the same point, and there is no problem about their simultaneity. Newton's third law, applied to cases where the action and reaction occur at the same point, does satisfy the postulate of relativity. As an example, let us consider the pressure in a fluid, which we have defined in Section 5.11 as the force per unit area in the fluid. Let us first choose a coordinate system, which we will designate by a superscript '0', in which the fluid at a particular point is at rest. Consider now a small surface element of area dA^0 in the fluid, and define a vector surface element $d\boldsymbol{A}^0$ whose magnitude is the area of the surface element and whose direction is perpendicular to the surface element. (See Fig. 14.2.) We can simplify the discussion without loss of generality by assuming that the vector $d\boldsymbol{A}^0$ lies in the x^0z^0-plane, making an angle θ^0 with the x^0-axis. The force exerted by the fluid at the back side of the surface element $d\boldsymbol{A}^0$ on the fluid at the front side is

$$\boldsymbol{F}^0 = p^0\, d\boldsymbol{A}^0, \tag{14.134}$$

according to the definition of pressure p^0. The components of force in the three directions are

$$\begin{aligned} F_x^0 &= p^0 \cos\theta^0 dA^0 = p^0 dA_x^0, \\ F_y^0 &= 0, \\ F_z^0 &= p^0 \sin\theta^0 dA^0 = p^0 dA_y^0. \end{aligned} \tag{14.135}$$

We now transform to a laboratory coordinate system in which the fluid is moving along the x-axis with a velocity $u = \beta c$. The velocity of the laboratory coordinate system relative to the rest-system is then $-\beta c$ in the x-direction. We may use Eqs. (14.83) to find the force components in the new coordinate system:

$$\begin{aligned} F_x &= F_x^0 = p^0 dA_x^0, \\ F_y &= \gamma^{-1} F_y^0 = 0, \\ F_z &= \gamma^{-1} F_z^0 = p^0 \gamma^{-1} dA_y^0, \end{aligned} \tag{14.136}$$

where we have used the fact that the velocity of the fluid on which the force acts is zero in the rest-system.

In order to find the expression for the element of area $d\boldsymbol{A}$ in the lab system, let us assume that the area element is in the form of a parallelogram bounded by a vector $\boldsymbol{a}$ in the xz-plane and a vector $\boldsymbol{b}$ in the y-direction (Fig. 14.2) so that

$$d\boldsymbol{A} = \boldsymbol{a} \times \boldsymbol{b}. \tag{14.137}$$

Since $\boldsymbol{b}$ is perpendicular to the direction of motion, it will have the same length in both coordinate systems:

$$\boldsymbol{b} = \boldsymbol{b}^0 = b^0 \hat{\boldsymbol{y}}. \tag{14.138}$$

Let the vector $\boldsymbol{a}$ be given in the form

$$\boldsymbol{a} = a_x \hat{\boldsymbol{x}} + a_z \hat{\boldsymbol{z}}. \tag{14.139}$$

Then because of the Lorentz contraction in the x-direction, we will have

$$a_x = \gamma^{-1} a_x^0, \qquad a_z = a_z^0. \tag{14.140}$$

We can now calculate the vector

$$\begin{aligned} d\boldsymbol{A} &= (a_x \hat{\boldsymbol{x}} + a_z \hat{\boldsymbol{z}}) \times b^0 \hat{\boldsymbol{y}} \\ &= b^0 a_x^0 \gamma^{-1} \hat{\boldsymbol{z}} - b^0 a_z^0 \hat{\boldsymbol{x}} \\ &= dA_x^0 \hat{\boldsymbol{x}} + \gamma^{-1} dA_z^0 \hat{\boldsymbol{z}}. \end{aligned} \tag{14.141}$$

This gives the transformation law for the area element:

$$dA_x = dA_x^0, \qquad dA_y = \gamma^{-1} dA_y^0, \qquad dA_z = \gamma^{-1} dA_z^0, \tag{14.142}$$

where we have included for completeness the y-component, which is zero in the present case. In the laboratory coordinate system, the pressure will be defined by the equation

$$\boldsymbol{F} = p\, d\boldsymbol{A}. \tag{14.143}$$

Substituting for the left-hand side from Eq. (14.136) and for the right-hand side from Eq. (14.142), we see that the force is given correctly in the laboratory coordinate system if we put

$$p = p^0, \tag{14.144}$$

from which we conclude that the pressure in a fluid is a 4-scalar. Newton's third law is of course automatically satisfied, since the reaction force is given by choosing the opposite area element $-d\boldsymbol{A}$. The law of force in this case will be the equation of state of the fluid, which expresses the pressure p^0 in terms of appropriate variables, for example the density and temperature of the fluid in the coordinate system in which the fluid is at rest.

In the case of forces acting between bodies at a distance from one another, the most satisfactory way of writing a force law which satisfies the postulates of relativity appears to be to introduce a force field which propagates from one body to another. In this way the difficulties in specifying a force which acts directly between bodies at a distance from each other are avoided. The conservation laws are satisfied by allowing the field itself to carry energy and momentum. Newton's third law is no longer required. As pointed out in Section 4.4, these remarks are also true in classical physics for a force field which propagates between bodies with a finite velocity. Although Newton's law of gravitation is the simplest of the fundamental force laws in classical physics, its relativistic generalization presents serious difficulties (see for example Problem 40). Einstein's general theory of relativity is a relativistic theory of gravitation; its mathematical development is however beyond the scope of this chapter. Maxwell's formulation of the laws of electrodynamics turns out to be already invariant under a Lorentz transformation as we shall now show.

It will be convenient to express the electric and magnetic fields in terms of the scalar and vector potentials. We assert here without proof that the electric and magnetic fields can be written in the following way in terms of a *scalar potential* ϕ and a *vector potential* $\boldsymbol{A}$:*

$$\begin{aligned} \boldsymbol{E} &= -\nabla\phi - \frac{1}{c}\frac{\partial \boldsymbol{A}}{\partial t}, \\ \left[\boldsymbol{E} \right. &= \left. -\nabla\phi - \frac{\partial \boldsymbol{A}}{\partial t}\right], \\ \boldsymbol{B} &= \nabla \times \boldsymbol{A}. \end{aligned} \tag{14.145}$$

We will use Gaussian units throughout this discussion of electrodynamics, as they bring out more clearly the structure of the equations from the point of view of the

*An understanding of electrodynamics is not required to follow the subsequent argument. The reader who wishes to pursue the subject further may consult any of the standard intermediate texts in electricity and magnetism, for example Panofsky and Phillips, *Classical Electricity and Magnetism*, Reading, Mass.: Addison-Wesley Publishing Company, 1955 (this text uses mks units); or J. D. Jackson, *Classical Electrodynamics*, New York: John Wiley and Sons, 1962 (this text uses Gaussian units).

theory of relativity. When we use Gaussian units, the only fundamental constant that appears in the equations of electrodynamics is the velocity of light c. As an aid to the reader who wishes to translate our discussion into mks units, we will include in square brackets the mks form of each equation which is different in the two systems of units.

We suspect that the 3-vector $\boldsymbol{A}$ and the 3-scalar ϕ together make up the components of a 4-vector. In order to check this hypothesis, we make use of the dynamical equations for the vector and scalar potentials, which we again quote without proof:

$$\nabla^2\phi - \frac{1}{c^2}\frac{\partial^2\phi}{\partial t^2} = -4\pi\rho,$$

$$\nabla^2\boldsymbol{A} - \frac{1}{c^2}\frac{\partial^2\boldsymbol{A}}{\partial t^2} = -\frac{4\pi}{c}\boldsymbol{j},$$

$$\left[\begin{aligned} \nabla^2\phi - \frac{1}{c^2}\frac{\partial^2\phi}{\partial t^2} &= -\frac{\rho}{\varepsilon_0} \\ \nabla^2 A - \frac{1}{c^2}\frac{\partial^2 A}{\partial t^2} &= -\mu_0\boldsymbol{j} \end{aligned}\right], \tag{14.146}$$

where ρ is the total electric charge density per unit volume and $\boldsymbol{j}$ is the total electric current density, defined by requiring that $\hat{\boldsymbol{n}}\cdot\boldsymbol{j}$ be the total charge per second per unit area crossing a surface in the direction of a unit vector $\hat{\boldsymbol{n}}$. In order for Eqs. (14.146) to hold, we also require the following equation, called the *Lorentz condition*:

$$\nabla\cdot\boldsymbol{A} + \frac{1}{c}\frac{\partial\phi}{\partial t} = 0,$$

$$\left[\nabla\cdot\boldsymbol{A} + \frac{1}{c^2}\frac{\partial\phi}{\partial t} = 0\right]. \tag{14.147}$$

Although we have quoted the above results from electrodynamics without proof, the reader familiar with Maxwell's equations will recognize that it follows from Eqs. (14.145) that the electric and magnetic fields satisfy the two homogeneous Maxwell equations

$$\nabla\cdot\boldsymbol{B} = 0, \tag{14.148}$$

$$\nabla\times\boldsymbol{E} = -\frac{1}{c}\frac{\partial\boldsymbol{B}}{\partial t}, \tag{14.149}$$

$$\left[\nabla\times E = -\frac{\partial B}{\partial t}\right],$$

and that Eqs. (14.146) then follow from Eq. (14.147) and the two remaining Maxwell equations

$$\nabla \times \boldsymbol{B} = \frac{1}{c}\frac{\partial \boldsymbol{E}}{\partial t} + \frac{4\pi}{c}\boldsymbol{j}, \tag{14.150}$$

$$\nabla \cdot \boldsymbol{E} = 4\pi\rho, \tag{14.151}$$

$$\left[\begin{aligned} \nabla \times \boldsymbol{B} &= \frac{1}{c^2}\frac{\partial \boldsymbol{E}}{\partial t} + \mu_0 \boldsymbol{j} \\ \nabla \cdot \boldsymbol{E} &= \frac{\rho}{\varepsilon_0} \end{aligned}\right].$$

We see from Eqs. (14.146) that in order to determine the way in which $\boldsymbol{A}$, ϕ transform between moving coordinate systems, we should determine how the quantities $\boldsymbol{j}$, ρ transform. To do this, let us imagine a situation in which at any point in space all electric charges are moving with the same velocity. For any given point, we may then choose a local rest coordinate system, in which the charges at that particular point are at rest. We designate quantities in the local rest frame by a superscript '0'. The charge density ρ^0 is given by the equation

$$\rho^0 = dQ^0/dV^0, \tag{14.152}$$

where dQ^0 is the total electric charge in a small volume element dV^0. Let us now return to any coordinate system of interest in which the charges at the point in question are moving with a velocity $\boldsymbol{u}$. The charge density is then given by the equation

$$\rho = dQ/dV, \tag{14.153}$$

where dQ is the total charge in a small volume element dV. Let us allow the volume element dV to move with the velocity of the charges which it encloses, so that it continues to contain the same charges. Since electric charge comes in multiples of a fundamental unit e (the charge on the electron being $-e$), the charge in the volume element dV can in principle be obtained simply by counting the total number of unit charges therein. Since the volume element dV moves along with the charges and contains always the same charges, the charges can be counted in any order and over any time period that we choose. No problems of simultaneity arise in counting the charges, and therefore the result must be the same in every coordinate system:

$$dQ = dQ^0, \tag{14.154}$$

so that the total electric charge in a volume moving with the charge is a 4-scalar. Because of the Lorentz contraction in the direction of motion, one of the three dimensions of the volume dV will be contracted by a factor γ, and we have therefore

$$dV = \gamma^{-1} dV^0. \tag{14.155}$$

We conclude from the preceding four equations that

$$\rho = \gamma \rho^0. \tag{14.156}$$

The current density in the present case is given by

$$\boldsymbol{j} = \rho \boldsymbol{u} = \gamma \boldsymbol{u} \rho^0. \tag{14.157}$$

The reader should be able to convince himself that this expression is correct in that $\hat{\boldsymbol{n}} \cdot \boldsymbol{j}$ is the proper expression for the total charge per second per unit area flowing across a surface perpendicular to $\hat{\boldsymbol{n}}$ (see for example the discussion of Fig. 8.6). Recalling formulas (14.26) for the 4-velocity, we now see that we can define a 4-vector, the *4-current*

$$J_\mu = \rho^0 U_\mu, \tag{14.158}$$

whose space and time parts are

$$J_i = j_i, \qquad J_0 = \rho c. \tag{14.159}$$

The densities $\rho c, \boldsymbol{j}$ therefore transform like the components of a 4-vector. Although the above derivation applies to the case when all the charges at any particular point in space are moving with the same velocity, we see that in fact it must apply in general. Since the sum of any number of 4-vectors is itself a 4-vector, in the case when there are a number of groups of charges at any point moving with different velocities, we may sum their contributions to the total 4-current and conclude that the total 4-current associated with all the charges is also a 4-vector.

We see now that if we define a set of four quantities A_μ by the equations

$$A_i = A_i, \qquad A_0 = \phi, \tag{14.160}$$

$$[A_i = A_i, \qquad A_0 = \phi/c]^*,$$

then we can write Eqs. (14.146) in the form

$$\Box^2 A_\mu = -\frac{4\pi}{c} J_\mu, \tag{14.161}$$

$$[\Box^2 A_\mu = -\mu_0 J_\mu].$$

Since the operator $\Box^2$ (Eq. 14.38) is a 4-scalar, we see that the quantities A_μ must transform as a 4-vector in order that Eq. (14.161) be relativistically covariant. The Lorentz condition (14.147) is then also properly covariant, since it has the form

$$(\Box_\mu, A_\mu) = 0, \tag{14.162}$$

as the reader may verify.

We are now ready to substitute Eq. (14.145) into the Lorentz force law given by the right-hand member of Eq. (14.103). We insert the result into the formulas

*Some authors use instead Φ_μ, defined in mks units by $\Phi_i = cA_i$, $\Phi_0 = \phi$.

(14.81) for the 4-force to obtain

$$\mathscr{F}_0 = -\frac{q\gamma}{c}\boldsymbol{u}\cdot\nabla\phi - \frac{q\gamma}{c}\boldsymbol{u}\cdot\frac{\partial \boldsymbol{A}}{\partial t},$$

$$\mathscr{F}_i = -q\gamma\frac{\partial\phi}{\partial x_i} - \frac{q\gamma}{c}\frac{\partial A_i}{\partial t} + \frac{q\gamma}{c}[\boldsymbol{u}\times(\nabla\times\boldsymbol{A})]_i.$$

It is a matter of straightforward algebra to verify that with definition (14.160) these equations may be written in the form

$$\mathscr{F}_\mu = \frac{q}{c}\Sigma_\nu\, g_\nu U_\nu\left(g_\mu\frac{\partial A_\nu}{\partial x_\mu} - g_\nu\frac{\partial A_\mu}{\partial x_\nu}\right), \tag{14.163}$$

or, equivalently,

$$\mathscr{F}_\mu = \frac{q}{c}\Box_\mu\,(U_\nu, A_\nu) - \frac{q}{c}(U_\nu, \Box_\nu)\,A_\mu. \tag{14.164}$$

[In mks units, omit the c in formulas (14.107), (14.163), and (14.164).] This form makes it clear that $\mathscr{F}_\mu$ is indeed a 4-vector, so that the Lorentz formula (14.107) for the electromagnetic force does have the correct transformation properties. The reader will readily verify that this formula for the force also satisfies Eq. (14.82), as it necessarily must since we used Eq. (14.81) in deriving it.

14.7 TENSOR ALGEBRA IN THE 4-DIMENSIONAL SPACE

The reader who has studied Chapter 10 may be interested in the possibility of generalizing tensor algebra to the 4-dimensional space. We will give in this section a brief summary of this development. We will omit most of the derivations, which are not very difficult and which the reader should be able to supply for himself. Chapter 10 was concerned only with second rank tensors; we will generalize this notion to a tensor of arbitrary rank in 4-dimensional space.

A *4-tensor of rank n* is defined as a set of 4^n quantities $T_{\mu\nu\cdots\lambda}$, labeled by $n_{\mu\nu,\ldots,\lambda}$, each taking on the values 0, 1, 2, 3, and transforming according to the rule

$$T^*_{\mu\nu\cdots\lambda} = \Sigma_{\alpha,\beta,\ldots,\gamma}\, a_{\mu\alpha}\, a_{\nu\beta}\ldots a_{\lambda\gamma}\, T_{\alpha\beta\cdots\gamma}. \tag{14.165}$$

A particular example of an nth rank tensor is an n-ad formed by taking the products $A_\mu B_\nu \cdots C_\lambda$ of the components of n vectors taken in all possible combinations, one component from each vector. The *sum*

$$T_{\mu\nu\cdots\lambda} + H_{\mu\nu\cdots\lambda} \tag{14.166}$$

of two tensors of the same rank, formed by adding corresponding components of each, is itself a tensor of the same rank. The $(m+n)$-*ad product*

$$T_{\mu\nu\cdots\lambda}\, H_{\alpha\beta\cdots\gamma} \tag{14.167}$$

of an mth rank tensor $T_{\mu\nu\cdots\lambda}$ and an nth rank tensor $H_{\alpha\beta\cdots\gamma}$ formed by taking all possible products of a component $T_{\mu\nu\cdots\lambda}$ and a component $H_{\alpha\beta\cdots\gamma}$ is a tensor of rank $m+n$. Given a tensor $T_{\mu\nu\cdots\nu}$, of rank greater than or equal to 2, we can form a tensor of rank $(n-2)$ by *contracting* in the following manner:

$$\Sigma_\alpha g_\alpha T_{\mu\nu\cdots\alpha\cdots\alpha\cdots\lambda}, \tag{14.168}$$

where we have chosen any two of the indices of $T_{\mu\nu\cdots\lambda}$, and have formed sums of four components in which those two indices have the same value, multiplied by the factor g_α corresponding to each particular value of the index.

As an example of the above operations, we note that scalar product (14.20) of two 4-vectors may be obtained by forming first the dyad product $A_\mu B_\nu$ according to the prescription (14.167) and then contracting the two indices according to the prescription (14.168). Since there is then no index left, the result is a 4-scalar.

A tensor may have the property that the value of its components does not change when the values of two particular indices are interchanged:

$$T_{\mu\nu\cdots\alpha\cdots\beta\cdots\lambda} = T_{\mu\nu\cdots\beta\cdots\alpha\cdots\lambda}. \tag{14.169}$$

Such a tensor is said to be *symmetric* in those two indices. Likewise a tensor is said to be *antisymmetric* in two particular indices if the value of its components changes sign when those two particular indices are interchanged:

$$T_{\mu\nu\cdots\alpha\cdots\beta\cdots\lambda} = -T_{\mu\nu\cdots\beta\cdots\alpha\cdots\lambda}. \tag{14.170}$$

The property of being symmetric or antisymmetric in a particular pair of indices is preserved under a Lorentz transformation (14.165). If a tensor is antisymmetric in two particular indices, then evidently those components for which those two indices have the same value must be zero. A contraction (14.168) performed on two antisymmetric indices evidently gives the result zero, that is, it gives a tensor all of whose components are zero. We saw that an antisymmetric tensor in three dimensions has three independent components. The reader may verify that an antisymmetric second rank tensor in four dimensions has 6 independent components. In four dimensions, an antisymmetric tensor cannot be placed in correspondence with any 4-vector, since it has the wrong number of components.

As an application of the algebra of 4-tensors, let us consider the *electromagnetic field tensor*, of second rank, defined by the equation

$$F_{\mu\nu} = \Box_\mu A_\nu - \Box_\nu A_\mu. \tag{14.171}$$

The tensor $F_{\mu\nu}$ is evidently antisymmetric, and has therefore six independent components which are just the components of the vectors $\boldsymbol{E}$, $\boldsymbol{B}$ which describe the electromagnetic field. The components of $F_{\mu\nu}$ are given according to Eq.

(14.145) by the following scheme as the reader may verify:

$$F_{\mu\nu} = \begin{array}{c|cccc} \mu\backslash\nu & 0 & 1 & 2 & 3 \\ \hline 0 & 0 & E_x & E_y & E_z \\ 1 & -E_x & 0 & B_z & -B_y \\ 2 & -E_y & -B_z & 0 & B_x \\ 3 & -E_z & B_y & -B_x & 0 \end{array}. \tag{14.172}$$

[In mks units, replace E_x, E_y, E_z by E_x/c, E_y/c, E_z/c.]* From this result, together with the transformation law (14.165), the reader can derive the rule for transforming the electric and magnetic fields from one coordinate system to another (see Problem 52). The electromagnetic force on a charged particle is obtained by contracting the triad product of the 4-velocity with the electromagnetic field tensor:

$$\mathscr{F}_\mu = \Sigma_\nu \frac{q}{c} g_\nu U_\nu F_{\mu\nu}. \tag{14.173}$$

Many further examples of the application of the algebra of 4-tensors would arise if we had space here to extend the discussion of Chapter 8 to the relativistic mechanics of a continuous medium. (See Problems 53, 54, 55.)

14.8 THE GENERAL THEORY OF RELATIVITY

We will conclude this chapter with a few qualitative remarks about the general theory of relativity. As we noted in the first section, it would be intuitively plausible to try to postulate quite generally that the laws of physics should be independent of the motion of the coordinate system. This postulate however seems immediately to be in conflict with our experience. While it may be difficult or even impossible to detect whether one is in uniform motion or standing still, anyone who has ridden in an automobile on a bumpy road or going around a rapid curve knows that acceleration of the coordinate system is readily detectable by simple experiments. A way out of the dilemma was suggested by the physicist-philosopher Ernst Mach, who proposed that the effects which we observe are not due to acceleration relative to absolute space, a notion which he regarded as meaningless, but may rather be ascribed to the acceleration of our local coordinate system relative to the rest of the matter in the universe. According to Mach, an inertial coordinate system is one in which the matter in the universe is not accelerated on the average. Einstein attempted to incorporate Mach's ideas into his general theory of relativity.

We saw in Chapter 7 that we can write Newton's law of motion in an ac-

---

*When $\Phi_\mu = cA_\mu$ is used instead as the 4-potential, $F_{\mu\nu}$ in Eq. (14.172) is also multiplied by c. See footnote following Eq. (14.160).

celerated coordinate system in the same form that it has in an inertial system, if we transpose the terms arising from the acceleration of the coordinate system to the other side of the equation, so that they appear as additional forces. These "inertial" forces which act on any body in an accelerated coordinate system are proportional to the mass of the body. According to Newton's law of gravitation, the gravitational force acting on the body is also proportional to its mass. Einstein proposed that this is not a coincidence, but that the inertial forces should be regarded as gravitational forces, on the same footing as the forces given by Newton's law of gravitation. The law of gravitation should then be modified so that the gravitational forces depend on the positions and the velocities of other bodies in such a way that the inertial forces arise from accelerations of the rest of the mass of the universe relative to our coordinate system. We see from Eq. (7.37) that the coriolis force on a body depends upon its velocity, so that the generalized law for the gravitational force on a body must contain such terms; they will be analogous to the term giving the magnetic force on a charged body, which also depends upon its velocity.

In the special case of a uniformly accelerated coordinate system, we see from Eq. (7.10) that the equation of motion is the same as in a uniform gravitational field, where the gravitational acceleration $\mathbf{g}$ is the negative of the acceleration of the coordinate system. Einstein proposed to base his general theory of relativity on the principle that this is true not only of the law of motion but of all the laws of physics. He therefore postulated the *principle of equivalence:*

The laws of physics in a uniform gravitational field $\mathbf{g}$ *are the same as they would be in the absence of a gravitational field in a uniformly accelerated coordinate system with acceleration* $-\mathbf{g}$. (14.174)

In the neighborhood of a gravitating mass such as the earth, the acceleration of gravity is not uniform, but varies with the distance from the earth, and also varies in direction depending upon the position relative to the earth. Such a gravitational field cannot be eliminated by introducing an accelerated coordinate system. In order to include nonuniform gravitational fields, we can formulate the principle of equivalence in the following more satisfactory form:

In a small, freely falling laboratory, the laws of physics are the same as the laws of the special theory of relativity without any gravitational field. (14.175)

By a freely falling laboratory, we mean a physical system subject to no external forces except gravitational forces. An example would be a space ship coasting in a vacuum, together with its contents. By "small" laboratory, we mean that it is to be small enough so that differences in the magnitude and direction of the gravitational field are negligible over the dimensions of the laboratory, for those experiments which are to be performed. From the mathematical point of view, we may formulate the principle by asserting that in the neighborhood of any point in space-time, we may always introduce a coordinate system whose origin is falling freely in that region, and such that in the neighborhood of the origin the laws of special relativity hold without any gravitational field.

An immediate consequence of the principle of equivalence is that all bodies fall at the same rate in a gravitational field, or equivalently that the gravitational force is exactly proportional to the mass. This is not a new effect. It was first reported by Galileo, and was the effect which suggested the principle of equivalence to Einstein. It was tested in 1890 by Eötvös, essentially by comparing ratios of the gravitational forces and the centrifugal forces due to the Earth's rotation on a number of bodies. His measurements showed the proportionality of gravitational force and mass within one part in 10^8 for all the bodies he tested. The experiment was repeated to a part in 10^{11} by R. H. Dicke in 1964. This is by far the greatest precision with which a prediction of the general theory of relativity has been tested. It is adequate to show that the mass associated with internal nuclear and chemical energy is acted on also by the gravitational field.

The principle of equivalence provides us only with the laws of physics in a local, freely falling coordinate system. In the presence of gravitating masses, it is evidently not possible to introduce a single coordinate system which is everywhere freely falling. We are therefore left with the problem of formulating the laws of physics so that they can be applied not just to a local infinitesimal region, but globally (or rather, universally). Without attempting this problem, which was solved by Einstein, we can draw a number of immediate conclusions from the principle of equivalence.

It follows from the principle of equivalence that the propagation of light is affected by a gravitational field. According to the special theory of relativity, and therefore in a local, freely falling coordinate system, light travels with the constant velocity c in a straight line. It follows that light must also experience the acceleration of gravity in the presence of a gravitational field. In particular, we would expect starlight passing near the sun to be bent in an orbit about the sun, so that the apparent position of a star behind the sun would be shifted by a slight angle. This was first predicted by Einstein. In the absence of a suitable global formulation of the law of propagation of light, we cannot make a numerical prediction. If, however, we calculate the deflection on the simple assumption that the light at a distance from the sun is traveling with the velocity c and that it is then accelerated according to Newton's law of gravitation in the gravitational field of the sun, we can arrive at a formula for the deflection (see Problem 56), which turns out to be just one-half of the actual deflection predicted by the general theory of relativity. The predicted deflection is so small that it is difficult to measure very precisely, but within the accuracy of measurements made during eclipses, the observed deflection of starlight agrees with the value predicted by Einstein.

Let us now consider a pair of clocks A and B, separated in height by a distance h in a uniform gravitational field $\boldsymbol{g}$. (See Fig. 14.3.) Let the gravitational acceleration be in the direction from clock A toward clock B. Let us assume that the clocks are physically identical and that a light signal of frequency f_B as measured

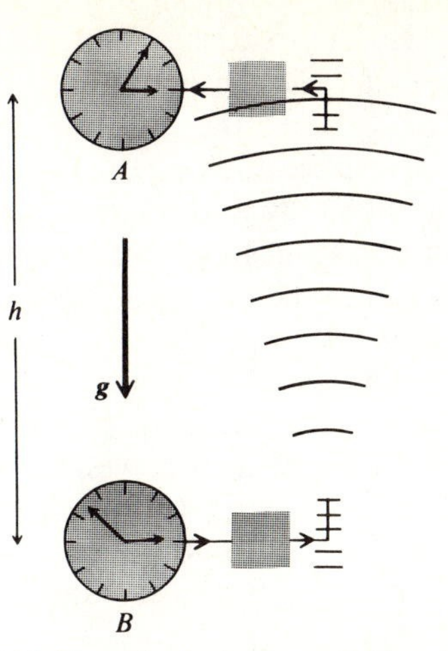

Fig. 14.3 Two clocks in a gravitational field.

by clock B is sent out from B. We wish to calculate the frequency f_A of the signal as measured with the aid of clock A when it is received at the position of A. To do so, let us introduce a freely falling coordinate system falling with the acceleration $\mathbf{g}$. In this coordinate system, both clocks A and B are accelerated upward in the direction from B toward A. Let us choose the coordinate system so that at the moment when clock B sends its signal, it is at rest in the coordinate system. We assume also that the separation h is not too large, so that the velocities of both clocks remain small during the time of interest. There are then no special relativistic effects to take into account in our freely falling coordinate system. The light signal leaving B at time $t = 0$ arrives at A at time $t = h/c$. At that time, clock A is moving in a direction away from B with the velocity $v = gt = gh/c$. (Clock B by that time is also of course traveling with the same velocity.) The frequency f_A is therefore reduced by the Doppler effect by a factor $[1-(v/c)]$. We conclude that

$$f_A = f_B[1-(gh/c^2)],$$

or

$$\frac{\Delta f}{f} = \frac{f_B - f_A}{f_B} = \frac{gh}{c^2}. \tag{14.176}$$

As viewed from clock A, clock B appears to run slowly; the shift in frequency $\Delta f/f$ is given by the above equation in terms of (gh), the difference in gravitational potential between A and B. This is the *gravitational red shift*. The effect is very small, except in large gravitational fields g and at large distances h. An appreciable shift in the frequencies of spectral lines coming from near the surface of white dwarf stars is expected due to this effect. Some such effect seems to be observed, although a precise numerical check is difficult to make. If we use the acceleration of gravity at the surface of the earth, and take for the height h a distance of 10 meters, we conclude that the shift in frequency should be about one part in 10^{15}. Although this is a very small effect, it can be measured utilizing the Mössbauer effect. This conclusion of the general theory of relativity has therefore been verified by experiments at the surface of the earth.

We can use the above result to discuss the twin paradox (Section 13.6) from the point of view of the traveling twin. In general relativity, we are permitted to use a coordinate system in which the astronaut is always at rest. Recall that, relative to the outward moving astronaut, the earth-clock was reading 7.5 years when he arrived at the distant planet, but that shortly afterward, when he had landed on the planet, the earth clock was reading 10 years relative to the then stationary astronaut. In a coordinate system in which the astronaut is fixed, there is a gravitational field pointing from Earth toward the distant planet during the time the rocket is decelerating (relative to Earth). Hence in the astronaut's (accelerated) coordinate system, the clock at Earth runs fast relative to one located on the planet. Although formula (14.176) was derived only to lowest order in gh/c^2 and therefore cannot legitimately be applied to this example where $gh/c^2 \gg 1$, a correct general relativistic analysis of this problem leads to the conclusion that in the accelerated coordinate system, the earth-clock does indeed gain 2.5 years during the landing.

It should be emphasized again that all physical processes go on at rates determined by the proper time as read on local co-moving clocks. Any physical process whose rate is known may be used as a clock. Differences in the rates of clocks, and hence of all physical processes, due to differences in velocities or in gravitational potential, are predicted by the special and the general theory of relativity. These differences in rate are real, in the sense that we must take them into account if we wish to correlate correctly the time sequences of events in distant or rapidly moving systems. On the other hand, these differences are relative, in the sense that we may choose a coordinate system in which any particular clock is at rest and then it is to be regarded as running at a fixed standard rate; peculiarities in the matching of time sequences with other systems are then ascribed to variations in the rates of their clocks. Relative to the astronaut, there is a sudden speedup of the clocks back on Earth during the deceleration process. The twin who stays on Earth observes no such speedup; he accounts for the entire sequence of events in terms of the time dilation exhibited by the astronaut's clock.

A third crucial experiment suggested by Einstein in his original paper on the general theory of relativity involves the motion of the perihelion of the planet Mercury. We do not have the machinery available here to make any quantitative estimate of this effect, except to point out, as noted in Chapter 3, that any deviation from the Newtonian law of gravitation or the Newtonian law of motion will be expected to lead to a precession of the major axis of the elliptical orbit. The predicted precession is 43 seconds of arc per century. Our measurements and understanding of planetary orbits is now sufficiently precise so that this precession can be measured and agrees with the figure predicted by Einstein within about 5%, that is, within the experimental accuracy.

One of Einstein's motivations in seeking a general theory of relativity was the suggestion by Mach that the inertial coordinate systems are those which are at rest, or at least not accelerated, relative to the mean motion of the matter in the universe. One might therefore expect that a nearby rapidly rotating mass would produce some slight rotation in an inertial coordinate system. The theory of relativity does indeed predict such an effect. For practically obtainable rotations of reasonable masses, the effect is extremely small. However, experiments are underway which it is hoped may be sufficiently precise to detect this effect.

Since the principle of relativity requires that gravitational effects cannot be transmitted instantaneously from one mass to another but must propagate through the intervening space, it is perhaps not surprising that the theory of relativity predicts the existence of gravitational waves traveling with the velocity c. Again the effects of such waves are expected to be very slight. However, by 1969, experiments by J. Weber appeared to have detected such waves.*

An interesting and important conclusion of the general theory of relativity is that gravitational fields can affect the geometry of space. In a gravitational field the postulates of Euclidean geometry will not be quite satisfied. From the operational point of view, the concepts of geometry, straight line, angles, etc., are to be defined in terms of measurements made with physical instruments. The assertion that geometry is non-Euclidean therefore amounts to the assertion that gravitational fields affect the behavior of measuring instruments in certain ways. In order to see how this comes about, let us consider an inertial coordinate system, which we will call the fixed system, in which we assume that the ordinary laws of physics (in particular special relativity and Euclidean geometry) hold. Let us imagine a coordinate system rotating with respect to the fixed system; we will call it the rotating system. Let us now consider a circle whose center is on the axis of rotation and let us assume that the radius and circumference of the circle are measured by means of meter sticks at rest in the fixed system and also by means of meter sticks at rest in the rotating system. We will assume that the radius of the circle and the rate of rotation are such that a point on the circumference in the rotating system travels with a velocity less than c, but sufficiently great so that the factor γ is measureably different from unity. It is clear that the concept of a rigidly rotating coordinate system cannot be extended out to radii for which $\omega r > c$ without running into difficulty with the postulates of relativity. We consider first measurements of the radius of the circle. Since the meter sticks in the rotating system are moving perpendicular to the length which they are measuring, according to our previous discussions the measurements of the radius of the circle in the moving and the fixed system should agree. Let us assume that a number of meter sticks in the rotating coordinate system are laid out so that they completely fill the circumference of the circle. If we consider these meter sticks from the point of view of the fixed coordinate system, we are entitled to use the laws of special relativity, according to which these sticks will be contracted by a factor

*J. Weber, *Physical Review Letters*, **22**, 1320 (1969).

$\gamma = [1-(\omega r/c)^2]^{-1/2}$; hence the number of such meter sticks which can be fitted into the circumference of the circle will exceed the number which can be laid around the circle in the fixed coordinate system by the factor γ. Since, in the moving system, we may measure the circumference of the circle simply by counting the meter sticks, we conclude that the measured circumference C^* will be greater than the circumference C, and in fact $C^* = \gamma C$. Since $r^* = r$, we conclude that in the rotating system the circumference of the circle is given by

$$C^* = 2\pi\gamma r^* = 2\pi r^* [1-(\omega r^*/c)^2]^{-1/2}. \tag{14.177}$$

The circumference of a circle in a rotating coordinate system is larger than the value that would be given by the usual Euclidean formula by a factor which depends upon the radius of the circle. The effect is larger for larger circles. For sufficiently small circles, the formula approaches the Euclidean value.

Since the distortion of the geometry is associated with the acceleration of the coordinate system, since this acceleration also produces gravitational effects (centrifugal and coriolis force fields), and since according to the principle of equivalence these gravitational effects are physically indistinguishable from the gravitational effects due to gravitating masses, we conclude that we expect to find deviations from Euclidean geometry associated in general with gravitational fields. In the rotating coordinate system, we have a centrifugal gravitational field (as well as a coriolis force field), and associated with it a distortion in the geometry in which the circumference of a circle is larger in proportion to its radius than would be given by the Euclidean formula. In the neighborhood of a gravitating mass, we have a centripetal gravitational field. There is no rigid acceleration of the coordinate system which will produce such a gravitational field, and therefore no way of deducing directly from the principle of equivalence the nature of the geometry produced by such a field. Although we cannot make it plausible by any simple argument, the reader will perhaps not be surprised that the Einstein theory, when it is worked out, predicts that the centripetal gravitational field around a gravitating mass is associated with a distortion of the geometry in which the circumference of a circle is less than $2\pi r$. Likewise the area of a sphere around a gravitating mass is less than the Euclidean value $(4\pi/3)r^2$, and increases less rapidly with the radius. One consequence of this distortion of the geometry is to produce an additional bending of a light ray over and above what one might calculate from the gravitational acceleration alone. Since a light ray traveling in empty space follows a path which is the shortest distance between two points along it, one can see that because of the distortion of the geometry, a light ray passing near the sun will find that it can travel a shorter distance by following an apparently curved path which carries it farther from the sun at its point of closest approach than would be the case if it followed a straight line in ordinary Euclidean geometry. This effect alone gives a deflection of a light ray which is equal to the deflection that would be calculated from the gravitational acceleration, and the total deflection predicted by the general theory of relativity then comes out to be twice what one would calculate simply from the acceleration of gravity. Actual measurements on the bending of starlight agree with the Einstein prediction within the accuracy of the measurements (about 10%). A similar effect which has more recently been measured is the delay in arrival time of radio signals coming either from astronomical sources or from artificial satellites circling the sun. The length of a line from a distant object to the earth which passes near the sun is greater than it would be if the sun were not present, and this causes a predicted delay in a radio signal which is confirmed by the measurements.

One fascinating feature of the general theory of relativity is that it is a theory which can be applied to the entire universe and to the entire history of the universe. There are a large variety of solutions to Einstein's gravitational field equations. There are solutions in which the density of matter throughout the universe is uniform and is such that the gravitational effects produce a closed space in which the total volume is finite. The volume of a sphere about any given point increases with the radius less rapidly than would be given by Euclid's formula, and eventually at a certain maximum radius the volume becomes equal to the total volume of space. A light ray or a high-speed particle would eventually return to the neighborhood of its origin, if it continued to travel freely without being deflected or absorbed. This is a situation which could in principle be verified by astronomical observations. If for example it should turn out that the universe is homogeneous in the sense that the density of galaxies is the same in all parts of it, then by counting the number of galaxies which can be seen as a function of the distance to the galaxy, one could effectively measure the volume of a large sphere as a function of its radius. Unfortunately the necessary observations are very difficult to make and to interpret, and furthermore it appears that the universe may very well not be homogeneous in the sense defined.

It turns out that static solutions of Einstein's gravitational field equations are unstable, so that the theory effectively predicts that the universe should either be expanding or contracting. The observations agree with this prediction in that it appears that the universe is indeed expanding. Whether it will expand indefinitely, or will reach a maximum size and then contract again, depends according to the theory upon the mean density of matter, which has not yet been measured with sufficient precision to determine the answer to this question.

If the geometry of space around gravitating bodies is non-Euclidean, then Cartesian space-coordinate systems x, y, z of the type that we have used in our discussion of the special theory of relativity will not in general exist. A similar difficulty arises with the time coordinate. In the presence of gravitational fields, clocks in different places may run at different rates even if they are at rest in some particular coordinate system. We may see this either by including clocks in our discussion of a rotating coordinate system, or from the discussion of the gravitational red shift. There is then no reasonable way of synchronizing clocks so that they stay synchronized. If we define a time coordinate t at each point (x, y, z) by correcting for the travel time of light signals from a standard clock at the origin, then this time will not agree (except perhaps at one instant) with the time read by a local clock at the point (x, y, z).

It turns out to be convenient in the general theory of relativity to give up the attempt to set up any standard system of coordinates based on standardized systems of measured lengths and times to locate events. Instead, we simply assign four coordinates, say q_0, q_1, q_2, q_3, to events in space-time in any arbitrary (continuous) way. The coordinate system in itself therefore contains no information about either physics or geometry, and any arbitrary change to any other set of coordinates $q_0^*, q_1^*, q_2^*, q_3^*$ would be permissible. The geometry is then defined by specifying the results of measurements of times and distances between the events.

Near any particular event in space-time the principle of equivalence guarantees that we may set up a local inertial coordinate system x_λ by means of measurements made with freely falling clocks and meter sticks. The distance ds or time interval $d\tau$ measured between two nearby events is then given, according to Eq. (14.4), by

$$ds^2 = -c^2 d\tau^2 = \Sigma_\lambda g_\lambda (dx_\lambda)^2, \tag{14.178}$$

where dx_λ are the coordinate differences between the two events, and Formula (14.178) holds only within the small region [in the sense of the principle of equivalence (14.175)] within which our inertial system x_λ may be set up. If we now return to an arbitrary coordinate system q_μ, the local coordinates x_λ may be expressed in terms of q_μ by some set of functions

$$x_\lambda = x_\lambda (q_0, q_1, q_2, q_3). \tag{14.179}$$

We substitute in Eq. (14.178) to obtain a formula for the interval between two nearby events in terms of the general coordinates q_μ:

$$ds^2 = -c^2 d\tau^2 = \Sigma_{\mu\nu}\, g_{\mu\nu}\, dq_\mu\, dq_\nu, \tag{14.180}$$

where

$$g_{\mu\nu} = \Sigma_\lambda g_\lambda \frac{\partial x_\lambda}{\partial q_\mu} \frac{\partial x_\lambda}{\partial q_\nu}. \tag{14.181}$$

Formula (14.180) gives the desired connection between the coordinates q_μ, which may be chosen arbitrarily, and physical measurements of distances and times between nearby events. The geometrical (physical) relationships between nearby events are characterized by the quantities $g_{\mu\nu}$, which are called the components of the *metric tensor*. Since $g_{\mu\nu} = g_{\nu\mu}$, only ten of the $g_{\mu\nu}$ are independent. If $ds^2 > 0$, then the interval between the two nearby points is space-like and ds is the result of measuring the distance in a coordinate system in which the two events occur simultaneously. If $ds^2 < 0$, then the two events are separated by a time-like interval and $d\tau$ is the time interval measured by a clock which passes through both events.

The metric tensor $g_{\mu\nu}$ describes on the one hand the geometry of space-time, and on the other hand the gravitational field in the coordinate system q_μ. We may regard Eq. (14.181) as an operational definition of $g_{\mu\nu}$; given an arbitrarily chosen coordinate system q_μ, set up near each space-time point a local inertial system x_λ by means of measurements with freely falling clocks and meter sticks, and calculate $g_{\mu\nu}$ from Eq. (14.181). Conversely, given $g_{\mu\nu}$ in a particular region in space-time, we may work backward from Eq. (14.180) and ask for a set of functions $x_\lambda\ (q_\mu)$ which convert ds^2 into the form (14.178). [The reader who has studied Chapter 10 will recognize that if the region in question is sufficiently small so that $g_{\mu\nu}$ may be regarded as constant, we have to solve the problem of diagonalizing a symmetric tensor in the four-dimensional space.] The quantities $g_{\mu\nu}$ therefore implicitly define the freely falling coordinate systems x_λ relative to q_μ; by the principle of equivalence, they therefore characterize the gravitational field in the system q_μ.

Formula (14.180) only holds in the limit when the events are sufficiently close together so that the interval between them may be regarded as infinitesimal. In the case of events with a finite separation, the interval between them must be defined by an integration of Formula (14.180).

Unfortunately there is not space here to pursue the mathematical development of the theory. We have therefore had to content ourselves with these qualitative remarks. The reader who wishes to pursue the subject further may consult any one of a number of excellent texts on relativity (see the list in the Bibliography).

PROBLEMS

1. Check that relations (14.13) hold for the special cases (14.7) and (14.8).

2. Given two events E^1, E^2, write the quantity S^{21} given by Eq. (14.4) in a starred coordinate system, make the substitution (14.6), and verify that if Eqs. (14.13) hold, S^{21} reduces to the same form (14.4) in unstarred coordinates.

3. Prove that the scalar product (14.20) is invariant under a Lorentz transformation if A_μ, B_μ are 4-vectors.

4. Prove theorem (14.21). [*Hint:* If $\Sigma_\mu K_\mu B_\mu = \Sigma_\mu L_\mu B_\mu$ for arbitrary B_μ, then $K_\mu = L_\mu$.]

5. Derive formula (13.47), which gives the relativistic Doppler effect by making a Lorentz transformation on the wave vector k_μ.

6. A starred coordinate system moves with a velocity $\boldsymbol{v}$ relative to an unstarred system. The velocity $\boldsymbol{v}$ lies in the xy-plane at an angle α with the x-axis, measured counterclockwise relative to the unstarred system. The velocity $\boldsymbol{v}$ makes an angle α^* with the x^*-axis, measured counterclockwise in the starred system. The z- and z^*-axes are parallel. The origins O, O^* coincide at $t = t^* = 0$. Find the coefficients $a_{\mu\nu}$ for the transformation (14.6) for this case. [*Hint:* Make three successive transformations of the forms (14.7) and (14.8) and use Eq. (14.12).] Check your algebra by showing that the origin O moves correctly in the starred coordinate system.

Show that, unless $\alpha = 0$, it is impossible to choose α^* so that both the x^*- and y^*-axes are parallel respectively to the x- and y-axes. In other words, the new axes are not at right angles in the old coordinate system (and vice versa). Given $\alpha \neq 0$, what value of α^* makes the x^*-axis parallel to the x-axis? [*Hint:* At $t = 0$, the x^*-axis can be found in the unstarred system by setting $y^* = z^* = 0$.] If $\alpha = \alpha^*$, what is the angle between the x^*- and y^*-axes as measured in the unstarred system?

Show that if $\beta \ll 1$, the starred axes are nearly at right angles relative to the unstarred system.

7. Show that, if p_0 is positive in any inertial coordinate system, it is positive in all inertial systems. (Note the analogy with the absolute meaning of the future light-cone in space-time.)

Show that the rest mass m of a particle remains constant in magnitude and sign under a Lorentz transformation of p_μ.

8. Calculate the group velocity, whose components are

$$v_{gi} = \partial\omega/\partial k_i,$$

as a function of $\boldsymbol{k}$, for the dispersion relation (14.59). Show that the relation (14.58) then implies that the velocity $\boldsymbol{u}$ of a particle is equal to the group velocity $\boldsymbol{v}_g$ of its associated wave in relativistic quantum mechanics.

9. Given a set of 4-vectors $p_{j\mu}$, all time-like or light-like, show that their sum is time-like or light-like. Show that the sum is light-like only if all the $p_{j\mu}$ are light-like and all the space parts $\boldsymbol{p}_j$ are parallel.

10. It has been proposed that particles may exist whose 4-momentum p_μ is space-like. Define a suitable parameter for such a particle analogous to the mass of an ordinary particle, and write equations analogous to Eqs. (14.46) expressing p_μ in terms of the velocity $\boldsymbol{u}$. Since $u > c$, such particles are called *tachyons*. Find approximate formulas for E and $\boldsymbol{p}$ for a tachyon in the ultrarelativistic limit $u \gg c$. What happens in the limit $u \to c$? Could a tachyon carry any signal from one place to another?

11. A horizontal beam of antineutrons of kinetic energy 100 MeV enters a hydrogen bubble chamber. A reaction is observed in which two positive pions of momenta 141 MeV/c and 920 MeV/c leave tracks in the vertical plane at angles of 45° with the direction of the incident beam, one above and one below it. A third track of a negative pion is seen moving horizontally at right angles with the incident beam with a momentum of 145 MeV/c. All energies and momenta are measured to an accuracy of about 1%. Assume that the target particle was a proton. The mass of a nucleon (neutron or proton) is 938 MeV/c^2, and the mass of a pion is 140 MeV/c^2. (It is convenient to measure energies of high energy particles in MeV, masses in MeV/c^2, and momenta in MeV/c. Kinematical calculations can then be carried out without any messy conversion of units.)

a) Show that at least one unseen (therefore uncharged) particle must have been produced in the reaction, and that it was almost surely a single neutral pion. What was the magnitude and direction of its momentum?

b) It is suspected that an ω^0 meson may have been produced in the reaction. The ω^0 is a very short-lived particle (10^{-20} sec) with a mass of 783 MeV/c^2 and zero electric charge, which decays into three pions. Show that it is very likely that an ω^0 was produced, and find its kinetic energy and direction of motion.

12. Express the quantity T given by Eq. (14.67) in terms of the masses m_a, m_b, m_c, m_d, given by Eq. (14.64) and the invariants M, S given by Eq. (14.65) and (14.66).

13. Given a system of particles with total 3-momentum $\boldsymbol{P}$ and total energy $E = P_0 c$, find a starred coordinate system in which $\boldsymbol{P}^* = \mathbf{0}$, and show that $P_0^* = Mc$ (Eq. 14.65). [*Hint:* Take the x-axis in the direction of $\boldsymbol{P}$.]

14. A proton beam of kinetic energy T enters a hydrogen bubble chamber. Find the threshold energy T_t for producing antiprotons in the reaction

$$p^+ + p^+ \to p^+ + p^+ + p^+ + \bar{p}^-,$$

where the target proton is assumed to be at rest. The rest energy of the proton and of the antiproton is 938 MeV.

15. A pion of kinetic energy 1200 MeV is incident on a proton at rest, producing a number of pions in the reaction

$$\pi + p \to p + n\pi.$$

What is the maximum number n of pions that can be produced in the reaction? The rest energy of the proton is 938 MeV and that of the pion 140 MeV. [*Hint:* Use the center-of-mass coordinate system.]

16. A pion of kinetic energy 1000 MeV is scattered through an angle of 45° from its original direction of motion in a collision with a proton at rest. What is the scattering angle of the pion in the center-of-mass coordinate system? The masses of the pion and the proton are given in Problem 15.

17. An incident particle (a) reacts with a particle (b) at rest, producing particles (c) and (d). Show that if the masses, the invariants M and S, and the angle θ_{ac} between $\mathbf{p}_a$ and $\mathbf{p}_c$ are given, all energies and momenta and the angle θ_{ad} can be determined.

18. Find an expression for the scalar product $(P_{a\mu}, P_{d\mu})$ of the 4-momenta of particles (a) and (d) in the reaction shown in Fig. (14.1), in terms of the masses of the four particles and the invariants M, S defined by Eqs. (14.65) and (14.66).

19. Write the invariant S in a particular coordinate system in terms of the energies E_a, E_c, the masses m_a, m_c, and the angle θ_{ac} between the directions of motion of the two particles.

20. a) Show that the four-dimensional volume element $dV\,dt = dx\,dy\,dz\,dt$ is a 4-scalar. [*Hint:* Show that the Jacobian $\partial(x^*, y^*, z^*, t^*)/\partial(x, y, z, t) = 1$ for transformations of the type (14.7) and (14.8), and hence conclude that the Jacobian is unity for a transformation between any two inertial coordinate systems.]

b) Show that the same argument leads to the conclusion that the volume element $dp_x\,dp_y\,dp_z\,dE$ in a four-dimensional energy-momentum space is also invariant under a Lorentz transformation. (Note the connection with Problems 7 and 19 of Chapter 13.)

21. Show by direct calculation that the quantity $\gamma_a\gamma_b|u_a-u_b|$, where the velocities u_a, u_b are directed along the x-axis, is invariant under a special Lorentz transformation of the type (14.7). The velocities u_a, u_b are taken positive or negative according to the directions of motion along the x-axis.

22. Show that the Lorentz-invariant quantity

$$F_{ab} = [(p_{a\mu}, p_{b\mu})^2 - m_a^2 m_b^2 c^4]^{1/2}$$

has the value

$$F_{ab} = m_a m_b c^2 \gamma_a \gamma_b |u_a \mp u_b|$$

if $\boldsymbol{u}_a$ is parallel or antiparallel to $\boldsymbol{u}_b$. Using this result, write a generalization of Eq. (14.72) which defines a cross section σ which is a scalar under arbitrary Lorentz transformations.

23. Two particles (a) and (b) with 4-momenta $p_{a\mu}$, $p_{b\mu}$ have given masses m_a, m_b and a given total 4-momentum $P_\mu = p_{a\mu} + p_{b\mu}$. Show that the quantity

$$\frac{p_a^3\, d\Omega_a}{\boldsymbol{p}_a \cdot (E_b \boldsymbol{p}_a - E_a \boldsymbol{p}_b)}$$

is invariant under a Lorentz transformation, where $\boldsymbol{p}_a$ is the 3-momentum of particle (a), and its direction lies in the solid angle $d\Omega_a = \sin\theta_a\, d\theta_a\, d\phi_a$. (Suggestion: Start from the invariance of the volume element dp_{a0}, $dp_{ax}\, dp_{ay}\, dp_{az}\, dp_{b0}\, dp_{bx}\, dp_{by}\, dp_{bz}$ in 8-space. First make a change of variables from p_{ax}, p_{ay}, p_{az} to spherical coordinates p_a, θ_a, ϕ_a in $\boldsymbol{p}_a$-space. Then make the change of variables p_{b0}, p_{bx}, p_{by}, p_{bz}, p_{a0}, p_a, θ_a, $\phi_a \rightarrow P_0$, P_x, P_y, P_z, $(m_a^2c^2)$, $(m_b^2c^2)$, θ_a, ϕ_a. The 8×8 Jacobian determinant for the latter transformation is not too difficult to evaluate.)

Using the above result, show that the cross section for elastic scattering of particle (a) by particle (b) into a solid angle $d\Omega_a$ can be written in the form

$$d\sigma_a = \frac{F(M, S)p_{aF}^3\, d\Omega_a}{\boldsymbol{p}_{aF}\cdot(E_{bF}\boldsymbol{p}_{aF} - E_{aF}\boldsymbol{p}_{bF})},$$

where $\boldsymbol{p}_{aF}$, E_{aF}, $\boldsymbol{p}_{bF}$, E_{bF} are the final energies and momenta of the two particles after scattering, $S = (p_{aF\mu} - p_{aI\mu}, p_{aF\mu} - p_{aI\mu})$, and the subscript '$I$' denotes initial values.

24. Evaluate the cross section $d\sigma_a$ given in Problem 23 in the center-of-mass coordinate system for the elastic scattering of two particles of equal mass m. Express the result in terms of the mass m, the invariant mass M, and the angle Θ through which particle (a) is scattered. Express the invariant S in terms of m, M, and Θ. Show that if, for some center-of-mass energy M_0c^2, the scattering is isotropic, i.e., if the cross section per unit solid angle $d\sigma_a/d\Omega_a$ is independent of Θ, then the function $F(M_0, S)$ is independent of S at that center-of-mass energy.

Given $F(M_0, S) = F_0$, independent of S, evaluate $d\sigma_a/d\Omega_a$ in the laboratory coordinate system (where particle (b) is stationary), and hence find the angular distribution in the laboratory of a scattering process which is isotropic in the center-of-mass.

25. Verify Eqs. (14.83). Check that all four components $\mathscr{F}_u^*$ come out correctly when $\boldsymbol{F}^*$ is given by Eqs. (14.83).

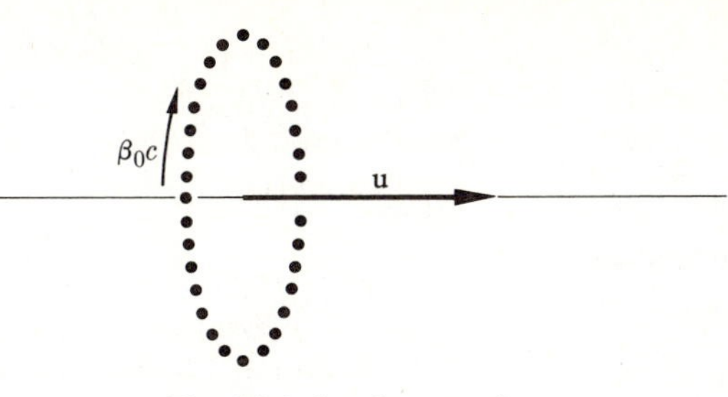

Fig. 14.4 An electron ring

26. A particle starts from the origin with momentum $\boldsymbol{p}_0$ directed along the y-axis. It is subject to a constant force $\boldsymbol{F}$ in the x-direction. Find its motion. How does its orbit compare with the parabola that would result if the classical equations of motion were used?

27. In an electron-ring accelerator, a circular ring of electrons travels with velocity $\boldsymbol{u}$ along the symmetry axis of the ring (Fig. 14.4.). In a coordinate system in which the ring is at rest, it contains a large number N of electrons uniformly distributed around the circular ring of radius a moving around the ring with energy $\gamma_0 mc^2$. Show that in the accelerator coordinate system, its energy and momentum are the same as that of a single particle of mass $N\gamma_0 mc^2$ moving with velocity $\boldsymbol{u}$ down the symmetry axis.

28. A particle of mass m moving along the x-axis is subject to a force $F_0 \cos \omega t$ in the x-direction.

a) Given the initial velocity v_0 of the particle, find $v(t)$.
b) Given that the relativistic corrections are small, find the position $x(t)$ to first order in c^{-2}.
c) Given that F_0 is small, but $\gamma_0 \gg 1$, find $x(t)$ to first order in F_0.
d) Given that $v_0 = 0$, but $F_0 \gg \omega mc$, sketch $v(t)$ and describe the motion.

29. A particle moving along the x-axis is subject to a linear restoring force $F = -kx$. Describe its motion. Carry the solution by the energy method as far as you can. Under what conditions does the motion agree with that of a classical harmonic oscillator? Find the lowest order relativistic corrections to the frequency of oscillation and to the motion $x(t)$.

30. A particle moves along the x-axis under a potential energy

$$V(x) = (Ax^2 - B)e^{-\alpha x^2}.$$

If it starts from the origin with velocity v_0, use the energy integral (14.92) to discuss the motion. Describe the kinds of motion which may occur and give the range of initial velocities v_0 for each. If the particle can escape, give its final velocity in terms of v_0.

31. Devise a force function $F(x)$ for which the relativistic equations of motion have an exactly sinusoidal solution $x = A \sin \omega t$ for a given amplitude A.

32. A light high-speed particle of mass m, charge $-e$ is scattered by a heavy fixed charge Ze which attracts it with the coulomb force $-Ze^2/r^2$. Given the initial velocity u_0 and the impact parameters (Section 3.16), find the distance of closest approach. Assume the fixed charge remains at rest. Show that if $m\gamma_0 \ll M$, where M is the mass of the fixed particle, then it is legitimate to neglect the motion of the latter. (If the fixed charge remains at rest, then it is also correct to use the ordinary coulomb law for the force.)

33. Solve Problem 29 by using the equivalent potential V_{rel} (Eq. 14.101). Show that this method gives the same result.

34. Use theorem (14.102) and the result of Problem 67 in Chapter 3 to find the lowest order relativistic correction to the Rutherford scattering cross section formula (3.276).

35. Show that in the limit $E_c \gg V$, the relativistic orbit of a particle of mass m is approximately the same as the classical orbit of a particle of mass $m(mc^2/E)$ moving in the same potential $V(x)$ with the same energy $E_c = E - mc^2$.

36. Generalize the theorem (14.102) to the case of a charged particle subject to an electrostatic field $\boldsymbol{E} = -\nabla\phi$ and a static magnetic field $\boldsymbol{B}$. Apply your theorem to solve Problem 75, Chapter 3, for the relativistic case.

37. A particle of charge q moves in uniform crossed electric and magnetic fields $\boldsymbol{E} = E_0\hat{\boldsymbol{x}}$, $\boldsymbol{B} = B_0\hat{\boldsymbol{y}}$. Show that if $E_0 < B_0$, the projection of its motion on the xz-plane is an ellipse of eccentricity E_0/B_0, moving in the z-direction with an average velocity $cE_0/\gamma_0 B_0$, and that it traverses the ellipse with an angular frequency ω_1 given by

$$\omega_1 = q(B_0^2 - E_0^2)^{1/2}/\gamma_0 mc,$$

where $\gamma_0 mc^2$ is the particle energy when it is at the end of a major diameter of the ellipse. What is the nature of the y-motion? [*Hint:* Use the proper time as independent variable. The corresponding problem in classical mechanics was solved in Section 3.17.]

38. Show by a direct calculation that the four quantities

$$B_\mu = \Sigma_i\, a_{\mu i} V_i,$$

used in the definition (14.126) of the 4-thrust of a rocket engine, transform like the components of a 4-vector. Remember that V_i are the components of the exhaust velocity relative to the rocket and therefore do not transform, and that $a_{\mu\nu}$ are the coefficients of the transformation from the rest frame of the rocket to the coordinate system in which B_μ is calculated [*Hint:* Don't be afraid of this problem. Just calculate B_μ^* and it will come right out.].

39. A rocket ship of total mass $M(t)$ exhausts rest mass at a rate $dM_e/d\tau$ with a velocity $\boldsymbol{V}$ relative to the rocket. In addition, waste material is thrown overboard at zero velocity relative to the rocket at a rate $dM_w/d\tau$.

a) Show that the equation of motion (14.128) still applies, but the thrust $\boldsymbol{T}$ must be defined in terms of $dM_e/d\tau$.

b) Show that, given V and $dM_e/d\tau$, the maximum final velocity of a rocket accelerated from rest will be obtained if any unneeded materials are thrown overboard as soon as possible.

c) Show that for any given fuel, an optimum exhaust velocity exists in the following sense. Let the properties of the fuel be such that it is possible to convert a fraction α of its rest mass into kinetic energy, so that $dE_k = \alpha c^2 dM_f$, where dM_f is the mass of fuel used up in time $d\tau$.

This energy is used to accelerate part of the spent fuel of rest mass $dM_e = -\eta\, dM_f$ to whatever velocity corresponds to kinetic energy dE_k. Any remaining spent fuel $(1-\alpha-\eta)\, dM_f$ is thrown overboard. Show that the maximum thrust is obtained by taking the maximum possible value of η, that is putting all spent fuel into the exhaust. What is the optimum value of V? Assume that the rocket is to be accelerated in a straight line so that $\boldsymbol{V}$ is in the direction of $-\boldsymbol{u}$.

40. Equations (6.16) and (6.33) suggest a possible relativistic law of gravitation in which the force on a particle of mass m is given in terms of a 4-scalar potential Φ by

$$\mathscr{F}_{m\mu} = m\Box_\mu \Phi.$$

Write a covariant equation to be solved for Φ which reduces in the static case (all masses at rest, $\partial\Phi/\partial t = 0$) to Eq. (6.33). Show that unfortunately this simple theory does not satisfy Eq. (14.82), and hence that the rest mass of a particle does not remain constant. Show that, according to this theory,

$$\frac{dm}{dt} = -m\frac{d\Phi}{dt},$$

where $d\Phi/dt$ is the rate of change of Φ at the position of the moving particle.

41. A vector $d\boldsymbol{A}$ represents a surface element of constant area and orientation which moves with velocity $\boldsymbol{u}$ in an unstarred coordinate system. Using the fact that pressure is a scalar, find the vector $d\boldsymbol{A}^*$ which represents the same surface element in a starred coordinate system whose axes are parallel and which moves with velocity βc along the x-axis relative to the unstarred system.

42. Verify that Eq. (14.163) is a correct 4-vector transcription of the preceding equations for $\mathscr{F}_0, \mathscr{F}_i$.

43. Verify that the electromagnetic 4-force given by Eq. (14.164) satisfies Eq. (14.82). (The proof requires only two or three lines.)

44. Show that the Lagrangian function

$$L = -mc^2(1-u^2/c^2)^{1/2} - q\phi + q\boldsymbol{u}\cdot\boldsymbol{A}/c$$

gives the correct relativistic equation of motion (14.79) for a charged particle.

45. It can be shown that if $L\,dt$ is a 4-scalar, then the Lagrangian function L will lead to equations of motion which are covariant under a Lorentz transformation. Show that for the Lagrangian L given by Problem 44, $L\,dt$ is a 4-scalar.

46. Show that the Lagrangian function L given in Problem 44 yields the Hamiltonian function

$$H = mc^2\gamma + q\phi.$$

Express H in terms of coordinates and momenta.

47. Prove that sums, products, and contractions formed according to the formulas (14.166), (14.167), and (14.168) transform like 4-tensors of appropriate rank.

48. Prove that if a 4-tensor is symmetric or antisymmetric in two particular indices, this property is preserved by a Lorentz-transformation.

49. Count the number of independent components of a completely antisymmetric 4-tensor of various ranks, that is, one antisymmetric in every pair of indices. Show that no nontrivial completely antisymmetric 4-tensor exists of rank greater than four. Show that a completely antisymmetric 4-tensor of rank four has only one independent component. Does it transform as a 4-scalar?

50. How many independent components has a symmetric 4-tensor of second rank? How many has a completely symmetric 4-tensor of third rank, that is, symmetric in every pair of indices?

51. Verify that the electromagnetic field tensor defined by Eq. (14.171) is correctly given in terms of $\boldsymbol{E}$ and $\boldsymbol{B}$ by Eq. (14.172).

52. Find the components of $\boldsymbol{E}^*$, $\boldsymbol{B}^*$ in a starred coordinate system in terms of $\boldsymbol{E}$, $\boldsymbol{B}$ in an unstarred system related to it by the transformation coefficients (14.7). (Use the relation 14.172.)

53. a) A fluid moving with velocity $\boldsymbol{u}$ has a mass density ρ_0 in a coordinate system in which the fluid is at rest. Show that the density of rest mass when the fluid moves with velocity $\boldsymbol{u}$ is $\gamma\rho_0$, and that the total energy density is $\gamma^2\rho_0 c^2$.

b) The kinetic energy-momentum tensor for the fluid is defined to be

$$T_{\mu\nu} = \rho_0 U_\mu U_\nu.$$

Show that T_{00} is the energy density, that T_{0i} is the energy current in the i-direction, that $c^{-1}T_{0i}$ is the density of i-component of momentum, and that T_{ij} is the momentum current, i.e. the current of i-component of momentum in the j-direction.

54. a) Given that $\boldsymbol{f}$ is the body force density (3-force per unit volume), show that the four quantities $f_0 = \boldsymbol{u}\cdot\boldsymbol{f}_c$ and f_i are the components of a 4-vector, where $\boldsymbol{u}$ is the velocity of the matter on which the force acts.

b) Using the above result, and that of Problem 53, show that the conservation of energy and momentum in a moving fluid subject to a body force density $\boldsymbol{f}$ may be expressed in the form

$$\frac{\partial T_{\mu\nu}}{\partial x_\nu} = f_\mu.$$

***55.** Comparison of the above result with that of Problem 35 of Chapter 10 suggests a conservation law for a medium containing stresses in the form

$$\frac{\partial}{\partial x_\nu}(T_{\mu\nu} + P_{\mu\nu}) = f_\mu,$$

where $P_{\mu\nu}$ is the stress energy-momentum tensor. Assume that in a fluid at rest the only nonzero components of $P_{\mu\nu}$ are $P_{11} = P_{22} = P_{33} = p$, where p is the pressure. Find $P_{\mu\nu}$ in a coordinate system in which the fluid moves with velocity $\boldsymbol{u}$ along the x-axis. Show that P_{0i} is the energy current due to pressure, and that P_{ij} is the momentum current due to pressure.

56. Assume that a photon moving with speed c when it is far from the sun is accelerated by the gravitational field of the sun, the acceleration being given by Newton's law of gravitation. Find the deflection angle α in the direction of a light ray passing near the sun, as a function of the impact parameter s.

57. According to quantum mechanics, a photon of frequency ν has energy $h\nu$, and therefore according to the special theory of relativity, it has a (kinetic) mass $h\nu/c^2$. A photon moves upward against a uniform gravitational field whose acceleration is g. Assume that the total energy, including gravitational potential energy, is conserved and find the frequency ν' of the photon when it has climbed a distance s. Show that your result agrees with eq. (14.176).

58. A rocket ship leaves Earth and travels a distance L at a speed βc, $\beta \ll 1$, to arrive at a planet at rest relative to the Earth. Clocks on Earth, at the planet, and in the rocket are synchronized to read zero at the moment of departure. Find the reading t_p of the clock at the planet at the moment of arrival of the rocket. Find the reading t_R of the rocket clock. Find, relative to a coordinate system in which the rocket is fixed, the time t_E read on the Earth clock when the rocket arrives at the planet. The rocket ship now decelerates with acceleration $-g$ for a time Δt_R until it is at rest relative to the planet. Using formula (14.176) in a decelerating coordinate system in which the rocket remains always at rest, calculate (to first order in β) the time $\Delta t_E - \Delta t_R$ gained by the Earth clock relative to the rocket clock during the deceleration. Show that the time gained is just enough to bring the Earth clock back in synchronism with the planet clock.

59. A clock A at rest on the axis of a rotating coordinate system sends out a signal of frequency f_A as measured by clock A. The signal arrives at clock B, also at rest in the rotating system at a distance r from the axis.

a) Using the laws of special relativity, in a nonrotating system, show that the frequency f_B of the signal, as measured by clock B, can be written in the form

$$f_B = f_A(1 - 2\mathcal{G}/c^2)^{-1/2}$$

where $\mathcal{G}$ is the gravitational potential difference between A and B, defined by Eq. (6.17), for the centrifugal force field in the rotating system.

b) Show that this same formula agrees to first order in $\mathcal{G}/c^2$ with formula (14.176) derived for a uniform gravitational field. (Formula (14.176) was only derived to first order in gh/c^2.)

60. a) Show from the result (14.177) that the distance ds measured in a rotating coordinate system between two nearby points may be written in the form

$$(ds)^2 = (dr^*)^2 + r^{*2}[1 - 2\mathcal{G}/c^2]^{-1}(d\theta)^2,$$

where r^*, θ are polar coordinates in the rotating system, and $\mathcal{G}$ is the gravitational potential defined by Eq. (6.17) due to the centrifugal force field in the rotating system. Take the reference radius where $\mathcal{G} = 0$ to be $r^* = 0$.

b) Assume that the above result applies also, at least approximately, to the gravitational field around the sun, but with $\mathcal{G}$ now defined to vanish at $r = \infty$, so that far from the sun Euclidean geometry is valid. Consider a "straight line" passing within a distance h from the sun. Find to first order in $\mathcal{G}$ the increase in length of this line due to the gravitational field of the sun. What is the corresponding increase in travel time of a radio signal passing near the sun?

c) Assuming the result of Problem 59 applies near the sun, and recalling that the speed of light c is measured by using local clocks, find the increase in travel time of a radio signal passing within a distance h of the sun, due to the slower rate of local clocks near the sun. Show that this effect gives a time delay equal to that calculated in part (b), so that the total time delay is doubled.

BIBLIOGRAPHY

The following is a list, by no means complete, of books related to the subject matter of this text which the reader may find helpful.

ELEMENTARY MECHANICS TEXTS

1. J. W. Campbell, *An Introduction to Mechanics*. New York: Pitman, 1947.
2. R. A. Millikan, D. Roller, and E. C. Watson, *Mechanics, Molecular Physics, Heat, and Sound*. Boston: Ginn and Co., 1937.

INTERMEDIATE MECHANICS TEXTS

3. W. Arthur, and S. K. Fenster, *Mechanics*. New York: Holt, Rinehart, and Winston, 1969.
4. R. A. Becker, *Introduction to Theoretical Mechanics*. New York: McGraw-Hill, 1954.
5. R. B. Lindsay, *Physical Mechanics*, 2nd ed. New York: D. Van Nostrand, 1950.
6. W. D. MacMillan, *Theoretical Mechanics*. New York: McGraw-Hill. Vol. 1: *Statics and Dynamics of a Particle*, 1927. Vol. 3: *Dynamics of Rigid Bodies*, 1936.
7. W. F. Osgood, *Mechanics*. New York: Macmillan, 1937.
8. M. Scott, *Mechanics, Statics and Dynamics*. New York: McGraw-Hill, 1949.
9. R. J. Stephenson, *Mechanics and Properties of Matter*. New York: John Wiley & Sons, 1952.
10. J. L. Synge, and B. A. Griffith, *Principles of Mechanics*, 3rd ed. New York: McGraw-Hill, 1959.

ADVANCED MECHANICS TEXTS

11. H. C. Corbin, and P. Stehle, *Classical Mechanics*. New York: John Wiley & Sons, 1950.
12. H. Goldstein, *Classical Mechanics*. Reading, Mass.: Addison-Wesley, 1950.
13. H. Lamb, *Hydrodynamics*, 6th ed. Cambridge: Cambridge University Press, 1932. (New York: Dover Publications, 1945.)
14. L. D. Landau, and E. M. Lifshitz, *Mechanics*. London: Pergamon Press, 1960. (Reading, Mass.: Addison-Wesley, 1960.)
15. L. D. Landau, and E. M. Lifshitz, *Fluid Mechanics*. London: Pergamon Press, 1959. (Reading, Mass.: Addison-Wesley, 1959.)
16. L. D. Landau, and E. M. Lifshitz, *Theory of Elasticity*. London: Pergamon Press, 1959. (Reading, Mass.: Addison-Wesley, 1959.)
17. Lord Rayleigh, *The Theory of Sound* (2 vols.), 2nd ed. London: Macmillan, 1894–96. (New York: Dover Publications, 1945.)
18. E. J. Routh, *Dynamics of a System of Rigid Bodies*, Advanced Part, 6th ed. London: Macmillan, 1905. (New York: Dover Publications, 1955.)
19. J. C. Slater, and N. H. Frank, *Mechanics*. New York: McGraw-Hill, 1947.
20. A. G. Webster, *The Dynamics of Particles and of Rigid, Elastic, and Fluid Bodies*. Leipzig: B. G. Teubner, 1904.

21. E. T. Whittaker, *A Treatise on the Analytical Dynamics of Particles and Rigid Bodies*, 4th ed. Cambridge: Cambridge University Press, 1937. (New York: Dover Publications, 1944.)
22. A. Wintner, *The Analytical Foundations of Celestial Mechanics*. Princeton: Princeton University Press, 1941.

TEXTS ON ELECTRICITY AND MAGNETISM

23. R. G. Fowler, *Introduction to Electric Theory*. Reading, Mass.: Addison-Wesley, 1953.
24. N. H. Frank, *Introduction to Electricity and Optics*, 2nd ed. New York: McGraw-Hill, 1950.
25. G. P. Harnwell, *Principles of Electricity and Magnetism*, 2nd ed. New York: McGraw-Hill, 1949.
26. A. F. Kip, *Fundamentals of Electricity and Magnetism*. New York: McGraw-Hill, 1969.
27. L. Page, and N. I. Adams, *Principles of Electricity*. New York: D. Van Nostrand, 1931.
28. W. T. Scott, *Physics of Electricity and Magnetism*. New York: John Wiley & Sons, 1966.
29. J. C. Slater, and N. H. Frank, *Electromagnetism*. New York: McGraw-Hill, 1947.

ADVANCED TEXTS ON ELECTROMAGNETIC THEORY

30. J. D. Jackson, *Classical Electrodynamics*. New York: John Wiley & Sons, 1962.
31. W. K. H. Panofsky, and M. Phillips, *Classical Electricity and Magnetism*. Reading, Mass.: Addison-Wesley, 1962.

BOOKS AND ELEMENTARY TEXTS ON RELATIVITY

32. A. Einstein, and L. Infeld, *The Evolution of Physics*. New York: Simon & Schuster, 1938. An excellent popular account.
33. A. Einstein, H. A. Lorentz, H. Weyl, H. Minkowski, *The Principle of Relativity*. New York: Dover Publications, 1952. (Originally published London: Methuen, 1923.) A collection of translations of the original papers by the authors, remarkably readable.
34. D. Bohm, *The Special Theory of Relativity*. New York: W. A. Benjamin, 1965.
35. M. Born, *Einstein's Theory of Relativity*. New York: Dover Publications, 1962.
36. R. B. Lindsay, and H. Margenau, *Foundations of Physics*. New York: John Wiley & Sons, 1936. Contains two chapters on the special and general theories of relativity.
37. N. D. Mermin, *Space and Time in Special Relativity*. New York: McGraw-Hill, 1968.
38. R. D. Sard, *Relativistic Mechanics*. New York: W. A. Benjamin, 1970.
39. J. H. Smith, *Introduction to Special Relativity*. New York: W. A. Benjamin, 1965.
40. E. F. Taylor, and J. A. Wheeler, *Spacetime Physics*. San Francisco: W. H. Freeman, 1963.

ADVANCED TEXTS ON RELATIVITY

41. P. G. Bergmann, *An Introduction to the Theory of Relativity*. New York: Prentice-Hall, 1946.
42. R. C. Tolman, *Relativity, Thermodynamics, and Cosmology*. Oxford: Oxford University Press, 1934.

WORKS ON QUANTUM MECHANICS

43. W. Heisenberg, *The Physical Principles of the Quantum Theory*, trans. by C. Eckart and F. C. Hoyt. Chicago: University of Chicago Press, 1930. (New York: Dover Publications, 1949.)

44. D. Bohm, *Quantum Theory*. New York: Prentice-Hall, 1951.
45. M. Born, *Atomic Physics*, trans. by John Dougall, 4th ed. New York: Hafner, 1946.
46. L. D. Landau, and E. M. Lifshitz, *Quantum Mechanics—Non-relativistic Theory*. London: Pergamon Press, 1958. (Reading, Mass.: Addison-Wesley, 1958.)
47. R. B. Lindsay, and H. Margenau, *Foundations of Physics*. New York: John Wiley & Sons, 1936. Contains a chapter on quantum mechanics.

TEXTS AND TREATISES ON MATHEMATICAL TOPICS

48. R. Bellman, *Stability Theory of Differential Equations*. New York: McGraw-Hill, 1953.
49. R. V. Churchill, *Fourier Series and Boundary Value Problems*. New York: McGraw-Hill, 1941.
50. R. Courant, *Differential and Integral Calculus*, trans. by E. F. McShane. London: Blackie & Son, 1934.
51. L. Hopf, *Introduction to the Differential Equations of Physics*, trans. by Walter Nef. New York: Dover Publications, 1948.
52. D. Jackson, *Fourier Series and Orthogonal Polynomials*. Menasha, Wisc.: George Banta Publishing Co., 1941.
53. T. von Karman, and M. A. Biot, *Mathematical Methods in Engineering*. New York: McGraw-Hill, 1940.
54. W. Kaplan, *Advanced Calculus*. Reading, Mass.: Addison-Wesley, 1952.
55. O. D. Kellogg, *Foundations of Potential Theory*. Berlin: J. Springer, 1929.
56. M. S. Knebelman, and T. Y. Thomas, *Principles of College Algebra*. New York: Prentice-Hall, 1942.
57. W. Leighton, *An Introduction to the Theory of Differential Equations*. New York: McGraw-Hill, 1952.
58. H. Levy, and E. A. Baggott, *Numerical Solutions of Differential Equations*, New York: Dover Publications, 1950.
59. W. E. Milne, *Numerical Calculus*. Princeton: Princeton University Press, 1949.
60. W. F. Osgood, *Introduction to Calculus*. New York: Macmillan, 1922.
61. W. F. Osgood, *Advanced Calculus*. New York: Macmillan, 1925.
62. W. F. Osgood, and W. C. Graustein, *Plane and Solid Analytic Geometry*. New York: Macmillan, 1938.
63. B. O. Peirce, *Elements of the Theory of the Newtonian Potential Function*, 3rd ed. Boston: Ginn & Co., 1902.
64. B. O. Peirce, *A Short Table of Integrals*, 3rd ed. Boston: Ginn & Co., 1929.
65. H. B. Phillips, *Vector Analysis*. New York: John Wiley & Sons, 1933.
66. E. T. Whittaker, and G. Robinson, *The Calculus of Observations*. New York: Van Nostrand, 1924.
67. A. P. Wills, *Vector Analysis, with an Introduction to Tensor Analysis*. New York: Prentice-Hall, 1931.
68. E. B. Wilson, *Advanced Calculus*. Boston: Ginn & Co., 1912.
69. D. R. Wylie, Jr., *Advanced Engineering Mathematics*. New York: McGraw-Hill, 1951.

ANSWERS TO ODD-NUMBERED PROBLEMS

ANSWERS TO ODD-NUMBERED PROBLEMS

Chapter 1

1. 4.06×10^{-42} dyne; 9.22×10^{-3} dyne.
7. b) $\mu mg/(\sin\theta - \mu\cos\theta)$.
9. $t = (v_0/g)[(\sin\theta + \mu\cos\theta)^{-1} + (\sin^2\theta - \mu^2\cos^2\theta)^{-1/2}]$.
11. 2.20×10^{27} tons.
13. 1.4×10^{11} sun masses.

Chapter 2

1. a) 160 hp, 801 hp, 2403 hp.
   b) 3011 lb-wt, 602 lb-wt, 201 lb-wt.
3. $p = p_0$. $v_\infty = v_0 + (p_0/m)$.
5. b) $v = (F_0/4\omega m)(2\omega t - \sin 2\omega t)$, $x = (F_0/8\omega^2 m)(2\omega^2 t^2 - 1 + \cos 2\omega t)$.
7. a) $v = v_0$, $x = v_0 t$, when $t < t_0$,
   $v = v_0 + p_0(t - t_0)/m\delta t$, $x = v_0 t + p_0(t - t_0)^2/2m\delta t$,
   when $t_0 \le t \le t_0 + \delta t$,
   $v = v_0 + p_0/m$, $x = v_0 t - p_0\delta t/2m + p_0(t - t_0)/m$,
   when $t > t_0 + \delta t$.
9. $x = 200t - 2000(1 - e^{-t/20})(3 - e^{-t/20})$, ($x$ in ft, t in sec), $v_\infty = 200$ ft/sec. Assumption F independent of v.
11. b) $t_s = m(1 - e^{-\alpha v_0})/(\alpha b)$, $x_s = [m/(\alpha^2 b)][1 - e^{-\alpha v_0} - \alpha v_0 e^{-\alpha v_0}]$.
13. $v = \sqrt{F_0/b}\tanh(\sqrt{bF_0}\, t/m)$.
15. $v = \sqrt{P/b}\,(1 - e^{-2bt/m})$.
17. $v = [v_0^{(1-n)} - (1 - n)(bt/m)]^{1/(1-n)}$;
    $x = m\{v_0^{(2-n)} - [v_0^{(1-n)} - (1 - n)(bt/m)]^{(2-n)/(1-n)}\}/(2 - n)b$;
    $t_s = mv_0^{(1-n)}/(1 - n)b$, $(n < 1)$;
    $x_s = mv_0^{(2-n)}/(2 - n)b$, $(n < 2)$.
19. $m\ddot{x} = k/x^3$, $x = [x_0^2 + (kt^2/mx_0^2)]^{1/2}$.
21. a) $F = -2ax + 3bx^2$.
    b) $v_c^2 = 8ma^3/27b^2$.

23. a) $x^2 = \dfrac{E}{k} + \dfrac{\sqrt{E^2 - ka}}{k}\cos\left(2\sqrt{\dfrac{k}{m}}\,t + \theta_0\right)$.

b) $x \doteq \sqrt{\dfrac{a}{2E}} + \sqrt{\dfrac{2E}{k}}\left|\cos\left(\sqrt{\dfrac{k}{m}}\,t + \theta_0/2\right)\right|$; (interpret).

25. c) $(2b/a)^{1/6}$, $(2\pi/3)(m^3b^4/4a^7)^{1/6}$.

27. b) $x_{eq} = 0, \pm\sqrt{2}\,a$; at $\pm\sqrt{2}\,a$, $\omega = (v_0/3ma^2)^{1/2}$.

c) $\alpha > [1 + (4V_0/9mv_0^2)]^{-1}$, $\alpha > [1 + (7\,V_0/36mv_0^2)]^{-1}$,
$x = a, (7/2)^{1/2}a$.

31. $x = \dfrac{m}{2b}\ln\left(1 + \dfrac{bv_0^2}{mg}\right) + \dfrac{m}{b}\ln\cos\left[\sqrt{\dfrac{bg}{m}}\,(t_0 - t)\right]$, $(0 < t < t_0)$,

$= \dfrac{m}{2b}\ln\left(1 + \dfrac{bv_0^2}{mg}\right) - \dfrac{m}{b}\ln\cosh\left[\sqrt{\dfrac{bg}{m}}\,(t - t_0)\right]$, $(t > t_0)$,

where $t_0 = \sqrt{m/bg}\tan^{-1}(\sqrt{b/mg}\,v_0)$.

33. $x = (x_0^{3/2} + t\sqrt{9MG/2})^{2/3}$.

37. $C_1 = A\cos\theta$, $C_2 = -\omega_1 A\sin\theta$.

39. $x = x_0e^{-\gamma t}[\cos\omega_1 t + (\gamma/\omega_1)\sin\omega_1 t]$,
$x = x_0(1 + \gamma t)e^{-\gamma t}$, $x = x_0(\gamma_1 - \gamma_2)^{-1}(\gamma_1 e^{-\gamma_2 t} - \gamma_2 e^{-\gamma_1 t})$.

41. $x = [(\gamma_1 x_0 + v_0)e^{-\gamma_2 t} - (\gamma_2 x_0 + v_0)e^{-\gamma_1 t}]/(\gamma_1 - \gamma_2)$, $(v_0 < 0)$.

43. a) $k = 4.9 \times 10^4$ kgm sec^{-2}, $b = 7.07 \times 10^4$ kgm sec^{-1}.
b) 0.076 sec.

45. $x = (F_0/k) + Ae^{-\gamma t}\cos(\omega_1 t + \theta)$.

47. $x = [F_0/m\omega_0(\omega_0^2 - \omega^2)](\omega_0\sin\omega t - \omega\sin\omega_0 t)$.

49. $x = \dfrac{F_0}{m(\gamma^2 + \omega^2)^2}\left[(\gamma^2 - \omega^2)\cos\omega t + 2\gamma\omega\sin\omega t - (\gamma^2 - \omega^2 + \gamma^3 t - 3\gamma\omega^2 t)e^{-\gamma t}\right]$
$+ (x_0 + v_0 t + \gamma x_0 t)e^{-\gamma t}$.

53. $x = (v_0/\omega_0)\sin\omega_0 t$, $t \leq 3\pi/2\omega_0$;
$x = (B/m)(\omega_0^2 - \omega^2)^{-1}[\cos(\omega t + \theta) + \cos\alpha\sin\omega_0 t + (\omega/\omega_0)\sin\alpha\cos\omega_0 t]$
$+ (v_0/\omega_0)\sin\omega_0 t$, $\alpha = (3\pi\omega/2\omega_0) + \theta$, $t \geq 3\pi/2\omega_0$.

55. a) $x = 0$ if $t < t_0$, $x = (p_0/k\,\delta t)[1 - \cos\omega_0(t - t_0)]$, if $t_0 \leq t \leq t_0 + \delta t$,
$x = (2p_0/k\,\delta t)\sin(\frac{1}{2}\omega_0\,\delta t)\sin\omega_0(t - t_0 - \frac{1}{2}\delta t)$, if $t > t_0 + \delta t$.

57. a) $m\omega_0^2 x = (\frac{3}{2}A + \frac{1}{34}B)e^{-\omega_0 t/3}\cos(\frac{2}{3}\sqrt{2}\,\omega_0 t)$
$+ (\frac{3}{8}\sqrt{2}\,A + \frac{37}{136}\sqrt{2}\,B)e^{-\omega_0 t/3}\sin(\frac{2}{3}\sqrt{2}\,\omega_0 t)$
$- \frac{3}{2}A\cos\omega_0 t - \frac{1}{34}B\cos 3\omega_0 t - \frac{2}{17}B\sin 3\omega_0 t$.

59. $x = (F_0/ma^2)[1 - (1 + at + \frac{1}{2}a^2t^2)e^{-at}]$.

61. $x = (2F_0/\pi\,\omega_0^2)\left[1 + 2\sum_{n=1}^{\infty}\cos 2n\omega_0 t\right]$.

63. $$x = \begin{cases} 0, & \text{if } t < t_0, \\ \dfrac{p_0}{m\omega_0^2\,\delta t}\left[1 - e^{-\gamma(t-t_0)}\cos\omega_1(t-t_0) - \dfrac{\gamma}{\omega_1}e^{-\gamma(t-t_0)}\sin\omega_1(t-t_0)\right], & \text{if } t_0 \le t \le t_0 + \delta t \\ \dfrac{p_0}{m\omega_0^2\,\delta t}\,e^{-\gamma(t-t_0)}\left[\left(e^{\gamma\delta t}\cos\omega_1\,\delta t - 1 - \dfrac{\gamma}{\omega_1}e^{\gamma\delta t}\sin\omega_1\,\delta t\right)\cos\omega_1(t-t_0) \right. \\ \quad \left. + \left(e^{\gamma\delta t}\sin\omega_1\delta t + \dfrac{\gamma}{\omega_1}e^{\gamma\delta t}\cos\omega_1\,\delta t - \dfrac{\gamma}{\omega_1}\right)\sin\omega_1(t-t_0)\right], & \text{if } t > t_0. \end{cases}$$

Chapter 3

11. F_0R.

13. a) $k[1 - \sqrt{5} + \ln\frac{1}{2}(1+\sqrt{5})]$; (b) $-k\pi/\sqrt{3}$.

15. $(\ddot{A}_\rho - 2\dot{A}_\varphi\dot{\varphi} - A_\varphi\ddot{\varphi} - A_\rho\dot{\varphi}^2)\hat{\boldsymbol{\rho}}$
$+ (\ddot{A}_\varphi + 2\dot{A}_\rho\dot{\varphi} + A_\rho\ddot{\varphi} - A_\varphi\dot{\varphi}^2)\hat{\boldsymbol{\varphi}} + \ddot{A}_z\hat{\boldsymbol{z}}$.

17. b) $\dfrac{\partial\hat{\boldsymbol{f}}}{\partial f} = -\dfrac{1}{2}\left(\dfrac{h}{f}\right)^{1/2}\dfrac{\hat{\boldsymbol{h}}}{(f+h)}$, $\dfrac{\partial\hat{\boldsymbol{f}}}{\partial h} = \dfrac{1}{2}\left(\dfrac{f}{h}\right)^{1/2}\dfrac{\hat{\boldsymbol{h}}}{f+h}$,
$\dfrac{\partial\hat{\boldsymbol{h}}}{\partial f} = \dfrac{1}{2}\left(\dfrac{h}{f}\right)^{1/2}\dfrac{\hat{\boldsymbol{f}}}{f+h}$, $\dfrac{\partial\hat{\boldsymbol{h}}}{\partial h} = -\dfrac{1}{2}\left(\dfrac{f}{h}\right)^{1/2}\dfrac{\hat{\boldsymbol{f}}}{f+h}$;
$\dot{\boldsymbol{r}} = (f+h)^{1/2}\left(\dfrac{\dot{f}}{f^{1/2}}\hat{\boldsymbol{f}} + \dfrac{\dot{h}}{h^{1/2}}\hat{\boldsymbol{h}}\right)$.

21. $\left(\dfrac{1}{\rho}\dfrac{\partial A_z}{\partial\varphi} - \dfrac{\partial A_\varphi}{\partial z}\right)\hat{\boldsymbol{\rho}} + \left(\dfrac{\partial A_\rho}{\partial z} - \dfrac{\partial A_z}{\partial\rho}\right)\hat{\boldsymbol{\varphi}} + \left(\dfrac{\partial A_\varphi}{\partial\rho} - \dfrac{1}{\rho}\dfrac{\partial A_\rho}{\partial\varphi} + \dfrac{A_\varphi}{\rho}\right)\hat{\boldsymbol{z}}$.

23. a) $F_x = \frac{1}{2}F_0 + \left(\dfrac{mv^2}{R} - \frac{1}{2}F_0\right)\cos\dfrac{vt}{R}$, $F_y = \left(\frac{1}{2}F_0 - \dfrac{mv^2}{R}\right)\sin\dfrac{vt}{R}$.
b) $\dfrac{\pi F_0 R}{2v}\hat{\boldsymbol{x}} + \left(\dfrac{F_0R}{v} - 2mv\right)\hat{\boldsymbol{y}}$.

25. $\boldsymbol{L} = (-2mbct^3,\ mact^2 - mcx_0,\ mabt^4 + 3mbx_0t^2)$,
$\boldsymbol{F} = (2ma,\ 6mbt,\ 0)$,
$\boldsymbol{N} = (-6mbct^2,\ 2mact,\ 4mabt^3 + 6mbx_0t)$.

29. $\delta = \exp(-bv_{z0}/mg)$.

33. a) $\alpha_0 = \sin^{-1}(gx_0/v_0^2)$.
b) Angle of elevation α_0 should be increased by $4bv_0\cos\alpha_0/3mg(\cot^2\alpha_0 - 1)$.

35. a) $5bx^4y^2 - 6abxyz^3$.
c) $-\int_{x_s}^{x} F_x\,dx - \int_{y_s}^{y} F_y\,dy - \int_{z_s}^{z} F_z\,dz$.

37. a) $\frac{1}{2}e^{-R}$.
b) $-\int^{s=\boldsymbol{A}\cdot\boldsymbol{r}} f(s)\,ds$.

41. $F_x = -ae^2(r_1^{-3} + r_2^{-3}) - xe^2(r_1^{-3} + r_2^{-3})$,
$F_y = -ye^2(r_1^{-3} - r_2^{-3})$, $\quad F_z = -ze^2(r_1^{-3} - r_2^{-3})$.

43. $\dot{\theta} = (k/m)^{1/2}$, $\omega_r = 2(k/m)^{1/2}$.

45. $kr^2 = E + (E^2 - \omega^2 L^2)^{1/2} \cos(2\omega t + 2\alpha_0)$, $\omega = (k/m)^{1/2}$,
$\tan(\theta - \theta_0) = (\omega L)^{-1}[E - (E^2 - \omega^2 L^2)^{1/2}] \tan(\omega t + \alpha_0)$,
θ_0 = angle at aphelion. (This is a Lissajous figure with $\omega_x = \omega_y$, i.e., an ellipse.)

47. a) $F = (1 + \alpha r)Ke^{-\alpha r}/r^2$.
d) $L^2 = -mKa(1 + \alpha a)e^{-\alpha a}$, $E = (1 - \alpha a)Ke^{-\alpha a}/2a$.
e) $\tau_c = 2\pi[-K(1 + \alpha a)e^{-\alpha a}/ma^3]^{-1/2}$,
$\tau_r = 2\pi[-K(1 + \alpha a - \alpha^2 a^2)e^{-\alpha a}/ma^3]^{-1/2}$.
[Stable circular motion is not possible if $\alpha a \geq \frac{1}{2}(1 + \sqrt{5})$.]

49. b) Opposite direction, 1.2×10^{-7} gm m^{-3}.

51. c) Ellipse precesses $2\pi(1 - \alpha)/\alpha$ radians per revolution, in same direction as $\dot{\theta}$ if $\alpha < 1$, in opposite direction if $\alpha > 1$, where $\alpha^2 = 1 + (mK'/L^2)$.

53. 1301 km.

55. $\omega_p \doteq \frac{3}{5}\eta(R/r)^2(MG/r^3)^{1/2}$, 0.88 degrees per revolution.

57. $\omega_p \doteq \frac{1}{2}K^{3/2}m^{-3/2}r^{-5/2}c^{-2}$.

59. 2.8 km s^{-1}, 12.3 km s^{-1}.

61. $v_1 = (2\pi r_1/Y_1)\{[2r_2/(r_2 + r_1)]^{1/2} - 1\}$,
$v_2 = (2\pi r_1/Y_1)(r_1/r_2)^{1/2}\{[2r_1/(r_2 + r_1)]^{1/2} - 1\}$.
Venus: -5700 mi hr^{-1}; Mars: 6700 mi hr^{-1}.

63. Perigee at point of maximum $\dot{\theta}$; $a^3 = gR^2\tau^2/4\pi^2$;
$\epsilon = (\lambda - 1)/(\lambda + 1)$, λ = ratio of maximum to minimum $\dot{\theta}$. (There are many other possible answers.)

73. $x = (E_0/2m\omega^2)(\sin \omega t - \omega t \cos \omega t)$,
$y = (E_0/2m\omega^2)(2 \cos \omega t - 2 + \omega t \sin \omega t)$,
$z = 0$.

75. d) If $\dot{z}_0 = \dot{\rho}_0 = 0$, $\dot{\varphi}_0 = -(qB/2mc) \pm [(qB/2mc)^2 - (qa/m\rho_0^2)]^{1/2}$.
e) $\omega_\rho = 2[(qB/2mc)^2 - (qa/2m\rho_0^2)]^{1/2}$.

Chapter 4

3. $2[1 + (M/m)](gl)^{1/2} \sin \frac{1}{2}\theta$.

5. $\cos^{-1}[1 - 0.293m_1^2/(m_1 + m_2)^2]$.

7. 371 sec.

9. $M_1 = 39{,}800$ kgm, $M_2 = 1946$ kgm (fuel + rocket, excluding payload).

13. $r = r_0[1 + (5\pi)^{-1}(a^2/r_0^2)(\omega_0 - \omega)Y_0]^2$, Y_0 = length of present year. 49 miles. Less if moon were included.

17. $m_1(1 + \alpha^2 + 2\alpha \cos \theta_1)/(1 - \alpha^2)$, $\alpha = p_{1F}/p_{1I}$. Measure θ_2.

21. $(1 + \gamma)p_{1F} = (p_{1I} - p_{2I}) \cos \vartheta_1 \pm [(\gamma p_{1I} + p_{2I})^2 - (p_{1I} - p_{2I})^2 \sin^2 \vartheta_1]^{1/2}$, $\gamma = m_2/m_1$.

23. $Q = \frac{p_1^2}{2m_1}\left[\frac{(m_1/m_3)\sin^2\vartheta_4 + (m_1/m_4)\sin^2\vartheta_3}{\sin^2(\vartheta_3+\vartheta_4)} - 1\right]$.

29. $2\gamma mV^2(1+\gamma+\gamma^2)^{-1}[\gamma - (v_0/V)(1+\gamma)^{1/2}]$, where $\gamma = MG/R(v_0^2+V^2)$.

39. $x_1 = x_2 = Ae^{-\gamma t}\cos(\omega_1 t + \theta)$, $\gamma = b_1/2m_1$, $\omega_1^2 = -\gamma^2 + (k_1' + k_3)/m_1$;
and $x_1 = -x_2 = Ae^{-\gamma t}\cos(\omega_2 t + \theta)$, $\omega_2^2 = -\gamma^2 + (k_1' - k_3)/m_1$.

41. $x_1 = \frac{(m_2\omega^2 - k_2')F_0\cos\omega t}{k_3^2 - (m_1\omega^2 - k_1')(m_2\omega^2 - k_2')}$, $\quad x_2 = \frac{k_3F_0\cos\omega t}{k_3^2 - (m_1\omega^2 - k_1')(m_2\omega^2 - k_2')}$.

Chapter 5

5. $\omega_0^2k^2(1+\mu^2)^{1/2}/4\pi\mu ga$ turns.

7. $\theta = \theta_0 + (N_0/b)t + [\alpha N_0/(I^2\omega_0^3 + b^2\omega_0)][b\sin\omega_0 t - I\omega_0\cos\omega_0 t]$.

11. a) 60° above the horizontal!
    b) $\frac{1}{3}\pi(l/g)^{1/2}$, compared with $\frac{1}{2}\pi(l/g)^{1/2}$.

13. $g = [4\pi^2(h+h')/\tau^2][1 + 2h'\delta/(h-h')]$.

15. $x_G = 0$, $y_G = -4a/9\pi$; $I_{Ox} = I_{Oy} = I_{Gy} = 3\pi a^4\sigma/8$, $I_{Oz} = 3\pi a^4\sigma/4$,
$I_{Gz} = (81\pi^2 - 32)a^4\sigma/108\pi$, $I_{Gx} = (81\pi^2 - 64)a^4\sigma/216\pi$.

17. $2112\rho cm^5$, $2491\rho cm^5$, $1169\rho cm^5$.

19. 30 yards.

23. a) $l\sin(\alpha\sqrt{3})$; b) $(5/36)Ml^2$.

25. $2\sqrt{2}$ kgm-wt, acting at a point on the third side extended 0.75 m beyond the corner. Equilibrant direction is 135° clockwise from 3-kgm-wt force.

27. a) $\mathbf{F}_0 = (0, -6\text{ lb}, -14\text{ lb})$ at center, $\mathbf{F}_c = (0, -3\text{ lb}, -8\text{ lb})$ at any front corner, $-\mathbf{F}_c$ at adjacent rear corner. (x-axis outward, y-axis horizontal to right, z-axis vertical, origin at center of cube.)
    b) $\mathbf{F}_1 = (0, 0, 2\text{ lb})$ at center, $\mathbf{F}_2 = (0, -6\text{ lb}, -16\text{ lb})$ at center of front face. (There are other correct answers.)
    c) $\mathbf{F}_0$ [part (a)] at the point (65/116 ft, 0, 0), $\mathbf{N} = (0, 9/58\text{ lb-ft}, 21/58\text{ lb-ft})$.

29. $\beta_{opt} = \frac{1}{2}\alpha$, $\alpha_{opt} = 60°$.

31. $A = (100W/Y)e^{100wz/Y}$.

33. a) C is 8 ft from A in direction 22°50′ below the horizontal, 235 lb-wt, 463 lb-wt.
    b) C is 8.283 ft from A in direction 28°45′ below the horizontal, 223.4 lb-wt, 438.7 lb-wt.

35. $y = w(x^2 - \frac{1}{4}D^2)(4\tau_0^2 - w^2D^2)^{-1/2}$.

37. $L = \sqrt{3}\,l$, $2WL^3/Yl^4$, $2WL^2/Yl^4$.

39. $-\rho L^4/192Y(b^2+a^2) - \rho L^2/8n$.

41. 1.55×10^5 lb in^{-2}.

43. $p = p_0(1-\alpha z)^{Mg/RT_0}$, where $T = T_0(1-\alpha z)$.

Chapter 6

3. $x = \alpha a[\alpha^2 + (1-\alpha)^2(a^2/b^2)]^{-3/4}$, $\quad y = (1-\alpha)b[(1-\alpha)^2 + \alpha^2(b^2/a^2)]^{-3/4}$.

5. $\boldsymbol{g} = -(MG/r^2)(\boldsymbol{r}/r),\ (r \geq a),\ = -(MG\boldsymbol{r}/a^3),\ (r \leq a),$
$\mathcal{G} = (MG/r),\ (r \geq a),\ = (MG/2a^3)(3a^2 - r^2),\ (r \leq a).$

9. $T\dfrac{d}{dr}\left(\dfrac{R^2Tr^2}{A^2p}\dfrac{dp}{dr}\right) = -4\pi r^2 Gp$, arbitrary constants determined by

$$M = \int_0^\infty \frac{4\pi A p r^2}{RT}\,dr, \quad p \to 0 \text{ as } r \to \infty.$$ A is molecular weight.

11. $$p = \frac{M^2G}{4\pi a^4}\left[\ln\left(\frac{r^2 + a^2}{r^2}\right) - \frac{a^4}{2(r^2 + a^2)^2} - \frac{a^2}{r^2 + a^2}\right],$$
$$T = \frac{AMGr}{2a^2R}\left[\frac{(r^2 + a^2)^2}{a^4}\ln\left(\frac{r^2 + a^2}{r^2}\right) - \frac{3}{2} - \frac{r^2}{a^2}\right].$$

13. a) $g - g_0 = (MG/a^2)(2 - \sqrt{2})$; (b) $g - g_0 = -(\frac{3}{4})(MG/a^2)$.

15. a) $(MG/r) + (MGa^2/4r^3)(1 - 3\cos^2\theta)$;
b) $g_r = -(MG/r^2) - (3MGa^2/4r^4)(1 - 3\cos^2\theta)$,
$g_\theta = (3MGa^2/4r^4)\sin 2\theta$.

19. a) $2\pi\sigma G$, toward sheet. (b) It is half the field outside the shell.

Chapter 7

1. b) $m\boldsymbol{a}^* = -b\boldsymbol{v}^* - bgt$.

5. $2\rho\omega v\cos\theta$, ω = angular velocity of earth's rotation, θ = colatitude; approximately 0.0003 lb $\text{in}^{-2}\cdot\text{mile}^{-1}$.

7. a) $2m\omega gt\sin\theta$, eastward b) $(8\omega^2h^3/9g)^{1/2}\sin\theta$.

9. 5.25×10^{-4} radians.

13. $\omega = (k/m)^{1/2}$; in rotating system, m moves with angular velocity -2ω in circle of arbitrary radius with arbitrary center.

17. 0.73×10^6 radians sec^{-1}. Decrease ω if the electron circles in the positive sense relative to $\boldsymbol{B}$; otherwise, increase ω.

21. $$(2MG)^{1/2}\left[\frac{1}{R} + \frac{1}{a - R} - \frac{4}{a} + \frac{(a - 2R)^2}{4a^3}\right]^{1/2}.$$

Chapter 8

1. b) $u = A\sin(n\pi x/2l)\cos(n\pi ct/2l) + B\sin(n\pi x/2l)\sin(n\pi ct/2l)$,
$n = 1, 3, 5, \ldots$

3. $$u = \frac{4l}{5\pi^2}\left(\sin\frac{\pi x}{l}\cos\frac{\pi ct}{l} - \frac{1}{9}\sin\frac{3\pi x}{l}\cos\frac{3\pi ct}{l} + \frac{1}{25}\sin\frac{5\pi x}{l}\cos\frac{5\pi ct}{l} - \cdots\right).$$

5. $u = A\sin\omega t(\cos kx - \operatorname{ctn} kl\sin kx)$, $k = \omega/c$.

7. $u = Ae^{-bt/2\sigma}\sin(n\pi x/l)\cos\{[(n^2\pi^2c^2/l^2) - (b^2/4\sigma^2)]^{1/2}t + \theta\}$,
$n = 1, 2, 3, \ldots$

9. $$\left(b\frac{\partial u}{\partial t} = -\tau\frac{\partial u}{\partial x}\right)_{x=0};\quad g(\eta) = \frac{\tau - bc}{\tau + bc}f(\xi),\ \xi = -\eta.$$

11. $u = f(x - ct) + g(x + ct)$, where $f(\xi) = g(\xi)$ = function obtained by joining by straight lines the points $f = (-1)^n l/20$, $\xi = (n + \frac{1}{2})l$.

17. $u = (RT/M) \ln (p/p_0)$;
v is a solution of $v_0^2 - v^2 + (2RT/M) \ln (vS/v_0S_0) = 2gh$;
$p = p_0v_0S_0/vS$.

19. $\boldsymbol{v} = -(a/r^2)\hat{\boldsymbol{r}}$.

23. $v_x = -(l\pi A/\omega\rho_0 L_x) \sin (l\pi x/L_x) \cos (m\pi y/L_y) \sin (k_z z - \omega t)$,
$v_y = -(m\pi A/\omega\rho_0 L_y) \cos (l\pi x/L_x) \sin (m\pi y/L_y) \sin (k_z z - \omega t)$,
$v_z = (k_z A/\omega\rho_0) \cos (l\pi x/L_x) \cos (m\pi y/L_y) \cos (k_z z - \omega t)$.

25. $A \cos (l\pi x/L_x) \cos (m\pi y/L_y) \cos (k_z z + \omega t)$.

27. $v_g = \left(\frac{\tau h}{m}\right)^{1/2} \cos \frac{kh}{2}, \; c = \left(\frac{\tau h}{m}\right)^{1/2} \frac{2}{kh} \sin \frac{kh}{2}$.

29. $v = (3I/2\rho l^3)(l^2 - 4x^2)$, $dp/dz = 12\eta I/\rho l^3$, where x is the distance from a plane midway between the walls.

Chapter 9

1. $T = \frac{1}{2}ma^2(\dot{w}^2 \cos^2 \zeta + \dot{u}^2 \sin^2 \zeta) \exp (2w \cos^2 \zeta + 2u \sin^2 \zeta)$,
$Q_u = a \sin \zeta (F_r \sin \zeta - F_\theta \cos \zeta) \exp (w \cos^2 \zeta + u \sin^2 \zeta)$,
$Q_w = a \cos \zeta (F_r \cos \zeta + F_\theta \sin \zeta) \exp (w \cos^2 \zeta + u \sin^2 \zeta)$;
$Q_u = -2m\dot{s}^2 \sin^2 \zeta, \quad Q_w = -m\dot{s}^2 \cos^2 \zeta$.

3. a) $T = \frac{1}{2}m(f + h)\left(\frac{\dot{f}^2}{f} + \frac{\dot{h}^2}{h}\right), \; p_f = \frac{m(f + h)}{f}\dot{f}$,
$p_h = \frac{m(f + h)}{h}\dot{h}$.

5. c) 'Q_r' = 'F_r' = $mr\omega^2 \sin^2 \theta + 2mr\omega\dot{\varphi} \sin^2 \theta$,
'Q_θ' = r'F_θ' = $mr^2\omega^2 \sin \theta \cos \theta + 2mr^2\omega\dot{\varphi} \sin \theta \cos \theta$,
'Q_φ' = $r \sin \theta$'F_φ' = $-2mr\dot{r}\omega \sin^2 \theta - 2mr^2\omega\dot{\theta} \sin \theta \cos \theta$.

11. $\omega^2 = [2g/\mu(l_1 + l_2)][1 \pm (1 - \mu)^{1/2}], \qquad \mu = [m_1/(m_1 + m_2)][4l_1l_2/(l_1 + l_2)^2]$.

15. b) $\theta \doteq [a\omega^2/(g - l\omega^2)] \cos \omega t$.

17. $\cos \theta_1 = 2E/3mgR$ if $p_\varphi^2 < \frac{2}{3}mR^2E - 8E^3/27mg^2$; otherwise the string will not collapse.

19. b) $z = 2l - 2(m + M)g/(m\omega^2)$,
$\omega_z^2 = (m + M)g \sin^2 \theta/(m + 2M \sin^2 \theta)\, l \cos \theta$,
where $\cos \theta = 1 - (z/2l)$.

21. b) $U = -\frac{1}{2}m\omega^2(x^2 + y^2) - m\omega(x\dot{y} - y\dot{x})$
$= -\frac{1}{2}m\omega^2 r^2 \sin^2 \theta - m\omega r^2 \dot{\varphi} \sin^2 \theta$.

29. $H = c\left[m^2c^2 + \left(p_x - \frac{q}{c}A_x\right)^2 + \left(p_y - \frac{q}{c}A_y\right)^2 + \left(p_z - \frac{q}{c}A_z\right)^2\right]^{1/2} + q\phi$.

31. $H = \frac{p_x^2 + p_y^2 + p_z^2}{2(m_1 + m_2)} + \frac{p_r^2}{2\mu} + \frac{p_\theta^2}{2\mu r^2} + \frac{p_\varphi^2}{2\mu r^2 \sin^2 \theta} - (m_1 + m_2)gZ - \frac{m_1m_2G}{r}$,
$V_Z = -(m_1 + m_2)gZ, \; V_r = \frac{p_\theta^2}{2\mu r^2} + \frac{p_\varphi^2}{2\mu r^2 \sin^2 \theta} - \frac{m_1m_2G}{r}$.

Chapter 10

5. $T'_{11} = -40, T'_{12} = 15, T'_{13} = 35\sqrt{2},$
$T'_{22} = 10, T'_{23} = 15\sqrt{2}, T'_{33} = 40.$

13. $T'_1 = 4, T'_2 = 10, T'_3 = -8, \hat{e}'_1 = (\sqrt{3}\,\hat{e}_1 + \sqrt{2}\,\hat{e}_2 - \hat{e}_3)/\sqrt{6},$
$\hat{e}'_2 = (-\sqrt{3}\,\hat{e}_1 + \sqrt{2}\,\hat{e}_2 - \hat{e}_3)/\sqrt{6}, \hat{e}'_3 = (\hat{e}_2 + \sqrt{2}\,\hat{e}_3)/\sqrt{3}.$

21. Eigenvalues: $+1, e^{\pm i\alpha}$; $\cos\alpha = \frac{1}{2}(\cos\psi + \cos\theta + \cos\psi\cos\theta)$.

23. $\mathbf{I} = (5/36)Ml^2\mathbf{1}, M = 6m.$

25. $0.15\, mh^2 \sin^2\alpha(6 + \tan^2\alpha).$

27. $I_{xx} = \dfrac{M}{12}\left(\dfrac{a^4 + b^4}{a^2 + b^2} + c^2\right), \quad I_{xy} = \dfrac{M}{12}\left(\dfrac{ab^3 - a^3b}{a^2 + b^2}\right),$

$I_{xz} = 0, \quad I_{yy} = \dfrac{M}{12}\left(\dfrac{2a^2b^2}{a^2 + b^2} + c^2\right), \quad I_{yz} = 0, I_{zz} = \dfrac{M}{12}(a^2 + b^2).$

29. a) $I_1 = mb^2/12, I_2 = ma^2/12, I_3 = m(a^2 + b^2)/12; \hat{e}_1 \parallel a, \hat{e}_2 \parallel b.$

b) $I_1 = \frac{1}{6}ma^2 + \left(\dfrac{16}{15} - \dfrac{\sqrt{3}}{5}\right)mb^2, \quad I_2 = \frac{3}{4}ma^2 + \left(\dfrac{49}{60} - \dfrac{\sqrt{3}}{5}\right)mb^2,$
$I_3 = \frac{11}{12}ma^2 + \frac{1}{4}mb^2$; ($\hat{e}_3$ vertical).

31. $(w^2 + h^2)x^2 + (h^2 + l^2)y^2 + (l^2 + w^2)z^2 = 5a^2/M.$

37. $\mathbf{P} = [p_0 - (\Delta pz/l)]\mathbf{1} + (r\,\Delta p/2l)(\hat{\boldsymbol{\rho}}\hat{z} + \hat{z}\hat{\boldsymbol{\rho}}).$

39. c) $2\eta|(\nabla\boldsymbol{v})_{ts}|^2 + \eta'(\nabla\cdot\boldsymbol{v})^2$

$$= (\tfrac{4}{3}\eta + \eta')\sum_i\left(\frac{\partial v_i}{\partial x_i}\right)^2 - (\tfrac{4}{3}\eta - 2\eta')\sum_{i>j}\frac{\partial v_i}{\partial x_i}\frac{\partial v_j}{\partial x_j}$$
$$+ \eta\sum_{i\neq j}\left(\frac{\partial v_i}{\partial x_j}\right)^2 + 2\eta\sum_{i>j}\frac{\partial v_i}{\partial x_j}\frac{\partial v_j}{\partial x_i}, \quad \text{where } |\mathbf{T}|^2 = \sum_{i,j=1}^{3} T_{ij}^2.$$

41. $$\mathbf{P} = c_1\hat{z}\hat{z}\frac{\partial\zeta}{\partial z} + c_2\left(\hat{z}\frac{\partial\boldsymbol{\rho}_\perp}{\partial z} + \frac{\partial\boldsymbol{\rho}_\perp}{\partial z}\hat{z}\right) + c_3(\hat{z}\nabla_\perp\zeta + \nabla_\perp\zeta\hat{z})$$
$$+ c_4[\nabla_\perp\boldsymbol{\rho}_\perp + (\nabla_\perp\boldsymbol{\rho}_\perp)^t] + \tfrac{1}{2}(c_5 - c_4)\nabla_\perp\cdot\boldsymbol{\rho}_\perp(\hat{x}\hat{x} + \hat{y}\hat{y}),$$
where z is the axis of symmetry, and
$$\nabla_\perp = \hat{x}\frac{\partial}{\partial x} + \hat{y}\frac{\partial}{\partial y}, \quad \boldsymbol{\rho}_\perp = \hat{x}\xi + \hat{y}\eta.$$

Chapter 11

5. $\omega_3 = N_3(t + t_0)/I_3, \omega_1 = \omega_{10}\cos[\alpha(t + t_0)^2] - \omega_{20}\sin[\alpha(t + t_0)^2],$
$\omega_2 = \omega_{10}\sin[\alpha(t + t_0)^2] + \omega_{20}\cos[\alpha(t + t_0)^2],$
$\alpha = N_3(I_3 - I_1)/(2I_3I_1), t_0 = \omega_{30}I_3/N_3.$

9. $$\begin{matrix} \cos\psi\cos\phi - \cos\theta\sin\phi\sin\psi & \cos\psi\sin\phi + \cos\theta\cos\phi\sin\psi & \sin\psi\sin\theta \\ -\sin\psi\cos\phi - \cos\theta\sin\phi\cos\psi & -\sin\psi\sin\phi + \cos\theta\cos\phi\cos\psi & \cos\psi\sin\theta \\ \sin\theta\sin\phi & -\sin\theta\cos\phi & \cos\theta \end{matrix}$$

13. $L = \frac{5}{4}Ma^2\dot{\theta}^2 + \frac{5}{4}Ma^2\dot{\phi}^2 \sin^2\theta + \frac{1}{4}Ma^2(\dot{\psi} + \dot{\phi}\cos\theta)^2$
$+\frac{3}{8}Ma^2(\dot{\psi}' + \dot{\phi}\cos\theta)^2 - \frac{17}{14}Mag\cos\theta,$

where ψ, ψ' refer to disk and rings, respectively. Angular velocities ω_3, ω_3' of disk and rings are separately constant. Precession and nutation as for top in Section 11–5, but with p_ψ replaced by $p_\psi + p_\psi'$, I_1 by $\frac{5}{2}Ma^2$, I_3 by $\frac{1}{2}Ma^2$, ω_3 by $\omega_3 + \frac{9}{2}\omega_3'$.

15. Top rises in time $t \doteq (r^2\omega_{30}\theta_1)/(\mu gl)$ after $\theta_1/(2\pi\mu)$ revolutions of precession; center of mass moves with angular velocity $(gl/r^3\omega_{30})$ in circle of radius $(gl^2/\mu\omega_{30}^2)^{1/3}$, 90° out of phase with precession. Top wobbles after $t \doteq (r^2\omega_{30})/(2\mu ga)$.

17. $L = \frac{1}{2}I_1\dot{\theta}^2 + \frac{1}{2}I_1\dot{\phi}^2\sin^2\theta + \frac{1}{2}I_3(\dot{\psi} + \dot{\phi}\cos\theta)^2 + \frac{1}{2}(M+m)(\dot{r}^2 + r^2\dot{\alpha}^2)$
$-(M+m)M'G/r - mM'Ga^2/(4r^3)[1 - 3\sin^2\theta\cos^2(\phi - \alpha)].$

19. $M = 6.0 \times 10^{24}$ kgm, $m = 1.6 \times 10^{22}$ kgm, $a = 6400$ km., 24,000 years.

23. $(2\pi)^{-1}[(\omega_0\cos\theta_0)(I_3\omega_3 - I_1\omega_0\cos\theta_0)/I_1]^{1/2}$.

Chapter 12

1. $x_1 = \kappa^2[m_1 S\,\Delta\omega^2]^{-1/2}q_1 - \frac{1}{2}[m_1 S/\Delta\omega^2]^{-1/2}q_2,$
$x_2 = \frac{1}{2}[m_2 S/\Delta\omega^2]^{-1/2}q_1 + \kappa^2[m_2 S\,\Delta\omega^2]^{-1/2}q_2,$
$S = \frac{1}{2}\Delta\omega^2 + \frac{1}{2}(\omega_{10}^2 - \omega_{20}^2)$, in the notation of Section 4–10.

3. $\omega_j^2 = \dfrac{5V_0}{ma^2}e^{-77},\quad \dfrac{V_0}{2ma^2}e^{-77}(13 + \sqrt{73}),\quad \dfrac{V_0}{2ma^2}e^{-77}(13 - \sqrt{73}).$

5. $x_1 = a + A_1\cos(\omega_1 t + \theta_1) + A_2\cos(\omega_2 t + \theta_2),$
$x_2 = a + A_1\cos(\omega_1 t + \theta_1) - A_2\cos(\omega_2 t + \theta_2),$
$y_1 = A_3\cos(\omega_3 t + \theta_3) + A_4\cos(\omega_4 t + \theta_4),$
$y_2 = -A_3\cos(\omega_3 t + \theta_3) + A_4\cos(\omega_4 t + \theta_4),$
$z_1 = A_5\cos(\omega_5 t + \theta_5) + A_6\cos(\omega_6 t + \theta_6),$
$z_2 = -A_5\cos(\omega_5 t + \theta_5) + A_6\cos(\omega_6 t + \theta_6),$
$\omega_1^2 = \omega_3^2 = \omega_5^2 = k/m,\ \omega_2^2 = k(l + 6a)/m(l + 2a),\qquad \omega_4^2 = \omega_6^2 = kl/m(l + 2a),$
where x_1, y_1, z_1, and x_2, y_2, z_2 are measured from A, B respectively, the x-axis is parallel to $\overline{AB}$, and a is the positive root of $ka(l + 2a)^2 - q^2 = 0$.

7. Both ions oscillate in phase parallel to the electric field with amplitude $(qE_0/m)/(\omega_0^2 - \omega^2)$.

11. $x_k^0 = \Delta_k'/\Delta_0$, where $\Delta_0 = |K_{lm}^0|$, and Δ_k' is Δ_0 with K_{lk}^0 replaced by

$$(\partial V'/\partial x_l)_0,\ K_{kl}' = \left(\frac{\partial^2 V'}{\partial x_k\,\partial x_l}\right)_0 + \sum_m \frac{\Delta_m'}{\Delta_0}\left(\frac{\partial^3 V^0}{\partial x_k\,\partial x_l\,\partial x_m}\right)_0.$$

13. $$W_1' \doteq W_{11}' + \sum_{r=2}^{f}\frac{W_{1r}'W_{r1}'}{W_1^0 - W_r^0} + \sum_{r,l=2}^{f}\frac{W_{1r}'W_{rl}'W_{l1}'}{(W_1^0 - W_r^0)(W_1^0 - W_l^0)}.$$

15. $$\omega_1^2 = \frac{g}{l}\left(1 + 4\frac{m}{M}\right),\quad \omega_2^2 = \frac{\sqrt{2}}{\sqrt{2}+1}\,\frac{g}{l}\left[1 - (\sqrt{2}+1)\frac{m}{M}\right],$$

$$\omega_3^2 = \frac{\sqrt{2}}{\sqrt{2}-1}\,\frac{g}{l}\left[1 + (\sqrt{2}-1)\frac{m}{M}\right].$$

17. $x_1 = x_4 = (1+\sqrt{5})A\cos(\omega_1 t+\theta)$, $x_2 = x_3 = -2A\cos(\omega_1 t+\theta)$,
$\omega_1 = [k(3+\sqrt{5})/2m]^{1/2}$;
$x_1 = -x_4 = [(1+\sqrt{5}) + 2(k'/k\sqrt{5})]A\cos(\omega_2 t+\theta)$,
$x_2 = -x_3 = -[2 - 2(1+\sqrt{5})(k'/k\sqrt{5})]A\cos(\omega_2 t+\theta)$,
$\omega_2 = \omega_1[1 + (2k'/k)/(3+\sqrt{5})]$;
$x_1 = x_4 = 2A\cos(\omega_3 t+\theta)$, $x_2 = x_3 = (1+\sqrt{5})A\cos(\omega_3 t+\theta)$,
$\omega_3 = [k(3-\sqrt{5})/2m]^{1/2}$;
$x_1 = -x_4 = [2 + (1+\sqrt{5})(k'/k\sqrt{5})]A\cos(\omega_4 t+\theta)$,
$x_2 = -x_3 = [(1+\sqrt{5}) - 2(k'/k\sqrt{5})]A\cos(\omega_4 t+\theta)$,
$\omega_4 = \omega_3[1 + (2k'/k)/(3-\sqrt{5})]$.

By symmetry, modes 1 and 3 will be exactly as given above for any k'; hence two roots ω_1, ω_3 are known, and the secular equation can be factored.

19. $$\omega_1^2 = \frac{\pi^2}{4}\frac{\pi}{l^2\sigma_0}\left[1 - \frac{8a}{3\pi\sigma_0} + \frac{64a^2}{9\pi^2\sigma_0} + \frac{64a^2}{\pi^2\sigma_0}\sum_{j=3,5,7,\ldots}\frac{1}{j^2(j-1)(j^2-4)}\right],$$
$$u = A\cos(\omega_1 t+\theta)\left[\sin\frac{\pi x}{l} - \frac{8a}{\pi\sigma_0}\sum_{j=3,5,7,\ldots}\frac{j^{1/2}}{(j-1)(j^2-4)}\right].$$

21. Steady motions: $z = 0$, $r = r_0$, $\dot{\theta} = \omega_0$. (Cylindrical coordinates with $+Ze$ at $z = \pm a$, $r = 0$.) Normal vibrations: $r = r_0 + A\cos(\omega_1 t+\theta)$, $z = 0$; $r = 0$, $z = A\cos(\omega_2 t+\theta)$, if $r > \sqrt{2}a$, otherwise unstable.
$$\omega_0^2 = \frac{2Ze^2}{ma^3}\left(1+\frac{r_0^2}{a^2}\right)^{-3/2}, \quad \omega_1^2 = \omega_0^2\left(\frac{r_0^2}{a^2}+4\right)\bigg/\left(\frac{r_0^2}{a^2}+1\right),$$
$$\omega_2^2 = \omega_0^2\left(\frac{r_0^2}{a^2}-2\right)\bigg/\left(\frac{r_0^2}{a^2}+1\right).$$

23. Steady motions:
$r = r_0$, $\theta = \theta_0$, $\dot{\alpha}^2 = Mg/(mr_0\cos\theta_0)$, $\dot{\varphi}^2 = g/[(l-r_0)\cos\theta_0]$.
$$\omega^2 = \dot{\varphi}^2(A \pm B), \quad A = \frac{3M}{M+m}\left(\frac{l}{r_0} - \cos^2\theta_0\right) + 1 + 3\cos^2\theta_0,$$
$$B^2 = A^2 + \frac{3M}{M+m}\left[4\cos^2\theta_0 - \frac{l}{r_0}(1 - 3\cos^2\theta_0)\right].$$
[Lower mode unstable if $\cos^2\theta_0 < \frac{1}{3}$ and $r_0 < \frac{1}{4}l(\sec^2\theta_0 - 3)$].

25. Steady motions: $\theta = \theta_0$, $\phi = 0$, $\omega_3 =$ constant.
Normal vibrations: $\theta = \theta_0 + A\lambda\omega_3\cos(\omega t+\beta)$,
$\phi = (\omega/\sin\theta_0)A\sin(\omega t+\beta)$, $\omega^2 = \lambda^2\omega_3^2 + (\omega_0^2/\sin^2\theta_0)$, $\lambda = I_3/I_1$;
$\theta = \theta_0 + A + (\omega_0^2 Bt/\lambda\omega_3\sin\theta_0)$, $\phi = B$, (unstable mode with $\omega^2 = 0$).

27. Steady motions: $r = r_0$, $\alpha = \pi$, α-mode always unstable;
$r = r_0$, $\alpha = 0$, possible only if $l < 3r_0$, both modes always stable.

33. $X_1 = X_2 = 0$, $E_1 = E_2 = A$. Guess $x_1 = x_2 = B$, $\epsilon_1 = \epsilon_2 = \dot{\alpha}t$; checks if $\dot{\alpha} = -3\omega B/2a$.

Chapter 13

3. a sent out 1 microsecond later than b.

Chapter 14

11. a) $P_x = -305$ MeV/c, $p_y = -145$ MeV/c, $p_z = -553$ MeV/c, where the incident beam is in the x-direction.
    b) 260 MeV, $p_x = -205$ MeV/c, $p_y = 0$, $p_z = -653$ MeV/c.
15. Six.
19. $S = (m_a^2 + m_c^2)c^2 - 2E_aE_c/c^2 + 2p_ap_c \cos\theta_{ac}$.
    $[p = (E^2 - m^2c^4)^{1/2}/c.]$
29. $ct = \int \{1 - [1 + (T - \frac{1}{2}kx^2)/mc^2]^{-2}\}^{-1/2}\, dx,\ T = E - mc^2,$
    $\omega = (k/m)^{1/2}[1 - (3T/8mc^2) + \cdots],$
    $x = (2T/k)^{1/2}(1 + 3T/32mc^2)\sin(\omega t + \theta) - (2T/k)^{1/2}(3T/32mc^2)$
    $\sin 3(\omega t + \theta) + \cdots$
31. $F = -\partial V/\partial x,\ V = -mc^2[1 - \omega^2(A^2 - x^2)/c^2]^{-1/2}$.
39. c) $V = c[1 - (1 - \alpha)^2]^{1/2}$.
41. Replace F by dA in Eq. (14.83).
49. Six of rank two, four of rank three, one of rank four.

INDEX OF SYMBOLS

INDEX OF SYMBOLS

The following list is not intended to be complete, but includes important symbols and those which might give rise to ambiguity. In general, standard mathematical symbols, and symbols used in a specialized sense occurring only once, are omitted. To facilitate reference, the page on which the symbol first occurs is listed immediately after the definition of the symbol. When use of a symbol in a particular sense is restricted to one or two sections or chapters, this is indicated by chapter or section numbers in parentheses following the definition.

Scalar quantities are designated in the text by italics. Vector quantities are designated by boldface italic letters, beginning in Chapter 3. An italic letter is used for the magnitude of the vector represented by the same letter in boldface. An italic letter with subscripts is used to denote components of the vector represented by the same letter in boldface. A caret over a vector symbol denotes a vector of unit length; a caret over a coordinate (boldface) denotes a unit vector in the direction in which that coordinate increases. In Chapter 2, boldface roman letters are used for some complex quantities. Tensors are represented by sans-serif boldface capitals, beginning in Chapter 10. The same letter in italics with a double subscript designates a tensor component, and with a prime or single subscript, an eigenvalue. A dot over a letter indicates differentiation with respect to time. Single quotes are used to mark quantities associated with fictitious forces which arise in moving coordinate systems.

Latin Letters

A amplitude of vibration, 32

A area, 224

A constant coefficient, 59

AA' plane perpendicular to beam, 241 (Section 5.10)

$\boldsymbol{A}$ vector potential, 386 (Sections 9.8, 12.7, 14.6)

A_μ 4-potential, 573 (Chapter 14)

a constrained generalized coordinate, 371 (Section 9.4)

a distance from focus to directrix in parabola, 133 (Chapter 3)

a semimajor axis of ellipse or hyperbola, 131 (Chapter 3)

a, $\boldsymbol{a}$ acceleration, 5, 93

a_{ij} coordinate transformation coefficients, 410

$a_{\mu\nu}$ Lorentz transformation coefficients, 539

L_O, $\mathbf{L}_O$	angular momentum about point O, 105, 106
l	length, 12
l_0	rest length, 522
M	Mach number, 344 (Sections 8.14, 8.15)
M	mass, usually total mass of a body or system of particles, 10
M	molecular weight, 251 (Section 5.11)
M	invariant mass, 554 (Section 14.3)
m	mass, usually of a particle, 6, rest mass, 548
N	bending moment, 241 (Section 5.10)
N, $\mathbf{N}$	torque, 106, 107
N	total number of particles, 160 (Chapters 4, 9)
N_O, $\mathbf{N}_O$	torque about point O, 84, 85
n	shear modulus, 237 (Chapter 5, Section 10.6)
$\hat{\mathbf{n}}$	unit vector normal to surface, outward normal from closed surface, 100
O	point in space, usually the origin, 91
OO'	line through centroid of beam cross section, 243 (Section 5.10)
P, P'	points in space, 229
P	power, 303 (Chapter 8)
P	power per unit area, 333 (Section 8.10)
P_x	component of dipole moment per unit volume, 27 (Chapter 2)
$\mathbf{P}$	total linear momentum, 161
P	stress tensor, 432
p	complex coefficient in exponential time dependence (e^{pt}), 44
p	generalized momentum (usually with subscript), 360
p	pressure, 248
p, $\mathbf{p}$	linear momentum, 8, 103
p_μ	4-momentum, 548
p'	excess pressure, 311 (Chapter 8)
Q	energy produced in inelastic collision, 180 (Chapters 4, 14)
Q	generalized force (usually with subscript) 361
Q	point in space, 163 (Chapter 4)
Q	source density, 323 (Chapter 8)
q	electric charge, 81
q	generalized coordinate (usually with subscript), 354
R	gas constant, 251 (Section 5.11)
R	Reynolds number, 348 (Section 8.15)

Greek Letters

Other Symbols

Subscripts

Superscripts

INDEX

INDEX